Lecture Notes in Computer Science 13828

The series Lecture Notes in Computer Science (LNCS), including its subseries Lecture Notes in Artificial Intelligence (LNAI) and Lecture Notes in Bioinformatics (LNBI), has established itself as a medium for the publication of new developments in computer science and information technology research, teaching, and education.

LNCS enjoys close cooperation with the computer science R & D community, the series counts many renowned academics among its volume editors and paper authors, and collaborates with prestigious societies. Its mission is to serve this international community by providing an invaluable service, mainly focused on the publication of conference and workshop proceedings and postproceedings. LNCS commenced publication in 1973.

Meikang Qiu · Zhihui Lu · Cheng Zhang
Editors

Smart Computing and Communication

7th International Conference, SmartCom 2022 New York City, NY, USA, November 18–20, 2022 Proceedings

Editors
Meikang Qiu
Dakota State University
Madison, SD, USA

Zhihui Lu
Fudan University
Shanghai, China

Cheng Zhang
Ibaraki University
Ibaraki, Japan

ISSN 0302-9743 ISSN 1611-3349 (electronic)
Lecture Notes in Computer Science
ISBN 978-3-031-28123-5 ISBN 978-3-031-28124-2 (eBook)
https://doi.org/10.1007/978-3-031-28124-2

This Springer imprint is published by the registered company Springer Nature Switzerland AG
The registered company address is: Gewerbestrasse 11, 6330 Cham, Switzerland

Preface

This volume contains the papers presented at SmartCom 2022: the Seventh International Conference on Smart Computing and Communication held during November 18–20, 2022, in New York City, USA.

There were 312 submissions. Each submission was reviewed by at least three reviewers, and on average 3.5 Program Committee members. The committee decided to accept 64 regular papers. Thanks to the hard work of the program chairs for the strict peer review process and the quality responsibility they took on. Without this, the proceedings would not have been possible. Thanks to the general chairs for their time and efforts to organize this hybrid conference with great success.

Recent booming developments in Web-based technologies and mobile applications have facilitated a dramatic growth in the implementation of new techniques, such as cloud computing, edge computing, big data, pervasive computing, Internet of Things, security and privacy, blockchain, Web 3.0, and social cyber-physical systems. Enabling a smart life has become a popular research topic with an urgent demand. Therefore, SmartCom 2022 focused on both smart computing and communications fields and aimed to collect recent academic work to improve the research and practical application in the field.

The scope of SmartCom 2022 was broad, from smart data to smart communications, from smart cloud computing to smart security. The conference gathered high-quality research/industrial papers related to smart computing and communications and aimed at proposing a reference guideline for further research. SmartCom 2022 was held in New York City and its conference proceedings publisher is Springer.

SmartCom 2022 continued in the series of successful academic get togethers, following SmartCom 2021 (virtual), SmartCom 2020 (Paris, France), SmartCom 2019 (Birmingham, UK), SmartCom 2018 (Tokyo, Japan), SmartCom 2017 (Shenzhen, China) and SmartCom 2016 (Shenzhen, China).

We would like to thank the conference sponsors: Springer, North America Chinese Talents Association, and Longxiang High Tech Group Inc.

November 2022

Meikang Qiu
Zhihui Lu
Cheng Zhang

Organization

General Chairs

Yu Huang	Peking University, China
Yonghao Wang	Birmingham City University, UK
Yongxin Zhu	Chinese Academy of Sciences, China

Program Chairs

Meikang Qiu	Dakota State University, USA
Zhihui Lu	Fudan University, China
Cheng Zhang	Ibaraki University, Japan

Local Chairs

Keke Gai	Beijing Institute of Technology, China
Xiangyu Gao	New York University, USA
Tao Ren	Beihang University, China

Publicity Chairs

Han Qiu	Tsinghua University, China
Lixin Tao	Pace University, USA
Zhenyu Guan	Beihang University, China

Technical Committee

Yue Hu	Louisiana State University, USA
Aniello Castiglione	University of Salerno, Italy
Maribel Fernandez	King's College, University of London, UK
Hao Hu	Nanjing University, China
Oluwaseyi Oginni	Birmingham City University, UK
Thomas Austin	San Jose State University, USA
Zhiyuan Tan	Edinburgh Napier University, UK

Xiang He	Birmingham City University, UK
Peter Bull	QA Ltd., UK
Hendri Murfi	Universitas Indonesia, Indonesia
Bo Du	ZF Friedrichshafen AG, Germany
Cefang Guo	Imperial College London, UK
Zengpeng Li	Lancaster University, UK
Paul Kearney	Birmingham City University, UK
Wenbo Shi	Inha University, South Korea
Pietro Ferrara	JuliaSoft SRL, Italy
Ke Miao	Mitacs Inc., Canada
Xiaohu Zhou	Birmingham City University, UK
Guangxia Xu	Chongqing University of Posts and Telecommunications, China
Kui Zhang	Birmingham City University, UK
Matthew Roach	Swansea University, UK
Chunhua Deng	Wuhan University of Science and Technology, China
Katie Cover	Pennsylvania State University, USA
Jinguang Gu	Wuhan University of Science and Technology, China
Wei Cai	Chinese University of Hong Kong, Shenzhen, China
Haikuo Zhang	China Internet Network Information Center (CNNIC), China
Junwei Zhang	Xidian University, China
Vitor Jesus	Birmingham City University, UK
Jue Wang	SCCAS, China
Hao Tang	City University of New York, USA
Md Ali	Rider University, USA
Li Bo	Beihang University, China
Zijian Zhang	Beijing Institute of Technology, China
Ian A. Williams	Birmingham City University, UK
Meng Ma	Peking University, China
Cheng Zhang	Waseda University, Japan
Fausto Spoto	Dipartimento di Informatica, Verona, Italy
Song Yang	Beijing Institute of Technology, China
Lixin Tao	Pace University, USA
Alan Dolhasz	Birmingham City University, UK
Shaojing Fu	National University of Defense Technology, China
Agostino Cortesi	Università Ca' Foscari, Italy
Yunxia Liu	Huazhong University of Science and Technology, China

Yongxin Zhu	Chinese Academy of Sciences, China
Rehan Bhana	Birmingham City University, UK
Songmao Zhang	Chinese Academy of Sciences, China
Dawei Li	Beihang University, China
Jongpil Jeong	Sungkyunkwan University, South Korea
Shuangyin Ren	Chinese Academy of Military Science, China
Ding Wang	Peking University, China
Wenjia Li	New York Institute of Technology, USA
Peng Zhang	Stony Brook University, USA
Wayne Collymore	Birmingham City University, UK
Zehua Guo	Beijing Institute of Technology, China
Jeroen van den Bos	Netherlands Forensic Institute, The Netherlands
Haibo Zhang	University of Otago, New Zealand
Zhiqiang Lin	University of Texas at Dallas, USA
Xingfu Wu	Texas A&M University, USA
Dalei Wu	University of Tennessee at Chattanooga, USA
Jun Zheng	New Mexico Tech, USA
Minzhou Pan	Virginia Tech, USA
Suman Kumar	Troy University, USA
Shui Yu	Deakin University, Australia
Jeremy Foss	Birmingham City University, UK
Cham Athwal	Birmingham City University, UK
Petr Matousek	Brno University of Technology, Czechia
Javier Lopez	University of Malaga, Spain
Andrew Aftelak	Birmingham City University, UK
Yi Zheng	Virginia Tech, USA
Dong Dai	Texas Tech University, USA
Hiroyuki Sato	University of Tokyo, Japan
Bo Luo	University of Kansas, USA
Syed Rizvi	Pennsylvania State University, USA
Paul Rad	Rackspace, USA
Ruisheng Shi	Beijing University of Posts and Telecommunications, China
Fuji Ren	University of Tokushima, Japan
Jian Zhang	Institute of Software, Chinese Academy of Sciences, China
Chungsik Song	San Jose State University, USA
Sang-Yoon Chang	Advanced Digital Science Center, Singapore
Mohan Muppidi	UTSA, USA
Kan Zhang	Tsinghua University, China
Malik Awan	Cardiff University, UK
Ming Xu	Hangzhou Dianzi University, China

Allan Tomlinson	University of London, UK
Long Fei	Google, USA
Emmanuel Bernardez	IBM Research, USA

Contents

Design and Implementation of Deep Learning Real-Time Streaming Video Data Processing System

Qiming Zhao, Jing Wu(✉), Xiang Wu, Jie Fan, Linhao Wang, and Fengling Wu

College of Computer Science, Wuhan University of Science and Technology, Wuhan, China
{kenlig,jacobevans,wujingecs}@wust.edu.cn

Abstract. With the arrival of big data era and the rapid development of artificial intelligence, deep learning has made breakthroughs in many fields. However, although it has been widely used in many fields, there are still many challenges in itself, such as the slow reference speed of neural networks. In stream processing scenarios that integrate deep learning algorithms, data is often massive and generated with high-speed, requiring the system to respond in seconds or even milliseconds. If the response speed is too slow, the actual application requirements may not be met, and the user experience cannot be guaranteed. How to use stream processing technology to improve the speed and throughput of such systems has become an urgent problem to be solved. This paper used the popular real-time stream processing framework Flink to implement a complete data stream processing program, and integrated three algorithms of face detection, facial key points detection and face mosaic into the processing logic. Setting the operator parallelism realized parallel processing of video data, which improved the system throughput. The user can choose which algorithm to perform on the video, and can also choose the parallelism according to the performance of the machine. The system implemented the Flink framework to process video in parallel, achieved the effect of improving processing efficiency, and completely implemented the front-end interface and back-end program.

Keywords: Flink · Big data · Deep learning · Image processing

1 Introduction

With the advent of the era of big data [1,2] and the rapid development of computer hardware [3,4] and network infrastructure [5–7], deep learning [8,9] has made breakthroughs in many fields. Deep learning has made great achievements in human-computer communication, computer vision, natural language processing, autonomous driving, etc. For example, it can be applied to image

This work was supported by Hubei Innovation and Entrepreneurship Training Program for University Students under Grant NO. S202210488057.

M. Qiu et al. (Eds.): SmartCom 2022, LNCS 13828, pp. 1–10, 2023.
https://doi.org/10.1007/978-3-031-28124-2_1

synthesis and image classification by *Generative Adversarial Networks* (GAN) [10]; autonomous driving by integrating target detection and target tracking into vehicles, etc. However, despite having been widely used in several fields, deep learning is still challenged by various aspects, such as the speed of neural network inference.

The inference speed of neural networks is very important, especially in stream processing scenarios. In such scenarios, there are usually huge amounts of data generated at a very high speed [14]. This data is imported into the stream processing system and requires the system to respond in a very short time in order to ensure the availability of the system and the user experience. However, a large amount of computation [11–13] is usually involved in neural networks. And, when the neural network model is larger and the model is more complex, the amount of computation is greater and its performance requirements on the machine are higher [15]. Therefore, in practical applications, we have to use stream processing techniques to improve the processing speed and throughput of such systems.

The application of stream processing techniques is carried out through the stream processing framework. We can choose Apache Hadoop, Apache Spark, or Apache Flink. Apache Flink framework is suitable for batch and stream data processing and has the advantages of low latency, high throughput, fault tolerance, while being faster than Apache Hadoop. By comparison, we choose the Apache Flink framework [16]. Flink is a distributed processing engine for streaming and batch data with high throughput, low latency, and high performance. Flink uses parallel and streaming computing and is very efficient in processing datasets.

This paper was organized as the follows. Section 2 introduces background and related work. Section 3 introduces the system design. Section 4 introduces the system experiments. Finally, in Sect. 5, we discuss our conclusions.

2 Related Work

Several related studies exist in the world for this topic. In terms of big data processing, Cecilia Calavaro used Apache Flink to process market data in real time and improved the performance of big data processing by parallel processing [17]. Baolin Xu developed an information intelligence system using Apache Flink, using Flink streaming technology as the processing center, Kafka as the message queue, and MySQL as the storage system, which greatly improves the processing efficiency of big data real-time streaming data [18]. A large amount of data may cause data competition and reduce processing efficiency when parallel processing is performed, and the resource competition problem associated with automatic resource control mechanisms can be solved by Flink and processing across distributed clusters [19].

Of course, Flink and deep learning are also closely connected. Tae Wook Ha process multivariate data streams through Apache Flink, train an LSTM encoder-decoder model to reconstruct multivariate input sequences, and develop a detection algorithm that can use the reconstruction error between the input and reconstructed sequences [20]. Yuansheng Dong building a deep learning-based intrusion detection system for implemented networks using Flink with an accuracy of 94.32% [21]. Not only that, Flink can also be used with GPU arithmetic, memory management of Flink, and Flink clusters to form a set of distributed neural net-work frameworks that can be flexible to build their desired network architecture [22]. In deep learning, we can introduce hardware to improve efficiency [23]. We can also use Flink in this case.

Currently, deep learning requires extremely large training and prediction sets [24]. After the model is trained, data processing of the prediction set is a tricky challenge [25]. When the size of the prediction set is large, the model has limited speed to process the prediction set, and a performance bottleneck arises. Flink can handle big data and improve the performance of big data processing by parallel processing. Flink can also be combined with deep learning to form a flexible network architecture. So, we can build Flink with deep learning and big data processing in one system, which can improve the speed of processing results in that system. Not only that, it can also improve the quality of the data processed by deep learning.

3 System Design

3.1 Overall System Design

The system mainly contains four modules: video upload module, video processing module, result generation module, video download module, and provides a user-friendly Web interface (see Fig. 1).

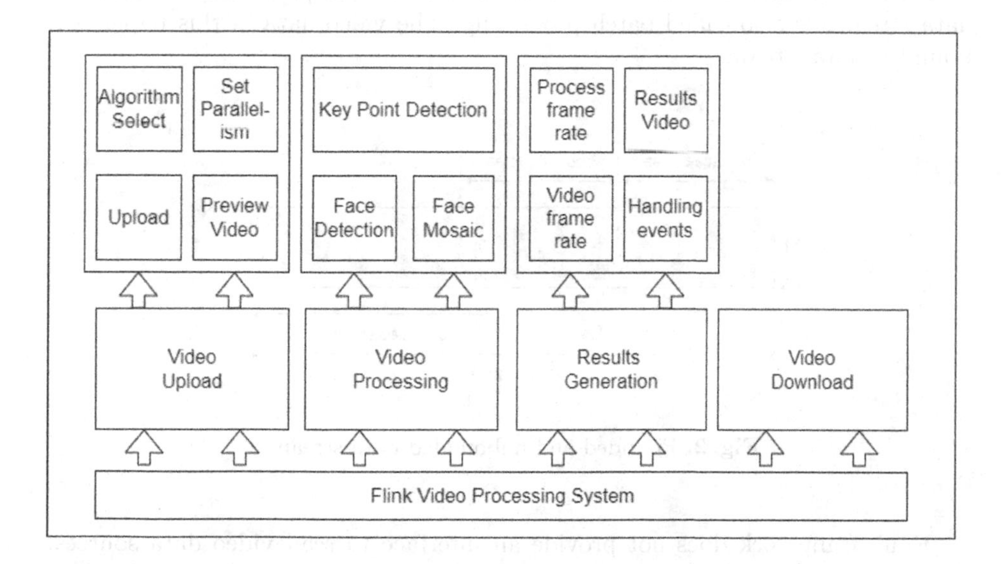

Fig. 1. System function module diagram

- Video upload module. Users can upload videos in the web interface, select the processing to be performed on the video, the parallelism of the processing, and also preview the video.
- Video processing module. The server side processes the video accordingly according to the user's choice, including face detection, key point detection, face mosaic.
- Result generation module. The server stitches the frames in order according to the result frames to generate the result video, and at the same time will calculate the number of frames of the video, the events processed, the frame rate processed, and return the result to the front-end.
- Video download module. Users can download video results to local.

3.2 Video Processing Module

The video processing module is the core part of the system design in this paper. The basic process of Flink framework is to read the data source by Source operator, to operate Transform operator on the data source, and to output the data to the external connector by Sink operator.

Source Operator Reads the Data Source. The Source operator defines the source of Flink data streams. In Flink, there are two types of data streams: unbounded data streams, and bounded data streams (see Fig. 2). Unbounded data flow means that there is a beginning of defined flow and no end of defined flow, which will generate data endlessly. Bounded data flow means that there is both the beginning and the end of a defined flow. The processing of bounded data streams is also called batch processing. The video data in this paper is a bounded data stream.

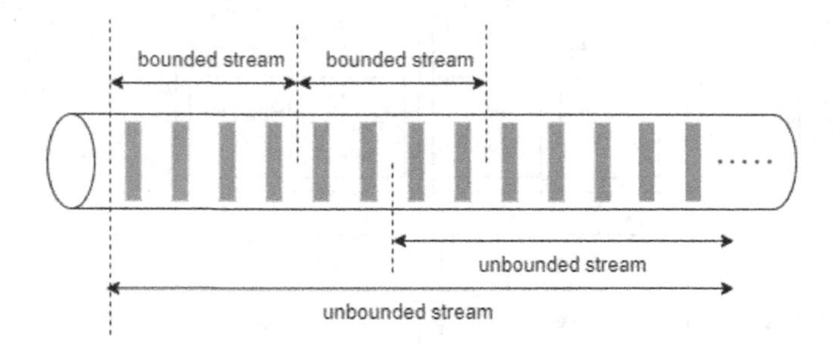

Fig. 2. Bounded and unbounded data stream

Flink framework does not provide an interface to read video data sources, so we designed the Source operator to read the data. In this paper, the data is video. Source arithmetic improves the speed of data reading compared to traditional video data reading using streams, and it can split the video source into video frames to be provided to the arithmetic afterwards. Source arithmetic provides two functions: generating the data source and defining the operations to be performed when the data source stops. Of course, the Source operator is tightly integrated with the Flink framework and uses the interface provided by the Flink framework to send the processed data, i.e. video frames, to the Transform operator.

Transform Operator for Video Frames. The Transform operator refers to the operation that needs to be performed on each element of the data source. Flink provides predefined operators, but these operators cannot perform the operations we need. In this article, we implement the Transform operator on top of Flink's own filter, window, and keyBy operators, and the Transform operator can accept data passed by the Source operator. The Transform operator combines the filter, window, and keyBy operators to filter and aggregate the data to form a window data stream, and partition the data according to a certain attribute, and the data with the same attribute value will be partitioned into the same Data with the same attribute will be partitioned into the same partition.

For face detection and keypoint detection, we can propose our own face detection and keypoint detection solutions based on the OpenIMAJ software library. In this paper, we propose a more efficient solution. We call the OpenIMAJ software library with one object corresponding to each keypoint detected. When the detection is finished, we draw the detection result on the original video frame with all objects. The face part is drawn as a rectangle and each key point is drawn as a dot, both of them are distinguished by different colors set by the col parameter (see Fig. 3).

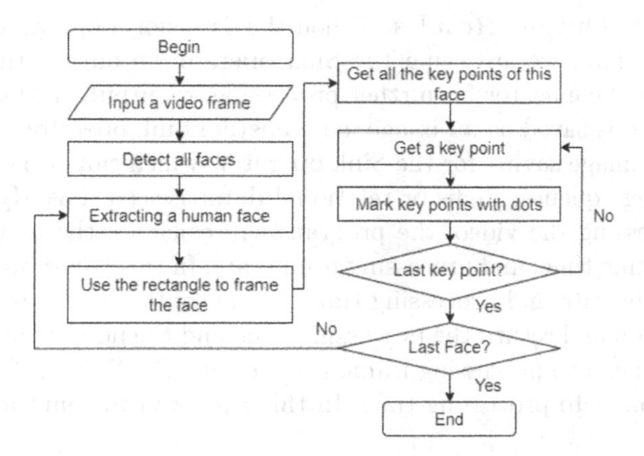

Fig. 3. Face recognition and key point detection algorithm flow chart

For the face mosaic function, this paper customizes the algorithm to implement. The mosaic area is the rectangle corresponding to the face area, and the size of the mosaic block is set to 40. The algorithm first calculates the number of mosaics drawn on the x-axis and y-axis, then iterates through each block in the area and draws the mosaic in turn, and finally returns the finished result to the Transform operator. The mosaic is drawn by taking the color of the center point of each block and then applying the color to the entire block. The flowchart of the algorithm is shown (see Fig. 4).

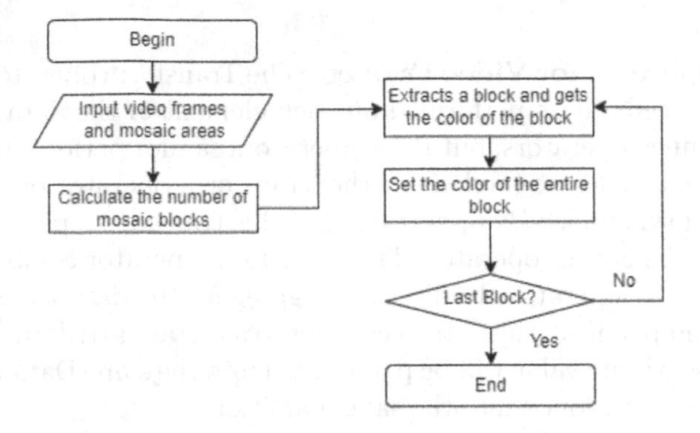

Fig. 4. Mosaic algorithm flow chart

In addition, the parallelism of the Map operator needs to be set according to the parallelism degree selected by the user. In Flink, multiple levels of parallelism can be set. The higher level of parallelism will override the lower level of parallelism. Higher parallelism means more threads are opened, and the program will be executed more efficiently.

Sink Operator Output Results. When the Transform operator has finished processing the data, we are using the Sink operator to output the result to a specified external connector for further processing or output. In this paper, each processed frame is saved as an image, so a custom Sink operator is needed. We do the custom image saving for the Sink operator, which can also connect to file systems, message queues, Redis or relational databases such as MySQL.

After processing the video, the program will calculate the number of video frames, processing time, and processing frame rate. In this paper, we combine the processing frame rate and processing time. We make the Flink program process the whole stream and return the processing time and the number of video frames at the same time. The processing frame rate (Frame Per Second, FPS), which is the ratio of frames to processing time. In this way, we can count all the data.

4 System Experiments

4.1 Experimental Setup

The software development environment includes the compiler, development framework, programming environment, interface debugging tools used in development. The hardware development environment mainly refers to the hardware parameters configuration of the computer. We have conducted experiments using the environment as in Table 1 and obtained the expected results.

Table 1. Software and hardware development environment.

Category	Details
Compiler	IntelliJ IDEA, visual studio code
Framework	Spring boot, Apache Flink, Vue, OpenIMAJ
Environment	JDK 11, Nodejs 16.15.0
APITool	APIFox
System	Windows 11
CPU	Intel®Core^{TM}i5-11320H @ 3.20 GHz
Memory size	16 GB
GPU	Intel®Iris®Xe Graphics

4.2 Experimental Result

To visualize Flink's processing of streaming data, we will set the Global Parallelism and Transform Parallelism for raw data. Global parallelism and Transform parallelism are the things that can be set in the Flink framework.

First, we set the global parallelism to 1 and the Transform parallelism to 2. Then we can see the physical execution of the Flink program (see Fig. 5). It can be seen that the data of the whole program is divided into two data streams, and the results of the two data streams are processed in parallel, which improves the running efficiency of the program. Not only that, the data stream is split into two and finally merged. The merging of multiple data streams can significantly improve efficiency when processing in parallel.

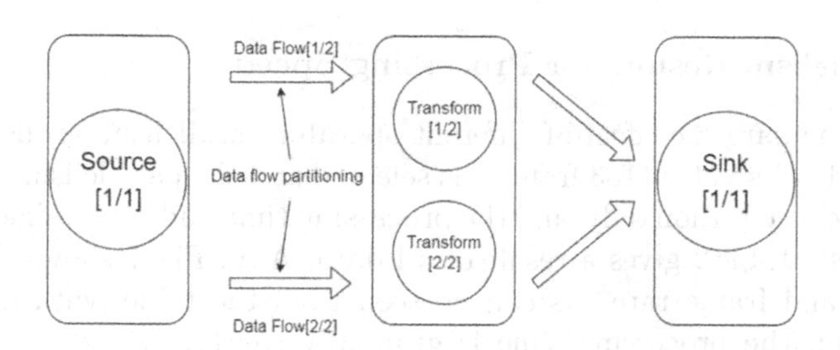

Fig. 5. Physical execution diagram when transform is 2

In the actual program run, we set the Transform parallelism of the program to 8 and show the task IDs being processed and the frame IDs being processed (see Fig. 6). It can be seen that the parallel tasks are called at different times, while each task processes one frame, and each task corresponds to one frame ID.

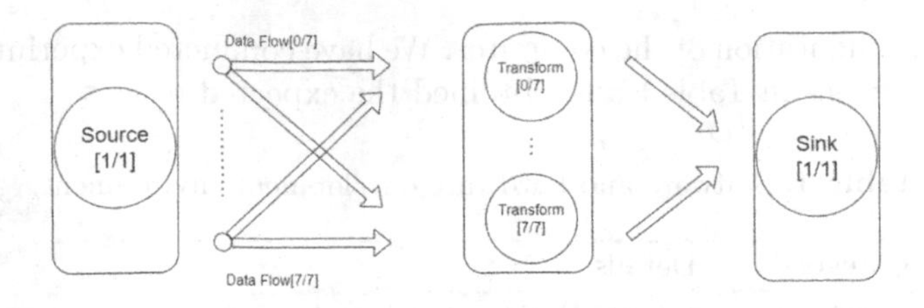

Fig. 6. Physical execution diagram when transform is 8

Table 2. Processing time and frame rate with parallelism table.

Parallelism	Processing time	Processing frame rate
2	34s	4
4	23s	6
6	15s	10

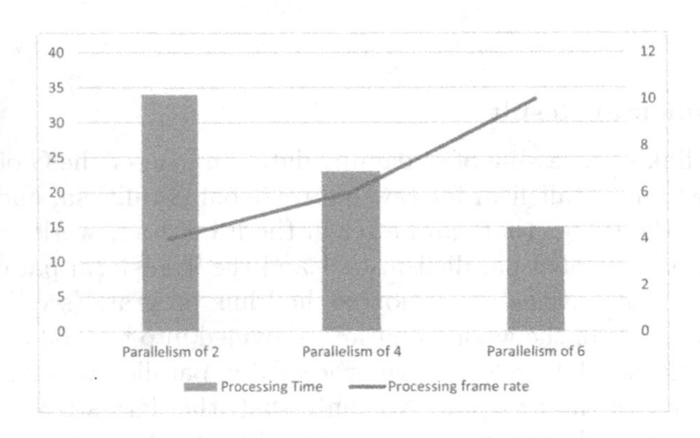

Fig. 7. Processing time and frame rate with parallelism

4.3 Parallelism Results for Processing Speed

In order to compare the effect of different operator parallelism on the video processing speed, a 5 s video (153 frames) is selected and the parallelism is set to 2, 4 and 6 for face recognition only, and the processing time and processing frame rate are compared. Table 2 gives a result of all data. Also, Fig. 7 shows the changes of the time and frame rate. As can be seen from the table, with the increase of parallelism, the processing time is gradually shortened and the processing frame rate is gradually increased, which basically achieves the requirement of increasing the system processing speed.

5 Conclusion

This paper showed our design of a parallel video processing system used to solve the problem of slow inference speed of deep learning algorithms by using Flink framework and several operators. We set the global parallelism degree and the Transform operator parallelism degree, so that the system completed the parallel processing of offline video data. Also, we compared the effect of different parallelism degrees on the processing speed and improved the system throughput. Finally, after running our system on the testing platform and adjusting the parallelism degree, our system performed even better.

In the future, for this topic, we can consider other deep learning algorithms and how to further improve the throughput of the system. Specifically, we can do further research and improvements. In this paper, we only improve the throughput of the system by setting the parallelism of the Flink operator. In a real application scenario, the real-time stream processing system is deployed to a cluster. The next step of the study focuses on how to build a cluster for a distributed processing system to further improve the throughput of the system. In the meantime, we will test the performance of the program at higher frame rates and higher parallelism, including the performance of the cluster, and consider deploying it to a real application. Also, we will work on demonstrating the time complexity of parallel processing algorithms.

References

1. Li, Y., Gai, K., et al.: Intercrossed access controls for secure financial services on multimedia big data in cloud systems. ACM Trans. Multimed. Comp. Comm. App. **12**(4s), 1–18 (2016)
2. Qiu, M., Chen, Z., Ming, Z., Qin, X., Niu, J.: Energy-aware data allocation with hybrid memory for mobile cloud systems. IEEE Syst. J. **11**(2), 813–822 (2014)
3. Qiu, M., Li, H., Sha, E.: Heterogeneous real-time embedded software optimization considering hardware platform. ACM Sym. Appl. Comp. 1637–1641 (2009)
4. Qiu, M., Xue, C., Shao, Z., Sha, E.: Energy minimization with soft real-time and DVS for uniprocessor and multiprocessor embedded systems. In: IEEE DATE Conference, pp. 1–6 (2007)
5. Niu, J., Gao, Y., et al.: Selecting proper wireless network interfaces for user experience enhancement with guaranteed probability. JPDC **72**(12), 1565–1575 (2012)
6. Qiu, M., Xue, C., Shao, Z., et al.: Efficient algorithm of energy minimization for heterogeneous wireless sensor network. In: IEEE EUC, pp. 25–34 (2006)
7. Gai, K., Qiu, M., Elnagdy, S.: A novel secure big data cyber incident analytics framework for cloud-based cybersecurity insurance. In: IEEE BigData Security (2016)
8. Qiu, H., Dong, T., Zhang, T., et al.: Adversarial attacks against network intrusion detection in IoT systems. IEEE IoT J. **8**(13), 10327–10335 (2020)
9. Qiu, M., Qiu, H.: Review on image processing based adversarial example defenses in computer vision. In: IEEE Conference on BigData Security, pp. 94–99, Baltimore, MD, USA (2020)

10. Creswell, A., White, T., Dumoulin, V., Arulkumaran, K., Sengupta, B., Bharath, A.A.: Generative adversarial networks: an overview. IEEE Signal Proc. Mag. **35**(1), 53–65 (2018)
11. Li, J., Ming, Z., et al.: Resource allocation robustness in multi-core embedded systems with inaccurate information. J. Syst. Arch. **57**(9), 840–849 (2011)
12. Qiu, M., Jia, Z., et al.: Voltage assignment with guaranteed probability satisfying timing constraint for real-time multiproceesor DSP. J. Signal Proc. Sys. **46**, 55–73 (2007)
13. Qiu, M., Yang, L., Shao, Z., Sha, E.: Dynamic and leakage energy minimization with soft real-time loop scheduling and voltage assignment. IEEE TVLSI **18**(3), 501–504 (2009)
14. Tantalaki, N., Souravlas, S., Roumeliotis, M.: A review on big data real-time stream processing and its scheduling techniques. Int. J. Parallel Emerg. Distrib. Syst. **35**(5), 571–601 (2020)
15. Memeti, S., Pllana, S., Binotto, A., et al.: Using meta-heuristics and machine learning for software optimization of parallel computing systems: a systematic literature review. Computing **101**, 893–936 (2019)
16. Isah, H., Abughofa, T., Mahfuz, S., et al.: A survey of distributed data stream processing frameworks. IEEE Access **7**, 154300–154316 (2019)
17. Calavaro, C., Russo, G.R., Cardellini, V.: Real-time analysis of market data leveraging Apache Flink. In: 16th ACM International Conference on Distributed and Event-Based Systems (DEBS), New York, NY, USA, pp. 162–165 (2022)
18. Xu, B., Jiang, J., Ye, J.: Information intelligence system solution based on Big Data Flink technology. In: 2022 ACM 4th International Conference on Big Data Engineering (BDE), New York, NY, USA, pp. 21–26 (2022)
19. HoseinyFarahabady, M.R., Jannesari, A., et al.: Q-Flink: A QoS-Aware Controller for Apache Flink. In: 20th IEEE/ACM Symposium on Cluster, Cloud and Internet Computing (CCGRID), pp. 629–638 (2020)
20. Ha, T.W., Kang, J.M., Kim, M.H.: Real-time deep learning-based anomaly detection approach for multivariate data streams with Apache Flink. In: Bakaev, M., Ko, I.-Y., Mrissa, M., Pautasso, C., Srivastava, A. (eds.) ICWE 2021. CCIS, vol. 1508, pp. 39–49. Springer, Cham (2022). https://doi.org/10.1007/978-3-030-92231-3_4
21. Dong, Y., Wang, R., He, J.: Real-time network intrusion detection system based on deep learning. In: IEEE 10th International Conference on Software Engineering and Service Science (ICSESS), pp. 1–4 (2019)
22. Del Monte, B., Prodan, R.: A scalable GPU-enabled framework for training deep neural networks. In: 2nd International Conference on Green High Performance Computing (ICGHPC), pp. 1–8 (2016)
23. Hu, F., Lakdawala, S., et al.: Low-power, intelligent sensor hardware interface for medical data preprocessing. IEEE Trans Inf. Tech. Biomed. **13**(4), 656–663 (2009)
24. Qiu, H., Zheng, Q., et al.: Topological graph convolutional network-based urban traffic flow and density prediction. IEEE Trans. Intel. Trans. Sys. **22**(7), 4560–4569 (2021)
25. Li, Y., Song, Y., et al.: Intelligent fault diagnosis by fusing domain adversarial training and maximum mean discrepancy via ensemble learning. IEEE Trans. Indus. Inform. **17**(4), 2833–2841 (2021)

GenGLAD: A Generated Graph Based Log Anomaly Detection Framework

Haolei Wang, Yong Chen, Chao Zhang, Jian Li, Chun Gan, Yinxian Zhang, and Xiao Chen(✉)

State Grid Zhoushan Electric Power Supply Company of Zhejiang Power Corporation, Zhoushan 316021, China
18768070482@163.com

Abstract. Information systems record the current states and the access records in logs, so logs become the data basis for detecting anomalies of system security. To realize log anomaly detection, frameworks based on text, sequence, and graph are applied. However, the existing frameworks could not extract the complex associations in logs, which leads to low accuracy. To meet the requirements of the hyperautomation framework for log analysis, this paper proposes GenGLAD, a generated graph based log anomaly detection framework. The generated graph is used to express the log associations, and the node embedding of the generated graph is obtained based on random walk and word2vec. Finally, we use clustering to realize unsupervised anomaly detection. Experiments verify the detection effect of GenGLAD. Compared with the existing detection frameworks, GenGLAD achieves the highest accuracy and improves the comprehensive detection effect.

Keywords: Log anomaly detection · Graph learning · Hyperautomation

1 Introduction

Information technology [14,31] and computer capability [30,36,37] promoted the machine learning [11,34,39] and intelligent development [26,27] of various industries. Many enterprises and institutions rely on open environments to carry out business. To deal with the automatic and diversified network attacks [24,28, 35] in open environments, a series of network security devices and systems are deployed on the network boundary to ensure the security of the business system within the boundary. The security system [6,8,15] prompts the administrator for network attacks, SQL injection, and other abnormal or malicious behaviors in the network environment through access logs and accompanying alarm labels.

With the proposal of the hyperautomation framework [2], automatic analysis of logs has become a major demand in the industry. However, traditional log analysis frameworks rely on manual analysis. Through simple statistics of the log content, the key information such as the source IP or source user and the

M. Qiu et al. (Eds.): SmartCom 2022, LNCS 13828, pp. 11–22, 2023.
https://doi.org/10.1007/978-3-031-28124-2_2

attack duration in the log is extracted, and the authenticity of each logline is determined by combining the prior knowledge such as the sensitivity of the source IP.

These frameworks face a series of problems in real-world scenes, including 1) the volume of logs is huge [5,29,46]. There are many kinds of logs in the existing system, but most kinds of log files contain a large number of lines [43], which exceeds the capacity limit of manual analysis; 2) Most of the logs are low-level and useless [22], which are not triggered by real malicious behaviors. Malicious behaviors [32,33] are hidden in a large number of messy logs and invalid alarms, resulting in great security vulnerabilities [7,9]. Therefore, how to detect anomaly logs that represent malicious behaviors is the key to maintaining network security.

To meet the need for hyperautomation, a series of representation learning technologies based on machine learning or deep learning, such as TCN [1] and LSTM [17], have been proposed and applied to the detection of various logs [3,41] or optimization for security systems [38]. On this basis, the detection methods based on graph models such as log2vec [18] are proposed. The complex associations in original logs are characterized by the graph structure. After obtaining appropriate representations of loglines or network entities, the detection of anomaly logs could be realized through a relatively simple classification method.

Therefore, we propose GenGLAD, a log anomaly detection framework based on the generated graph. We use the generated graph model to characterize the original log associations, optimizing the existing generated graph construction method. The initial attribute of the generated graph node is determined by setting the key features of the log, and the detection of anomaly logs is realized based on clustering. Through experiments on public simulation datasets, the availability and high accuracy of GenGLAD are proved.

The main contributions of this paper are as follows:

- The construction method of the generated graph is optimized, and the number of edges is reduced. as a result, the speed of model training is improved.
- GenGLAD, a novel generated graph based anomaly detection framework for logs is proposed, which can effectively detect the anomaly logs out of a large number of logs.
- The detection effect of GenGLAD exceeds that of the popularly used methods.

The rest of this paper is organized as follows: Sects. 2 introduce the related work. Section 3 describes the framework of GenGLAD, while Sect. 4 shows the experiments. Finally, Sect. 5 summarizes the work and discusses future work.

2 Related Works

Our research belongs to the field of log anomaly detection. The existing methods could be divided into three categories: text-based methods, sequence-based methods, and graph-based methods.

2.1 Text Based Methods

Logs are semi-structured text data. Therefore, researchers have realized the anomaly detection of logs by migrating methods in *Natural Language Processing* (NLP), word embedding, and other fields. The most typical idea of the text-based methods is to analyze the keywords contained in the log and the related word frequency of the recorded access behaviors [12], considering the significantly different text features as anomalies. However, such methods could only directly use text features, lacking the ability to characterize the high-level features and deep associations contained in logs, so the detection accuracy is not as high as that of other kinds of methods.

2.2 Sequence Based Methods

Logs are real-time records of systems, so it is naturally a kind of time-series data. Deeplog [3] regards the normal logs as a sequence with a certain pattern, and learns the normal log sequence based on LSTM, to analyze the abnormal possibility when a new logline is recorded. On this basis, the technology migration of the GRU classifier and full connection layer further improves the detection index [42].

Transformer framework, which shows unparalleled sequence learning ability has also been applied in the field of log detection [44], and realizes feasible log detection. In addition, the generative adversarial network is also directly applied to the log detection scenario [4,13], and the method migration of attention mechanism realizes the effective detection of anomalies in logs. However, sequence based methods could only analyze the associations between loglines from the perspective of time series, and could not extract and analyze the complex associations between days, resulting in a decrease in accuracy.

2.3 Graph Based Methods

Thanks to the representation ability of graph structure, graph based frameworks are used to model original logs. Graph anomaly detection algorithms are migrated into the field of log anomaly detection. Log2vec [18] defines the node construction rules and edge link rules of the log generation graph, detecting anomalies with high accuracy through the direct definition of log associations. The use of a heterogeneous graph further improves the graph embedding effect of the algorithm.

On the other hand, the provenance graph based methods derived from network attack detection are also migrated into the field of log anomaly detection [10]. This kind of method constructs a provenance graph that completely describes the behaviors of each IP or user, and analyzes the behaviors based on the critical paths in the graph. The graph based methods could extract the complex association between logs, and effectively improve the accuracy of existing detection methods. However, existing graph construction methods would lead to an excessive number of nodes and edges. The accuracy of the relevant graph anomaly detection methods also needs to be further improved.

3 GenGLAD: Detection Framework

GenGLAD includes two main steps, as shown in Fig. 1. The first step is to construct a generated graph based on raw logs, and the second step is to detect anomaly nodes based on the generated graph, to detect the anomaly logs.

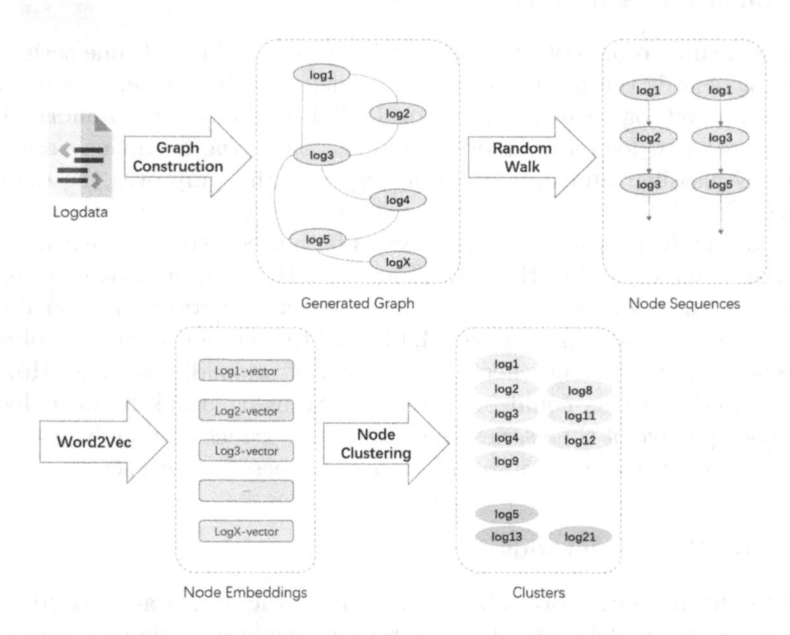

Fig. 1. The framework of GenGLAD. There are two main steps: 1) Construct the generated graph based on original logs; 2) Perform node anomaly detection, including node sequences calculation by random walk, embedding vectors calculated by word2vec, and unsupervised anomaly detection based on clustering.

3.1 Generated Grapg Construction

We use each node in the graph to represent a logline. Therefore, the edges between nodes represent the association between loglines. Through the construction method, we characterize the associations between logs on the graph.

Nodes. Each node corresponds to a logline raw logs, so if a node is detected as an anomaly node, its corresponding logline is an anomaly log. Based on this node definition, we can assign the attributes and label to each node.

The attributes of each node are the string of the corresponding logline, that is, the feathers of the access behavior recorded in the log. The most important attributes include 1) Source entity of the behavior, such as an IP address or user; 2) Destination entity of the behavior; 3) Type of the behavior; 4) Time when the behavior occurred.

The labels include normal and abnormal. GenGLAD is an unsupervised detection framework, so the initial tags are set to normal.

Edges. In anomaly detection scenarios, if behavior is more closely related to a known anomaly behavior, it is more likely to be abnormal [40]. Therefore, the edges in the graph should express the associations between loglines corresponding to the nodes. The existing method defines 10 connection rules between nodes to achieve this [18].

However, the existing rules are relatively complex and are not suitable for the single log style in real-world scenes. Most anomaly behaviors in real-world scenes represent potential attacks, so the destination is the core parameter. In addition, anomaly behaviors usually show concentration in time dimension [25].

Therefore, we improve the connection rules so that they could be applied to the anomaly detection tasks of most kinds of logs.

The connection rules we define are as follows. To make the expression more concise, the key information of the log line corresponding to a node is called the information of the node, such as time and source entity.

- *Rule1: Connections within one day.* Within the same day, all nodes are connected in chronological order, and nodes with the same source entity or action type are connected.
- *Rule2: Connections between days.* Daily node sequences are connected in chronological order, and sequences with at least one same source entity or action type are connected.
- *Rule3: Connections based on destination.* Nodes with the same destination entity are connected.

The simplified rules strengthen the portability of the method, and reduce the number of edges in the generated graph. Thereby reducing the time consumption of model training.

3.2 Node Anomaly Detection

Labels of unknown nodes could be obtained through the node anomaly detection framework on the generated graph, to infer the anomaly logs existing in the original data. We use the random walk algorithm on the graph to obtain the node sequence [45], and migrate the word2vec algorithm from the NLP domain to realize graph embedding [19,20]. Finally, we can obtain the abnormal node set by clustering the embedding vectors.

Random Walk. Most of the existing random walk methods [45] could be applied to heterogeneous graphs, heterogeneous information networks, and knowledge graphs. The generated graph is a static isomorphic graph, so the transition probability of random walk needs to be adjusted. Specifically, we adjust the transfer probability to:

$$P(t|v) = \begin{cases} \frac{1}{\boldsymbol{N}(v)}, & (t, v) \in E \\ 0, & otherwise \end{cases} \tag{1}$$

where $\boldsymbol{N}(v)$ denotes specific neighbor nodes of node v.

It is proven that the random walk sequence generated by focusing on only one kind of association could achieve the best effect in generating graph anomaly detection, the best practice walk sequence length is also provided [18].

Embedding Based on Word2vec. The word2vec algorithm, which is migrated into the graph learning field, is a coding method that embeds nodes into vectors and makes the embedding vectors obtained by nodes with similar attributes as close as possible [21]. It aims to maximize the probability of the neighbors conditioned on a node. For node n_v, in node list $n_{v-c}, \ldots, n_{v+c}$, The objective function to be maximized is:

$$\sum_{v=1}^{V} logP(n_{v-c}, \ldots, n_{v+c}|n_v) \tag{2}$$

We regard logging as independent and identically distributed events, so the probability in formula 2 could be converted into the product of a series of probabilities. In addition, softmax function is used to define function P. Therefore, the objective function could be calculated as:

$$\frac{e^{V_{n_v}^T V'_{n_{v+j}}}}{\sum_{i=1}^{V} e^{V_{n_v}^T V'_{n_i}}} \tag{3}$$

where V_{n_i} represents the input vector of node n_i, and V'_{n_i} represents the output vector of node n_i. In the process of i increasing from 1 to V, formula 3 calculates the embedding results of all nodes.

Anomaly Detection Method. Graph embedding based on random walk and word2vec makes the distribution of embedding vectors easy to distinguish. Therefore, the unsupervised clustering method could be used to detect anomaly nodes. Based on the satisfactory embedding results, we adopt a simple distance based clustering method. Specifically, we add all nodes to the initial set N_0, and let sets $N_1, N_2, \ldots$ be empty. Then, check whether each node meets condition 4 and condition 5.

$$\forall n_1 \in N_i, \forall n_2 \in N - N_i, dis(n_1, n_2) \geq d_0 \quad (4)$$

$$\forall n_1 \in N_i, \exists n_2 \in N_i \ s.t. \ dis(n_1, n_2) \leq d_0 \quad (5)$$

where N represents the set of all nodes, $dis()$ is the distance between embedding vectors of two nodes, and d_0 is a distance threshold. If a node does not meet the requirements of the conditions, move the node to an existing or new set to make it meet the requirements. Condition 4 makes the distance between nodes in different clusters relatively far, and condition 5 makes nodes in the same cluster have close embedding vectors so that the nodes in each cluster would have the same label.

4 Experiments and Results

4.1 Experimental Setup

Datasets. To verify the detection effect of GenGLAD, we use CERT [16], an open synthetic dataset for testing. CERT contains many different log files, describing more than 100 million behaviors of 4000 users. It covers the logs of device interaction, e-mail, file system, and so on. It covers the logs of device interaction, e-mail, file system, and other aspects. We select the device login and logout logs in version r4.2 to simulate the logs with insufficient data and features in real-world scenes. Specifically, we selected device interaction logs, recording the users' login actions on PCs for a span of 45 days. The main fields include device ID, user ID, action type, and time.

Baselines. We adopted three representative log anomaly detection methods as baselines: one-class *Support Vector Machine* (SVM), *Gaussian Mixture Model* (GMM), and Deeplog.

- *SVM* [23]. SVM trains a non-probabilistic binary linear classifier, which represents the logs as points in the space, and obtains the classification plane through the training process. Then SVM maps a new log to the same space and predicts its label based on which side of the classification plane it falls on.
- *GMM* [47]. GMM is one of the most widely used statistical methods. It uses maximum likelihood estimation to estimate the mean and variance of Gaussian distribution. Several Gaussian distributions are combined to represent the feature vectors and find out the anomalies.
- *Deeplog* [3]. Deeplog regards the logs as a sequence, calculates the type of each log through an analysis algorithm, and determines whether the newly generated log is an anomaly log based on a sequence learning framework, such as LSTM.

Parameters Selection. We use 40,000 device interaction logs, containing 1,154 anomaly ones. In the process of random walk, we choose a walk length of 60 and only focus on the edges generated by the same rule in each walking path. During node embedding, the dimension is set as 100 and the window length is 10. We set the threshold based on the average distance in the process of clustering.

Metrics. We use common metrics in the field of anomaly detection to measure the detection effect, including: $accuracy = \frac{TP+TN}{TP+TN+FP+FN}$, $recall = \frac{TP}{TP+FN}$, $precision = \frac{TP}{TP+FP}$. And F1 score for comprehensive evaluation.

$$F1 = 2 \cdot \frac{recall \cdot precision}{recall + precision} \tag{6}$$

where TP represents true positive, FP represents false positive, TN represents true negative, and FN represents false negative. We also use the *Receiver Operating Characteristic* (ROC) curve and *Area Under the Curve* (AUC) to evaluate the detection effect of GenGLAD. The closer the AUC value is to 1, the better the detection effect is.

4.2 Results

Figure 2 shows the ROC curve of GenGLAD.

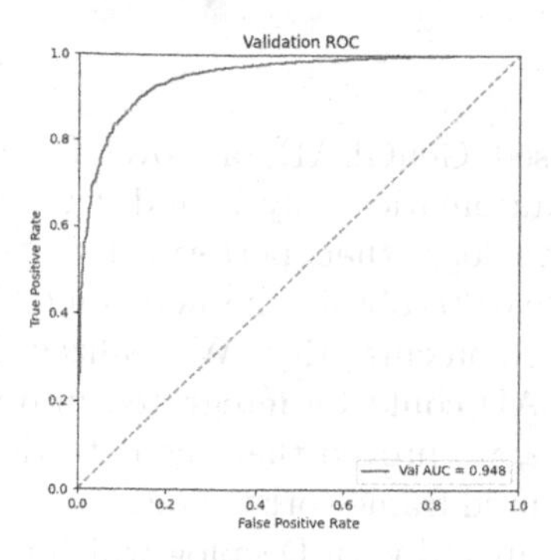

Fig. 2. ROC curve of the detection result of GenGLAD with AUC 0.948.

It is shown that the AUC value exceeds 0.948, indicating that GenGLAD has obtained effective detection results. However, in the anomaly detection scenario, the sample proportion of datasets is not balanced, and the detection effect could not be comprehensively evaluated only by the ROC curve. Therefore, it is necessary to compare various metrics in detail. Table 1 shows the detection results of GenGLAD and baselines on the CERT dataset.

Table 1. Detection results

Framework	Accuracy	Recall	Precision	F1
GenGLAD	**0.9273**	**0.6646**	0.7973	**0.7249**
SVM	0.6032	0.4474	**0.8095**	0.5763
GMM	0.5780	0.4168	0.1109	0.1752
Deeplog	0.9039	0.6310	0.6742	0.6519

It is shown that GenGLAD obtains the highest F1 score, which indicates that it has the best comprehensive detection effect. GenGLAD is superior to all baselines in accuracy and recall. However, for precision, SVM has better performance. This might be due to the more strict classification of SVM as a linear classifier, which makes it get a low false positive rate with low accuracy. As shown in Table 1, the accuracy of GMM is the lowest among all methods.

It is also shown that both GenGLAD and Deeplog have achieved relatively high accuracy, which means that deep learning helps to improve the effect of log anomaly detection. Among the baseline methods, GMM could only predict the distribution of normal logs, and SVM could only provide a linear classification surface, so these two methods could not achieve high detection accuracy. On the other hand, Deeplog regards logs as time series data, which could capture the correlation in the time dimension. In contrast, GenGLAD captures more correlations between logs, so it achieves the highest accuracy rate, although this results in higher time complexity.

5 Conclusion

In this paper, we proposed GenGLAD, a novel framework for log anomaly detection. To realize the automatic analysis and detection of logs, we first constructed a graph based on logs, then performed a random walk on the generated graph, using word2vec to obtain the node embedding vector, and finally detected anomaly logs based on clustering. We realized the automatic processing of logs and made GenGLAD could be integrated into a hyperautomation system. Through experiments, we proved that the detection effect of GenGLAD is better than the popularly used frameworks. The accuracy reached 0.927, and the AUC value was 0.948. Compared with Deeplog which is widely used, GenGLAD achieved an improvement of about 11% in F1 score. In further research, we plan to adjust relevant parameters to further improve the detection effect. In addition, we consider using deep neural network based methods like GCN to embed nodes in an attempt to obtain better embedding vectors, so that the clustering results could accurately reflect the distribution of anomaly nodes.

Acknowledgements. This work was supported by State Grid Zhoushan Electric Power Supply Company of Zhejiang Power Corporation under grant No. B311ZS220002 (Research on hyperautomation for information comprehensive inspection).

References

1. Bai, S., Kolter, J.Z., Koltun, V.: An empirical evaluation of generic convolutional and recurrent networks for sequence modeling. CoRR abs/1803.01271 (2018)
2. Bornet, P., Barkin, I., Wirtz, J.: Intelligent Automation: Welcome to the World of Hyperautomation - Learn How to Harness Artificial Intelligence to Boost Business & Make Our World More Human. WorldScientific (2021)
3. Du, M., Li, F., Zheng, G., Srikumar, V.: DeepLog: anomaly detection and diagnosis from system logs through deep learning. In: ACM SIGSAC Conference on Computer and Communication Security, pp. 1285–1298 (2017)
4. Duan, X., Ying, S., Yuan, W., Cheng, H., Yin, X.: A generative adversarial networks for log anomaly detection. Comput. Syst. Sci. Eng. **37**(1), 135–148 (2021)
5. Gai, K., Du, Z., et al.: Efficiency-aware workload optimizations of heterogeneous cloud computing for capacity planning in financial industry. In: IEEE 2nd CSCloud (2015)
6. Gai, K., Qiu, M., Elnagdy, S.: A novel secure big data cyber incident analytics framework for cloud-based cybersecurity insurance. In: IEEE BigData Security Conference (2016)
7. Gai, K., Zhang, Y., et al.: Blockchain-enabled service optimizations in supply chain digital twin. In: IEEE TSC (2022)
8. Gai, K., et al.: Electronic health record error prevention approach using ontology in big data. In: IEEE 17th HPCC (2015)
9. Gao, X., Qiu, M.: Energy-based learning for preventing backdoor attack. In: KSEM (3), pp. 706–721 (2022)
10. Han, X., Pasquier, T.F.J., Bates, A., Mickens, J., Seltzer, M.I.: Unicorn: runtime provenance-based detector for advanced persistent threats. In: 27th Network and Distributed System Security Symposium, NDSS 2020 (2020)

11. Hu, F., Lakdawala, S., et al.: Low-power, intelligent sensor hardware interface for medical data preprocessing. IEEE Trans. Inform. Tech. Biomed. **13**(4), 656–663 (2009)
12. Kent, A.: Cyber security data sources for dynamic network research. In: Dynamic Networks and Cyber-Security, pp. 37–65 (05 2016)
13. Kulyadi, S.P., Mohandas, P., et al.: Anomaly detection using generative adversarial networks on firewall log message data. In: 13th IEEE Conference on Electronics, Computers and Artificial Intelligence ECAI, pp. 1–6 (2021)
14. Li, J., Ming, Z., et al.: Resource allocation robustness in multi-core embedded systems with inaccurate information. J. Sys. Arch. **57**(9), 840–849 (2011)
15. Li, Y., Gai, K., et al.: Intercrossed access controls for secure financial services on multimedia big data in cloud systems. In: ACM TMCCA (2016)
16. Lindauer, B.: Insider threat test dataset (2020). https://kilthub.cmu.edu/articles/dataset/Insider_Threat_Test_Dataset/12841247
17. Lindemann, B., Maschler, B., Sahlab, N., Weyrich, M.: A survey on anomaly detection for technical systems using LSTM networks. Comput. Ind. **131**, 103498 (2021)
18. Liu, F., Wen, Y., et al.: Log2vec: a heterogeneous graph embedding based approach for detecting cyber threats within enterprise. In: ACM SIGSAC Conference on Computer and Communications Security, pp. 1777–1794 (2019)
19. Mikolov, T., Chen, K., Corrado, G., Dean, J.: Efficient estimation of word representations in vector space. In: Bengio, Y., LeCun, Y. (eds.) 1st International Conference on Learning Representations, ICLR 2013, Workshop Track Proceedings (2013)
20. Mikolov, T., Sutskever, I., et al.: Distributed representations of words and phrases and their compositionality. In: Advances in Neural Information Processing Systems, vol. 26. Curran Associates, Inc. (2013)
21. Moon, G.E., Newman-Griffis, D., et al.: Parallel data-local training for optimizing word2vec embeddings for word and graph embeddings. In: IEEE/ACM Workshop on Machine Learning in High Performance Computing Environment, MLHPC@SC, 2019, pp. 44–55 (2019)
22. Nehinbe, D.J.: A review of technical issues on ids and alerts. Global J. Comput. Sci. Technol. **17**, 55–62 (2018)
23. Nguyen, T.-B.-T., Liao, T.-L., Vu, T.-A.: Anomaly detection using one-class SVM for logs of juniper router devices. In: Duong, T.Q., Vo, N.-S., Nguyen, L.K., Vien, Q.-T., Nguyen, V.-D. (eds.) INISCOM 2019. LNICST, vol. 293, pp. 302–312. Springer, Cham (2019). https://doi.org/10.1007/978-3-030-30149-1_24
24. Niu, J., Gao, Y., et al.: Selecting proper wireless network interfaces for user experience enhancement with guaranteed probability. JPDC **72**(12), 1565–1575 (2012)
25. Pawlicki, M., Kozik, R., Choras, M.: A survey on neural networks for (cyber-) security and (cyber-) security of neural networks. Neurocomputing **500**, 1075–1087 (2022)
26. Qiu, H., Dong, T., et al.: Adversarial attacks against network intrusion detection in IoT systems. IEEE IoT J. **8**(13), 10327–10335 (2020)
27. Qiu, H., Zheng, Q., et al.: Topological graph convolutional network-based urban traffic flow and density prediction. IEEE Trans. ITS (2020)
28. Qiu, M., Chen, Z., et al.: Energy-aware data allocation with hybrid memory for mobile cloud systems. IEEE Sys. J. **11**(2), 813–822 (2014)
29. Qiu, M., Gai, K., Xiong, Z.: Privacy-preserving wireless communications using bipartite matching in social big data. FGCS **87**, 772–781 (2018)

30. Qiu, M., Jia, Z., et al.: Voltage assignment with guaranteed probability satisfying timing constraint for real-time multiproceesor DSP. J. Signal Proc. Sys. **46**, 55–73 (2007)
31. Qiu, M., Li, H., Sha, E.: Heterogeneous real-time embedded software optimization considering hardware platform. In: ACM Symposium on Applied Computing, pp. 1637–1641 (2009)
32. Qiu, M., Qiu, H.: Review on image processing based adversarial example defenses in computer vision. In: IEEE 6th International Conference on BigData Security, pp. 94–99 (2020)
33. Qiu, M., Qiu, H., et al.: Secure data sharing through untrusted clouds with blockchain-enabled key management. In: 3rd SmartBlock Conference, pp. 11–16 (2020)
34. Qiu, M., Sha, E., et al.: Energy minimization with loop fusion and multi-functional-unit scheduling for multidimensional DSP. JPDC **68**(4), 443–455 (2008)
35. Qiu, M., Xue, C., Shao, Z., et al.: Efficient algorithm of energy minimization for heterogeneous wireless sensor network. In: IEEE EUC Conference, pp. 25–34 (2006)
36. Qiu, M., Xue, C., et al.: Energy minimization with soft real-time and DVS for uniprocessor and multiprocessor embedded systems. In: IEEE DATE Conference, pp. 1–6 (2007)
37. Qiu, M., Yang, L., Shao, Z., Sha, E.: Dynamic and leakage energy minimization with soft real-time loop scheduling and voltage assignment. IEEE TVLSI **18**(3), 501–504 (2009)
38. Qiu, M., Zhang, L., Ming, Z., Chen, Z., Qin, X., Yang, L.T.: Security-aware optimization for ubiquitous computing systems with SEAT graph approach. J. Comput. Syst. Sci. **79**(5), 518–529 (2013)
39. Shao, Z., Wang, M., et al.: Real-time dynamic voltage loop scheduling for multi-core embedded systems. IEEE Trans. Circuits Syst. II **54**(5), 445–449 (2007)
40. Wang, S., Balarezo, J.F., Kandeepan, S., Al-Hourani, A., Chavez, K.G., Rubinstein, B.: Machine learning in network anomaly detection: a survey. IEEE Access **9**, 152379–152396 (2021)
41. Wang, Z., Tian, J., Fang, H., Chen, L., Qin, J.: Lightlog: a lightweight temporal convolutional network for log anomaly detection on the edge. Comput. Netw. **203**, 108616 (2022)
42. Xie, Y., Ji, L., Cheng, X.: An attention-based GRU network for anomaly detection from system logs. IEICE Trans. Inf. Syst. **103D**(8), 1916–1919 (2020)
43. Zeng, L., Xiao, Y., Chen, H., Sun, B., Han, W.: Computer operating system logging and security issues: a survey. Secur. Commun. Netw. **9**(17), 4804–4821 (2016)
44. Zhang, C., Wang, X., Zhang, H., Zhang, H., Han, P.: Log sequence anomaly detection based on local information extraction and globally sparse transformer model. IEEE Trans. Netw. Serv. Manag. **18**(4), 4119–4133 (2021)
45. Zhang, H., Duan, D., Zhang, Q.: RWREL: a fast training framework for random walk-based knowledge graph embedding. In: ACAI 2021: 4th International Conference on Algorithms, Computing and Artificial Intelligence, pp. 67:1–67:6. ACM (2021)
46. Zhang, L., Qiu, M., Tseng, W., Sha, E.: Variable partitioning and scheduling for MPSOC with virtually shared scratch pad memory. J. Signal Proc. Sys. **58**(2), 247–265 (2018)
47. Zhou, F., Qu, H.: A GMM-based anomaly IP detection model from security logs. In: Qiu, M. (ed.) SmartCom 2020. LNCS, vol. 12608, pp. 97–105. Springer, Cham (2021). https://doi.org/10.1007/978-3-030-74717-6_11

A New Digital-Currency Model Based on Certificates

Wei-Tek Tsai[1,2,3,4], Weijing Xiang[1], Shuai Wang[1](✉), and Enyan Deng[3]

[1] Digital Society and Blockchain Laboratory, Beihang University, Beijing, China
{xwj9712,sy2006125}@buaa.edu.cn
[2] Arizona State University, Tempe, AZ 85287, USA
[3] Beijing Tiande Technologies, Beijing, China
{tsai,deng}@tiandetech.com
[4] Andrew International Sandbox Institute, Qingdao, China

Abstract. This paper proposes a blockchain-based digital-currency model based on certificates, and uses stablecoins as an example to illustrate the model. A stablecoin is anchored on a fiat currency, and often backed by 100% fiat such as USD. Unlike traditional cryptocurrency models, the blockchain system using this model does not store the assets directly, but instead store the certificates of these assets. People can trade digital certificates like trading digital currencies, but the value of these certificates in stored in custodian banks. The transaction process is slightly different from cryptocurrency transactions. The advantage of trading certificates is that the loss of these digital certificate does not mean the loss of value. If a private key is lost, a client can recover her assets from the custodian bank with a replacement certificate. Due to this nature, using this model, the center of digital economy remains at banks, rather than the blockchain network, as suggested by Digital-Currency Areas (DCA).

Keywords: Blockchains · Digital certificate · Stablecoin · Digital currency

1 Introduction

Digital currency receives significant attention recently with a flood of cryptocurrencies, stablecoins, CBDC (Central Bank Digital Currency) projects in the last 8 years. While these are all digital currencies, but they differ significantly from working mechanisms, economic models, and financial market implications [1].

However, all these models store the value in the underlying blockchain systems, and owners hold only the private keys of these on-chain digital assets. An owner never stores the digital value in her digital wallet, only the private key of the on-chain assets. This is inconsistent with the current financial market practices where owners either hold cash in their own wallets, or deposit money in a bank and the bank returns a passbook. If the passbook is lost, the owner can still recover the money by asking the bank to re-issue the passbook. In other words, the passbook is merely a certificate rather than actual value. But in cryptocurrencies, the value is stored on the blockchain, but the value can be accessed only by the right private key. The loss of private key means the loss of value.

M. Qiu et al. (Eds.): SmartCom 2022, LNCS 13828, pp. 23–34, 2023.
https://doi.org/10.1007/978-3-031-28124-2_3

This paper proposes a new digital-currency model where the value is still stored in a custodian bank, rather than stored in the blockchain network. Instead, the certificate (or the passbook) is stored in the blockchain network, and the owner holds the private key of the certificate. In this way, if the owner loses the private key, the owner can request the bank to re-issue the digital certificate to recover the value. Another important consideration is that current digital currencies follow two main designs: the token-based model 1 and the account-based model. The new model can fit with both models.

The contributions of this paper as follows:

- Propose a new digital-currency model with transaction processes, recovery protocols, and settlement processes (Sect. 3); This paper uses stablecoin as an example to illustrate the model. But the model can be used for other kinds of digital currencies such as CBDC or digital assets.
- Analyze various digital-currency models and provide detailed comparisons (Sect. 4).

This paper is organized in as follows: Sect. 2 discusses the token model and the account models as they are related to our proposed model; Sect. 3 uses a stablecoin to illustrate the new model including various protocols; Sect. 4 compares various digital currency models; Sect. 5 concludes this paper.

2 Two Different Payment Systems

In 2009, a seminal paper by Kahn and Roberds 3 presents two different payment models: "account-based" and “store-of-value" systems. In their description, the dichotomy boils down to the type of verification each system requires: authentication is at the heart of the account system, and anti-counterfeiting protection is at the heart of the store of value system. This suggests that authentication is an essential difference between account-based payment systems and store-of-value payment systems.

The token-based model has strong anonymity, much like passing banknotes from one person to another. Ownership changes with actual possession. And banknotes are the instruments, one can irreversibly pass control of value to another person, both parties do not need to prove their identities and creditability, nor to any third party. Therefore, under the token-based model, users have a higher degree of freedom and better privacy, but compliance and supervision are the issues.

2.1 Bitcoin System

Bitcoin system is often viewed as the first and pure token-based system, but it still fits well with the definition of an account-based system. One can treat a Bitcoin address is an account [4] and its private key is the proof of identity required to transact the account. Every time a user wants to spend Bitcoin, that user verifies their identity using a private key, and importantly, the user must follow the protocol provided by the Bitcoin system to verify their established identity. The Bitcoin system uses a UTXO [5] model with transaction records only and without traditional account information, thus Bitcoin’s blockchain is called “half ledger model”, and a Bitcoin account is used only once.

Likewise, Bitcoin fits well with the token-based system. When someone wants to spend Bitcoin, the protocol verifies its validity by tracking its history. The current transaction history is used to verify the validity of the "object" being transferred, and it is valid only if it has not been used. The balance in a certain "account" is not represented by a number, but consists of all the UTXOs related to the current "account" in the blockchain network. When the balance in an address needs to be calculated, all relevant blocks in the entire network need to be traversed. Therefore, if the system has only transaction records and no account information, it takes effort to trace these transactions.

2.2 Ethereum System

Ethereum [7], has a native cryptocurrency, Ether (ETH), used to pay for all transactions processed by the network. However, Ether is not an ERC-20 token6; rather it was an intrinsic part of the blockchain platform before any ERC-20 token existed. The primary function of Ethereum is electronic record keeping.

In Ethereum, the accounts and associated balances on any ERC-20 ledgers are distributed across the network of participating nodes, so the only place where ERC-20 tokens exist is on this network. While the software used to control user balances for these tokens is called a digital wallet, such wallets do not contain anything in units of value. Instead, wallets hold private keys, allowing their holders to authorize transactions on the blockchain platform in a similar way to putting a signature on a check. While tokens can be viewed or controlled by wallet software as long as they "exist", it exists only in a replicated database maintained by the computing nodes of the blockchain platform and in the form of account balances, not digital objects in the wallet software itself.

The Ethereum system is a full ledger model [8], where one address corresponds to one account with a balance. When querying the account balance, there is no need to perform the complicated computation on UTXO, and read the balance information in that address.

2.3 Regulation Concerns

For regulators, the token-based model is not conducive for compliance as tracing transactions among individuals. Under the account-based model, the STRISA system [9] can be used to supervise the personal wallet of the personal account, which can effectively prevent the risk of criminals using digital currency for money laundering.

In terms of clearing and settlement, a token-based system often has instantaneous settlement, i.e., transactions and settlement are done together as a single step. This is possible only if the digital currency can be treated as cash. In 2000 US Treasury Department Office of Comptroller of Currency (OCC) states that instantaneous settlement is a game changer in the financial market, and this is one of reasons that OCC pushes for stablecoin projects in the US [10]. But the PFMI (Principles of Financial Market Infrastructure) [11], a financial system design guide developed after 2008 the world financial crisis, states that these two processes should be separated, this reduces financial risks associated with trading particularly related to money laundering.

3 Digital-Certificate Model

3.1 Participants and Overall Process

Most stablecoin projects require equal amount of fiat currency deposited into a custodian bank to ensure that the stablecoin can be considered as cash. In traditional case, the value is now placed in the blockchain network, but in our case, a digital certificate is stored in the network, and its value is kept in the custodian bank.

A client uses the digital certificate to complete transactions; the stablecoin issuer obtains the transaction results from the blockchain consensus protocols, generates a new digital certificate, completes the transfer of the stablecoin ownership, and completes the transaction. This model has three types of players, namely:

1) Stablecoin (or another digital asset) issuer;
2) Virtual Asset Service Providers (VASPs), i.e., those entities that involve in digital currencies or digital assets on behalf of clients, and;
3) Clients.

A client may be an enterprise or an individual, and the issuer has been authorized to issue digital certificates. The holder obtains the corresponding digital certificate to indicate ownership. All stablecoin transactions are completed through verification, generation and validation of the digital certificates on the blockchain. The blockchain also maintains a list of valid and invalid certificates. The specific issues related to the issuance, circulation and retrieval of stablecoins are as follows:

1) The stablecoin issuer maintains a stablecoin list with the currency type, value, certificate ID, and ownership including time of ownership. Each time a stablecoin is issued, the corresponding value is recorded so that it can be retrieved in case of a private key is lost. Every stablecoin transaction is also recorded.
2) The first step of a transaction verifies the digital certificate is true including ownership, and the amount stated in the certificate. Once the transaction is to be completed, the original certificate is invalidated as it has been used, and generate a new digital certificate for the recipient to represent the new ownership.
3) When the stablecoin is lost and recovered, the client submits a recovery application. The blockchain first verifies that the client is indeed of the certificate and the certificate is true. Once these are proven, the blockchain puts the certificate ID in the list of invalid certificates, and notifies the issuer to review and confirm the accuracy of the stablecoin ownership, and generate a new digital certificate and returns it to the client.
4) A VASP provides local supports for clients including buying, selling digital certificates. A VASP does this by connecting to the blockchain maintained by the issuer. If a transaction is done within the same VASP, the VASP may process this transaction locally.

3.2 Digital Certificate Data Structure

The stablecoin issuer is authorized to issue stablecoin digital certificates, and the issuer maintains an issuance list, which is shown in Table 1. The list of information can be used by VASP during transactions, helps in recovering the value if the private key is lost. The VASP uses the VASP ID as the financial institution identification code to apply to the issuer to exchange a certain amount of fiat currency for a certain amount of stablecoins. The stablecoin issuer verifies the application and issues the corresponding number of digital certificates to the VASP.

Table 1. VASP Digital Certificate Data Structures

Data type	Description
Certificate ID	The certificate unique identifier
VASP ID	The identification code of the financial institution
Currency value	The value of the currency represented
Currency type	The type represents the currency represented, such as USD
Blockchain ID	The ID of the blockchain to which it belongs
Previous owner ID	The ID of the source of the digital certificate is used to record the previous owner of the stablecoin
Timestamp	The date on which ownership of the stablecoin was acquired
Valid bits	The identification of whether the digital certificate is valid
Extended bits	1 indicates that the digital certificate is followed by a linked stablecoin certificate belonging to the same owner, 0 indicates that this is the owner's last said digital certificate
The next certificate ID	The field exists when the extension bit is 1, and does not exist when the extension bit is 0

Enterprises or individuals obtain stablecoins from VASPs, and the ownership of stablecoins is transferred. After receiving applications, a VASP generates digital certificates with different identities according to different roles, such as enterprise or personal digital certificate. The data structure of those are similar and thus omitted.

3.3 Stablecoin Transfer Process

The process of transferring stablecoin is as follows:

1) The client initiates a stablecoin transfer application, and this may include digital certificate information, timestamps, involved blockchain ID, transfer target ID.
2) The blockchain system receives the stablecoin transfer application, reviews the enterprise/personal digital certificate in the application, determines whether the digital certificate ID is in the invalid list, whether the digital certificate is valid, and records the amount and transfer of the stablecoin in the digital certificate. Address, after the audit results and transaction records are agreed, record them on the blockchain, and notify the VASP to check the transaction results and audit results;

3) The VASP checks whether the transaction results for compliance, and approves the generation of new digital certificates. At the same time, the nodes on the blockchain invalidate the sender's old digital certificates, including:

   ① Put the digital certificate ID into the list of invalid certificates;
   ② Set the valid position of the digital certificate to 0.

4) The blockchain generates a new digital certificate, sends a stablecoin ownership transfer response to the client, and submits an ownership modification application to the stablecoin issuer. All relevant information will be recorded including new certificate ID, timestamp, blockchain ID.
5) The issuer receives the application for stablecoin transfer, and the blockchain ensures the validity of the transfer amount through the digital certificate of the VASP in the application for modification of stablecoin ownership, and at the same time agrees to the modification of stablecoin ownership. The node consensus on the update result is uploaded to the chain, and a successful response to the stablecoin transfer is generated.
6) The node informs the VASP that the transaction is successful, and the VASP receives a successful response to the modification of the stablecoin ownership and notifies the client side.

Figure 1 illustrates this process.

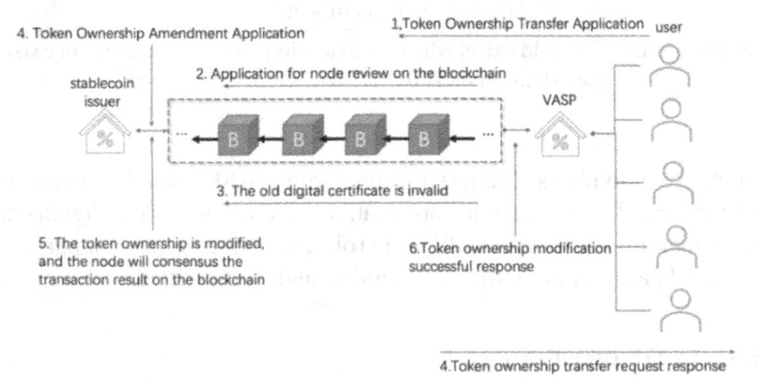

Fig. 1. Stablecoin transfer process

3.4 Stablecoin Recovering Process

The process of recovering the accidentally lost stablecoin digital certificate is as follows:

1) The enterprise/individual submits a digital certificate loss application to a VASP, finds the backup of the certificate ID in the wallet, and then generates a stablecoin certificate loss application. The data structure of the stablecoin certificate loss application includes business/personal ID, and Certificate ID.

2) After submitting the application for the loss of the stablecoin certificate, a two-step process is performed. The first step is to invalidate the original certificate, and the second step is to issue a new valid digital certificate. The processing of the original certificate invalidation includes:

   ① The VASP takes out the certificate ID from the application for lost digital certificate and asks the issuer to inquire whether the owner ID under the corresponding stablecoin is the enterprise/personal ID in the application. If so, the issuer agrees to grant the user to apply for a new digital ID the permission of the certificate;
   ② The blockchain obtains the processing result of the stablecoin issuer, and immediately puts the original digital certificate ID into the list of invalid certificates;
   ③ Return the successful result of invalidation processing to the client;
   ④ The client initiates a new-certificate generation request with relevant information such as IDs of involved parties.

3) The VASP issues new and valid certificates, including:

   ① The VASP receives a valid digital certificate generation request and confirms that a new digital certificate license can be obtained.
   ② The VASP broadcasts on the blockchain to notify the node to generate a new digital certificate;
   ③ The node on the blockchain records the new digital certificate generated this time as a transaction under the same user, and records the transaction result on the chain;
   ④ Return the newly generated digital certificate to the client.

   Figure 2 illustrates the process.

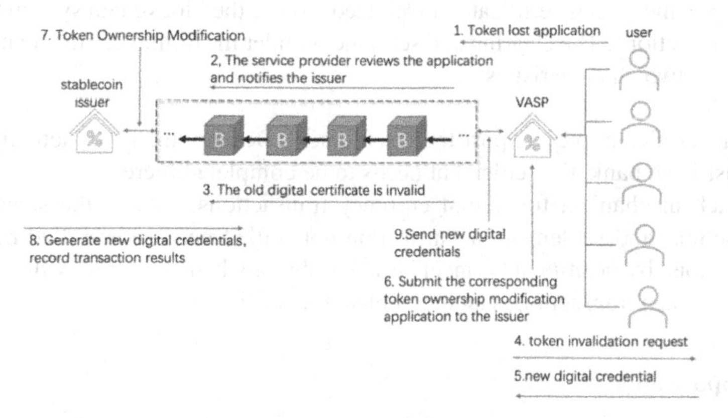

Fig. 2. Stablecoin recovery process

3.5 Digital Certificates for Transactions

VASP uses the VASP ID as the financial institution identification code to apply to the issuer to exchange a certain amount of legal currency for the same amount of stablecoins.

The stablecoin issuer verifies the application and issues a corresponding number of digital certificates to the VASP.

There are two situations in this certificate transaction, namely A → B and A → A. A → B is to transfer stablecoins from A to B. A → A is to apply for a new digital certificate after A loses the certificate. During the generation of a new digital certificate, if other customers obtain the old digital certificate, they will not be able to pass the verification process and will not be able to perform any transactions.

In the digital certificate model, there is a one-to-one correspondence between reserves and digital certificates. In fact, it is digital certificates that are traded on the blockchain. If there is an accident in the transaction, e.g., several nodes collapse at the same time, the issuer's system can facilitate this situation. The issuance record generates a new digital certificate, and the problem of stablecoin loss or invalidation will not occur.

3.6 Clearing and Settlement

In terms of settlement, whether the token is based on a token or an account-based system, the settlement can be achieved in one step. As the transaction can be settled immediately, the transaction can be completed quickly.

In the digital certificate model, settlement is divided into two steps: The first step is performed in the blockchain system, and the second step is completed in the custodian bank, and the second settlement is the final settlement. As all transactions are settled in banks, banks remain the center of financial market.

After settlement, the custodian may need to perform clearing possibly need to collaborate other custodian banks. These custodian banks may participate the blockchain associated with the issuer, or they may create and participate their own blockchain network to perform clearing.

Therefore, the digital certificate model needs to use the blockchain system to record all these transactions. The clearing and settlement under the digital certificate model has the following two characteristics:

1) Digital certificates require post-trade settlement. Because the funds actually exist in the custodian bank, the settlement needs to be completed there.
2) Rollback mechanism for digital currency transactions. Due to the separation of transactions and settlement, a transaction not settled need to be rolled back. This can be done by positing a "Cancel" trade to the blockchain system with indicating the concerned transaction is an incomplete transaction.

3.7 Comparison

The following table lists the differences between four models: token digital currency (such as Bitcoin), account digital currency, token digital certificate and account digital certificate (Table 2).

Table 2. Different digital-currency models

	Asset	Asset private key	Ledger	Features
Token digital currency	On the digital currency network	Token (private key) exists in the wallet	Token can only be used once in an account	Losing the private key, the assets on the network become orphans
Account digital currency	On the digital currency network	ID card identification account to identify asset owners by account	Accounts with full account books and complete history	Loss of ID card (such as biometric ID card) loss of assets
Token digital certificate	Real Assets in Custody Bank; Digital Certificates Left on Network	The digital certificate exists in the wallet	Accounts where digital certificates can only be used once	If the digital certificate is lost, the actual assets are still there. After the review, the lost handover will be invalidated, and then the digital certificate will be re-sent; the assets in the custodial bank will be taken out through the compliance process
Account digital certificate	Real Assets in Custody Bank; Digital Certificates Left on Network	Digital certificate transactions require digital certificates and ID cards	There is a full digital voucher ledger. Ledger with complete digital certificate history; Custodian bank maintains full account book and complete history of fiat currency	

4 Transactions at Exchanges

Section 3 discusses various aspects of the digital-certificate model including transactions at the blockchain network. However, most of the cryptocurrency trading now happen in exchanges, and the transaction process at these exchanges are different from the on-chain transaction processes. Table 3 compares trading at the blockchains and exchanges.

Table 3. Different trading methods

	Trading on the blockchain	Trading at exchanges
Bitcoin	Recognize token but not others, a token is verified by the Bitcoin network. If it passes, one can trade, the old token is no longer valid, and a new token is generated. Any Bitcoin assets still exist on the network, and the new token is stored in the wallet	Exchanges collect tokens, trade them themselves, and hand them back to customers after the last transaction on the Bitcoin network
Ethereum	Identify wallet accounts, transfers between wallets are verified by the Ethereum network, and if they pass, they can be traded. The balance of the transfer-outside decreased, and the balance of the transfer-in side increased	The exchange collects tokens, trades them several times between internal accounts, and finally returns them to the customer after a single transaction on the Ethereum network
Token digital currency	Wallet accounts are registered, and transfers between wallets are verified by the digital currency network. If they pass, transactions can be made. The balance of the transfer-outside decreased, and the balance of the transfer-in side increased	Tokens are collected by the exchange, exchanged several times between internal accounts, and finally returned to the customer after a single transaction on the digital currency network
Account digital currency	With an ID card linked to an account, transactions need to be verified by a third party. The balance of the transfer-outside decreased, and the balance of the transfer-in side increased	Exchanges collect identity information, make a number of trades between internal accounts, and periodically package the total results up the chain
Token digital certificate	Identify the wallet account, the transfer between wallets is verified by the digital voucher network, if passed, you can trade	The exchange collects the notes or identity information, makes sure they haven't been reported missing, makes a number of transactions between internal accounts, makes one transaction on the digital currency network, and continues to settle with the bank, which then sends the information to the customer

(*continued*)

Table 3. (*continued*)

	Trading on the blockchain	Trading at exchanges
Account digital certificate	With an ID card linked to an account, transactions need to be verified by a third party. The balance of the transfer-outside decreased, and the balance of the transfer-in side increased	

5 Conclusion

This paper presents a new digital-currency model based on digital certificates. Key advantages of this model is that assets are stored in custodian banks, and the loss of private key or tokens will not lose the asset value. This model fits better with current physical assets, e.g., real estate. As long as the real estate is still around, the value is not lost even if the corresponding private key is lost.

Another unique feature is related to regulation. Previously, all digital currency systems store value at the blockchain network, and it is necessary to manage and control each network individually. However, with the digital certificate model, the regulators can focus on custodian banks because ultimately, all transactions must be settled (and potentially also cleared) at these custodian banks. Thus, regulators can establish a firm foundation for regulation at the custodian banks while allowing financial technology innovations. This is not possible with traditional digital-currency models such as Bitcoin, Ethereum, and numerous other systems. In this way, this paper proposes a new system for regulating all the future digital currencies as long as they use the digital-certificate model.

Acknowledgment. This work is supported by Chinese Ministry of Science and Technology (Grant No. 2018YFB1402700).

References

1. Siahbidi Kermanshahi, S., Konari Zadeh, H.: An overview of digital currency regulation and its legal implications. Fares Law Res. **1**(1), 149–167 (2019)
2. Ferreira, A.: Emerging regulatory approaches to blockchain based token economy. J. Br. Blockchain Assoc. **3**(1), 1–9 (2020). https://doi.org/10.31585/jbba-3-1-(6)2020
3. Kahn, C.M., Roberds, W.: Why pay? An introduction to payments economics. J. Financ. Intermed. **18**(1), 1–23 (2009)
4. Nakamoto S. Bitcoin: A Peer-to-Peer Electronic Cash System. consulted
5. Delgado-Segura, S., PérezSolà, C., NavarroArribas, G., HerreraJoancomartí, J.: Analysis of the bitcoin utxo set. In: Zohar, A., et al. (eds.) Financial Cryptography and Data Security. LNCS, vol. 10958, pp. 78–91. Springer, Heidelberg (2019). https://doi.org/10.1007/978-3-662-58820-8_6
6. Liu, S.-H., Liu, X.F.: Co-investment Network of ERC-20 tokens: network structure versus market performance. Front. Phys. **9**, 631659 (2021). https://doi.org/10.3389/fphy.2021.631659

7. Buterin, V.: A next-generation smart contract and decentralized application platform. White Paper **3**(37), 2–1 (2014)
8. Wood, G.: Ethereum: a secure decentralised generalised transaction ledger. Ethereum Project Yellow Paper **151**(2014), 1–32 (2014)
9. Tsai, W.-T., Yang, D., Wang, R., Wang, K., Xiang, W., Deng, E.: STRISA: a new regulation architecture to enforce travel rule. In: Park, Y., Jadav, D., Austin, T. (eds.) Silicon Valley Cybersecurity Conference. CCIS, vol. 1383, pp. 49–67. Springer, Cham (2021). https://doi.org/10.1007/978-3-030-72725-3_4
10. Menand, L.: Why supervise banks? The foundations of the American monetary settlement. 74 VAND.L. REV. 951 (2021)
11. Mills, D.C., et al.: Distributed ledger technology in payments, clearing, and settlement. (2016)

Meter Location System Base on Jetson NX

Chengjun Yang[1], Ling zhou[1], and Ce Yang[2,3](✉)

[1] Hechi University, Hechi, Guangxi, China
05062@hcnu.edu.cn

[2] School of Computer Science and Engineering, Hunan University of Science and Technology, Xiangtan, Hunan, China
yangce@hnust.edu.cn

[3] Hunan Key Laboratory for Service computing and Novel Software Technology, Xiangtan, China

Abstract. Analog meter is still widely used due to their mechanical stability and electromagnetic impedance. Relying on humans to read mechanical meters in some industrial scenarios is time-consuming or dangerous, it is difficult for current meter reading robots to operate quickly and maintain high accuracy in edge computing devices. Computer vision-based meter reading systems can solve such dilemmas. We designed an SSD network-based meter image acquisition system that can run in real time in an NVIDIA Jetson NX development board. Moreover, the model can quickly classify meter types and locate meter coordinates in the presence of light changes, complex backgrounds, and camera angle deflection. Tested on NVIDIA Jetson NX using TensorRT acceleration, the inference speed and accuracy reached 9.238 FPS and 53.95 mAP, respectively.

Keywords: Analog meter · SSD Network · Computer vision · NVIDIA Jetson NX · NVIDIA TensorRT

1 Introduction

1.1 Background and Motivation

In recent years, with the development of computer [1–3] and cloud systems [4–6], the drive for automated monitoring and analysis is causing fully operational existing equipment to become obsolete, consuming significant economic and environmental costs due to the rapid development of industrial *Internet of Things* (IoT) [7–9] and big data technologies [9–11]. Digital instrumentation has replaced analog instrumentation in many scenarios. However, digital meters cannot be used in flammable, explosive, electromagnetic interference, high temperature, high pressure environments, etc. Due to the good mechanical stability and anti-electromagnetic interference performance of the analog pointer instrument, it is still widely used in petroleum, electric power, chemical, and other industries.

Supported by Hechi University.

M. Qiu et al. (Eds.): SmartCom 2022, LNCS 13828, pp. 35–44, 2023.
https://doi.org/10.1007/978-3-031-28124-2_4

The traditional manual method of collecting pointer-type meter readings [12–14] is not only inefficient and with low precision, but also labor-intensive. In addition, it is not real-time. In some extreme environments, e.g., radiation, high temperature, and high voltage, the stability of digital meters are poor and the manual reading of pointer meters is very inconvenient. Thus, it would be beneficial to have these analog systems connected to modern digital systems through an artificial intelligence-based meter reading system [15–17].

1.2 Prior Work and Limitations

The automatic reading system of computer vision can be roughly divided into three stages: identification and positioning, rotation correcting, and reading, in which most practices of the reading stage are calculated following the deflection angle of the pointer. For the first stages, various recognition and correction algorithms have been proposed for the problems of tilt, rotation, occlusion, distortion, and uneven illumination of the meter images collected in the field. However, the current algorithms generally fail to balance robustness, training cost, compensation correction, and reading accuracy. These methods fall into two main categories, one using traditional computer vision-based methods, such as hypergraph convolution [18–21], structural patterns [22–24], and self-attention [25–28]. The other using deep vision methods, such as dual channel [29–32], self-supervised graph [33–35], and hyperbolic hypergraph [36–39].

We consider the meter reading system with practical value meter reading system performance is mainly reflected in the following four aspects:

(1) **Accuracy.** Accuracy is mainly reflected in the key point positioning algorithm and reading algorithm design of the meter reading system, the meter reading system must be able to truly reflect the current system data, reflecting the real system situation. Previous researchers have done numerous studies on traditional vision-based meter reading systems, but a mass of engineering practice has proven that traditional vision cannot overcome environmental interference to achieve good accuracy in real application scenarios, and even with the use of image enhancement algorithms [40,41], e.g., Detection of meter pointers and scales in complex backgrounds. Compared with the very low robustness of traditional machine vision in real application scenarios, the deep vision applied to meter readings can effectively separate the meter from its natural environment.

(2) **Real-Time.** Factory meter location is relatively scattered, and the meter data is real-time changes, meter reading system as part of the overall project, if not real-time data reading and feedback, can not accurately reflect the overall project system situation, and can not be timely in the stress state for early warning. Traditional method-based methods are typically used to run more quickly, but traditional object detection and segmentation algorithms are less accurate and robust, susceptible to illumination and background clutter.

(3) **Stability.** Stability is reflected in both hardware design and software design. Hardware stability is mainly reflected in the ability to work stably in some extreme environments and can work stably for a long time under unattended conditions. Software stability is reflected in the visual recognition algorithm can adapt to various interference factors in the natural environment, to locate the key points of the instrument in a stable and accurate manner.
(4) **Compatibility.** Instrument types and specifications are diverse, if not compatible with a variety of instrument types, the system is difficult to popularize in practical applications. The table reading system using instance segmentation [38] has better generalization capability because of its larger parameter space, but its annotated dataset consumes more time [27,42] and is much more computationally expensive than the keypoint detection-based approach.

1.3 Our Contribution

To improve the training speed and inference accuracy of our analog meter reading system, We have designed a meter detection and localization system, which is based on a lightweight *Single-Shot MultiBox Detector* (SSD) [42], This system can pick keyframes for the gauge reading system in complex real-world environments and crop the dial area from the background according to bbox coordinates. Moreover, our system can run in real-time on edge computing devices, tested on NVIDIA Jetson NX using tensorRT [43] acceleration, the inference speed and accuracy reached 9.238 FPS and 53.95 mAP, respectively. And it maintains high accuracy even under light intensity change and angle deflection, etc.

1.4 Organization

This paper is organized as follows. Section 2 describes in detail the key technologies involved, Sect. 3 introduces our experimental environment and analyzes our experimental results, and finally in Sect. 4 we summarize our design and look forward to future research directions.

2 Main Methods

2.1 SSD-Based Meter Positioning

Compared with other single-stage object detection algorithms, the SSD model is used to classify and locate the meter. The SSD uses the VGG16 model as the backbone network, removes the Dropout layer and FC8 layer in the VGG16 model, and supplements the four convolutional layers with FC6 and FC7 as convolutional layers. The image or video is preprocessed to 300 × 300 × 3 images before being passed into the SSD network, the feature images of different levels are synthesized, the category and confidence of the default bounding box are calculated, and the target detection results are obtained by nonmaximal suppression. The image with <90% confidence was then selected as non-keyframe by the test result, and the high-definition picture was cropped before preprocessing following the bounding box coordinate scale to obtain a high-definition dial plate border image. Moreover, the unit and maximum scale values of the meter are obtained using the recognized meter type. The objective function used by the SSD network model is:

$$L(x,c,l,g) = \frac{1}{N}\left(L_{conf}(x,c) + \alpha L_{loc}(x,l,g)\right) \tag{1}$$

N is the number of target boxes; c is the index of the target class; x is an indicator function indicating whether the default bounding box matches the true bounding box; l and g indicate the prediction bounding box and the true bounding box, respectively; the positioning loss L_{loc} the loss between l and g is calculated using a smooth L_1 function; the confidence loss L_{conf} calculated by Softmax; and the α is a weight parameter. The specific definition of the L_{loc} and confidence loss L_{conf} is:

$$\mathrm{L}_{loc}(x,l,g) = \sum_{i\in Pos}^{N} \sum_{m\in loc_i} x_{ij}^{k} \operatorname{smooth}_{L1}\left(l_i^m - \hat{g}_i^m\right) \tag{2}$$

$$L_{\mathrm{conf}}(x,c) = -\sum_{i\in Pos}^{N} x_{ij}^{p} \log\left(\hat{c}_i^p\right) - \sum_{i\in Ne.g.} x_{ij}^{p} \log\left(\hat{c}_i^0\right), \hat{c}_i^p = \frac{\exp\left(c_i^p\right)}{\sum_p \exp\left(c_i^p\right)} \tag{3}$$

thereinto:

$$\hat{g}_j^{cx} = \frac{\left(g_j^{cx} - d_j^{cx}\right)}{d_i^w}, \hat{g}_j^{cy} = \frac{\left(g_j^{cy} - d_j^{cy}\right)}{d_i^h}, \hat{g}_j^{w} = \log\left(\frac{g_j^w}{d_i^w}\right), \hat{g}_j^{h} = \log\left(\frac{g_j^h}{d_i^h}\right)$$

d_j^{cx}, d_j^{cy}, d_i^w, and d_i^h contain the location information of the target; m represents the number of feature maps; and Pos,loc, and Neg represents the positive, negative, and bounding box coordinate position sets, respectively.

2.2 NVIDIA TensorRT

Even though the number of CUDA cores is the main constraint on the speed of model inference, a lot of time is also wasted on CUDA core startup and read/write operations for each layer of the input/output tensor, which results in memory bandwidth bottlenecks and wasted GPU resources. NVIDIA TensorRT makes a significant reduction in the number of layers by merging horizontally or vertically between layers, so that kernel launches and memory reads and writes can be reduced to some extent. Most deep learning frameworks use full 32-bit precision (FP32) for the tensor in the network when training neural networks. Once the network model is completed in training, it is entirely possible to reduce the data accuracy appropriately, for instance to FP16 or INT8 accuracy, during the deployment of inference as back propagation is not required. Lower data precision will result in lower memory usage and latency, and fewer model parameters.

INT8 has only 256 different values and using INT8 to represent FP32 precision values will definitely lose information and cause model accuracy degradation. TensorRT will provide a fully automated calibration process based on relatively entropic (also known as *Kullback-Leibler Divergence*, KLD) calculation that will reduce FP32 precision data to INT8 precision with the best matching performance, minimizing the performance loss of the model the KLD scatter is formulated as:

$$D_{KL}(P\|Q) = \sum_{i=1}^{n} P_i \log\left(\frac{P_i}{Q_i}\right) \tag{4}$$

The calculated KLDs are different for different network models when performing the conversion from FP32 to INT8. This requires a calibration dataset to pick the optimized activation value thresholds.

3 Experiment

3.1 Experimental Environment and Dataset

We conduct our experiments on a GPU Server equipped with NVIDIA 3090 GPU, 32 GB memory, and Intel I5-9400F Processor. The proposed model is implemented based on the PyTorch 1.8.0 and Python 3.8. In terms of model settings, the gradient descent algorithm for SSD model training, this paper selects the adaptive moment estimation algorithm (Adam), and the batch size is set to 16. The learning rate is multistep, the initial learning rate is 0.001, and the gamma value is set to 0.9.

The model was trained using a dataset consisting of 6 gauges, where the gauges Pressure-2.5bar, Pressure-1.5bar, Pressure-4.5bar, and Pressure-3bar from the video dataset provided by [27] were relabeled using the CVAT tool to obtain the dataset in coco format. These three meters have a small shooting camera deflection angle, the pointer deflection angle cannot contain the minimum scale to the maximum scale, and no overexposure and underexposure states. It is worth mentioning that the number of samples per class in our dataset is not balanced due to the fact that not all were taken by ourselves. There are 130 images for each of the three gauge types from [27], while our own collection of 1450 images of each of the three gauges (Oxygen-2.5bar, Nitrogen-2.5bar, and Propane-2.5bar) contains a variety of natural light and angular deflection scenarios. Total of 4745 images split into the training, validation, and test sets with a ratio of 80%, 20%, and 20%, respectively. The percentage of each class of our validation set is shown in Fig. 1.

3.2 Model Performance

We accelerated the whole training process by freezing the vgg16 network, the model was able to converge quickly and steadily, even with our small data sample. This illustrates that we only need a small amount of dataset to get the required model by transfer learning when we need to add a new instrument type to

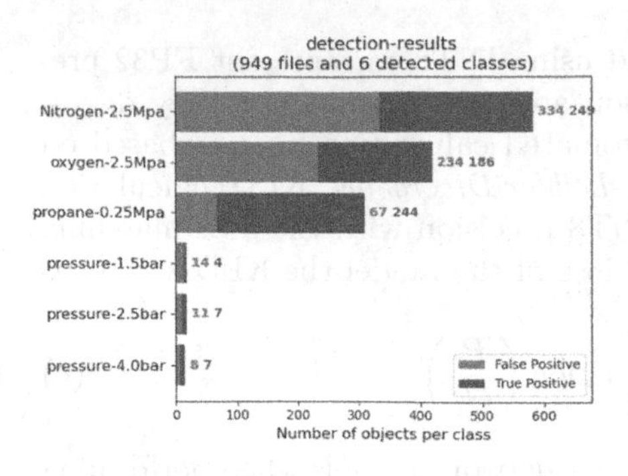

Fig. 1. Validation dataset composition

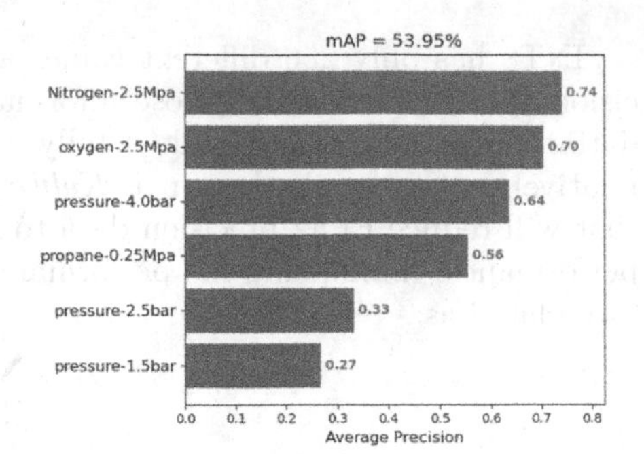

Fig. 2. mAP for each type of meter

the model. The accuracy of the three types of gauges collected from [27] is significantly lower due to the impact of unbalanced data distribution, as shown in Fig. 2. To further verify our point, we calculated the recall of various classes of meters, as shown in Fig. 3, where the sparsity of the positive samples leads to a lower recall for the rare three classes of meters with the same accuracy in explicit detail.

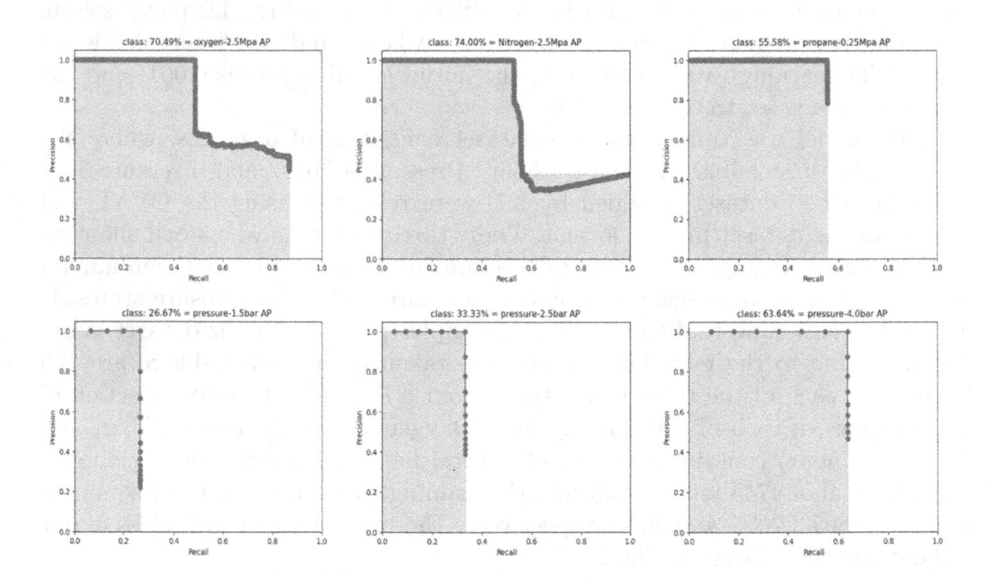

Fig. 3. Recall for each type of meter

We further deployed the model into NVIDIA Jetson NX to test the inference speed of the model in edge devices as well as the accuracy loss. The inference speed of the model stays above 9.238 FPS, which meets the demand of real-time detection, and the accuracy loss can be almost neglected by the acceleration optimization of the model through TensorRT.

3.3 Hyperparameter Sensitivity

Since our model is trained on an unbalanced dataset, because the 3 types of gauges are from the previous study so the samples are small, while we collected more samples of the 3 types of gauge types ourselves. [44] presents the theory shows that the SGD momentum is essentially a confounder in long-tailed classification. We believe it is necessary to perform erosion experiments on momentum, and we believe that the *learning rate* (lr) also directly affects the effect of momentum on the gradient descent process. Moreover, the computational complexity is high for deep neural network, one should consider the balance between model performance and the computation cost. We only selected two hyperparameters, learning rate and momentum decay coefficient, as the adjustment objects.

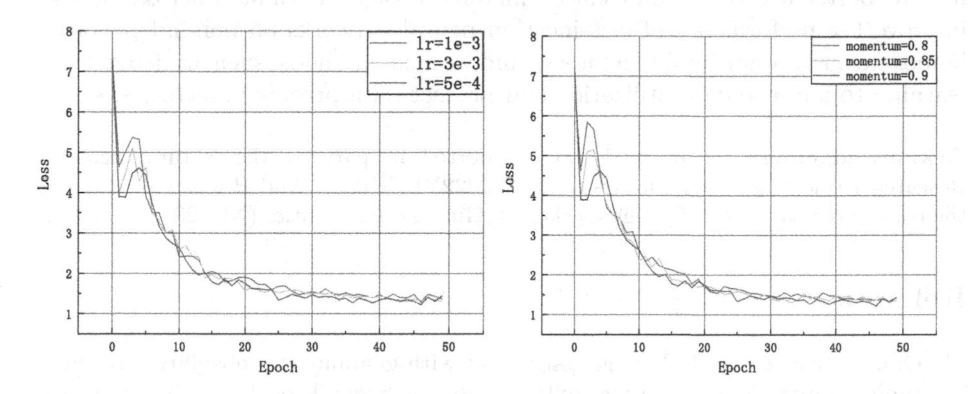

Fig. 4. The effect of learning rate on model convergence

Fig. 5. The effect of momentum on model convergence

First, we fixed momentum to 0.9 to test the effect of learning rate on the model training process, as shown in Fig. 4. The model converges fastest when learning rate takes the value of 0.0005 and we observe in the tensorboard that they have almost the same test loss. We further fixed the learning rate at 0.0005 to test the effect of momentum on the convergence effect of the model. As shown in Fig. 5, the model converges fastest when momentum takes the value of 0.9, the loss value swings slightly as it approaches 50 epochs. Experimental results show that our method has stable and fast convergence performance on unbalanced datasets.

4 Conclusions

We proposed a meter monitoring system based on SSD network to address the problem of slow speed and low accuracy of automatic meter reading systems on edge devices in the current industrial environment. Based on the NVIDIA TensorRT acceleration engine, the system is realized to run in real-time on the NVIDIA Jetson NX edge computing device, and the inference speed is preserved above 9.238 FPS. The average accuracy reaches 53.95 mAP in an unbalanced dataset.

5 Outlook

We realize that the collection of industrial meter datasets is a major hindrance to the development of visual meter reading systems, as these data are often trade secrets to companies. In our subsequent research, we will continue to delve into how to better utilize the unbalanced meter datasets distributed across plants, improve the performance of classification neural networks on non-independent homogeneously distributed datasets, and utilize methods such as federation learning to improve data utilization and protect data privacy in factories.

Acknowledgements. This work was supported in part by the Natural Science Research Project in Hechi University (NO. 2022YLXK003). And Project to improve the basic research ability of young teachers in Guangxi universities (NO. 2022KY0602).

References

1. Qiu, M., Jia, Z., et al.: Voltage assignment with guaranteed probability satisfying timing constraint for real-time multiproceesor. J. Signal Proc. Sys. DSP **46**, 55–73 (2007)
2. Qiu, M., Yang, L., Shao, Z., Sha, E.: Dynamic and leakage energy minimization with soft real-time loop scheduling and voltage assignment. IEEE TVLSI **18**(3), 501–504 (2009)
3. Qiu, M., Xue, C., et al.: Energy minimization with soft real-time and DVS for uniprocessor and multiprocessor embedded systems. In: IEEE DATE, pp. 1–6 (2007)
4. Qiu, M., Chen, Z., et al.: Energy-aware data allocation with hybrid memory for mobile cloud systems. IEEE Syst. J. **11**(2), 813–822 (2014)
5. Y. Li, K. Gai, et al. Intercrossed access controls for secure financial services on multimedia big data in cloud systems. In: ACM TMMCCA (2016)
6. Gai, K., Qiu, M., Elnagdy, S.: A novel secure big data cyber incident analytics framework for cloud-based cybersecurity insurance. In: IEEE BigData Security (2016)
7. Qiu, M., Liu, J., et al.: A novel energy-aware fault tolerance mechanism for wireless sensor networks. In: IEEE/ACM Conference on GCC (2011)
8. Niu, J., Gao, Y., et al.: Selecting proper wireless network interfaces for user experience enhancement with guaranteed probability. JPDC **72**(12), 1565–1575 (2012)

9. Qiu, M., Xue, C., Shao, Z., et al.: Efficient algorithm of energy minimization for heterogeneous wireless sensor network. In: IEEE EUC Conference, pp. 25–34 (2006)
10. Li, J., Ming, Z., et al.: Resource allocation robustness in multi-core embedded systems with inaccurate information. J. Syst. Arch. **57**(9), 840–849 (2011)
11. Hu, F., Lakdawala, S., et al.: Low-power, intelligent sensor hardware interface for medical data preprocessing. IEEE Trans. Inf. Technol. BioMed. **13**(4), 656–663 (2009)
12. Gai, K., Du, Z., et al.: Efficiency-aware workload optimizations of heterogeneous cloud computing for capacity planning in financial industry. In: IEEE CSCloud (2015)
13. Gai, K., Qiu, M., et al.: Electronic health record error prevention approach using ontology in big data. In: IEEE 17th HPCC (2015)
14. Zhang, L., Qiu, M., Tseng, W., Sha, E.: Variable partitioning and scheduling for MPSOC with virtually shared scratch pad memory. J. Signal Proc. Sys. **58**(2), 247–265 (2018)
15. Qiu, H., Dong, T., et al.: Adversarial attacks against network intrusion detection in IoT systems. IEEE IoT J. **8**(13), 10327–10335 (2020)
16. Qiu, H., Zheng, Q., et al.: Topological graph convolutional network-based urban traffic flow and density prediction. IEEE Trans. ITS (2020)
17. Qiu, H., Qiu, M., Lu, R.: Secure V2X communication network based on intelligent PKI and edge computing. IEEE Netw. **34**(2), 172–178 (2019)
18. Bao, H., Tan, Q., Liu, S., Miao, J.: Computer vision measurement of pointer meter readings based on inverse perspective mapping. Appl. Sci. **9**(18), 3729 (2019)
19. Sablatnig, R., Kropatsch , W.G.: Automatic reading of analog display instruments. In: Proceedings of 12th International Conference on Pattern Recognition, vol. 1, pp. 794–797. IEEE (1994)
20. Wang, J., Huang, J., Cheng, R.: Automatic reading system for analog instruments based on computer vision and inspection robot for power plant. In: 2018 10th International Conference on Modelling, Identification and Control (ICMIC), pp. 1–6. IEEE (2018)
21. Chen, Y.-S., Wang, J.-Y.: Computer vision-based approach for reading analog multimeter. Appl. Sci. **8**(8), 1268 (2018)
22. Mai, X., Li, W., Huang, Y., Yang, Y.: An automatic meter reading method based on one-dimensional measuring curve mapping. In: International Conference on Intelligent Robotics and Control Engineering (IRCE), pages 69–73 (2018)
23. Selvathai, T., Ramesh, S., Radhakrishnan, K.K., et al.: Automatic interpretation of analog dials in driver's instrumentation panel. In: 2017 Third International Conference on Advances in Electrical, Electronics, Information, Communication and Bio-Informatics (AEEICB), pp. 411–415. IEEE (2017)
24. Chi, J., Liu, L., Liu, J., Jiang, Z., Zhang, G.: Machine vision based automatic detection method of indicating values of a pointer gauge. In: Mathematical Problems in Engineering 2015 (2015)
25. Zheng, C., Wang, S., Zhang, Y., Zhang, P., Zhao, Y.: A robust and automatic recognition system of analog instruments in power system by using computer vision. Measurement **92**, 413–420 (2016)
26. Ma, Y., Jiang, Q.: A robust and high-precision automatic reading algorithm of pointer meters based on machine vision. Measur. Sci. Technol. **30**(1), 015401 (2018)
27. Lauridsen, J.S., Graasmé, J.A.G., et al.: Reading circular analogue gauges using digital image processing. In: 14th Conference Visigrapp, pp. 373–382 (2019)
28. Li, Z., Zhou, Y., Sheng, Q., Chen, K., Huang, J.: A high-robust automatic reading algorithm of pointer meters based on text detection. Sensors **20**(20), 5946 (2020)

29. Xuang, W., Shi, X., Jiang, Y., Gong, J.: A high-precision automatic pointer meter reading system in low-light environment. Sensors **21**(14), 4891 (2021)
30. Dumberger, S., Edlinger, R., Froschauer, R.: Autonomous real-time gauge reading in an industrial environment. In: 2020 25th IEEE International Conference on Emerging Technologies and Factory Automation (ETFA), vol. 1, pp. 1281–1284. IEEE (2020)
31. Huang, J., Wang, J., Tan, Y., Dongrui, W., Cao, Yu.: An automatic analog instrument reading system using computer vision and inspection robot. IEEE Trans. Instrum. Measure. **69**(9), 6322–6335 (2020)
32. Salomon, G., Laroca, R., Menotti, D.: Deep learning for image-based automatic dial meter reading: Dataset and baselines. In: 2020 International Joint Conference on Neural Networks (IJCNN), pp. 1–8. IEEE (2020)
33. Alexeev, A., Kukharev, G., et al.: A highly efficient neural network solution for automated detection of pointer meters with different analog scales operating in different conditions. Mathematics **8**(7), 1104 (2020)
34. Liu, Y., Liu, J., Ke, Y.: A detection and recognition system of pointer meters in substations based on computer vision. Measurement **152**, 107333 (2020)
35. Cai, W., Ma, B., Zhang, L., Han, Y.: A pointer meter recognition method based on virtual sample generation technology. Measurement **163**, 107962 (2020)
36. Lin, Y., Zhong, Q., Sun, H.: A pointer type instrument intelligent reading system design based on convolutional neural networks. Front. Phys. **8**, 618917 (2020)
37. Zhuo, H.-B., Bai, F.-Z., Xu, Y.-X.: Machine vision detection of pointer features in images of analog meter displays. Metrol. Measur. Syst. **27**, 589–599 (2020)
38. Zuo, L., He, P., Zhang, C., Zhang, Z.: A robust approach to reading recognition of pointer meters based on improved mask-RCNN. Neurocomputing **388**, 90–101 (2020)
39. Howells, B., Charles, J., Cipolla, R.: Real-time analogue gauge transcription on mobile phone. In: Proceedings of the IEEE/CVF Conference on Computer Vision and Pattern Recognition, pp. 2369–2377 (2021)
40. Liang, W., Long, J., Li, K.-C., Xu, J., Ma, N., Lei, X.: A fast defogging image recognition algorithm based on bilateral hybrid filtering. ACM Trans. Multimed. Comput. Commun. Applications (TOMM) **17**(2), 1–16 (2021)
41. Xiao, W., Tang, Z., Yang, C., Liang, W., Hsieh, M.-Y.: ASM-VoFDehaze: a real-time defogging method of zinc froth image. Connection Science **34**(1), 709–731 (2022)
42. Liu, W., et al.: SSD: single shot multibox detector. In: Leibe, B., Matas, J., Sebe, N., Welling, M. (eds.) ECCV 2016. LNCS, vol. 9905, pp. 21–37. Springer, Cham (2016). https://doi.org/10.1007/978-3-319-46448-0_2
43. NVIDIA. Nvidia Tensorrt. https://developer.nvidia.com/tensorrt. Accessed 10 July 2022
44. Tang, K., Huang, J., Zhang, H.: Long-tailed classification by keeping the good and removing the bad momentum causal effect. Adv. Neural Inf. Process. Syst. **33**, 1513–1524 (2020)

Research on Cross-Domain Heterogeneous Information Interaction System in Complex Environment Based on Blockchain Technology

BaoQuan Ma[1(✉)], YeJian Cheng[1,2], Ni Zhang[1], Peng Wang[1], XuHua Lei[1,2], XiaoYong Huai[1], JiaXin Li[1], ShuJuan Jia[1], and ChunXia Wang[1]

[1] National Computer System Engineering Research Institute of China, Beijing 102200, China
15501210877@163.com

[2] School of Computer Science and Technology, Xidian University, Xi'an 710071, China

Abstract. In the field of information interaction, when a project involves a large amount of cross-domain heterogeneous information, it is difficult to transmit and update the required information in a timely, accurate, reliable and safe manner in such a complex environment, thereby maintaining synchronization. The decentralization and immutability of blockchain technology provide new ideas for solving this problem. Therefore, this paper takes this as a starting point and proposes a cross-domain heterogeneous information exchange system based on blockchain, which calls smart contracts to realize data transmission and other functions. This paper states the design idea of the system, introduces the system software framework and network topology in detail, shows the flow chart of the system operation, and conducts simple software testing and verification. Finally, this work is summarized, and the development direction of future work is prospected, hoping to provide inspiration for the solution of such problems.

Keywords: Blockchain · Smart Contract · Heterogeneous · Information Interaction

1 Introduction

In recent years, with the rapid development of information technology, blockchain technology has been widely used in information interaction control scenarios and cross-domain collaboration due to its unique characteristics such as distributed multi-party accounting, data immutability, and data traceability [1]. For interactive control, the current business scenarios that require multi-organization collaboration (such as user management, task management, data analysis, etc.) are mostly processed based on the systems, processes, and file systems built by each organization [2]. The system has poor openness, and there is a lack of a secure and credible resource sharing, distribution and display mechanism, and it is difficult to integrate domain information. In addition, the system has various security levels, and it is necessary to dynamically grant fine-grained permissions to each participant [3]. Large amounts of data are processed during transmission and security requirements are required. In the case of high performance,

M. Qiu et al. (Eds.): SmartCom 2022, LNCS 13828, pp. 45–54, 2023.
https://doi.org/10.1007/978-3-031-28124-2_5

there are also strict requirements for real-time performance, which constitutes a complex environment with obvious characteristics.

In order to complete unified collaborative scheduling in complex environments, ensure that all parties reach a consensus on the content and order of information exchange, and ensure a reliable and immutable record of proven operation history, this paper takes a task-oriented cross-domain collaboration system as an example [4]and proposes a trusted data exchange and logic execution system between cross-domain heterogeneous systems based on blockchain technology. Committed to helping agencies achieve multi-agency mission collaboration on the basis of minimizing system modifications [5].

In view of the above situation, this paper takes task-oriented interactive control business as an example, and uses blockchain technology to achieve four goals among multiple cross-domain heterogeneous systems [6]:

First, based on the mapping between the blockchain user system and cross-domain application users, it realizes user permission control, user access management, and consensus mechanism joint execution.

The second is to build a trusted data exchange system between cross-domain applications based on the characteristics of blockchain data that cannot be tampered with.

The third is to build a credible cross-domain logic execution mechanism, coordinate multi-party process flow, and intelligently execute multi-party agreements based on blockchain-based smart contracts, consensus algorithms [7], and remote call protocol data transmission functions.

The fourth is to realize the interaction of data flow, task flow and message flow between cross-domain applications on the basis of the above three points.

The structure of the paper is as follows: Sect. 2 describes the Blockchain Architecture Design for Task Management Business; Sect. 3 describes the Software Design; Sect. 4 summarizes work.

2 Blockchain Architecture Design for Task Management Business

The overall framework designed in this paper is of cross-domain heterogeneous system information exchange software with layered encapsulation, bottom-up abstraction and top-level application-oriented. The entire information interaction software is divided into four layers from bottom to top, namely the task collaboration blockchain Fabric [5], the basic support layer, the smart contract layer and the application layer composed of functional modules corresponding to smart contracts. At the same time, strict authority control is implemented on the entire framework to enhance the overall security of the system. The overall framework is shown in Fig. 1. The proposed method implementation is based on a permissioned blockchain and the Hyperledger Fabric blockchain platform (HLF; www.hyperledger.org) [8].

2.1 Task Collaboration Blockchain Fabric Layer

The core part of this paper uses blockchain to achieve cross-domain heterogeneous information interaction. Fabric integrates three parts: data layer, network layer, and consensus

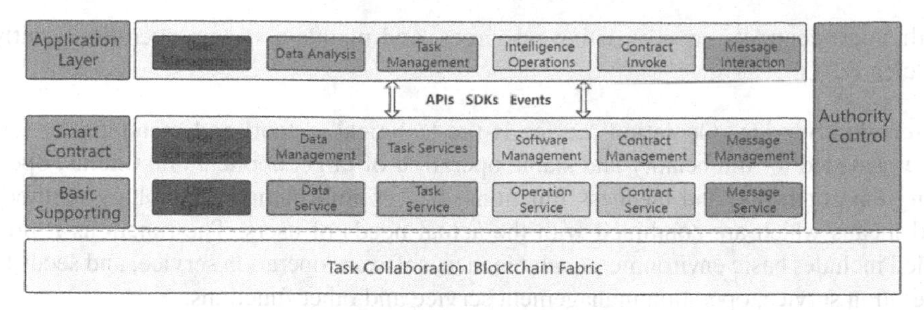

Fig. 1. System Framework

layer. The data layer of blockchain uses a chain structure to connect heterogeneous data blocks from different sources. In this paper, we define the corresponding data structure for tasks, message, etc. The data propagated on the chain is encrypted with a hash function, both of sender node and receiving node use asymmetric encryption at the same time to maintain data security; At the network layer, reflected in the P2P network architecture, the sender notifies the receiver to receive data through broadcast, and the receiver verifies the received data; The consensus layer is core of security and stability of the blockchain, the receiver determines whether the new module established by the sender can be accepted through a consensus algorithm. If the returned received information meets the specified standards, a consensus is reached, and the receiver adds the new module to its own library. The consensus mechanism is used to jointly maintain the stability of the blockchain.

2.2 Basic Support Layer

The basic support layer is the foundation that supports the entire software framework. It is divided into six modules: user service, data service, task service, operation service, contract service, and message services.

User Service. User information is stored in a separate database, and functions such as user registration and cancellation can be performed. Users can log in to a successfully registered account to use related functions.

Data Service. During the operation of the system, a large amount of initial data, process data and result data will be generated. Heterogeneous data from different systems, departments and even network domains need to be stored and processed in a timely, complete and secure manner. Data transmission and processing have high requirements on the reliability of network transmission. The data service combined with blockchain technology can avoid this problem well.

Task Service. With the help of the blockchain network, the task distribution can be notified in real time, and participants can also use it as a receiver to receive tasks in time. The distribution and execution of tasks, the allocation of appropriate resources and other necessary requirements for each task, each participant, etc., relies on the support of blockchain smart contract technology. All participants advance tasks in accordance with smart contracts, avoid wasting resources, and maintain system operation jointly efficiently.

Operation Service. Operation service is the technical support and maintenance service provided for the healthy and stable operation of target application systems, operating environments, and business functions. It uses information technology methods and means which are combined with the actual needs of users. The functions it provided includes basic environment maintenance, software operation service, and security operation service, operation management service and other functions.

Contract Service. Contract service is the technical support for the normal and effective operation of smart contracts. It supports a variety of smart contracts defined on the blockchain to make promises in a digital form, and stipulates the rights and obligations of each participant. Operations on the blockchain need to invoke the relevant smart contracts and operate according to the contract content, which solves the problem of mutual trust between the participants and make all participants jointly maintains the operation of the system.

Message Service. Message services provide the underlying logic to update the message system in real time. Message services need to process a large amount of information, such as regional weather information, equipment resource information, task execution information, etc. A variety of information can be transmitted across domain to achieve real-time updates, providing the required data sources for data analysis, intelligent operation, etc.

2.3 Smart Contract Layer

There are six major contracts in the smart contract layer, which correspond to the six major sectors provided by the basic support layer. The smart contract mechanism is the core key technology of blockchain technology, and the interaction standards on the chain are unified through this mechanism. By accessing the smart contract, the data is classified by different purpose modules, and the connection between the upper and lower layers is completed relatively quickly. It is the link between the layers and provides data and other support for the application layer.

2.4 Application Layer

The application layer is the software functions that are finally provided to users, such as User Management, Data Analysis, Task Management, Intelligence Operation, Contract Invoke, and Message Interaction. The six modules are independent of each other, coupled with each other, and support each other. Task management is the most commonly used and the most important function. Nodes can accept tasks. They can also formulate tasks and submit them to the blockchain through the analysis of initial message and the current needs of the department, coordinate task allocation, and promote tasks. The Message Interaction function allows users to grasp the latest developments in real time and realize real-time message updates within or across domains. The data analysis function can quickly and deeply analyze the latest intelligence, dig out more useful information, and allow users to proceed more rationally. Based on the existing operation and maintenance data, the intelligent operation system can analyze and summarize the rules from the data based on the existing operation data, and provide solutions, improve the operation and efficiency, and solve the problems that the traditional operation system cannot solve.

2.5 Rights Management

Fabric proposes the concept of Member Service Provider (MSP), which abstractly represents an authentication entity. MSP can be used to verify the authority of identity certificates for different resources. Specifically, users who join the system will get the initial minimum permission by default after the registration review. The relevant personnel of authority management can open the corresponding authority for specific registered users according to their computing power, network, equipment platform and other resources' access restrictions, as well as their actual levels. The permissions that users have are not fixed. The blockchain records various resource authorization operations in cross domain collaboration as a trusted authorization record to ensure the authenticity and reliability of authorization operations, and it serves as a trusted basis for restructuring information relationships. Managers have the right to adjust the permissions of relevant personnel according to changes in the situation, and increase or close the corresponding permissions. The strict implementation of authority management ensures the security of data and makes the system run smoothly and efficiently.

3 Software Design

3.1 Network Topology Design

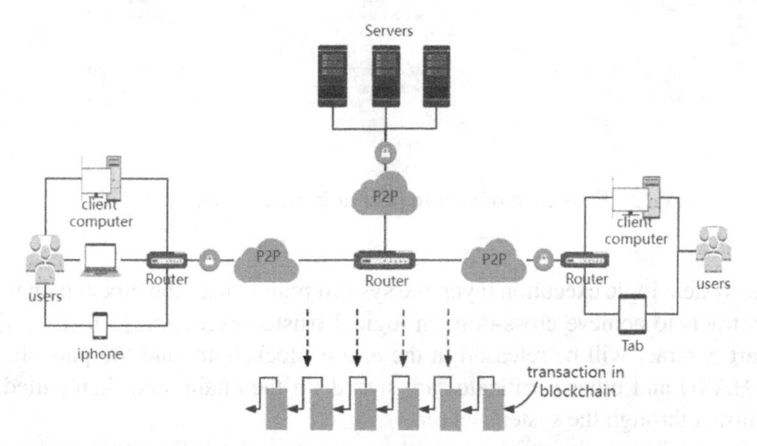

Fig. 2. Network topology design diagram of cross-domain heterogeneous information interaction

In the case of large-scale project tasks involving data processing and participants from different organizations, the environment is complex and changeable, which brings great difficulties to the execution of the task. The equipment systems used by each participant are heterogeneous and not unified, such as servers, desktop computers, portable computers, and mobile phones and other terminals (see Fig. 2), all of which are connected in their own network domains. Compared with devices in other network domains, it processes data in a cross-domain heterogeneous system. Therefore, we use blockchain technology to build distributed network structures in complex environments and connect them to P2P [9]. All transactions between them take place in the blockchain. Through the access to the blockchain and the scheduling of smart contracts, the policy of mutual data access by all parties is achieved.

3.2 Cross-Domain Information Interaction Process

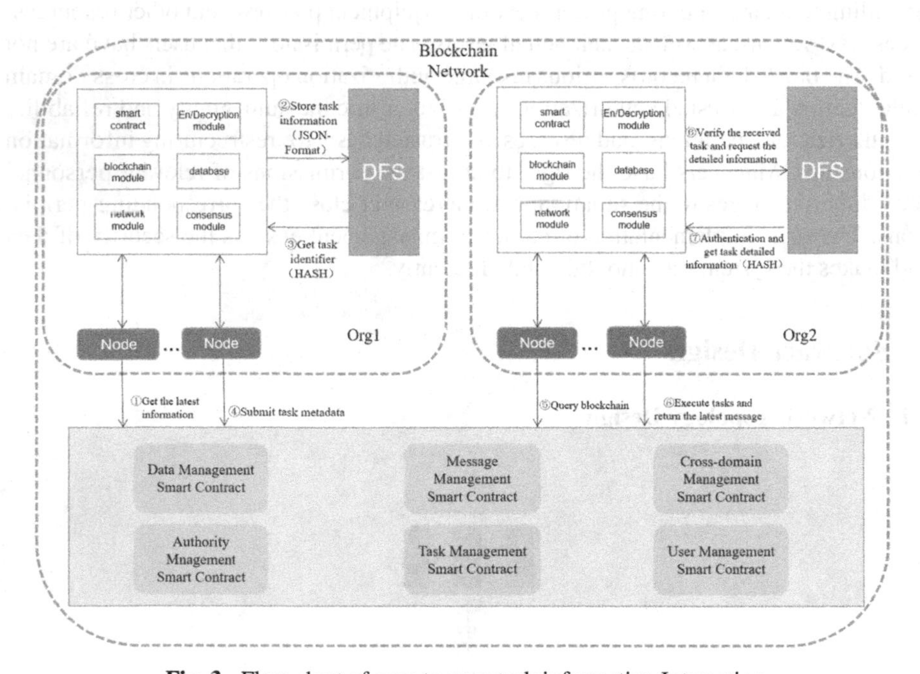

Fig. 3. Flow chart of peer-to-peer task information Interaction

At the system logic execution layer, the system mainly uses the blockchain to invoke smart contracts to achieve cross-domain logical trusted execution [10] (Fig. 3). First, each smart contract will be released in the entire blockchain, and the publisher, time, version, HASH and other certificates are stored on the chain, and then called by the smart contract through the system server.

In the cross-domain execution stage, the smart contract will write the cross-domain call protocol data into the blockchain, and after the cross-domain application obtains the cross-domain call protocol data from the blockchain, the entire cross-domain execution is completed according to the contract initiator program [11]. And write the execution status to the blockchain according to the cross-domain call contract. Finally, the system server will further invoke the smart contract according to the execution result obtained from the blockchain until the execution of the smart contract is completed. In the meantime, the system will provide an interface so that each participant can directly invoke the smart contract [12].

This paper designs a proof-of-concept model for decentralized data, focusing on smart contracts and query components, and provides a functional overview of the task system. The process is as follows:

- The task sender node obtains the latest message, analyzes the message, and then formulates tasks.
- Store detailed task information including data structures, user identities, etc. The task information is stored in the Distributed File Storage System (DFS).
- When task information is successfully stored in DFS, the file storage framework returns a storage identifier (HASH).
- The sender node must store the task metadata including the storage identifier and contract type into the blockchain and broadcast it to all other nodes.
- The consensus nodes use the PBFT consensus algorithm to check the block information, execute the smart contract and compare the results. It will not pass until they confirm that the transaction is the latest and the signature information is consistent. Other nodes can then query the blockchain after that.
- The receiver node verifies the received task and checks it against the information received by the DFS and blockchain.
- The receiver node obtains the task details through the storage identifier and DFS verification
- The receiver node completes the task and returns the latest message information of the blockchain.

3.3 Case Study

To describe the specific business processes within the framework of the HLF platform, many concepts are used, mainly assets, participants, transactions and events. In our case, the most important assets are data files in distributed storage, and their attributes are traceability metadata [13]. Based on the development status and best practices of international open data traceability, we have extracted the provenance metadata in relevant standards and specifications for project tasks. Participants are members of a task collaboration. They can own assets and make transaction requests, that is, when the assets (equipment, personnel, etc.) required by organization 1 are insufficient, they can make transaction requests to organization 2 to meet the resources required to complete the task by trading assets. Transactions are the mechanism for participants to interact with assets. All transactions take place in the blockchain. When all nodes reach a consensus and generate results, transactions cannot be tampered with. Every operation with data contains at least two types of transactions: one for client requests and one for server responses. Event messages can be sent by transaction processors to notify the changes of external software components in the blockchain. Applications can subscribe to receive event messages via HLF's API.

The task leader can publish specific task information through the interface, including task ID, task title, invited participants, etc. (see Fig. 4). The middle module is the task information interaction module, which will update the execution of the task and obtain other information which is collected or stored by the application system and called real-time intelligence information, in real time. The commander conducts data analysis according to all the data in the module through the data analysis function, and formulates new tasks and releases them to the participants. The nodes participating in the task can view the specific task information through the individual user's task list, and can also update the information in real time to the task information interaction module for all nodes to view. The task description template data structure is displayed (see Fig. 5).

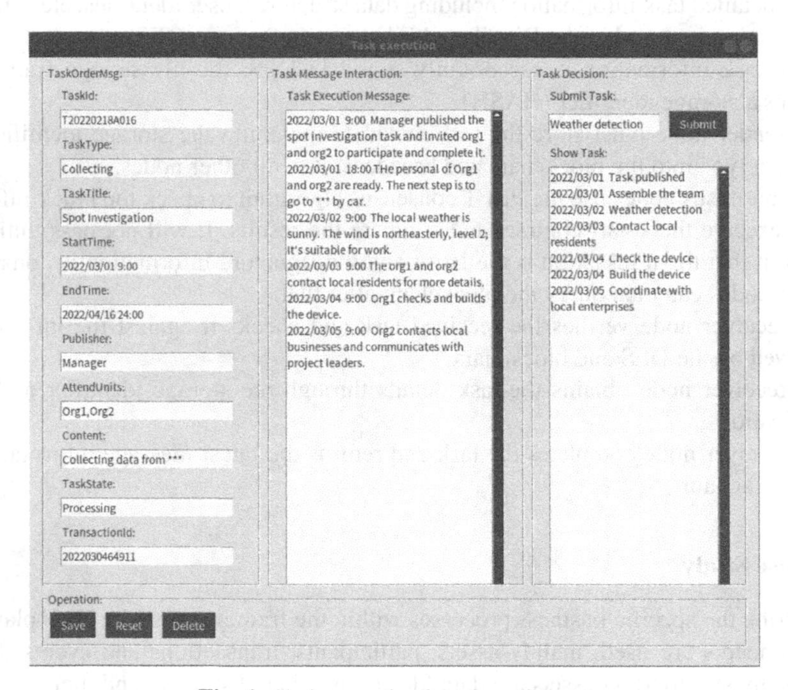

Fig. 4. Task execution interface diagram

3.4 Performance Analysis

In the traditional information interaction system, resource discovery, information interaction, etc. all need to follow multi-level processes. In this way, processes are solidified, and failure at any level in the middle may cause overall process failure, causing task interruption. Compared with the traditional way, the delay is longer and fragile. Table 1 lists the qualitative comparison between the performance of traditional hierarchical interaction and the performance of cross domain heterogeneous information interaction in a complex blockchain based environment in the case of task management-oriented business.

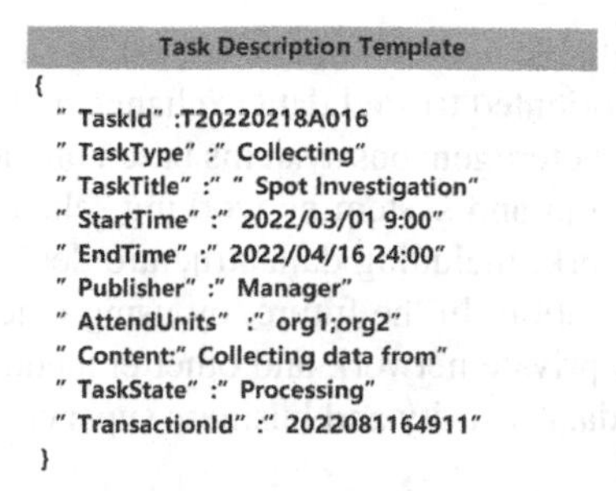
Task Description Template
{
" TaskId" :T20220218A016
" TaskType" :" Collecting"
" TaskTitle" :" " Spot Investigation"
" StartTime" :" 2022/03/01 9:00"
" EndTime" :" 2022/04/16 24:00"
" Publisher" :" Manager"
" AttendUnits" :" org1;org2"
" Content:" Collecting data from"
" TaskState" :" Processing"
" TransactionId" :" 2022081164911"
}

Fig. 5. Description of a task data struct in JSON format

Table 1. Qualitative comparison of information interaction system performance

Parameter	Traditional information interaction system	Cross domain heterogeneous information interaction system based on blockchain
Number of data forwarding	More	Less
Number of instructions forwarding	More	Less
Access control mode	Control circulation scope	Password technology authorization
Data request steps	More	Less
Emergency response speed	Slow	Fast
Single point decision speed	More	Less
Data storage space	Small	Large
Network traffic	Small	Large

For the performance of the cross domain heterogeneous information interaction mode of the blockchain, this paper makes a preliminary illustrate as follows.

Task instructions are recorded in the form of transactions as reliable regulatory records and objective basis. The tasks in the information interaction system are strictly standardized and distributed, and the feedback information of each task completed is also realized in the form of transaction. Under the scale of 1000 nodes, assuming that each node issues 30 instructions to each unit under its jurisdiction within 1 min, the TPS required to be provided by the blockchain system is 30/5 * 1000 * 2 = 1000 TPS. At present, the consensus mechanism blockchain system such as PBFT can fully meet such TPS speed requirements.

4 Summary

Aiming at the need to complete unified collaborative scheduling in complex environments, to ensure that all parties reach a consensus on the content and order of information exchange, and to ensure reliable and immutable recording of proven operation history, this paper proposes a task-oriented trusted data exchange and logic execution architecture between cross-domain heterogeneous systems based on blockchain technology and the system architecture design and system networking scheme are analyzed. We have shown some preliminary work, including data structure design, business interface and Proof-of-Concept Demonstration. In the future, by using emerging technologies, such as edge computing, virtual private network and other embedded blockchain platforms, the security guarantees of data security and business supervision in various application scenarios can be enhanced.

References

1. Lanlan, G., Yijing, L., Ran, L., Jian, L.: Design and application prospects of task-oriented cross-domain collaboration system. In: 2nd Information Technology and Internet Security, February **40**(2), pp. 19–23 (2021)
2. Qiu, M., Zhang, L., Ming, Z., Chen, Z., Qin, X., Yang, L.T.: Security-aware optimization for ubiquitous computing systems with SEAT graph approach. J. Comput. Syst. Sci. **79**(5), 518–529 (2013)
3. Qiu, H., Qiu, M., Lu, Z.: Selective encryption on ECG data in body sensor network based on supervised machine learning. Inf. Fusion **55**, 59–67 (2020). https://doi.org/10.1016/j.inffus.2019.07.012
4. Pathak, J., Peterson, K., Kanabar, M., & Schmeelk, S.: Electronic health records and blockchain interoperability requirements: a scoping review. JAMIA open. **5**(3), ooac068. (2022)
5. Qiu, M., Gai, K., Xiong, Z.: Privacy-preserving wireless communications using bipartite matching in social big data. Fut. Gener. Comput. Syst. **87**, 772–781 (2018). https://doi.org/10.1016/j.future.2017.08.004
6. Chen, C., Zhu, F., Liu, G., Zhou, J., Li, L., Wang, Y.: A unified framework for cross-domain and cross-system recommendations. IEEE Trans. Knowl. Data Eng. (2021)
7. Gilad, Y., Hemo, R., Micali, S., Vlachos, G., Zeldovich, N.: Algorand: Scaling byzantine agreements for cryptocurrencies. In: Proceedings of the 26th Symposium on Operating Systems Principles, pp. 51–68, October 2017
8. Androulaki, E., et al.: Hyperledger fabric: a distributed operating system for permissioned blockchains. In: Proceedings of the Thirteenth EuroSys Conference. Article No.30, 23–26, Porto, Portugal, April (2018)
9. Demichev, A., Kryukov, A.: Complete Decentralization of Distributed Data Storages Based on Blockchain Technology. arXiv preprint arXiv:2112.10693 (2021)
10. Wei, F., Ru, H., Tao H., Jiang, L., Shuo, W., Shiqin, Z.: A review of blockchain technology research: principles, progress and applications. J. Commun. (2020)
11. Ramachandran, G.S., Radhakrishnan, R., Krishnamachari, B.: Towards a decentralized data marketplace for smart cities. In: 2018 IEEE International Smart Cities Conference (ISC2), pp. 1–8. IEEE, September 2018
12. Raikwar, M., Gligoroski, D., Velinov, G.: Trends in development of databases and blockchain. In: 2020 Seventh International Conference on Software Defined Systems (SDS), pp. 177–182. IEEE, April 2020
13. Nagayama, R., Banno, R., Shudo, K.: Trail: a blockchain architecture for light nodes. In: 2020 IEEE Symposium on Computers and Communications (ISCC), pp. 1–7. IEEE, July 2020

A New Blockchain Design Decoupling Consensus Mechanisms from Transaction Management

Wei-Tek Tsai[1,2,3], Zimu Hu[1](✉), Rong Wang[1], Enyan Deng[3], and Dong Yang[1]

[1] Digital Society and Blockchain Laboratory, Beihang University, Beijing, China
{huzmu049,wangrong,yangdong2019}@buaa.edu.cn
[2] Arizona State University, Tempe, AZ 85287, USA
[3] Beijing Tiande Technologies, Beijing, China
{tsai,deng}@tiandetech.com

Abstract. In a traditional blockchain system, the consensus protocol and the transaction management are coupled, i.e., transactions are performed during the consensus process. When the consensus protocol completes its process, the transaction is considered as completed. Such an integrated approach challenges the system scalability and regulatory issues. And is not in line with the regulatory principles such as PFMI (Principles of Financial Market Infrastructures). This paper proposed a new decoupling mechanism that separates consensus protocols from transaction management, and further separate transactions from settlement with regulation compliance. Based on the new scheme, this paper proposed a new block-building protocol for blockchain systems. This protocol allows consensus steps, transaction management, settlement, and regulation compliance can be done concurrently.

Keywords: Blockchain · Consensus mechanism · Consensus and transaction decoupling · Regulation · Block building method

1 Introduction

Blockchain technology has been developed for more than ten years with numerous designs such as Bitcoin, Ethereum, Hyperledger, and Diem. With recent attention on digital assets and Web3, blockchain systems have received significant attention as they can be applied to digital currencies as well as digital assets. Not only support one-to-one digital currency transfers, but also complex data asset transactions using smart contracts. Therefore, a new generation of blockchains needs to not only make transactions, but also ensure that transactions are trusted, secure, and regulated.

Traditional blockchain systems such as Bitcoin, Ethereum, and Hyperledger has coupled with transaction management with the consensus protocols. Some drawbacks, such as the lack of decoupling between consensus and transaction, low efficiency, and inability to be regulated. Now blockchain is becoming a kind of financial system infrastructure, these are the problems that must be solved [10].

M. Qiu et al. (Eds.): SmartCom 2022, LNCS 13828, pp. 55–65, 2023.
https://doi.org/10.1007/978-3-031-28124-2_6

This paper is organized in as follows: Sect. 2 discusses the reasons and disadvantages of the coupling between consensus and transaction; Sect. 3 introduces the avalanche phenomenon caused by the coupling of consensus and transaction; Sect. 4 introduces the principle of decoupling between consensus and transaction in Diem, and analyzes the regulatory problems that still cannot be solved in Diem; Sect. 5 introduces how the new blockchain architecture represented by the ChainNet decouple consensus and transaction, and how to solve the regulatory problems in Diem; Sect. 6 proposed a new block building method of blockchain based on consensus and transaction decoupling; Sect. 7 shows the experimental result; Sect. 8 concludes this paper.

2 Coupling Consensus and Transaction Management

In a traditional blockchain system, if a transaction has failed during the consensus process, the entire block will be considered failed. Coupling the consensus protocol and transaction management incurs two problems:

Inefficiency: Any problem in a single transaction will cause the current round of the consensus process to fail including all the transactions within the same block. This will significantly slow down the system operation.

Without Independent Settlement and Compliance Verification: As the consensus protocol operates in real time, thus any regulation compliance mechanisms must be completed at the same pace. Furthermore, in most of these systems, transactions are settled when the consensus protocol completes. But this is not in line with the PFMI [1] where "settlement bank" is mentioned many times [5], meaning that another entity will be responsible for the settlement, providing independent transaction verification [9].

As blockchain has become an important infrastructure in financial markets, blockchain system not only execute trades, but also need to regulate these transactions.

3 Avalanche Phenomenon

If the consensus protocol and transaction management are coupled, the system has a chance to incur an avalanche phenomenon causing the entire system to break down. We discovered this phenomenon in our laboratory. It fails mainly for two reasons:

First, data on participating nodes are inconsistent with each other. As nodes have different data, the consensus protocol cannot reach an agreement. Second, data in blockchain nodes are consistent, but a specific transaction within the block is not proper.

During our seven years of laboratory experimentation, we have not identified and documented a single case for the first case. i.e., data in different nodes are inconsistent. One reason maybe we use CBFT (Concurrent Byzantine Fault Tolerance), and temporary data inconsistency are automatically handled, but we have observed the second case several times. We found that this phenomenon can happen with only failures of few transactions:

1) A single transaction fails in a block during the consensus protocol;
2) All the transactions within the same block are now considered as failed;
3) All the failed transactions need to put back into the transaction pool, but the transaction pool already has many transactions waiting;
4) Due to the large number of waiting transactions, the system will increase the block size to accommodate these transactions hoping to reduce the backlog;
5) Unfortunately, the consensus on the next block also failed, and all transactions in this new block had to go through the consensus protocol again;
6) As a result, the number of transactions waiting to go through the consensus process increased, and the block size continues to increase at the same time;
7) After few rounds, the system is jammed with data in their databases and caches. The blockchain system will enter an avalanche scenario and the whole system will stop.

The larger the block (the larger the number of transactions in it) is, the higher probability for this block to fail. Assume each transaction will fail with 0.01% probability, i.e., one failure in 10K transactions. If a block has over 5K transactions, the block has a 50% chance of failing. In our experiments, we encountered a block with 10K transactions or more. The faster the system performs, the higher probability that the system will encounter the avalanche phenomenon. The reason is simple, if a blockchain system can process 20K transactions per second, the likelihood of block failure is close to 1. Thus, a fast-processing system will have a higher probability encountering an avalanche event.

4 Decoupling Mechanisms in Diem

Diem [6] is a blockchain system proposed by Meta (Formerly Facebook) in June 2019. Diem aspires to build a decentralized blockchain and has multiple versions with significant improvement during these years. Diem proposed an innovative scheme to decouple consensus protocols and transaction management, Diem uses multiple Merkel trees to maintain accounts, transactions, and events separately.

4.1 Diem's Data Structure

The data in the Diem blockchain are stored in Merkel trees, thus any data changes are detected based on the hash value of the root node. Figure 1 shows the overall structure of the Diem blockchain. Unlike other blockchain systems that treat a blockchain as a collection of transaction blocks, Diem system has three Merkel trees:

First Tree for Storing Ledger History: This tree stores ledger history, where each leaf node is the ledger history generated during the consensus process. In each ledger history, three attributes are recorded.

1) The hash value of the Merkel root node of the ledger state;
2) All the transactions included in this block, and
3) The Merkel tree composed of events.

Second Tree for Storing Account Information: In this tree, account information is recorded, and the hash value is obtained and recorded in the Merkel tree, so as to prevent the data in the account from being tampered.

Third Tree for Storing Event Information: For each transaction execution, an event will be generated, and the event information is tracked by the Event Merkel tree.

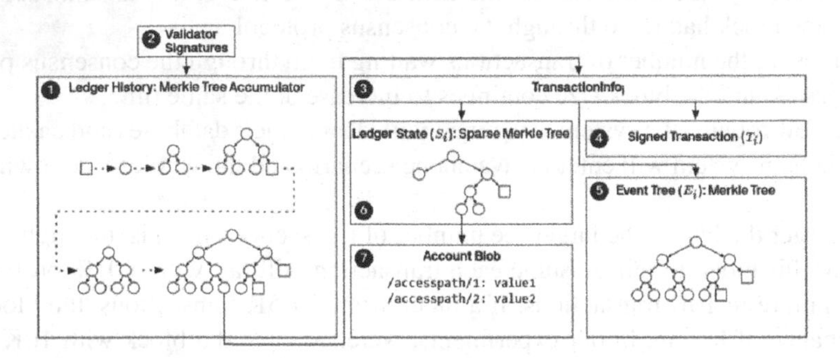

Fig. 1. Diem blockchain structure

4.2 Operation Processes in Diem

The Diem system has three separate Merkel trees, each is responsible for tracking its own data, its operation process is different from traditional blockchain systems. Thus, if some transactions within a block fails to complete, due to various reasons such as insufficient funds or lack of gas, the whole block will still go through the consensus protocol, with majority of transactions recorded as "Completed", and those failed transactions recorded "Failed". Regardless if there is any transaction failed in a block, the block will go through the consensus protocol. Thus, it is not possible for an avalanche event to happen, as no failed transaction can cause a block to fail in the consensus protocol. A consensus protocol will fail only if nodes within a blockchain contain different data, but this rarely happen in practice.

The Diem system also puts transactions in sequence to ensure that transactions will get the right results, regardless of the actual order of transaction execution. The consensus process needs only to ensure that the above three data are consistent, so that the consensus can be passed without guaranteeing that the transaction will be completed successfully.

4.3 Discussions

Diem's approach is similar but different from the ABC/TBC (Account Blockchain /Trading Blockchain) structure [12], ABC is responsible for maintaining account activities and information, and TBC for trading and storing trading information. But Diem keeps both account and trading information within the same system.

Diem has gone through significant changes from 2019 to 2022, and one of principal changes is to meet regulation requirements. In the second whitepaper published in 2020 [6], Diem mentioned that they will use an embedded compliance mechanism to ensure that transactions are regular, i.e., proper KYC and no money laundering. However, Diem has not released their compliance mechanisms other than mentioning VASPs (Virtual Assert Service Providers) need to comply with various regulations including domestic and international regulations.

5 New Decoupling Scheme

This section proposes a new blockchain architecture that 1) decouples consensus protocols and transaction management [11]; 2) decouples transactions and settlement; 3) performs compliance verification during the entire process; 4) decouples transaction information and account information.

5.1 Transaction Process with Multiple Rounds of Consensus

Figure 2 illustrates the transaction process with three consensus processes.

Pre-trading. This step identifies the involved parties for a given transaction. When a transaction is verified by KYC, it will go through the first round of consensus protocol. If a transaction fails the KYC examination, it will be rejected without execution.

Trading. The transaction will go through the execution, and at the same time, the second round of regulation compliance process will be carried out. If a transaction fails to complete or has been found to be irregular, it will be recorded as "Failed" in the blockchain; if a transition has passed both tests, it will be recorded as "Traded".

Settlement. In the third stage, those transactions that are completed will go through the settlement process. At the same time, each transaction will go through another round of regulation compliance checking. This may be the last chance for regulators to enforce any compliance before transactions are settled in banks, financial institutions, or VASP. Once a transaction is settled, it will be difficult to reverse the trade [4].

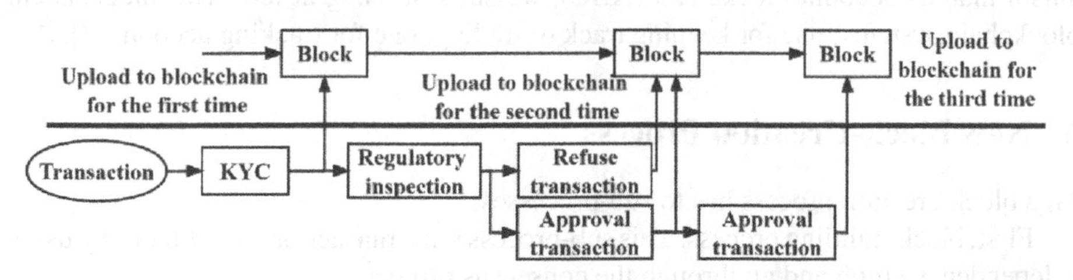

Fig. 2. Mechanisms consistent with PFMI principles

Each round of the consensus process can use various Byzantine fault tolerance protocols [8].

5.2 Characteristics of New Scheme

The new scheme has following key properties:

Separation of Transaction and Settlement: This scheme explicitly separates trading from settlement, thus providing compliance processes more time to complete their verification [4]. The time between the second stage and third stage can be hours (such as 2 h as suggested by Bank of England), or days, or seconds.

Consensus Process will Fail Only if Data are Inconsistent: Like the Diem system, any consensus protocol will fail only if data in various system are inconsistent. Transaction failure, settlement failures, regulation failures will not cause the consensus process to interrupt.

Continuously Regulation Monitoring with Adaptive Scheduling: At each stage, transactions are monitored for compliance. The time between the 2nd stage and 3rd stage can be dynamically adjusted. For suspicious trades, the system can adjust the time to allow more compliance verification. For example, FATF has reported that some money laundering events have taken more than 200 paths and thus it took a long time for regulators to identify the laundering path.

Linking to BigData Analytics: While blockchain systems are often with limited capabilities and databases, but the new scheme can be connected to a separate bigData platform with sophisticated machine learning capabilities so that even recent fraud cases can be learned and incorporated into the database.

Can be Adjusted for Similar Processes: The overall process can be divided into four stages: pre-transaction, transaction, settlement, clearing [3]. This scheme can be adjusted with four rounds of consensus processes. In practices, we have seen a process with 7 to 8 stages, or even 20 stages for credit card transactions.

Can Work with Multiple Blockchain Systems: The scheme can work with single blockchain systems, or multiple blockchain systems [2]. For example, each stage can have its own blockchain systems, or they can share a single blockchain system. The issue with a single blockchain system serving all the stages is that the single system may be jammed due to heavy loads. As a trading blockchain (TBC) has a different scaling mechanism than an account blockchain (ABC), we suggest using at least two independent blockchain systems, one for keeping track of trading, one for tracking account [7].

6 New Block-Creation Process

This block-creation process has to sub-processes:

First, block-building process. This sub-process puts transactions into blocks by using a dependency graph and go through the consensus protocol.

Second, transaction-verification process. This sub-process evaluates if a transaction within an accepted block (done by the block-consensus process) can be completed.

These two sub-processes can be carried out concurrently without interfering with each other. They can be placed in different processors for execution if necessary. Another important feature of this process the use if dependency graph.

6.1 Block-Building Process

If a transaction wants to proceed through the consensus process, it must first be incorporated into a block. Our block-building process has the following steps:

1) Each node puts the received transactions into the transaction pool. If necessary, a transaction can be broken up into multiple sub-transactions and the sub-transactions can be put into the transaction pool. The following is a discussion of intra-pool transactions, either the original trade or the decomposed sub-trade;
2) Establish a dependency graph for the transactions in the transaction pool;
3) The dependency graph is decomposed, and the transaction is sorted according to the dependency graph, and then blocks are built;
4) The block-building node broadcasts the block-building result to all nodes. Other nodes, upon receiving the block, verify that the transaction information in the block is correct (without verifying whether the transaction is complete). If yes, a vote will be cast based on the verification result, and the vote will be sent to the other nodes. There are different consensus protocols that can be used here, including the Byzantine Fault Tolerance protocol or other consensus protocols;
5) After receiving the voting results of other nodes, each node makes statistics and decides whether to accept the voting block through consensus analysis;
6) If accepted, the block is accepted by the blockchain system and this information is put into the blockchain;
7) If not, the block is rejected, and all transactions in the block remain in the transaction pool until the next block-building opportunity;
8) In all the intermediate processes, once a transaction is put into the pool, it stays in the pool, and ultimately a "Transaction verification process" decides when to leave.

Figure 3 shows the block-building process. This includes the whole process from the transaction being placed into the transaction pool, to the block eventually being added to the blockchain.

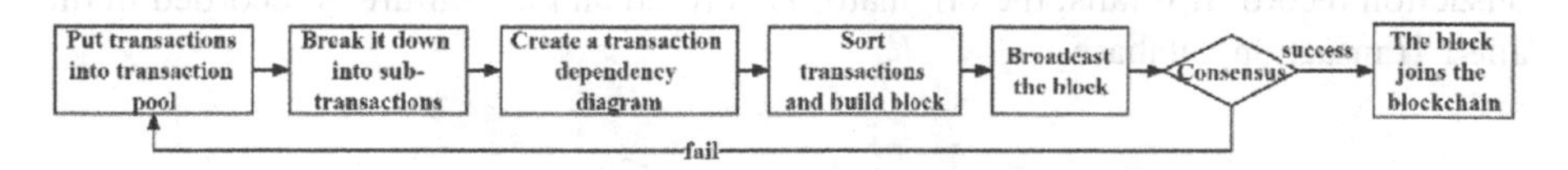

Fig. 3. Flowchart of block-building process

The first step puts transactions into the pool, each node does the same operation, and develops a dependency graph on the transactions in the pool. Where an edge in the graph represents a transaction, the node in the graph represents the user of the transaction, the direction represents the point from the transaction initiator to the transaction payee, and the edge attribute represents the transaction amount. The transaction dependency diagram is a directed graph.

In the process of establishing the transaction dependency graph, the degree of each node in the dependency graph is computed, and the new dependency graph is obtained by segmentation from the original dependency graph.

After the block is built, the participating node will sign and encrypt the block and broadcast it to all the voting nodes. After receiving the block, other nodes verify the transaction information in the block. The transaction information verified is only to check whether the information is correct, and it does not verify the transaction is completed. If all the transaction information is correct, vote yes, otherwise vote no, and send the result to other nodes.

After receiving the voting results of other nodes, each node uses the public key of the corresponding node to verify the signature. If the number of "agree" nodes exceeds the number of "disagree" nodes, the block is accepted. Once passed, all transactions in this block can be processed by the transaction verification process. If this block does not pass, this block will be rejected.

6.2 Transaction Verification Process

The transaction verification process includes the following steps:

1) When a block is accepted, each transaction in the block is subject to transaction verification. The verification checks whether the funds are real, sufficient, and legitimate;
2) If a transaction passes the verification process, the transaction is described as completed and removed from the pool;
3) If a transaction in the block fails the verification, check to see if the transaction may be completed later.
4) If the transaction has the potential to be completed in the future, the transaction will remain in the transaction pool, waiting the opportunity for the next block building;
5) If the transaction is deemed unlikely to succeed, the transaction is marked as a failure and is removed from the pool with notification notice to its sponsor.

Figure 4 shows the transaction verification process. A transaction in the pool means it has a chance to complete. If the transaction succeeds, it is recorded on the successful transaction record. If it fails, the originator is notified and the failure is recorded in the Failed Transaction database.

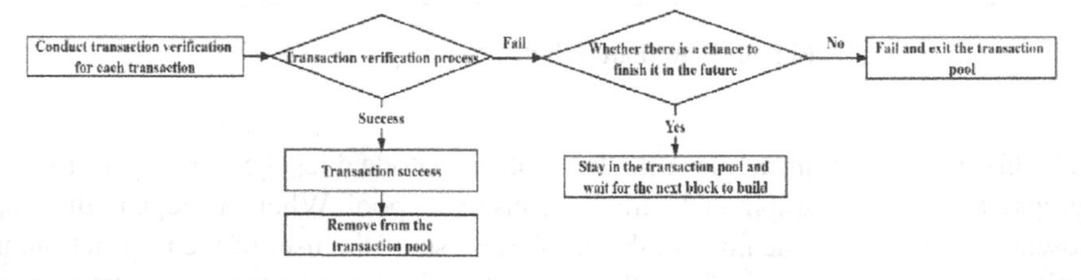

Fig. 4. Flow chart of transaction verification process

6.3 Concurrent Processes

As the consensus protocol and transaction management are now decoupled, these two processes can be carried out concurrently. Figure 5 shows transactions flows in the transaction pool. The transaction pool is divided into two parts. New transactions will first be put into the left side of the transaction pool. Block-building process will choose transactions from left side and put them into right side. Then transaction verification process checks the transactions in the right side. If the transaction is completed, it will be removed from the transaction pool. If the transaction cannot be completed this time, it will be put back to left side. Otherwise, the transaction will be marked as a failure and be removed from the pool. Multiple Transaction Verification processes can be carried out at the same time.

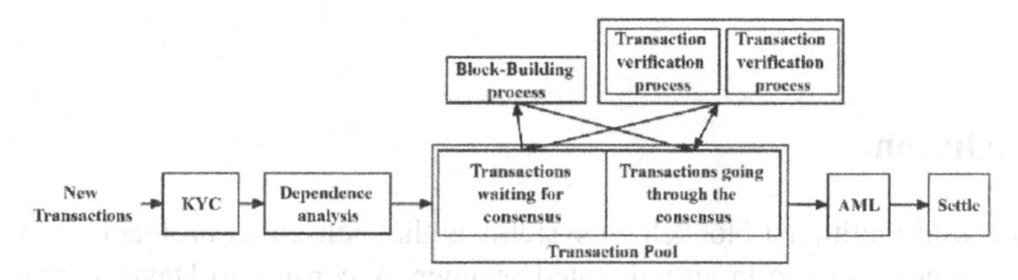

Fig. 5. Concurrent Processing with the Transaction Pool

Potentially, multiple Block Building processes can be active at the same time, e.g., they can work in a pipeline manner, or each can work on different groups of transactions with no dependency among the groups. Theoretically, this is possible, but in practice, it is difficult to implement such as lots of computation will be needed before any concurrent Block Building processes can be activated.

7 Experimental Verification

The experimental environment is shown in Fig. 6:

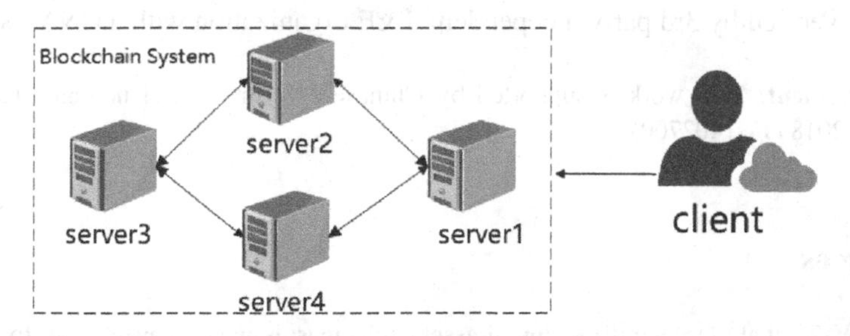

Fig. 6. Test Environment

The blockchain system under test consensus from trading. At the same time, we use a high-performance blockchain database as the underlying database. The client sends

200,000 transaction commands to the blockchain system, and the data are stored in the blockchain system after encryption, signature verification, and consensus. The actual test time is 7656 ms, and the transaction speed is 26,123 transactions/second. The test results are shown in Fig. 7.

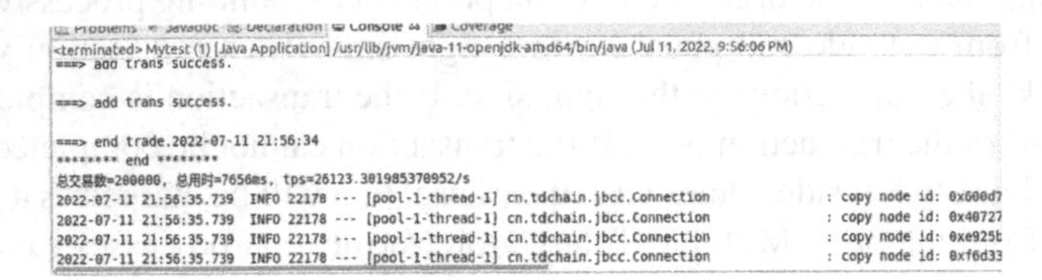

Fig. 7. The test results

8 Conclusion

One issue with traditional blockchain systems is that consensus protocol and transaction management is done in an integrated manner. According to David Parnas in his 1972 seminal paper [11], each module should perform one function only, a module that handles two or more functions will be difficult to maintain. Thus integrating consensus protocols with transaction management make the system hard to maintain and scale. In 2016, we proposed to classify blockchain systems into TBCs and ABCs, and each process only one specific function to reduce system complexity, improve system performance and scalability. The Diem system came with a different design that separate consensus protocol from transaction management. This paper proposed another mechanism to separate consensus protocol from transaction management, and further separate transaction from settlement, with a new block-creation process with two concurrent subprocesses. Another advantage is that compliance processes can be carried throughout the entire process with bigData and machine learning capabilities, and each step of our process can be carried in different processors if necessary to improve performance.

We have developed the system and the test results are shown in Sect. 7. The test results are verified by 3rd party independent T&E organization with a CNAS seal.

Acknowledgment. This work is supported by Chinese Ministry of Science and Technology (Grant No. 2018YFB1402700).

References

1. Tsai, W.-T., et al.: Decentralized digital-asset exchanges: issues and evaluation. In: 2020 3rd International Conference on Smart BlockChain (SmartBlock). IEEE (2020)
2. Wei-Tek, T., Mingding, L., Dong, Y.: DB-in-DB: a new blockchain database architecture. J. Inf. Secur. Res. **8**(5), 429 (2022)

3. Tsai, W.-T., Xiang, W., Wang, R., Deng, E.: LSO: a dynamic and scalable blockchain structuring framework. In: Gao, W., et al. (eds.) Intelligent Computing and Block Chain. CCIS, vol. 1385, pp. 219–238. Springer, Singapore (2021). https://doi.org/10.1007/978-981-16-1160-5_18
4. Tsai, W.-T., Yang, D., Wang, R., Wang, K., Xiang, W., Deng, E.: STRISA: a new regulation architecture to enforce travel rule. In: Park, Y., Jadav, D., Austin, T. (eds.) Silicon Valley Cybersecurity Conference. CCIS, vol. 1383, pp. 49–67. Springer, Cham (2021). https://doi.org/10.1007/978-3-030-72725-3_4
5. Bai, X., Tsai, W.-T., Jiang, X.: Blockchain design - a PFMI viewpoint. In: 2019 IEEE International Conference on Service-Oriented System Engineering (SOSE). IEEE (2019)
6. The Diem Association. The Diem Blockchain. https://developers.diem.com/docs/technical-papers/the-diem-blockchain-paper/. Accessed 2020
7. Tsai, W.-T., et al.: ChainNet. Oriental Publishing House, Beijing (2020)
8. Vukolic, M.: The quest for scalable blockchain fabric: proof-of-work vs. BFT replication. In: Camenisch, J., Kesdogan, D. (eds.) open problems in network security. LNCS, vol. 9591, pp. 112–125. Springer, Cham (2016). https://doi.org/10.1007/978-3-319-39028-4_9
9. Lin, L.X.: Deconstructing decentralized exchanges. Stan. J. Blockchain L. Pol'y. **2**, 58–77 (2019)
10. Payment Systems: Liquidity Saving Mechanisms in A Distributed Ledger Environment. European Central Bank and Bank of Japan (2017)
11. Parnas, D.L.: On the Criteria to Be Used in Decomposing Systems into Modules. In: Broy, M., Denert, E. (eds.) Pioneers and their contributions to software engineering, pp. 479–498. Springer Berlin Heidelberg, Heidelberg (2001). https://doi.org/10.1007/978-3-642-48354-7_20
12. Tsai, W.-T., et al.: A system view of financial blockchains. In: 2016 IEEE Symposium on Service-Oriented System Engineering (SOSE). IEEE (2016)

Adaptive Byzantine Fault-Tolerant ConsensusProtocol

Rong Wang[1], Wei-Tek Tsai[1], Feng Zhang[2](✉), Le Yu[2], Hongyang Zhang[2], and Yaowei Zhang[3,2]

[1] Digital Society and Blockchain Laboratory, Beihang University, Beijing, China
wangrong@buaa.edu.cn, tsai@tiandetech.com

[2] China Mobile Information Security Management and Operation Center, Beijing, China
{zhangfeng,yule,zhanghongyangxa,yaowei}@chinamobile.com

[3] Beijing Institute of Big Data Research, Beijing, China
yaowei@mail.ustc.edu.cn

Abstract. The existing blockchain consensus protocol has reached the level of availability in replicas in small-scale scenarios. However, if the blockchain system is composed of hundreds or even thousands of replicas, the throughput and delay will significantly decrease as the number of replicas increases, which makes it difficult to apply it in large-scale scenarios. This paper proposes an *Adaptive Byzantine Fault-Tolerant* (AdBFT) consensus protocol, which introduces the optimistic response assumption and proposes an adaptive approach to reach consensus with a latency of 2Δ in steady state, while providing the advantage of tolerating up to half of Byzantine failures, which can ensure security in a weaker synchronization model. Under the optimistic response condition, $O(\Delta)$ latency is achieved when the leader is honest and more than three-quarters of the replicas respond, and up to 1/3 of the Byzantine failures can be tolerated under synchronization.

Keywords: Blockchain · Byzantine Fault Tolerance · Synchronous Network · Optimistic Response · Parallel Pre-Built Block

1 Introduction

Blockchain is a distributed ledger system [1–3], in which each replica (usually called a node) contains a state machine [4–6], whose state must be consistent with that of all other nodes, and no node needs to know or trust other nodes in the system [7–9]. In order to reach a consensus, the consensus mechanism of each blockchain must be able to ensure the verification process of the current blockchain, help determine that the replica should create the next state (i.e., the next block), assist in verifying the next state, and ensure that all nodes agree to the next state. Blockchain is a distributed system. The immutable transaction history is stored in the distributed data structure on the peer-to-peer network. The nodes in each network contain replica of all data [10] and are organized into a block structure [11–14].

The *Practical Byzantine Fault Tolerance* (PBFT) [15] protocol forms the basis of most BFT consensus mechanisms, such as Tendermint [16], Streamlet [17] and HotStuff

M. Qiu et al. (Eds.): SmartCom 2022, LNCS 13828, pp. 66–75, 2023.
https://doi.org/10.1007/978-3-031-28124-2_7

[18]. The PBFT protocol builds on the Paxos [19] protocol and extends its crash failure to Byzantine fault tolerance to defend against adversarial participants who may arbitrarily deviate from the protocol. The PBFT protocol has Byzantine fault tolerance, which can ensure the correctness of the system when the number of evil nodes is less than one third. However, PBFT protocol has performance and availability problems. PBFT generates O (n^2) message complexity, which limits the scalability and performance of the consensus mechanism. Secondly, PBFT protocol utilizes a stable leader and changes it only when the leader is suspected to be Byzantine. Triggering leader change requires a slow, expensive, and error prone protocol, which is called the view change protocol. The complexity of secondary messaging prevents PBFT from building applications on a large scale [20]. In addition, if the failed client uses inconsistent verifiers in the request [21], client interaction may cause frequent changes to the system view without making progress.

In this paper, we study *Byzantine Fault Tolerance* (BFT) state machine replication under partial synchronization, and propose an *Adaptive Byzantine Fault Tolerance* (AdBFT) mechanism. This protocol is based on the leader's BFT protocol, reaching a consensus in a stable state with a delay of 2Δ (Δ represents the maximum network transmission delay), and at the same time, it can tolerate up to half Byzantine faults, which can ensure the security of the weak synchronization model.

Under the optimistic response condition, when the leader is honest and more than three-quarters of the replica respond, the $O(\delta)$ delay (δ represents the actual network transmission delay) is realized, and the Byzantine failure can be tolerated up to 1/3 under weak synchronization. In addition, the AdBFT protocol adopts the method of parallel pre-building blocks to realize parallel transaction processing, and adopts the identity-based distributed key generation threshold signature scheme to aggregate the signatures of multi-consensus nodes into one signature, thus reducing the communication complexity between nodes.

The main contributions of this paper are as follows:

(1) We propose an adaptive Byzantine fault-tolerant consensus protocol that achieves consensus with a latency of 2Δ in steady state, while being able to have the advantage of tolerating up to half of Byzantine failures, which ensures security in a weaker synchronization model. Under the optimistic response condition, $O(\delta)$ latency is achieved when the leader is honest and more than three-fourths of the replicas respond, and up to 1/3 of Byzantine failures can be tolerated under synchronization.
(2) We put forward a parallel pre-built block method. The traditional blockchain block building method is to wait until the previous block is completed before building a new block. In this paper, we use a parallel block building method and the consensus leader node performs the block building operation before proposing. The leader node proposes the pre-built block, broadcasts it to other consensus nodes in the network for verification, and forms a new block after verification, thus greatly improving the efficiency of building the block.
(3) We analyze theoretically the security of AdBFT protocol and demonstrate that this protocol can maintain the security of weak synchronization while achieving optimistic responses.

In this paper, the sections are organized as follows: Sect. 2 introduces the relevant theoretical background, Sect. 3 presents the protocol design of AdBFT consensus protocol, Sect. 4 elaborates the theoretical analysis of AdBFT security, and Sect. 5 summarizes the work of this paper and the prospects for future work.

2 Background

The Byzantine fault tolerance problem was proposed by Lamport, Shostak and Peases et al. [24] and put forward the solution to this problem. However, the solution implicitly assumes the use of synchronous networks, which means that messages between nodes are delivered within a fixed known time range. Compared with the asynchronous network, the message of this scheme can be delayed or even reordered at will, which is impossible to be used in the actual asynchronous network. The practical Byzantine fault tolerance was later proposed by Castro and Liskov [15], making it possible to reduce the time complexity to $O(n^2)$. In each PBFT consensus round (called view), the network is divided into two types of nodes (primary and backup), and blocks are added to the blockchain if more than one-third of the nodes reply using the same hash value. In each view, one node is the master node and the other nodes are the replicas nodes.

The Zyzzyva algorithm is a BFT protocol proposed by Kotla in 2007 [23] that introduces Speculation techniques to Byzantine protocols to reduce the cost of BFT replication systems. The Zyzzyva algorithm assumes that the server is in a normal state the vast majority of the time, so instead of executing every request after reaching consistency, it only needs to reach consistency after an error occurs. If there is inconsistency, the client discards the result and sends it back to the server to trigger the view replacement protocol to switch the master node. This may temporarily lead to system inconsistency, but does not affect the execution of non-Byzantine client requests. The Zyzzyva algorithm may improve system performance without too many Byzantine replica, but its consensus efficiency will be affected if more Byzantine replica are involved.

The *Concurrent Byzantine Fault Tolerance* (CBFT) consensus algorithm is proposed by Wei-Tek Tsai [25–27], which uses a four-communication BFT algorithm containing four stages: block determination, pre-prepare, prepare and commit, using parallel block building approach. It also uses a reputation mechanism to identify Byzantine fault nodes.

HotStuff is a base BFT-like protocol proposed by VMware Research [22] for a partial synchronization model, where the most important implementation is optimistic responsiveness while maintaining linear constraints on the complexity of the communication within each view. Once the network communication becomes synchronized, HotStuff enables the right leader to drive the protocol to consensus at the rate of the actual network delay, which is a property of responsiveness. And this network latency has a communication complexity that is linear in the number of replicas. In HotStuff, the leader of each view proposes a block, decides on a block after three consecutive blocks have been authenticated, and each view contains two rounds of communication. By adding another phase for each view, the leader replacement protocol is greatly simplified in exchange for the cost of reduced latency. However, the effectiveness of Hotstuff still requires implementing a protocol to ensure that the views are synchronized, i.e., that all replicas end up spending enough time in the same view with the correct leader. At the same time Hotstuff uses streamlined job execution in parallel to further improve throughput.

Sync HotStuff is a leader-based synchronous BFT solution, proposed by Abraham et al. [28]. Sync HotStuff has an optimistic responsiveness, for example, when less than a quarter of the replicas are not responding, it will move forward at network speed. By combining a number of practical solutions for partially synchronized BFT protocols, Sync HotStuff can tolerate up to half of Byzantine replicas without lock-step execution in steady state, it can handle weaker and more realistic synchronization models, and it was prototyped and proven to provide practical performance.

The HoneyBadgerBFT protocol, proposed by Miller et al. [29], is the first practical asynchronous BFT protocol that guarantees activity without making any temporal assumptions. The HoneyBadgerBFT uses gated public key encryption to prevent targeted censorship attacks and uses *Asynchronous Common Subset* (ACS) subcomponent for suboptional instantiation, and Reliable Broadcast (RBC) module with censoring code for transmitting each node's proposal to all other nodes.

The Dumbo protocol, proposed by Guo et al. [30, 31], is an asynchronous BFT protocol that asymptotically improves the runtime by providing two atomic broadcast protocols. It provides two atomic broadcast protocols, called Dumbo1 [30] and Dumbo2 [31]. Both protocols are based on the work of HoneyBadger BFT (as the first practical asynchronous atomic broadcast protocol), with some improvements on HoneyBadger BFT. The asynchronous public subset protocol of Dumbo1 runs only a small k (independent of n) instances of the *Asynchronous Binary Agreement* (ABA) protocol, while Dumbo2's further reduces it to a constant.

3 Adaptive Byzantine Fault-Tolerant Consensus Protocol

This section introduces the AdBFT protocol, which is based on a partially synchronous communication model, including the steady state AdBFT protocol and the optimistic state AdBFT protocol. This system is assumed to be partially synchronized, and this paper assumes that the adversary is able to delay the message for an arbitrary time, with an upper bound on the delay time of Δ. The actual message delay in the network is denoted by δ. After the *Global Stability Time* (GST), all messages between honest replica will arrive within Δ time. If the network environment satisfies the optimistic condition, blocks can be committed with a higher guarantee, i.e., full-speed network commit, and consensus can be reached with a delay of 2Δ after the optimistic response is not satisfied, with the advantage of being able to tolerate up to half of Byzantine failures, which ensures security in the weaker synchronization model.

3.1 Steady State AdBFT Protocol

In the steady state, the AdBFT protocol includes the Prepare phase, Commit phase, and Vote phase, and the process is shown in Fig. 1, and each phase is described in detail in this paper below.

Prepare phase: the leader node L proposes a block B_k: $= (b_k, H(B_{k-1}))$ by broadcasting $\langle$PREPARE, B_k, v, C_v $(B_{k-1})\rangle_L$, which proposal contains the view v arbitration certificate C_v of its predecessor block B_{k-1}, where the first view certificate (using $\perp$ denotes) is completed at system initialization time. If any replica receives a block containing a suspicious transaction, the replica is said to detect a possible Byzantine fault-tolerant failure of the leader.

Commit phase: each replica r, after receiving the above prepare phase message m, validates the transactions in block B_k, verifies whether block B_k is extended from B_{k-1}, and compares all the transactions in the block with the transactions in its own transaction buffer pool, if there is no problem with the validation then Bk is an acceptable proposal, otherwise it is considered that block B_k is unacceptable. When each replica r votes on the acceptance of block B_k, responding to the leader L by sending $\langle$COMMIT σ_c^i, B_k, v$\rangle_r$. The leader L starts a timer indicating the submission timer and, within a delay time of Δ, when leader L receives n-f proposals for the current block B_k votes, partial signatures are aggregated into a signature σ_c, and $\sigma_c \leftarrow Aggregation(\{\sigma_c^1, \sigma_c^2, ..., \sigma_c^{n-f}\})$, , Aggregation() is the aggregated signature function, and then the leader L sends $\langle$COMMIT, σ_c, B_k,v$\rangle_L$ L to the other replicas.

Vote phase: the replica r validates the received messages $\langle$COMMIT, σ_v, B_k, v, $C_v(B_{k-1})\rangle_L$ for validation, and once replica r votes for B_k, replica r starts a timer indicating the commit timer, and if replica r remains in view-v after Δ time (i.e., if r does not detect a leading modal blur or view change within Δ time), it commits the vote, responds to leader L by sending $\langle$VOTE, σ_v^i, B_k,v,$C_v(B_{k-1})\rangle_r$, and when leader L receives n-f votes for the current block B_k proposal, aggregates the partial signatures into one signature σ_v, compute $\sigma_v \leftarrow Aggregation(\{\sigma_v^1, \sigma_v^2, ..., \sigma_v^{n-f}\})$, Aggregation() is the aggregated signature function, and then the leader sends $\langle$REPLY, σ_v, B_k,v,$C_v(B_{k-1})\rangle_L$ to its client.

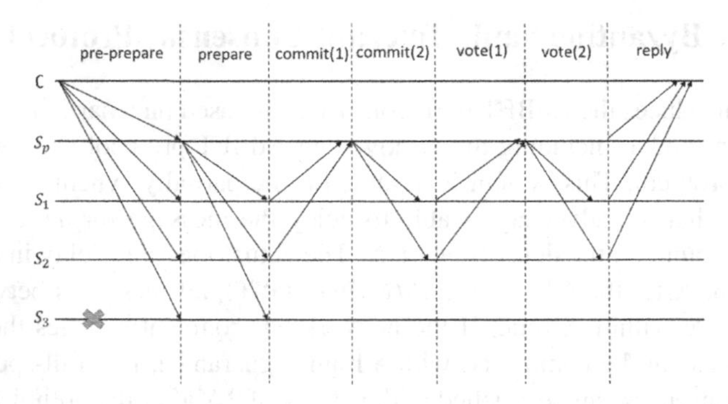

Fig. 1. Steady-state AdBFT protocol

The timer does not affect critical progress path commits, and the replica votes and starts the timer for subsequent heights without waiting to commit the previous height. In fact, replica r may have many previous heights with its commit timer still running. The view change protocol maintains security and ensures active status across views, with leader L proposing a block every 2Δ times, with one Δ for its proposal to reach other replicas and one Δ for the other replicas' vote to reach. If the leader is a Byzantine fault malicious node, i.e., sends a block proposal containing a tampered transaction or a replacement new view message proposal, it serves as a proof of Byzantine behavior and affects the reputation of the node. Once a replica r exits view v, it stops voting in that view and aborts all submitted timers for that view. The replica r then waits Δ time, selects a block with the highest authentication, locks it, reports the lock to L′, and then enters the new view. The Δ wait before locking ensures that each honest replica knows all blocks committed by all honest replicas in previous views before sending their lock status to the new leader, which maintains the security of all committed blocks for all honest replicas.

3.2 Optimistic State AdBFT Protocol

Under the optimistic response condition, the AdBFT protocol includes the Prepare phase and Commit phase, and the process is shown in Fig. 2, and each phase is described in detail in this paper below.

Prepare phase: the leader node L proposes a block B_k: $= (b_k, H (B_{k-1}))$ by broadcasting $\langle$PREPARE, B_k, v, C_v $(B_{k-1})\rangle_L$, which proposal contains the view v arbitration certificate C_v of its predecessor block B_{k-1}, where the first view certificate (using ⊥ denotes) is completed at system initialization time. If any replica receives a block containing a suspicious transaction, the replica is said to detect a possible Byzantine fault-tolerant failure of the leader.

Commit phase: each replica r, after receiving the above PREPARE message m, validates the transactions in block B_k, verifies whether block B_k is extended from B_{k-1}, and compares all the transactions in the block with the transactions in its own transaction buffer pool, if the validation is OK then B_k is an acceptable proposal, otherwise it is considered that block B_k is unacceptable. When each replica r votes on the acceptance of block B_k, respond to the leader L by sending $\langle$COMMIT σ_c^i, B_k, v, C_v $(B_{k-1})\rangle_r$. The leader L launches a timer indicating the submission timer, and within a delay time of δ, when the leader L receives 3n/4 votes for the current block B_k proposal, the partial signatures are aggregated into a signature σ_c, and $\sigma_c \leftarrow Aggregation(\{\sigma_c^1, \sigma_c^2, ..., \sigma_c^{n-f}\})$, „ Aggregation() is the aggregated signature function, and then the leader L sends $\langle$REPLY, σ_v, B_k, v, C_v $(B_{k-1})\rangle_L$ to its client.

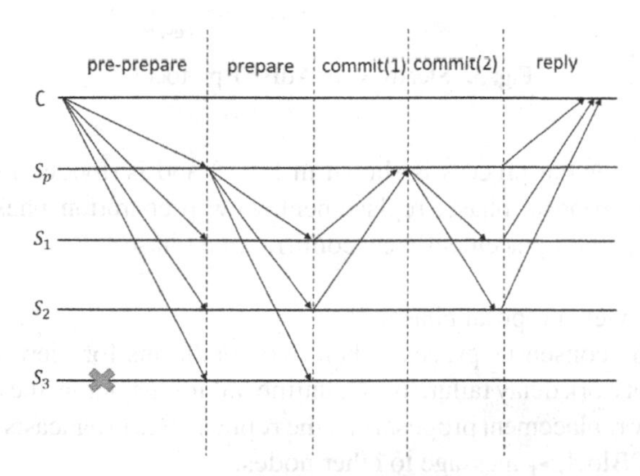

Fig. 2. Optimistic-state AdBFT protocol

Under the optimistic response condition, an O(δ) delay (δ denotes the actual network transmission delay) is achieved when the leader L is honest and, receives responses from more than three-quarters of the replicas, and to guarantee tolerance of up to 1/3 of Byzantine failures, the replica's Δ cycle after voting ensures two conditions: (1) each honest replica receives C_v (B_k) before going to the next view, and (2) no honest replica vote for a Byzantine fault block. To ensure condition (1), this paper requires that replicas start their timers C (B_{vk}) (included in the next proposal) from 3n/4 replicas after receiving. One of the replicas must be honest and prompted when the timer is started, so that all prompted replicas receive C (B_{vk}) in time. For condition (2), this paper requires all replicas to submit only after Δ time when no Byzantine faulted blocks are received.

3.3 View Change Protocol

The view change protocol maintains security and ensures active state across views, whenever a view fails due to a suspicious leader Byzantine failure or lack of progress, the replica uses the view change protocol to move to the next synchronized view of the leader. The two conditions for view replacement, i) network latency failure ii) suspicious Byzantine failure, defend against DDoS attacks and Byzantine fault tolerance failures, respectively.

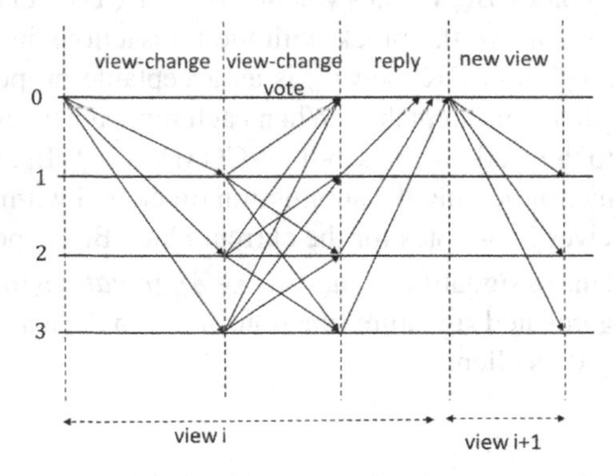

Fig. 3. Steady-state AdBFT protocol

The view replacement process is shown in Fig. 3 and is divided into four phases: replacement view proposal phase, replacement view preparation phase, replacement view voting phase, and replacement view confirmation phase.

(1) Replacement view proposal phase

During the consensus process when two conditions for view replacement are satisfied, i) network delay failure ii) Byzantine failure suspicion, the replica node can initiate a view replacement proposal and the replica node broadcasts <ViewChange, v, HighestQCBlock$>_r$ message to other nodes.

(2) Replacement view preparation phase

Other replica nodes verify the message at the stage of receiving the replacement view proposal, and the verification passes mutual broadcast of the replacement view <ViewChange, v, HighestQCBlock$>_r$ message.

(3) Replacement view voting phase

Other replica nodes validate the message at the stage of receiving the replacement view proposal, verify by broadcasting the replacement view voting ⟨NewView, v+1, $Cv'(B_k{}')$ ⟩$_r$ message to each other, replica wait for Δ time, and the replacement view proposal node collects the voting message.

(4) Replacement view acknowledgement phase

The replacement view proposal node receives n-f voting messages, synthesizes the view replacement certificate, selects a highest authenticated block, locks it, reports the lock to L′, and then enters the new view. The Δ-wait before locking ensures that each honest replica secures all blocks submitted by all honest replicas in the previous views before sending their lock status to the new leader. Each replica node starts a new round of consensus based on the received ⟨NewView, v + 1, $Cv'(B_k{}')$⟩$_L$ message.

4 Security Analysis

Due to the reputation mechanism and view replacement mechanism, Byzantine failure nodes must also submit the proposed block B_k within 2Δ time to avoid view changes and reputation being affected. However, due to the reputation mechanism and view replacement, view changes are also eventually triggered and there will be an honest leader. Once the leader node is an honest node, no view change will occur and all honest replica will keep submitting new blocks. At the beginning of a new view, the honest leader node will aggregate the latest PrepareQC certificates of the n-f replicas, according to the assumption that all honest nodes' replicas are located in the same view, so the leader node will pick the highest PrepareQC certificate as the highest highQC certificate, based on the synchronization assumption that all honest nodes' views are in a synchronized state, and since all replicas are located in the same view, so there exists bounded time T_f, so that all replicas complete the Prepare, Commit and Vote phases within T_f time.

The leader node may enter view v later than others and need to wait 2Δ time before it can propose. Other nodes need another Δ to receive the proposal. 2Δ waiting ensures that the honest leader receives locked blocks from all honest replica until that view starts. Thus, the block it proposes will extend the locked blocks of all honest replica and get votes from all honest replica. After that, an honest leader is able to propose a block every 2Δ times, one Δ for its proposed block to reach all honest replica and another Δ for the arrival of votes from all honest replica. Thus, an honest leader is able to vote for all other honest replica of the proposal in 2Δ time. Moreover, an honest leader does not have Byzantine failures. Therefore, no honest replica blames the honest leader, and all honest replica nodes are constantly submitting new blocks.

Suppose there exist two conflicting blocks B_k and $B_k{}'$ with different commits at height k. Assume that B_k is a direct commit in view v and $B_k{}'$ is a direct commit in view v′. If v < v′, then $B_k{}'$ is extended over Bk and there will be no conflict, and similarly if v > v′, then Bk′ is extended over B_k and there will be no conflict, and if v = v′, an honest replica node will not submit two conflicting blocks at the same time, so no two honest replica nodes submit different blocks at the same height. Assume that in the optimistic response state, there exist conflicting blocks B_k and $B_k{}'$, at this time block

B_k has been voted to support the PrepareQC certificate, at this time the size of the voting support replica group R_1 is $R_1 \geq 3/4n + 1$, and block $B_k{}'$ has been voted to support the PrepareQC certificate, at this time the size of the voting support replica group R_2 is $R_2 \geq 3/4n + 1$, there exists at least one honest replica replica j such that $j \in R_1 \cap R_2$, which contradicts the fact that an honest replica will only vote once at each stage in each view. Therefore, the assumptions in this paper are not valid. Thus, in the optimistic response state, the honest replica node will not submit two conflicting blocks at the same time.

5 Conclusion and Future Works

This paper proposed an adaptive Byzantine fault-tolerant consensus protocol AdBFT, a protocol that can maintain high throughput and scalability at hundreds of scale replica node consensus. The AdBFT protocol achieved consensus with a 2Δ delay in steady state, while being able to have the advantage of tolerating up to 1/2 Byzantine failures to ensure security in a weaker synchronization model. The AdBFT protocol can tolerate up to 1/3 Byzantine failures under synchronization by achieving $O(\Delta)$ latency under optimistic response conditions when the leader is honest and more than three-fourths of the replicas responded. In future work, we will perform formal verification of the AdBFT protocol and establish a general and scalable event logic-based framework for verifying the implementation of the AdBFT protocol using the Coq theorem provers.

Acknowledgment. This research was funded by the Chinese Ministry of Science and Technology (Grant No. 2018YFB1402700).

References

1. Gai, K., Zhang, Y., Qiu, M., Thuraisingham, B.: Blockchain-enabled service optimizations in supply chain digital twin IEEE Trans. Serv. Comput. 1–12 (2022).https://doi.org/10.1109/TSC.2022.3192166
2. Qiu, M., Qiu, H., Zhao, H., Liu, M., Thuraisingham, B.: Secure data sharing through untrusted clouds with blockchain-enabled key management. SmartBlock 2020, pp. 11–16 (2020)
3. Gai, K., Wu, Y., et al.: Privacy-preserving energy trading using consortium blockchain in smart grid. IEEE TII **15**(6), 3548–3558 (2019)
4. Tian, Z., Li, M., et al.: Block-DEF: a secure digital evidence framework using blockchain. Inf. Sci. **491**, 151–165 (2019)
5. Qiu, M., Qiu, H.: Review on image processing based adversarial example defenses in computer vision. In: IEEE 6th Intl Conf. BigData Security, pp. 94–99 (2020)
6. Qiu, H., Dong, T., Zhang, T., Lu, J., Memmi, G., Qiu, M.: Adversarial attacks against network intrusion detection in IoT systems. IEEE Internet Things J. **8**(13), 10327–10335 (2020)
7. Gao, X., Qiu, M.: Energy-based learning for preventing backdoor attack. Conf. KSEM. **3**, 706–721 (2022)
8. Qiu, H., Zheng, Q., et al.: Topological graph convolutional network-based urban traffic flow and density prediction. IEEE Trans. ITS (2020)
9. Li, Y., Gai, K., et al.: Intercrossed access controls for secure financial services on multimedia big data in cloud systems. ACM Trans. Multimed. Comp. Comm. App. (2016)
10. Natarajan, H., Krause, S., Gradstein, H.: Distributed ledger technology and blockchain (2017)

11. Dhumwad, S., Sukhadeve, M., Naik, C., Manjunath, K.N., Prabhu, S.: A peer to peer money transfer using SHA256 and Merkle tree. In: 2017 23RD Annual International Conference in Advanced Computing and Communications (ADCOM), September 2017
12. Meva, D.: Issues and challenges with blockchain: a survey. Int. J. Comput. Sci. Eng. **6**(12), 488–491 (2018)
13. Attaran, M., Gunasekaran, A.: Blockchain principles, qualities, and business applications. In: Applications of Blockchain Technology in Business (2019)
14. Bellini, E., Iraqi, Y., Damiani, E.: Blockchain-based distributed trust and reputation management systems: a survey. IEEE Access **8**, 21127–21151 (2020)
15. Castro, M., Liskov, B.: Practical byzantine fault tolerance. In: OsDI, vol. 99, No. 1999, pp. 173–186
16. Buchman, T.E.: Byzantine fault tolerance in the age of blockchains Doctoral dissertation, Master's thesis at University of Guelph, Ontario
17. Chan, B.Y., Shi, E.: Streamlet: textbook streamlined blockchains. In: Proceedings of the 2nd ACM Conference on Advances in Financial Technologies
18. Yin, M., Malkhi, D., Reiter, M.K., Gueta, G.G., Abraham, I.: Hotstuff: Bft consensus with linearity and responsiveness. In: 2019 ACM Symposium on Principles of Distributed Computing
19. Lamport, L.: Paxos made simple. In: ACM SIGACT News (Distributed Computing Column), vol. 32, no. 4, 51–58
20. Thakkar, P., Nathan, S., Viswanathan, B.: Performance benchmarking and optimizing hyperledger fabric blockchain platform. In: 2018 IEEE 26th Symposium on Modeling, Analysis, and Simulation of Computer and Telecommunication Systems (MASCOTS), pp. 264–276
21. Clement, A., Wong, E.L., Alvisi, L., Dahlin, M., Marchetti, M.: Making byzantine fault tolerant systems tolerate byzantine faults. In: NSDI, vol. 9, pp. 153–168
22. Aublin, P.L., Mokhtar, S.B., Quéma, V.: RBFT: redundant byzantine fault tolerance. In: 2013 IEEE 33rd International Conference on Distributed Computing Systems. IEEE
23. Kotla, R., Alvisi, L., Dahlin, M., Clement, A., Wong, E.: Zyzzyva: speculative byzantine fault tolerance. In: 21st ACM SIGOPS Symposium on Operating Systems Principles, pp. 45–58
24. Lamport, L., Shostak, R., Pease, M.: The Byzantine generals problem. In: Concurrency: the Works of Leslie Lamport
25. Tsai, W.T., Yu, L.: Lessons learned from developing permissioned blockchains. In: IEEE Conference on Software Quality, Reliability and Security Companion (QRS-C). IEEE, pp. 1–10 (2018)
26. Tsai, W.T., Blower, R., Zhu, Y., Yu, L.: A system view of financial blockchains. In: 2016 IEEE Symposium on Service-Oriented System Engineering, SOSE 2016 (2016)
27. Zhang, C., Wang, R., Tsai, W.T., He, J., Liu, C., Li, Q.: Actor-based model for concurrent byzantine fault-tolerant algorithm. In: IEEE International Conference on CNCI 2019 (2019)
28. Abraham, I., Malkhi, D., Nayak, K., Ren, L., Yin, M.: Sync hotstuff: simple and practical synchronous state machine replication. In: 2020 IEEE Symposium on Security and Privacy (SP) (2020)
29. Miller, A., Xia, Y., Croman, K., Shi, E., Song, D.: The honey badger of BFT protocols. In: 2016 ACM SIGSAC Conference on Computer and Communications Security (2016)
30. Guo, B., Lu, Z., Tang, Q., Xu, J., Zhang, Z.: Dumbo: faster asynchronous BFT protocols. In: 2020 ACM SIGSAC Conference on Computer and Communications Security (2020)
31. Lu, Y., Lu, Z., Tang, Q., Wang, G.: Dumbo-MVBA: optimal multi-valued validated asynchronous byzantine agreement, revisited. In: 39th Symposium on Principles of Distributed Computing

A Multi-level Corporate Wallet with Governance

Wei-Tek Tsai[1], Dong Yang[1], Zizheng Fan[1], Feng Zhang[2(✉)], Le Yu[2], Hongyang Zhang[2], and Yaowei Zhang[2]

[1] Digital Society and Blockchain Laboratory, Beihang University, Beijing 100191, China
tsai@tiandetech.com, yangdong2019@buaa.edu.cn

[2] China Mobile Information Security Management and Operation Center, Beijing 100000, China
{zhangfeng,yule,zhanghongyangxa,yaowei}@chinamobile.com

Abstract. This paper proposes a blockchain-based multi-level corporate digital wallet that meets financial regulatory requirements, including compliance with the Travel Rules, and identification of suspicious financial transactions. Unlike personal digital wallets, a corporate wallet is actually a large financial management system with internal blockchains, databases, and governing rules. Every transaction that the user operates on the wallet client will first be sent to the wallet server system. The wallet server will first check the compliance of the transaction, and then verify that the transaction meets the Travel Rules. At the same time, the digital wallet also checks if the transaction may be a financial fraud. In the Web3 era, many transactions will be performed using digital wallets, rather than bank accounts, a corporate wallet will play a significant role in the new digital world.

Keywords: Corporate wallet · Multi-level · Governance

1 Introduction

Blcockchain systems have received significant attention recently, especially as digital currency is one of the most important applications [1,2]. As of November 2021, the total market value of cryptocurrencies has exceeded 3 trillion, surpassing the size of the entire UK economy; at the same time, the total market value of Ethereum has also surpassed the market value of Chinese Internet giant Tencent [3]. It is estimated that in the near future, almost all economic activities such as clothing, food, housing and transportation, institutional transactions will rely on blockchains and digital currency.

However, traditional cryptocurrencies emphasize on privacy, people often do not reveal their identifies during the process. But this is against the financial regulations where the IDs of involved parties need to be known [4]. FATF (Financial Action Task Force) proposed Travel Rules to regulate cryptocurrency transaction, and the Travel Rules require involved parties to reveal their identifies.

M. Qiu et al. (Eds.): SmartCom 2022, LNCS 13828, pp. 76–84, 2023.
https://doi.org/10.1007/978-3-031-28124-2_8

Recently, US has further tightened the regulations concerning digital currency transactions, and many institutions start requiring enforcement of Travel Rules to digital wallets. At the same time, many financial institutions are exploring replacing accounts by digital wallets, and this will significant changes in the financial infrastructure systems [5].

This paper proposes a multi-level corporate digital wallet that can be used by a corporation to manage their digital assets including all kinds of digital currencies. Unlike personal digital wallets, a corporate wallet is a large financial management system with internal blockchains, databases, and governing rules.

2 Digital Wallets

A cryptocurrency transaction does not actually transfer any digital currency, but binds the ownership of a digital currency to the recipient's wallet address [6,7]. At present, many blockchain wallet products or design solutions have emerged, and their primary objective is to protect the security of private keys [8,9].

However, as more people and institutions use digital currencies, it is necessary to develop complex digital wallets to support corporate operations. As digital assets include digital real estate, digital futures, and digital stocks, and many of these assets are not personal assets, but institutional or corporate assets. Institution need to comply with their own rules, as well as regulation rules. An institution can have a few people to several million people [10] and the types of institutions are diverse, and can be government agencies, financial institutions, e-commerce companies, schools, courts, catering companies, travel companies and other types of corporations [11–13]. This means that current digital wallets need to be drastically expanded to better serve various institutions [11,14].

In traditional cryptocurrencies, values are stored on the blockchains. If the private key is lost, the value is considered as lost as it is no longer accessible. These orphan coins are unclaimed and unusable. The user's private key only means that they have the right to access and use the value [16].

We have proposed a new scheme to store value. The value is stored not on the blockchain system, but instead, in custodial banks. The blockchain system stores the certificate of the value, not like a passbook from a bank. A passbook represents values are kept in banks, but it is not the actual value. If a digital currency uses this model, even if the private key is lost, the value can be restored as no value is ever lost.

3 Multi-level Digital Wallets

As shown in the Fig. 1, each corporate wallet has one internal blockchain system, and an external blockchain system can be shared among multiple institutions. If an institution is small, it can use the external blockchain system to store its data, except all the data stored will be encrypted for security and privacy. In Fig. 1, accounting firm, tax agencies, auding firms, KYC units, AML organizations may participate in the external blockchain systems.

The difference between the corporate wallet and traditional wallet is that the asset management is divided into multiple levels, each level plays a certain role, and any use of a digital asset may require multiple levels of users cooperate with each other. For example, a large corporation has multiple departments, each department has multiple groups. In this case, the corporation has a large wallet, each department has a medium wallet, and each group has a small wallet. From the wallet point of view, a large wallet contains multiple medium wallet, and each medium wallet has multiple small wallets.

As shown in the Fig. 2, a large wallet controls the transfer assets in the medium wallets, and each medium wallet similarly also control transfer of funds among small wallets. In real life, the company's department may create new departments, abolish existing departments, merging several departments into one department, as well as other re-organization. Some corporations may have many levels, and not all branches will have the same heights. Thus, a corporate wallet may change its structure from time to time as the company changes its structure.

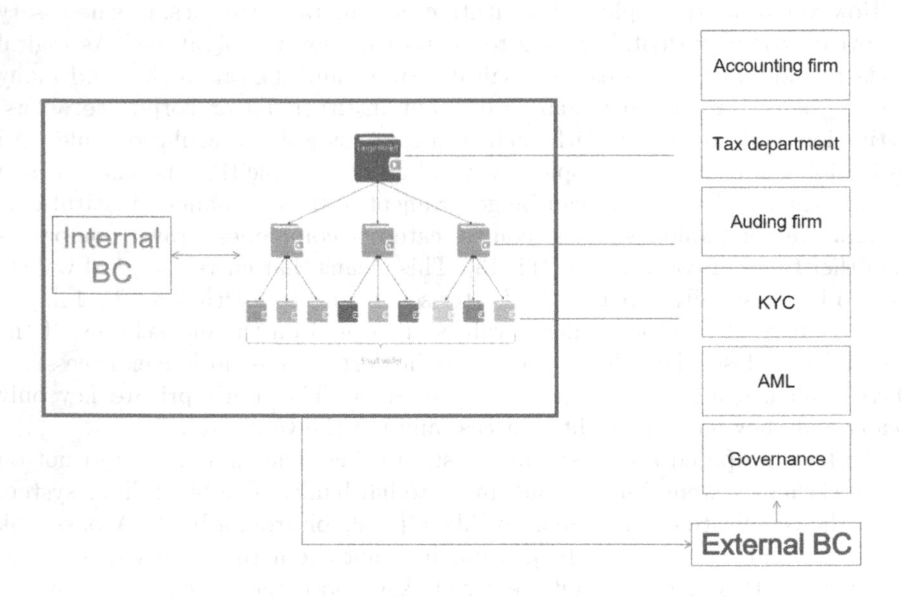

Fig. 1. Corporate wallet

4 Travel-Rule Interface Module

A corporate wallet is divided into two parts: wallet clients and wallet servers. A wallet client is responsible for keeping the user's private key, and the wallet server is responsible for supervising transactions, identifying suspicious-transactions and on-chain transactions. A wallet server is located between the wallet client and the blockchain system. Every transaction must go through

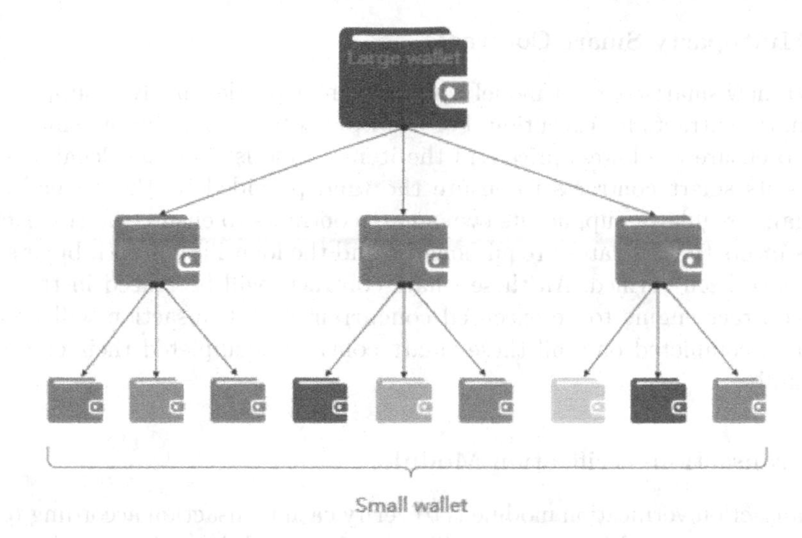

Fig. 2. Multi-level corporate wallet architecture

the wallet server system. After compliance verification, the transaction will be recorded in the blockchain system. The wallet server has five main modules: KYC module, Transaction Verification module, Travel-Rules Interface module, Suspicious-Transaction Identification module, and on-chain module.

The client and the server can have mutually agreed restrictions. For example, within a day, the client can only initiate 4 transactions, each with a maximum of 100 RMB, while the server can only initiate or accept at most one transaction related to this special client. The ledger data in the blockchain system can have various structure options:

1) There is only one ledger in the entire wallet, and all transactions go through this ledger, which can store different digital assets or digital asset certificates, such as digital RMB, or digital real estate certificates;
2) The server has a large ledger, while the client has its own (small) ledger, and the large ledger in the server also maintains all client small ledgers. The client needs to maintain ledger consistency with the server, such as using Byzantine Generals Protocol PBFT or CBFT or other consensus mechanisms.

In the first scheme, the entire wallet has only one ledger, which mainly supports the "full server control scheme", while in the second scheme, both the server and the client have ledger schemes, which mainly support the "client-side independent scheme" and " Client Partially Autonomous Scheme". Although the first scheme can also support "client-side autonomous scheme" and "clientside partial autonomy scheme", the process is more complicated.

4.1 Multi-party Smart Contracts

This is a new smart-contract model where different parties involved supply their own smart contracts for execution. For example, a buyer supplies its smart contracts to ensure the lowest price and the items trade is valid and legal, a seller supplies its smart contracts to ensure the fund provided by the buyer is real and legal, a regulator supplies its own smart contracts to ensure the transaction process meets the regulatory requirements, and the identifies of both buyers and sellers have been verified. All these smart contracts will be placed in the same smart-contract engine to be executed concurrently. A transaction will be considered as completed only all these smart contracts completed their execution successfully.

4.2 Transaction-Verification Module

The transaction verification module is to verify each transaction according to the level of the account. Whether the verification is passed depends not only on the current transaction itself, but also on some past transaction records. There are various rules for verification, such as the upper limit of the single-day transfer limit, the upper limit of the number of transfers per day, the upper limit of the annual transfer limit, and the upper limit of the number of annual transfers. Only after the requirements are met, the next step will be performed.

4.3 Travel-Rules Interface Module

This module is designed to comply with the Travel Rules requirements proposed by the FATF. Involved parties of a transaction need to show their identities. This module has a set of Procedures with three main steps: verification of transaction initiator identity, verification of transaction recipient identity, and verification of transaction information. The module needs to interface with external systems if the involved parties are not from the same corporation, and may need to interact with STRISA system to assist in identity verification.

4.4 Suspicious-Transaction Identification Module

This module detects if a transaction is a suspicious-transaction in real time. It has three steps: identification of suspicious transaction subjects (can be an entity or individual), outliers, and suspicious-transaction paths. Once a transaction is performed, subjects that participated in the transaction is linked in a graph model. A clustering algorithm is used to cluster various subjects, and those highly connected subjects will be linked together. Once a network of clusters is formed, those inter-cluster transactions will show those subjects that may have irregular transactions. Once identified, the system uses an improved CURE clustering to identify outliers. Once identified, the subjects involved in those outlier transactions are listed as high-risk subjects. Once the high-risk subjects are identified, the system searches all other subjects that traded with the suspicious subjects, and these may become high-risk subjects. Once high-risks are identified, their transaction paths are cross examined to identify money laundering paths. After thorough examination, the high-risk subjects may be cleared and removed from the high-risk list, or they may be reported for further investigation.

The system will import a lit of high-risk subjects from regulators, and they will be placed before the corporate wallet start operation so that it will not have a cold start. Whenever a transaction arrives, the system determines if the subjects involved in the transaction are on the list of high-risk subjects, if yes, the new transaction is suspicious and may be terminated in real time. The list of high-risk subjects will be continuously updated as the corporate wallet operates. Furthermore, new algorithms may be added into the wallet as they are available.

5 Method Implementation

5.1 Regulation Rules Implementation

Each transaction consists of multiple steps: First, construct a weighted undirected graph based on each account and its previous transactions. The nodes in the graph represent accounts, and the edges represent transactions between two accounts. The weight of the edges depends on the transaction amount and frequency. The final weight value is calculated as follows:

$$w = ln(total * fred) \tag{1}$$

If the cumulative transaction amount between the two subjects (or accounts) is small, it indicates that the relationship between these two is weak; and when the cumulative transaction amount is large and the frequency is high, it indicates that these two are closely related. and there is greater credibility that they belong to the same community. The weight will be a positive number.

The relationship between various subjects (and/or accounts) is saved in the subject database. The outlier database stores outliers information. Clustering algorithm can identify outliers in the system. in the database. This paper use an improved CURE clustering algorithms to identify outliers. In general, an otlier may be considered as noise, but often outliers in transaction may mean irregular transactions. Each subject will have connections with other subjects and this can be modeled a weighted graph, the weight of the link is calculated as follows:

$$L = -ln(w/w_{max}), \tag{2}$$

where $0 < w/w_{max} \leq 1$. Thus, if w_{max} is inverse proportional to L, when w_{max} is small, L is large.

Once the graph is computed, one can identify risks subjects in the database by using shortest path algorithms such as Dijkstra's algorithm. After determining the shortest path, calculate whether the shortest path exceeds the threshold k (the threshold is between 0 and 1). If the threshold is exceeded, the subjects will be added to the high-risk list. As shown in Figure Fig. 3, the suspicious-transaction identification module has two steps, an initialization phase and a determination phase.

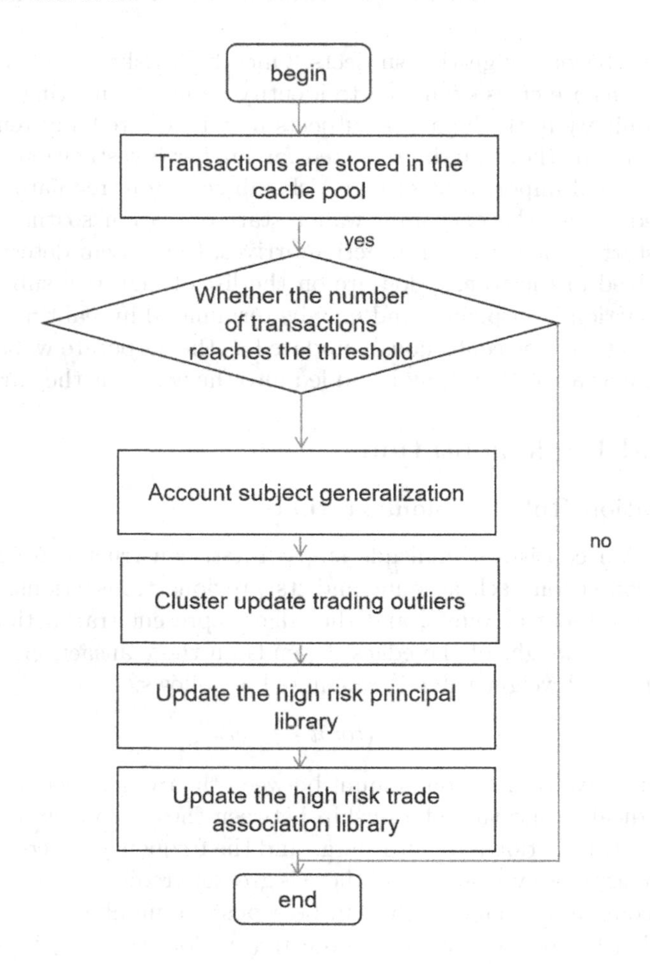

Fig. 3. Suspicious-transaction identification module

5.2 Multi-Level Wallet Architecture

A multi-level wallet includes the following parts: 1) corporate private keys, 2) virtual accounts, 3) threshold key generation, and 4) signature verification.

The corporate private key is the private key that controls the digital assets of the corporation. Each unit within the corporation has a virtual account with its own balance. Each corporation controls all its subordinate departments, each department controls each of its subordinate groups, and each group controls each of its subordinate employees. Using this structure, each unit, regardless of its level, has the corresponding virtual account with spending limit and balance. The spending limit is the amount the unit is allow to spend without consulting with its superior unit, and the balance is the amount that is available to spend. For example, a unit can have a balance of $300K and spending limit of $50K. If the unit wishes to spend $60K, its needs to consult with the superior unit for approval even though it has over $300K to spend.

Figure 4 shows an example, where a unit α controls 5 subordinate units A, B, C, D, E. All these units have their own virtual account with balance and spending limit. Furthermore, α also may change the rules and regulation of each subordinate account, e.g., by adding new funds into A's virtual accounts, or by removing funds from B's virtual account. In this way, the CEO of the corporation can have a transparent and real-time of funds within the corporation.

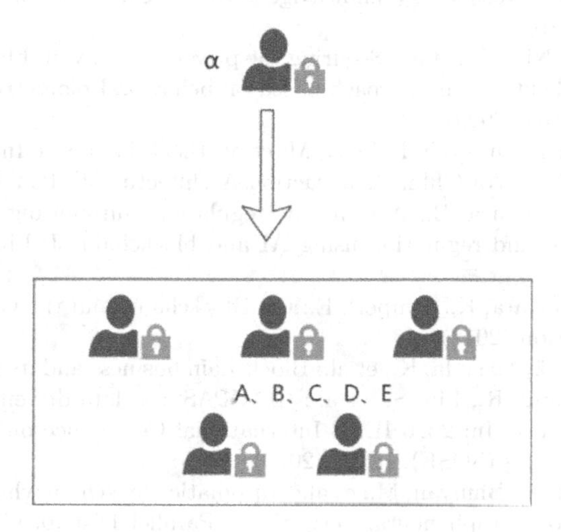

Fig. 4. Hierarchical Model

6 Conclusion

This paper proposes a blockchain-based multi-level corporate digital wallet. Each digital wallet can be used in a variety of organizations including enterprises, schools, and government agencies. Each wallet contains multiple virtual accounts (can be considered as sub-wallets) with its own governance rules and regulations. Each corporation can design and deploy different rules and regulations to meet its needs. Each wallet also has automated compliance mechanisms to assist regulators as well as management to manage various transactions.

In May 2022, JP Morgan discussed the possibility of completely eliminating bank accounts and replacing them with digital wallets instead. While this possibility may take years to take place, but the importance of digital wallets in the Web3 era cannot be underestimated.

Acknowledgments. The research was funded by National Key Research and Development Program of China (No. 2018YFB1402700).

References

1. Li, G., You, L.: A consortium blockchain wallet scheme based on dual-threshold key sharing. Symmetry **13**(8), 1444 (2021)
2. Han, J., Song, M., Eom, H., et al.: An Efficient Multi-Signature Wallet In Blockchain Using Bloom Filter (2021)
3. Ryou, J.K., Pan, Y.H.: A study on the user experience design of blockchain wallet considering users' technology knowledge level. J. Digital Contents Soc. **21**(12), 2073–2081 (2020)
4. Hu, T., Liu, X., Niu, W., et al.: Securing the private key in your blockchain wallet: a continuous authentication approach based on behavioral biometric. J. Phys. Conf. Ser. **1631**, 012104 (2020)
5. Igboanusi I S, Dirgantoro K P, Lee J M, et al. Blockchain Side Implementation Of Pure Wallet (PW): An Offline Transaction Architecture[J]. ICT Express, 2021
6. Treleaven, P., Batrinca, B.: Algorithmic regulation: automating financial compliance monitoring and regulation using AI and blockchain. J. Financ. Transf. **45**, 14–21 (2017)
7. Ibanez, L.D., O'Hara, K., Simperl, E.: On Blockchains and the General Data Protection Regulation (2018)
8. Yano, M., Dai, C., Masuda, K., et al.: Blockchain business and its regulation (2020)
9. Tsai, W.T., Wang, R., Liu, S., et al.: COMPASS: a data-driven blockchain evaluation framwework. In: 2020 IEEE International Conference on Service-Oriented System Engineering (SOSE). IEEE (2020)
10. Liu, L., Tsai, W.T., Bhuiyan, M., et al.: Automatic blockchain whitepapers analysis via heterogeneous graph neural network. J. Parallel Distrib. Comput. **145**, 1–2 (2020)
11. Tsai, W.T., He, J., Wang, R., et al.: Decentralized digital-asset exchanges: issues and evaluation. In: 2020 3rd International Conference on Smart BlockChain (SmartBlock) (2020)
12. Boubeta-Puig, J., Rosa-Bilbao, J., Mendling, J.: CEPchain: a graphical model-driven solution for integrating complex event processing and blockchain. Expert Syst. Appl. **184**, 115578 (2021)
13. Gao, J., Pattabhiraman, P., Bai, X., et al.: SaaS performance and scalability evaluation in clouds. In: Proceedings of 2011 IEEE 6th International Symposium on Service Oriented System (SOSE) (2011)
14. Lin, L., Wtta, B., Mzab C., et al.: Blockchain-Enabled Fraud Discovery Through Abnormal Smart Contract Detection On Ethereum - ScienceDirect (2022)
15. Lian, Y.U., Tsai, W.T.: State synchronization in process-oriented Chaincode. Front. Comput. Sci. (print) **13**(006), 1166–1181 (2019)
16. Hentschel, A., Hassanzadeh-Nazarabadi, Y., Seraj, R., et al.: Flow: Separating Consensus And Compute - Block Formation and Execution (2020)
17. Min, X., Li, Q., Lei, L., et al.: A permissioned blockchain framework for supporting instant transaction and dynamic block size. In: 2016 IEEE Trustcom/BigDataSE/ISPA. IEEE (2017)
18. Kang, J., Xiong, Z., Niyato, D., et al.: Incentivizing consensus propagation in proof-of-stake based consortium blockchain networks. IEEE Wirel. Commun. Lett. **8**, 157–160 (2018)

Blockchains with Five Merkle Trees to Support Financial Transactions

Wei-Tek Tsai[1], Dong Yang[1], Feng Zhang[2(✉)], Le Yu[2], Hongyang Zhang[2], and Yaowei Zhang[2]

[1] Digital Society and Blockchain Laboratory, Beihang University, Beijing 100191, China
tsai@tiandetech.com, yangdong2019@buaa.edu.cn
[2] China Mobile Information Security Management and Operation Center, Beijing 100000, China
{zhangfeng,yule,zhanghongyangxa,yaowei}@chinamobile.com

Abstract. Traditional blockchain systems the consensus protocol and transaction management is performed together. In other words, the completion of the consensus protocol means the completion of a transaction. Furthermore, the transaction is considered as settled. This scheme is not compatible with modern financial transaction rules where transaction is distinct from settlement, and thorough regulation compliance should be performed to ensure that no money laundering. This paper proposes using five Merkle trees to maintain data in a blockchain system to separate consensus protocol from transaction management.

Keywords: Consensus protocol · Transaction management · Merkle trees

1 Introduction

Traditional blockchain systems use single Merkle trees to maintain their state information, and the consensus protocol not only manages data consistency, but also performs transaction management. Thus, when the consensus protocol for a block is completed, the corresponding transactions within the block are also completed. Furthermore, these transactions are also settled at the same time [1]. This kind of one-step completion process is convenient but inconsistent with modern financial market practices, for example, this is against the PFMI (Principles of Financial Market Infrastructure). According to PFMI, transaction processes are separated from settlement processes.

One of the main issues is that if a transaction is a fraud, but it cannot be detected immediately, involved parties may lose money, and the money cannot be recovered. Thus, in traditional financial market, a transaction is settled later, giving regulation mechanisms more time to complete analysis.

In addition to the risk of money laundering, the reliability is another major issue. If the number of transactions per second is large, the probability of block

M. Qiu et al. (Eds.): SmartCom 2022, LNCS 13828, pp. 85–94, 2023.
https://doi.org/10.1007/978-3-031-28124-2_9

failure will be high. This is due to traditional blockchain design where all the transactions within a block will be considered as completed or failure together. For example, if a block has 5000 transactions, even if only one of them cannot be completed, the rest of 4999 transactions are also considered failed. All of them need to go through the consensus protocol again. Assuming each transaction will fail with a small probability of 0.01%, a block of 10K transactions will fail with 63% of probability respectively. That means a lot of blocks will fail during the consensus process. This problem becomes more acute with increasing block size, bu a large block size often means better performance.

Another major issue is that as governments such as US government start enforcing various regulations on digital assets and digital currencies (such as stablecoins), these regulation mechanisms need to be integrated into the blockchain design.

One way to address this problem is to reduce the number of transactions in a block, e.g., limit the block size to only hundreds of transaction only. In this case, the probability of the block to fail in a consensus protocol will remain small (less than 0.1% with a transaction failure rate of 0.01%). However, this may not be the optimal way. For example, Federal Reserve and MIT have released a Project Hamilton CBDC (Central Bank Digital Currency) report in Feb. 2022, but no blockchain system is used in this project. One reason cited is that they did not find a blockchain system with sufficient performance to meet the CBDC requirements.

Thus, currently one faces a dilemma here: if one wants high-performance blockchain systems, the reliability can be an issue; but if one chooses reliability as the primary objective, the performance can be an issue. But this is the issue of the traditional blockchain design, and this can be addressed by decoupling existing functionalities in the blockchain, and allocating these functionalities to different sub-systems to handle. In this way,

- One can address performance and reliability at the same time;
- One can separate the consensus protocol from transaction management;
- One can separate transaction management from transaction settlement;
- One can integrate regulation mechanisms into the blockchain architecture.

To do these, this paper proposes using five Merkle trees to manage blockchain data:

- Consensus Event Merkle Tree (CEMT);
- Transaction Event Merkle Tree (TEMT)
- Temporary Settlement Merkle Tree (TSMT);
- AML (anti-money laundering) Data Merkle Tree (ADMT);
- Bank Settlement Merkle Tree (BSMT).

This paper also proposes a new state temporary settlement or pre-settlement into the blockchain process. This is the state between trade completion and final settlement. In this state, a trade is considered temporarily settled, with all the data recorded in various blockchain systems, but the trade is not finally settled [2].

This new design has significant implication to financial market operations including scalability, reliability, and regulation, blockchain design, blockchain architecture, and blockchain networks.

The Facebook (now renamed as Meta) Diem project uses three Merkle trees to store 1) ledger, 2) account; 3) event information. It also released a settlement system FastPay, the system does not use a blockchain, but uses protocols commonly used in blockchain systems [3,5–7]. This paper proposes five Merkle trees storing different data.

2 Five Merkle Trees

The five trees are maintained independently by sub-systems, but reference each other including:

- The CEMT is related to the TEMT: the transaction data in the TEMT must be found in the CEMT;
- The TEMT is related the TSMT as only successful transactions in the TEMT can be temporarily settled and stored in the TSMT;
- The TEMT (and TSMT) is related the BSMT as the transaction settled in a bank must have been successfully gone through the consensus protocol (and the temporary settlement process) [8];
- The TEMT (and TSMT) is related the ADMT as a transaction that needs to perform AML analysis must have gone through the consensus protocol (and the temporary settlement process);
- Once a transaction is accepted by the AML process, the bank system reads data from the TSMT, and settle these transactions in banks, possibly with multiple banks with inter-bank transaction processing, and settlement data will be stored in the BSMT.

2.1 Consensus Event Merkle Tree (CEMT)

The CEMT is maintained by the consensus protocol, and records the results of the consensus protocol. It has nothing to do with transactions stored in the block that goes through the consensus protocol. The consensus protocol is responsible to make sure all the participating nodes have consistent data, and nothing else. It does not care if any transactions within the block can be successfully completed or not.

The CEMT includes the consensus event IDs, timestamps, voting results, IDs of participation node and the leader node, and the first associated consensus IDs [9]; Consensus event labeling includes blockchain IDs, the associated consensus IDs includes the consensus IDs of pre-transaction, in-transaction, pre-settlement, AML, and settlement; Where the data block unique identifier is the unique IDs of this transaction data that requires consensus; and consensus status includes consensus result (success or failure); All consensus events over time were stored into the CEMT.

2.2 Transaction Event Merkle Tree (TEMT)

The TEMT is maintained by the transaction manager of the blockchain system. It records all the transactions that gone through the consensus protocol. Once a block of transaction has been accepted by the consensus protocol, the transactions within the block can now be verified if they can be completed. Regardless if a transaction can complete its process, its resulting data will be stored in the TEMT with relevant information such as the consensus IDs, transaction IDs, participating subjects, timestamps. All successful transactions will be handled by the TSMT, and all failed transaction will be reported back to the sender, but all the transaction data are stored in the TEMT.

2.3 Temporary Settlement Merkle Tree (TSMT)

The TSMT is maintained by the temporary settlement system, and this is a new sub-system that manages only successful transactions. It records the transaction and settlement information. The TSMT reads the successful transactions in the TEMT for settlement. The TSMT can settle accounts in real time if necessary [3]. For example, if all the funds and items to be traded are cleared before the transaction, a completed transaction can be considered as settled too in futures markets [10].

2.4 AML Data Merkle Tree (ADMT)

The ADMT is maintained by the AML analysis system, and it reads the data from the TSMT for AML analysis. The AML data includes: AML event IDs, timestamps, related money laundering transaction collection, related account collection, related pre-settlement IDs, related transaction IDs and related consensus IDs. The associated consensus IDs includes the consensus IDs of pre-transaction, during transaction, pre-settlement and settlement; Each period of temporary money laundering information constitutes an AML event [11].

2.5 Bank Settlement Merkle Tree (BSMT)

The BSMT is maintained by another sub-system that responsible for settling accounts in banks or financial institutions. Sometimes, banks or financial institutions may take days (or hours) to settle balances in accounts, and modern payment systems can settle balance in seconds. But regardless the time needed to settle balance, the BSMT records the event that banks finally settle the balance in the right accounts. The data structure of the bank settlement event includes: settlement event IDs, timestamps, pre-settlement balance, post-settlement balance, associated money laundering transaction collection, associated banks including inter-bank accounts, relevant pre-settlement IDs, relevant transaction identity, and relevant consensus IDs, where the relevant consensus IDs includes pre-transaction, in-transaction, pre-settlement, settlement consensus IDs [12]. The banking system data may also include inter-bank transactions, and these can be carried out concurrently [13].

3 Method Implementation

With availability of Merkle trees operated by independent sub-systems, the blockchain systems now have enough capabilities to support a variety of trading methods, including real-time trading with delayed settlement, real-time trading with instantons settlement, trading with separate settlement and clearing processes. These can be realized by using a subset of five Merkle trees, or by adding new Merkle trees to support additional processes, such as clearing.

3.1 Separating Consensus Protocol from Transaction Management

The CEMT consists of several parts shown in Table 1, including consensus event IDs, timestamps, consensus participation node and leader node, and first associated consensus IDs; blockchain identity card, associated consensus IDs such as pre-transaction, transaction, pre-settlement, AML, settlement; where the unique identifier is the unique IDs of this transaction data requiring consensus, including consensus success and consensus failure.

Table 1. Consensus event structure

Consensus event marking (blockchain IDs card, etc.)
Timestamp
Consensus participating nodes, leading nodes
Associated consensus IDs (such as pre-transaction, mid-transaction, pre-settlement, anti-money laundering, and settlement consensus identifiers)

Based on the five Merkle-tree structure [14], the consensus mechanism is responsible for the data consistency between each blockchain node only, the rest of functionalities are allocated to other sub-systems to handle. Due to this decoupling, the consensus mechanism can simultaneously support multiple transactions mechanisms, temporary settlement mechanisms, AML mechanisms, and settlement mechanisms. For example, a trading system can have one consensus protocol, but six AML sub-systems, each AML sub-system is supplied by independent regulatory agencies including international, national, and local regulatory agencies. All these AML sub-systems can be active running on top of a bigData platforms at the same time. Potentially, some transactions can be settled instantaneously, while some must have delayed settlement, but the choice is made in real time based on the specific transactions at hand. In the case of instantaneous settlement, the TSMT will be automatically updated to reflect this situation [15,16].

3.2 Separating Transaction Management from Temporary Settlement

The TEMT is maintained by the transaction sub-system. The transaction subsystem can use bigData platform to process all consensus successful data blocks concurrently to determine each transaction can be completed. The transaction event structure is shown in Table 2, including transaction event labeling, transaction data processing start time, transaction data processing settlement time, successful transaction collection, failed transaction collection, failure reason and association consensus IDs, such as pre-transaction, transaction, pre-settlement, AML and settlement consensus IDs [19].

Table 2. Transaction event structure

Transaction event flag
Transaction data processing start time
Transaction data processing settlement time
Collection of successful transactions set of failed transactions Reason for failure
Associated consensus IDs (such as pre-transaction, mid-transaction, pre-settlement, anti-money laundering, and settlement consensus identifiers)

The structure of the TSMT is shown in Table 3 including: temporary settlement event ID, account unique identifier, related party transaction collection, balance before settlement, post-settlement balance, timestamps, related party transaction ID, and related party transaction consensus ID. The associated consensus ID, such as the pre-transaction, transaction, pre-settlement, AML and settlement.

The TSMT stores those transactions that can be settled but still subject to the AML analysis before final settlement. While the AML analysis is being done, those transaction data will be kept in the blockchain to ensure that no data will be tampered with. This feature enhances the reliability and reputation of the trading system. As multiple AML sub-systems may be operating at the same time, the system needs to wait until all the AML sub-systems provide green light before proceeding to final settlement.

3.3 Separation of Temporary Settlement from AML Mechanisms

The ADMT is maintained by the AML analysis systems, and it reads the data in the temporary settlement Merkle tree for AML analysis. Various AML algorithms can be used. The AML data structure is shown in Table 4, including AML events, timestamps, related money-laundering transactions, related accounts,

Table 3. Temporary settlement event

Consensus event marking (blockchain IDs card, etc.)
Temporarily settle the event IDs Account unique identifier
Collection of related party transactions
Balance before settlement Balance after settlement Timestamp
Related Party Transaction IDs
Associated consensus IDs (e.g., pre-transaction, pre-settlement, anti-money laundering, settlement consensus identifiers)

related pre-settlement IDs, related transaction IDs and related consensus protocol identity, such as the consensus IDs of pre-transaction, transaction, pre-settlement and settlement. The information of temporary money-laundering constitutes an AML event, and multiple AML events constitute an ADMT.

Table 4. Anti-money laundering event structure

Transaction event flag
Settlement event IDs Timestamp
Balance before settlement Balance after settlement
Associated collection of money laundering transactions
A collection of linked bank accounts, including cross-line
The relevant pre-settlement IDs Correlation transaction IDs
Relevant consensus IDs (e.g. pre-transaction, in-transaction, pre-settlement, settlement consensus identifiers)

3.4 Various Analysis can be Performed

Figure 1 shows the dependency diagram of the five trees. With these five trees, numerous analysis can be performed including:

- **Integrity Analysis**: One can perform integrity analysis by checking completeness of relevant data, e.g., each temporary settlement event requires a corresponding transaction event;
- **Consistency Analysis**: One can use five trees to make sure that data in three trees are consistent with each otehr as these trees contain redundant data. For example, a settlement event can only correspond to one transaction event, not tw transaction events;
- **Transaction-Ordering Analysis**: One can verify the execution order of various transactions, e.g., temporary settlement events must occur before the final settlement;

- **AML analysis**: Multiple parties can perform different set of AML analysis using different algorithms and systems, and they can run on platforms outside of the blockchain systems. This can support real-time monitoring and enforcement;
- **Timing Analysis**: One can determine the time needed to complete various step during the transaction-settlement process.

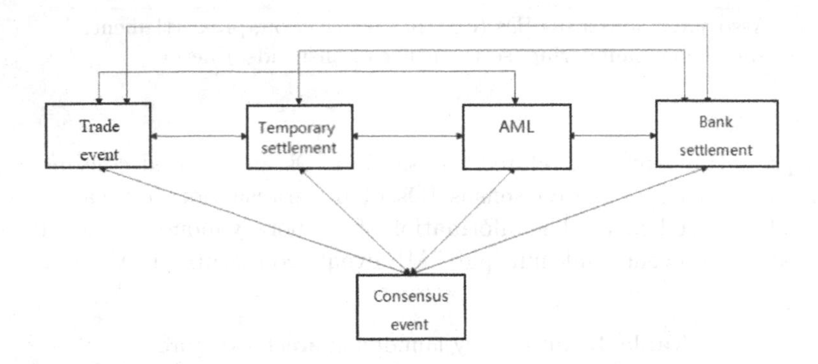

Fig. 1. Inter-Dependency among Merkle Trees

Most of these analysis can be performed concurrently in different sub-systems for optimal performance. The system can be equipped with one blockchain systems or multiple blockchain systems. They can be done on demand, periodically, or both.

The five-trees structures can support many different blockchain-based transaction processes, including large international cross-border payments, multinational financial institutions, multiple blockchain systems, multiple AML enforcement agencies, and complex transaction mechanisms. For example, JPMorgan bank's digital currency system that includes 20 large banks, 400 financial institutions, and spread over 78 countries. This large digital currency platform requires a large number of transactions, audit, AML, settlement mechanisms, and numerous cooperating processes.

4 Conclusion

This paper addresses the problems of potential money laundering and reliability in the blockchain system by proposing a complete process from consensus protocol to transaction management to settlement in banks with five Merkle trees. This scheme separates transaction management from consensus protocol, transaction from temporary settlement, temporary settlement from final settlement. With this structure, a blockchain system will be easier to manage and scale.

Acknowledgments. The research was funded by National Key Research and Development Program of China (No. 2018YFB1402700).

References

1. Tsai, W.T., Wang, R., Liu, S., et al.: COMPASS: a data-driven blockchain evaluation framwework. In: 2020 IEEE International Conference on Service-Oriented System Engineering (SOSE). IEEE (2020)
2. Liu, L., Tsai, W.T., Bhuiyan, M., et al.: Automatic blockchain whitepapers analysis via heterogeneous graph neural network. J. Parallel Distrib. Comput. **145**, 1–2 (2020)
3. Tsai, W.T., He, J., Wang, R., et al.: Decentralized digital-asset exchanges: issues and evaluation. In: 2020 3rd International Conference on Smart BlockChain (SmartBlock) (2020)
4. Boubeta-Puig, J., Rosa-Bilbao, J., Mendling, J.: CEPchain: a graphical model-driven solution for integrating complex event processing and blockchain. Expert Syst. Appl. **184**, 115578 (2021)
5. Gao, J., Pattabhiraman, P., Bai, X., et al.: SaaS performance and scalability evaluation in clouds. In: Proceedings of 2011 IEEE 6th International Symposium on Service Oriented System (SOSE) (2011)
6. Tsai, W.T., Chen, Y., Paul, R.A., et al.: Cooperative and group testing in verification of dynamic composite web services. Int. Comput. Softw. Appl. Conf. IEEE (2004)
7. Tsai, W.T., Chen, Y., Paul, R.A., et al.: Services-oriented dynamic reconfiguration framework for dependable distributed computing. In: 28th International Computer Software and Applications Conference (COMPSAC 2004), Design and Assessment of Trustworthy Software-Based Systems, 27–30 September 2004, Hong Kong, China, Proceedings. IEEE Computer Society (2004)
8. Lin, L., Wtta, B., Mzab, C., et al.: Blockchain-enabled Fraud Discovery through Abnormal Smart Contract Detection on Ethereum - ScienceDirect (2022)
9. Tsai, W.T., Luo, Y., Deng, E., et al.: Blockchain systems for trade clearing. J. Risk Financ. **21**(5), 469–492 (2020)
10. Tsai, W.T., Blower, R., Yan, Z., et al.: A system view of financial blockchains. In: 2016 IEEE Symposium on Service-Oriented System Engineering (SOSE). IEEE (2016)
11. Lian, Y.U., Tsai, W.T.: State synchronization in process-oriented Chaincode. Front. Comput. Sci. (print) **13**(006), 1166–1181 (2019)
12. Tsai, W.T., Zhao, Z., Chi, Z., et al.: A multi-chain model for CBDC. In: 2018 5th International Conference on Dependable Systems and Their Applications (DSA) (2018)
13. Erwu, L.A., et al.: Blockchain technologies and applications. **16**(6), 3 (2019)
14. Wei-Tek, T., Lian, Y.U., Yuan, B., et al.: Big data-oriented blockchain for clearing system. Big Data Res. (2018)
15. Sun, Y., Xue, R., Zhang, R., et al.: RTChain: a reputation system with transaction and consensus incentives for e-commerce blockchain. ACM Trans. Internet Technol. **21**(1), 1–24 (2020)
16. Bissas, G., Levine, B.N., Ozisik, A.P., et al.: An analysis of attacks on blockchain consensus (2016)
17. Hentschel, A., Hassanzadeh-Nazarabadi, Y., Seraj, R., et al.: Flow: separating consensus and compute - block formation and execution (2020)

18. Min, X., Li, Q., Lei, L., et al.: A permissioned blockchain framework for supporting instant transaction and dynamic block size. In: 2016 IEEE Trustcom/BigDataSE/ISPA. IEEE (2017)
19. Kang, J., Xiong, Z., Niyato, D., et al.: Incentivizing consensus propagation in proof-of-stake based consortium blockchain networks. IEEE Wirel. Commun. Lett. (2018)

A Survey of Recommender Systems Based on Hypergraph Neural Networks

Canwei Liu[1,3], Tingqin He[1,3](✉), Hangyu Zhu[1,2,3], Yanlu Li[1,3], Songyou Xie[4], and Osama Hosam[5]

[1] School of Computer Science and Engineering, Hunan University of Science and Technology, Xiangtan 411201, China
hetingqin@hnust.edu.cn

[2] Guangdong Financial High-tech Zone "Blockchain +" Fintech Research Institute, Guangzhou 510623, China

[3] Hunan Key Laboratory for Service computing and Novel Software Technology, Xiangtan 411201, China

[4] College of Computer Science and Electronic Engineering, Hunan University, Changsha 410082, China

[5] Higher Colleges of Technology, Abu Dhabi, United Arab Emirates

Abstract. Unlike highly purposeful search, a recommender system tends to uncover the user's potential interests and is a personalized information filtering system. Recently, the performance of hypergraph neural networks in classification tasks has attracted much attention. Compared with traditional recommender systems, hypergraph neural network-based recommender systems have better mining higher-order associations, accurate modeling of multivariate relationships, handling of multimodal and heterogeneous data, and clustering advantages. This fact drives the development of recommendation algorithms based on hypergraph neural networks. To this end, we 1) define generic links of recommender systems, and systematically analyze the challenges of hypergraph neural network-based recommender systems in different research directions. 2) present some new perspectives on existing weaknesses and future developments.

Keywords: Collaborative Filtering Algorithms · Graph Neural Networks · Hypergraph Neural Networks · Recommender Systems · Social Recommendation

1 Introduction

1.1 Introduction to Recommender Systems

With the rapid development of the IoT [1,2], storage technology [3], deep learning technology [4–6] and etc., the amount of data has also increased exponentially [7–9]. However, each person's life and upbringing are extremely different, so everyone

M. Qiu et al. (Eds.): SmartCom 2022, LNCS 13828, pp. 95–106, 2023.
https://doi.org/10.1007/978-3-031-28124-2_10

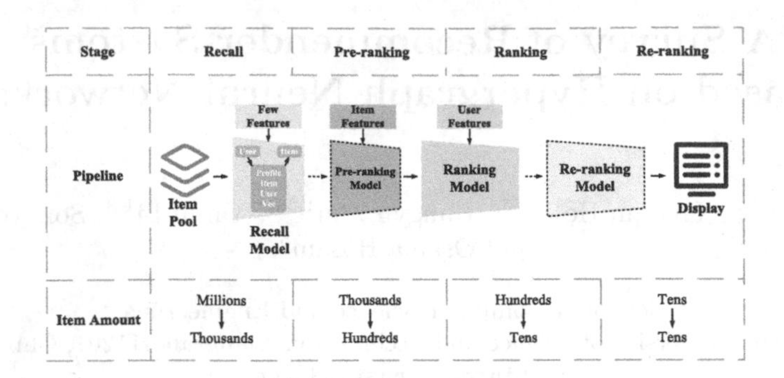

Fig. 1. Schematic diagram of recommender system structure

has different preferences and tastes. To handle massive information [10–12] and meet the various individual needs of users, the recommender system was born.

So far, the general framework of the recommender system has been relatively stable, and its full chain is generally divided into three stages: recall, pre-ranking, and ranking. Sometimes, to maximize the efficiency of the page, it also goes through re-ranking to reasonably distribute the flow, and these three stages are described below.

Recall. Recall uses a small number of features and a simple model to quickly filter out some of the data from a large amount of data for use in the subsequent sorting stage. That is, recall can reduce the scale of millions or hundreds of millions of items to thousands of candidates, which greatly reduces the computing time in the subsequent sorting stage. The size of the input data in this phase is so large that the model in this phase is usually designed to be simpler, considering the computational speed of the recall phase. In addition to improving the computational speed of the subsequent sorting stage that uses complex models, adding the recall stage can make up for the lack of mainly considering a single target in the sorting stage.

Pre-ranking. Pre-ranking is located between recall and ranking. In most cases, the structure of the pre-ranking model is similar to that of the ranking or recall model, so pre-ranking is not a necessary part of the recommender system. Pre-ranking has a very strict time requirement, usually, a few hundred data are selected from a candidate set of several thousand data in a very short time and sent to the ranking model for calculation. Generally speaking, pre-ranking does not consider whether it matches user characteristics, as long as the quality of the content is good, it will be ranked first.

Ranking. In the ranking stage, the candidate set has only a few hundred data. Therefore, more sophisticated models and finer strategies can be used than pre-ranking, and the most accurate predictions can be output. The models used in the ranking stage combine a large number of user features and are used to calculate the interest level of a particular user in certain content, thus enabling accurate personalized ranking of items (Fig. 1).

The main development stages of recommender systems are expert systems, neural network-based recommender systems, graph neural network-based recommender systems, and the recently emerged hypergraph neural network-based recommender systems.

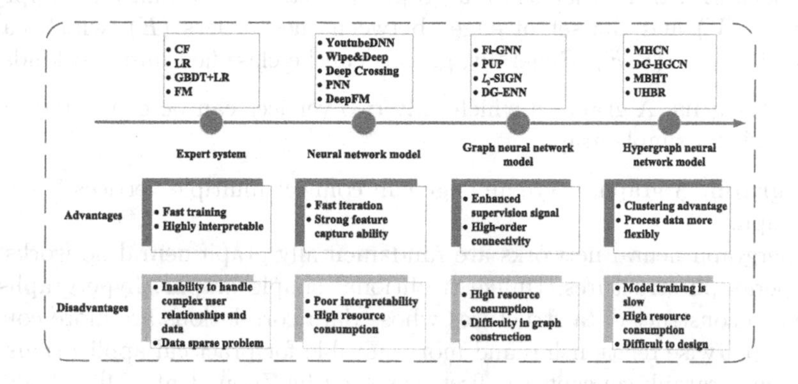

Fig. 2. History of Recommender System Development

Recommender systems started from the GroupLens research group at the University of Minnesota in 1994, which introduced the GroupLens system [13]. This article first proposed the idea of collaborative filtering algorithm to complete the recommendation task, which is also the User-based CF [14] widely studied by scholars later. Subsequently, Item-based CF [15], MF-based CF [16], and other collaborative filtering algorithms have been proposed one after another. Meanwhile, other non- collaborative filtering recommendation algorithms, such as PLOY2 [17], and FM [18] are also being developed. The above approaches belong to the expert system stage. Although the models at this stage are more interpretable, they face great challenges in dealing with complex user relationships and data as well as data sparsity problems. Therefore, neural network-based models are proposed.

The input of Deep Crossing [19] is a set of individual features, which can be dense or sparse, so it can better solve the data sparsity problem faced by expert systems. Meanwhile, Deep Crossing uses Multi-Layer Perceptron (MLP) to extend the capacity to better handle complex user relationships and data. Similarly, PNN networks [20] and DeepFM [21] are also combined with MLP. These approaches, however, are clumsy in modeling multi-hop information and difficult to encode higher-order information of the data to capture complex user preferences implicit in sequential behaviors. The high-order connectivity and other properties of graph neural networks can be a good solution to these problems, so recommender systems based on graph neural networks are proposed (Fig. 2).

1.2 Introduction to Hypergraph Neural Networks

Graph Neural Networks, a deep learning method based on graph structures [22, 23], is essentially a method for processing graph-structured data with the help of neural networks. In this section, we first introduce these two different graph structures, then explain the difference between hypergraph neural networks and graph neural networks, and finally summarize the current status of hypergraph neural networks. In mathematics, a graph consists of an infinite nonempty set of vertices (V) and the set of edges between the vertices (E), which can be expressed as $G = (V, E)$. Generally, graphs can be classified into two kinds.

Simple Graphs. A graph in which only two vertices can be connected by one edge is called a simple graph.

Hypergraph. A graph where an edge can connect multiple vertices is called a hypergraph.

Hypergraph neural networks are fundamentally graph neural networks that use hypergraph structures. Unlike traditional simple graphs, hypergraphs can be used to construct data structures whose data correlations are more complex than the pairwise relationship and more suitable for practical applications.

The hypergraph concept was first proposed by Zhou et al. [24] in 2006, but the computational complexity and high storage cost of hypergraphs at that stage prevented them from being widely used. With the development of deep learning, several studies combined GNN methods with hypergraphs to enhance representation learning. HGNN [25] was the first spectral model that used hypergraph convolution operations to process complex data, and hypergraph neural network models have grown rapidly since then. Bai et al. [26] introduced hypergraph attention to hypergraph convolutional networks to increase their capacity; Vijaikumar et al. [27] proposed HyperTenet in 2021 to add the next item to a user-generated list in a personalized manner; Jo et al. [28] proposed a hyper clustering and hyper drop approach overcome the limitations of the node-based pooling approach of existing graph neural network methods. The combination of a hypergraph and deep learning techniques solves the technical problems that existing graph neural network methods are difficult to directly model higher-order complex associations and the low learning efficiency of the models and promotes the development of deep learning techniques.

1.3 Advantages of Hypergraph Neural Network Applied to Recommender System

With the development of social media, recommender systems have gradually shifted from individual to group modeling. Group preferences are a mixture of various preferences of individuals, so group-oriented recommender systems need to consider the correlation among group members when modeling. Most of the previous approaches mainly focus on pairwise connections between individuals within a group and often ignore the complex higher-order interactions between groups, i.e., the influence of user interactions inside and outside the group.

While the hypergraph structure has a stronger ability to portray and mine nonlinear higher-order associations between data samples compared to the general graph structure [25] and can model multivariate relationships more accurately [29] and is more flexible in dealing with multimodal and heterogeneous data [30]. It is also more advantageous in the clustering process [31]. Therefore, compared with ordinary graph neural networks, hypergraph neural networks can better solve many problems in recommender systems in practical applications (Fig. 3).

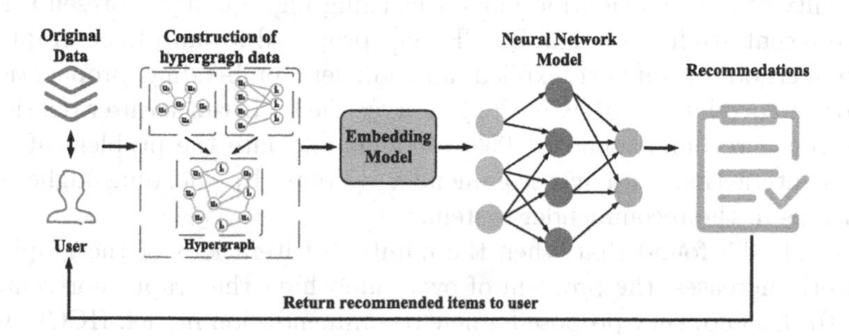

Fig. 3. Schematic diagram of the application of hypergraph neural network in recommender system

In the remainder of this paper, the structure is organized as follows: Sect. 2 mentions the model development of recommender systems; Sect. 3 describes the main research directions of hypergraph neural networks in recommender systems. Next, we discuss the challenges to be faced by hypergraph neural networks for recommender systems and provide an outlook on future work in Sect. 4. Finally, in Sect. 5, we conclude our work.

2 The Development of Recommender System Models

2.1 Traditional Model

Traditional recommendation algorithms mainly refer to collaborative filtering. Among them, Item-based CF [15] uses users' historical behaviors to make real-time recommendations to users, which is more convincing to users than User-based CF [14]. MF [16] is used to solve the problems such as the weak ability of the collaborative filtering algorithm to deal with sparse matrices and the difficulty of maintaining the similarity matrix. But the traditional MF algorithm, the reusability is not good. After that, the researchers proposed the FM [18], which is essentially an LR and cross-term feature combination.

And then, with the development of deep learning techniques, more and more recommendation models based on neural networks or graph neural networks have been proposed, such as neural network-based Deep Crossing [19], Wide&Deep [32] etc., graph neural network-based Fi-GNN [33], L_0-SIGN [34] etc.

2.2 Hypergraph Neural Network Model

Although existing graph neural network-based recommendation models have achieved great success in various recommendation tasks, there are still problems that need to be solved.

Most graph neural network-based recommendation models are supervised learning. In the real recommendation scenario, the user-commodity interactions are usually sparse and unbalanced, and the inevitable wrong clicks between users and commodities, etc., make the supervised signals noisy. All of these will lead to the difficulty of recommendation models learning high-quality representations.

Some recent studies, such as DHCF [35], proposed a jump hypergraph convolution method to support explicit and efficient embedding propagation of higher-order correlations; MHCN [36] leverages the higher-order user relations to enhance social recommendation. These models alleviate the problem of sparse user-item interaction data in recommender systems by encoding higher-order relationships in the recommender system.

Xia et al. [37] found that when the number of iterations of the graph neural network increases, the problem of over-smoothing the graph representation occurs. To this end, they proposed a new recommendation model, HCCF, which combines global hypergraph structure learning with local co-relational encoders, while designing a hypergraph comparison learning method to solve the problem of over-smoothing of graph representation (Table 1).

Table 1. Representative models for each developmental stage.

Stage	Model	Venue	Year
Expert System	POLY2 [17]	JMLR	2010
	FM [18]	ICDM	2010
Neural Network Model	Deep&Cross [19]	KDD	2016
	Wide&Deep [32]	DLRS	2016
Graph Neural Network Model	Fi-GNN [33]	CIKM	2019
	L_0-SIGN [34]	AAAI	2021
	S-MBRec [38]	IJCAI	2022
Hypergraph Neural Network Model	DHCF [35]	KDD	2020
	MHCN [36]	WWW	2021
	HCCF [37]	SIGIR	2022
	DH-HGCN [39]	SIGIR	2022
	MBHT [40]	KDD	2022
	H^2SeqRec [41]	CIKM	2021

3 The Main Research Directions of Hypergraph Neural Networks in Recommender Systems

This section details several major research directions of hypergraph neural networks in recommender systems, including social recommendation, sequential recommendation, bundled recommendation, and session-based recommendation, and illustrates what challenges these research directions to have, and the corresponding solutions.

3.1 Social Recommendation

In the past decade or so, the development of social media has dramatically changed people's lifestyles and thinking habits. A study [42] confirmed that people tend to connect with people who have similar preferences. At the same time, people's behavior is influenced by their friends. Inspired by these studies, social connections are often integrated into recommender systems to alleviate the data sparsity problem.

Hypergraphs give a way to model complex higher-order relationships between users. For example, MHCN [36] generates better social recommendation results by aggregating multiple user embeddings learned through multiple channels to obtain an integrated user representation. And the recently proposed DH-HGCN [39], advocates modeling dual homogeneity in social relationships and item connections to obtain higher-order correlations between users and items.

The privacy issues of online social networks have become a matter of great concern for many people [43], too. Hypergraphs can help address privacy issues. zhang et al. [44], modeled online social networks as hypergraphs and proposed a KFR scheme to address the problem of relationship privacy leakage in social recommendations.

3.2 Sequence Recommendation

In sequential recommendation scenarios, it needs to model user preferences over time through item sequences.

Although it is effective to use graph neural networks as a backbone model to capture item dependencies in sequential recommendation scenarios, existing approaches have been focused on sequential representations of items with a single interaction type and are therefore limited to capturing the dynamic heterogeneous structure of relationships between users and items [40].

Hypergraphs are a natural way to model higher-order relationships between users and items, and MBHT [40] incorporates global multi-behavioral dependencies into a hypergraph neural network architecture to capture both short-term and long-term cross-type behavioral dependencies; H^2SeqRec [41] addresses most of the existing general sparsity problem of sequence recommendations based on graphs and the lack of utilization of hidden hypergraphs among users.

3.3 Bundle Recommendation

Bundled recommendation aim to recommend a bundle of items to users so that they are consumed as a whole. It not only improves the user experience, but also increases sales.

A bundle consists of multiple items, if users are to be satisfied with most of the items in the bundle, the interaction between the user and the bundle will be more sparse. Chen et al. [45] proposed a multitasking framework for modeling user bundle interaction and user-item interaction simultaneously to alleviate the sparsity of user bundle interaction. However, this approach is difficult to balance the weight between primary and secondary tasks.

A recent study proposed a hypergraph framework, UHBR [46], for bundle recommendation, which unifies multiple relationships among users, bundles, and items into a hypergraph, which better solves the above problem.

3.4 Session-Based Recommendation

Session-based recommendation (SBR) focuses on the next prediction at a point in time. The main difference between *Sequence Based Recommendation* (SR) and Session Based Recommendation is the anonymity and the length of the sequence, i.e., in the SBR scenario, we do not know long-term information about the user, so the length of the sequence in SBR is shorter than the length of the sequence in SR.

In existing neural network-based recommendation methods, the session-based data are usually modeled as one-way sequences, and thus these data may be time-dependent. In contrast, in graph neural network-based recommendation methods, the session-based data are generally modeled as directed subgraphs and item transitions are treated as pairwise relationships, which can slightly alleviate the temporal dependence among consecutive items. However, in practical application scenarios, item transitions are co-triggered by clicks on previous items, and there are many-to-many and higher-order relationships between items. Hypergraphs are the best choice to describe such relationships. Xia et al. [47] then used hypergraphs to model the higher-order relationships in session items for capturing the higher-order correlations prevalent between items.

4 Challenges and Future Prospects

Although the application of hypergraph neural networks to recommender systems can be a good way to improve the performance of recommender systems, there are various challenges. We will list a few challenges and give an outlook on the future.

Hypergraph Construction Problem. The first step of applying hypergraph neural networks to recommender systems is obviously to construct a suitable hypergraph according to the actual situation, and the quality of the hypergraph directly affects the quality of relationship modeling. However, the construction of hypergraphs requires consideration of higher order relationships in the data than the construction of simple graphs. And the computational complexity of existing recommendation algorithms based on hypergraph neural networks is large, especially when updating the hypergraph structure, which exacerbates the computational complexity and makes it difficult to scale to large-scale data. Therefore, building a hypergraph that can handle the task well is not a minor challenge.

Model Collaboration Problem. A good recommender system is composed of multiple models, which requires efficient collaboration between the various modules of the system, so that users have a better subjective perception of the recommendation results. In addition, due to the complex data link of the recommender system, the recommendation results should be interpretable and diagnosable to facilitate system optimization.

Cold Start Problem. In a new domain, neither new users nor new items have relevant behavioral information, which is a common cold start problem in recommender systems. And the cross-domain recommendation is an effective approach to alleviate the cold-start [48] and data sparsity problems [49]. By discerning the relationship between items in several different domains, a recommender system can use the user's preferences in a known domain to recommend items to the user in a new domain. At this stage, hypergraph neural networks have been less studied in the cross-domain recommendation. We believe that the structure of hypergraphs can model and integrate information from different domains well, and the potential of hypergraph neural networks in this research direction is yet to be developed.

Noise Problem. In practical applications, noise and malicious attacks can have a relatively large impact on the recommendation results, and it is a challenge to ensure the quality of the training data. The self-supervised learning approach is proven to alleviate the noise problem, and the self-supervised hypergraph neural network learning approach may become a future research direction.

5 Conclusion

In recent years, hypergraph neural network models have been rapidly developed in the research field of recommender systems. In this paper, we summarized and analyzed the development history and the latest progress based on the existing literature. Traditional recommender systems often have problems such as data sparsity and noise, which can be well mitigated by using hypergraph neural network methods with high-order association mining capabilities. This paper analyzed the advantages of hypergraph neural network methods applied in recommender systems and summarizes the challenges, methods, and future directions in different research areas. Of course, the application of hypergraph neural networks to recommender systems is a new area of research that started in recent years, and there are still many open directions waiting to be explored.

References

1. Liang, W., Long, J., Li, K.C., Xu, J., Ma, N., Lei, X.: A fast defogging image recognition algorithm based on bilateral hybrid filtering. ACM TOMM **17**(2), 1–16 (2021)
2. Xu, Z., Liang, W., Li, K.C., Xu, J., Zomaya, A.Y., Zhang, J.: A time-sensitive token-based anonymous authentication and dynamic group key agreement scheme for industry 5.0. IEEE TII **18**(10), 7118–7127 (2021)
3. Wang, J., Luo, W., Liang, W., Liu, X., Dong, X.: Locally minimum storage regenerating codes in distributed cloud storage systems. China Commun. **14**(11), 82–91 (2017)
4. Liang, W., Li, Y., Xu, J., Qin, Z., Li, K.C.: Qos prediction and adversarial attack protection for distributed services under dlaas. IEEE Trans. Comput. **2021**, 1–14 (2021)
5. Diao, C., Zhang, D., Liang, W., Li, K.C., Hong, Y., Gaudiot, J.L.: A novel spatial-temporal multi-scale alignment graph neural network security model for vehicles prediction. IEEE Trans. Intell. Trans Syst. **24**, 904–914 (2022)
6. Peng, L., Peng, M., Liao, B., Huang, G., Liang, W., Li, K.: Improved low-rank matrix recovery method for predicting mirna-disease association. Sci. Rep. **7**(1), 6007 (2017)
7. Qiu, M., Chen, Z., et al.: Energy-aware data allocation with hybrid memory for mobile cloud systems. IEEE Syst. J. **11**(2), 813–822 (2014)
8. Li, Y., Gai, K., et al.: Intercrossed access controls for secure financial services on multimedia big data in cloud systems. ACM TMMCCA **12**(4s), 1–18 (2016)
9. Li, J., Ming, Z., et al.: Resource allocation robustness in multi-core embedded systems with inaccurate information. J. Syst. Arch. **57**(9), 840–849 (2011)
10. Gai, K., Qiu, M., Elnagdy, S.: A novel secure big data cyber incident analytics framework for cloud-based cybersecurity insurance. In: IEEE BigDataSecurity Conference (2016)
11. Hu, F., Lakdawala, S., et al.: Low-power, intelligent sensor hardware interface for medical data preprocessing. IEEE Trans. Infor. Tech. Biomed. **13**(4), 656–663 (2009)
12. Qiu, M., Xue, C, Shao, Z, et al.: Efficient algorithm of energy minimization for heterogeneous wireless sensor network. In: IEEE EUC Conference, pp. 25–34 (2006)
13. Resnick, P., Iacovou, N., et al.: Grouplens: an open architecture for collaborative filtering of netnews. In: ACM Conference on Computer Supported Cooperative Work, pp. 175–186 (1994)
14. Zhao, Z.D., Shang, M.S.: User-based collaborative-filtering recommendation algorithms on hadoop. In: IEEE 3rd International Conference on Knowledge Discovery and Data Mining, pp. 478–481 (2010)
15. Sarwar, B., Karypis, G., Konstan, J., Riedl, J.: Item-based collaborative filtering recommendation algorithms. In: 10th International Conference on World Wide Web (2001)
16. Koren, Y., Bell, R., Volinsky, C.: Matrix factorization techniques for recommender systems. Computer **42**(8), 30–37 (2009)
17. Chang, Y.-W., Hsieh, C.-J., et al.: Training and testing low-degree polynomial data mappings via linear SVM. J. Mach. Learn. Res. **11**, 1471–1490 (2010)
18. Rendle, S.: Factorization machines. In: 2010 IEEE ICDM (2010)
19. Shan, Y., Hoens, T.R., Jiao, J., et al.: Deep crossing: web-scale modeling without manually crafted combinatorial features. In: 22nd ACM SIGKDD (2016)

20. Qu, Y., et al.: Product-based neural networks for user response prediction. In: 16th IEEE ICDM (2016)
21. Guo, H., Tang, R., et al.: Deepfm: a factorization-machine based neural network for CTR prediction. In: 26th Conference on Artificial Intelligence, pp. 1725–1731 (2017)
22. Qiu, H., Dong, T., et al.: Adversarial attacks against network intrusion detection in IoT systems. IEEE IoT J. **8**(13), 10327–10335 (2020)
23. Qiu, H., Zheng, Q., et al.: Topological graph convolutional network-based urban traffic flow and density prediction. IEEE Trans. ITS **22**(7), 4560–4569 (2020)
24. Zhou, D., Huang, J., Schölkopf, B.: Learning with hypergraphs: clustering, classification, and embedding. In: Advances in Neural Information Processing Systems, vol. 19 (2016)
25. Feng, Y., You, H., Zhang, Z., et al.: Hypergraph neural networks. In: AAAI Conference on Artificial Intelligence (2019)
26. Bai, S., Zhang, F., Torr, P.H.L: Hypergraph convolution and hypergraph attention. Pattern Recognit. **110**, 107637 (2021)
27. Vijaikumar, M., Hada, D., Shevade, S.: Hypertenet: hypergraph and transformer-based neural network for personalized list continuation. In: IEEE ICDM, pp. 1210–1215 (2021)
28. Jo, J., Baek, J., Lee, S., et al.: Edge representation learning with hypergraphs. In: Advances in Neural Information Processing Systems (2021)
29. Do, M.T., Yoon, S.E., Hooi, B., Shin, K.: Structural patterns and generative models of real-world hypergraphs. In: 26th ACM SIGKDD (2020)
30. Kim, E.-S., Kang, W.Y., et al.: Hypergraph attention networks for multimodal learning. In: EEE/CVF Conference on Computer Vision and Pattern Recognition (2020)
31. Zhang, R., Zou, Y., Ma., J.: Hyper-sagnn: a self-attention based graph neural network for hypergraphs. In: ICLR (2020)
32. Cheng, H.-T., Koc, L., et al.: Wide & deep learning for recommender systems. In: 1st Workshop on Deep Learning for Recommender Systems (2016)
33. Li, Z., Cui, Z., Wu, S., et al.: Fi-gnn: modeling feature interactions via graph neural networks for CTR prediction. In: 28th ACM Conference on Information and Knowledge Management (2019)
34. Su, Y., Zhang, R., Erfani, S., Xu, Z.: Detecting beneficial feature interactions for recommender systems. In AAAI Conference on Artificial Intelligence (2021)
35. Ji, S., Feng, Y., et al.: Dual channel hypergraph collaborative filtering. In: 26th ACM SIGKDD (2020)
36. Yu, J., Yin, H., et al.: Self-supervised multi-channel hypergraph convolutional network for social recommendation. In: Web Conference (2021)
37. Xia, L., Huang, C., Xu, Y., et al.: Hypergraph contrastive collaborative filtering. In: 45th ACM SIGIR Conference on Research and Development in Information Retrieval (2022)
38. Gu, S., Wang, X., Shi, C., Xiao, D.: Self-supervised graph neural networks for multi-behavior recommendation (2022)
39. Han, J., Tao, Q., et al.: DH-HGCN: dual homogeneity hypergraph convolutional network for multiple social recommendations. In: 45th ACM SIGIR Conference on Research and Development in Information Retrieval, pp. 2190–2194 (2022)
40. Yang, Y., Huang, C., Xia, L., et al.: Multi-behavior hypergraph-enhanced transformer for sequential recommendation. In: 28th ACM SIGKDD (2022)

41. Li, Y., Chen, H., et al.: Hyperbolic hypergraphs for sequential recommendation. In: 30th ACM I Conference on Information Knowledge Management, pp. 988–997 (2021)
42. McPherson, M., Smith-Lovin, L., Cook, J.M.: Birds of a feather: homophily in social networks. Ann. Rev. Soc. **27**, 415–444 (2001)
43. Zhang, S., Yao, T., et al.: A novel blockchain-based privacy-preserving framework for online social networks. Connection Sci. **33**(3), 555–575 (2021)
44. Zhang, S., Li, X., et al.: A privacy-preserving friend recommendation scheme in online social networks. Sustain. Cities Soc. **38**, 275–285 (2018)
45. Chen, L., Liu, Y., et al.: Matching user with item set: Collaborative bundle recommendation with deep attention network. In: IJCAI (2019)
46. Yu, Z., Li, J., Chen, L., Zheng, Z.: Unifying multi-associations through hypergraph for bundle recommendation. Knowl.-Based Syst. **255**, 109755 (2022)
47. Xia, X., Yin, H., et al.: Self-supervised hypergraph convolutional networks for session-based recommendation. In: AAAI Conference on Artificial Intelligence (2021)
48. Abel, F., Herder, E., et al.: Cross-system user modeling and personalization on the social web. User Model. User-Adap. Inter. **23**, 169–209 (2013). https://doi.org/10.1007/s11257-012-9131-2
49. Pan, W., et al.: Transfer learning in collaborative filtering for sparsity reduction. In: AAAI Conference on Artificial Intelligence (2010)

Equipment Health Assessment Based on Node Embedding

Jian Li, Xiao Chen, Chao Zhang, Hao Wu, Xin Yu, Shiqi Liu, and Haolei Wang(✉)

State Grid Zhoushan Electric Power Supply Company of Zhejiang Power Corporation, Zhejiang 316021, Zhoushan, China
151491628@qq.com

Abstract. Equipment health assessment is a fundamental task in predictive equipment maintenance practice, which aims to predict the health of equipment based on information about the equipment and its operation, thus avoiding unexpected equipment failures. In the current context, equipment health assessment based on sequential deep learning methods is becoming more and more popular, however, such methods ignore the inter-device correlations, leading to their lack of readiness for health assessment of a large number of devices. To address this problem, this paper proposes a node-embedding-based device health assessment method, which creatively introduces a graph model for device health assessment and effectively improves the performance of health assessment. Firstly, this paper proposes a way to define equipment association graphs. Secondly, we introduce the node embedding technique to extract graph information. Finally, an equipment health assessment method based on the equipment association graph is proposed. Experiments show that the proposed method outperforms the existing prevailing methods.

Keywords: Health Assessment · Node Embedding · Association Graph

1 Introduction

With the improvement of information equipment automation [15,29] and integration [28,33,34] technology, the classic equipment asset management method can no longer satisfy the requirements of current equipment management. A large number of basic information data [?,?,?] and operation status data [?,?,?] derived from routers, switches [22,27,32] and professional production equipment are beyond the analysis capability of traditional expert experience, and there is an urgent need for intelligent analysis methods [24,26] to migrate and apply.

Most of the existing equipment health assessment methods are based on the *Reliability-Centered Maintenance* (RCM) concept, which describes historical failure data through quantitative modeling, combined with expert evaluation to determine the life and reliability of equipment, so as to make preventive

M. Qiu et al. (Eds.): SmartCom 2022, LNCS 13828, pp. 107–119, 2023.
https://doi.org/10.1007/978-3-031-28124-2_11

maintenance decisions to reduce potential downtime losses [1,35,36,45]. Among these approaches, traditional methods are generally based on statistic definition and regression analysis techniques, among which the relative healthiness model and its improvement models are typical [2,5,21,23]. However, such methods are unable to effectively extract high-level features from the data, and their classification or prediction capabilities are insufficient, resulting in poor health assessment accuracy and limited guidance for maintenance of equipment in production environments. The development of machine learning technology has introduced new ways of approaching equipment health assessment. The machine learning-based methods improve the evaluation accuracy to a certain extent [3,38,44,46], but it also relies on the introduction of expert knowledge and has insufficient migration capability for different application scenarios and different device states [6,19,39,43]. In recent years, deep learning-based methods have also been applied [4,13,18]. Some methods use sequential models to predict the future health of equipment [17,20,42] , however, these methods are only applicable to a single device and consume a large amount of computational resources, making it difficult to land applications in real scenarios with a large number of equipment [7,41,47].

To address these problems, this paper aims to provide a graph structure that can characterize the association between equipment operation information and equipment, and propose a method for equipment health assessment based on equipment association graphs, so that equipment health assessment can be free from the reliance on expert knowledge. Specifically, this paper first proposes the definition of a device association graph model and defines the node features in the device association graph by feature extraction; subsequently, the graph features are extracted based on the node embedding method; finally, the labels of unknown labeled nodes are predicted based on the perceptron and the information of known labeled nodes.

The main contributions of this paper can be summarized as follows:

1. This paper proposes a new equipment association graph definition and construction method. The traditional method tends to focus on the historical operation data [8,11,16] of a single equipment, and the analysis of the association between equipment is limited to the similarity of weights in the regression equation brought by the association of basic information such as the same manufacturer, without obtaining the influence of the association such as the physical location of the equipment. In contrast, the proposed equipment association graph can effectively characterize the complex associations between equipment and can more accurately reflect the effect of the influencing factors on equipment health, thus obtaining more accurate health values.
2. In this paper, we introduce node embedding based on random walk and Word2Vec into the field of equipment health assessment, which brings a new perspective to the research and development of this field. Compared with the existing methods based on statistics and machine learning, the node embedding method reduces the dimensionality of the feature vector, which reduces the complexity for the subsequent calculation; on the other hand, the vector

value of a device after embedding is influenced by the devices with which it has a strong association, which can extract higher-level features and achieve a more accurate health assessment.
3. This method is based on the graph structure for equipment health assessment, and is able to predict the health of all unknown devices through a single uniform node embedding, which solves the shortcomings of existing deep learning-based equipment health assessment methods that focus on single device health prediction.
4. We define the equipment characteristics through the most intuitive basic information and operation information of the equipment, and get rid of the dependence of the existing method on expert knowledge. For different types of equipment, the method can operate properly without the exclusive characteristics defined by expert knowledge, and accurately achieve equipment health assessment, reducing the threshold of personnel and data completeness for applying the method.

2 Related Work

There are many studies on equipment health assessment models, including traditional statistical models, machine learning models, and deep learning models. Most approaches are based on RCM concept and assess equipment health centered around remaining useful life.

2.1 Statistical Models

Earlier approaches modeled equipment health assessment based on expert knowledge defining statistics under application scenarios, by such as equipment operation indicators, equipment temperature, relevant product technical indicators, etc.; and then implemented statistical techniques such as multiple regression and entropy correction to calculate the weights of the statistics, and finally used the obtained relative health model to predict the health of equipment.

For example, statistical methods including hypothesis testing [21], extreme value theory [5] and maximum-likelihood estimation [2,23] are widely used in the field of equipment health assessment [37,40]. However, such methods rely on manual feature construction and have difficulty in obtaining complex fusion features, which leads to a strong dependence on feature construction for their accuracy, further affecting their accuracy and usability.

2.2 Machine Learning Models

Existing machine learning-based methods are also based on expert knowledge to define key equipment features, such as basic equipment information, operating indicators, etc., followed by feature modeling using machine learning algorithms such as XGBoost [14] and clustering [40], and training with large amounts of data [9,10,31] to obtain a good classification or prediction model.

Besides, other commonly used machine learning methods include support vector machines [3,46], Gaussian regression [38,44], the gamma process [43], least squares regression [6], hidden Markov model [19], and the Wiener processes [1,39]. Compared with statistical-based methods, this type of method improves the model's ability to fit the data, thus enhancing the evaluation accuracy, but it still relies on expert knowledge and feature selection, and its automatic feature extraction capability still needs further improvement.

2.3 Deep Learning Models

Deep learning techniques are also applied in this field, but limited by the amount of data [12,25,30] and the number of labels required for deep learning models, the migration of related technologies is still at a preliminary stage, and some researchers have used sequential models to predict the future health of a single device [4,13,18], but the related accuracy rate needs to be improved [17,20,42,47]. For example, [20] proposes a competition learning-based method for predicting long-term machine health status and [42] combines multiple sensor signals and *Long Short-Term Memory* (LSTM) models for modeling. In addition, there are also many approaches based on combining GAN models with sequence models to obtain better performance [17,47]. In addition, other network structures, such as *Convolutional Neural Networks* (CNN), are gradually applied to equipment health assessment [7,41]. For example, [7] combines CNN and LSTM to improve the accuracy of equipment remaining useful life estimation. However, as an important part of deep learning, deep graph models have been rarely applied in device health assessment. In particular, the graph node embedding-based approach has not yet been migrated to the field. This makes existing methods applicable only to a single device, ignoring practical application scenarios with a large number of devices.

3 Method

The health assessment method proposed in this paper is divided into two stages. First, a graph structure is defined to characterize the association between equipment operation information and equipment in order to free the health assessment method from the reliance on expert knowledge, and node features in the equipment association graph are defined by feature extraction. Second, the equipment association graph is embedded based on the node embedding method and the health level of the equipment to be evaluated is assessed. In this section, we first define and explain the concepts and graph structure related to equipment association graphs, then we explain how equipment association graphs are constructed, and finally we discuss the methods for equipment health assessment based on equipment association graphs. The complete flow of the proposed method is shown in Fig. 1.

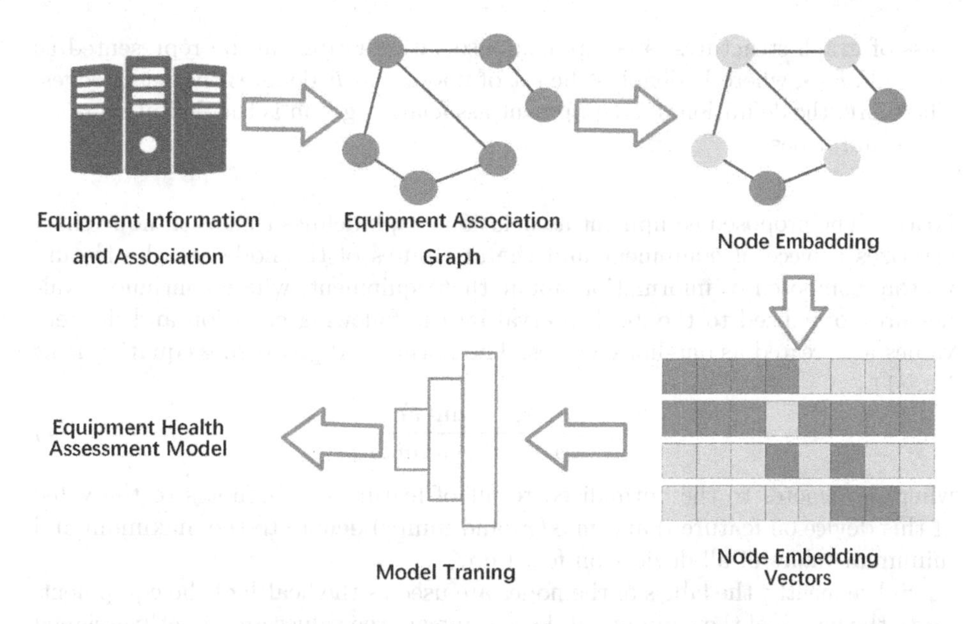

Fig. 1. Flow chart of the proposed method

3.1 Definition of Concepts

Equipment Information. Equipment information includes basic equipment information, equipment usage information and other information for specific equipment types. Among them, the basic information of equipment includes manufacturer, factory time, equipment type, etc.; equipment usage information includes physical location of equipment, average daily working time, average daily failure times, average daily temperature, etc.; other information for specific equipment type refers to the working information based on equipment type, for example, network switch includes average daily forwarding volume, average daily fan speed, etc. Based on specific scenarios and equipment, equipment information can be added without upper limit, thus forming a more complete device characteristic.

Equipment Association. Equipment association refers to the association between equipment information. If a piece of information of two equipment is the same, it is considered that there is an association between two equipment. In the actual production environment, the stronger the association, the more similar the health of the equipment. For example, equipment of the same batch, or equipment running at the same temperature.

3.2 Definition of Equipment Association Graph

To effectively describe the information of equipment and the association between the equipment, a equipment association graph needs to be constructed. As a class of graph structures, a equipment association graph can be represented as $G =< V, E >$, where V denotes the set of nodes and E denotes the set of edges. Therefore, the definition of a equipment association graph is the definition of its edges and nodes.

Node. The proposed equipment association graph defines that each node characterizes a piece of equipment and the attributes of the node are the feature vectors composed of information about that equipment, where continuous values are normalized to the $[0, 1]$ interval by the following equation and discrete values are treated as one-hot encodes. The normalized processing equation is as Eq. (1).

$$y_i = \frac{x_i - \min(x)}{\max(x) - \min(x)} \tag{1}$$

where y_i denotes to the normalized result of feature i, x_i denotes to the value of this device on feature i, and $\max(x)$ and $\min(x)$ denote to the maximum and minimum values of all devices on feature i.

Subsequently, the labels of the nodes are used as the health of the equipment. Since the health of the training set data is known, the values are directly assigned to the corresponding nodes as labels.

Edge. Each edge in the proposed equipment association graph links two nodes, and the edges have no direction but have a weight. The construction of edges follows the following flow.

1. Let the weight of the edge between any two nodes be 0.
2. For any two nodes, information about their corresponding equipment is examined. For each identical field in the equipment information, the weight of the edge between these two nodes is increased by 1.
3. Generate edges between nodes with ownership greater than 0.

Algorithm 1. Equipment Association Graph Construction

Input: Node Set $V\{v_i\}_{i=1}^n$ where n denotes the size of nodes. Equipment information Set $I = \{I_i\}_{i=1}^n$ where I_i denotes the equipment information of nodes v_i. Size T of equipment information I_i.

Output: Equipment Association Graph $G = (V, E)$.

1: **for** $v_i \in V, v_j \in V$ **do**
2: $\quad E_{ij} = 0$
3: $\quad$ **for** $t \in 0, \cdots, T$ **do**
4: $\quad\quad$ **if** $I_i^t = I_j^t$ **then**
5: $\quad\quad\quad E_{ij} \leftarrow E_{ij} + 1$
6: $G = (V, E)$
7: **return** G

Specifically, to generate the equipment association graph, we first process the raw data and, for each equipment, generate its equipment information vector as equipment characteristics; subsequently, we construct nodes for each equipment, add the equipment characteristics vector as attributes, or as labels if the health degree is known; finally, we calculate the weights between the nodes two by two and generate edges with corresponding weights greater than 0. The equipment association graph construction algorithm is summarized as Algorithm 1.

3.3 Equipment Health Assessment Based on Node Embedding

To perform equipment health assessment based on the equipment association graph, we first a) perform a random walk on the graph to obtain node sequences based on the equipment association graph; subsequently b) compute node embedding vectors based on the node sequences using the Woed2Vec algorithm; and finally c) predict node labels for all labeled locations based on a three-layer perceptron with the node embedding vectors and known labels as inputs.

First, most of the existing random walk methods can be applied to structures such as heterogeneous graphs and heterogeneous information networks. Equipment association graphs are a static class of homogeneous graphs, so the transfer probability of random walk needs to be adjusted. Specifically, we adjust the probability of being currently at node v, which will be transferred to node t in the next step, as Eqs. (2) and (3).

$$P(t \mid v)=\begin{cases}\frac{weight(t,v)}{N_w(v)},(t,v)\in E\\ 0, otherwise\end{cases} \tag{2}$$

$$N_w(v)=\sum_{t_i} weight\left(t_i, v\right) \tag{3}$$

where $weight(t, v)$ denotes the weight of the edge between node t and node v, $N_w(v)$ denotes the sum of weights of edges between node v and all neighboring nodes. We select the number of nodes in the path obtained by the random walk to be 10.

Second, the Word2vec algorithm is an encoding approach and we adapt it to a graph node embedding algorithm that embeds nodes into vectors and makes the embedding vectors obtained by nodes with similar attributes as close as possible.

For any node v in $n_{v-c}, ..., n_{v+c}$ of the paths obtained by random walks in the previous step, the objective function to be maximized by word2vec is

$$\sum_{v=1}^{v} \log P\left(n_{v-c}, \ldots, n_{v-1}, n_{v+1}, \ldots, n_{v+c}\right) \tag{4}$$

The probability in the Eq. (4) can be transformed into a product of a series of probabilities, and the final objective function can be transformed into

$$\frac{e^{V_{n_v}^{\top} V'_{n_{v+j}}}}{\sum_{i=1}^{V} e^{V_{n_v}^{T} V'_{n_v}}} \tag{5}$$

where V_{n_i} denotes the input vector of node n_i (i.e., its attributes), V'_{n_i} denotes the output vector of node n_i (i.e., its embedding vector), and V denotes the number of all nodes. During the calculation of i growth to V, the above equation calculates the embedding vector of all the nodes.

Finally, we use the embedding vectors of the nodes corresponding to all equipment with known health as the input to the three-layer perceptron, and their health values are used as the labels to be fitted to train the perceptron model. After the training is completed the embedding vectors of the nodes with unknown labels are fed into the perceptron model and the obtained output is the predicted health of the corresponding nodes, i.e., the corresponding equipment.

Each layer in the three-layer perceptron is a fully connected layer, and each neuron obeys the following formula.

$$output = f(net - \theta) \tag{6}$$

$$net = \sum z_i \cdot v_i \tag{7}$$

where z_i denotes the output value of the ith neuron in the previous layer, v_i is the weight of the ith neuron linking this neuron in the previous layer. θ is the deviation value of this neuron, which we set to θ. $f(x)$ is the activation function, and we set the activation function which is the sigmoid function.

4 Experiments

4.1 Dataset

The dataset is the equipment information and equipment association information of servers, disk arrays, network routers, network switches, firewalls, IPS, IDS, WAF, etc. from an enterprise in operation in China, and the comprehensive evaluation is carried out based on the relevant information.

We use equipment information as equipment characteristics and equipment association information as the basis for constructing equipment association diagrams, and experts are invited to evaluate the health of the equipment in the dataset in terms of years in operation, failure conditions, and product support periods, and use the health as the dataset label.

We finally constructed a dataset consisting of 1952 devices, which were randomly divided into training, validation, and test sets in the ratio of 8:1:1. Subsequently, all the data are used to construct an equipment association graph according to their relationships as input data for the proposed equipment health assessment method. Besides equipment information, we construct features such as years in operation, defect level, cumulative failures, percentage of failures in the most recent year, business system data loss, average trouble-free operation time, product support period, and repeated maintenance.

4.2 Evaluation

We compare our approach with the mainstream machine learning and deep learning methods. All methods use raw numerical features for normalization and category features for one-hot encoding as input features. We not only use RMSE and MAE as indicators of health assessment error, but also discretize health judgments into healthy and unhealthy (with a cut-off of whether health is greater than 0.5) to compare the accuracy of health trend assessment.

Among them, RMSE can be expressed as:

$$\mathrm{RMSE}(X,h)=\sqrt{\frac{1}{m}\sum_{\mathrm{i=1}}^{m}(h(x_i)-y_i)^2} \tag{8}$$

where X denotes the test dataset, m denotes the test dataset size, h denotes the health assessment model, $h(x_i)$ denotes the result of the ith test data predicted by the model, and y_i denotes the label of the ith test data. MAE can be expressed as:

$$\mathrm{MAE}(X,h)=\frac{1}{m}\sum_{i=1}^{m}|h(x_i)-y_i|\,. \tag{9}$$

The experimental results are shown in the Table 1.As shown in the table, the proposed method has reduced 6.1% and 2.2% in RMSE and MAE of health prediction and improved 2.3% in accuracy compared to recent deep learning methods [7]. The results show that the proposed method can effectively improve the performance of equipment health assessment and is closer to the expert assessment results than previous methods.

Table 1. Experimental results of comparison with prevailing methods (%)

Method	RMSE	MAE	Accuracy
SVM	42.3	33.2	78.0
XGBoost	34.8	25.1	85.2
CNN	35.6	27.9	81.8
LSTM	30.9	23.7	86.7
CNN+LSTM [7]	28.4	21.0	88.3
Proposed method	**22.3**	**18.8**	**90.6**

4.3 Ablation

To verify the validity of the proposed method, we compared the experimental results of the proposed method with the results of the equipment health assessment without node embedding.

The experimental results are displayed in Table 2, and it can be seen that the proposed method is effective in enhancing the final assessment results. Since the same model is used for training in both methods, the results namely show that the proposed method can effectively improve the feature representation without over-relying on expert knowledge.

Table 2. Experimental results of comparison with the methods without node embedding (%)

Method	RMSE	MAE	Accuracy
Proposed method	**22.3**	**18.8**	**90.6**
without node embedding	38.5	26.6	80.2

4.4 Hyperparameters and Model Selection

As mentioned earlier, we divided a portion of the training data as the validation set. In the training, we use MSE loss as the loss function, and compare the loss on the validation set for models trained with different combinations of hyperparameters to select the model parameters. Specifically, we select stochastic gradient descent as the optimizer, the number of walking steps from 1, 2, 3, 4, 5, the learning rate and weight decay from 0.0001, 0.0005, 0.001, 0.005, 0.01,0.05, 0.1, 0.5, and the number of epochs from 100, 150, 200, 250. The final parameters are shown in Table 3.

Table 3. Selected hyperparameter value

Hyperparameter	Value
Walking Steps	4
Learning Rate	0.005
Weight Decay	0.001
Epochs	200

5 Conclusion

In this paper, we propose a node-embedding based equipment health assessment method that introduces a graph model in the equipment health assessment task, which significantly reduces the RMSE and MAE of equipment health assessment and improves the task accuracy. Compared with previous methods, although the proposed method has been decoupled from expert knowledge to a large extent, it still requires a certain amount of expert annotation. In the next stage, combining the method with semi-supervised and unsupervised methods to further reduce the reliance on expert annotation may help to reduce the cost of the equipment health assessment to further enhance its application value.

Acknowledgements. This work was supported by State Grid Zhoushan Electric Power Supply Company of Zhejiang Power Corporation under grant No. B311ZS220002 (Research on hyperautomation for information comprehensive inspection).

References

1. An, D., Kim, N.H., Choi, J.H.: Practical options for selecting data-driven or physics-based prognostics algorithms with reviews. Reliab. Eng. Sys. Saf. **133**, 223–236 (2015)
2. Awate, S.P.: Adaptive, nonparametric Markov models and information-theoretic methods for image restoration and segmentation. Ph.D. thesis, School of Computing, University of Utah (2006)
3. Benkedjouh, T., Medjaher, K., Zerhouni, N., Rechak, S.: Health assessment and life prediction of cutting tools based on support vector regression. J. Intell. Manuf. **26**(2), 213–223 (2015)
4. Chen, C., Liu, Y., Sun, X., Di Cairano-Gilfedder, C., Titmus, S.: An integrated deep learning-based approach for automobile maintenance prediction with gis data. Reliab. Eng. Syst. Saf. **216**, 107919 (2021)
5. Clifton, D.A., Clifton, L.A., Bannister, P.R., Tarassenko, L.: Automated novelty detection in industrial systems. In: Advances of Computational Intelligence in Industrial Systems, pp. 269–296. Springer, Berlin (2008). https://doi.org/10.1007/978-3-540-78297-1_13
6. Coppe, A., Haftka, R.T., Kim, N.H.: Uncertainty Identification of Damage Growth Parameters Using Nonlinear Regression. AIAA J. **49**(12), 2818–2821 (2011). https://doi.org/10.2514/1.J051268
7. Fu, H., Liu, Y.: A deep learning-based approach for electrical equipment remaining useful life prediction. Auton. Intell. Syst. **2**(1), 1–12 (2022)
8. Gai, K., Qiu, M., Elnagdy, S.: A novel secure big data cyber incident analytics framework for cloud-based cybersecurity insurance. In: IEEE BigDataSecurity Conference (2016)
9. Gai, K., Qiu, M., Liu, M., Xiong, Z.: In-memory big data analytics under space constraints using dynamic programming. Fut. Gen. Comput. Syst. **83**, 219–227 (2018). https://doi.org/10.1016/j.future.2017.12.033
10. Gai, K., Zhang, Y., Qiu, M., Thuraisingham, B.: Blockchain-enabled service optimizations in supply chain digital twin. IEEE Trans. Serv. Comput , Early Access 1–12 (2022). https://doi.org/10.1109/TSC.2022.3192166
11. Gai, K., et al.: Electronic health record error prevention approach using ontology in big data. In: IEEE 17th HPCC (2015)
12. Gao, X., Qiu, M.: Energy-based learning for preventing backdoor attack. In: Memmi, G., Yang, B., Kong, L., Zhang, T., Qiu, M. (eds.) Knowledge Science, Engineering and Management: 15th International Conference, KSEM 2022, Singapore, August 6–8, 2022, Proceedings, Part III, pp. 706–721. Springer International Publishing, Cham (2022). https://doi.org/10.1007/978-3-031-10989-8_56
13. Hashemian, H.M.: State-of-the-art predictive maintenance techniques. IEEE Trans. Instrum. Meas. **60**(1), 226–236 (2011). https://doi.org/10.1109/TIM.2010.2047662
14. Jia, Z., Xiao, Z., Shi, Y.: Remaining useful life prediction of equipment based on xgboost. In: The 5th International Conference on Computer Science and Application Engineering. pp. 1–6 (2021)
15. Li, J., Ming, Z., Qiu, M., Quan, G., Qin, Xiao, C.: Tianzhou: Resource allocation robustness in multi-core embedded systems with inaccurate information. J. Syst. Archi. **57**(9), 840–849 (2011). https://doi.org/10.1016/j.sysarc.2011.03.005
16. Li, Y., Gai, K., Ming, Z., Zhao, H., Qiu, M.: Intercrossed access controls for secure financial services on multimedia big data in cloud systems. ACM Trans. Multim. Comput. Commun. Appl. **12**(4s), 1–18 (2016). https://doi.org/10.1145/2978575

17. Liu, C., Tang, D., Zhu, H., Nie, Q.: A novel predictive maintenance method based on deep adversarial learning in the intelligent manufacturing system. IEEE Access **9**, 49557–49575 (2021). https://doi.org/10.1109/ACCESS.2021.3069256
18. Liu, J., Pan, C., Lei, F., Hu, D., Zuo, H.: Fault prediction of bearings based on LSTM and statistical process analysis. Reliab Eng. Syst. Saf. **214**, 107646 (2021)
19. Liu, Q., Dong, M., Peng, Y.: A novel method for online health prognosis of equipment based on hidden semi-markov model using sequential monte carlo methods. Mech. Syst. Signal Process. **32**, 331–348 (2012)
20. Malhi, A., Yan, R., Gao, R.X.: Prognosis of defect propagation based on recurrent neural networks. IEEE Trans. Instrum. Meas. **60**(3), 703–711 (2011)
21. Markou, M., Singh, S.: Novelty detection: a review-part 1: statistical approaches. Signal Process. **83**(12), 2481–2497 (2003)
22. Niu, J., Gao, Y., et al.: Selecting proper wireless network interfaces for user experience enhancement with guaranteed probability. J. Paralell. Distrib. Comput. **72**(12), 1565–1575 (2012)
23. Pecht, M.: Prognostics and health management of electronics. In: Encyclopedia of Structural Health Monitoring. Wiley (2009)
24. Qiu, H., Dong, T., et al.: Adversarial attacks against network intrusion detection in IoT systems. IEEE IoT J. **8**(13), 10327–10335 (2020)
25. Qiu, H., Kapusta, K., et al.: All-or-nothing data protection for ubiquitous communication: challenges and perspectives. Inf. Sci. **502**, 434–445 (2019)
26. Qiu, H., Zheng, Q., et al.: Topological graph convolutional network-based urban traffic flow and density prediction. IEEE Trans. Intell. Transp. Syst. **22**, 4560–4569 (2020)
27. Qiu, M., Chen, Z., et al.: Energy-aware data allocation with hybrid memory for mobile cloud systems. IEEE Sys. J. **11**(2), 813–822 (2014)
28. Qiu, M., Jia, Z., et al.: Voltage assignment with guaranteed probability satisfying timing constraint for real-time multiproceesor DSP. J. of Signal Proc, Syst. **46**, 55–73 ((2007)
29. Qiu, M., Li, H., Sha, E.: Heterogeneous real-time embedded software optimization considering hardware platform. In: ACM Symposium on Applied Computing, pp. 1637–1641 (2009)
30. Qiu, M., Qiu, H.: Review on image processing based adversarial example defenses in computer vision. In: IEEE 6th International Conference on BigDataSecurity, pp. 94–99 (2020)
31. Qiu, M., Qiu, H., et al.: Secure data sharing through untrusted clouds with blockchain-enabled key management. In: 3rd SmartBlock Conference on Smart BlockChain (SmartBlock), pp. 11–16 (2020)
32. Qiu, M., Xue, C., Shao, Z., et al.: Efficient algorithm of energy minimization for heterogeneous wireless sensor network. In: 2021 IEEE 19th International Conference on Embedded and Ubiquitous Computing (EUC), pp. 25–34 (2006)
33. Qiu, M., Xue, C., et al.: Energy minimization with soft real-time and DVS for uniprocessor and multiprocessor embedded systems. In: IEEE Design, Automation Exhibition Conference, pp. 1–6 (2007)
34. Qiu, M., Yang, L., Shao, Z., Sha, E.: Dynamic and leakage energy minimization with soft real-time loop scheduling and voltage assignment. IEEE Trans. Very Large Scale Inetegr. **18**(3), 501–504 (2009)
35. Qiu, M., et al.: RNA nanotechnology for computer design and in vivo computation. Philos. Trans. R.Soc. A: Math., Phy. Eng. Sci. **371**(2000), 20120310 (2013)

36. Qiu, M., Xue, C., Shao, Z., Sha, E.H.M.: Energy minimization with soft real-time and DVS for uniprocessor and multiprocessor embedded systems. In: 2007 Design, Automation & Test in Europe Conference & Exhibition, pp. 1–6. IEEE (2007)
37. Roberts, S.J.: Novelty detection using extreme value statistics. IEE Proc. Vis. Image Signal Process. **146**(3), 124–129 (1999)
38. Seeger, M.: Gaussian processes for machine learning. Int. J. Neural Syst. **14**(02), 69–106 (2004)
39. Si, X.S., Wang, W., Hu, C.H., Chen, M.Y., Zhou, D.H.: A wiener-process-based degradation model with a recursive filter algorithm for remaining useful life estimation. Mech. Syst. Signal Process. **35**(1–2), 219–237 (2013)
40. Sutharssan, T., Stoyanov, S., Bailey, C., Yin, C.: Prognostic and health management for engineering systems: a review of the data-driven approach and algorithms. J. Eng. **2015**(7), 215–222 (2015)
41. Wang, B., Lei, Y., Li, N., Yan, T.: Deep separable convolutional network for remaining useful life prediction of machinery. Mech. Syst. Signal Process. **134**, 106330 (2019)
42. Wang, D., Liu, K., Zhang, X.: A generic indirect deep learning approach for multisensor degradation modeling. IEEE Trans. Autom. Sci. Eng. **19**, 1924–1940 (2021)
43. Wang, X., Balakrishnan, N., Guo, B., Jiang, P.: Residual life estimation based on bivariate non-stationary gamma degradation process. J. Stat. Comput. Simul. **85**(2), 405–421 (2015)
44. Wilson, A., Adams, R.: Gaussian process kernels for pattern discovery and extrapolation. In: International Conference on Machine Learning. pp. 1067–1075. PMLR (2013)
45. Wu, G., Zhang, H., Qiu, M., Ming, Z., Li, J., Qin, X.: A decentralized approach for mining event correlations in distributed system monitoring. J. Parallel Distrib Compu. **73**(3), 330–340 (2013)
46. Yan, J., Liu, Y., Han, S., Qiu, M.: Wind power grouping forecasts and its uncertainty analysis using optimized relevance vector machine. Renew Sustain Energy Rev. **27**, 613–621 (2013)
47. Zhang, L., Lin, J., Liu, B., Zhang, Z., Yan, X., Wei, M.: A review on deep learning applications in prognostics and health management. IEEE Access **7**, 162415–162438 (2019)

Improvement of ERP Cost Accounting System with Big Data

Jie Wan[1](✉) and Yiren Qi[2]

[1] Shanghai Publishing and Printing College, Shanghai 200093, China
wj20202@sppc.edu.cn

[2] School of International and Public Affairs, Columbia University, New York, USA
yq2337@Columbia.edu

Abstract. With the rapid development of computer technology, ERP system arises at the right time as an advanced cost management tool. ERP is mainly implemented in enterprises with the purpose of optimizing resource allocation and giving full play to its functions. ERP system is an enterprise resource planning system with strategic management thought, which tries to reduce the consumption of resources. Cost management is an important part of enterprise management. ERP system provides a tool for the cost management of enterprises. It greatly enhances the comprehensive management ability of enterprises. This paper discusses the improvement of ERP system to enterprise cost. We proposes to combine enterprise cost management theory with big data, use digital technology to adjust the strategic cost plan of enterprises, and improve the cost accounting system of ERP + MES + SCADA industry finance deep integration. We build lean production management and control platform of cost information sharing and use digital technology to deeply integrate performance evaluation index.

Keywords: Big data · Cost accounting · ERP system · Improvement · Innovation

1 Introduction

ERP (*Enterprise Resource Planning*) is established on the basis of information technology, using the advanced management ideas of modern enterprises, to provide enterprises with decision-making, planning, control and management performance evaluation of an all-round, systematic management platform. It integrates the functions of all aspects of enterprise management, including sales, production, finance, quality management, etc. It is an enterprise management information system centered on financial management. ERP is an advanced enterprise management mode in the world today. Its main purpose is to balance and optimize the comprehensive management of human, financial, material and other comprehensive data owned by the enterprise, so that the enterprise can exert sufficient ability in an all-round way in the fierce market competition and obtain better economic benefits.

With the rapid development of computer network technology, especially the rapid arrival of Internet + and big data era. Enterprise cost management is more and more

M. Qiu et al. (Eds.): SmartCom 2022, LNCS 13828, pp. 120–130, 2023.
https://doi.org/10.1007/978-3-031-28124-2_12

considered as a complex and huge system project. In order to meet the requirements of the times, enterprises begin to use computer technology for cost management. ERP cost accounting system is a good choice, it can help enterprises to realize the cost data is linkage, real-time update. Business personnel and managers of functional departments can grasp the real-time cost information at any time, which is conducive to scientific decision-making, timely and reasonable allocation of resources, and strengthening the control of product cost.

This paper first analyzes the development and current situation of ERP software products. Then it constructs the enterprise cost accounting system based on ERP environment improvement. We analyze the realization of cost accounting module and put forward the innovation path of cost management in enterprises with big data.

The paper structure is as the following. Section 2 gives the definitions. Section 3 builds the ERP system. Various novel approaches have been proposed in Sect. 4. Final, Sect. 5 concludes the paper.

2 Definition

2.1 Online Public Opinion

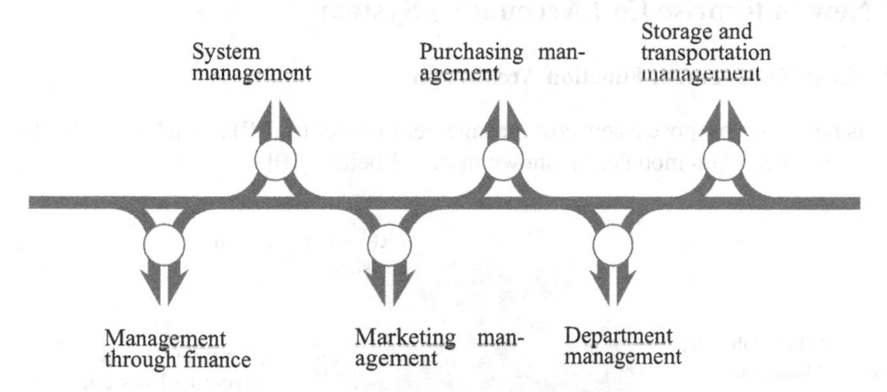

Fig. 1. The composition of ERP software

ERP is a modern enterprise management ideology and method developed in the early 1990s, which is based on MRP II, market and customer demand-oriented, with the goal of optimizing the allocation of internal and external resources, eliminating all ineffective labor and resources in the production and operation process, achieving organic integration of information flow, logistics and capital flow, and improving customer satisfaction, with planning and control as the main line. It is a modern enterprise management concept and method with network and information technology as the platform, integrating finance, logistics, production and other functions.

Due to the different design ideas and styles of ERP manufacturers, the modular structure of their ERP products also varies greatly. However, as far as the functions of ERP software are concerned, there is still a great deal of similarity in its modules, which are mainly composed of several parts as shown in Fig. 1.

2.2 Development and Current Status of ERP Software Products

ERP software is developing very rapidly, and the global ERP market revenue is growing at an alarming rate. In the face of the huge market, some companies have joined the ERP development and development team, and hundreds of companies specializing in the development, sales and consulting of MRP, MRP II and ERP products have emerged in the world [1]. In recent years, in order of ERP software and services revenue, the top companies include SAP, Oracle, J.D. Edwards, People soft, Ba an, SSA companies [2, 9].

ERP system has experienced five stages of development, namely, inventory management, material plan management, manufacturing resource planning and management, enterprise resource planning management, the collectivization of the remote control, is to make full use of the Internet, big data and cloud technology and a series of technology to realize the large-scale, industry, business and regional remote integration controls, It is the integrated control of multiple single and independent ERP "units". After several rounds of management innovation, ERP system becomes easy to learn, easy to use, low cost, driving the development of the whole industrial chain, has become a sharp tool in enterprise business management.

3 New Enterprise Cost Accounting System

3.1 Cost Management Function Architecture

In this paper, we propose a new cost management model for ERP, which consists of the following three main modules, as shown in Fig. 2 below [10].

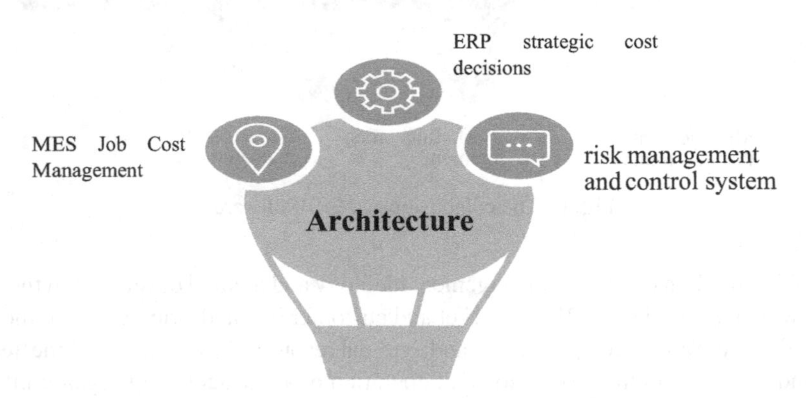

Fig. 2. Cost management model of ERP

First, strategic cost decision system. According to the strategic planning of the enterprise, the rules of strategic cost management are formulated, and the cost drivers are analyzed according to the external environment and internal resources of the enterprise, including production scope, production scale, production technology, utilization of production capacity, efficiency of plant layout, and supplier and customer relationships.

Describe the enterprise value chain and determine the optimal target cost for the enterprise based on the value chain, which is given to the bottom execution system. It receives the cost reports from the bottom layer and adjusts the enterprise value chain to reduce the non-value-added links in the value chain [7].

Second, the operation cost management center: it is the core of the whole cost control system, which consist of the operation cost control and management module, auditing system and data collection module in the production management system, and the operation cost. There are several modules in the management center:

Job Decomposition Center. It is realized in the job cost control and management module to distinguish the job centers and complete the job distinction of the whole production and operation process on the basis of job centers, and construct the cost motive base in each job center, and finally form the complete job chain and the corresponding motive base.

Cost Control Center. Realized in the operation cost control and management module, adopting the principle of comprehensiveness, target management and the principle of combining debt and rights, the cost of production and operation is controlled comprehensively, including the formula design before production, process flow arrangement, etc.

Job Cost Planning. Implemented in the job cost control and management module, the cost is apportioned to the whole job chain in monetary terms based on the job center's cost driver library, and the cost of the corresponding job in the corresponding job chain is specified in advance to form the standard cost. Corresponding to the job costing [8], the strategic costs are decomposed and the target costs are implemented. The data for this part is provided by the data collection module.

Job Costing. It is an important part of the auditing system, which collects and processes the actual costs collected according to the job costing method, and forms the cost report of actual costs.

Job Cost Analysis. It is implemented in the job cost control and management module, which analyzes the costing results and forms a cost report for decision making, including the consumption difference between the actual cost and the target cost, and the value-added of the jobs in the job chain. This information is provided to the ERP layer and is important information for decision making.

Third, risk management and control system. The internal control will be related to the ERP system, and the risk management and control will be implemented in real combat. Supply and marketing linkage, according to the single lock material, prevent material price fluctuations brought by the income risk, quotation is more scientific and reasonable, make quotation risk controllable. Scheme selection can set various quotation. Enterprise business contract closed loop management and evaluation: establish business credit assessment and credit management system, establish contract gross margin forecast and control early warning, alteration, cancellation of on-line examination and approval system, debt collection early warning, tracking, automatic reminders, penalty system, the sales price linkage purchase order quick response system, The operational risk can be visible and controllable, so as to ensure the profit of the enterprise, realize the closed-loop supervision of the whole process of the contract, and finally ensure the profit of the enterprise. The multi-dimensional data analysis and decision-making basis can be provided for enterprise managers.

3.2 Implementation of the Cost Accounting Module

Cost accounting is divided into two main accounting items: accounting for funds occupied by work in progress and raw material cost summary. The former reflects the funds occupied by the products in process, while the latter reflects the various costs of the finished products. The cost accounting statistics query is the core of the module, the interface uses a variety of query methods, while taking into account the user is accustomed to the use of EXCEL, in the query results provide a filtering function [10], and the program provides the query results into the EXCEL function, which greatly facilitates the user to further processing of the query results.

4 Innovation Path of ERP Cost Management

4.1 Use Digital Technology for Strategic Cost Planning of Enterprises

Strategic cost planning is an improvement and upgrade on the basis of the traditional cost planning concept. It takes the long-term development of the enterprise as the starting point and uses the strategic concept to carry out the cost planning for the enterprise. Due to the large number of products, complex production processes, unconcentrated distribution of workshops and factories, and high uncertainty factors of external environment and policies, it is difficult for the management of enterprises to make use of complex cost data elements for planning. In the context of digital economy, enterprises can use digital technology to adjust and upgrade the strategic cost planning, so as to make the cost planning more accurate [11].

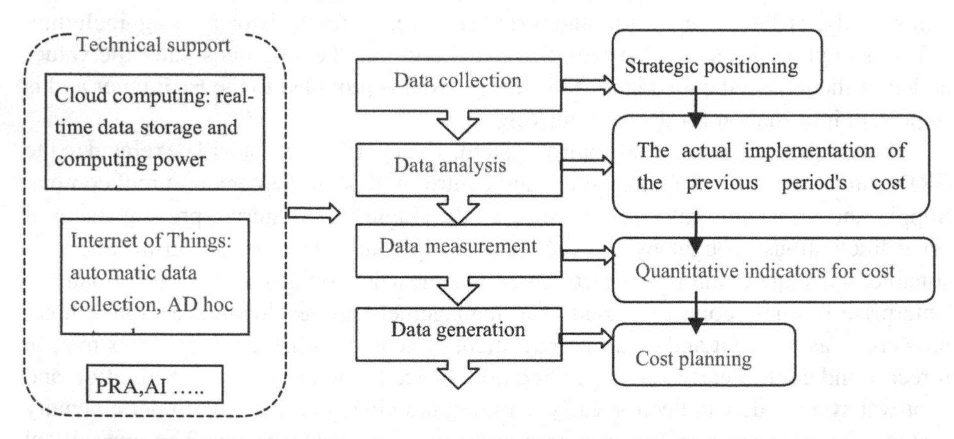

Fig. 3. Strategic cost planning process of enterprise under digital technology

As shown in Fig. 3,first of all, from the perspective of the enterprise value chain, manufacturing companies rely on big data, AI technology to build the Internet of things, in the heart of the information network to obtain higher quality of external and internal data resources, including external information industry development data, competitive enterprise information and policy support, the internal information including its own resources, technical ability and the development scale, Grasp the internal and external environment of enterprises in an all-round way, generate effective decision-making data information, that is, make clear the strategy of enterprises. On the platform of enterprise digital system, the integration process of cost data collection, cleaning and analysis of enterprises can be implemented. The massive cost data resources provide intelligent and networked decision-making for enterprises to define strategic positioning. Secondly, enterprises can integrate blockchain, cloud computing, AI recognition and other new generation of digital technologies, in-depth analysis of the actual implementation of the previous period of cost planning, provide cost prediction analysis and real-time visualization tools for enterprise cost planning, and build a cost data-driven planning and decision model. Thirdly, it calculates the quantitative indicators of cost planning reduction for enterprises for many times, mainly starting with the indicators of production expense, manufacturing expense, unit product cost and sub-product cost to upgrade the whole business process of cost planning for manufacturing enterprises [12, 13]. Finally, information technologies such as cognitive technology, RPA, visualization and cloud technology can also strengthen the ability of cost resource allocation of enterprises, enhance the decision support ability of cost planning of enterprises on the whole, and generate more effective and scientific strategic cost planning.

4.2 Improve the Cost Accounting System of ERP + MES + SCADA

At present, enterprise ERP system has completed many functions such as procurement, sales and production integration, and in the process of cost accounting,, because of the large number of manufacturers products order product variety is more, from the cost of production line element quality is not entirely accurate, at the end of production inventory is difficult to achieve [11, 14, 15].

In the context of digital economy, a cost accounting system integrating MES (manufacturing execution system) and SCADA (data acquisition and Monitoring system) can be built on the basis of enterprise ERP system. Cost data elements can be integrated through digital technology, and accurate accounting of cost data across account sets, modules and accounting periods can be unified. The complex business logic of cost accounting is integrated into the information platform to improve the cost accounting efficiency of enterprises.

As shown in Fig. 4, firstly, multi-dimensional basic data are obtained by relying on the IMS, OA office system, EPMS and other general systems of enterprises. Second, in the center of the DBMS, HDFS data acquisition based on enterprise labor costs, raw materials, manufacturing cost and other cost data, through the butt end of MES and ERP data, the standard cost of the enterprise and the product production order smallest unit step down to each worker, combination of MES and SCADA digital technology to grab the most accurate cost factors, Realize real-time data acquisition of the whole production process from raw materials to finished products, and directly control the equipment layer; Thirdly, digital technology is used to adjust the multi-caliber cost-related data, extract, clean and load the cost data elements, and load them into the data mart of ERP + MES

+ SCADA. The support center of ODS and OLAP is used to realize the difference analysis between the actual cost and the standard cost, dig out the abnormality of the cost data deeply, and get more quantitative data analysis. Finally, according to the cost accounting situation of enterprises to form data reports of different dimensions, such as cost information table, cost budget information table [16], etc.

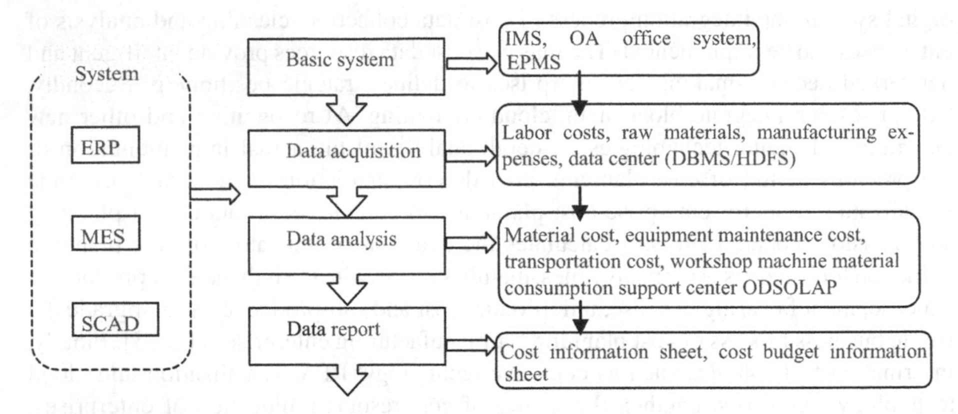

Fig. 4. ERP + MES + SCADA industry finance integration cost accounting system

4.3 Build a Lean Production Management and Control Platform

The core concept of lean production in enterprise cost control is to make use of system process reengineering, organizational structure upgrade and business standardization, so as to realize the enterprise system's response to external dynamic changes in the first time, so as to fully pursue production zero pause, product zero inventory and machine zero failure on the basis of ensuring product quality. At present, enterprise cost control typically involve multiple functions of system, such as ERP, CRM, inventory management systems, etc., but these do not match the data interface function system, the system is not each other, between enterprise inventory of raw materials, products market demand and production conditions to real-time matching, the enterprise cost control information lag problem. In the era of digital economy, data has become the production factor. Therefore, it is an inevitable choice for enterprises to build a lean production management and control platform based on digital technology to share cost information and improve the processing efficiency of cost management and control.

As can be seen from Fig. 5, the platform is divided into front-end data collection, central control data analysis and background technical support. Among them, the front key acquisition enterprises in purchasing, production, inventory and sales link of cost data, the enterprise the actual cost of, space complex flow of digital collection, intelligent decision support costs data optimization, reducing artificial intervention, safeguard data elements in the cost management value maximization. Control on cost accounting and cost control links, using ERP + MES + SCADA fetching production line real-time cost data elements, using the digital technology such as AI, cloud computing, Internet of things on the whole value chain of the enterprise cost of massive amounts of data for self-diagnosis and big data mining, unified cost under centralized data statistics and analysis, to generate real-time cost report, So that the enterprise management to check the cost difference, capacity difference, improve production efficiency [17, 18]. In the background, the intelligent and networked digital technology of cost information accounting and control is embedded to integrate the cost information, provide the actual basis for performance evaluation, realize visual and intelligent financial display, and provide strategic support for enterprise decision makers to make the next year's cost planning.

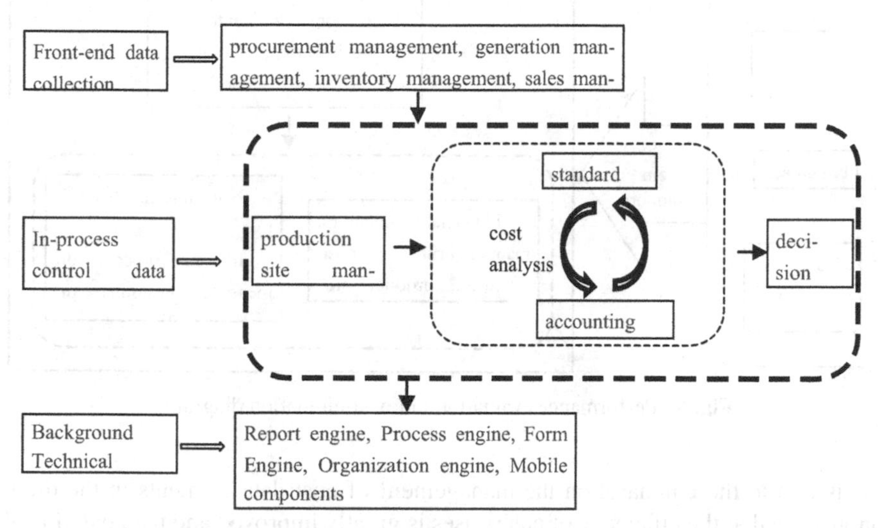

Fig. 5. Schematic diagram of lean production management and control platform

Through the lean production management and control platform of cost information sharing, multiple basic systems of the enterprise are integrated to effectively connect the functions of bidding and inquiry, order purchase, contract approval, fund payment, etc., and realize the intelligent verification and accounting of supplier invoices by the cost system. The cost information of various departments such as shared projects, contracts, suppliers and other basic information can also be centralized to solve the problem of "island" and information barrier of some departments' cost information, realize cross-system, cross-department and cross-level interconnection, and give full play to the advantages of cost data sharing.

4.4 Use Digital Techniques to Deeply Integrate Performance Indicator

Relying on digital technologies such as Internet of Things platform, IT hardware facilities and data management architecture, enterprises can build a more effective performance evaluation system, and use digital technology and data elements to deeply integrate financial indicators and non-financial indicators of performance evaluation.

This is shown in Fig. 6, first of all, in the responsibility center management mode, the responsibility center can be divided into income center, investment center and cost center, etc., the responsibility center of cost information in different categories is remote transmission to the composed of block chain, cloud computing and other digital infrastructure network, depends on the digital technology to realize standardization of performance evaluation of action and result evaluation standard. Of course, performance evaluation indicators such as controllable cost, sales revenue, pre-tax profit, return on investment and other financial data rely on digital technology, which can be obtained anytime and anywhere.

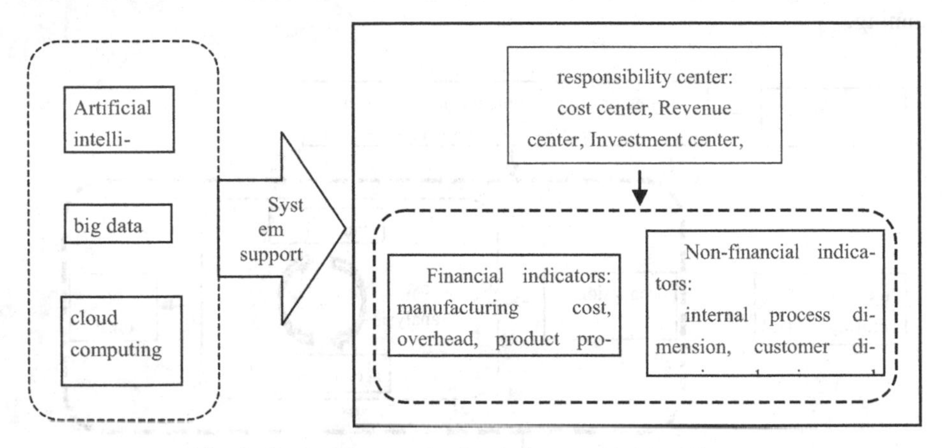

Fig. 6. Performance evaluation index combination diagram

It is due to the emphasis on the management of cost data elements in the digital economy era that the efficiency of enterprises is greatly improved and the goal of profit maximization is promoted. Secondly, the non-financial performance evaluation system is based on the balanced scorecard, and also includes financial dimension, internal process dimension, customer dimension, learning and growth dimension. In the context of digital economy, data + computing power + algorithms drive the mining of massive non-financial data, and the volume of non-financial data elements acquired by deep learning and AI technology supports the extension of the four dimensions. It is necessary to make full use of digital technology to promote the deep integration of financial performance evaluation system and non-financial performance evaluation system, so that enterprises can build a closed-loop evaluation system that is consistent with the product value chain, so as to achieve the purpose of motivating employee cost behavior and promoting human capital creativity.

For modern enterprises, with the great advance in computer hardware [20–22] and network techniques [23–25], big data [26–28] and machine learning [29, 30] have become the major factors in the competition between enterprises. Modern information technology [31, 32] and enterprise financial management highly integrated, improve the efficiency and quality of financial accounting. Enterprise financial accounting [33, 34], as an indispensable department in the process of enterprise development, must keep up with the trend of the era of big data and apply information technology to enterprise financial accounting to improve enterprise management level.

5 Conclusion

This paper analyzed the influence of modern information technology on enterprise financial accounting and put forward corresponding measures. Constructive suggestions were put forward that manufacturing enterprise cost management based on the background of digital economy. In order to promote the modern information technology to the enterprise financial accounting positive impact, we do the fol1owings: 1) use digital technology to adjust the strategic cost planning of manufacturing enterprises, 2) improve the cost accounting system with deep integration of ERP + MES + SCADA industry finance, 3) build a lean production management and control platform with cost information sharing, 4) use digital technology to deeply integrate performance evaluation indicators.

References

1. M. Liu, S. Zhang, et al., "H-infinite State Estimation for Discrete-Time Chaotic Systems Based on a Unified Model," IEEE Trans. on Systems, Man, and Cybernetics (B), 2012
2. Lai, J.: ERP system, ownership structure and earnings quality. Account. Res. **5**, 59–66 (2013)
3. Sun, J., Yuan, R., Wang, B.: ERP implementation can really improve business performance? China Soft Sci. **8**, 121–132 (2017)
4. Lu, Z., Wang, N., Wu, J., Qiu, M.: IoTDeM: an IoT big data-oriented MapReduce performance prediction extended model in multiple edge clouds. JPDC **118**, 316–327 (2018)
5. Zeng, J., Wang, L., Xu, H.: The implementation of ERP system vs agency cost – based on the evidence of ERP introduction period in China. Nankai Manage. Rev. Theor. **3**, 131–138 (2012)
6. Qiu, M., Cao, D., et al.: Data transfer minimization for financial derivative pricing using monte Carlo simulation with GPU in 5G. J. Commun. Syst. **29**(16), 2364–2374 (2016)
7. Liu, M., Huang, H., Tong, C., Liu, K.: Research on the basic framework and construction idea of intelligent finance. Account. Res. **409**(12), 194–194 (2020)
8. Qiu, M., Chen, Z., Liu, M.: Low-power low-latency data allocation for hybrid scratch-pad memory. IEEE Embed. Syst. Lett. **6**(4), 69–72 (2014)
9. Liu, P., Wang, J., Xia, H.: The application research of ERP in enterprise information management. Appl. Mech. Mater. **3744**(713–715), 2356–2359 (2015)
10. Chiang, C., Kou, T., Koo, T.: Systematic literature review of the IT-based supply chain management system: towards a sustainable supply chain management model. Sustainability **13**(5), 2547 (2021)
11. Tu, C., Zhang, L., Peng, Y.: Research on enterprise logistics cost accounting method based on project accounting. Financ. Account. Commun. **16**, 133–135 (2011)
12. Shi, Z., Zhang, S.: Review and prospect of neural marketing ERP research. Manage. World **38**(04), 2035–2054 (2022)
13. Feng, J.: Analysis of manufacturing enterprise cost management under ERP system. Account Learn. **36**, 153–154 (2019). Manage. China Mark. **18**, 169–174 (2016)
14. Fang, J.: Research on the difficulties and solutions of manufacturing cost accounting. Enterp. Reform Manage. **16**, 10–16 (2021)
15. Tang, X.: The impact of ERP system on manufacturing enterprise cost management and its application analysis. Admin. Bus. Assets Finances Serv. **22**, 85–86 (2019)
16. Ma, Y.: Current situation and prospect of ERP development in China. China Manage. Inf. **23**(13), 96–98 (2020)
17. Guo, K.: On the reform of manufacturing cost management in the era of smart manufacturing in China. Modernization Shopping Malls **21**, 135–136 (2018)

18. Qin, W.: Research on strategy + finance management system based on big data technology – taking electric power enterprises as an example. Account. Commun. **18**, 148–153 (2022)
19. Liu, K., Xu, J.: Development of strategic management accounting in the context of big data path analysis. Finan. Account. Commun. **22**, 43–46 (2019)
20. Qiu, M., Yang, L., Shao, Z., Sha, E.: Dynamic and leakage energy minimization with soft real-time loop scheduling and voltage assignment. IEEE Trans. Very Large Scale Integr. (VLSI) Systems **18**(3), 501–504 (2009)
21. Qiu, M., Li, H., Sha, E.: Heterogeneous real-time embedded software optimization considering hardware platform. In: ACM Symposium on Applied Computing, pp. 1637–1641 (2009)
22. Qiu, M., Jia, Z., et al.: Voltage assignment with guaranteed probability satisfying timing constraint for real-time multiproceesor DSP. J. Sig. Proc. Syst. **46**, 55–73 (2007)
23. Qiu, M., Chen, Z., Ming, Z., Qin, X., Niu, J.: Energy-aware data allocation with hybrid memory for mobile cloud systems. IEEE Syst. J. **11**(2), 813–822 (2014)
24. Niu, J., Gao, Y., Qiu, M., Ming, Z.: Selecting proper wireless network interfaces for user experience enhancement with guaranteed probability. JPDC **72**(12), 1565–1575 (2012)
25. Qiu, M., Xue, C., Shao, Z., Zhuge, Q., Liu, M., Sha, E.H.M.: Efficient algorithm of energy minimization for heterogeneous wireless sensor network. In: Sha, E., Han, S.K., Xu, C.Z., Kim, M.H., Yang, L.T., Xiao, B. (eds.) Embedded and Ubiquitous Computing. EUC 2006. Lecture Notes in Computer Science, vol. 4096, pp. 25–34. Springer, Berlin (2006). https://doi.org/10.1007/11802167_5
26. Hu, F., Lakdawala, S., et al.: Low-power, intelligent sensor hardware interface for medical data preprocessing. IEEE Trans. Inf. Tech. Biomed. **13**(4), 656–663, 2009
27. Qiu, M., Qiu, H., et al.: secure data sharing through untrusted clouds with blockchain-enabled key management. SmartBlock **2020**, 11–16 (2020)
28. Li, J., Ming, Z., et al.: Resource allocation robustness in multi-core embedded systems with inaccurate information. J. Sys. Arch. **57**(9), 840–849 (2011)
29. K. Gai, Y. Zhang, et al., "Blockchain-enabled Service Optimizations in Supply Chain Digital Twin", IEEE Transactions on Service Computing, 2022
30. Qiu, H., Zheng, Q., et al.: Topological graph convolutional network-based urban traffic flow and density prediction. IEEE Trans. ITS **22**, 4560–2569 (2020)
31. Gao, X., Qiu, M.: Energy-based learning for preventing backdoor attack. KSEM **3**, 706–721 (2022)
32. Qiu, H., Dong, T., Zhang, T., et al.: Adversarial attacks against network intrusion detection in IoT systems. IEEE IoT J. **8**(13), 10327–10335 (2020)
33. Gai, K., Qiu, M., Elnagdy, S.: A novel secure big data cyber incident analytics framework for cloud-based cybersecurity insurance. In: IEEE BigDataSecurity (2016)
34. Li, Y., Gai, K., et al.: Intercrossed access controls for secure financial services on multimedia big data in cloud systems. ACM Trans. Multi. Comput. Commun. Appl. **12**, 1–18 (2016)

OpenVenus: An Open Service Interface for HPC Environment Based on SLURM

Meng Wan[1], Rongqiang Cao[1,2(✉)], Yangang Wang[1,2], Jue Wang[1,2], Kai Li[1], Xiaoguang Wang[1], and Qinmeng Yang[1]

[1] Computer Network Information Center, Chinese Academy of Sciences, Beijing 100083, China
{wanmengdamo,kai.li,wangxg}@cnic.cn, {caorq,wangyg,wangjue}@sccas.cn
[2] University of Chinese Academy of Sciences, Beijing 100049, China

Abstract. With the emergence of more and more "AI + Field + HPC" applications, it is urgent to solve the problem of scheduling and management of *High-Performance Computing* (HPC) resources, as well as the fast and efficient "cloud service" of HPC applications. This engineering problem is particularly critical because it affects the progress of scientific research, the development period of the research platform, and the learning cost of scientists. To solve the problem, a set of reusable life cycle processes for HPC resources are designed. Based on the life cycle, we propose an open service interface based on HPC, which reduces the startup time under multiple refreshes and abnormal retries by using the mode of contention lock. The active interruption of users is a typical scenario in the startup phase. Furthermore, a read-write strategy with an overlay based on Singularity is implemented to save storage space and improve running speed. In order to evaluate the serviceability and performance of the proposed interface, we deploy the service on the Venus platform and make a startup comparison experiment. In addition, the reduction of storage for 100 users is also tested. The experimental results show that under the HPC environment with SLURM, the proposed open-service interface can effectively shorten 46% startup time of applications and services and reduce 25% storage at least for each user of the Venus platform.

Keywords: Open service · HPC · Cloud · Container · Singularity · Cloud application

1 Introduction

With the combination of various research fields and *High-Performance Computing* (HPC) and networks [1,2], more and more "AI + Field + HPC" applications are emerging. The rapidly changing application scenario puts forward new requirements for HPC platforms [3]. Many HPC platforms and applications provide services to various users in a cloud-like operation [4]. In the field of materials, MatCloud, as a first principle molecular dynamics computing platform, provides materials computing cloud services for researchers [5]. In the field

M. Qiu et al. (Eds.): SmartCom 2022, LNCS 13828, pp. 131–141, 2023.
https://doi.org/10.1007/978-3-031-28124-2_13

of life sciences, latchBio offers a series of bio information reasoning cloud services, including alphafold2 [6]. Furthermore, the Rainbird platform releases cloud services that automate decision-making and run deep learning tasks in a workflow manner [7].

The "cloud service" of the scheduling and management of HPC resources can enable users to operate HPC-based platforms and applications in a more familiar way and with a better experience, especially for some scientists in the field who are not familiar with the HPC command line [8]. Currently, most HPC centers use job scheduling tools such as SLURM and PBS [9]. However, the repetitive development work for each platform and application makes the entire development cycle lengthy and inconsistent, and users have high learning costs [10]. More importantly, cloud services are all based on docker containers. Singularity, as the primary container technology of major HPC centers, of which the related scheduling, and management tools are not as mature as Docker's ecology [11].

To optimize various problems in the cloud service process of the HPC center based on the SLURM job scheduling system, we propose an Open Service Interface for HPC Environment based on SLURM, which has three main contributions:

- Reasonable service life cycle design: which can quickly convert the SLURM-based HPC scheduling service into a container service that can be called by Web applications.
- Efficient and reusable service startup technology and read-only & writable strategy for overlay storage.
- A set of universal open interfaces, including unified standard open interface design, distribution, release, forwarding, etc. The WEB code can be embedded without invasion through loose coupling and pluggable mode.

This paper's second section mainly introduces the relevant work and technology. The third section presents the overall architecture of the system and the life cycle of the open service interface. The fourth section describes the key technical points to improve startup efficiency. The fifth section compares development and startup efficiency and summarizes the advantages of the open service interface.

2 Related Work

With the rapid demand for computing power [12–14] in most research fields, HPC has inexorably penetrated scientific research [15]. Cloud service [16–18] of computing power platforms and applications is an essential technical support for future scientific development. Rafael et al. used HPC-based cloud technology for virtual screening in drug discovery [19]. Wu et al. proposed an optimization algorithm to greatly reduce the time consumption of the computing life cycle [20]. Li et al. studied an HPC cloud architecture to reduce the complexity of HPC workflow in the containerization phase to provide more scalability, and user convenience for HPC resources [21]. Sawa et al. have developed a platform named LincoSim, which enables users to analyze virtual cupping [22] automatically.

Baidu's PaddlePaddle AI platform from the Internet field uses cloud services to open up core capabilities such as computing resources to users on demand, including CPU, GPU, and memory [23]. In order to improve the parallel efficiency of computing tasks, Qiu et al. tried to speed up the computation through the underlying hardware by RNA nanotechnology [4]. In this paper, considering HPC resources' scalability, management efficiency, and ease of use, we have developed a class of open service interfaces for various HPC platforms.

3 System Architecture

3.1 Overall Process

In this paper, the open service interface is built on the HPC cluster with SLURM as the scheduling system, which can quickly enable the entire HPC cluster to have cloud service capabilities. With administrator privileges, software, applications and computing resources deployed in HPC environment can be quickly released to users through port forwarding and scheduling tools, as shown in Fig. 1.

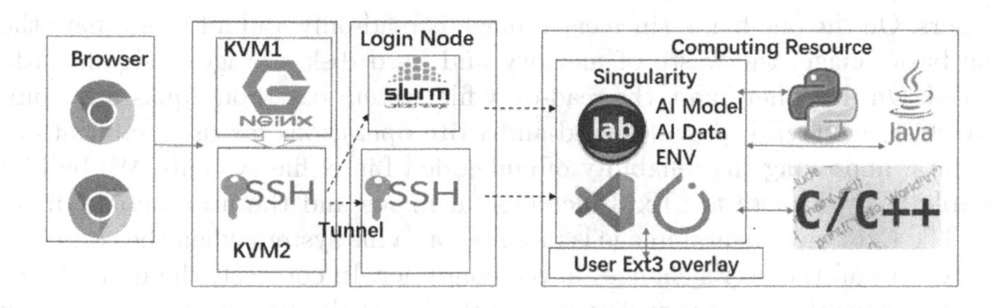

Fig. 1. The process of user access

This set of open service interfaces has been deployed on the Venus AI platform and officially launched for use. As shown in Fig. 2: the whole architecture design adopts hierarchical and modular modes to ensure system scalability. From the perspective of hierarchical division, the lowest layer is infrastructure, including proprietary server clusters and HPC clusters. The scheduling layer uses a SLURM job scheduling system and V-Slurm customized plug-ins. The middle layer realizes the Singularity container solution of separating reading and writing through various tools and technologies at the operating system level. The service layer provides various interfaces that are decoupled from each other, forming a set of open service interfaces that are systematic and complete, accelerating the development process of cloud services based on HPC. The main content of this paper is mainly reflected in the blue part of the container solution and the yellow part of the interface services.

The container scheme is implemented based on Singularity with a high affinity for HPC. We use Ubuntu and CentOS images to build a Singularity fundamental image repository. Any of which can be used as a read-only system image

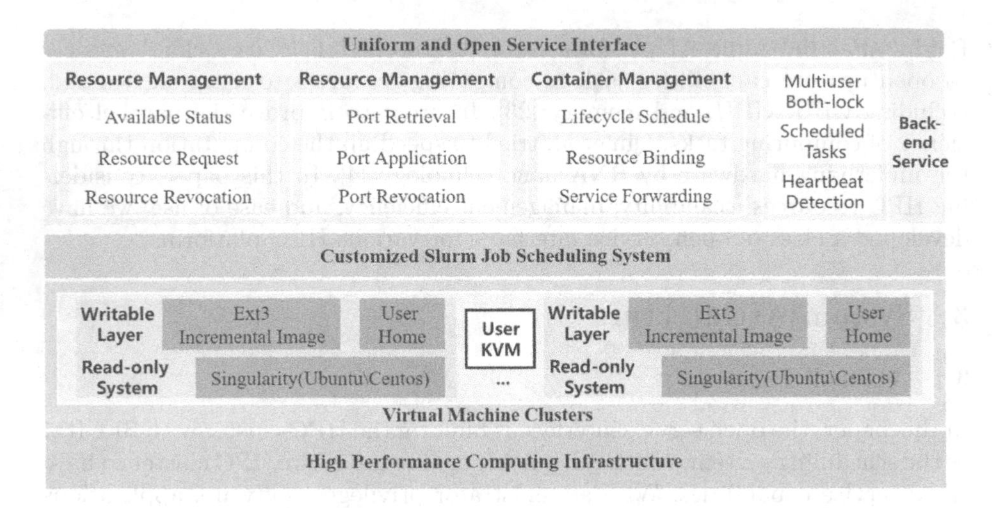

Fig. 2. The architecture of the main technology.

for users. On the one hand, since each image is read-only and all users share the same basic image, the waste of memory and hard disk storage is significantly reduced. On the other hand, the read-only file system based on SquashFS, combined with overlayFS, provides read-and-write operations for user applications and data, improving the reliability of embedded Linux file systems. We build a writable layer through the Ext3 incremental image and the user's home directory. The Ext3 incremental image is mounted as a file system when the container runs to save all the user's changes in the container. In contrast, the user's home directory maps the user's user directory in the KVM through shared storage and prevents the. bashrc file from serving as a bridge between the user's file transfer and the recovery environment. It is worth mentioning that by modifying the start script. sh and 99-base.in. syntax Scripts such as sh can customize different startup environments, such as initializing Conda, running the Jupyter service, and loading the current path configuration. The interface layer uses Python's Flask framework to build four modules as APIs: resource management, interface management, container management, and back-end services.

3.2 Lifecycle

For the transformation of applications or HPC resources into available cloud services, a life cycle diagram is roughly formed as shown in Fig. 3.

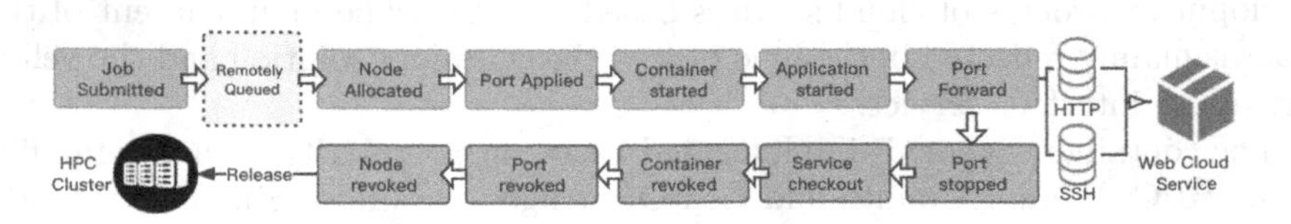

Fig. 3. The lifecycle of application service

The user triggers the startup action through the WEB browser, and the task will allocate resources according to the user's startup conditions (memory size, number of GPUs, etc.). If the resources are insufficient, enter the queue to wait. Otherwise, directly apply to the corresponding node in the cluster. When the node is successfully allocated, V-Slurm initializes the network card and system environment for the current node, and the node's IP is obtained. In order to obtain the available ports of the currently available springboard machine, it is necessary to retrieve the memory database with the bitmap structure. With the above parameter preparation, the container startup process begins. Most parameters are directly used through the Singularity startup command, and a small number of application parameters are passed in through environment variables. After the container is started, The startup script of an application or service will be automatically triggered under Singularity. Subsequently, the services related to port forwarding are started, and the complete chain port tunnel using HTTP and SSH protocols is opened. So far, the service startup process is completed.

The resource recycling process is relatively simple: The port forwarding process is recalled one by one. Before stopping the application service, it checks whether the service checkout has been saved from overlay. If it is saved successfully, continue the following operations: container stop, port database recycle, node resource release, etc. All physical and virtual resources occupied by users are restored. Due to incremental images, the user can directly restore the existing environment at the next boot.

4 Efficient Startup Technology

4.1 Serial Process with Multiuser Both-Lock

Serial Process with Multiuser Both lock mainly solves the problem of repeated startup and startup efficiency under multiple user requests. An application queuing or network card initialization timeout is a common problem caused by excessive resource load. In this case, certain stages in the life cycle can be reused to save time. In other words, a unique entity is generated for each user. Each user will increase their resource lock in a specific time window (currently 120 s) when the same function is called multiple times. Only the first call is actually executed, and subsequent operations can only obtain results from the cache. In order to complete this process, two essential conditions are required.

- Basic condition 1: Each user entity consists of a key or nameid identifier representing a unique instance. In the port, container, and tunnel objects, the jobId and other attributes ensure uniqueness and calculate md5 to generate a static, encrypted string. The nameid is the only user ID. One nameid can correspond to multiple instances, but only the latest instance is allowed to be retained during the inspection process. The job is identified by its name because the job number exists only after the job is successfully created. The uniqueness of the job can also be guaranteed when the lock is acquired first and then created. (If created from outside the controlled system, there will be multiple jobs with the same name because the SLURM system allows jobs to have duplicate names).

- Basic condition 2: The instance status can be queried in real-time according to the entity ID, and reliable status maintenance and retrieval service need to be built. To speed up retrieval time and reduce the time cost of deploying open interfaces. Bitmap based on memory database is used as the state service storage database. Each operation on entity and identifier are written to log records and memory database. At regular intervals, heartbeat detection is used to update the status of various entities for unfinished jobs.

The following Fig. 4 mainly shows the detailed Serial Process with Multiuser Both-lock. It uses a single instance process that provides fine-grained locks for single write and multiple read to solve the problem of multi-thread resource competition and reduce the difficulty of API development. All R requests perform the same operation. In the T-all window, only R2 executes, and the rest will be cached. After the T-all window, there will be other requests to acquire the lock again. However, since the entity still exists, the "create the entity" operation will be skipped to avoid resource waste and improve resource utilization efficiency. When applying for a job, if the job startup process is not finished and the API is called again (equivalent to refreshing the web page), a new resource application process will be started again. The original resources will not be released (or the resources can only be released after a long period through complex monitoring indicators). In the current mode, only one entity can be started for a period to avoid resource waste; To ensure the idempotence of functions and ease of programming, the write operations of the five entities shown in the figure below are idempotent within the lock adequate time, and repeated operations will not change the state of the entity. Call the upper layer applications of these APIs as long as they are called on demand without paying attention to the problem of repeated calls.

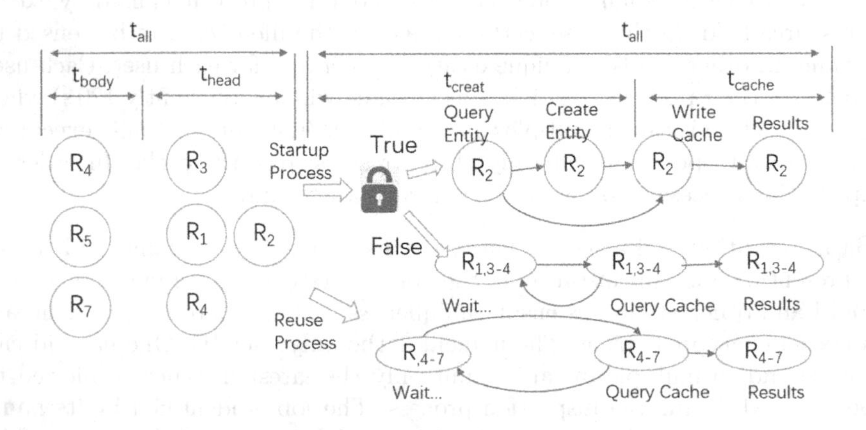

Fig. 4. The startup procedure of Multiuser Both-lock

Generally, this method makes every API call a simple request and response mode, which can be called many times until a long operation is completed. For example, the job takes a long time to start. When a request calls the API, the API starts to execute. Due to the long execution time, the front-end call ends before the execution process ends. The execution result can be obtained at the next call.

4.2 Scalable Overlay Storage

The Read-only&Writable Strategy for scalable overlay storage effectively ensures the container startup phase's efficiency in considering factors such as environment recovery and user-defined space. As shown in the Fig. 5 below, the basic image includes the Conda package, system software, and user software. The entire file system or a single directory is compressed using SquashFS, stored in a KVM disk file, and saved as a base image Sif file. Each user selects different project codes and code-dependent environments when starting the environment. The working directory is mounted through a shared folder, and the user directory goes directly to the container to form the root directory.

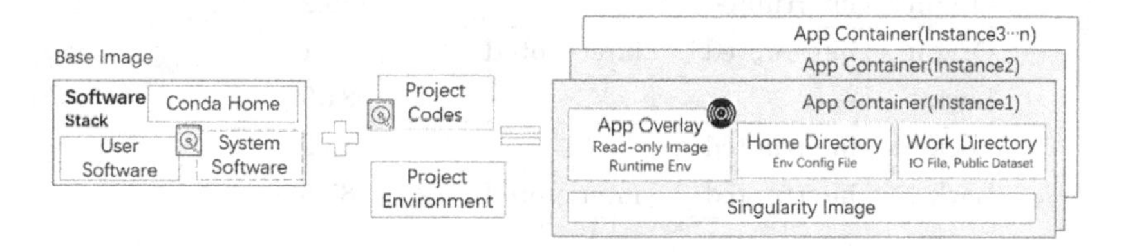

Fig. 5. The design of scalable overlay storage

An ordinary user can mount multiple read-only image files and one writable image file. The final startup command is as follows:

```
singularity exec \
—Nv Mount Nvida smi Driver
—Home home Running or development directory\
—Bind task working directory\
—Overlay read—only image 1_ Application overlay\
—Overlay read—only mirror 2_ Common Dataset\
—Overlay writable image (optional, temporary file)\
—env PORT & UUID.
```

5 Evaluation

Experimental Environment. The open service interface is deployed on the AI special HPC cluster. The host environment includes CentOS 7.6, 380 Nvidia P100 GPUs, and Intel Snapdragon CPUs. Each host is configured with 256 GB memory, and 50T shared storage by default.

5.1 Experiments A

In this paper, experiments A are designed to evaluate the time saved by using the serial process with Multiuser Both-lock technology. In order to verify the accurate time reduced by the technology, four groups of comparative experiments are designed. The first group of experiments ran the service startup interface process without external intervention. The second group of experiments was interrupted once in the startup phase and requested again when the lock technology was not used. In the third group of experiments, when using the lock technology, it was interrupted once in the startup phase and requested again. In the fourth group of experiments, when using the lock technology, the system interrupts once and requests again in the startup and forwarding phases, respectively. The statistics of reasoning results are shown in Table 1.

Table 1. Results of time consuming in different group

Strategy	Startup Phase	Forwarding phase	Time (Seconds)
Origin	–	–	81.7
Origin	interrupted	–	126.2
Origin	interrupted	interrupted	159.1
Lock	-	-	81.9
Lock	interrupted	-	83.2
Lock	interrupted	interrupted	85.5

Experimental Result. The first, second, and third groups of experiments show that the use of the lock contention strategy can significantly reduce the startup time of users under repeated refreshes and exceptional failure retries, which is a prevalent scenario in the process of user use. The third and fourth groups of experiments confirmed that interrupting the interface at any stage has little impact on the overall process and can quickly restore to the last state.

5.2 Experiments B

Experiments B are designed to evaluate the storage saved by using the read-only & writable strategy for scalable overlay storage technology. To test the accurate storage reduced by the technology, four groups of comparative experiments are designed. Each group represents the storage usage of a specific configured image with 100 users. In addition, the personal space of each user in this experiment is allocated to 10 GB. The statistics of the results are shown in Table 2.

Table 2. Results of 100 users' storage test

Image	Image Size (GB)	100 Users with Origin (TB)	100 Users with Overlay (TB)	Reduction (%))
CenOS7.5	4.7	1.47	1.0050	31.7
CentOS7.6 with Pytorch	8.5	1.85	1.0088	45.5
Ubuntu20.04	3.6	1.36	1.0037	26.2
Ubuntu18.04 with Tensorflow	7.2	1.72	1.0073	41.4

Experimental Result. The experiment assumes that there are 100 real users using the platform. In this case, even if the Ubuntu empty image with the smallest storage space is used, the overlay technology can reduce the total storage capacity by 26.2%.

5.3 Application

Venus AI Platform. OpenVenus makes it possible to quickly build a platform based on HPC and release the application and service capabilities. It has been integrated and applied on the Venus AI Platform [24]. As of the date of issuance, five applications are open to the public on Venus platform, the number of HPC users is up to 120, the number of successfully running tasks is up to 1200, and the total time of running GPU cards is 9000 h hours.

Application Service. We have integrated three application services through this open service interface: JupyterLab, Code Server, and Baihai IDP. These three services are currently open to all users on the platform and are highly praised by most users. In the construction phase of the initial version of the platform, it will take at least 1–2 months for a team of 10 people to integrate an application service based on HPC and open it to users. With our interface, the time in this phase is shortened to less than a week. Developers familiar with the interface can even go online and release a cloud service or application based on HPC in one day.

6 Conclusion

In order to improve the construction speed and universality of various cloud platforms, cloud services, and cloud applications based on HPC, we proposed an open service interface considering the scalability, management efficiency, and ease of use of HPC resources. The complete life cycle design makes the whole interface calling process flexible and easy to use and also solves the scheduling problems such as automatic application and allocation. The Serial Process with Multiuser Both lock technology improves the rapid release of application and service capabilities and significantly improves the startup speed of applications. Read-only and Writable strategy combines various technologies to reduce the occupation of hard disk space and solve the problem of reloading the user environment. Finally, the open service interface follows the design idea of loose coupling and pluggable in software engineering and is placed in the GitHub repository and open source for any developer.

Acknowledgments. This work was supported by the National Key R&D Program of China(No. 2020AAA0105202).

References

1. Niu, J., Gao, Y., et al.: Selecting proper wireless network interfaces for user experience enhancement with guaranteed probability. JPDC **72**(12), 1565–1575 (2012)
2. Qiu, M., Xue, C., et al.: Efficient algorithm of energy minimization for heterogeneous wireless sensor network. In: IEEE EUC Conference, pp. 25–34 (2006)
3. Jiang, Z., et al.: HPC AI500: a benchmark suite for HPC AI systems. In: Zheng, C., Zhan, J. (eds.) Bench 2018. LNCS, vol. 11459, pp. 10–22. Springer, Cham (2019). https://doi.org/10.1007/978-3-030-32813-9_2
4. Qiu, M., Khisamutdinov, E., et al.: RNA nanotechnology for computer design and in vivo computation. Philos. Trans. Royal Soc. A: Math. Phys. Eng. Sci. **371**(2000), 20120310 (2013)
5. Yang, X., Wang, Z., et al.: Matcloud: a high-throughput computational infrastructure for integrated management of materials simulation, data and resources. Comput. Mater. Sci. **146**, 319–333 (2018)
6. De Laurentiis, L., De Santis, D., et al.: A new user oriented platform to develop AI for the estimation of bio-geophysical parameters from EO data. In: IEEE International Geoscience and Remote Sensing Symposium, IGARSS, pp. 262–265 (2021)
7. Collins, R.A., Trauzzi, G., et al.: Meta-fish-lib: a generalised, dynamic DNA reference library pipeline for metabarcoding of fishes. J. Fish Biol. **99**(4), 1446–1454 (2021)
8. Qiu, M., et al.: Energy minimization with soft real-time and DVS for uniprocessor and multiprocessor embedded systems. In: IEEE Date, pp. 1–6 (2007)
9. Ahn, D.H., Garlick, J., Grondona, M., et al.: Flux: a next-generation resource management framework for large HPC centers. In: 43rd IEEE Conference on Parallel Processing Workshops, pp. 9–17 (2014)
10. Asatiani, A.: Why cloud?-a review of cloud adoption determinants in organizations. In: European Conference on Information Systems (2015)
11. Saha, P., Beltre, A., et al.: Evaluation of docker containers for scientific workloads in the cloud. In: Practice and Experience on Advanced Research Computing, pp. 1–8 (2018)
12. Qiu, M., Yang, L., et al.: Dynamic and leakage energy minimization with soft real-time loop scheduling and voltage assignment. IEEE TVLSI **18**(3), 501–504 (2009)

13. Qiu, M., Jia, Z., Xue, C. et al. Voltage assignment with guaranteed probability satisfying timing constraint for real-time multiproceesor DSP. J VLSI Sign. Process. Syst. Sign Image Video Technol. **46**, 55–73 (2007). https://doi.org/10.1007/s11265-006-0002-0
14. Li, J., Ming, Z., et al.: Resource allocation robustness in multi-core embedded systems with inaccurate information. J. Syst. Arch. **57**(9), 840–849 (2011)
15. Cieslak, W.R., Westrich, H.R.: Ldrd impacts
16. Zhao, H., Chen, M., et al.: A novel pre-cache schema for high performance android system. FGCS **56**, 766–772 (2016)
17. Gao, Y., et al.: Performance and power analysis of high-density multi-GPGPU architectures: a preliminary case study. In: IEEE 17th HPCC, pp. 29–35 (2015)
18. Gai, K., Qiu, M., Elnagdy, S.: A novel secure big data cyber incident analytics framework for cloud-based cybersecurity insurance. In: IEEE BigDataSecurity Conference (2016)
19. Dolezal, R., Sobeslav, V., Hornig, O., Balik, L., Korabecny, J., Kuca, K.: HPC cloud technologies for virtual screening in drug discovery. In: Nguyen, N.T., Trawiński, B., Kosala, R. (eds.) ACIIDS 2015. LNCS (LNAI), vol. 9012, pp. 440–449. Springer, Cham (2015). https://doi.org/10.1007/978-3-319-15705-4_43
20. Wu, G., Zhang, H., et al.: A decentralized approach for mining event correlations in distributed system monitoring. JPDC **73**(3), 330–340 (2013)
21. Li, G., Woo, J., Lim, S.B.: HPC cloud architecture to reduce HPC workflow complexity in containerized environments. Applied Sci. **11**(3), 923 (2021)
22. Salvadore, F., Ponzini, R.: Lincosim: a web based HPC-cloud platform for automatic virtual towing tank analysis. J. Grid Comp. **17**(4), 771–795 (2019)
23. Ma, Y., Yu, D., et al.: Paddlepaddle: an open-source deep learning platform from industrial practice. Front. Data Domputing **1**(1), 105–115 (2019)
24. Yao, T., Wang, J., Wan, M., et al.: Venusai: an artificial intelligence platform for scientific discovery on supercomputers. J. Syst. Arch. **128**, 102550 (2022)

Design and Analysis of Two Efficient Socialist Millionaires' Protocols for Privacy Protection

Xin Liu[1,2], Xiaomeng Liu[1(✉)], Xiaofen Tu[1], and Neal Xiong[2,3]

[1] School of Information Engineering, Inner Mongolia University of Science and Technology, Baotou 014010, Inner Mongolia, China
15552086354@163.com

[2] National Engineering Research Center for E-Learning, Central China Normal University, Wuhan 430079, China

[3] Department of Computer Science and Mathematics, Sul Ross State University, Alpine, TX 79830, USA

Abstract. Yao's Millionaires' problem has led to the emergence of secure multi-party computation. As an important tool for privacy protection in cryptography, secure multi-party computation has attracted more and more scholars to study it. The socialist millionaires' problem is the basic module of the secure multiparty computing protocol. Designing secure and efficient solutions for the socialist millionaires' problem can be effectively applied to the secret ballot, electronic auction, and so on. Based on the vector encoding method, the Paillier encryption scheme, and the Goldwasser-Micali encryption scheme, two efficient socialist millionaires' protocols are proposed and the protocols are analyzed. The correctness analysis, security proof, performance analysis, and experimental simulation show that the efficiency of the two protocols is superior to the existing schemes.

Keywords: Secure multi-party computation · Socialist millionaires' problem · Vector encoding method · Encryption scheme · Experimental simulation

1 Introduction

Yao's Millionaire Problem- was raised by Mr. Yao [1], a Turing Award winner in 1982. This problem is described as follows: Alice and Bob are two millionaires, they want to compare who is richer, but neither of them wants to disclose how much money they have. This problem can be seen everywhere in real life. For example, Alice wants to buy something from Bob at a price x, but Bob wants to sell it at a price y. They are not willing to disclose their bottom price, so they want to keep it secret. In order to solve these problems, Secure Multiparty Calculation is needed. In 1987, Goldreich proposed an MPC protocol based on a cryptographic security model [2], which can compute arbitrary functions. It is theoretically proved that all MPC protocols can be implemented by using Garbled Circuit. In 1998, Goldreich proposed the definition of security for MPC, and it has more security [3]. In addition, scholars at home and abroad have proposed theoretical definitions of MPC [4–6], applications [7–9], security proof methods [10–12], and MPC protocols for some problems [13–15], which are of great significance

M. Qiu et al. (Eds.): SmartCom 2022, LNCS 13828, pp. 142–151, 2023.
https://doi.org/10.1007/978-3-031-28124-2_14

for the development of cryptographic services [16–18] and efficiency [19, 20] in many fields of society [21–23].

In Yao's Millionaire Problem, two data comparison problems are studied, $x \geq y$ or $x < y$. But in real life, many situations are to judge whether the data are equal, that is $x = y$, which is the problem of socialist millionaires.

Some scholars have proposed solutions to the problem of socialist millionaires. In the semi-honest model, Liu Wen [24] and others have proposed a protocol to solve the problem of socialist millionaires by means of sliding windows and exchange encryption functions. Boudot et al. [25] have proposed a protocol based on *discrete logarithm* (DL), *Diffie-Hellman* (DH), and *Decision Diffie-Hellman* (DDH) hypothesis and zero-knowledge proof Socialist Millionaire protocol, but the computational complexity is high. Qin Jing et al. [26] proposed a new protocol for socialistic millionaires without information leakage based on ϕ-hiding hypothesis and semantic security, which needs the participation of a blank third party. The solution to Yao's Millionaire Problem [1], efficiency is exponential. If the data is comparatively large, the protocol is unrealistic. Protocols in references [27] and [28] can only judge two data ">, ≤". It is impossible to judge whether the data are equal.

Although some socialist millionaire protocols have been proposed, there are also some defects as described in the preceding paragraph. It is necessary to further study more secure and efficient socialist millionaire Protocols. Therefore, this paper proposes two socialist millionaire Protocols. The main contributions are as follows:

(1) Based on a special encoding method-vector encoding, the data to be compared is transformed into a vector, and the problems of data comparison and judgment are transformed into the problem of selecting vector elements.
(2) Based on the additive homomorphism of the Paillier encryption scheme, a socialist millionaire protocol is proposed, which can compare the two data at one time. The computational complexity of the protocol is $2(s + 2)\lg N$ modular multiplications (vector encoding dimension).
(3) Based on the difference or homomorphism of the *Goldwasser-Micali* (GM) encryption scheme, a social attention millionaire protocol is proposed. Its computational complexity is $6s+4$ modular multiplications, which further improves computational efficiency.

Section 2 mainly introduces the relevant knowledge used in this paper, including simulation example proof security, Paillier encryption scheme, GM encryption scheme and vector coding method. Section 3 proposes two socialist millionaire protocols, and analyses the correctness and security of the protocols. Examples are given. Section 4 analyses the performance of the protocol and simulates it experimentally. Section 5 summarizes the whole paper.

2 Relevant Knowledge

2.1 Simulation Example Proof

The simulation example proposed by Goldreich [9] is widely used in security proof of secure multi-party computation protocols. It can simulate the process of protocol execution by protocol participants. Its proof principle is that each participant simulates the protocol separately with its own input and output. The implementation process of the protocol and any information he can get from the secure multi-party computation protocol show that participants cannot get more information from the actual secure multi-party computation protocol than from the ideal secure multi-party computation protocol, that is to say, the protocol is secure.

2.2 Paillier Encryption Scheme

Paillier's encryption scheme is an encryption algorithm based on high order residual class problem proposed by Paillier et al. [29]. in 1999. It has additive homomorphism.

Get ready: set up $N = pq$, p and q is two large prime numbers: $\lambda(N) = lcm(p-1, q-1)$; $B = \{x|x^{N\mu} \mod N^2 = 1, \mu \in \{1, 2 \cdots, \lambda\}\}$; $S_N = \{u < N^2 | \mu \equiv 1 \mod N\}$; $L(u) = \frac{u-1}{N}(\forall u \in S_N)$; $g \in B$ is public key; λ is private key.

Encryption: Select a random number $r < N$, plaintext $m < N$, the encryption process is $c = E(m) = g^m r^N \mod N^2$.

Decryption: ciphertext $c < N^2$, decryption calculation as $m = \frac{L(c^\lambda \mod N^2)}{L(g^\lambda \mod N^2)} \mod N$.

Paillier encryption algorithm has additive homomorphism, that is $E(m_1) = g^{m_1} r^N \mod N^2$, then $E(m_1)E(m_2) = g^{m_1} r^N \mod N^2 \cdot g^{m_2} r^N \mod N^2 = g^{m_1+m_2} r^N \mod N^2 \equiv E(m_1+m_2)$.

2.3 Goldwasser-Micali Encryption Scheme

Goldwasser-Micali encryption scheme (GM) was proposed by Goldwasser and Micali in 1984 based on the difficulty of quadratic residue [30].

The scheme is described as follows:

Prepare: Set a security parameter k, select two k-bit prime numbers p and q, and calculate $n = pq$. Select the quadratic non-residual $t \in Z_n^1$ of module n, where Z_n^1 is a subset of Jacobi elements containing Z_n^*. The public key is (n, t) and the private key is (p, q).

Encryption: plaintext $m = m_1 m_2 \cdots m_s$, $m_i \in \{0, 1\}$ in binary representation, public key (n, t), random number r, encrypted message m_i:

$$E(m_i) = t^{m_i} r_i^2 \mod n = \begin{cases} t r_i^2 \mod n, \ m_i = 1 \\ r_i^2 \mod n, \ m_i = 0 \end{cases}.$$

Decryption: Using the private key (p, q), the decryption process is as follows:

$$m_i = \begin{cases} 0 \ (\frac{E(m_i)}{p}) = (\frac{E(m_i)}{q}) = 1 \\ 1 (\frac{E(m_i)}{p}) = (\frac{E(m_i)}{q}) = -1 \end{cases}.$$

Among them, $(\frac{a}{p})$ is the Jacobian symbol is defined as follows:

$$(\frac{a}{p}) = \begin{cases} 1 \ \text{P cannot divide a, P is the second residue of a ;} \\ -1 \ \text{P is not divisible by a, P is a quadratic non - residue of a ;} \\ 0 \ \text{P can divide a.} \end{cases}$$

GM encryption algorithm is bit-by-bit encryption. GM encryption algorithm can encrypt messages with two opposing properties, such as 0 and 1, right and wrong, profit and non-profit, complete and incomplete. In addition, the encryption algorithm is different or homomorphic.

$$E(m_i)E(m_j) = \begin{cases} r_i^2 r_j^2 \bmod n & m_i = 0, m_j = 0 \\ t r_i^2 r_j^2 \bmod n & m_i = 0, m_j = 1 \\ t^2 r_i^2 r_j^2 \bmod n & m_i = 1, m_j = 1 \\ t r_i^2 r_j^2 \bmod n & m_i = 1, m_j = 0 \end{cases}$$

2.4 Vector Coding

Vector encoding has the following advantages in designing secure data comparison protocols:

(1) Vector coding is simple and easy to implement.
(2) The comparative data can be converted into a vector, and the form of problem-solving can be converted into the form of easy calculation.
(3) As long as the number of elements of the vector is limited, the amount of computation can be reduced. This section introduces a vector coding method, which can encode a number k into a vector v: $v = (v_1, v_2, \cdots, v_i, \cdots, v_n)$, where $v_i = \begin{cases} \alpha, & 1 \le i < k \\ \beta, & i \ge k \end{cases}, \alpha \ne \beta$.

3 Efficient Protocol of Socialist Millionaires' Problem

Based on the vector coding method, two socialist millionaire protocols are proposed in this paper. One is vector encryption using the Paillier encryption scheme, the other is vector encryption using the GM encryption scheme. By selecting the encrypted vector elements, we can judge whether the two data are equal.

3.1 Protocol of Socialist Millionaires' Problem Based on Paillier Algorithm

Alice has data x, Bob has data y. Alice encodes his data into vector X by vector encoding, and sends $E(X)$ to Bob using Paillier encryption scheme to encrypt vector X. Alice selects the Y element from it, and uses Paillier additive homomorphism to encrypt the selected y element to generate another ciphertext and send the ciphertext. Decrypt Alice. Alice and tell Bob the result. The protocol is described as follows:

Protocol 1: Secure Data Judgment Protocol Based on Paillier Encryption System.

Input: Alice's input is x, Bob's input is y, $x, y \in U = \{u_1, \cdots, u_s\}$.

Output: $x = y$, $x \neq y$.

Step 1: Using Paillier encryption scheme, Alice generates public key (g, N) and private key λ and sends public key (g, N) to Bob.

Step 2: Using vector coding method, Alice encodes data x into vector X, as follows:$X = (m_1, \cdots, m_i, \cdots, m_s)$, Among them, $m_i = \begin{cases} \alpha, u_i \neq x \\ \beta, u_i = x \end{cases}, \alpha \neq \beta$.

Step 3: Alice chooses s random numbers $r_1, \cdots r_s$ and encrypts vector X with Paillier encryption scheme to get $E(X) = (E(m_1, r_1), \cdots, E(m_s, r_s))$, where $E(m_i, r_i) = g^{m_i} r_i^N \mod N^2, (i = 1, \cdots, s)$.

Step 4: Alice sends $E(X)$ to Bob.

Step 5: Bob selects a random number r_b. He selects $E(m_i, r_i)$ from $E(X)$ to make $i = y$ and calculates $E(m_i, r_i) \times E(0, r_b) = g^{m_y} r_y^N \mod N^2 \times g^0 r_b^N \mod N^2 \to E(\mu)$.

Bob sends $E(\mu)$ to Alice.

Step 6: Alice decrypts $D(E(\mu))$ and tells Bob the result $P(x, y)$:

If $\mu = \alpha$, then $x \neq y$; If $\mu = \beta$, then $x = y$.

Correctness Analysis:

(1) (1)Alice decryption $D(E(\mu)) = \mu$, that is, $u = m_y$.Since $m_y \in \{m_1, \cdots m_{x-1}, m_x, m_{x+1}, ..., m_s\} = \{\alpha, \cdots, \alpha, \beta, \alpha, \cdots, \alpha\}$, if $y = x$,$y \neq x$; if $m_y = \beta, y = x$.
(2) When Alice receives the data $E(\mu)$ transmitted by Bob, she does not know how to calculate $E(\mu)$,because she does not know the random number r_b, so $E(m_y, r_y)$ is confidential, and y is confidential.
(3) When Bob knows the result $E(m_y, r_y)$, Bob doesn't know which α is equal to m_y, so Bob can't know the value of x. When Bob knows $x = y$, it doesn't mean that he leaks the information of x, because it's the same under the ideal secure multi-party computation model.

3.2 Protocol on Socialist Millionaires' Problem Based on GM Algorithm

Alice has data x, Bob has data y, Alice encodes x into vector X composed of 0–1 codes by vector encoding method. Alice uses GM encryption scheme to encrypt X to get $E(X)$, and sends $E(X)$ to Bob. Bob selects the y element, and returns it to Alice after encrypting and decrypting Alice, then we can know whether $x = y$.

Protocol 2: Secure Data Judgment Protocol Based on GM XOR Homomorphism

Input: Alice's input is x, Bob's input is y, $x, y \in U = \{u_1, \cdots, u_s\}$.

Output: $x = y$, $x \neq y$.

The protocol is basically the same as Protocol 1, but the difference lies in step 2.

Step 1: According to GM encryption scheme, Alice generates public key (n, t) and private key (p, q), and chooses random number $r_1, \cdots, r_s$.

Step 2: Using the vector coding scheme, Alice encodes x into a 0–1 vector X: $(m_1, \cdots, m_i, \cdots m_s)$. Among them, $m_i = \begin{cases} 0, i \neq x \\ 1, i = x \end{cases}$.

Step 3: Alice encrypts X with GM encryption scheme:$E(X) = (E(m_1, r_i), \cdots, E(m_i, r_i), \cdots, E(m_L, r_L))$. Among them, $E(m_i, r_i) = \begin{cases} tr_i^2 \bmod n, & m_i = 1 \\ r_i^2 \bmod n, & m_i = 0 \end{cases}$.

Step 4: Alice sends $E(X)$ to Bob;

Step 5: Bob selects the y element from $E(X)$. He selects a random number r_b and calculates it: $E(m_y, r_y) \times E(0, r_b) = E(m_y, r_y) \times r_b^2 \bmod n \rightarrow e'_y$;

Step 6: Bob sends e'_y to Alice;

Step 7: Alice decryption e'_y:

If $(\frac{e'_y}{p}) = (\frac{e'_y}{q}) = 1$, then $D(e'_y) = 0$, so $x \neq y$; If $(\frac{e'_y}{p}) = (\frac{e'_y}{q}) = -1$, then $D(e'_y) = 1$, so $x = y$;

Step 8: Alice tells Bob the result $P(x, y)$.

The Protocol is over.

Correctness Analysis:

(1) In the above protocol, step 5 is based on the difference or homomorphism of GM encryption algorithm, that is to say:$E(m_y, r_y) \times E(0, r_b) = E(m_y, r_y) \times r_b^2 \bmod n = E(m_y \oplus 0)$.
If $m_y = 0$, then $D(E(m_y \oplus 0)) = 0$, so in the protocol $x \neq y$. If $m_y = 1$, then $D(E(m_y \oplus 0)) = 1$, so in the protocol $x = y$.
(2) Because the GM encryption scheme is probabilistic, the same plaintext is encrypted and the ciphertext is different. So Bob will not find its rule when he receives $E(X)$.
(3) Both Alice's random number r_i and Bob's random number r_b are confidential. They don't know each other's random number, so Alice can't calculate $E(0, r_b)$ and Bob can't calculate $E(m_i, r_i)$.
(4) Bob chooses ciphertext $E(m_y, r_y)$ and encrypts $E(m_y, r_y)$, so Alice does not know which ciphertext Bob chooses.
(5) p and q are private keys, Bob doesn't know, so Bob can't decrypt them.

Security Proof:

Theorem 3.2 Protocol 3.2 is secure under the semi-honest model.

Proof: We prove the security of the protocol by constructing simulators S_1 and S_2.

(1) According to $P(x, y)$ and S_1,y' is selected to make $P(x, y) = P(x, y')$ and S_1 use (x, y') to simulate the whole process of protocol. S_1 encodes x into $X = (m_1, \cdots, m_i, \cdots m_s)$ by vector coding method.
(2) Using GM encryption scheme, S_1 chooses different random numbers to encrypt X, and obtains: $E(X) = (E(m_1, r_i), \cdots, E(m_i, r_i), \cdots, E(m_s, r_s))$.
(3) S_1 chooses a random number r' and calculates it. $E(m_{y'}, r_{y'}) \times E(0, r') = E(m_{y'}, r_{y'}) \times r'^2 \bmod n \rightarrow e_{y'}$.
(4) S_1 decrypts $D(e_{y'})$ and gets $P(x, y')$. In the Protocol, $view_1^{\pi}(x, y) = \{X, E(X), e'_y, P(x, y)\}$. Let $\{S_1(x, P(x, y))\} = \{X, E(X), e_{y'}, P(x, y')\}$, because $P(x, y) = P(x, y')$, $e_{y'} \stackrel{c}{\equiv} e'_y$, therefore, $\{(S_1(x, f_1(x, y)), P(x, y))\}_{x,y} \stackrel{c}{\equiv} \{(view_1^{\pi}(x, y), output_2^{\pi}(x, y))\}_{x,y}$.

By using the same method, S_2 can be constructed to obtain:
$\{(f_1(x, y), S_2(y, f_2(x, y)))\}_{x,y} \stackrel{c}{\equiv} \{(output_1^{\pi}(x, y), view_2^{\pi}(x, y))\}_{x,y}$.
The proof is complete.

4 Performance Analysis and Experimental Simulation

4.1 Computational Complexity Analysis

For the solution of Yao's Millionaire Problem [1], its efficiency is exponential. If the data is comparatively large, the protocol is unrealistic. In reference [27], Lin uses ElGamal. The multiplication homomorphism of Al is used to judge $x > y$ or $x \le y$, and its computational overhead is $(5b \lg N + 4b - 6)$ times modular multiplication. For Blake et al. [28]. Who proposed the protocol, only $x > y$ or $x \le y$ can be judged, and its computational overhead is $((4b + 1) \lg N + 6b)$ times modular multiplication.

Liu Wen et al. [24] proposed a protocol to solve the problem of socialist millionaires. The computational complexity of the protocol is $O(log^2 N)$. Boudot et al. 125]. Proposed a socialist millionaire protocol, the complexity is $O(N^2)$-modular multiplication.

In protocol 3.1, Alice encodes her data as an s-dimensional vector. She calculates s Paillier encryption and one Paillier decryption. Bob uses one Paillier encryption. The computational cost of each Paillier encryption and decryption is $2 \log N$ modular multiplication. Therefore, the total computational cost of protocol 3.1 is $(2(s + 2) \log N)$ times modular multiplication. In protocol 3.2, we use GM encryption scheme to encrypt vector X. The computational cost of GM encryption algorithm is three times modular multiplication. Therefore, encryption vector X needs $3s$ times modular multiplication and decryption e'_y needs 2 times modular multiplication. Therefore, the two protocols need $(6s+4)$ times modular multiplication at most. Table 1 compares the computational complexity of each protocol.

Table 1. Computational complexity of data comparison protocols

Protocol	Judgement result	Modular multiplication
Yao[1]	$>, \le$	Exponential level
L.T. [27]	$>, \le$	$5b \lg N + 4b - 6(4b + 1) \lg N + 6b$
B.K. [28]	$>, \le$	$O(\log^2 N)$
Liu [24]	$=$	
Boudot [25]	$=$	$O(N^2)$
Protocol 3.1	$=$	$2(s + 2) \lg N$
Protocol 3.2	$=$	$6s + 4$

b: Bit Number of Input Information, N: Modules of Public Key Encryption Scheme, s: Vector Coding Dimension.

4.2 Communication Complexity Analysis

Communication complexity is another important index to measure the efficiency of secure multi-party computation protocols, which is usually measured by the number of rounds of interaction. Reference [24] needs $O(r)$-round interaction (where r is the data bit value), and document [25] needs n-round communication, but protocol 3.1 and protocol 3.2 only need one round of communication, which is more efficient.

4.3 Experimental Simulation

Experimental environment: Windows 10 (64-bit) operating system, Intel (R) Core (TM) i5-6600 CPU @3.30 GHz processor, memory 8.00 GB, run on MyEclipse with Java language.

The running time of protocol 3.1, 3.2 and references [24, 25] are compared by simulation. The number of bits of input data is set to $b = 1, 2, \cdots, 20$ respectively, and each bit value of b is tested 2000 times, ignoring the preprocessing time in the protocol, and the average execution time of the protocol is counted. Figure 1 compares the execution time of the protocols.

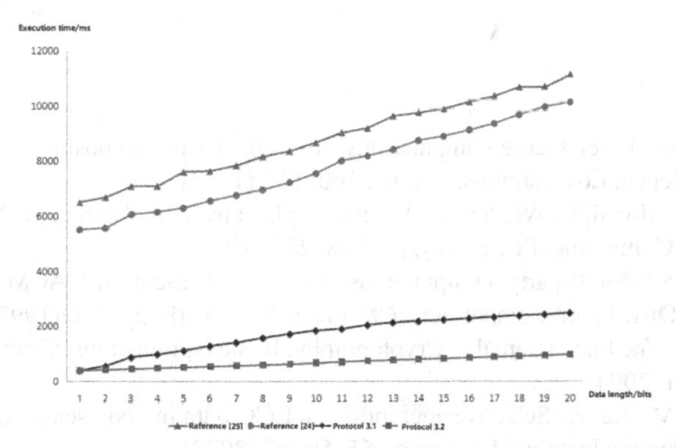

Fig. 1. The Rule of the Implementation Time of the Socialist Millionaire Problem Protocol with the Number of Input Bits.

From the experimental data, it can be seen that the execution time of protocol 3.1 varies slightly with the increase in the number of bits in the input data, and the execution time of protocol 3.2 does not change much with the increase in the number of bits in the input data. From the test results, the computational efficiency of protocol 3.1 is higher.

5 Conclusion

This paper has proposed two efficient solutions to the problem of socialist millionaires based on vector coding. One was the socialist millionaire protocol using the Paillier encryption algorithm, the other was the socialist millionaire protocol using the Goldwasser-Micali encryption algorithm. The correctness and security of the two protocols were analyzed. It was proved that the efficiency of the two protocols was high through performance analysis and experimental simulation. The protocol could be applied in electronic auction, secret voting, electronic commerce, and other fields.

Funding. This work is supported by the National Natural Science Foundation of China: Big Data Analysis based on Software Defined Networking Architecture, grant numbers 62177019 and F0701; NSFC, grant numbers 62271070, 72293583, and 61962009; Inner Mongolia Natural Science Foundation, grant number 2021MS06006; 2023 Inner Mongolia Young Science and Technology Talents Support Project, grant number NJYT23106; 2022 Fund Project of Central Government Guiding Local Science and Technology Development, grant number 2022ZY0024; 2022 Basic Scientific Research Project of Direct Universities of Inner Mongolia, grant number 20220101; 2022 "Western Light" Talent Training Program "Western Young Scholars" Project; the 14th Five Year Plan of Education and Science of Inner Mongolia, grant number NGJGH2021167; 2023 Open Project of the State Key Laboratory of Network and Exchange Technology; 2022 Inner Mongolia Postgraduate Education and Teaching Reform Project, grant number 20220213; the 2022 Ministry of Education Central and Western China Young Backbone Teachers and Domestic Visiting Scholars Program, grant number 2022015; Inner Mongolia Discipline Inspection and Supervision Big Data Laboratory Open Project Fund, grant number IMDBD202020; Baotou Kundulun District Science and Technology Plan Project, grant number YF2020013; Inner Mongolia Science and Technology Major Project, grant number 2019ZD025.

References

1. Yao. A.: Protocols for secure computations. In: 23th IEEE Symposium on Foundations of Computer Science, Los Alamitos, CA, pp. 160–164 (1982)
2. Goldreich, O., Micali, S., Wigderson, A.: How to play any mental game, In: ACM Conference on Theory of Computing. Piscataway, pp. 218–229 (1987)
3. Goldwasser, S.: Multi-party computations: past and present. In: ACM Symposium on Principles of Distributed Computing, ACM Press, New York, pp. 1–6 (1997)
4. Goldreich, O.: The Fundamental of Cryptography: Basic Applications. Cambridge University Press, London (2004)
5. Qiu, H., Qiu, M., Lu, Z.: Selective encryption on ECG data in body sensor network based on supervised machine learning. Inf. Fusion **55,** 59–67 (2020)
6. Qiu, M., Zhang, L., et al.: Security-aware optimization for ubiquitous computing systems with SEAT graph approach. J. Com. Sys. Sci. **79**(5), 518–529 (2013)
7. Xia, F., Hao, R., et al.: Adaptive GTS allocation in IEEE 802.15. 4 for real-time wireless sensor networks. J. Syst. Arch. **59**(10), 1231–1242 (2013)a
8. Kumar, P., Kumar, R., et al.: PPSF: a privacy-preserving and secure framework using blockchain-based machine-learning for IoT-driven smart cities. IEEE Trans. Netw. Sci. Eng. **8**(3), 2326–2341 (2021)
9. Wu, C., Luo, C., et al.: A greedy deep learning method for medical disease analysis. IEEE Access **6**, 20021–20030 (2018)
10. Cheng, H., et al.: Multi-step data prediction in wireless sensor networks based on one-dimensional CNN and bidirectional LSTM. IEEE Access **7**, 117883–117896 (2019)
11. Yao, Y., Xiong, N., et al.: Privacy-preserving max/min query in two-tiered wireless sensor networks. Comput. Math. Appl. **65**(9), 1318–1325 (2013)
12. Zhao, J., et al.: An effective exponential-based trust and reputation evaluation system in wireless sensor networks. IEEE Access **7**, 33859–33869 (2019)

13. Gao, Y., Xiang, X., et al.: Human action monitoring for healthcare based on deep learning. IEEE Access **6**, 52277–52285 (2018)
14. Wu, C., Ju, B., et al.: UAV autonomous target search based on deep reinforcement learning in complex disaster scene. IEEE Access **7**, 117227–117245 (2019)
15. Fu, A., Zhang, X., et al.: VFL: a verifiable federated learning with privacy-preserving for big data in industrial IoT. IEEE Trans. Indust. Inform. 18, 3316–3326 (2020)
16. Zhang, W., Zhu, S., Tang, J., Xiong, N.: A novel trust management scheme based on Dempster–Shafer evidence theory for malicious nodes detection in wireless sensor networks. J. Supercomput. **74**(4), 1779–1801 (2018)
17. Chen, Y., Zhou, L., et al.: KNN-BLOCK DBSCAN: fast clustering for large-scale data, IEEE Trans. Syst. Man Cybernet. Syst. **51**(6), 3939–3953 (2019)
18. Huang, S., Zeng, Z., Ota, K., Dong, M., Wang, T., Xiong, N.N.: An intelligent collaboration trust interconnections system for mobile information control in ubiquitous 5G networks, IEEE Trans. Netw. Sci. Eng. **8**(1), 347–365 (2020)
19. Qiu, M., Xue, C., Shao, Z., et al.: Efficient algorithm of energy minimization for heterogeneous wireless sensor network. In: IEEE EUC, pp. 25–34 (2006)
20. Niu, J., et al.: Selecting proper wireless network interfaces for user experience enhancement with guaranteed probability. J. Parallel Distrib. Comput. **72**(12), 1565–1575 (2012)
21. Qiu, H., Dong, T., et al.: Adversarial attacks against network intrusion detection in IoT systems. IEEE IoT J. **8**(13), 10327–10335 (2020)
22. Li, Y., Gai, K., et al.: Intercrossed access controls for secure financial services on multimedia big data in cloud systems. ACM Trans. Multim. Comput. Commun. Appl. **12**(4) (2016)
23. Qiu, H., Zheng, Q., et al.: Topological graph convolutional network-based urban traffic flow and density prediction. IEEE Trans. Intell. Transp. Syst. **22**, 4560–4569 (2020)
24. Liu, W., Luo, S., Chen, P.: A new solution to SMP based on sliding window and exchange encryption function. Comput. Eng. **33**(22), 163–171 (2007)
25. Boudot, F., Schoenmakers, B., Traore, J.: A fair and efficient solution to the socialist millionaires. Problem. Discrete Appl. Math. **111**, 23–36 (2003)
26. Qin, J., Zhang, Z., Feng, D., et al.: Comparisons without information leakage. J. Softw. **15**(3), 421–427 (2004)
27. Lin, H.Y., Tzeng, W.G.: An efficient solution to the millionaires' problem based on homomorphic encryption. In: Third International Conference on Applied Cryptography and Network Security (ACNS), pp. 456–466New York, USA (2005)
28. Blake, I.F., Kolesnikov, V.: Strong conditional oblivious transfer and computing on intervals. In: Advances in Cryptology-AISACRYPT 2004, pp. 515–529 (2004)
29. Paillier, P.: Public-key cryptosystems based on composite degree residuosity classes. In: Stern, J. (eds.) Advances in Cryptology — EUROCRYPT 1999. EUROCRYPT 1999, LNCS, vol. 1592, pp. 223–238. Springer, Berlin (1999). https://doi.org/10.1007/3-540-48910-X_16
30. Goldwasser, S., Micali, S.: Probabilistic encryption. J. Comput. Syst. Sci. **28**(2), 270–299 (1984)

ABODE-Net: An Attention-based Deep Learning Model for Non-intrusive Building Occupancy Detection Using Smart Meter Data

Zhirui Luo[1], Ruobin Qi[1], Qingqing Li[1], Jun Zheng[1](✉), and Sihua Shao[2]

[1] Department of Computer Science and Engineering, New Mexico Institute of Mining and Technology, Socorro, NM 87801, USA
jun.zheng@nmt.edu

[2] Department of Electrical Engineering, New Mexico Institute of Mining and Technology, Socorro, NM 87801, USA

Abstract. Occupancy information is useful for efficient energy management in the building sector. The massive high-resolution electrical power consumption data collected by smart meters in the advanced metering infrastructure (AMI) network make it possible to infer buildings' occupancy status in a non-intrusive way. In this paper, we propose a deep leaning model called ABODE-Net which employs a novel Parallel Attention (PA) block for building occupancy detection using smart meter data. The PA block combines the temporal, variable, and channel attention modules in a parallel way to signify important features for occupancy detection. We adopt two smart meter datasets widely used for building occupancy detection in our performance evaluation. A set of state-of-the-art shallow machine learning and deep learning models are included for performance comparison. The results show that ABODE-Net significantly outperforms other models in all experimental cases, which proves its validity as a solution for non-intrusive building occupancy detection.

Keywords: Building occupancy detection · Smart meter · Deep learning · Machine learning · Attention

1 Introduction

Recently, efficient energy management of buildings has attracted a lot of attention because of the significant potential for energy reduction. Building occupancy detection has many applications in this area such as improving the energy saving of building appliances and providing demand-response services for smart grids. The occupancy-based control of indoor Heating, Ventilation, and Air Conditioning (HVAC) systems and lighting in the buildings can lead up to 40% reduction of the power consumption of the buildings and 76% reduction of the power used for lighting, respectively [22]. Moreover, occupancy status information benefits

M. Qiu et al. (Eds.): SmartCom 2022, LNCS 13828, pp. 152–164, 2023.
https://doi.org/10.1007/978-3-031-28124-2_15

demand-response services of smart grids by (i) determining users' peak demand periods [6], (ii) anticipating the willingness of deferring their consumption to off-peak hours [2], and (iii) jointly optimizing the occupancy-based demand response and the thermal comfort for occupants in microgrids [15].

Due to the wide deployment of Advanced Metering Infrastructures (AMI) globally, the massive high-resolution electrical power consumption data make non-intrusive building occupancy detection possible. With smart meters installed in customers' buildings, the building electricity usage data can be recorded and transmitted distantly in real-time for occupancy detection which does not require additional in-door sensors (e.g., environment measurement sensors or surveillance cameras). By using smart meter data, a number of studies were conducted that used data-driven machine learning models for building occupancy detection, such as support vector machine (SVM) [4,14], hidden Markov model (HMM) [14], k-nearest neighbors (kNN) [1,14], etc. Furthermore, a deep learning-based method proposed recently in [6] sequentially stacks a convolutional neural network (CNN) and a bidirectional long short-term memory network (BiLSTM) to capture spatial and temporal patterns in the smart grid data, which outperforms other state-of-the-art methods. However, all of the aforementioned studies use a set of features manually extracted from the raw power consumption data. The goal of this paper is to develop an end-to-end deep learning model to automatically capture the discriminative information in the raw smart meter data to infer the building occupancy status.

Building occupancy detection based on raw smart meter data can be considered a time series classification (TSC) problem. In recent years, there were a considerable amount of studies that used deep learning to solve the challenging TSC problem [11]. Specifically, CNN has demonstrated its powerful capability to solve multivariate TSC problems in many areas [5,13,17,19]. Although recurrent neural networks (RNN), e.g. long short-term memory (LSTM), is good at capturing the time dependency, the lack of the parallel training ability caused by the recurrent calculation costs more computational power when scaling up. CNN can better utilize GPU parallelism since it does not have the recurrent structure like RNN. With the attention mechanism, the capability of a CNN-based model on capturing temporal and spatial patterns can be enhanced while keeping its advantage of parallel training.

In this paper, we propose an **A**ttention-based **B**uilding **O**ccupancy **De**tection Deep Neural **Net**work (ABODE-Net) which uses raw smart meter data as input. ABODE-Net utilizes a Fully Convolutional Network (FCN) block and a new Parallel-Attention (PA) block to learn both temporal and spacial patterns from smart meter data. The FCN uses three CNN layers to extract features automatically from power consumption readings and corresponding time information. The PA block combines temporal attention (TA), variable attention (VA), and Squeeze-and-Excitation (SE) modules to focus on important features by blending temporal, variable, and cross-channel information. The contributions of this paper are summarized as follows: (1) we propose an attention-based deep learning model called ABODE-Net for building occupancy detection in an end-to-end manner using raw smart meter data and corresponding time information; (2) we propose a novel lightweight PA block as a key component of ABODE-Net to capture discriminative information for building occupancy detection; and (3) we compare the performance of ABODE-Net with a set of state-of-the-art baseline methods using two popular smart meter datasets and prove that ABODE-Net is a viable solution for non-intrusive building occupancy detection using smart meter data.

This paper is organized as follows. In Sect. 2, we define the building occupancy detection problem. The proposed ABODE-Net for non-intrusive building occupancy detection using smart meter data is described in Sect. 3. Section 4 presents the performance evaluation experiments and results. Finally, we conclude the paper in Sect. 5.

2 Problem Definition

The building occupancy detection problem targeted in this paper is to infer the real-time occupancy status, $\mathbf{y} = \{y_i \mid i \in [1, N]\}$, of a building from its historical power consumption data, $\mathbf{X} = \{X_i \in \mathbb{R}^{1 \times F \times T} \mid i \in [1, N]\}$, where N, F and T are the number of samples, the number of features in each time step, and the number of time steps in the time window of a sample, respectively. Specifically, we consider power consumption data and periodical time information as our features. The occupancy status y_i of a building corresponding to the sample X_i collected from the building is either vacant or occupied. We seek to learn a set of trainable parameters, $\mathbf{W}$, of a deep neural network M which predicts the building occupancy status $\hat{y}_i$ for the sample X_i. The prediction model for building occupancy detection can be written as:

$$\hat{y}_i = M(\mathbf{W}, X_i) \tag{1}$$

3 Proposed Method

The architecture of the proposed ABODE-Net is shown in Fig. 1. ABODE-Net consists of three sequentially connected components: (i) an FCN block, (ii) a PA block, and (iii) a classification block. The FCN block serves as the function of automatic feature extraction. The PA block applies the attention mechanism to focus on good features and suppress poor features extracted from the FCN block. The classification block consists of a global pooling layer followed by a fully connected layer to generate the predicted possibility of occupancy for a given input. The symbols used in this section and their descriptions are listed in Table 1.

3.1 FCN Block

FCN has been proven to be an efficient feature extractor for many TSC tasks [9,12,13]. In ABODE-Net, however, we use a different set of hyperparameters

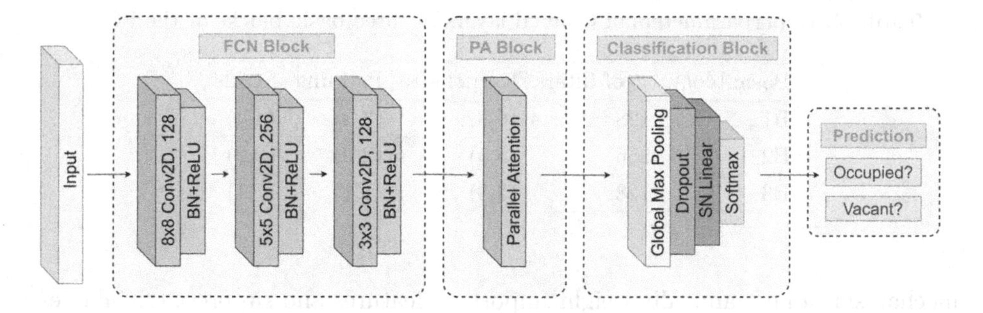

Fig. 1. The architecture of ABODE-Net

Table 1. Symbols and their descriptions

Symbol	Description	Symbol	Description
W	Learnable weight parameters	Softmax	Softmax activation
FC	Fully-connected layer	δ	ReLU activation
Conv2d	Convolution with 2D kernel	BN	Batch normalization
Conv1x1	Conv2d with 1×1 kernel	MLP	Multilayer perceptron
GAP	Global average pooling	$\oplus$	Element-wise matrix addition
GMP	Global max pooling	$\otimes$	Batch matrix multiplication
Sigmoid	Sigmoid activation	$\odot$	Element-wise multiplication
tanh	tanh activation	TRS	Tensor transpose operation

tailored for the building occupancy detection problem. For TSC tasks, FCN is usually constructed by using a large kernel size at the beginning to achieve enough perception fields, but with a shorter network depth compared to VGG-like networks [23] and no pooling layer between convolution layers. In ABODE-Net, the FCN component consists of three sequentially connected basic blocks, B1 to B3. Each basic block sequentially stacks a Conv2d layer, a BN layer, and a ReLU activation function which can be written as:

$$h_l = B_l(x_l) = \sigma(\mathtt{BN}(\mathtt{Conv2d}(x_l))), l \in \{1, 2, 3\} \tag{2}$$

where B_l indicates the l-th basic block, x_l and h_l are the input and output of B_l, respectively. Note that $x_1 \in \mathbb{R}^{1 \times F \times T}$ is the input of the network which consists of the raw power consumption data and corresponding time information. The hyperparameters of Conv2d layers in the three basic blocks are listed in Table 2.

3.2 PA Block

In the past decade, the attention mechanisms have played an increasingly important role in different deep learning applications such as computer vision [7,10,26], natural language processing (NLP) [18,20,24], and TSC [12,13,25]. An attention

Table 2. Hyperparameters of Conv2d layers in three basic blocks of the FCN

Basic block	# of filters	Kernel size	Padding	Stride
B1	128	(8,8)	(3,3)	(4,4)
B2	256	(5,5)	(2,2)	(2,2)
B3	128	(3,3)	(1,1)	(1,1)

mechanism can dynamically weight important features and suppress trivial ones based on the input. The general form of the attention mechanism, $Attention(x)$, can be written as:

$$Attention(x) = F(A(x), x) \tag{3}$$

where $A(x)$ generates attention weights based on the input x and $F(A(x), x)$ applies the attention weights $A(x)$ on the corresponding input x to attend critical regions.

A recent study classifies attention mechanisms into six categories: (i) channel attention, (ii) spatial attention, (iii) temporal attention, (iv) branch channel, (v) channel & spatial attention, and (vi) spatial & temporal attention [8]. On the other hand, a mechanism combining channel attention, variable attention, and temporal attention still remains unexplored in the TSC research area. We propose a novel PA block in ABODE-Net as shown in Fig. 2 to combine the three kinds of attention modules in a parallel way to signify important features.

The three attention modules in the PA block are the SE module, the VA module, and the TA module which are for cross-channel attention, variable attention, and temporal attention, respectively. Given an intermediate feature map $H \in \mathbb{R}^{C \times F \times T}$ as input, PA parallelly infers a channel attention map $O_{\mathtt{CA}} \in \mathbb{R}^{C \times F \times T}$, a variable attention map $O_{\mathtt{VA}} \in \mathbb{R}^{C \times F \times T}$, and a temporal attention map $O_{\mathtt{TA}} \in \mathbb{R}^{C \times F \times T}$. Once all attention maps are generated, they are combined according to Eq. (4) to obtain the final attended feature map $M_{\mathtt{PA}} \in \mathbb{R}^{C \times F \times T}$.

$$M_{\mathtt{PA}} = (H \oplus O_{\mathtt{TA}} \oplus O_{\mathtt{VA}}) \odot O_{\mathtt{CA}} \tag{4}$$

Channel Attention Module. To capture the cross-channel relationship, we adopt the SE block [10] as the channel attention module. SE exploits the channel dependencies by considering the channel-wise statistics from squeezed global variable and temporal information. The *squeeze* operation $F_{sq}(\cdot)$ is defined by GAP. The *excitation* operation $F_{ex}(\cdot, W)$ uses a simple gating mechanism with a sigmoid activation, where W is the weights of a MLP. Finally, the output of the block is re-scaled back via the *scale* operation F_{scale}. The SE module is formulated as follows:

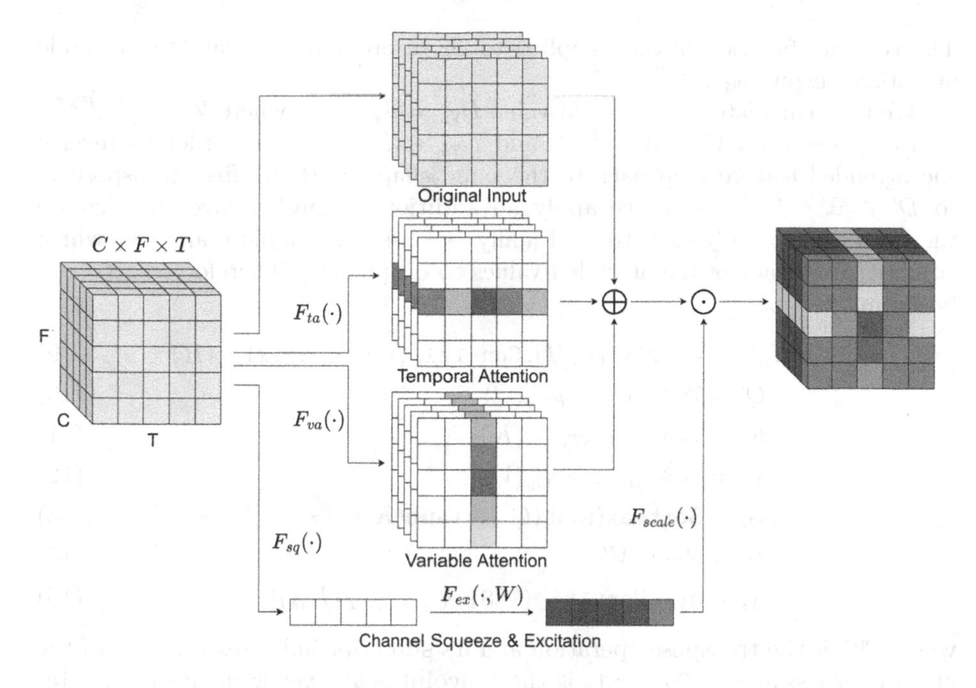

Fig. 2. Illustration of the PA block

$$
\begin{aligned}
H_{sq} &= F_{sq}(H) = GAP(H) && (5)\\
H_{ex} &= F_{ex}(H_{sq}, W) = \texttt{Sigmoid}(MLP(H_{sq})) && (6)\\
&= \texttt{Sigmoid}(W_2\delta(W_1(GAP(H_{sq})))) && (7)\\
O_{\texttt{CA}} &= F_{scale}(H_{ex}) && (8)
\end{aligned}
$$

where $W_1 \in \mathbb{R}^{C\times\frac{C}{r}}$ and $W_2 \in \mathbb{R}^{\frac{C}{r}\times C}$ are the trainable weights of two FC layers, $H_{sq} \in \mathbb{R}^C$ is the output of the *squeeze* operation, $H_{ex} \in \mathbb{R}^C$ is the output of the *excitation* operation, and $O_{\texttt{CA}} \in \mathbb{R}^{C\times F\times T}$ is the output of the *scale* operation. Same as [10], we set the reduction ratio r to 16.

Variable Attention Module. The variable attention module in the PA block is implemented with the VA operation $F_{va}(\cdot)$. Given a local feature map $H \in \mathbb{R}^{C\times F\times T}$, the VA module first applies three convolutional layers with 1×1 filters (Conv1x1) on H to generate three feature maps Q, K, and V, respectively, where $\{Q, K\} \in \mathbb{R}^{C_1\times F\times T}$ and $\{V\} \in \mathbb{R}^{C_2\times F\times T}$. C_1 and C_2 are the number of channels, which are less than C for dimension reduction. We set C_1 to $C/8$ and C_2 to $C/2$.

We transpose Q and K before the attention weights are generated, where the transpose operation on tensor input swaps the axes of a given tensor. For the feature map $Q \in \mathbb{R}^{C_1\times F\times T}$, we transpose it into $Q' \in \mathbb{R}^{T\times F\times C_1}$. We transpose $K \in \mathbb{R}^{C_1\times F\times T}$ into $K' \in \mathbb{R}^{T\times C_1\times F}$. Following this, we apply a tanh activation function to Q' and K' separately, where we get Q'' and K''. For each time step $t \in T$, we multiply Q''_t by K''_t which results in a intermediate matrix $S_t \in \mathbb{R}^{F\times F}$.

The softmax function is then applied to S_t in order to generate the variable attention weight $A_{\mathtt{VA}} \in \mathbb{R}^{T\times F\times F}$.

We then calculate the attended value $D_{\mathtt{VA}} = A_{\mathtt{VA}} \otimes V'$, where $V' \in \mathbb{R}^{T\times F\times C_2}$ is transposed from $V \in \mathbb{R}^{C_2\times F\times T}$ and $D_{\mathtt{VA}} \in \mathbb{R}^{T\times F\times C_2}$. In order to reverse the attended feature map back to the same shape of H, we first transpose D to $D' \in \mathbb{R}^{C_2\times F\times T}$. Next, we apply an additional Conv1x1 layer to increase the channels from C_2 back to C. Finally, we use a trainable scalar $\sigma_{\mathtt{VA}}$ which adjusts the efficacy of the attended values to output $O_{\mathtt{VA}}$. Therefore, $F_{va}(\cdot)$ can be formulated as:

$$Q, K, V = \mathtt{Conv1x1}_{C_1}(H), \mathtt{Conv1x1}_{C_1}(H), \mathtt{Conv1x1}_{C_2}(H) \tag{9}$$
$$Q' = \mathtt{TRS}_{C_1FT\rightarrow TFC_1}(Q) \tag{10}$$
$$K' = \mathtt{TRS}_{C_1FT\rightarrow TC_1F}(K) \tag{11}$$
$$V' = \mathtt{TRS}_{C_2FT\rightarrow TFC_2}(V) \tag{12}$$
$$A_{\mathtt{VA}} = \mathtt{Softmax}(\tanh(Q') \otimes \tanh(K')) \tag{13}$$
$$D_{\mathtt{VA}} = A_{\mathtt{VA}} \otimes V' \tag{14}$$
$$O_{\mathtt{VA}} = \sigma_{\mathtt{VA}} \cdot \mathtt{Conv1x1}_C(\mathtt{TRS}_{TFC_2\rightarrow C_2FT}(D_{\mathtt{VA}})) \tag{15}$$

where TRS is the transpose operation and its subscript indicates the axes of the given tensor swapped. $\mathtt{Conv1x1}_c$ is the convolutional layer with the 1×1 filter and c output channels.

Temporal Attention Module. The temporal dependencies are captured via the TA operation $F_{ta}(\cdot)$. The TA module is similar to the VA module with some differences in transposing Q, K, V feature maps. The resulting attention weight of TA is $A_{\mathtt{TA}} \in \mathbb{R}^{F\times T\times T}$ while the attention weight for VA is $A_{\mathtt{VA}} \in \mathbb{R}^{T\times F\times F}$. $F_{ta}(\cdot)$ is formulated as follows:

$$Q, K, V = \mathtt{Conv1x1}_{C_1}(H), \mathtt{Conv1x1}_{C_1}(H), \mathtt{Conv1x1}_{C_2}(H) \tag{16}$$
$$Q' = \mathtt{TRS}_{C_1FT\rightarrow FTC_1}(Q) \tag{17}$$
$$K' = \mathtt{TRS}_{C_1FT\rightarrow FC_1T}(K) \tag{18}$$
$$V' = \mathtt{TRS}_{C_2FT\rightarrow FTC_2}(V) \tag{19}$$
$$A_{\mathtt{TA}} = \mathtt{Softmax}(\tanh(Q') \otimes \tanh(K')) \tag{20}$$
$$D_{\mathtt{TA}} = A_{\mathtt{TA}} \otimes V' \tag{21}$$
$$O_{\mathtt{TA}} = \sigma_{\mathtt{TA}} \cdot \mathtt{Conv1x1}_C(\mathtt{TRS}_{FTC_2\rightarrow C_2FT}(D_{\mathtt{TA}})) \tag{22}$$

3.3 Classification Block

Given the attended feature map $M_{\mathtt{PA}}$, we use a GMP layer followed by an FC layer and a softmax function to predict the occupancy status. The classification block can be formulated as:

$$h_{fc} = \mathtt{SN}(W_c(\mathtt{GMP}(M_{\mathtt{PA}})) + b_c) \tag{23}$$
$$\hat{\mathtt{y}} = \mathtt{Softmax}(h_{fc}) \tag{24}$$

where $W_c \in \mathbb{R}^{128 \times N_c}$ and b_c are the weight and bias of the FC layer, respectively. The output of the FC layer is $h_{fc} \in \mathbb{R}^{N_c}$ and $\hat{\mathrm{y}} \in \mathbb{R}^{N_c}$ is the vector of predicted probabilities, where N_c is the number of classes. The spectral normalization (SN) is utilized to stabilize the FC layer, which is a normalization technique proposed in [21] to stabilize the training of the discriminator in generative adversarial networks (GANs). Unlike input based regularizations, e.g. BN, SN does not depend on the space of data distribution, instead it normalizes the weight matrices.

3.4 Model Training

The three trainable components of ABODE-Net are trained together. Since the building occupancy detection task is a classification problem, we use the negative log likelihood loss ($\mathtt{NLL}_{loss}$) shown in Eq. (25) for training:

$$\mathtt{NLL}_{loss}(y, \hat{y}) = -\sum_{i=1}^{N_c} y_i \ln \hat{y}_i \tag{25}$$

where N_c is the number of classes, y is the one-hot encoding of the ground truth label, and y_i and $\hat{y}_i$ are the actual and predicted probabilities of the label being ith class, respectively. The predicted occupancy status is the class whose predicted probability is the highest among all classes.

We use Adam with a learning rate of 1e–3 as the optimizer to learn all trainable parameters of ABODE-Net. To prevent over-fitting, weight decay is set to 5e–4. The max training epoch is 100. We use a scheduler with a learning rate that decreases following the value of cosine function between the initial learning rate and 0 with a warm-up period of 7 epochs.

4 Performance Evaluation and Results

4.1 Smart Meter Datasets

To evaluate the performance of ABODE-Net, two smart meter datasets widely used for non-intrusive building occupancy detection, the Electricity Consumption and Occupancy (ECO) dataset [3] and the Non-Intrusive Occupancy Monitoring (NIOM) dataset [4], are adopted in our study. The ECO dataset was collected from five houses in summer and winter. The NIOM dataset was collected from two houses, Home-A in both spring and summer and Home-B only in summer. We use the power consumption readings, i.e. the *powerallphases* of the ECO dataset and the power usage trace of the NIOM dataset, and their corresponding occupancy statuses for our experiments. The occupancy status of the NIOM dataset is for two occupants. However, we consider a home to be occupied when at least one occupant is at home. The sampling rates of the two datasets are different. The ECO dataset has a sampling rate of one sample per second, while the NIOM dataset has a lower sampling rate of one sample per minute. Thus, we aggregate the samples of every 60s of the ECO dataset into one sample via averaging.

In addition to power consumption information, time information is also important for occupancy detection because the life style of a household usually depends on the time of a day and the day of a week. We add time of the day P_{time} and day of the week P_{day} to model the temporal dependency and life-style cycle. P_{time} is the timestamp ranging from 0 to 1439 corresponding to each minute between 00:00 to 23:59. P_{day} is the day of a week ranging from 0 to 6 corresponding to Monday to Sunday. For both datasets, the input features of a sample are generated from a 60-min power consumption readings and their corresponding time information, P_{time} and P_{day}, resulting in a 3×60 feature matrix. We add a dummy dimension to the feature matrix to form the input of the FCN of shape $1 \times 3 \times 60$. The label of an input is the majority class (occupied or vacant) of samples in the 60-min segment.

4.2 Data Preparation

For a fair comparison, the datasets are quality-controlled with two criteria [6]: (1) the data length should be more than 900 samples, and (2) the samples of each class should be more than 10% of all samples. Based on the criteria, four periods of the three houses in the ECO dataset and all periods of the two houses in the NIOM dataset are qualified which are shown in Table 3. Each qualified period of a house is used as a case for performance evaluation.

Table 3. Qualified cases of the two datasets

Case	Household-period	Occupied/Vacant	Total # of samples
ECO-1	01-Summer	769/168	937
ECO-2	01-Winter	834/270	1104
ECO-3	02-Winter	769/311	1080
ECO-4	03-Summer	1038/330	1368
NIOM-1	HomeA-Spring	125/43	168
NIOM-2	HomeA-Summer	142/26	168
NIOM-3	HomeB-Summer	125/43	168

As shown in Table 3, all cases have significantly more occupied samples than vacant samples which result in imbalanced datasets for performance evaluation. Therefore, we use the random oversampling method implemented by `Imbalanced-learn` [16] to over-sample the minority vacant class. The random oversampling method randomly select samples in the minority class with replacement.

In an evaluation experiment, the data of each case are randomly divided into training, validation, and test sets by a ratio of 3:1:1. The experiment is repeated 10 times randomly by using different random seeds. We apply the min-max normalization to scale each feature of the data into the range between zero to one. After the normalization, we oversample the minority class of the training set with the validation and test sets untouched.

4.3 Baseline Models

We compare ABODE-Net with a collection of state-of-the-art shallow ML and DL models which have been applied for building occupancy detection. All models are implemented using Python 3.8. The DL models and ABODE-Net are implemented with the PyTorch framework. The included shallow ML models are: kNN [1,14], GMM [14], and SVM [4,14], which are implemented with the `scikit-learn` library. We use the grid search to find the best hyperparameters of those models.

The DL models included in our experiments are CNN, LSTM, and CNN-BiLSTM of [6]. We re-implement CNN-BiLSTM which uses the same features as ABODE-Net. The CNN model has the same architecture as the CNN block of CNN-BiLSTM. The LSTM model consists of two LSTM layers with 50 hidden neurons and bias weights, and its classification block contains a dropout layer with 40% dropping rate followed by spectral-normalized FC layer. We train baseline DL models the same way as we train ABODE-Net which is discussed in Sect. 3.4. Specifically, the hyperparameters of a deep model achieving the highest F1 score on the validation phase are used for testing.

4.4 Performance Metrics

Two popular metrics, accuracy and F1 score, are used for evaluating the performance of all models. By denoting the two occupancy statuses, occupied and vacant, as positive and negative, respectively, the two performance metrics are defined as:

$$Accuracy = \frac{T_p + T_n}{T_p + T_n + F_p + F_n} \tag{26}$$

$$Precision = \frac{T_p}{T_p + F_p} \tag{27}$$

$$Recall = \frac{T_p}{T_p + T_n} \tag{28}$$

$$F_1 = 2 \cdot \frac{precision \cdot recall}{precision + recall} \tag{29}$$

where T_p, T_n, F_p, and F_n are true positives, true negatives, false positives, and false negatives, respectively.

4.5 Results

The performance evaluation results in term of averaging accuracy and F1 score over 10 trials are shown in Tables 4 and 5 for the cases of the ECO and NIOM datasets, respectively. As shown in Table 4, ABODE-Net outperforms all baseline models for every case of the ECO dataset, whose average accuracy and F1 score over all cases are 0.8649 and 0.8198, respectively. It can be seen that the three shallow ML models have significantly worse performance than DL models. This shows that DL models have better capability of capturing time dependencies and spatial patterns in time series data. Another reason is that these shallow ML models are affected by the curse of dimensionality due to the high dimensional data. Similar results can be observed in Table 5 for the NIOM dataset. ABODE-Net achieves the best average accuracy and F1 score over all cases among all models. It only has a slightly lower F1 score than LSTM for case 3. On the other hand, LSTM has much worse performance than ABODE-Net in other two cases. Overall, the performance evaluation results demonstrate that ABODE-Net is more capable than the state-of-the-art baseline models in terms of utilizing the temporal and spatial information in smart meter data to detect building occupancy.

Table 4. Performance evaluation results for the ECO dataset

Case	Metric	Model						
		ABODE-Net	kNN	GMM	SVM	CNN	LSTM	CNN_BiLSTM
1	ACC	**0.8585**	0.7766	0.7133	0.7851	0.8426	0.8154	0.8324
	F1	**0.7832**	0.7153	0.6483	0.7217	0.7562	0.7509	0.7621
2	ACC	**0.8208**	0.5900	0.7262	0.6701	0.8131	0.8190	0.7584
	F1	**0.7719**	0.5553	0.7024	0.6414	0.7654	0.7650	0.5147
3	ACC	**0.9218**	0.8292	0.7588	0.8537	0.9194	0.9185	0.9111
	F1	**0.9050**	0.8081	0.6287	0.8349	0.9004	0.9015	0.8672
4	ACC	**0.8584**	0.6818	0.6562	0.6569	0.8248	0.7482	0.8036
	F1	**0.8191**	0.6497	0.6367	0.6362	0.7923	0.7181	0.7655
Average	ACC	**0.8649**	0.7194	0.7136	0.7415	0.8500	0.8253	0.8264
	F1	**0.8198**	0.6821	0.6540	0.7085	0.8036	0.7839	0.7274

Table 5. Performance evaluation results for the NIOM dataset

Case	Metric	Model						
		ABODE-Net	kNN	GMM	SVM	CNN	LSTM	CNN_BiLSTM
1	ACC	**0.9206**	0.8500	0.8029	0.8529	0.8912	0.7529	0.9000
	F1	**0.9007**	0.8248	0.6300	0.8207	0.8601	0.7202	0.8633
2	ACC	**0.9000**	0.8235	0.8824	0.8353	0.8706	0.7412	0.8294
	F1	**0.7730**	0.7388	0.6267	0.7276	0.7513	0.6198	0.7269
3	ACC	**0.9324**	0.9029	0.8353	0.9294	0.9206	**0.9324**	0.9265
	F1	0.9190	0.8878	0.7041	0.9162	0.9047	**0.9194**	0.9091
Average	ACC	**0.9176**	0.8588	0.8402	0.8725	0.8941	0.8088	0.8853
	F1	**0.8643**	0.8171	0.6536	0.8215	0.8387	0.7531	0.8331

5 Conclusion

In this paper, we propose an attention-based deep learning model called ABODE-Net to infer the building occupancy status in an end-to-end manner by using the raw smart meter data and corresponding time information. ABODE-Net consists of a three-layer FCN block, a novel light-weight PA block, and a spectral-normalized classification block. The proposed PA block captures discriminative information for occupancy detection through a combination of temporal, variable, and channel attentions. Two smart meter datasets widely used for building occupancy detection are adopted for performance evaluation. The experimental results demonstrate that ABODE-Net achieves significantly better performance than the state-of-the-art baseline methods.

Acknowledgements. This work was supported in part by the National Science Foundation under EPSCoR Cooperative Agreement OIA-1757207 and in part by the Institute for Complex Additive Systems Analysis (ICASA) of New Mexico Institute of Mining and Technology.

References

1. Akbar, A., Nati, M., Carrez, F., Moessner, K.: Contextual occupancy detection for smart office by pattern recognition of electricity consumption data. In: 2015 IEEE International Conference on Communications (ICC), pp. 561–566 (2015)
2. Albert, A., Rajagopal, R.: Smart meter driven segmentation: what your consumption says about you. IEEE Trans. Power Syst. **28**(4), 4019–4030 (2013)
3. Beckel, C., Kleiminger, W., Cicchetti, R., Staake, T., Santini, S.: The ECO data set and the performance of non-intrusive load monitoring algorithms. In: Proceedings of the 1st ACM Conference on Embedded Systems for Energy-efficient Buildings (BuildSys 2014), pp. 80–89 (2014)
4. Chen, D., Barker, S., Subbaswamy, A., Irwin, D., Shenoy, P.: Non-intrusive occupancy monitoring using smart meters. In: Proceedings of the 5th ACM Workshop on Embedded Systems for Energy-Efficient Buildings (BuildSys 2013), pp. 1–8 (2013)
5. Chen, W., Shi, K.: Multi-scale attention convolutional neural network for time series classification. Neural Netw. **136**, 126–140 (2021)
6. Feng, C., Mehmani, A., Zhang, J.: Deep learning-based real-time building occupancy detection using AMI data. IEEE Trans. Smart Grid **11**(5), 4490–4501 (2020)
7. Fu, J., et al.: Dual attention network for scene segmentation. In: Proceedings of the 2019 IEEE/CVF Conference on Computer Vision and Pattern Recognition (CVPR), pp. 3146–3154 (2019)
8. Guo, M.-H., et al.: Attention mechanisms in computer vision: a survey. Comput. Vis. Media 1–38 (2022). https://doi.org/10.1007/s41095-022-0271-y
9. Hao, Y., Cao, H.: A new attention mechanism to classify multivariate time series. In: Proceedings of the Twenty-Ninth International Joint Conference on Artificial Intelligence, pp. 1999–2005 (2020)
10. Hu, J., Shen, L., Sun, G.: Squeeze-and-excitation networks. In: Proceedings of the 2018 IEEE/CVF Conference on Computer Vision and Pattern Recognition (CVPR), pp. 7132–7141 (2018)

11. Ismail Fawaz, H., Forestier, G., Weber, J., Idoumghar, L., Muller, P.-A.: Deep learning for time series classification: a review. Data Min. Knowl. Discovery **33**(4), 917–963 (2019). https://doi.org/10.1007/s10618-019-00619-1
12. Karim, F., Majumdar, S., Darabi, H., Chen, S.: LSTM fully convolutional networks for time series classification. IEEE Access **6**, 1662–1669 (2017)
13. Karim, F., Majumdar, S., Darabi, H., Harford, S.: Multivariate LSTM-FCNs for time series classification. Neural Netw. **116**, 237–245 (2019)
14. Kleiminger, W., Beckel, C., Santini, S.: Household occupancy monitoring using electricity meters. In: UbiComp 2015 - Proceedings of the 2015 ACM International Joint Conference on Pervasive and Ubiquitous Computing, pp. 975–986, September 2015
15. Korkas, C.D., Baldi, S., Michailidis, I., Kosmatopoulos, E.B.: Occupancy-based demand response and thermal comfort optimization in microgrids with renewable energy sources and energy storage. Appl. Energy **163**, 93–104 (2016)
16. Lemaître, G., Nogueira, F., Aridas, C.K.: Imbalanced-learn: a python toolbox to tackle the curse of imbalanced datasets in machine learning. J. Mach. Learn. Res. **18**(17), 1–5 (2017)
17. Li, Q., Luo, Z., Zheng, J.: A new deep anomaly detection-based method for user authentication using multichannel surface EMG signals of hand gestures. IEEE Trans. Instrum. Meas. **71**, 1–11 (2022)
18. Liu, G., Guo, J.: Bidirectional LSTM with attention mechanism and convolutional layer for text classification. Neurocomputing **337**, 325–338 (2019)
19. Luo, Z., Li, Q., Zheng, J.: Deep feature fusion for rumor detection on twitter. IEEE Access **9**, 126065–126074 (2021)
20. Luong, M.T., Pham, H., Manning, C.D.: Effective approaches to attention-based neural machine translation. arXiv preprint arXiv:1508.04025 (2015)
21. Miyato, T., Kataoka, T., Koyama, M., Yoshida, Y.: Spectral normalization for generative adversarial networks. arXiv preprint arXiv:1802.05957 (2018)
22. Razavi, R., Gharipour, A., Fleury, M., Akpan, I.J.: Occupancy detection of residential buildings using smart meter data: a large-scale study. Energy Buildings **183**, 195–208 (2019)
23. Simonyan, K., Zisserman, A.: Very deep convolutional networks for large-scale image recognition. arXiv preprint arXiv:1409.1556 (2014)
24. Sutskever, I., Vinyals, O., Le, Q.V.: Sequence to sequence learning with neural networks. In: Advances in Neural Information Processing Systems, vol. 27 (2014)
25. Tang, Y., Xu, J., Matsumoto, K., Ono, C.: Sequence-to-sequence model with attention for time series classification. In: 2016 IEEE 16th International Conference on Data Mining Workshops (ICDMW), pp. 503–510. IEEE (2016)
26. Woo, S., Park, J., Lee, J.Y., Kweon, I.S.: Cbam: convolutional block attention module. In: Proceedings of the 2018 European Conference on Computer Vision (ECCV), pp. 3–19 (2018)

Confidentially Computing DNA Matching Against Malicious Adversaries

Xiaofen Tu[1], Xin Liu[1,2](✉), Xiangyu Hu[1], Baoshan Li[1], and Neal N. Xiong[3,4]

[1] School of Information Engineering, Inner Mongolia University of Science and Technology, Baotou 014010, Inner Mongolia, China
lx2001.lx@163.com, libaoshan@imust.edu.cn

[2] School of Computer Science, Shanxi Normal University, Xi'an 710062, China

[3] National Engineering Research Center for E-Learning, Central China Normal University, Wuhan 430079, China

[4] Department of Computer Science and Mathematics, Sul Ross State University, Alpine, TX 79830, USA

Abstract. DNA is one of the most important information in every living thing. The DNA matching experiment is helpful for the study of paternity testing, species identification, gene mutation, suspect determination, and so on. How to study the DNA matching in the case of privacy protection has become the inevitable problems in the research of information security. The Hamming distance can reflect the similarity degree of two DNA sequences. The smaller the Hamming distance is, the more similar the two DNA sequences are. In this paper, the DNA sequence with l length is encoded with a 0–1 string with $3l$ length, and the protocol of confidentially computing Hamming distance is designed, which calculated the matching degree of two DNA under the premise of protecting DNA privacy. In addition, in view of the criminal suspect DNA matching problem, we design a secure computation protocol against malicious adversaries using the zero-knowledge proof and the cut-choose method to prevent or find malicious behaviors, which can resist malicious attacks.

Keywords: DNA matching · Secure multiparty computation · Hamming distance · Malicious model · Cut-choose method

1 Introduction

Currently, DNA sequence coding has been widely studied. In medicine, it is often necessary to compare the similarity degree of DNA sequences to carry out the study of paternity tests, species identification, gene mutation, and other medical issues [1, 2]. Hamming distance refers to the number of different characters in two strings X and Y of the same length, which is denoted as $HMD(X, Y)$. The Hamming distance can reflect the similarity degree of two DNA sequences.

In real life, people usually don't want to disclose their DNA sequences to each other. Aiming at this problem, with the help of the secure multiparty computation,

M. Qiu et al. (Eds.): SmartCom 2022, LNCS 13828, pp. 165–174, 2023.
https://doi.org/10.1007/978-3-031-28124-2_16

DNA information can be compared in secret. Secure *Multi-Party Computation* (MPC) is first proposed by Yao in 1982, and then Goldreich, Cramer, and other scholars make in-depth research on MPC, which has become a research hotspot, including wireless sensor networks [3–5], data mining [6–8], secure computational geometry [9, 10], secure scientific computing [11, 12], smart computing [13–15], etc. These researches promote the development of MPC and solve many practical problems.

Reference [16] studies the Hamming distance calculation of DNA sequences for privacy protection. Based on the reference, we make improvements to shorten the coding length of DNA sequence, and uses the elliptic curve cryptography (ECC) [17] to design a MPC protocol under the semi-honest model [18], which greatly improves the computing efficiency [19–21]. In view of the possible security issues [22–24], we design another MPC protocol of DNA comparison under the malicious model [25], which can prevent or detect malicious attacks, and analyzes the probability of successfully attacking by malicious adversaries. The contributes are as follows:

(1) A new 0–1 coding rule of DNA sequence is proposed.
(2) This paper designs a MPC Hamming distance protocol under the semi-honest model.
(3) Based on the zero-knowledge proof [26] and cut-choose method [25], a secure judgment protocol of two DNAs' equality under the malicious model is designed to resist the possible malicious attacks.

2 The 0–1 Coding for the DNA Sequence

Let a DNA sequence with length l be $D = d_1, d_2, \cdots, d_l \in \{A, G, T, C\}$, then the 0–1 codes of the corresponding DNA sequence is coded as $D' = d'_1, d'_2, \cdots, d'_{3l} \in \{0, 1\}$. The specific coding rule is: $A \to 111$, $G \to 100$, $T \to 010$, $C \to 001$. As shown in Fig. 1.

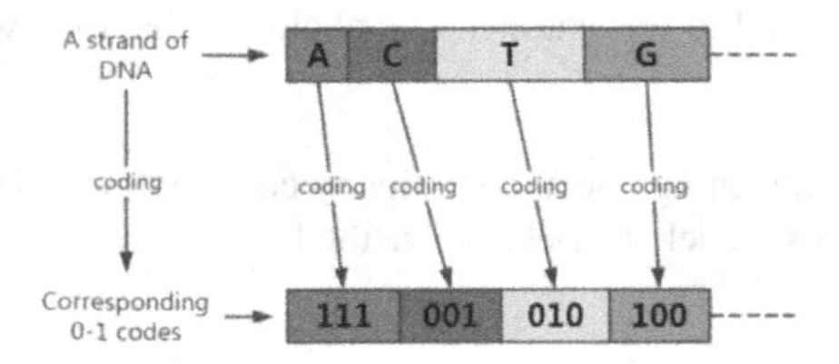

Fig. 1. DNA sequence 0–1 coding rules

We use the exhaustive method to prove that the Hamming distance of two DNA sequences after 0–1 coding is twice that of the original two DNA sequences. Table 1 shows the comparison of two Hamming distances in six cases:

3 The MPC Protocol for DNA Similarity Under the Semi-honest Model

Assuming that Steven and Tom separately have a DNA sequence of length l, Steven and Tom respectively encode their DNA sequences into the 0–1 codes, that is $X =$

Table 1. Hamming distances for all unequal cases

DNA comparison	Hamming distance of DNA sequence	0–1 code comparison	Hamming distance of 0–1 codes
A vs G	1	111 vs 100	2
A vs T	1	111 vs 010	2
A vs C	1	111 vs 001	2
G vs T	1	100 vs 010	2
G vs C	1	100 vs 001	2
T vs C	1	010 vs 001	2

$(a_1, a_2, \cdots, a_{3l})$ and $Y = (b_1, b_2, \cdots, b_{3l})$. Steven and Tom secretly calculate the Hamming distance $HMD(X, Y)$ of 0–1 codes, so that the Hamming distance of their DNA sequence is $HMD(X, Y)/2$.

Protocol 1. The MPC protocol for the DNA similarity under the semi-honest model.

Input: Steven's sequence X, Tom's sequence Y.

Output: $HMD(X, Y)$.

Preparation Stage: The generator of ECC is G. Steven chooses the private key k, then calculates $K = k \cdot G$ to get the public key K, and sends the public key (K, G) to Tom.

(1) Steven selects $3l$ random numbers r_i $(i = 1, 2, \cdots 3l)$, and uses the random number r_i and the public key K to encrypt each element of the sequence X bit by bit to obtain the encryption vector $E_r(X) = (E_{r_1}(a_1), E_{r_2}(a_2), \cdots, E_{r_i}(a_{3l}))$. The encryption process is as follows:

$$E_{r_i}(a_i) = \begin{cases} G + r_i \cdot K & a_i = 1 \\ r_r \cdot K & a_i = 0 \end{cases}$$

Then he calculates the identifier C_i corresponding to each ciphertext $E_{r_i}(a_i)$, $C_i = r_i \cdot G$. Finally, Steven sends Tom $E_r(X)$ and $3l$ identifiers C_i, namely $(E_r(X), C_i)$.

(2) After Tom receives $(E_r(X), C_i)$, he performs the following steps:

(2.1) At first, $3l$ random numbers s_i are selected, where $i = 1, 2, \cdots 3l$, and each element in the sequence Y is encrypted bit by bit by using s_i and Steven's public key K to obtain an encryption vector $E_s(Y) = (E_{s_1}(b_1), E_{s_2}(b_2), \cdots, E_{s_i}(b_{3l}))$, as follows:

$$E_{s_i}(b_i) = \begin{cases} G + s_i \cdot K & b_i = 1 \\ s_r \cdot K & b_i = 0 \end{cases}$$

(2.2) Add the elements in the same position of the two encryption vectors $E_r(X)$ and $E_s(Y)$ to get the encryption vector of length $3l$, as follows:

$$E_r(X) + E_s(Y) = (E_{r_1}(a_1) + E_{s_1}(b_1), E_{r_2}(a_2) + E_{s_2}(b_2), \cdots, E_{r_i}(a_{3l}) + E_{s_i}(b_{3l}))$$

At the same time, the corresponding $3l$ C_i' are calculated, that is $C_i' = C_i + s_i \cdot G$.

(2.3) The secret permutation T in set $\{1, 2, \cdots, 3l\}$ is randomly selected, and the substitution of $E_r(X) + E_s(Y)$ and C_i' is carried out to obtain $T(E_r(X) + E_s(Y))$ and $T(C_i')$, where:

$$T(E_r(X) + E_s(Y)) = (E_{r_1}(a_{T(1)}) + E_{s_1}(b_{T(1)}), E_{r_2}(a_{T(2)}) + E_{s_2}(b_{T(2)}), \cdots, E_{r_i}(a_{T(3l)}) + E_{s_i}(b_{T(3l)})),$$

$$T(C_i') = (C'_{T(1)}, C'_{T(2)}, \cdots, C'_{T(3l)})$$

Then Tom sends $T(E_r(X) + E_s(Y))$ and $T(C_i')$ to Steven.

(3) After obtaining $T(E_r(X) + E_s(Y))$ and $T(C_i')$, Steven decrypts each element in $T(E_r(X) + E_s(Y))$ successively by using the decryption character $T(C_i')$ and his private key k, and obtains:

$$T(X + Y) = (a_{T(1)} + b_{T(1)}, a_{T(2)} + b_{T(2)}, \cdots, a_{T(3l)} + b_{T(3l)})$$

Then, the value of $HMD(X, Y)$ can be obtained by adding up the number of G elements in $T(X + Y)$, which is denoted as *Sum*. Steven tells Tom the calculation result *Sum*.

The protocol ends.

4 The MPC Protocol for DNA Similarity Under the Malicious Model

What malicious acts cannot be prevented in the ideal model protocol [25]: (1) The participants refuse to participate in the protocol; (2) Participants provide false inputs; (3) Stop the protocol halfway.

We designed the malicious model protocol through finding malicious behaviors:

(1) It is possible that the party holding the private key tells the other party the wrong result. The solution is that both parties have the public and private keys.
(2) When Steven or Tom encrypts their own data, they provide false ciphertexts. The solution is to verify the correctness of the ciphertext by using the cut-choose method.
(3) In the last step of Protocol 1, Steven and Tom can send the wrong decrypted data m' to each other. The solution is to use the zero-knowledge proof.

4.1 The MPC Protocol for DNA Equality Under the Malicious Model

Suppose Steven and Tom each has a DNA sequence of length l. Steven and Tom encode their DNA sequences to 0–1 codes, and get decimal numbers x and y, and then encode x and y onto the elliptic curve to get two points M_1 and M_2. Steven and Tom secretly compare two points under the malicious model to see if two DNA are equal.

Protocol 2. The MPC protocol of two DNA equality under the malicious model.

Input: Steven has M_1, Tom has M_2.

Output: Whether M_1 is equal to M_2.

Preparation Stage: Steven and Tom jointly select the generator G of ECC, and they respectively select their private keys k_1 and k_2 of ECC, and calculate the public keys

$K_1 = k_1 \cdot G$ and $K_2 = k_2 \cdot G$. Then they respectively select the random numbers s and t, and calculate $u = s \cdot K_1$ and $u = s \cdot K_1$, and publish (K_1, G, u) and (K_2, G, v).

(1) Steven and Tom respectively selected m random numbers s_i and t_i $(i = 1, 2, \cdots, m)$, and calculated

$$(c_{1s}^i, c_{2s}^i) = (s_i \cdot M_1 + K_1, M_1 + s_i \cdot M_1 + s \cdot G)$$

$$(c_{1t}^i, c_{2t}^i) = (t_i \cdot M_2 + K_2, M_2 + t_i \cdot M_2 + t \cdot G)$$

Both parties publish (c_{1s}^i, c_{2s}^i), (c_{1t}^i, c_{2t}^i).

(2) Using the idea of the cut-choose method, Steven chooses $m/2$ groups of (c_{1t}^i, c_{2t}^i) from m group (c_{1t}^i, c_{2t}^i) and asks Tom to publish the corresponding $t_i \cdot M_2$. Steven uses Tom's public key K_2 to verify $(t_i \cdot M_2 + K_2 = c_{1t}^i)$. If the verification passes, the next step is executed, otherwise the protocol is terminated. Tom verifies the same. Tom implements the same process as Steven.

(3) They randomly select a (c_{1t}^j, c_{2t}^j) and (c_{1s}^i, c_{2s}^i) from the remaining (c_{1t}^i, c_{2t}^i) and (c_{1s}^i, c_{2s}^i) respectively. Meanwhile, Steven secretly selects two random numbers a and p_1, while Tom secretly selects two random numbers b and p_2.

Steven calculates
$c_t = a \cdot (c_{2t}^j - c_{1t}^j - M_1 + K_2) = a \cdot (M_2 - M_1) + a \cdot t \cdot G, P_1 = p_1 \cdot G, \lambda_t = p_1 \cdot K_2,$
Tom calculates
$c_s = b \cdot (c_{2s}^j - c_{1s}^j - M_2 + K_1) = b \cdot (M_1 - M_2) + b \cdot s \cdot G, P_2 = p_2 \cdot G, \lambda_s = p_2 \cdot K_1,$
then Steven and Tom send $c_t + P_1$ and $c_s + P_2$ to each other.

(4) Steven calculates $\omega_s = k_1 \cdot (c_s + P_2)$ and sends it to Tom, and Tom calculates $\omega_t = k_2 \cdot (c_t + P_1)$ and sends it to Steven.

(5) At this time, Steven and Tom send c_t and c_s to each other.

(6) Steven calculates $m_s = k_1 \cdot c_s$ and sends it to Tom. Tom calculates $m_t = k_2 \cdot c_t$ and sends it to Steven.

(7) Steven uses the zero-knowledge proof to verify that the m_t sent by (Steven) is correct, that is, to prove that (Steven) indeed has the m_t obtained by multiplying his private key k_2 with c_t. The proof method is to verify whether $(m_t = \omega_t - \lambda_t)$ is valid. Tom verifies the same. The proof method is to verify whether $(m_s = \omega_s - \lambda_s)$ is valid. If either party does not pass, then the party who does not pass is malicious.

(8) If both parties pass, Tom can get $k_1 \cdot b \cdot (M_1 - M_2)$ by calculating $m_s - b \cdot u$, so as to judge whether M_1 and M_2 are equal, but he may still be cheated by Steven. (It will be analyzed later that the probability of success of Steven's cheating is close to zero.) Steven calculates similarly.

The protocol ends.

4.2 Security Proof

(1) In the first two steps, after Tom publishes the encrypted data (c_{1t}^i, c_{2t}^i) and $t_i \cdot M_2$, Steven cannot deduce the value of M_2 because he does not know the t_i of $c_{1t}^i = t_i \cdot M_2 + K_2$ and he does not know the t of $c_{2t}^i = M_2 + t_i \cdot M_2 + t \cdot G$.

(2) step (5), because Steven does not know the random number b selected by Tom in $c_s = b \cdot (M_1 - M_2) + b \cdot s \cdot G$, so he could not calculate the data $b \cdot (M_1 - M_2)$ to get the output results in advance.
(3) In Protocol 2, the only way for each party to cheat is to provide a false ciphertext in step (1), the other party passes the cut-choose method verification, and the wrong encrypted data is selected in step (3) so that both parties cannot reach the correct conclusion. The probability of success is analyzed as follows:

Assuming that Steven wants to cheat, then the optimal choice is that only one group of data (c_{1s}^i, c_{2s}^i) in m group is wrong, and then the probability of success of deception is $1/m$. If there are n groups that do not meet the requirements, where $n < m/2$, then the probability of successful cheating is $(\frac{C_{m-n}^{m/2}}{C_m^{m/2}}) \times \frac{n}{m/2}$.If m is larger, the probability of success will approach to zero.

Next, the ideal/real model paradigm [18] is used to prove the security of the protocol.

Theorem 1. Protocol 2 (denoted $\prod$) is secure.

Proof. If $\prod$ is safe, it is necessary to convert the acceptable policy $\overline{A} = (A_1, A_2)$ pair when $\prod$ is executed into the corresponding policy pair $\overline{B} = (B_1, B_2)$ in the ideal model protocol and make the output of A_1 and A_2 in $\prod$ computationally indistinguishable from the output of the ideal model protocol B_1 and B_2.

LEt's Start with the First Case, A_1 is Honest (that is, A_2 is Dishonest). In this case, B_1 will execute the protocol according to the protocol (without causing termination of $\prod$), and output according to the protocol $\prod$. We need to invoke A_2, because the adversary B_2 in the ideal model does not know how A_2 should make a decision in face of some particular problems.

When A_1 is honest and $\prod$ is executed, then

$REAL_{\Pi,\bar{A}(M_1,M_2)} = \{F(M_1, A_2(M_2)), A_2((C_{1s}^i, C_{2s}^i), m_s, S)\}$,

where $i = 1, \cdots, m$ and S is the data generated by A_2 in the zero-knowledge proof process. If we can find the policy pair $\overline{B} = (B_1, B_2)$ under the ideal model, and their outputs are computationally indistinguishable from the $REAL_{\prod,\bar{A}(M_1,M_2)}$, it can be proved the execution protocol $\prod$ is secure as follows.

(1) In the ideal model, B_1 would send the real M_1 to TTP. The message dishonest B_2 sends to the TTP depends on its policy, and B_2's policy should be the same as A_2's policy, so A_2 needs to decide what message to send to the TTP. B_2 sends M_2 to A_2 and gets $A_2(M_2)$ from A_2. $A_2(M_2)$ is the private data used by A_2 in the actual model. B_2 sends $A_2(M_2)$ to TTP and gets $F(M_1, A_2(M_2))$ from TTP. (B_1 will also get $F(M_1, A_2(M_2))$, because B_2 can only get a result if B_1 gets a result and the protocol is not terminated).
(2) B_2 gets $F(M_1, A_2(M_2))$ from TTP and uses this result to try to get a $view_{B_2}^F(M_1, A_2(M_2))$ that should be computationally indistinguishable from the $view_{A_2}^{\prod}(M_1, A_2(M_2))$ obtained by A_2, and passes $view_{B_2}^F(M_1, A_2(M_2))$ to A_2 and outputs the output of A_2.

- B_2 randomly selects M_1' to make $F(M_1', A_2(M_2)) = F(M_1, A_2(M_2))$, and B_2 uses M_1' to simulate the protocol, uses M_1' to generate the corresponding $(C_{1s}^{i'}, C_{2s}^{i'})$ and publishes it.
- B_2 publishes the information which A_2 requires A_1 to publish in step (2) of the protocol.
- B_2 and A_2 execute the rest of the protocol. B_2 gets the corresponding m_s' and proves to A_2 that the private key used in m_s' calculation is correct with the zero-knowledge proof.

(3) B_2 calls A_2 with $((C_{1s}^{i'}, C_{2s}^{i'}), m_s', S')$ and output $A_2((C_{1s}^{i'}, C_{2s}^{i'}), m_s', S')$ to get:

$$IDEAL_{F,\overline{B}}(M_1, M_2) = \{F(M_1, A_2(M_2)), A_2((C_{1s}^{i'}, C_{2s}^{i'}), m_s', S')\}$$

For A_2, in the elliptic curve encryption system, $C_{1s}^{i} \overset{c}{\equiv} C_{1s}^{i'}$, $C_{2s}^{i} \overset{c}{\equiv} C_{2s}^{i'}$, while the random numbers S and S' in m_s and m_s' are computations indistinguishable, so $m_s \overset{c}{\equiv} m_s'$. In addition, the zero-knowledge proof system guarantees $S \overset{c}{\equiv} S'$, then

$$\{IDEAL_{F,\overline{B}}(M_1, M_2)\} \overset{c}{\equiv} \{REAL_{\prod,\overline{A}}(M_1, M_2)\}$$

In the Second Case, A_2 is Honest (A_1 is Dishonest). In this case, only convert the opponent A_1 in the actual model to the opponent B_1 in the ideal model. Due to space limitations, it is similar to the above proof process, so the second case proof process is omitted. Thus,

$$\{REAL_{\prod,\overline{A}}(M_1, M_2)\} \overset{c}{\equiv} \{IDEAL_{F,\overline{B}}(M_1, M_2)\}$$

To sum up, for any acceptable probabilistic polynomial time policy pair $\overline{A} = (A_1, A_2)$ in the actual protocol, there will be an acceptable probabilistic polynomial time policy pair $\overline{B} = (B_1, B_2)$ in the ideal model, satisfying

$$\{IDEAL_{F,\overline{B}}(M_1, M_2)\} \overset{c}{\equiv} \{REAL_{\prod,\overline{A}}(M_1, M_2)\}$$

Therefore, Protocol 2 is secure under the malicious model.

5 Performance Analysis and Comparison

Performance Comparison of Protocol 1: In this paper, a comparison is made between Protocol 1 and Reference [16], and the DNA sequence length is assumed to be l. In Reference [16], the GM encryption algorithm is used with a computational complexity of $16l$ modular exponential operations, and a communication complexity is 1 round. In Protocol 1 of this paper, the ECC algorithm is adopted, the computational complexity is $15l$ multiplication operations, and the communication complexity is 1 round.

Performance Comparison of Protocol 2: This paper compares Reference [25] with Protocol 2, assuming that both parties encrypt m-group ciphertext. In Reference [25], the Paillier encryption algorithm is used with a computational complexity of $5m + 12$ modular exponential operations, and the communication complexity is 3 rounds. In this paper, the ECC algorithm is adopted, the computational complexity is $2m + 14$ multiplication operations, and the communication complexity is 3 rounds.

Through comparison, it can be seen that the efficiency of Protocol 1 is higher than Reference [16], and that of Protocol 2 is higher than Reference [25], shown in Table 2.

Table 2. Performance comparison

Protocol	Computational complexity	Communication complexity	Resist the malicious attacks
Protocol 1	$15l$ elliptic curve multiplication	1 round	×
Reference [16]	$16l$ modular exponential operations	1 round	×
Protocol 2	$2m + 14$ elliptic curve multiplication	3 rounds	√
Reference [25]	$5m + 12$ modular exponential operations	3 rounds	√

*l: DNA sequence length; m: number of groups

Comparing Protocol 1 and Protocol 2, Protocol 2 under the malicious model uses the zero-knowledge proof, the cut-choose method, so under the condition of the same input and output Protocol 2's computational complexity and communication complexity are slightly higher than Protocol 1, but Protocol 2 can resist the malicious attacks.

Note: Because the MPC protocols under the malicious model require the use of the zero-knowledge proof, and cut-choose method, so Protocol 2's computational complexity and communication complexity are slightly higher than Protocol 1, we can use the way of pre-processing or service outsourcing to improve efficiency.

6 Conclusions

DNA matching plays an important role in practical scenarios such as paternity tests, species identification, etc. In this paper, we design two MPC protocols based on ECC for DNA matching securely. Protocol 1 is designed to calculate the DNA Hamming distance protocol confidentially; Protocol 2 secretly compares DNA equivalence under the malicious model. In this paper, the cut-choose method and zero-knowledge proof method are used to construct the MPC protocol under the malicious model, which has more practical value and provides a solution to the MPC of DNA matching problem.

Acknowledgement. This work is supported by National Natural Science Foundation of China: Big Data Analysis based on Software Defined Networking Architecture (No. 62177019; F0701); Inner Mongolia Natural Science Foundation (2021MS06006); 2022 Basic Scientific Research Project of Direct Universities of Inner Mongolia (20220101); 2022 Fund Project of Central Government Guiding Local Science and Technology Development (20220175); 2022 "Western Light" Talent Training Program "Western Young Scholars" Project; Inner Mongolia Discipline Inspection and Supervision Big Data Laboratory Open Project Fund (IMDBD202020);Baotou Kundulun District Science and Technology Plan Project (YF2020013); the 14th Five Year Plan of Education and Science of Inner Mongolia (NGJGH2021167); Inner Mongolia Science and Technology Major Project (2019ZD025); 2022 Inner Mongolia Postgraduate Education and Teaching Reform Project (20220213); the 2022 Ministry of Education Central and Western China Young Backbone Teachers and Domestic Visiting Scholars Program (20220393); Basic Scientific Research Business Fee Project of Beijing Municipal Commission of Education (110052972027); Research Startup Fund Project of North China University of Technology (110051360002).

References

1. Wu, C., Luo, C., et al.: A greedy deep learning method for medical disease analysis. IEEE Access **6**, 20021–20030 (2018)
2. Gao, Y., Xiang, X., et al.: Human action monitoring for healthcare based on deep learning. IEEE Access **6**, 52277–52285 (2018)
3. Fu, A., Zhang, X., et al.: VFL: a verifiable federated learning with privacy-preserving for big data in industrial IoT, IEEE Trans. Indust. Inform. **18**, 3316–3326 (2020)
4. Xia, F., Hao, R., et al.: Adaptive GTS allocation in IEEE 802.15. 4 for real-time wireless sensor networks. J. Syst. Archit. **59**(10), 1231–1242 (2013)
5. Kumar, P., Kumar, R., et al.: PPSF: a privacy-preserving and secure framework using blockchain-based machine-learning for IoT-driven smart cities. IEEE Trans. Netw. Sci. Eng. **8**(3), 2326–2341 (2021)
6. Cheng, H., Xie, Z., et al.: Multi-step data prediction in wireless sensor networks based on one-dimensional CNN and bidirectional LSTM. IEEE Access **7**, 117883–117896 (2019)
7. Yao, Y., Xiong, N., et al.: Privacy-preserving max/min query in two-tiered wireless sensor networks. Comput. Math. Appl. **65**(9), 1318–1325 (2013)
8. Zhao, J., Huang, J., et al.: An effective exponential-based trust and reputation evaluation system in wireless sensor networks. IEEE Access **7**, 33859–33869 (2019)
9. Wu, C., Ju, B., et al.: UAV autonomous target search based on deep reinforcement learning in complex disaster scene. IEEE Access **7**, 117227–117245 (2019)
10. Zhang, W., Zhu, S., Tang, J., Xiong, N.: A novel trust management scheme based on Dempster–Shafer evidence theory for malicious nodes detection in wireless sensor networks, J. Supercomput. **74**(4), 1779–1801 (2018)
11. Chen, Y., Zhou, L., et al.: KNN-BLOCK DBSCAN: fast clustering for large-scale data. IEEE Trans. Syst. Man Cybernet. Syst. **51**(6), 3939–3953 (2019)
12. Huang, S., Zeng, Z., Ota, K., et al.: An intelligent collaboration trust interconnections system for mobile information control in ubiquitous 5G networks. IEEE Trans. Netw. Sci. Eng. **8**(1), 347–365 (2020)
13. Qiu, M., Gai, K., Xiong, Z.: Privacy-preserving wireless communications using bipartite matching in social big data. Futur. Gener. Comput. Syst. **87**, 772–781 (2018)
14. Qiu, H., Qiu, M., Lu, Z.: Selective encryption on ECG data in body sensor network based on supervised machine learning. Inf. Fusion **55**, 59–67 (2020)

15. Qiu, M., Zhang, L., Ming, Z., et al.: Security-aware optimization for ubiquitous computing systems with SEAT graph approach. J. Comput. Syst. Sci. **79**(5), 518–529 (2013)
16. Ma, M.Y., Xu, Y., Liu, Z.: Privacy preserving Hamming distance computing problem of DNA sequences. J. Comput. Appl. **39**(09), 2636–2640 (2019)
17. Liu, Z., Seo, H., Kim, H.et al.: Memory-efficient implementation of elliptic curve cryptography for the Internet-of-Things. IEEE Trans. Depend. Secur. Comput. **16**(3), 521–529 (2019)
18. Goldreich, O.: The Fundamental of Crytography: Basic Application. Cambridge University Press, London (2004)
19. Qiu, M., Xue, C., Shao, Z., et al.: Efficient algorithm of energy minimization for heterogeneous wireless sensor network. In: IEEE EUC, pp. 25–34 (2006)
20. Qiu, M., Chen, Z., Ming, Z., Qin, X., Niu, J.: Energy-aware data allocation with hybrid memory for mobile cloud systems. IEEE Syst. J. **11**(2), 813–822 (2014)
21. Niu, J., Gao, Y., Qiu, M., Ming, Z.: Selecting proper wireless network interfaces for user experience enhancement with guaranteed probability. J. Parallel Distrib. Comput. **72**(12), 1565–1575 (2012)
22. Gai, K., Qiu, M., Elnagdy, S.: A novel secure big data cyber incident analytics framework for cloud-based cybersecurity insurance. In: IEEE BigDataSecurity (2016)
23. Qiu, H., Zheng, Q., et al.: Topological graph convolutional network-based urban traffic flow and density prediction. IEEE Trans. Intell. Transp. Syst. **22**, 4560–4569 (2020)
24. Qiu, H., Dong, T., Zhang, T., et al.: Adversarial attacks against network intrusion detection in IoT systems. IEEE Internet Things J. **8**(13), 10327–10335 (2020)
25. Li, S., Wang, W., Du, R.: Protocol for millionaires' problem in malicious models. Sci. Sin. Inform. **51**(1), 75 (2021)
26. Goldwasser, S., Micali, S., et al.: The knowledge complexity of interactive proof systems. SIAM J. Comput. **18**(1), 186–208 (1989)

Research on Action Recognition Based on Zero-shot Learning

Hui Zhao[1], Jiacheng Tan[2], and Jiajia Duan[3(✉)]

[1] Educational Information Technology Lab., Henan University, Kaifeng 475000, China
zhh@henu.edu.cn
[2] Software College, Henan University, Kaifeng 475000, China
[3] Henan International Joint Laboratory of Theories and Key Technologies on Intelligence Networks, Henan University, Kaifeng 475000, China
robots7@163.com

Abstract. At present, the research on human action recognition has achieved remarkable results and is widely used in various industries. Among them, human action recognition based on deep learning has developed rapidly. With sufficient labeled data, supervised learning methods can achieve satisfactory recognition performance. However, the diversification of motion types and the complexity of the video background make the annotation of human motion videos a lot of labor costs. This severely restricts the application of supervised human action recognition methods in practical scenarios. Since the zero-shot learning method can realize the recognition of unseen action categories without relying on a large amount of labeled data. In recent years, action recognition methods based on zero-shot learning have received great attention from researchers. In this paper, we propose an attention-based zero-shot action recognition model ADZSAR. We design a novel attention-based mechanism feature extraction method that introduces the current state-of-the-art semantic embedding model (Word2Vec). Experiments show that this method performs the best among similar zero-shot action recognition methods based on spatio-temporal features.

Keywords: Deep learning · Zero-shot learning · Attention mechanism · Semantic embedding

1 Introduction

With the rapid advance in computer hardware [1–3], system infrastructure [4–6], and the popularization of 5G technology, the explosive growth of network data [7–9] shows that human society has already entered the era of big data [10–12]. Correctly understanding and analyzing video content has become an

M. Qiu et al. (Eds.): SmartCom 2022, LNCS 13828, pp. 175–187, 2023.
https://doi.org/10.1007/978-3-031-28124-2_17

important task in the field of computer vision [13,14]. In recent years, human action recognition has achieved remarkable results and has been widely used in various industries, such as video surveillance, smart home [15–17], autonomous driving [18,19], and human-computer interaction [20,21].

Human action recognition processes and analyzes pre-segmented time series. It analyzes people's actions and behaviors and establishes the mapping relationship between video content and action categories, so that the computer can correctly understand the video. The generalization performance of human action recognition methods based on traditional features is weak. And it is difficult to deal with the problems of illumination changes and occlusions in the process of data collection [22,23]. With the vigorous development of deep learning and the proposal of large-scale datasets, action recognition methods based on deep learning have achieved great results [24]. The method is trained in an end-to-end manner, with the network autonomously learning behavioral representations in videos to complete the classification.

The deep learning method has good recognition performance due to the powerful computing resources, the complexity of the neural network and the huge data set. However, today's era of explosive growth of video data, it is extremely time-consuming and labor-intensive to finely label massive video data [25–27]. This severely restricts the scalability of deep learning-based action recognition methods. Therefore, how to enable a machine learning system to effectively capture discriminative action visual features from a small number of samples or action semantic descriptions has become a blueprint that many machine learning researchers are eager to achieve. This is also a research difficulty in video action recognition.

The proposal of *Zero-shot Learning* (ZSL) provides a solution to the problem of high cost caused by a large amount of labeled data in deep learning [28]. This has great research significance for the exploration of the generalization ability of deep learning network models. There are many researches on zero-shot learning in images, and certain results have been achieved. However, the research on video has only emerged in recent years, and the research results are still relatively few. Human action recognition based on zero-shot learning learns never-before-seen video action categories by establishing a mapping relationship between visual space and semantic space. Zero-shot learning can reduce the gap between artificial intelligence and humans in action recognition, and is the only way to develop general types. Zero-shot learning does not rely on a large amount of labeled data in training, which greatly improves the generalization and transfer capabilities of the model. It enables machines to also have the ability to reason, and is an important step towards true artificial intelligence.

In this paper, we propose an attention-based zero-shot action recognition model ADZSAR. This model proposes a novel feature extraction method based on attention mechanism, which can adaptively extract salience information in features for different kinds of actions to construct a non-redundant visual space. This model introduces the current state-of-the-art semantic embedding model (Word2Vec) to effectively extract semantic embedding information of class labels. First, we introduce the research status of action recognition technology, as shown in Sect. 2. Second, we introduce ADZSAR, an action recognition model based on zero-shot learning, as shown in Sect. 3. Then, we conduct experiments and compare the experimental results, as shown in Sect. 4.

2 Related Work

Human action recognition technology can be divided into traditional feature-based methods and deep learning-based methods. Among them, deep learning-based methods are widely used, mainly including two-stream convolution-based methods and three-dimensional convolution-based methods.

In terms of two-stream convolution based, Simonyan et al. [29] proposed a new two-stream architecture including spatial convolutional flow and temporal convolutional flow. However, this model has limitations in time acquisition. In order to extract long-term dependencies, Wang et al. [30] proposed a method to extract multi-dimensional convolutional feature maps using a two-stream convolutional network, and based on this, they proposed a *Temporal Segmentation Network* (TSN) [31]. Diba et al. [32] proposed a new modeling and characterization method to extract motion information of the whole video. And embed the extracted information into the traditional CNN. It can simultaneously capture the long-term structure between video frames and encode it into a compact representation.

In terms of 3D convolution, Ji et al. [33] proposed a 3D convolution model to extract temporal and spatial features and obtain continuous action information between frames. On this basis, Tran et al. [34] proposed the C3D (Convolution 3D) network architecture, which is simple and effective. It uses a large number of supervised data sets to train deep 3D convolutional networks. On this basis, Joao et al. [35] proposed a three-dimensional convolutional I3D (Two-Stream Inflated 3D ConvNets) model, which is expanded from a 2D convolutional network and adds optical flow to the input. The network will process based on RGB video and optical flow separately to get the result of fusion of the two streams.

The concept of *Zero-shot Learning* (ZSL) was first proposed by Palatucci et al. [28], which can use the attribute information of semantic labels to identify new sample categories without training samples. Zero-shot learning is a branch of Transfer Learning [36], which draws on the learning and reasoning ability of humans to make computers also have the ability to transfer and reason. Zeynep et al. [37] proposed to establish a mapping relationship between visual space and semantic space. Through the mapping relationship between the two to find the most appropriate relationship. Then attribute-based image classification is performed. In 2015, Alata et al. [38] proposed a joint embedding space containing attributes, textual and hierarchical relations, and learned the visual and semantic mapping relations. Gan et al. [39] directly exploit the semantic relationship between action classes and then perform label transfer learning. Wang et al. [40] proposed a staged bidirectional latent embedding framework for the subsequent two learning stages of zero-shot visual recognition. Wang et al. used the idea of joint embedding for the first time to study the multi-label zero-shot action recognition problem. Gao et al. [41] proposed a *two-stream graph convolutional network* (TS-GCN). Zhang et al. [42] proposed an end-to-end deep embedding for zero-shot work. And they pointed out that the visual feature space works better as the embedding space than the semantic space as the embedding space.

3 Model Design

In this section, we introduce the framework design of the model ADZSAR. The overall frame diagram of ADZSAR is shown in Fig. 1, which mainly includes embedding modules (semantic space and visual space) and related modules.

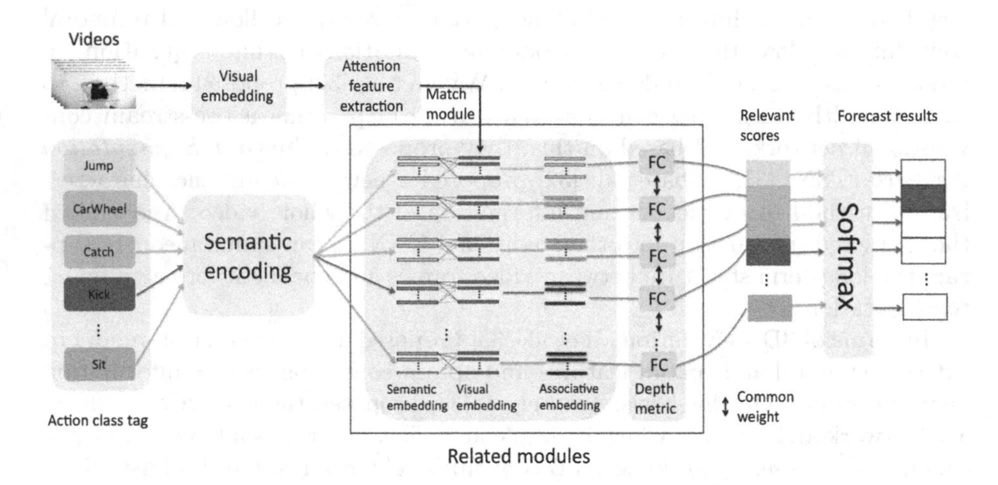

Fig. 1. Zero-Shot Action Recognition Model ADZSAR.

3.1 Visual Embedding Module

The visual embedding module inputs the graph of each frame of video and the corresponding optical flow graph into the C3D model and the TSN model, respectively. Through training, the spatio-temporal features and optical flow features based on can be obtained. The salience behavior information is further captured by the fully connected layer, and the spatio-temporal feature embedding and optical flow feature embedding of each video are generated after stacking in temporal order. Then, these two features are fused to obtain visual embeddings. An attention mechanism is added after obtaining the fused visual embedding. Deep metric learning is performed to obtain more salient visual features, as shown in Fig. 2.

3.2 Semantic Encoding Module

The Skip-gram model is used in the ADZSAR model to obtain word vectors for all labels. The model obtains the final semantic embedding representation using two fully connected layer modules, as shown in Fig. 3. For action class labels, the Skip-Gram model is used to generate semantic word vectors for action class labels. Specifically, the Skip-gram model uses the input action class labels to predict the words surrounding such labels, and its prediction probability is $P = (\omega_i|\omega_j)$, where $j - l \leq i \leq j + l, i \neq j$ and l is the window size. The larger the l, the higher the accuracy. For a language model of a set of words $\omega_1, \omega_2 ... \omega_j$, the optimization objective is:

$$\frac{1}{J}\sum_{J}^{j=1} \sum_{-1 \leq t \leq 1, t \neq 0} logp\left(\omega_{i+t}|\omega_j\right) \tag{1}$$

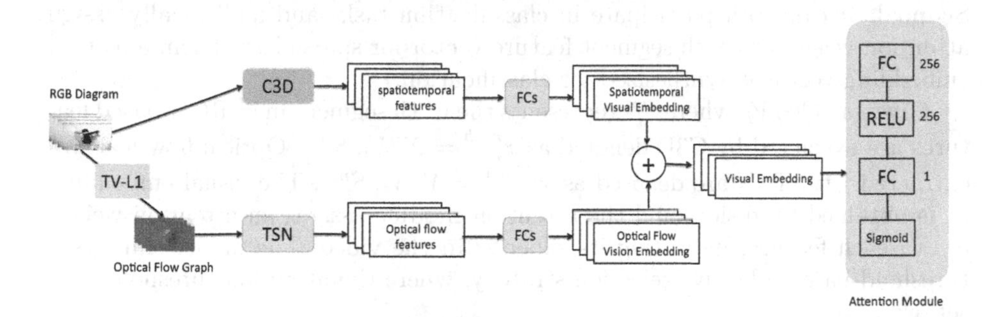

Fig. 2. ADZSAR Model Vision Space.

After obtaining the semantic representation u_y of the class label through the word vector model, it is embedded into the joint embedding space. Embed it into the joint embedding space. The specific operation is to learn through a fully connected layer alignment, and then pass through an embedding layer to generate a semantic embedding representation:

$$g(u_c) = \sigma\left(\phi_c \cdot u_c\right) \tag{2}$$

where σ is the activation function ReLU and ϕ_c is all the parameters involved in the semantic model.

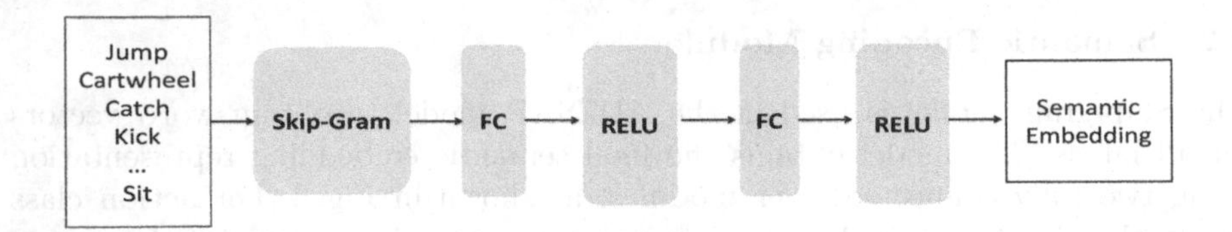

Fig. 3. ADZSAR Model Semantic Space.

3.3 Feature Extraction for Attention Mechanism

In the ADZSAR model, an attention mechanism is added after the visual embedding obtained by the two-stream structure of the video embedding module. Adding attention mechanism can learn to measure the importance of each segment. The structure diagram of the attention mechanism module is shown in Fig. 4. The attention module consists of two *fully connected* layers (FC) and Sigmoid. It can both participate in classification tasks and additionally assign attention weights to each segment feature vector for sparse loss. Then, a feature embedding vector is synthesized for classification.

Given a video V_i, where S_i^m represents the m-th segment in it. RGB-based features are extracted by C3D denoted as $x_i^{rgb} = X(V_i; S_i^m)$. Optical flow features extracted with TSN are denoted as $x_i^{flow} = Y(V_i; S_i^m)$. The visual embedding x_i is obtained by fusion, and the attention module assigns an attention weight a_m^i to each feature, indicating its saliency in the video. The model employs a threshold-based adaptive selection strategy, where the attention threshold a_i^t is set as:

$$a_i^t = \frac{1}{M} \sum_{M}^{m=1} a_i^m \quad (3)$$

If the attention weight of the video clip is greater than the threshold. Then focus on detecting the corresponding segment. And filter out the video clips below the threshold in the video. Finally, these features with large amount of information are generated into visual embedding representations. The loss of the saliency detection module is:

$$L_A = L_c + \alpha \|\alpha\|_1 \quad (4)$$

where L_c is the binary cross-entropy loss between the prediction result and the class label. α is a constant that balances the classification loss and regularization loss of attention weights.

3.4 Related Modules

The related modules include a matching module and a depth metric module. The matching module compares visual embeddings and semantic embeddings. Then, the results of the comparison are fed back to the deep neural network and output the relationship score. Finally, The depth metric module uses the Softmax layer to map relational scores to a probability distribution on the sample class.

The matching module takes as input the visual embedding V obtained from the visual space (representing the embedding of the t-th video clip) and the semantic embedding Y (representing the embedding of the y-th class label) obtained from the semantic space. The match score is calculated according to the formula [43]:

$$S_i^{t\,y} = \left(v_i^t - u_y\right) \Box \left(v_i^t - u_y\right) \tag{5}$$

where $\Box$ represents element-wise multiplication. i is the index of the video and t is the index of the video segment.

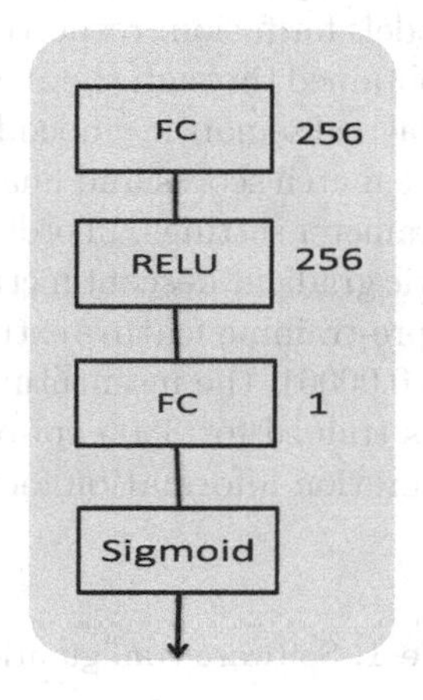

Fig. 4. Attention module structure.

The deep metric module uses deep neural networks to perform deep metric learning. Input the obtained matching scores into the deep neural network. In a deep neural network, a fully connected layer (FC) is used to aggregate all query video clips to generate the final relation score. Finally, the relation scores are fed back to the Softmax layer to be mapped to the probability distribution of sample classes. The overall model is trained end-to-end. Use binary cross-entropy loss as the loss function across the batch:

$$L_c = -\frac{1}{N}\sum_{i=0}^{N}\left(T_i log S_i^j + (1 - T_i)\, log\left(1 - S_i^j\right)\right) \tag{6}$$

where N is the number of videos of unseen classes. T_i is the target relationship score and S_i^j is the predicted relationship score.

4 Evaluation

In this section, the relevant experimental setup and experimental results are analyzed and discussed. We verify the effectiveness of the proposed ADZSAR model.

4.1 Datasets and Experimental Settings

In zero-shot action recognition, we use the HMDB51 [44] and UCF101 [45] datasets to evaluate the ADZSAR model. Using class-based data segmentation, each dataset is equally divided into visible and invisible classes. Additionally, for each class label in the HMDB51 and UCF101 datasets, we employ Word2Vec to extract the semantic embedding representation of the class label. The Semantic Embedding Autoencoder consists of 6 fully connected layers that take as input the semantic embeddings of action labels and object names. On datasets HMDB51 and UCF101, we extract spatio-temporal features and optical flow features from pre-trained C3D and TSN models for fusion, respectively. Then the salient visual embedding representation is obtained through the attention mechanism. We use a matching module to align visual and semantic embeddings. Then, it is input to the depth metric module to get the match scores and final classification results of the matches. In terms of hyperparameter settings, in order to speed up the training of the model, we use the stochastic gradient descent method to optimize the network parameters when performing pre-training feature extraction on both datasets. We set the initial learning rate to 0.00001, the momentum to 0.9, and the decay step size to 1000. The network was trained for 3000 episodes and evaluated every 30 episodes. The software configuration information for the experiment is shown in Table 1.

Table 1. Software configuration

Software Name	Description
Python3.6	Data preprocessing and algorithm framework building.
Pytorch	Deep Learning Computing Platform
OpenCV	Video Preprocessing Tools
Gensim	Semantic encoder
Matplotlib	Drawing tool

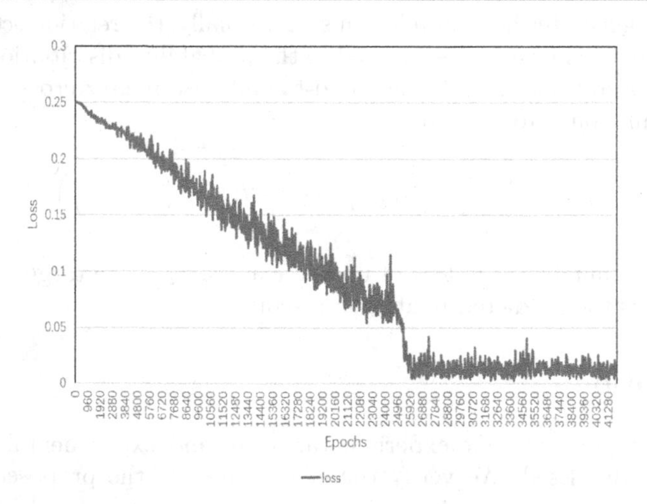

Fig. 5. The change curve of the objective function value as the Epochs changes.

4.2 Experimental Results and Analysis

We select five typical methods from previous zero-shot action recognition methods for comparison,as shown in Table 2. First, we compared the ADZSAR model with the methods in Table 2 on the datasets HMDB51 and UCF101.

Table 2. Zero-Shot Action Recognition Methods

Method	Introduction
ZSECOC [40]	CVPR 2017.
ESZSL [46]	ICML 2015.
BiDiLEL [24]	IJCV 2017.
GMM [47]	IEEE 2018.
TARN [48]	EMVC 2019.

Experimental results are shown in Table 3. All methods in the table have the same zero-sample evaluation criteria. And they all use spatiotemporal features as visual embedding representations and word vectors as semantic representations. From the comparative experimental data in Table 3, it can be seen that the ADZSAR model has at least 2.1% improvement on the dataset HMDB51 compared to the current best method. There is at least a 2.2% improvement on the dataset UCF101. There are good recognition results on both datasets.

Table 3. Comparative experimental results on HMDB51 and UCF101 datasets.

Model	Accuracy on HMDB51.(%)	Accuracy on UCF101.(%)
ZSECOC	16.5 ± 3.9	13.7 ± 0.5
ESZSL	18.5 ± 2.0	15.0 ± 1.3
BiDiLEL	18.6 ± 0.7	18.9 ± 0.4
GMM	19.3 ± 2.1	17.3 ± 1.1
TARN	19.5 ± 4.2	19.0 ± 2.3
ADZSAR	$\mathbf{21.8 \pm 1.6}$	$\mathbf{21.2 \pm 1.2}$

In order to achieve the final result, the ADZSAR model optimizes the model parameters through the iteration of the objective function. Take the dataset HMDB51 as an example. The specific method is to randomly select a Split in the data set to record the convergence of the model. Figure 5 shows the change curve of the Loss value of the objective function with different iteration cycles. From Fig. 5, it can be seen that the model reaches a state of convergence around Epoch 25920.

Finally, we performed ablation experiments on the ADZSAR model to analyze the role of each module. From the overall structure, the ADZSAR model mainly includes visual space, semantic space and related modules. When extracting visual features, a new feature extraction method based on attention mechanism is mainly used. On dataset HMDB51, videos are preprocessed with OpenCV. The experimental results of different methods in extracting visual features are given respectively. They are: C3D, TSN, C3D+TSN, C3D+TSN+Attention. From the experimental results in Table 4, it can be seen that the feature extraction method based on the attention mechanism adopted by the ADZSAR model in this paper has a certain improvement in the effect of action recognition.

Table 4. Ablation Experiments—Classification results on the HMDB51 dataset.

Model	Accuracy on HMDB51.(%)
C3D	19.0 ± 2.3
TSN	17.3 ± 1.1
C3D+TSN	20.2 ± 0.8
C3D+TSN+Attention(ADZSAR model)	**21.8** ± **1.6**

5 Conclusion

In this paper, we proposed an attention-based zero-shot action recognition model ADZSAR. First, the model proposes a novel attention-based feature extraction method, which can adaptively extract salience information in features for different kinds of actions to construct a non-redundant visual space. Second, a current state-of-the-art semantic embedding model (Word2Vec) is introduced to effectively extract the semantic embedding information of class labels. Extensive experiments showed that our method can outperform existing methods in zero-shot action recognition methods based on spatio-temporal features.

References

1. Qiu, M., Li, H., Sha, E.: Heterogeneous real-time embedded software optimization considering hardware platform. In: ACM Symposium on Applied Computing, pp. 1637–1641 (2009)
2. Qiu, M., Jia, Z., Xue, C., Shao, Z., Sha, E.: Voltage assignment with guaranteed probability satisfying timing constraint for real-time multiproceesor DSP. J. VLSI Signal Proc. Syst. **46**, 55–73 (2007)
3. Qiu, M., Yang, L., Shao, Z., Sha, E.: Dynamic and leakage energy minimization with soft real-time loop scheduling and voltage assignment. IEEE TVLSI **18**(3), 501–504 (2009)
4. Qiu, M., Xue, C., Shao, Z., Sha, E.: Energy minimization with soft real-time and DVS for uniprocessor and multiprocessor embedded systems. In: IEEE DATE Conference, pp. 1–6 (2007)

5. Shao, Z., Wang, M., et al.: Real-time dynamic voltage loop scheduling for multi-core embedded systems. IEEE Trans. Circuits Syst. II **54**(5), 445–449 (2007)
6. Qi, D., Liu, M., et al.: Exponential synchronization of general discrete-time chaotic neural networks with or without time delays. IEEE Trans. Neural Netw. **21**(8), 1358–1365 (2010)
7. Niu, J., Gao, Y., et al.: Selecting proper wireless network interfaces for user experience enhancement with guaranteed probability. JPDC **72**(12), 1565–1575 (2012)
8. Qiu, M., Xue, C., Shao, Z., et al.: Efficient algorithm of energy minimization for heterogeneous wireless sensor network. In: IEEE EUC Conference, pp. 25–34 (2006)
9. Qiu, M., Chen, Z., Ming, Z., Qin, X., Niu, J.: Energy-aware data allocation with hybrid memory for mobile cloud systems. IEEE Syst. J. **11**(2), 813–822 (2014)
10. Hu, F., Lakdawala, S., et al.: Low-power, intelligent sensor hardware interface for medical data preprocessing. IEEE Trans. Inf. Tech. Biomed. **13**(4), 656–663 (2009)
11. Gai, K., Qiu, M., Elnagdy, S.: A novel secure big data cyber incident analytics framework for cloud-based cybersecurity insurance. In: IEEE BigDataSecurity Conference (2016)
12. Li, J., Ming, Z., et al.: Resource allocation robustness in multi-core embedded systems with inaccurate information. J. Syst. Arch. **57**(9), 840–849 (2011)
13. Qiu, H., Dong, T., et al.: Adversarial attacks against network intrusion detection in IoT systems. IEEE Internet Things J. **8**(13), 10327–10335 (2020)
14. Qiu, H., Zheng, Q., et al.: Topological graph convolutional network-based urban traffic flow and density prediction. IEEE Trans. ITS **22**(7), 4560–4569 (2020)
15. Qiu, M., Khisamutdinov, E., et al.: Rna nanotechnology for computer design and in vivo computation. Philosophical Trans. R. Soc. A **371**(2000), 20120310 (2013)
16. Qiu, M., Su, H., Chen, M., Ming, Z., Yang, L.: Balance of security strength and energy for a PMU monitoring system in smart grid. IEEE Comm. Magazine **58**(5), 142–149 (2012)
17. Huang, H., Chaturvedi, V., et al.: Throughput maximization for periodic real-time systems under the maximal temperature constraint. ACM Trans. Embedded Comput. Syst. (TECS) **13**(2s), 1–22 (2014)
18. Li, J., Qiu, M., et al.: Thermal-aware task scheduling in 3D chip multiprocessor with real-time constrained workloads. ACM Trans. Embed. Comp. Sys. (TECS) **12**(2), 1–22 (2013)
19. Qiu, M., Ming, Z., et al.: Three-phase time-aware energy minimization with DVFS and unrolling for chip multiprocessors. J. Syst. Archit. **58**(10), 439–445 (2012)
20. Zhang, K., Kong, J., et al.: Multimedia layout adaptation through grammatical specifications. Multimedia Syst. **10**(3), 245–260 (2005)
21. Tao, L., Golikov, S., Gai, K., Qiu, M., A reusable software component for integrated syntax and semantic validation for services computing. In: IEEE Symposium on Service-Oriented System Engineering, pp. 127–132 (2015)
22. Gai, K., Du, Z., et al.: Efficiency-aware workload optimizations of heterogeneous cloud computing for capacity planning in financial industry. In: IEEE 2nd CSCloud (2015)
23. Zhang, L., Qiu, M., Tseng, W., Sha, E.: Variable partitioning and scheduling for mpsoc with virtually shared scratch pad memory. J. Sig. Proc. Syst. **58**(2), 247–265 (2018)
24. Wang, L., et al.: Temporal segment networks for action recognition in videos. IEEE Trans. Pattern Anal. Mach. Intell. **41**(11), 2740–2755 (2018)
25. Li, Y., Gai, K., et al.: Intercrossed access controls for secure financial services on multimedia big data in cloud systems. ACM Trans. Multimedia Comput. Commun. Appl. **12**(4s), 1–18 (2016)

26. Qiu, M., Sha, E., Liu, M., Lin, M., Hua, S., Yang, L.: Energy minimization with loop fusion and multi-functional-unit scheduling for multidimensional DSP. JPDC **68**(4), 443–455 (2008)
27. Gai, K., Qiu, M., Chen, L., Liu, M.: Electronic health record error prevention approach using ontology in big data. In: IEEE 17th HPCC (2015)
28. Palatucci, M., Pomerleau, D., Hinton, G.E., Mitchell, T.M.: Zero-shot learning with semantic output codes. In: International Conference on Neural Information Processing Systems (2009)
29. Simonyan, K., Zisserman, A., Two-stream convolutional networks for action recognition in videos. In: Advances in Neural Information Processing Systems, vol. 1 (2014)
30. Wang, L., Qiao, Y., Tang, X.: Action recognition with trajectory-pooled deep-convolutional descriptors. In: Computer Vision & Pattern Recognition (2015)
31. Wang, L., et al.: Temporal segment networks: towards good practices for deep action recognition, arXiv e-prints (2016)
32. Diba, A., Sharma, V., Gool, L.V.: Deep temporal linear encoding networks. In: Computer Vision and Pattern Recognition (2017)
33. From, M., Related, S., Breast, M., Related, S., Breast, M.: 3d convolutional neural networks for human action recognition. [44]
34. Tran, D., Bourdev, L., Fergus, R., Torresani, L., Paluri, M.: Learning spatiotemporal features with 3d convolutional networks. In: IEEE International Conference on Computer Vision (2015)
35. Carreira, J., Zisserman, A.: Quo vadis, action recognition? a new model and the kinetics dataset. In: 2017 IEEE Conference on Computer Vision and Pattern Recognition (CVPR) (2017)
36. Pan, S.J., Qiang, Y.: A survey on transfer learning. IEEE Trans. Knowl. Data Eng. **22**(10), 1339–1345 (2010)
37. Akata, Z., Perronnin, F., Harchaoui, Z., Schmid, C.: Label-embedding for image classification. IEEE Trans. Pattern Anal. Mach. Intell. **38**(7), 1425–1438 (2016)
38. Akata, Z., Reed, S., Walter, D., Lee, H., Schiele, B.: Evaluation of output embeddings for fine-grained image classification. In: IEEE Computer Vision and Pattern Recognition (2015)
39. Gan, C., Lin, M., Yang, Y., Zhuang, Y., Ha, G.: Exploring semantic inter-class relationships (sir) for zero-shot action recognition. In: National Conference on Artificial Intelligence (2015)
40. Wang, Q., Chen, K.: Zero-shot visual recognition via bidirectional latent embedding. Int. J. Comput. Vis. **124**(3), 356–383 (2017). https://doi.org/10.1007/s11263-017-1027-5
41. Gao, J., Zhang, T., Xu, C.: I know the relationships: Zero-shot action recognition via two-stream graph. In: Convolutional Networks and Knowledge Graphs, pp. 8303–8311 (2019)
42. Li, Z., Tao, X., Gong, S.: Learning a deep embedding model for zero-shot learning, IEEE (2017)
43. Wang, S., Jing, J.: A compare-aggregate model for matching text sequences (2016)
44. Kuehne, H., Jhuang, H., Garrote, E., Poggio, T., Serre, T.: Hmdb: a large video database for human motion recognition. In: IEEE International Conference on Computer Vision (2011)
45. Soomro, K., Zamir, A.R., Shah, M.: Ucf101: a dataset of 101 human actions classes from videos in the wild. Computer Science (2012)

46. Romera-Paredes, B., Torr, P.H.S.: An embarrassingly simple approach to zero-shot learning. In: Proceedings of the 32nd International Conference on Machine Learning (ICML 2015) (2015)
47. Mishra, A., Verma, V.K., Reddy, M., Arulkumar, S., Mittal, A.: A generative approach to zero-shot and few-shot action recognition. In: WACV, vol. 2018 (2018)
48. Bishay, M., Zoumpourlis, G., Patras, I.: Tarn: temporal attentive relation network for few-shot and zero-shot action recognition (2019)

DPSD: Dynamic Private Spatial Decomposition Based on Spatial and Temporal Correlations

Taisho Sasada[1,2](✉), Yuzo Taenaka[1], and Youki Kadobayashi[1]

[1] Nara Institute of Science and Technology, Nara, Japan
{sasada.taisho.su0,yuzo,youki-k}@is.naist.jp
[2] Japan Society for the Promotion of Science, Nara, Japan

Abstract. IoT data collected in a physical space have intrinsic values, such as the spatial and temporal correlation of people's activities. As such spatio-temporal data inevitably includes private information like trajectory and waypoints, privacy exposure becomes a serious problem. Local Differential Privacy (LDP) has been gaining attention as a privacy protection procedure on a device collecting spatio-temporal data. However, LDP cannot retain spatial and temporal properties which are essential for cyber-physical systems. The is because LDP makes each data indistinguishable and inevitably removes spatial and temporal properties as well. In this paper, we propose a method enabling LDP to keep spatial and temporal properties on privacy protection process. Our method dynamically changes the strength of privacy protection (called privacy budget) for each of device groups who has resemble spatial and temporal behavior. This makes data of each device in a group indistinguishable within the group but a set of data made by a group distinguishable between groups in terms of spatial and temporal domains. As the whole data merged in a data store will consists of modified data with wide variety of privacy budgets, we arrange every privacy budgets so that merged data keeps particular strength of privacy protection. We call this process as Dynamic Private Spatial Decomposition (DPSD). The experimental results show that our LDP preserves the data utility while maintaining the privacy protection of the entire client because of DPSD.

Keywords: Local Differential Privacy · Private Spatial Decomposition · Spatio-Temporal Data · Geospatial Clustering · Edge Computing · Database Security

1 Introduction

Spatio-temporal data collected by mobile devices has intrinsic value for driving society. People's trajectory and waypoints are one of the important spatio-temporal data because the spatial and temporal correlation of such data could be useful for urban planning, epidemiology, etc. However, such spatio-temporal data could also bring serious privacy exposure. It allows others to identify the

M. Qiu et al. (Eds.): SmartCom 2022, LNCS 13828, pp. 188–202, 2023.
https://doi.org/10.1007/978-3-031-28124-2_18

home, workplace, or other sensitive activities of a particular person. That is why we needs privacy protection maintaining spatial and temporal correlation even after the process of privacy protection.

As privacy protection in data collection, Local Differential Privacy (LDP) [14] is regarded as one of the promising technologies. Specially, LDP adds numerical noise to each value of data so as to make data indistinguishable. The amount of noise is determined based on the degree of the statistical distribution of original data and a specified strength of privacy protection (called privacy budget). For example, if many data are very similar values, the amount of noise will be a little to satisfy the privacy budget because the differentiation of data is very difficult in the first place. If many data are very distributed (or the amount of data is a little), a much amount of noise satisfying the privacy budget is chosen to make data differentiation difficult. Each device adds noise to a value of data at collecting it and only sends the processed data to a data store. The combination of such privacy-protection processing and on-device mechanism significantly prevents both the exposure of original (non-processed) data outside a device and identification of private information from processed data. However, LDP is fundamentally unable to maintain spatial and temporal properties in data after adding noise because the noise itself conceals these properties. Furthermore, it is also difficult to keep adding appropriate noise for satisfying a certain privacy budget because the trend of data changes time by time depending on people's movements.

To protect privacy while preserving both the spatial and temporal properties in waypoints and trajectory, Cormode et al. [7] have proposed Private Spatial Decomposition (PSD). PSD divides data into several pieces having spatial and temporal similarities and chooses/adds noise to each piece, This gives distinguishability of data between that with different amount of noises, and keeping spatial and temporal properties. All original data must be aggregated into a single data store like a cloud server, for computing the amount of noise for privacy protection by PSD, and the client itself cannot perform privacy protection processing on the device.

In this study, we propose a novel privacy-protection scheme applying LDP to spatio-temporal data while keeping spatial/temporal properties. We assign privacy budgets to spatially and temporally similar groups of clients. The scheme preserves spatial and temporal correlations by making data within groups indistinguishable but leaving features between groups. Then, the method requires a mechanism to ensure that data processed with different privacy budgets satisfy an arbitrary privacy protection strength for the aggregated data as a whole. Although data must be aggregated for proper grouping to satisfy an arbitrary privacy protection strength, aggregation in the data store is not as reliable from the client as in the case of PSD.

We therefore utilize a Local Edge Server (LES) to prevent the original data from being aggregated into a data store, and let the processing to satisfy arbitrary privacy budgets while maintaining spatial/temporal features. LES arranges adaptively changing privacy budget in the managing area and also contributes to avoiding that the data concentrates on a single data store. We define this process of changing the privacy budget according to the spatio-temporal correlation dynamic of all clients as Dynamic Private Spatial Decomposition (DPSD). To preserve the spatial/temporal correlation (intrinsic value) of the data, DPSD makes clusters including clients having similar characteristics and chooses/added noise to each separately. We execute DPSD in a certain interval to follow the changes in the statistical distribution of clients. Through this process, we can minimize the loss of intrinsic values as a whole data while protecting privacy. In this scheme, dividing clients into too many clusters can lead to excessive privacy protection and loss of intrinsic information in the data. The privacy protection method to de-identify depends on the number of clients. If the local edge server detects small clusters and sets the protection level, it increases the noise amount across clients and removes intrinsic information inside data. To address this problem, we exploratively vary the parameters of waypoints, trajectory, and time interval to detect clusters and design an optimal scheme to control protection levels based on spatio-temporal correlations for each dataset. The contributions of this work are as follows:

- Design algorithms to allocate the privacy protection strength for the preservation of temporal and spatial properties.
- Dynamic noise addition based on similarity of waypoints and trajectory (total amount of noise is the same, thereby privacy protection strength is constant).
- Clustering independent of transportation method (walking or driving speed) by tuning parameters.

The structure of this paper is as follows: First, we give the related work about existing protection for spatio-temporal data in Sect. 2. In Sect. 3, we define problem between spatio-temporal data and LDP. In Sect. 4, we describe the content of our proposals to address the existing problems. In Sect. 5, we explain the contents of experiments to evaluate balancing privacy protection and preserving intrinsic values in data and a discussion of the results. Finally, in Sect. 6, we summarize this study and refer to future work.

2 Related Work

2.1 Anonymization for Trajectories

Never Walk Alone (NWA) [3] is a proposal for trajectory anonymization to preserve spatio-temporal properties. NWA generalizes similar trajectories to non-identical ones so that satisfy (k, δ)-anonymity. This generalization ensures that adversaries cannot identify trajectories from each other with a probability $1/k$. To preserve the correlation of trajectories, Abul et al. [4] proposed to Wait for Me (W4M) as an extension of NWA. In addition to the location distance, W4M also considers the timestamp distance for anonymization, thus protecting privacy in terms of both waypoints and trajectory. However, clustering in NWA and W4M is based on the already collected trajectories. This approach cannot continue to anonymize trajectories where the data volume increases spatially and temporally over time. Moreover, anonymization that satisfies privacy metrics such as k-anonymity requires assumptions about the background knowledge of the adversary. If there is an unassumed adversary, they can identify the client and expose its privacy. In spatio-temporal data cases, it is hard for data stores to make assumptions about the adversary's knowledge. In other words, data stores must choose methods to protect client's privacy under any conditions.

2.2 DP/LDP for Trajectories

As anonymization without assuming the adversary's knowledge, Differential Privacy (DP) is typical privacy protection. As trajectory protection using DP, Cormode et al. [7] proposed Private Spatial Decomposition (PSD) to spatially and temporally control the protection level to preserve the properties of waypoints and trajectories. PSD optimally adjusts the privacy budget ϵ which is the protection level in DP depending on the distribution of spatio-temporal data. PSD can optimally control the protection level by adopting small ϵ for sensitive regions and large ϵ for relatively safe regions. This privacy protection guarantees client de-identification while preserving spatio-temporal information. LDP not only does not require any knowledge assumption by the adversary but also does not require the client to trust the data store. Therefore, there are higher expectations for LDP than for DP.

PSD can only be applied after data collection because PSD controls the protection level ϵ according to the distribution of spatio-temporal data. In LDP, clients sanitize data on their devices before data collection. Clients cannot know each other's waypoints and trajectories and cannot adjust the protection level appropriately. If clients accidentally set an excessively small privacy budget ϵ, privacy protection deprives the intrinsic information of trajectories. On the contrary, inadequate protection levels (large ϵ) lead to the leakage of sensitive information. For clients to preserve the intrinsic information of trajectories over LDP, they should continue to control the protection level while grasping the properties of location and trajectory.

3 Problem Statement

In this section, we explain the waypoints and trajectory correlations that should be preserved and the problems in applying LDP to spatio-temporal data.

3.1 Spatial and Temporal Properties Necessary for LBS

There is a spatio-temporal query for both waypoints and trajectories, and the data store must return meaningful results for both queries. Here we describe in detail the characteristics of waypoints and trajectories that are necessary for keeping spatio-temporal correlations.

In pattern mining of waypoints, by applying various clustering methods to the spatio-temporal data consisting of latitude, longitude, and timestamp, we can extract the degree of visitation by area and the Point of Interest (POI), which represents the location that the user visited with interest [19]. In the real world, it is mainly useful for understanding congestion caused by events, and for understanding the clusters of COVID-19 [6,16]. Since the data store wants to extract the locations where the client stayed, it is important to preserve the density of waypoints, the occurrence points, and the shapes of clusters. On the other hand, waypoints clustering also causes privacy exposure because waypoints clustering identify clients who do not belong to a cluster.

Pattern mining of trajectories is performed by applying a clustering method to trajectories generated through moving a geographical space [11,19]. The trajectory is a network structure consisting of a series of nodes connected in the order of their timestamps and is represented by a set of spatio-temporal locations. Trajectory pattern mining can be used to predict road congestion based mainly on similar routes and to analyze the propagation routes of user-health conditions and infectious diseases caused by movement. Therefore, clients must correctly preserve the continuity and temporal correlation of the trajectories. However, an adversary can infer home address, workplace, and even health status from the trajectory and threaten privacy if the client's trajectory exposed.

As summary, applying LDP to spatio-temporal data requires the preservation of spatial correlations and temporal correlations. By perturbing the data while preserving the intrinsic information, we can achieve to protect privacy while suppressing the utility loss in spatio-temporal data.

3.2 LDP Adaptation Mismatch to Spatio-Temporal Data

Unfortunately, LDP is clumsy to apply to spatio-temporal data. As described in Sect. 3.1, the client must protect privacy while preserving the density of waypoints, trajectory continuity, and temporal correlation because queries in spatio-temporal data are issued for both waypoints and trajectories. Due to this, clients should dynamically and voluntarily set protection strengths to preserve the density of waypoints, trajectory continuity, and temporal correlation. However, since the clients do not know each other's spatial and temporal distances, they cannot set protection strength appropriately. To preserve useful waypoints and trajectories while protecting privacy under all conditions in spatio-temporal data, we have to design a novel method that optimally controls the strength of protection so that clients can grasp and preserve the properties of each other's locations and trajectories.

4 Proposed Scheme

To design a novel scheme enabling LDP to address problems described in Sect. 3, we propose how to grasp spatio-temporal correlation on client's device and how to dynamically control the protection strength based on these features.

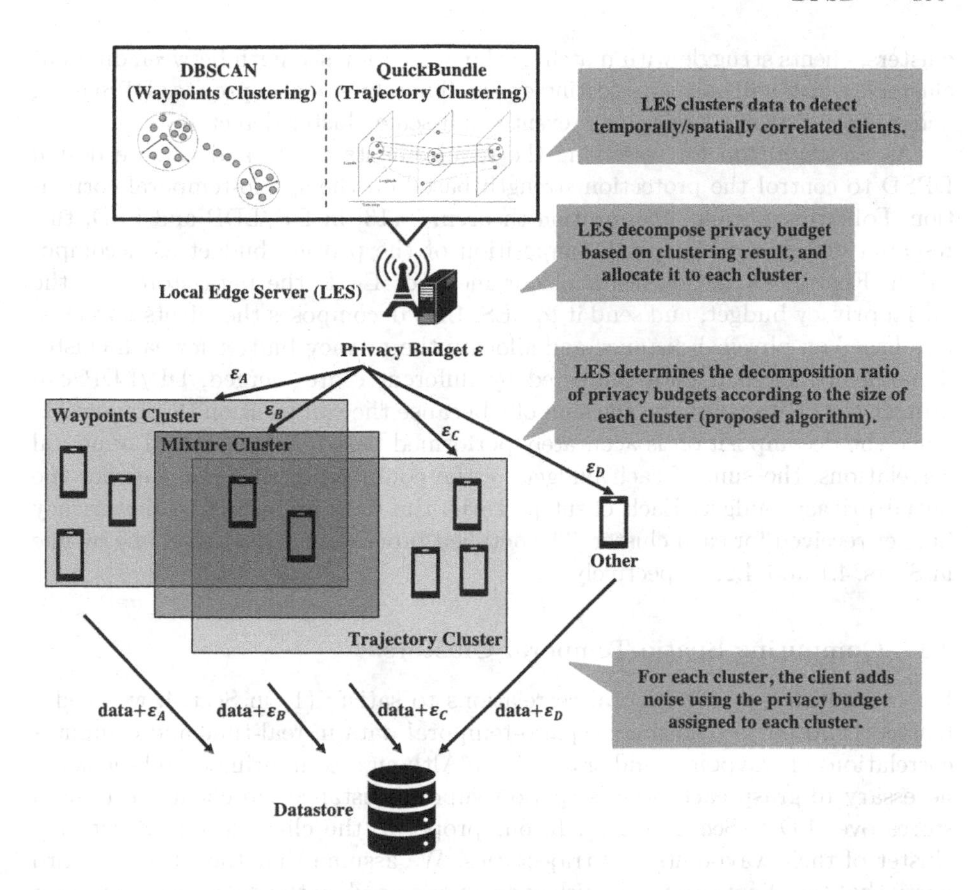

Fig. 1. The overview of our proposal. This figure show the flow of allocating privacy budget ϵ based on spatial/Temporal correlation.

In order to apply LDP in spatio-temporal data, there are two essential functions for the design. (1) To preserve intrinsic information in spatio-temporal data, we need clustering and indexing clients based on spatial and temporal correlation. Since the client cannot trust other clients or data stores, clients have to communicate with trusted third-party and let them cluster their data. For this reason, we first propose a method of clustering and indexing by a trustworthy local edge server (LES). Clients basically cannot trust third parties, but they use services provided by a telecommunication company. Then, it can be said that clients trust the telecommunication companies. Since Telecommunication companies such as Deutsche Telekom AG in Germany and KDDI Corporation in Japan collect spatio-temporal data [2,9], the use of LES like the proposed method is practically feasible. (2) To protect privacy based on spatio-temporal correlation, we also demand a scheme enabling LDP for clients to control protection strengths dynamically. Dynamically controlling the protection strength cluster-by-cluster let the clients remain useful waypoints and trajectory information while protecting their privacy under any conditions. However, if the LES detects small-scale clusters, clients struggle with matching the protection strength between different clusters, which will cause protecting errors. Thereby, we also propose DPSD and detection parameter tuning to prevent small-scale cluster detection.

As an algorithm for satisfying the requirements of (1) and (2), we design DPSD to control the protection strength based on the spatio-temporal correlation. Following parallel composition theorem [8,13] in DP/LDP and PSD, this research defines direct sum decomposition of the privacy budget as decomposition. Figure 1 is the overview of our method. First, the data store sets the initial privacy budget, and send it to LES. LES decomposes the clients into clusters based on physical distance and allocate the privacy budget for each cluster dynamically. Even if data protected by different ϵ are merged, DP/LDP can protect the privacy only by the sum of ϵ because the composition theorem work. Since the decomposition is accurately performed based on spatial and temporal correlations, the sum of each budget by the composition theorem satisfies the initial privacy budget. Each client perturbs the data according to the privacy budget received for each cluster. The detailed processing is described line by line in Sects. 4.1 and 4.2, respectively.

4.1 Computing Spatio-Temporal Clusters

For computing spatio-temporal correlations to satisfy (1) in Sect. 4, we need a trusted third party who shares spatio-temporal data in real-time and computes correlations of waypoints and trajectories. Although computing correlations are necessary to grasp each other's spatio-temporal distance, no client trusts data stores over LDP (See Sect. 2.2). In our proposal, the client lets LES detect a cluster of their waypoints and trajectories. We assume that the LES is a third party that the client can trust, such as a server owned by the device manufacturer or the device's communication carrier. Since the original spatio-temporal data is not communicated between clients, there is no need to worry about privacy exposure, and clients can securely grasp the spatial and temporal distance of each other.

An overview of the proposed algorithm is given in Algorithm 1. In the setup phase of Algorithm 1, data store $\mathcal{S}$ is responsible for setting the parameters required for DPSD and deciding which clients will participate in PSD. In line 2, $\mathcal{S}$ set the initial privacy budget ϵ according to the strength of protection that $\mathcal{S}$ wants to set. In line 3, $\mathcal{C}$ selects the trustworthy LES $\mathcal{L}$ that decomposes the privacy budget and clients that receive the decomposed privacy budget. In lines 5–7, $\mathcal{L}$ computes spatial and temporal correlations between each client. In practice, DPSD computes cluster by DBSCAN [10] and QuickBundle [12], and calculate the number of clusters. By using DBSCAN and QuickBundle, which are unsupervised learning that do not require specifying the number of clusters in advance, we can detect multiple clusters based on the similarity of waypoints and trajectories. When the LES computes spatio-temporal correlations of waypoints and trajectories, there are three parameters exist: the first parameter defines the correlation between waypoints, the second one defines the correlation between trajectory and the last one defines the computation interval.

Algorithm 1. Dynamic Private Spatial Decomposition (DPSD)

1: ***(1) Setup Phase.***
2: *$\mathcal{S}$ sets privacy budget $\epsilon(\epsilon > 0)$ and sends ϵ to $\mathcal{C}$.*
3: $\mathcal{C}$ picks trustworthy LES $\mathcal{L}$.
4: ***(2) Clustering and Decomposing Phase.***
5: **for** $i \in [d]$ **do**
6: $\quad\mathcal{L}$ compute spatial correlation set $a_i \leftarrow$ DBSCAN(d_i), temporal correlation set $s_i \leftarrow$ QuickBundle(d_i)
7: $\quad\mathcal{L}$ compute all cluster count $N_i \leftarrow max([a_i], [s_i])$.
8: $\quad$**for** $j \in [N_i]$ **do**
9: $\quad\quad\mathcal{L}$ create indexes $m_{\text{idx}}^{(j)} \leftarrow a_i^{(j)} \cap s_i^{(j)}$ on device.
10: $\quad\quad\mathcal{L}$ create indexes $a_{\text{idx}}^{(j)}, s_{\text{idx}}^{(j)} \leftarrow a_i^{(j)} \setminus s_i^{(j)}, a_i^{(j)} \setminus s_i^{(j)}$ on device.
11: $\quad\quad\mathcal{L}$ create indexes $o_{\text{idx}}^{(j)}, \leftarrow d_i^{(j)} \setminus (s_i^{(j)} \cup a_i^{(j)})$ on device.
12: $\quad$**end for**
13: $\quad\mathcal{L}$ decompose budget $\sum \epsilon = \epsilon_{\text{adj}} + \epsilon_{\text{sim}} + \epsilon_{\text{mlt}} + \epsilon_{\text{other}}$
14: $\quad$**if** $a_{\text{idx}}^{(j)} \leq m_{\text{idx}}^{(j)}$ **then**
15: $\quad\quad\mathcal{L}$ dispersively decompose $\epsilon_{\text{mlt}}^{(j)} \leftarrow \epsilon_{\text{mlt}}/[m_{\text{idx}}^{(j)}]$
16: $\quad$**else**
17: $\quad\quad\mathcal{L}$ dispersively decompose $\epsilon_{\text{adj}}^{(j)} \leftarrow \epsilon_{\text{adj}}/[a_{\text{idx}}^{(j)}]$
18: $\quad$**end if**
19: $\quad\mathcal{L}$ send $\epsilon_{\text{adj}}, \epsilon_{\text{sim}}, \epsilon_{\text{mlt}}, \epsilon_{\text{other}}$ to $\mathcal{C}$.
20: **end for**
21: ***(3) Perturbation Phase.***
22: **for** $i \in n$ **do**
23: $\quad c_i$ set θ unif. in $[0, 2\pi)$
24: $\quad c_i$ draw p unif. in $[0, 1)$
25: $\quad c_i$ set $r \leftarrow C_{\epsilon_i}^{-1}(p) = -\frac{1}{\epsilon_i}(W_{-1}(\frac{p-1}{\epsilon_i}) + 1)$
26: $\quad c_i$ pertubate $d_i' \leftarrow d_i + (r\cos(\theta), r\sin(\theta))$
27: $\quad c_i$ send LDP-processed data d' to $\mathcal{S}$
28: **end for**

Depending on these parameters, the result of the spatio-temporal correlation will change. Let Θ_{wypt} and Θ_{traj} be the parameters that determine how many waypoints and trajectories are included in the cluster to control the privacy budget, respectively. This parameter Θ_{wypt} and Θ_{traj} are very important: if the clients are correlated in waypoints and trajectory, protecting them with a small privacy budget that does not deteriorate the data utility. On the contrary, if they assign a large privacy budget to areas where the clients are isolated and distributed, inadequate protection exposes client's privacy. In addition, as time passes, the strength of the assigned privacy budget becomes insufficient. By recalculating according to this discrepancy, optimal privacy protection can always be provided to the terminal. The ideal time interval Θ_{tmsp} is expected to vary depending on the type of terminal movement (walking, bicycling, driving, flying, etc.), we thereby need an analysis with varying time intervals as a parameter. For these reasons, we exploratively search the parameters with grid-search [15] and analyze which thresholds are best correlated spatially and temporally to control the protection strength.

4.2 Decomposition of Privacy Budget

If clients set extremely small ϵ or ignore privacy (clients set $\epsilon = \infty$), the data store does not confirm how much the privacy budget is. Instead of clients that cannot properly set, a trustworthy LES allocates the privacy budget. If there are numerous small-scale clusters, it is hard to match the privacy budget of all clusters. To prevent LES from detecting small-scale clusters, we must satisfy condition (2) and dynamically changes the privacy protection strength for each temporally and spatially correlated cluster.

Depending on the computed clustering results, $\mathcal{L}$ creates the index set which clusters each client belongs to decompose and allocate the privacy budget (lines 8–12). If waypoints and trajectories belong to more than one cluster, excessive noise will be added. To handle it, we index the waypoints index set a_{idx}, the trajectory index set a_{idx}, the mixture index set m_{idx}, and other index set o_{idx}, separately. Based on this index, $\mathcal{L}$ decomposes the privacy budget and shares it with clients (lines 13–20). Specifically, $\mathcal{L}$ assigns privacy budgets based on the ratio of the number of elements in each cluster to the total number of clients. In the perturbation phase, all clients add noise to their data to satisfy the LDP. As a privacy mechanism to satisfy LDP, we use the Planar Laplace Mechanism [5]. This mechanism guarantees indistinguishability between a waypoint d and a waypoint d' contained in a circle of radius r centered on it. Given each waypoint as input, we get the probability density distribution of the possible output points by adding noise according to a planar laplace function. If the difference of the planar laplace probability density distributions of the two waypoints is small, anyone cannot identify and clients data satisfy DP/LDP (Geo-Indistinguishability [5]). Each c_i defines the angle and radius in lines 22–25 and adds noise in line 25–26 (If c_i belongs to a_{idx}, then c_i add planar laplace noise with ϵ_{adj}). Finally, c_i send their perturbed data to $\mathcal{S}$ in line 27.

Moreover, the allocation of the privacy budget decomposed by PSD may not be sufficient because clients are always on the move in physical spaces. To re-allocate the privacy budget, we perform the DPSD again at a fixed time interval Θ_{tstp}. We use Algorithm 1 to decompose and allocate the privacy budget again, depending on the locality of the location and trajectory that have moved over time. Over time, the strength of the decomposed and allocated privacy budget becomes insufficient. By recalculating according to this discrepancy, we can always provide clients with optimal privacy protection. Since the ideal time interval is expected to vary depending on the client's means of transportation (e.g., walking, bicycling, driving, flying, etc.), we do not fix the time interval but vary it as a parameter and analyze the appropriate value exploratory.

5 Experiment

In Sect. 4, we proposed a novel scheme that controls the protection strength so that the client preserves the intrinsic information of waypoints and trajectories by grasping the spatio-temporal correlation. The noise amount that our proposal can reduce depends on the definition of the minimum waypoints in client's

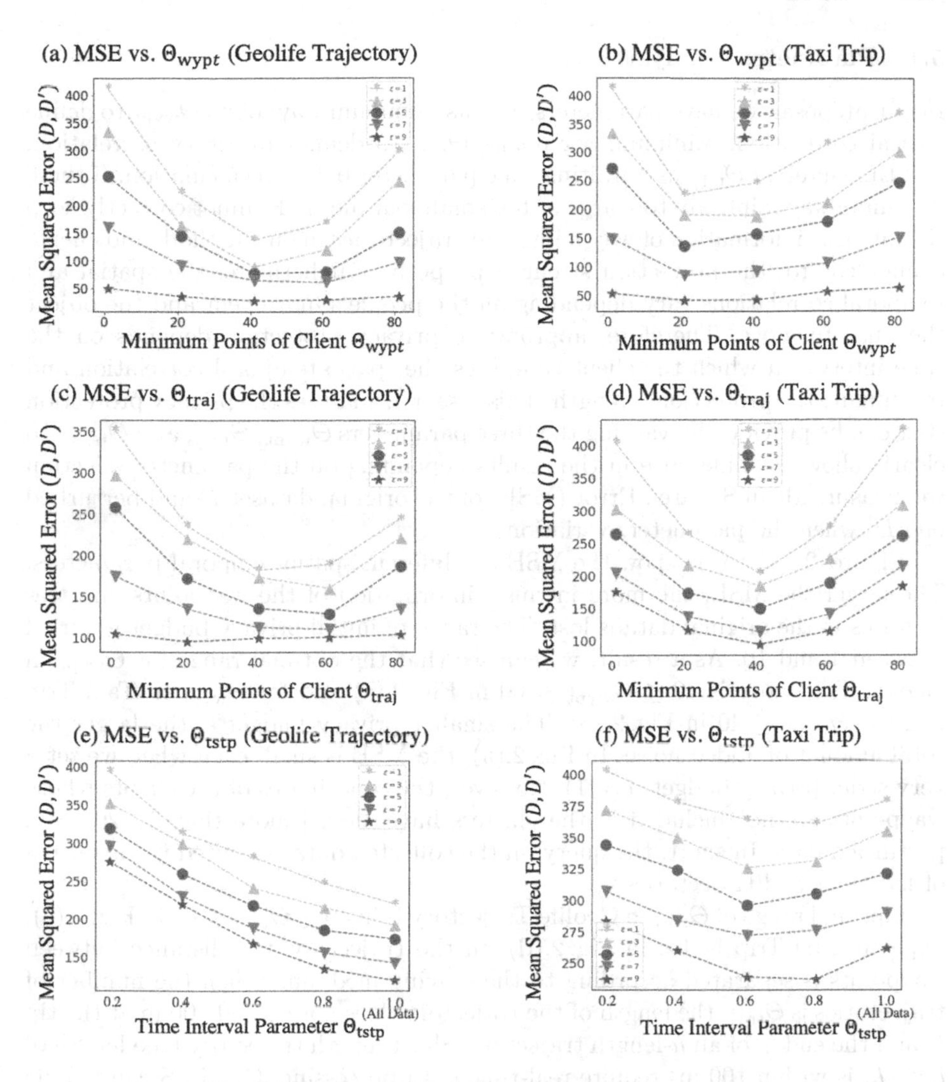

Fig. 2. MSE vs. Each Spatio-Temporal Parameter

spatio-temporal correlation. To evaluate this effect, we analyze the parameters used in computing spatio-temporal correlations and evaluate the average noise amount when using the parameters obtained from the analysis in Sect. 5. As the experimental environment, we perform all experiments with ASRockRack 3U8G+/C621E workstation, CPU is 40-core Intel Xeon Gold 6230 Processor @ 2.10 GHz (2 threads, 27.5 MB Cache), 262 GB RAM, and the host OS is Ubuntu 18.04 LTS. As datasets, we adapt Geolife GPS trajectory dataset [20], and ECML/PKDD 15 taxi trip time prediction dataset [1].

5.1 Parameter Analysis

In our proposal, we have parameters such as minimum waypoints Θ_{wypt} to define spatial correlation, minimum waypoints Θ_{traj} to define trajectory correlation, and time interval Θ_{tstp} as a waiting time parameter before recalculation. If these minimum waypoints are too large or too small, our method cannot correctly keep the intrinsic information of waypoints and trajectories in our method, and clients cannot control the protection strength properly. Furthermore, the spatial and temporal correlations vary depending on the people's movement and the object they use to move. Therefore, appropriate privacy protection depends on the time interval in which the client computes the spatio-temporal correlation and reconfigure the protection strength. This research analyzes the privacy protection that can be provided by varying the three parameters Θ_{wypt}, Θ_{traj}, and Θ_{tstp}. To clearly show the difference in the results depending on the parameter selection we measure Mean Squared Error (MSE) of the original dataset D and perturbed one D' when the parameters variation.

Figures 2. (a) to (f) show the MSE for different spatio-temporal parameters. The larger the MSE, the more intrinsic information of the waypoints and trajectories in the original data is lost. The range of initial privacy budget is varied between 1 and 10. As a result, we can see that the optimal range for Θ_{wypt} in Geolife Trajectory is $40 \leq \Theta_{wypt} \leq 60$ in Fig. 2.(a) and for Θ_{wypt} in Taxi Trip is $20 \leq \Theta_{wypt} \leq 40$ in Fig. 2.(b). The smaller privacy budget ϵ, the larger the total amount of added noise. In Fig. 2.(a), the MSE is small even when we set a very strict privacy budget ($\epsilon = 1$). However, the coordinates of the clients whose waypoints are not included in the clusters have moved more than 10 20km in perturbed data. In short, the query on the collected data is limited to clustering of locations and trajectories.

Optimal range of Θ_{traj} in Geolife Trajectory is in $40 \leq \Theta_{traj} \leq 60$ in Fig. 2.(c), Θ_{traj} in Taxi Trip is 40s in Fig. 2.(d). In the trajectory, the distance between waypoints is separated according to the moving medium. When the number of trajectories is Θ_{traj}, the length of the trajectory does not exceed 100 m at the tip l_1 and the end l_n of an n-length trajectory. Short-length trajectory (the length of l_1 to l_n is within 100 m) require real-time data processing. Our LES can handle data by that speed because we implemented the LES part by using PySpark. Without a framework for real-time processing, performance of LES must be overhead about clustering. Also, the trajectory clustering is not only affected by Minimum Points, but also by other parameters that should be considered more strictly, such as the angle of the direction of movement and the speed of movement. This may be the reason for the large scatter in the MSE between different privacy budgets.

The time interval Θ_{tstp} is very different from the other two parameters. The longer the time interval Θ_{tstp} is, the less noise is added. A large number of client data correlate position and trajectory in physical space in proportion to Θ_{tstp}. Therefore, from the point of view of the nature of differential privacy, it is natural that the MSE becomes smaller because the amount of data becomes larger and harder to identify. However, in the Taxi dataset, where clients move around

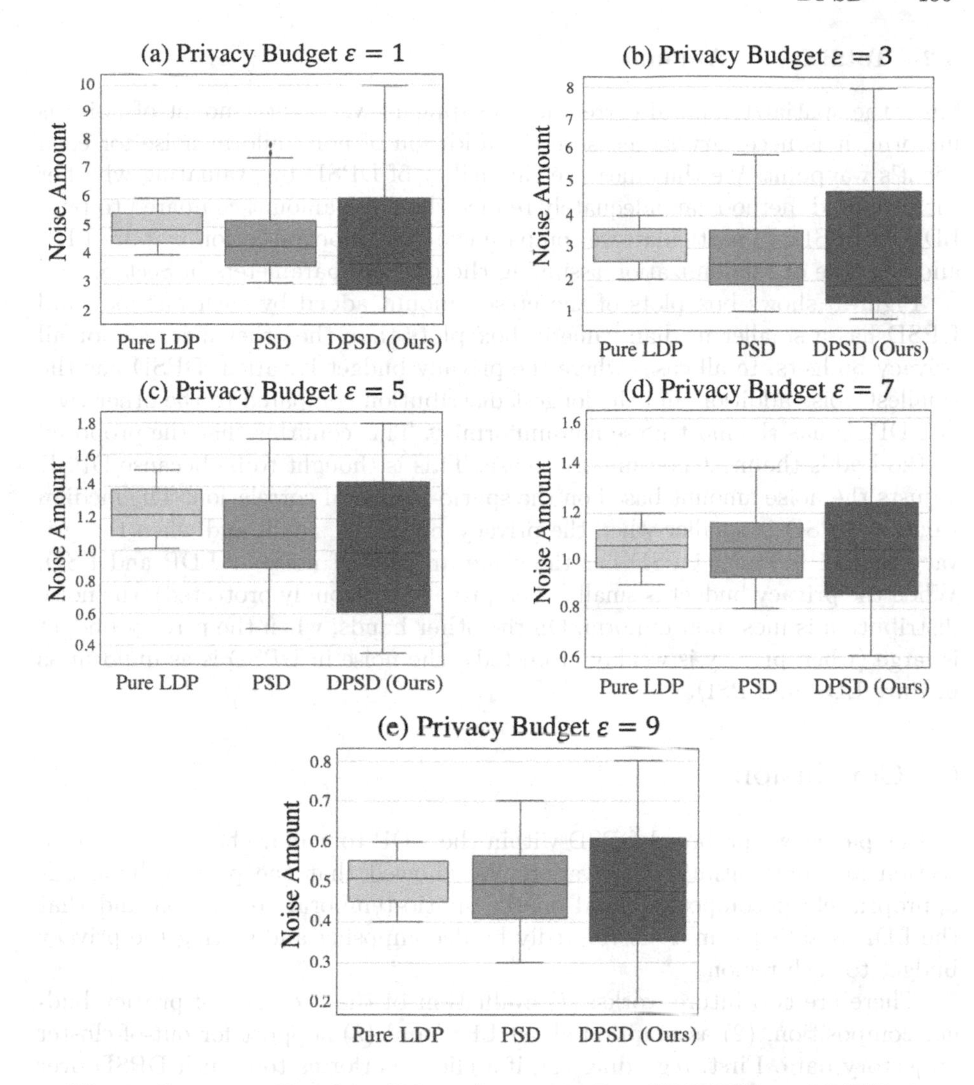

Fig. 3. Comparing Boxplot of Noise Amount in Pure LDP, PSD, and DPSD.

a lot due to too long time intervals, the MSE is larger. Lastly, we describe the differences among the datasets. The parameters of each dataset are very different. While pedestrians are not restricted by physical objects (buildings and obstacles) in walking, cars are restricted by geographic feature, speed, traffic rules, and many other factors. As a result, there is a large difference in the definition of Θ_{traj} that can reduce the amount of noise added between the Geolife GPS trajectory data and the ECML/PKDD 15 Taxi Trip Time Prediction dataset.

5.2 Total Noise Amount

Since the spatial/temporal correlation disappears when the amount of noise is uniform, it is necessary to measure the addition of non-uniform noise for each client's waypoint. We thus measure the utility of DPSD by evaluating whether the proposed method can adequately reduce the noise amount compared to pure LDP and PSD. In particular, we compare the case of perturbation by pure LDP and the case of randomization assuming the optimal parameters in Sect. 5.1.

Figure 3 shows box plots of the noise amount added by each method, and DPSD has a smaller median value in box plots than the other methods for all privacy budgets. In all cases where the privacy budget is varied, DPSD has the smallest noise addition and the longest distribution compared to the other two, i.e., DPSD has the most noise non-uniformity. This confirms that the proposed method adds the most non-uniform noise. This is thought to be because DPSD adjusts the noise amount based on the spatio-temporal correlation. The median value of DPSD is smaller when the privacy budget is small, and when the privacy budget is large, DPSD has the same amount of noise as LDP and PSD. When the privacy budget is small (when privacy is strongly protected), the noise distribution is most non-uniform. On the other hands, when the privacy budget is large (when privacy is weakly protected), the noise in DPSD is as uniform as in Pure LDP and PSD.

6 Conclusion

In this paper, we proposed DPSD within the LDP to achieve both privacy protection and data utility preservation. We showed that the privacy budget is appropriately decomposed based on the spatio-temporal correlation and that the LDP is satisfied in a distributedly by decomposing and setting the privacy budget to each region.

There are two future works: (1) evaluation of the accuracy of privacy budget composition, (2) assumption about LES, and (3) support for out-of-cluster trajectory data. First, regarding (1), if a client performs too much DPSD over LDP, the decomposed privacy budget may not satisfy the appropriate value due to overflow or underflow of the decimal calculation. In order to grasp the limits of our proposal, it is necessary to compose the privacy budgets of each client and evaluate the accuracy of composition. As for (2), our proposed method assumes a trusted third-party such as device manufacturers or a telecommunication companies to set a privacy budget. Even in Intel SGX and AMD SEV, which protect privacy at the hardware level, the manufacturer is assumed to be trusted [18], and clients using communication devices can be said to have a certain level of trust in the manufacturer or telecommunication company. However, in LDP, it is preferable not to provide a data and privacy budget to a third party because if the device manufacturer or carrier has malicious intentions, it may lead to a privacy exposure. Regarding the last future work (3), we applied a large amount of noise to trajectory data that has no temporal or spatial correlation, as it has little intrinsic information value. However, there are a few cases where trajectories that do not resemble other trajectories have value. For example, trajectory data that are outliers may be useful in the design of deviant behavior and evacuation routes [17]. As future work, we will extend and reexamine the proposed method to solve the above three problems.

Acknowledgements. This work was supported in part by Information-technology Promotion Agency (IPA)'s ICS-CoE Core Human Resources Development Program and Japan Society for the Promotion of Science (JSPS)'s KAKENHI Grant Number JP22J23910.

References

1. Kaggle ECML/PKDD 15: taxi trip time prediction II. https://www.kaggle.com/c/pkdd-15-taxi-trip-time-prediction-ii. Accessed 14-Jul 2022
2. KDDI location analyzer. https://k-locationanalyzer.com/en/. Accessed 14 Jul 2022
3. Abul, O., Bonchi, F., Nanni, M.: Never walk alone: uncertainty for anonymity in moving objects databases. In: 2008 IEEE 24th International Conference on Data Engineering, pp. 376–385. IEEE (2008)
4. Abul, O., Bonchi, F., Nanni, M.: Anonymization of moving objects databases by clustering and perturbation. Inf. Syst. **35**(8), 884–910 (2010)
5. Andrés, M.E., Bordenabe, N.E., Chatzikokolakis, K., Palamidessi, C.: Geo-indistinguishability: differential privacy for location-based systems. In Proceedings of the 2013 ACM SIGSAC Conference on Computer & Communications Security, pp. 901–914 (2013)
6. Jie Bao, Yu., Zheng, D.W., Mokbel, M.: Recommendations in Location-Based Social Networks: a Survey. GeoInformatica **19**(3), 525–565 (2015)
7. Cormode, G., Procopiuc, C., Srivastava, D., Shen, E., Yu, T.: Differentially private spatial decompositions. In: 2012 IEEE 28th International Conference on Data Engineering, pp. 20–31. IEEE (2012)
8. Dwork, C., Lei, J.: Differential privacy and robust statistics. In: Proceedings of the Forty-First Annual ACM Symposium on Theory of Computing, pp. 371–380 (2009)
9. Eichler, G., Pohlink, C., Kurz, W.: The telecommunication data cockpit – full control for the household community. In: Rautaray, S.S., Eichler, G., Erfurth, C., Fahrnberger, G. (eds.) I4CS 2020. CCIS, vol. 1139, pp. 3–22. Springer, Cham (2020). https://doi.org/10.1007/978-3-030-37484-6_1
10. Ester, M., Kriegel, H.-P., Sander, J., Xiaowei, X., et al.: A density-based algorithm for discovering clusters in large spatial databases with noise. In kdd **96**, 226–231 (1996)
11. Feng, Z., Zhu, Y.: A survey on trajectory data mining: techniques and applications. IEEE Access **4**, 2056–2067 (2016)
12. Garyfallidis, E., et al.: Quickbundles, a method for tractography simplification. Front. Neurosci. **6**, 175 (2012)
13. Kairouz, P., Oh, S., Viswanath, P.: The composition theorem for differential privacy. In: International Conference on Machine Learning, pp. 1376–1385. PMLR (2015)

14. Kasiviswanathan, S.P., Lee, H.K., Nissim, K., Raskhodnikova, S., Smith, A.: What can we learn privately? SIAM J. Comput. **40**(3), 793–826 (2011)
15. Lerman, P.M.: Fitting segmented regression models by grid search. J. R. Stat. Soc. Ser. C **29**(1), 77–84 (1980)
16. Lyu, H., Chen, L., Wang, Y., Luo, J.: Sense and sensibility: characterizing social media users regarding the use of controversial terms for Covid-19. IEEE Trans. Big Data **7**(6), 952–960 (2020)
17. Meng, F., Yuan, G., Lv, S., Wang, Z., Xia, S.: An overview on trajectory outlier detection. Artif. Intell. Rev. **52**(4), 2437–2456 (2019)
18. Mofrad, S., Zhang, F., Lu, S., Shi, W.: A comparison study of intel SGX and AMD memory encryption technology. In: Proceedings of the 7th International Workshop on Hardware and Architectural Support for Security and Privacy, pp. 1–8 (2018)
19. Zheng, Yu.: Trajectory data mining: an overview. ACM Trans. Intell. Syst. Technol. **6**(3), 1–41 (2015)
20. Zheng, Yu., Xie, X., Ma, W.-Y.: Geolife: a collaborative social networking service among user, location and trajectory. IEEE Data Eng. Bull. **33**(2), 32–39 (2010)

Scheduling Algorithm for Low Energy Consumable Parallel Task Application Based on DVFS

Xun Liu[1](✉) and Hui Zhao[2]

[1] College of Computer Science and Technology, Wuhan University of Science and Technology, Hubei Province Key Laboratory of Intelligent Information Processing and Real-Time Industrial System, Wuhan, China
liuxunecs@gmail.com
[2] Henan University, Kaifeng 475000, China
zhh@henu.edu.cn

Abstract. With the continuous improvement of various high-performance computing systems, various data centers had also been fully expanded. Energy consumption and actual performance measurement were very important indicators, which were also key issues in how to judge parallel calls for some tasks in high-performance computer systems. Modern processors were basically equipped with software control functions such as DVFS (Dynamic Voltage Frequency Scaling), in the actual system operation to ensure that the system could ensure the reasonable operation of the system while reducing energy consumption indicators. This paper considered how the designed scheduling algorithm first divides tasks reasonably to ensure that the maximum completion time and energy consumption of the processor were sufficiently reduced when the directed acyclic graph was executed. Then considered making reasonable adjustments to the processor frequency using DVFS technology to adapt to the task while ensuring the critical path of the task. At the end of the article, make sure that the experimental verification algorithm could ensure that the task was completed and could reduce the energy consumption during task execution as much as possible.

Keywords: Slacking phase · DVFS technique · Low-energy-consumption · Parallel applications · Task classification

1 Introduction

The current survey could be named that the actual operation of modern processors had a large amount of energy consumption that cannot be ignored, which had become a key direction that must be paid attention to in high-performance distributed computing systems. A recent study showed that nearly 2% of the global energy was consumed by data centers [1]. Such huge consumption could be attributed to multiple levels, such as the rapid spread of distributed computing platforms, the rapid development of computer cloud. In addition, previous studies had shown that about 52% of data center energy was

M. Qiu et al. (Eds.): SmartCom 2022, LNCS 13828, pp. 203–212, 2023.
https://doi.org/10.1007/978-3-031-28124-2_19

consumed by the actual operation of computing system servers, while the corresponding supported systems consume a lot of energy [2].

The computer center must strike a balance between energy consumption and performance, thereby reducing completion time and energy consumption while improving throughput. Therefore, reducing energy consumption was essential to ensure the future sustainable growth of the computing research center. The performance metrics and operational consumption of various servers and AC networks in data centers could be directly affected by the higher priority parallel applications designed and developed [3, 4].

At the same time, the processor of the computer center basically supported the application of DVFS technology. Moreover, since today 's data center processors basically had voltage regulators, the voltage of various frequencies could be easily switched in actual operation, which made the voltage switching in actual operation low consumption and low working hours [5, 6]. This also made the current various different areas the server could be relatively simple to take the DVFS strategy. This provided a prerequisite for reasonable and efficient task scheduling.

DVFS technology could also be applied to a variety of different domain models such as cloud computing distributed computing platform, virtual machine scheduling, high performance computing communication phase [7–9]. For different task models, there were also commonly used prediction algorithms and energy-aware scheduling algorithms [10]. At the same time, through cooperation with users as a certain basis, in the basic environment of ensuring service level agreement negotiation [11, 12], it could also achieve a balance between the relative maximum completion time of tasks and different energy consumption [13, 14].

And variety of processor models adjusted different voltage frequencies and adopted corresponding reasonable operation methods, which could more easily adapt to these different applications subject to parallel priority constraints [15–17]. In addition, the corresponding tasks for tasks could be distinguished to divide the server, allocate the idle time period and reasonably realize the scaling of voltage frequency [18].

Effective low power DVFS algorithm for heterogeneous multicore tasks Processing characteristics played a key role and could reasonably reduce system power Consumption, improve the overall performance of the system. How to implement in heterogeneous multi-core systems Now reasonable low power algorithm to achieve the purpose of reducing system power consumption. Become the main direction of current research [18–20].

The main structure of this paper was as follows: The second part introduces the DVFS usage strategy; the third part mainly analyzed the system model; the fourth part described the specific algorithm implementation strategy of DVFS for task scheduling. The fifth part of the experimental setup and results; the sixth section drew a conclusion.

2 DVFS Strategy and Task Partition

DVFS technology was achieved by dynamically changing voltage and frequency. The purpose of saving electricity. From the processor dynamic power consumption model could be seen. Reducing the power supply voltage and clock frequency of the chip could reduce the system [21]. Dynamic power consumption. DVFS technology includes both voltage and frequency tuning. Whole Voltage adjustment was achieved by a regulator (DC-DC), and Frequency adjustment was realized by programmable clock generator, but there was Some systems only realize voltage regulation (DVFS) or frequency regulation (DVFS). Now the regulator had been implemented Integrated in the processor chip, greatly reduced the voltage conversion time and Power consumption promotes the development of DVFS technology. For multinuclear System, DVFS could be divided into global DVFS and single-core DVFS. By The task model in the system was sometimes known and sometimes unknown [22]. Prediction algorithms were often used for unknown task models and for known tasks Energy-aware scheduling algorithm was often used in the model [23].

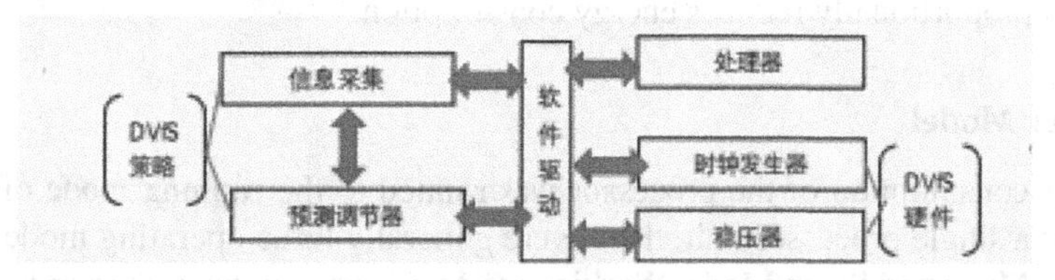

Fig. 1. Operational framework

Firstly, in order to reasonably reduce communication energy and increase throughput, the technology of minimizing and eliminating the expensive communication overhead in the process of data transmission between tasks by assigning tasks to the same processor. If applied correctly, the technology could reduce the maximum completion time and energy consumption, maximize throughput and minimize the number of active processors for task scheduling (Fig. 1).

At the same time,the communication cost berween processors assigned to tasks in communication was prevented by creating data regions.For DAG with larger CCR, this technology could more effectively reduce the maximum completion time and energy consumption and to meet maximum completion time,throughput and energy constraints [24].Task clustering reduced task completion time by zeroing the edge of high communication time and appropriate strategies [25].Task replication reduced communication overhead by reducing the allocation of certain tasks to multiple proceisors,thereby reducing energy consumption [27-29]The proposed approaches can also be used in data-intensive applications [29–31] and security-aware applications [32–34].

3 System Model

3.1 Processor Model

This paper set the server to P, and the chip of the processor was integrated with a voltage stabilizer. It supported DVFS strategy and could freely adjust different voltage frequencies. The voltage of the processor was defined as V, and the clock frequency was defined as f. It ensured that the clock frequency of the processor always has a level corresponding to the power supply voltage. In addition, for the adjustment of voltage frequency in this paper, λ was designed as the scaling factor of processor frequency, which was the ratio of the theoretical maximum frequency of the processor to the current frequency in the real-time operation of the processor.

This paper converts the scheduled task set into a DAG graph reasonably. In order to ensure the completion time of the task, the critical path and non-critical path of the task were determined first before scheduling. The critical path of the scheduled task graph was used as the execution time of the task from the earliest task to the last task and the time slot set of data exchange. After confirming the completion time of the task, the scheduling algorithm would not consider scaling the corresponding voltage frequency of the tasks on the critical path to ensure the completion of the task. At the same time, the scaling of the corresponding processor voltage frequency would be applied to the tasks on the non-critical path in the task scheduling graph. Adjusting the voltage frequency of these tasks proportionally reduces energy consumption.

3.2 Power Model

The power consumption of the processor was related to the running mode of the processor. For a single processor unit, there were generally three operating modes: Active Mode, Idle Mode and Sleep Mode. Working mode means that the processor is currently executing a task. The overall processor power consumption considered in this paper consists of three parts, namely dynamic power consumption, static power consumption and inherent power consumption. The overall total power consumption was defined as P_{total}. And used $P_{dynamic}$ to represent dynamic power consumption, P_{static} to represent static power consumption, and P_{on} to represent inherent power consumption.

$$P_{total} = P_{dynamic} + P_{static} + P_{on} \tag{1}$$

$$P_{dynamic} = a \times C \times f \times V^2 \tag{2}$$

$$P_{static} = I \times \mathrm{V} \tag{3}$$

where a was the switching coefficient of the gate circuit, C was the equivalent load capacitance, f was the clock frequency, V was the power supply voltage, and I was the leakage current. Dynamic power consumption was caused by circuit state switching.

In the normal task scheduling process, it could be found that the static energy consumption of the processor in a relatively static state was much weaker than the dynamic energy consumption generated in the actual running state of the processor, so the static power consumption of the task would not be considered in this paper. At the same time, this paper preferentially determines that when the processor adjusted the use mode of the processor to the idle mode due to the idle task, the voltage frequency of the processor maintains a relatively low state to ensure the reduction of task energy consumption, and the idle time period under the task running state was less than the running time period. This paper also mainly considered the reduction of energy consumption during the operation of the processor. Therefore, the main control of energy consumption in this paper focused on dynamic power consumption $P_{dynamic}$, and the optimization method of inherent power consumption P_{on} was not considered. Therefore:

$$P_{total} = P_{static} = k \times V^2 \cdot f \tag{4}$$

In the above equation, K was the operating constant used in the processor running device, and V was the voltage level of the power supply, which could be further deduced.

$$V = \beta \cdot f \tag{5}$$

In the above equation, β was a constant that runs during the actual scheduling of the processor. The processor power consumption considered in this paper is the energy consumption generated during the task scheduling process and execution process. Therefore, the overall energy consumption of this paper could be regarded as $E_{new} = P_{total} \cdot \Delta t$.

3.3 Task Model

In this paper, the task scheduling graph and related data would be reasonably transformed into DAG, so as to determine the task dependency and scheduling, simply expressed as G (T, D). Where T was composed of a set of tasks in the processor, and D was composed of a set of directed edges between tasks in the processor. For all tasks in the overall DAG graph $\forall t_i \in$ T were part of each node of the DAG in the running algorithm. At the same time, each task was running on different processor systems, thus $d_{ij} \in$ D was used to determine the priority relationship between tasks and determine the prerequisite constraints of the edge. Indicates that tasks t_i and t_j were separated and satisfies the condition that t_j must start running after task t_i. In addition, suc(t_i) was used to represent the successor set of task t_i, , and pre(t_i) was used to represent the predecessor set of task t_i. No previous task enters the processor and starts to run, which was regarded as task (t_{entry}). No subsequent task would exit and run on the processor, which is regarded as task (t_{exit}). At the same time, since the task would run on one or more processors, only when all task inputs are determined to be ready, the task began to execute, resulting in corresponding output results.

4 Efficient Energy Scheduling Algorithm

After the task was divided reasonably, it was necessary to determine the critical path and non-critical path to further reduce the energy consumption of the processor. Determine the maximum completion time by considering the critical path on the DAG as the longest path from task start to task end. It consisted of a set of tasks, which itself determines the maximum completion time of the task. When the delay occurs, the corresponding entire task framework would be delayed together with this. Therefore, whose LST was equal to their EST, it constitutes a key path.

Algorithm slack time ()

1. Confirm idle slots for tasks based on the completion time of each task
2. Calculate EST (t_i), EFT (t_i), LST (t_i), LFT (t_i)
3. end for
4. For each task t_i in sort do
5. Calculate Slack time of t_i
6. if Slack time task t_i = 0 then
7. Add task t_i to Critical Path List end if
8. else
9. Add task t_i to Non-Critical Path List
10. end if
11. end for

The non-critical tasks in DAG had corresponding relaxation time. Slack time is the amount of time that delays the start of an activity without delaying completion time. The key task in the whole DAG was not to consider any slack time, and the corresponding non-critical tasks generally had slack time. For each task, calculated the relaxation time by the following equation:

$$t_i = LST(t_i) - EST(t_i) \; or \; LST(t_i) - EST(t_i) \tag{6}$$

Specific time data for tasks should be derived from the following equation:

$$EST(t_i) = \left\{ \begin{array}{cc} 0 & Ifpre(t_i) = \varphi \\ MAX(EFT(t_i), MAX(EFT(t_k) + (d_{ki}))) & Otherwise \end{array} \right\} \tag{7}$$

The earliest completion time EFT of the task could be derived from the following equation by calculating the EST value:

$$EFT(t_i) = EST(t_i) + et(t_i, pm) \tag{8}$$

At the same time, the value of the final completion time (LFT) should be calculated. It was worth calculating that the task was completed from the last task to the end of the first task, which should be obtained by the following equation: d_{ik}

$$LFT(t_i) = \left\{ \begin{array}{cc} EFT(t_i) or makespan & EFT(t_i) or makespan \\ MIN(LST(t_i), MIN(LST(t_k) + (d_{ik}))) & Otherwise \end{array} \right\} \tag{9}$$

Correspondingly, the task ' s recent start time (LST) could be calculated by the following equation:

$$LST(t_i) = LFT(t_i) - et(t_i, pm) \tag{10}$$

After confirming the need to use DVFS technology to adjust the voltage/frequency task and the idle period of the task, it was necessary to determine the reasonable adjustment range of voltage/frequency and confirm the ideal voltage frequency f'n as much as possible.

The key path and non-key path were classified in different ways. The processor performs the key task, and the frequency was amplified to the highest level. Priority was given to ensuring that the key task was not delayed so as to avoid affecting the subsequent task progress. When facing the non-key task, the voltage/frequency level should be adjusted to f_n:

$$f_n = \text{freq}_{\text{highest}} \times \frac{\text{et}\left(t_i, P_M\left(v_{\text{highest}}, f_{\text{highest}}\right)\right)}{\text{slack time for task } t_i} \tag{11}$$

It could be determined from the above that the actual running time of some tasks could be adjusted and the voltage frequency could be reasonably adjusted within a certain range. In addition, the specific allocation of idle time periods for tasks did not take into account the idle time periods of task scheduling and processor frequency adjustment, but adjusts the task allocation in idle time.

The algorithm in this paper gave priority to the energy consumption of the task under the precondition of ensuring that the performance of the task scheduling cycle is satisfied. The main task would be arranged in advance of the task classification, through reasonable methods such as task clustering, task replication, so that the task before the corresponding module to run. Then, the tasks on the DAG were divided, the tasks that could be adjusted by DVFS technology were confirmed, and the idle time of the tasks was determined, and a reasonable scheduling scheme is realized.

5 Experimental Results and Discussion

The algorithm proposed in this paper will be tested in the following. In this experiment, because specific tasks could not be identified in advance, priority needs to be given. Since there were many situations that need to be considered for the random generation of graphs, this experiment would determine some different situations from several basic characteristics of DAG graphs, including the specific number of tasks in the DAG graph, the communication computation ratio CCR (which was the ratio between the average communication time and the average computation time of the application DAG) and the level number of the application DAG.

In this simulation experiment, we would mainly consider the different number of tasks, and set it from 30 tasks to 150 tasks. At the same time, we would consider the ratio of different CCR changes, and set it from CCR = 0.1 to 1 and even to 5. Generally speaking, for CCR = 5 and CCR = 0.1, it was an analysis that could clearly determine the different situations of the calculation-intensive DAG graph and the peer-intensive DAG graph. In order to carry out the corresponding simulation experiment, would use Lenovo ' s y7000P model computer running, using Intel Courier i7 11800 CPU, 2.3 GHz, eight cores, 16 GB memory running, through the calculation simulation tool MATLAB R2020b to simulate multiple data centers for experiments, while ensuring that these simulated processor types supported the use of DVFS technology. Simulation compares the proposed algorithm with other algorithms, namely RASD, HEFT.

The results were shown in Fig. 2. For the analysis of a graph sample, the algorithm proposed in this paper was under sufficient constraint conditions, in the environment of

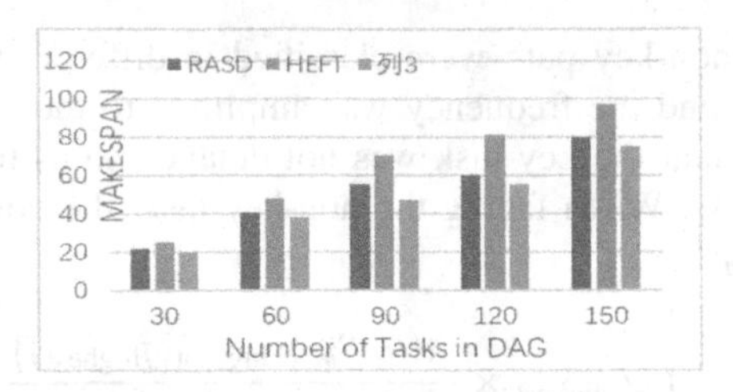

Fig. 2. Comparison of makespan between devised method and RASD, HEFT for CCR = 0.1

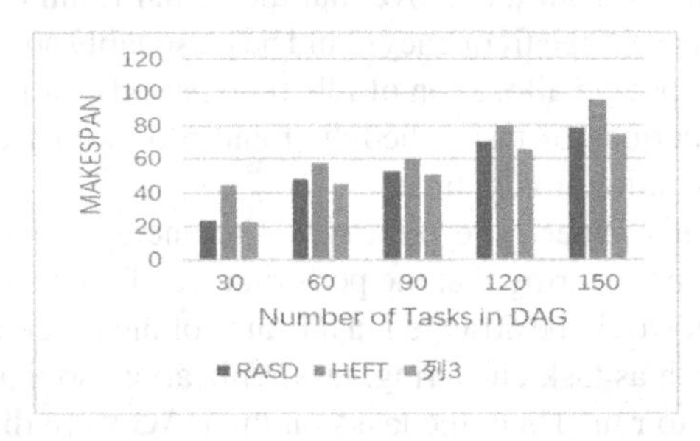

Fig. 3. Comparison of makespan between devised method and RASD, HEFT for CCR = 1

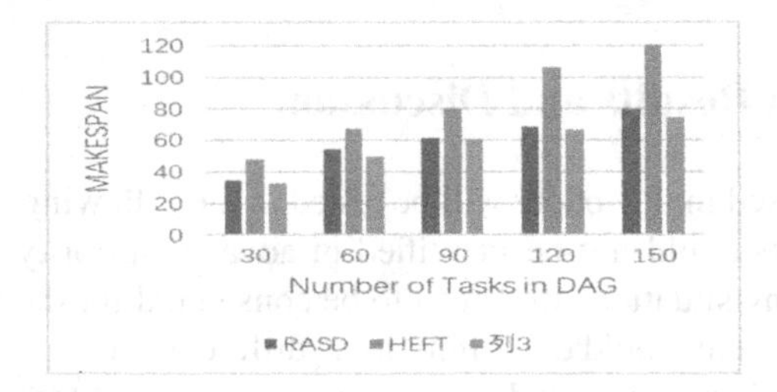

Fig. 4. Comparison of makespan between devised method and RASD, HEFT for CCR = 5

CCR = 0.1, and in the face of the DAG graph generated under different numbers of conditions, the overall makespan value was basically smaller than that of other algorithms. Through the following Fig. 3 and 4, it could be concluded that the overall advantage still exists when the CCR value reached 1 and 5. However, when the number of tasks in the DAG graph continues to rise backward and the CCR value continues to increase, the overall completion time was also significantly increased.

At the same time, with the number of tasks reaching 200 and beyond, the algorithm proposed in this paper would increase significantly. But with CCR worth changing, DAG applications, both computationally and communicationally intensive, still performed well.

6 Conclusion and Future Work

This paper considered the design of a reasonable algorithm to adapt to the data center composed of processors that support the use of DVFS technology through the voltage/frequency adjustment of processors. It could meet the constraints as much as possible through task replication and cluster strategy, and design a reasonable task planning in advance. At the same time, in the process of actual scheduling of tasks, it was necessary to ensure the rapid division of different types of tasks, and determine the idle time period of tasks to make reasonable adjustments to reduce energy consumption.

The future work outlook and how to adapt to more complex different types of DAG graphs. Under the premise of meeting the constraints, while facing more tasks in the task model, how to adapt, improve the efficiency of task completion, and reduced energy consumption as much as possible. For example, a model was proposed in some practical application samples to alleviate the energy consumption in task scheduling of different components.

References

1. Li, J., Qiu, M., Niu, J., et al.: Thermal-aware task scheduling in 3D chip multiprocessor with real-time constrained workloads. ACM TECS **12**(2), 1–22 (2013)
2. Berenjian, G., Motameni, H., et al.: Distribution slack allocation algorithm for energy aware task scheduling in cloud datacenters. J. Intell. Fuzzy Syst. **41**, 251–272 (2021)
3. Zhou, J., Yan, J., et al.: Thermal-aware correlated two-level scheduling of real-time tasks with reduced processor energy on heterogeneous MPSoCs. J. Syst. Archi. **82**, 1 (2018)
4. Yu, K., Han, D., Youn, C., Hwang, S., Lee, J.: Power-aware task scheduling for big LITTLE mobile processor. In: IEEE International SoC Design Conference, pp. 208–212 (2013)
5. Cheng, Z., Shaoheng, L.: Noise-aware DVFS for efficient transitions on battery-powered IoT devices. In: IEEE TCAD (2020)
6. Nielsen, L.S., Niessen, C., Sparso, J., et al.: Low-power operation using self-timed circuits and adaptive scaling of the supply voltage. IEEE TVLSI **2**(4), 391–397 (1994)
7. Qiu, M., Liu, J., Li, J., et al.: A novel energy-aware fault tolerance mechanism for wireless sensor networks. In: IEEE/ACM International Conference on GCC (2011)
8. Niu, J., Gao, Y., Qiu, M., Ming, Z.: Selecting proper wireless network interfaces for user experience enhancement with guaranteed probability. JPDC **72**(12), 1565–1575 (2012)
9. Hu, F., Lakdawala, S., et al.: Low-power, intelligent sensor hardware interface for medical data preprocessing, IEEE Trans. Info. Tech. Biomed. **13**(4), 656–663 (2009)
10. Huang, J., Raabe, A., Buckl, C., Knoll, A.: A workflow for run time adaptive task allocation on heterogeneous MPSoCs. In: Design, Automation and Test in Europe (2011)
11. Cheng, D., Zhou, X., Lama, P., et al.: Energy efficiency aware task assignment with DVFS in heterogeneous Hadoop clusters. IEEE TPDC **29**(1), 70–82 (2018)
12. Jia-Li, X., Hui, C., Bing, Y.: A real-time tasks scheduling algorithm based on dynamic priority. Chinese J. Comput. **35**(12), 2685–2695 (2012)
13. Smanchat, S., Viriyapant, K.: Taxonomies of workflow scheduling problem and techniques, in the cloud. Futur. Gener. Comput. Syst. **52**, 1–12 (2015)
14. Qiu, M., Xue, C., Shao, Z., Sha, E.: Energy minimization with soft real-time and DVS for uniprocessor and multiprocessor embedded systems. In: IEEE DATE Conference, pp. 1–6 (2007)

15. Qiu, M., Guo, M., Liu, M., et al.: Loop scheduling and bank type assignment for heterogeneous multi-bank memory. JPDC **69**(6), 546–558 (2009)
16. Qiu, M., Sha, E., et al.: Energy minimization with loop fusion and multi-functional-unit scheduling for multidimensional DSP. JPDC **68**(4), 443–455 (2008)
17. Jun, X., Shuai, Y., Yi, Y.: Scheduling algorithm for periodic tasks with low energy consumption based on heterogeneous multicore platforms. J. Comput. App. **39**(10), 2980–2984 (2019)
18. Hu, W., Ma, T., Wang, Y., Xu, F., Reiss, J.: TDCS: a new scheduling framework for real-time multimedia OS. Int. J. Parallel Emerg. Distrib. Syst. **35**(3), 396–411 (2020)
19. Anderson, J.H., Erijckson, J.P., Devi, U.C., et al.: Optimal semi- partitioned scheduling in soft real-time systems. J. Signal Process. Syst. **84**(1), 3–23 (2016)
20. Wang, J., Qiu, M., Guo, B., Zong, Z.: Phase-reconfigurable shuffle optimization for Hadoop MapReduce. IEEE Trans. Cloud Comput. **8**(2), 418–431 (2020)
21. Qiu, M., Xue, C., Shao, Z., et al.: Efficient algorithm of energy minimization for heterogeneous wireless sensor network. In: IEEE EUC, pp. 25–34 (2006)
22. V. Berten, C. Chang, and T. Kuo, "Managing Imprecise Worst Case Execution Times on DVFS Platforms," RTCSA, 2009, pp. 181–190
23. Venkatachalam, V., Franz, M.: Power reduction techniques for microprocessor systems. ACM Comput. Surv. **37**(3), 195–237 (2005)
24. Qiu, M., Li, H., Sha, E.: Heterogeneous real-time embedded software optimization considering hardware platform. In: ACM Symposium Applied Computing, pp. 1637–1641 (2009)
25. Qiu, M., Chen, M., et al.: Online energy-saving algorithm for sensor networks in dynamic changing environments. J. Embed. Comput. **3**(4), 289–298 (2009)
26. Barzegar, B., Motameni, H., Movaghar, A.: EATSDCD: a green energyaware scheduling algorithm for parallel task-based application using clustering, duplication and DVFS technique in cloud datacenters. J. Intell. Fuzzy Syst. **36**, 5135-5152 (2019)
27. Qiu, M., Jia, Z., et al.: Voltage assignment with guaranteed probability satisfying timing constraint for real-time multiproceesor DSP. J. Signal Process. Syst. **46**, 55–73 (2007)
28. Qiu, M., Yang, L., et al.: Dynamic and leakage energy minimization with soft real-time loop scheduling and voltage assignment. IEEE Trans. VLSI **18**(3), 501–504 (2009)
29. Li, J., Ming, Z., et al.: Resource allocation robustness in multi-core embedded systems with inaccurate information. J. Syst. Architect. **57**(9), 840–849 (2011)
30. Qiu, M., Chen, Z., Ming, Z., Qin, X., Niu, J.: Energy-aware data allocation with hybrid memory for mobile cloud systems. IEEE Syst. J. **11**(2), 813–822 (2014)
31. Qiu, H., Zheng, Q., et al.: Topological graph convolutional network-based urban traffic flow and density prediction. IEEE Trans. ITS **22**(7), 4560–4569 (2020)
32. Li, Y., Gai, K., et al.: Intercrossed access controls for secure financial services on multimedia big data in cloud systems. ACM Trans. MCCA **12**, 1–18 (2016)
33. Gao, X., Qiu, M.; Energy-based learning for preventing backdoor attack. KSEM **3**, 706–721 (2022)
34. Qiu, M., Qiu, H., et al.; Secure data sharing through untrusted clouds with blockchainenabled key management. In: The 3rd SmartBlock Conference, China, pp. 11–16 (2020)
35. Gai, K., Zhang, Y., et al.: Blockchain-enabled service optimizations in supply chain digital twin. IEEE Trans. Serv. Comput. (2022)

Research of Duplicate Requirement Detection Method

Dansheng Rao[1](✉), Lingfeng Bian[2], and Hui Zhao[3]

[1] COMAC Shanghai Aircraft and Research Institute, Shanghai 201210, China
raodansheng@comac.cc
[2] Fudan University, Shanghai 200433, China
lfbian20@fudan.edu.cn
[3] Henan University, Kaifeng 475000, China
zhh@henu.edu.cn

Abstract. With the rapid development of the software industry, the number of requirements included in the software is also increasing . In particular, in a complex software system such as an avionics software system, dozens of subsystems are usually included, and each subsystem includes hundreds of requirements, so the entire software system will include thousands of requirements. Furthermore, during software development, requirements will undergo frequent changes. When the number of requirements is huge and there are many people from different backgrounds participating in requirements development, it is easy to duplicate requirements. Therefore, in order to prevent redundant development, in the requirements analysis stage, requirements analysts usually need to detect all duplicate requirements in the requirements document and eliminate them, thereby improving the quality of software requirements and saving software development costs. However, when the number of requests is large, manually detecting duplicate requests can be a time-consuming and error-prone task. Aiming at this problem, this paper studies the method of duplicate requirements detection in English requirements.

Keywords: Duplicate Requirement · Detection Method · Avionics software Requirement · Requirement Engineering

1 Introduction

Requirement Engineering (RE) refers to a research field that studies how to use effective methods and techniques to help users analyze requirements and clarify all the characteristics that the system needs to have. Currently, requirements engineering has been considered by most people as an important success factor in software development [1]. The main purpose of this stage is to check out the errors in the requirements document, and correct all errors in time to ensure that the subsequent software development will not be affected, resulting in problems such as project overdue, over budget, and failure to meet user needs. In addition, in the process of requirement analysis, detecting duplicate requirements in requirement documents can further improve software reusability and save development costs [2–4]. Therefore, duplicate requirement detection is an important task in requirement engineering.

M. Qiu et al. (Eds.): SmartCom 2022, LNCS 13828, pp. 213–225, 2023.
https://doi.org/10.1007/978-3-031-28124-2_20

2 Related Work

With the rapid development of computer capability [5–7] and network techniques [8–10], software development becomes more and more important. New big data [11–13] processing algorithms [14–16] and machine learning techniques [17–20] have greatly reduced the cost of software development costs. As a basic problem in requirements engineering, duplicate requirement detection is of great significance for improving the quality of software requirements. Scholars at home and abroad have carried out many studies on this issue, and obtained many research valuable results.

Dag et al. [21–23] first proposed to use some common methods in the field of information retrieval to calculate the similarity of requirements, and judge whether the requirements are repeated according to it. The method first calculates the similarity coefficient between requirements according to the number of the same words in different requirements. Among them, the commonly used similarity coefficients mainly include cosine coefficient [24], Dice coefficient and Jaccard coefficient [25].

Hayes et al. [26] proposed a requirement similarity calculation method based on Term Frequency-Inverse Document Frequency (TF-IDF), and proposed a combination calculation method of TF-IDF and the thesaurus dictionary. D. FALESSI et al. [27–29] developed a tool (Proactive Reuse OpportUnity Discovery, PROUD) based on natural language processing technology, and proposed a requirement similarity calculation method. Manel Mezghani [30] et al. developed a tool, SEMIOS, based on K-means clustering algorithm, which can detect duplicate requirement.

To sum up, for the previous research on repeated requirement detection, most of the methods proposed by researchers are based on the co-occurrence frequency of words in different requirements. These methods only consider the similarity of words in requirements, and do not consider the similarity of word order and syntax in requirements, so there are limitations. Aiming at this problem, this paper proposes a method that considers the features of lexical, syntactic and semantic aspects to calculate the similarity of requirements and combines support vector machine to detect repeated requirements. In addition, with the development of deep learning technology, many valuable results have been achieved in the field of natural language processing. Therefore, this paper explores and proposes a deep learning-based model for repetitive requirement detection.

3 Duplicate Requirement Detection Based on Multi-feature Fusion

3.1 Detection Method

This chapter not only considers the similarity features of different requirement sentences at the lexical level, but also extracts multiple similarity features at the syntactic and semantic levels. Then, use all the extracted features together as an overall measure of requirement similarity. In order to realize the detection of repeated requirements, this paper first calculates the similarity features of each sample at the lexical level, syntactic level and semantic level on the training set, and obtains the feature vector corresponding to the sample. Then, the feature vector corresponding to each sample is input into the SVM for training. Considering that the main purpose of this paper is to detect duplicate requirements, the two categories output by the classifier are duplicate requirement pairs and irrelevant requirement pairs. Finally, when the training process is completed, SVM can be used to predict whether the input requirement pair is a duplicate requirement pair or an irrelevant requirement pair, so as to realize the detection of duplicate requirement.

3.2 Data Collection and Processing

At present, most of the requirements engineering researches use unpublished requirements data to conduct experiments, which also makes it impossible for other researchers to reproduce relevant experiments and obtain corresponding results. To solve this problem, Ferrari [31] et al. collected and organized a public requirement dataset PURE (PUblic REquirements dataset) from the Internet. The dataset contains 79 requirements documents described in natural language, with a total of 34268 sentences in these documents. The author reviewed all requirements documents and added some identification to each requirement document, including name, subject, format, number of pages, number of pictures or tables, etc. At the same time, the author exposes the requirement dataset on his website, and displays the identification information of each requirement document in the form of a table.

Since the author did not annotate the requirement data in PURE, this paper needs to annotate the data to construct a dataset suitable for repeated requirement detection tasks. The process of constructing the dataset in this paper is as follows:

(1) First, download all the XML format requirement documents in PURE, and select 24 requirement documents with relatively standardized language description. At the same time, 2560 requirement statements are extracted from the XML format document according to the rules.
(2) Then, this paper filters out the semantically repeated requirement pairs and the semantically irrelevant requirement pairs from the requirement sentences, and sets the corresponding tags for them. When annotating data in this paper, the samples with semantically repeated demand pairs are regarded as positive samples, and the corresponding label is 1; the semantically irrelevant demand pairs are regarded as negative samples, and the corresponding label is 0.
(3) Finally, after analysis and screening, this paper extracts a total of 102 positive samples and 1240 negative samples from PURE.

If only these data are used to construct a repeated demand detection data set, the proportion of positive and negative samples contained in the data set will be extremely unbalanced, which will affect the experimental evaluation results and lead to inaccurate evaluation indicators. Therefore, this paper considers increasing the number of positive samples through data augmentation. Data augmentation is a commonly used method in natural language processing, and data augmentation methods can generally be divided into methods based on text generation and methods based on word replacement. These two data augmentation methods essentially use various techniques to obtain text with the same semantics as the original text. At the same time, this paper attempts to use word replacement and back translation to generate demand pairs with the same semantics, and compares the data generated by different data enhancement methods. Therefore, this paper uses back translation to construct 1138 positive samples. The specific process of constructing positive samples is as follows:

(1) First, in each requirement document, about half of the requirements are selected, and a total of 1138 requirements are extracted.
(2) Second, after obtaining the requirements described in other languages by machine translation, machine translation is used again to translate the requirements expressed in other languages into English requirements. Since the machine translation effect is not guaranteed to be completely correct, the back-translation results will be checked and corrected manually, and the corrected results and the original requirements will form a positive sample.
(3) Finally, this paper constructs a repetitive demand detection task dataset containing 2480 samples, of which there are 1240 positive and negative samples. The data sample format of the positive and negative samples in the duplicate demand detection task data set is shown in Table 1 below:

Table 1. Positive and negative sample data samples in the duplicate demand detection data set

Type	Content	Label
Positive sample	A system administrator shall have unrestricted access to all aspects of the system	1
	The system administrator will have all access authorities of the system	
Negative sample	A system administrator shall have unrestricted access to all aspects of the system	0
	All systems and application source code shall be available to or on the systems that execute it	

3.3 Evaluation Indicators

This section mainly introduces the evaluation metrics used in the experiments of the duplicate demand detection method. Since the duplicate requirement detection problem is a classification problem, this paper considers the accuracy and F1 value as performance metrics to evaluate the method proposed in this paper.

The sample set constructed in this paper contains two categories of samples, namely positive samples and negative samples. At this point, the relationship between the results predicted by the model and the real situation will appear in four cases. The first case is that the result predicted by the model based on the input sample is the same as the real sample and both are positive examples. In this case, this situation is called *True Positive* (TP). The second case is also that the prediction result is consistent with the real situation, but both are negative examples, and the situation at this time is called

True Negative (TN). The third situation is that the predicted result is different from the real situation, and the predicted result at this time is a negative example, but the real situation is a positive example, so it is called a false negative (False Negative, FN). The last case is predicted as a positive example, but the real situation is a negative example, this situation is also called false positive (False Positive, FP). According to the above analysis, it is not difficult to draw the confusion table shown in Table 2.

Table 2. Confusion table

Reality	Predict result	
	Positive	Negative
Positive	TP	FN
Negative	FP	TN

According to the above table, the indicators used in this paper can be further obtained. Among them, the calculation formula of the accuracy rate (Accuracy) is shown in the following formula (1)

$$Accuracy = \frac{TP + TN}{TP + TN + FP + FN} \tag{1}$$

The F1 value is determined by the precision and recall. Among them, the precision rate and recall rate can be calculated according to formulas (2) and (3):

$$Precision = \frac{TP}{TP + FP} \tag{2}$$

$$Recall = \frac{TP}{TP + FN} \tag{3}$$

The F1 value can be calculated according to formula (4):

$$F1 = 2 * \frac{Precision * Recall}{Precision + Recall} \tag{4}$$

3.4 Experimental Method

In order to further evaluate the method in this chapter, this paper divides the constructed duplicate requirement detection task dataset into ten parts, seven of which are used for training, and the remaining three are used for testing. Therefore, the training set used in this experiment contains 1736 samples, and the test set contains 744 samples. At the same time, this paper sets the ratio of positive samples and negative samples in the data of the two sets to 1:1.

During training, this paper firstly input all the features extracted from the lexical level, syntactic level and semantic level of the required sentence as the original feature input to the classifier. Then, in the process of parameter setting of SVM, this paper uses the accuracy rate as the scoring standard, and uses grid search and ten-fold cross-validation for parameter selection, and the final penalty coefficient C is 0.9.

In the experiment process, we will first use the calculation method proposed in this chapter to extract multiple similarity features from the lexical level, syntactic level and semantic level. Among them, the lexical level extracts three similarity features of word form, word frequency and word order, and the syntactic level extracts the semantic similarity features of three syntactic topics, namely, syntactic structure similarity, sentence subject, predicate and direct object. At the semantic level, TF-IDF is used to calculate the semantic similarity features of all words in the sentence.

When all features are extracted, combine them into a feature vector. Then, take these vectors as training data for the SVM and train it. Finally, the input is predicted according to the trained classifier, and the prediction results corresponding to all samples in the test set are counted. Calculate the accuracy rate and F1 value according to the real labels corresponding to the samples, and use them as the evaluation criteria for the repeated demand detection method based on multi-feature fusion proposed in this chapter.

At the same time, this experiment uses the VSM-based demand similarity calculation method proposed in [28] as the baseline model. Then, it is compared with the multi-feature fusion method proposed in this chapter. When testing the baseline model, this paper first calculates the similarity between each requirement pair in the test set sample based on the model. Then, by setting a certain threshold, the similarity of the two requirements is compared to it. If the similarity between the two requirements in the sample exceeds this threshold, the input requirement pair will be classified as a duplicate requirement pair; otherwise, it is classified as an unrelated requirement pair. Finally, calculate the accuracy and F1 value obtained by the model on the test set under different similarity thresholds, and compare the highest accuracy and F1 value obtained with the method proposed in this chapter.

3.5 Experimental Results and Analysis

According to the experimental method introduced in the previous section, the experimental results shown in Table 3 are further obtained:

Table 3. Experimental results

Model	Accuracy	F1
VSM	60.3	67.8
Methods in this chapter	67.5	72.4

According to the experimental results, it can be seen that the method based on the combination of multi-feature and SVM proposed in this paper achieves higher accuracy and F1 value than the method based on the VSM model proposed in the literature [28], which is increased by 4.6% and 7.2% respectively. This further shows that the method proposed in this chapter can achieve good results in repeated requirement detection. This paper analyzes that the reason why the method based on multi-feature fusion proposed in this chapter can achieve higher accuracy is that it can obtain more semantic information in the demand sentence, including lexical, syntactic and semantic information. However, the method based on VSM only considers the co-occurrence frequency of words in different requirements, and can obtain the semantic information taken is relatively small, so the accuracy rate is not high. For example, for some requirements with complex syntactic structure, the method based on the VSM model cannot obtain the similarity at the level of syntactic structure. The method based on multi-feature fusion can obtain a certain similarity of syntactic structure, thereby improving the accuracy of repeated demand detection.

4 Duplicate Demand Detection Based on Deep Learning

4.1 Model Structure

This section proposes a duplicate requirement detection model based on BERT and Tree-LSTM. The model adopts the Siamese network architecture in the architecture. The structure of the model is generally composed of four layers: input layer, encoding layer, feature fusion layer and classification layer. The architecture diagram of this model is shown in Fig. 1. The specific content of each level is described below:

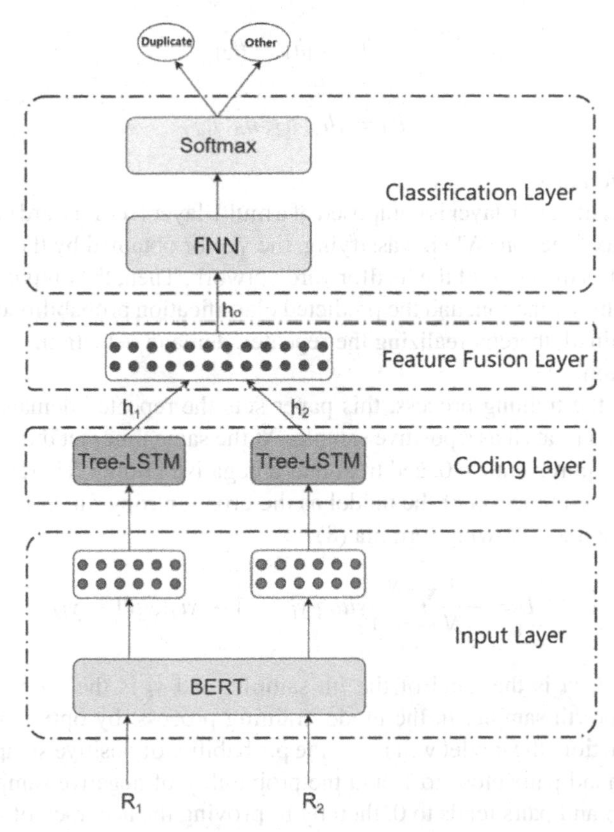

Fig. 1. Risk index calculation

(1) Input layer

Because BERT has strong language representation ability, this model uses the BERT model in the input layer to obtain the word vector representation of each word in the input demand sentence.

(2) Coding layer

Tree-LSTM is a tree-shaped long short-term memory network that can extract rich syntactic structure features, so this model uses Tree-LSTM in the coding layer to obtain the corresponding syntactic structure features. In the encoding process, first, the dependency syntax analysis is used to construct the corresponding dependency syntax tree. Then, a Child-Sum Tree-LSTM is used to represent the dependency syntax tree. Finally, after the Tree-LSTM model is calculated, the hidden state value of the root node is taken as the demand feature vector output by the encoding layer.

(3) Feature fusion layer

Assume that the inputs are the requirements $R1$ and $R2$, and the feature vectors obtained after the encoding layer are $h1$ and $h2$, respectively. In this paper, we consider the distance and angle of $h1$ and $h2$ as new features, and concatenate them with $h1$ and $h2$ to form a new feature vector h*o*. This process can be represented by the following Eqs. (5) to (6):

$$h_d = h_1 \odot h_2 \tag{5}$$

$$h_a = |h_1 - h_2| \tag{6}$$

$$h_O = (h_1, h_2, h_a, h_d) \tag{7}$$

(4) Classification layer

The classification layer is composed of a multi-layer feedforward neural network and Softmax function. When classifying, the vector obtained by the previous layer is first used as the input of the feedforward network. Then, the output is normalized by the Softmax function, and the predicted classification probability of the demand pair is obtained, thereby realizing the repeated demand detection.

(5) Loss function

During the training process, this paper sets the repeated demand pair sample label to 1 and treats it as a positive sample. At the same time, set the sample label of irrelevant requirements to 0, and treat it as a negative sample. Therefore, this paper defines the loss function of the model as the cross entropy function, which can be expressed by the following formula (8):

$$L = -\frac{1}{N}\sum\nolimits_{i=1}^{N} yilog\,\widehat{y_i} + (1 - yi)\log(1 - \widehat{y_i}) \tag{8}$$

Among them, *yi* is the label of the *i*th sample, and $\widehat{y_i}$ is the prediction result of the model for the *i*th sample. In the model training process, by optimizing the cross-entropy loss function, the model will make the probability of positive samples predicted as repeated demand pairs close to 1, and the probability of negative samples predicted as irrelevant demand pairs tends to 0, thereby improving the accuracy of the model. To sum up, for a given demand $R1$ and $R2$, the process of this model predicting whether the demand $R1$ and $R2$ is a repeated demand is as follows: First, $R1$ and $R2$ are respectively input into the BERT model, and the word vector representation of all the words in the required sentence is obtained. Secondly, use the dependency syntax analysis to construct the corresponding dependency syntax tree, and construct the corresponding Tree-LSTM according to the dependency syntax tree. Then, the word vector of the word corresponding to each node in the dependency tree is used as the input of each unit in the Tree-LSTM. After the Tree-LSTM model is calculated, the hidden states $h1$ and $h2$ corresponding to the root node are taken as the features of $R1$ and $R2$ respectively vector representation. At the same time, according to formula (5) and formula (6), the distance and angle between $h1$ and $h2$ are calculated respectively, and they are combined as new features with $h1$ and $h2$ as a new feature vector ho. Finally, take ho as the input of the feedforward network and combine the Softmax function to normalize the output of the feedforward neural network to output the classification result of the demand pair.

4.2 Dataset

This chapter uses the requirements dataset introduced in the previous chapter for the repetitive requirements detection task for the experimental evaluation. The dataset contains a total of 2480 samples, including 1240 positive and negative samples. At the same time, the data in the dataset is equally divided into ten parts, each containing 248 samples. Then, based on these data, a training set and a test set are constructed. Among them, the training set accounts for seven tenths, that is, contains 1736 samples. At this point, the remaining samples constitute the test set, which contains 744 samples. In addition, this paper sets the ratio of positive and negative samples in both parts to 1:1.

4.3 Experimental Method

During the experiment, the model adopts the parameters shown in Table 4:

Table 4. Model training parameter setting table

Parameter	Value
BERT version	BERT-base-uncased
Word vector dimension	768
Learning rate	5e−5
Batch_size	8
Epoch	10
Dropout	0.4
Loss function	Cross entropy
Optimization	Adam

The two models selected in this paper for comparison are Tree-LSTM [33] and BERT-Avg [32]. The experimental process of these two models is briefly described below:

(1) The Tree-LSTM model first obtains a word vector with a fixed dimension of 300 dimensions through the Glove [34] model, and then inputs the generated word vector into the Tree-LSTM for calculation, and obtains the feature vector representation of the demand statement. Next, the feature fusion proposed in this chapter is performed on the two feature vectors. Finally, the fused feature vector is used as the input of the multi-layer feed-forward neural network and the output of the feed-forward neural network is normalized by Softmax function to obtain final classification result.
(2) The BERT-Avg model first generates the word vector representation of all the words in the input requirement according to the BERT model, and sums and averages them to obtain the feature vector representation of the requirement sentence. Then, feature vectors are generated using the same feature fusion method as in the model in this chapter. At the same time, the feature vector is input into the same classification layer to get the final classification result.

4.4 Experimental Results and Analysis

According to the experimental method in the previous section, the experimental results shown in Table 5 are obtained in this paper:

Table 5 Experimental Results

Model	Accuracy	F1
Tree-LSTM	68.2	80.1
BERT-Avg	69.3	82.4
Our Model	75.1	87.6

As can be seen from Table 5 above, the BERT-Avg model is slightly better than the Tree-LSTM in the repetitive demand detection task, and the accuracy and F1 value are increased by 1.1% and 2.3% respectively, which also shows that the BERT model has better performance. Strong language representation ability. At the same time, the model in this chapter achieves better results than the BERT-Avg model in the detection of repeated requirements. Among them, the accuracy and F1 value are improved by 5.8% and 5.2% respectively compared with the BERT-Avg model. The results also show that the model can achieve good results in repeated demand detection. This paper analyzes that this model can achieve better results mainly because this model integrates the respective advantages of the other two models, and captures richer syntactic structure information while acquiring rich semantics. Therefore, when this model predicts the relationship between some demand pairs with complex syntactic structure and a large number of words, The feature vector of the demand statement obtained by this model at the coding layer can contain richer semantics, which can improve the accuracy and F1 value of the model.

The main research work of this paper is as follows:

(1) The method of duplicate demand detection based on multi-feature fusion is studied. Among the current duplicate requirement detection methods, most of the methods calculate the semantic similarity of requirements based on the co-occurrence frequency of words in different requirements, and then judge whether they are duplicated according to the semantic similarity between different requirements. Since the overall meaning of requirements includes many aspects such as lexical, syntactic, and semantic aspects, the previous detection methods for repeated requirements only consider the co-occurrence frequency of words in requirements from the perspective of statistical theory, which has certain limitations. Therefore, this paper considers the similarity features of multiple aspects in the lexical layer, syntactic layer and semantic layer, and combines SVM to predict whether the demand pair is repeated and realizes the detection of repeated demand.
(2) Research on the duplicate demand detection model based on deep learning

Compared with traditional feature engineering, the neural network model can more easily extract the semantic features of each dimension of the sentence and obtain richer semantic information. Therefore, based on the technology of neural network and deep learning, this paper proposes a repeated demand detection model combining BERT and Tree-LSTM, and through experiments, it is verified that the model can achieve good results in repeated demand detection.

5 Conclusion

Duplicate requirement detection is a basic task in requirement analysis, which is beneficial to reduce redundant development, save software development cost and improve software requirement quality. With the rapid development of the software industry, the number of requirements contained in the software is also increasing. At this time, the requirement analyst will consume a lot of energy and be prone to errors in the process of repeated requirement detection, which will affect subsequent software development. Therefore, this paper summarized the duplicate demand detection methods proposed by the predecessors and combined natural language processing and deep learning methods to propose two different duplicate demand detection methods. At the same time, this paper designed and implemented a duplicate demand detection system, which can assist humans to quickly and accurately complete the repetitive demand detection task.

References

1. Hofmann, H.F., Lehner, F.: Requirements engineering as a success factor in software projects. IEEE Softw. **18**(4), 58–66 (2001)
2. Qiu, M., Zhang, K., Huang, M.: Usability in mobile interface browsing. Web Intell. Agent Syst. J. **4**(1), 43–59 (2006)

3. Qiu, M., Li, H., Sha, E.: Heterogeneous real-time embedded software optimization considering hardware platform. ACM Symposium on Applied Computing, pp. 1637–1641 (2009)
4. Gai, K., Qiu, M., Chen, L., Liu, M.: Electronic health record error prevention approach using ontology in big data. In: IEEE 17th HPCC (2015)
5. Qiu, M., Yang, L., Shao, Z., Sha, E.: Dynamic and leakage energy minimization with soft real-time loop scheduling and voltage assignment. IEEE TVLSI **18**(3), 501–504 (2009)
6. Qiu, M., Jia, Z., et al.: Voltage assignment with guaranteed probability satisfying timing constraint for real-time multiproceesor DSP. J. Sig. Process. Syst. (2007)
7. Qiu, M., Xue, C., Shao, Z., Sha, E.: Energy minimization with soft real-time and DVS for uniprocessor and multiprocessor embedded systems. In: IEEE DATE Conference, pp. 1–6 (2007)
8. Niu, J., Gao, Y., Qiu, M., Ming, Z.: Selecting proper wireless network interfaces for user experience enhancement with guaranteed probability. JPDC **72**(12), 1565–1575 (2012)
9. Qiu, H., Dong, T., Zhang, T., Lu, J., Memmi, G., Qiu, M.: Adversarial attacks against network intrusion detection in IoT systems. IEEE IoT J. **8**(13), 10327–10335 (2020)
10. Qiu, M., Xue, C., Shao, Z., et al.: Efficient algorithm of energy minimization for heterogeneous wireless sensor network. IEEE EUC, pp. 25–34 (2006)
11. Gai, K., Qiu, M., Elnagdy, S.: A novel secure big data cyber incident analytics framework for cloud-based cybersecurity insurance. IEEE BigDataSecurity (2016)
12. Li, Y., Gai, K., et al.: Intercrossed access controls for secure financial services on multimedia big data in cloud systems. ACM Trans. Multimedia Comput. Commun. Appl. **12**(4s), 1–18 (2016). https://doi.org/10.1145/2978575
13. Qiu, M., Chen, Z., Ming, Z., Qin, X., Niu, J.: Energy-aware data allocation with hybrid memory for mobile cloud systems. IEEE Syst. J. **11**(2), 813–822 (2014)
14. Hu, F., Lakdawala, S., et al.: Low-power, intelligent sensor hardware interface for medical data preprocessing. IEEE Trans Inform. Technol. Biomed. **13**(4), 656–663 (2009). https://doi.org/10.1109/TITB.2009.2023116
15. Li, J., Ming, Z., et al.: Resource allocation robustness in multi-core embedded systems with inaccurate information. J. Syst. Architect. **57**(9), 840–849 (2011)
16. Qiu, M., Qiu, H.: Review on image processing based adversarial example defenses in computer vision. IEEE 6th Conference on BigDataSecurity, pp. 94–99. USA (2020)
17. Qiu, M., Qiu, H., et al.: Secure data sharing through untrusted clouds with blockchain-enabled key management. IEEE SmartBlock 2020, pp. 11–16. China (2020)
18. Gai, K., Zhang, Y., Qiu, M., Thuraisingham, B.: Blockchain-enabled service optimizations in supply chain digital twin. IEEE Trans. Serv. Comput. 1–12 (2022). https://doi.org/10.1109/TSC.2022.3192166
19. Qiu, H., Zheng, Q., et al.: Topological graph convolutional network-based urban traffic flow and density prediction. IEEE Trans. Intell. Transport. Syst. **22**(7), 4560–4569 (2021). https://doi.org/10.1109/TITS.2020.3032882
20. Gao, X., Qiu, M.: Energy-based learning for preventing backdoor attack. In: Memmi, G., Yang, B., Kong, L., Zhang, T., Qiu, M. (eds.) Knowledge Science, Engineering and Management: 15th International Conference, KSEM 2022, Singapore, August 6–8, 2022, Proceedings, Part III, pp. 706–721. Springer International Publishing, Cham (2022). https://doi.org/10.1007/978-3-031-10989-8_56
21. Dag, J.N., Regnell, B., Carlshamre, P., et al.: A feasibility study of automated natural language requirements analysis in market-driven development. Requirem Eng. **7**(1), 20–33 (2002)
22. Dag, J.N., Regnell, B., Gervasi, V., et al.: A linguistic-engineering approach to large-scale requirements management. IEEE Softw. **22**(1), 32–39 (2005)

23. Dag, J.N., Thelin, T., Regnell, B.: An experiment on linguistic tool support for consolidation of requirements from multiple sources in market-driven product development. Empir. Softw. Eng. **11**(2), 303–329 (2006)
24. Salton, G.: Automatic text processing: the transformation, analysis, and retrieval of, p. 16. Addison-Wesley, Reading (1989)
25. Niwattanakul, S., Singthongchai, J., Naenudorn, E., et al.: Using of Jaccard coefficient for keywords similarity. Intl. Conf. Eng. Comput. Sci. **1**(6), 380–384 (2013)
26. Hayes, J.H., Dekhtyar, A., Sundaram, S.K.: Advancing Candidate Link Generation for Requirements Tracing: The Study of Methods. IEEE Press (2006)
27. Falessi, D., Briand, L.C., Cantone, G.: The impact of automated support for linking equivalent requirements based on similarity measures, Simula Research Labartery. Technical Report (2009)
28. Falessi, D., Cantone, G., Canfora, G.: A comprehensive characterization of NLP techniques for identifying equivalent requirements. In: Proceedings of the 2010 ACM-IEEE (2010)
29. Falessi, D., Cantone, G., Canfora, G.: Empirical principles and an industrial case study in retrieving equivalent requirements via natural language processing techniques. IEEE Trans. Software Eng. **39**(1), 18–44 (2011)
30. Mezghani, M., Kang, J., Sèdes, F.: Industrial requirements classification for redundancy and inconsistency detection in SEMIOS. In: 2018 IEEE 26th International Requirements Engineering Conference (RE), pp. 297–303 (2018)
31. Ferrari, A., Spagnolo, G.O., Gnesi, S.: Pure: a dataset of public requirements documents. In: IEEE 25th International Requirements Engineering Conf. (RE), pp. 502–505 (2017)
32. Reimers, N., Gurevych, I.: Sentence-bert: Sentence embeddings using siamese bert-networks. arXiv preprint arXiv:1908.10084 (2019)
33. Tai, K.S., Socher, R., Manning, C.D.: Improved semantic representations from tree-structured long short-term memory network. arXiv preprint arXiv:1503.00075 (2015)
34. Pennington, J., Socher, R., Manning, C.D.: Glove: Global vectors for word representation. In: Conference on Empirical Methods in Natural Language Processing (EMNLP), pp. 1532–1543 (2014)

Bayesian Causal Mediation Analysis with Longitudinal Data

Yu Zhang[1], Lintao Yang[1], Fuhao Liu[1(✉)], Lei Zhang[2], Jingjing Zheng[2], and Chongxi Zhao[2]

[1] Central China Normal University, Wuhan 430079, China
ltaoyang@mail.ccnu.edu.cn, 1634820263@qq.com
[2] Wuhan Fiberhome Technical Services Co., Ltd., Wuhan 430073, China
{leizhang,jjzheng,chxzhao}@fiberhome.com

Abstract. Mediation analysis was concerned with the decomposition of the total effect of exposure on the outcome into the indirect effects and the remaining indirect effects, through a given mediation. However, when longitudinal data including time varying exposure and mediator variables, the estimated causal effects are affected by time varying confounders. Standard generalized linear equations did not give unbiased estimates. In this paper, we introduced inverse probability weighting technique to adjust such time varying confounders. Considering that the amount of data may be small and the distribution is not uniform, we decide to use Bayesian Inference to estimate the *Structural Equation Model* (SEM) parameters, and finally estimates the causal effect through counterfactual thought. This paper summarized the relevant theoretical knowledge of this method, verified the feasibility of this method by using the simulated data, and compared the performance of different methods.

Keywords: Bayesian · Longitudinal data · Time-varying causal mediation analysis · Structural equation model

1 Introduction

With the rapid development of information technology [1–3] and cloud computing [4–6], large amounts of data [7–9] had been generated and processed. To get the information efficiently [10], accurately [11], and safely [12–14], new mechanisms for big data analytics are urgently needed. Machine learning and artificial intelligence (AI) [15–17] have become one of the hot areas of research, which can advance many related applications such as image processing, natural language processing, and autonomous driving. Mediation analysis is one of the important topics of big data analysis and prediction [18–20].

Exposure often acts directly or indirectly on the outcome of interest, mediated by some mediators. Identification and quantification of these two effects contribute to further understanding of potential causal mechanisms. In the literature on causal inference, many advances have been made in causal mediation analysis in non-longitudinal settings [21–23]. However, the above literature does not allow exposure and mediator to change over

M. Qiu et al. (Eds.): SmartCom 2022, LNCS 13828, pp. 226–235, 2023.
https://doi.org/10.1007/978-3-031-28124-2_21

time. The main reason is that in the presence of time varying confounders, it is difficult to identify causal effects, especially the confounder of mediator-outcome relationships affected by exposure [24]. These confusions cannot be treated by standard methods, but can be adjusted by using inverse probability weighting or g-formula [25]. However, there are other problems besides time-varying confounding. For example, standard linear regression and other machine learning model learning parameters are derived from data [26]. When the data volume is small or the distribution is not uniform, the degree of estimation bias may be higher. Because the above method cannot make causal inference for data more accurately based on prior information (such as model parameter distribution and data obedience distribution). At the same time, for this problem, there is a lack of literature on causal mediation analysis under longitudinal setting.

In this paper, we extend the bayesian model to longitudinal causal mediation analysis, adjust the time-varying confounders with the inverse probability weighting method, conduct path analysis with the structural equation model, and estimate the direct, indirect and interactive effects of time varying exposures on the results under the counterfactual framework. At the same time, we also test experiments to analyze the above problems. This paper provides a simple and general procedure for causal mediation analysis of longitudinal data, suitable for researchers who are not familiar with causal mediation analysis, and for problems with small data sets and uneven distribution. We believe that bayesian mediation analysis will be the basis for the development of time varying exposures and mediation evaluation.

The rest of this paper is organized as follows; in Sect. 2, we present structural equation model, bayesian model and the estimation of direct, indirect, and interactive effect. Section 3, we present our we present our experiment and results. At last, in Sect. 4 we conclude the paper and propose our future work.

2 Method

2.1 Structural Equation Model

Structural equation model (SEM) is a powerful multivariate technique used to examine and evaluate multivariate causal relationships. At the beginning, SEM was mainly applied to build causal model and do path analysis [27]. Path analysis aims to quantify causality among multiple variables. In this section, I will introduce the basic principle and application of structural equation model by using simple SEM composed of nonparametric linear equations.

In order to extend SEM methods to cover all variable types, linear and nonlinear cases, we need to separate the concept of "effect" from the coefficient in the equation and redefine "effect" as the general ability to transfer changes between variables. In Fig. 1, for example, there are two variables Y and X, each corresponding to linear function variables, including causal influence the direction of the arrows indicate the assumption that the image coding (direct) possible causal impact, and there is no causal impact, coding the quantitative relationship between the relevant variables and equations, will determine from the data. The "path coefficient" β quantifies the (direct) causal effect on Y. The coefficient β refers to that if you increase the value of X by one, you increase the value of X by one.

The variables U_X and U_Y are called "exogenous"; They represent observed or unobserved influence factors, factors that the researchers cannot explain, and factors that are constantly changing but not affected by other variables in the model. At the same time, the exogenous variables not observed in SEM, sometimes called "disturbance" or "error", are fundamentally different from the residual term in the regression equation. Residual error ϵ_x in regression equation is the error term obtained by analysis and is not related to other variables in regression. Exogenous variables such as U_Y and U_X are shaped by physical realities (e.g., genetic factors, socioeconomic conditions) rather than by artificial analysis. They can be defined as any other variable, and although we cannot measure them, we must acknowledge their existence and evaluate them qualitatively in relation to other variables in the system. Generally, dashed lines are used to show the relationship between exogenous variables and explanatory variables. In this case, U_X and U_Y are independent and subject to arbitrary distribution. Each of the linear functions (1) represents a causal process (or mechanism) that can determine the value of the right-hand variable (output) from the left-hand variable (input).

When reading a pathway map, parents, children, ancestors, and offspring. For example, the arrow in $X \rightarrow Y$ specifies the parent of Y, and a child of X. "Path" is a sequence of any connected edges represented by solid or dashed lines. For example, in Fig. 1, there are two paths between X and Y that can be expressed in the form of a set, $\{X - Y, X - U_X - U_Y - Y\}$, so that the path is not oriented.

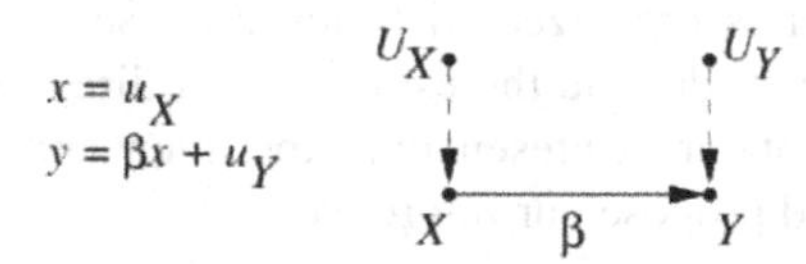

Fig. 1. A causal diagram with time-varying confounders affected by exposures

2.2 Bayesian Model

Under Bayes' formula, statistical inference of unknown parameters is based on a posterior distribution, which is obtained by Bayes' principle as follows:

$$\mathrm{pr}(\beta|data) \propto \mathrm{pr}(\beta)\mathrm{pr}(data|\beta) \tag{1}$$

Where, $\mathrm{pr}(\beta)$ is the prior distribution of parameter β. And $\mathrm{p}(data|\beta)$ is the likelihood function of observed data. The prior distribution $\mathrm{pr}(\beta)$ quantifies our prior knowledge of the distribution β [28]. The posterior distribution $\mathrm{pr}(\beta|data)$ summarizes the latest information of the data pairs we observe.

In the previous section, we illustrate the relevant principles of the SEM. Figure 1, where a causal effect on Y. Suppose the datasets D $= (x_i, y_i), i = 1, 2, \ldots, n$, represents a sample, and Y obey the following formula relationship:

$$y = \beta \cdot x + \varepsilon + U_Y \tag{2}$$

The parts of the formula have been introduced in the previous section and not much elaborated. Here, U_Y is assumed to obey a normal distribution.

In statistical inference using Bayesian method, the most important thing is to determine the prior distribution. Prior distribution is both the starting point and the core problem to ensure accurate statistical inference. Prior distribution is both the starting point and the core problem to ensure accurate statistical inference. Prior distribution pr(β) is the recognition of the possible value of parameter β before sampling the sample. After obtaining the sample, since the sample contains information of β, the recognition of β changes. After adjusting the value of, the posterior distribution p(β) of parameter β is obtained. According to The Bayesian formula, the prior distribution of β can be determined, assuming that it obeys the normal distribution, i. e.

$$\mathrm{pr}(\beta) = \frac{1}{\sqrt{2\pi}\sigma_\beta} \cdot \exp\left(\frac{\beta^T\beta}{2\sigma_\beta^2}\right) \tag{3}$$

After determining the prior distribution, we need to know the likelihood function of parameter β. Likelihood function refers to the probability of data set D given parameter $\mathrm{p}(D|\beta)$, that is, $\mathrm{p}(D|\beta)$ can also be written as $\mathrm{p}(Y|X, \beta)$. Since U_Y obeys the normal distribution, the likelihood here also obeys the normal distribution, and the expression is:

$$\mathrm{pr}(Y \mid X, \beta) = \prod\nolimits_{i=1}^{N} \mathrm{pr}(y_i \mid x_i, \beta) \tag{4}$$

Based on the prior distribution and likelihood function, the posterior distribution expression of parameter β can be obtained by Bayesian formula:

$$\mathrm{pr}(\beta|X, Y) \propto \prod\nolimits_{i=1}^{N} \mathrm{pr}(y_i \mid x_i, \beta) \cdot \frac{1}{\sqrt{2\pi}\sigma_\beta} \cdot \exp\left(\frac{\beta^T\beta}{2\sigma_\beta^2}\right) \tag{5}$$

The template is used to format your paper and style the text. All margins, column widths, line spaces, and text fonts are prescribed; please do not alter them. You may note peculiarities. For example, the head margin in this template measures proportionately more than is customary. This measurement and others are deliberate, using specifications that anticipate your paper as one part of the entire proceedings, and not as an independent document. Please do not revise any of the current designations.

2.3 Estimation of Direct, Indirect, and Interactive Effects

Suppose now that the exposure, mediators and possible confounders vary over time. $\overline{A}(t) = \{A(1), A(2), ..., A(t)\}$, $\overline{M}(t) = \{M(1), M(2), ..., M(t)\}$, $\overline{L}(t) = \{L(1), L(2), ..., L(t)\}$ denote the variable of interest measured at time points 0,1,…,T. The relationships between the variables are given in Fig. 2. The effect of observed confounders on exposure variables can be removed by inverse probability weighting (IPTW). However, when the sample size of some exposure or mediator values is small, the weight estimation of exposure variables or mediator variables will be very large, which will affect the final analysis results. To solve this problem, stable weights (SW)

are used to estimate the weights of the mediator variable and the outcome variable [29]. Our proposed algorithm mainly includes the following three steps: 1) The IPTW method was used to assign stability weights to the data samples and adjust the influence of time-varying confounding variables; 2) SEM was constructed and the parameters of SEM were estimated by Bayesian model; 3) causal effects were estimated by the effect estimation formula, including direct, indirect and interactive effects.

Step 1. Firstly weighted the sample data, and estimated the stability weights (*SW*) of the mediator and outcome by using Eqs. (6) and (7), so as to adjust the influence of time-varying confounding.

To estimate the *SW* of the outcome variable Y:

$$sw_y = \frac{pr\left(M(t)|\overline{A}(t), \overline{M}(t-1)\right)}{pr\left(M(t)|\overline{A}(t), \overline{M}(t-1), \overline{L}(t-1), V\right)} \times \frac{pr\left(A(t)|\overline{A}(t-1), \overline{M}(t-1)\right)}{pr\left(A(t)|\overline{A}(t-1), \overline{M}(t-1), \overline{L}(t-1), V\right)} \tag{6}$$

To estimate the *SW* of the mediator variable M

$$sw_m = \frac{\mathrm{pr}\left(A(t)|\overline{\mathrm{A}}(\mathrm{t}-1)\right)}{\mathrm{pr}\left(A(t)|\overline{\mathrm{A}}(\mathrm{t}-1), \overline{\mathrm{M}}(\mathrm{t}-1), \overrightarrow{L}(\mathrm{t}-1), \mathrm{V}\right)} \tag{7}$$

The estimated *SW* can be used for parameter estimation of conditional probability by logistic regression.

Step 2. After processing the time varying confounders with inverse probability weighting, the structural equation model can be used to construct the basic function for the outcome *Y* and the intermediate *M*. Considering that *Y* is a continuous type, we assume that the model for *Y* is:

$$E\left[Y_{\overline{AM}}\right] = \lambda_0'\overline{A}(t) + \lambda_1'\overline{M}(t) + \lambda_2\mathrm{cum}\left(\overline{A}(t)\right)\cdot\mathrm{cum}\left(\overline{M}(t)\right) + \lambda_3'\overline{L}(t) + \lambda_4'V + \lambda_5 \tag{8}$$

$\mathrm{cum}\left(\overline{M}(t)\right) = \sum_{t=1}^{T} M(t)$ and $\mathrm{cum}\left(\overline{A}(t)\right) = \sum_{t=1}^{T} A(t)$ refer to are the cumulative totals of $\overline{A}(t)$ and $\overline{M}(t)$ respectively, T is the maximum follow-up time. Construct a SEM for the mediators *M*,

$$\mathrm{logit}\left(E\left[M_{\overline{A}}(t)\right]\right) = \gamma_0(t)'\overline{A}(t) + \gamma_1(t)'\overline{M}(t-1) + \gamma_2(t)'\overline{L}(t) + \gamma_3(t)'\overline{V} + \gamma_4(t) \tag{9}$$

We can use the Bayesian model mentioned in Section Bayesian model to fit the formulas 8 and 9. Then the potential outcome $E\left[Y_{\overline{a}\overline{G_a}}\right]$ of exposure $\overline{A}(t)$ can be obtained by substituting $E\left[M_{\overline{A}}(t)\right]$ in $E\left[Y_{\overline{AM}}\right]$. Direct effects, indirect effects, interaction effects, and total effects can be estimated using the formula of causal effect estimation [30].

Step 3. To avoid variable type inconsistency, we assumed that the outcome variable is continuous and exposures is binary. The causal effect referred to in this paper includes direct effect, indirect effect, interaction effect and total effect. The formula of the causal effect is shown below, where $\overline{a}^*$ represents exposure $\overline{A}(t)$ with a value set to 0 and $\overline{a}$ represents exposure $\overline{A}(t)$ with a value set to 1.

Control direct effect: $\text{CDE}(\overline{m}^*) = E(Y_{\overline{a}\overline{m}^*} - Y_{\overline{a}^*\overline{m}^*})$, is equal to a function of $\overline{m}^*$;
Reference interaction effect: $\text{INT}_{\text{ref}}(\overline{m}^*) = \text{E}\Big[\Big(Y_{\overline{a}G_{\overline{a}^*}} - Y_{\overline{a}^*G_{\overline{a}^*}}\Big) - -(Y_{\overline{a}\overline{m}^*} - Y_{\overline{a}^*\overline{m}^*})\Big]$;
Mediated interaction effect: $\text{INT}_{\text{med}} = E\Big(Y_{\overline{a}G_{\overline{a}}} - Y_{\overline{a}G_{\overline{a}^*}} - Y_{\overline{a}^*G_{\overline{a}}} + Y_{\overline{a}^*G_{\overline{a}^*}}\Big)$;
Pure indirect effect: $\text{PIE} = (Y_{\overline{a}^*G_{\overline{a}}} - Y_{\overline{a}^*G_{\overline{a}^*}})$;
Total effect: $\text{TE} = (Y_{\overline{a}G_{\overline{a}}} - Y_{\overline{a}^*G_{\overline{a}^*}})$.

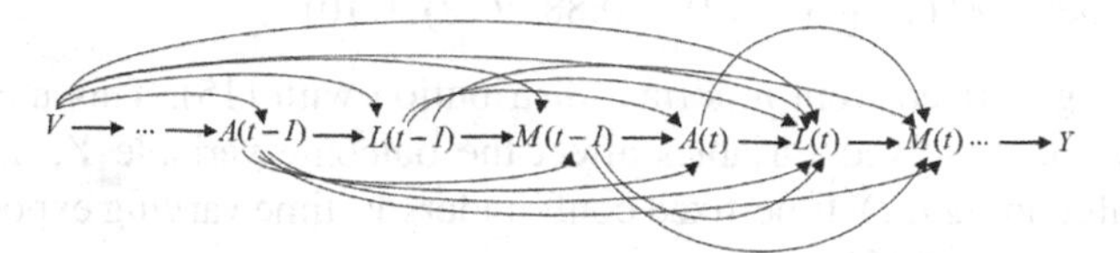

Fig. 2. A causal diagram with Time varying variables $\overline{A}(t)$, $\overline{L}(t)$, $\overline{M}(t)$

3 Experiments and Results

3.1 Simulated Data

The volume of simulated data sample is N, including time-varying exposures $A(t)$, time-varying mediators variable $M(t)$, time-varying confounders $L(t)$, baseline confounders V and outcome variable Y. The measurement times of time-varying exposure, time-varying intermediary and time-varying confounder are T = 2. Assuming no unobserved confounding variables, we simulate data on a sample of N subjects as follows:

- $V = \{V_0, V_1, V_2\}$ is drawn from Bernoulli distribution or Normal distribution. That is, baseline confounder is randomized and not affected by other variables. Where $V_0 \sim B(0.5)$, $V_1 \sim B(0.43)$, $V_2 \sim Normal(15, 5)$;
- Then conditional on $V, A(1)$ is drawn from a Bernoulli distribution with formula (1). That is, the exposure variable $A(1)$ is affected by baseline confounding, including time-varying and time-invariant.

$$B(0.05 \cdot V_0 + 2.42 \cdot V_1 + 1.41 \cdot V_2 - 20) \tag{10}$$

- Conditional on V and $A(1)$, $L(1)$ is drawn from a Bernoulli distribution with (11).

$$Normal(2.88 \cdot V_0 + 0.92 \cdot V_1 + 0.31 \cdot V_2 + 2.99 \cdot A(1) + 14, 1) \tag{11}$$

- Conditional on V, $A(1)$ and $L(1)$, $M(1)$ is drawn from a Bernoulli distribution with (12).

$$B(1.19 \cdot V_0 - 2.17 \cdot V_1 - 1.24 \cdot V_2 + 2.5 \cdot A(1) + 0.13 \cdot L(1) + 13) \tag{12}$$

- Conditional on V, $A(1)$, $L(1)$ and $M(1)$, $A(2)$ is drawn from a Bernoulli distribution with (13).

$$B(0.5 \cdot V_0 + 0.17 \cdot V_1 - 2.62 \cdot V_2 + 1.71 \cdot A(1) + 2.33 \cdot L(1) - 0.62 \cdot M(1) - 9) \tag{13}$$

- Conditional on V, $A(1)$, $L(1)$, $M(1)$ and $A(2)$, $L(2)$ is drawn from a Bernoulli distribution with (14).

$$Normal(-1.42 \cdot V_0 - 0.34 \cdot V_1 + 1.75 \cdot V_2 + 0.71 \cdot A(1) - 0.23 \cdot L(1)) \quad (14)$$

- Conditional on $V, A(1), L(1), M(1)$ $A(2)$ and $L(2)$, $M(2)$ is drawn from a Bernoulli distribution with (15).

$$Normal(0.50 \cdot V_0 - 2.49 \cdot V_1 + 0.03 \cdot V_2 + 2 \cdot A(1) - 0.64 \cdot L(1) - 0.64 \cdot M(1) + 3 \cdot A(2) + 0.88 \cdot L(2) + 10) \quad (15)$$

- Finally, Y is generated from a normal distribution with (15). This means that in a causal DAG, all observed variables affect the outcome variable Y, including time varying confounding $L(t)$, time-fixed confounders V, time varying exposure $A(t)$ and time varying mediator $M(t)$.

$$Normal(3.53 \cdot V_0 - 0.89 \cdot V_1 - 0.24 \cdot V_2 + 2.1 \cdot A(1) - 2.31 \cdot L(1) - 1.3 \cdot M(1) + 2.1 \cdot A(2) - 0.97 \cdot L(2) + 2.5 \cdot M(2) + 16, 0.32) \quad (16)$$

Figure 1 represents the data generating process for the simulation dataset. According to the causal relationship between the variables indicated by the arrow in Fig. 1, we simulated the data generation process with Python code.

3.2 Experiments and Results

The simulated data used in this section are as described in Sect. 3.1. It is assumed that there are no missing items in the data set, and the causal effect estimation is performed on the simulated data using the method proposed in this paper.

The method used in Sect. 2.3 in this paper is based on A counterfactual framework to estimate the causal effect of time-varying exposure A (t) on result Y. The direct effect is *control direct effect* (CDE), and the indirect effect is *pure indirect effects* (PIE). Interaction effects include mediated interaction effect (INT_{med}) and Reference interaction effect (INT_{ref}). Here, the mathematical meaning of causal effect is risk difference, also known as RD value. In general, the significance of the RD value can be summarized as follows (assuming the outcome is a continuous variable): RD = 0, there is no association between exposure and outcome; RD > 0, the risk of exposure to the outcome increases and exposure is positively correlated with the outcome; RD < 0, the risk of exposure to the outcome decreases and exposure is negatively correlated with the outcome.

The experimental part mainly includes two parts. The first part tests for different sample sizes and compares the accuracy of the estimated effect values of the proposed method under different sample sizes. Sample sizes were set as 200,600 and 1000, respectively. The second part is to compare the difference of estimated effect values between different models, including linear regression model and random forest model. There are many types of machine learning models, but most tend to predict large data samples and ignore the ability to interpret the results. We compute causal effects by counterfactual, which requires machine learning models with strong interpretability. Here we choose the linear regression model with strong interpretability and the random forest model with weak interpretability to make a comparison. The experimental test process is as follows:

1) Computing the true value. The true value in the experiment is directly calculated according to the formula of causal effect in Sect. 2.3, including direct effect, indirect effect, interactive effect and total effect.
2) To estimate the effect value and generate simulation data set, we use different models to estimate the causal effect. Experiment repeats 50 times independently, using the formula bias $= |\frac{1}{100}\sum_{n=1}^{100} CE^n{}_{est} - CE_{true}|$ and true value to get the average effect of bias.$CE^n{}_{est}$ refers to the estimate for the Nth experiment (Table 1)

Table 1. The results of the simulated dataset with the method proposed

		N = 200		N = 600		N = 1000	
Causal effect	Method	CE_{est}	bias	CE_{est}	bias	CE_{est}	bias
CDE	Linear model	4.18	0.02	4.19	0.01	4.19	0.01
	Random forest	5.68	1.48	4.89	0.69	4.38	0.18
	Bayesian model	4.20	0.00	4.20	0.00	4.20	0.00
INT_{ref}	Linear model	1.45	1.26	0.83	0.64	0.62	0.43
	Random forest	2.61	2.42	2.6	2.41	2.81	2.62
	Bayesian model	0.52	0.29	0.31	0.12	0.28	0.09
INT_{med}	Linear model	1.04	1.67	1.90	0.80	2.17	0.53
	Random forest	0.83	1.88	1.45	1.26	1.82	0.89
	Bayesian model	2.42	0.29	2.70	0.00	2.70	0.00
PIE	Linear model	0.89	1.53	1.66	0.75	1.95	0.47
	Random forest	0.03	2.38	0.09	2.32	0.06	2.35
	Bayesian model	2.20	0.21	2.42	0.01	2.45	0.03
TE	Linear model	7.55	1.96	8.58	0.92	8.93	0.58
	Random forest	9.15	0.36	9.03	0.48	9.07	0.44
	Bayesian model	9.30	0.21	9.64	0.13	9.63	0.12

4 Conclusion

Duplicate requirement detection is a basic task in requirement analysis, which is beneficial to reduce redundant development, save software development cost and improve software requirement quality. With the rapid development of the software industry, the number of requirements contained in the software is also increasing. At this time, the requirement analyst will consume a lot of energy and be prone to errors in the process of repeated requirement detection, which will affect subsequent software development. Therefore, this paper summarized the duplicate demand detection methods proposed by the predecessors and combined natural language processing and deep learning methods to propose two different duplicate demand detection methods. At the same time, this paper designed and implemented a duplicate demand detection system, which can assist humans to quickly and accurately complete the repetitive demand detection task.

Acknowledgment. This work was supported by the key research and development project of Hubei Province "Research and Application of key Technologies of Intelligent Operation and maintenance and data security for 5G Micro data Center", project number: 2020BAA001. We were grateful for the participation of all researchers and thanked project funding.

References

1. Qiu, M., Jia, Z., et al.: Voltage assignment with guaranteed probability satisfying timing constraint for real-time multiprocessor DSP. J. Signal Proc. Systems (2007)
2. Qiu, M., Yang, L., et al.: Dynamic and leakage energy minimization with soft real-time loop scheduling and voltage assignment. IEEE Trans. VLSI **18**(3), 501–504 (2009)
3. Qiu, M., Li, H., Sha, E.: Heterogeneous real-time embedded software optimization considering hardware platform. ACM Symposium on Applied Computing, pp. 1637–1641 (2009)
4. Niu, J., Gao, Y., et al.: Selecting proper wireless network interfaces for user experience enhancement with guaranteed probability. JPDC **72**(12), 1565–1575 (2012)
5. Qiu, M., Chen, Z., Ming, Z., Qin, X., Niu, J.: Energy-aware data allocation with hybrid memory for mobile cloud systems. IEEE Syst. J. **11**(2), 813–822 (2014)
6. Qiu, M., Xue, C., Shao, Z., et al.: Efficient algorithm of energy minimization for heterogeneous wireless sensor network. IEEE EUC, pp. 25–34 (2006)
7. Gai, K., Qiu, M., Chen, L., Liu, M.: Electronic health record error prevention approach using ontology in big data. In: IEEE 17th HPCC (2015)
8. Gai, K., Qiu, M., Elnagdy, S.: A novel secure big data cyber incident analytics framework for cloud-based cybersecurity insurance. IEEE BigDataSecurity (2016)
9. Zhang, K., Kong, J., et al.: Multimedia layout adaptation through grammatical specifications. Multimedia Syst. **10**(3), 245–260 (2005)
10. Qiu, M., Xue, C., Shao, Z., Sha, E.: Energy minimization with soft real-time and DVS for uniprocessor and multiprocessor embedded systems. IEEE DATE Conference, pp. 1–6 (2007)
11. Li, J., Ming, Z., et al.: Resource allocation robustness in multi-core embedded systems with inaccurate information. J. Syst. Architect. **57**(9), 840–849 (2011)
12. Gai, K., Zhang, Y., et al.: Blockchain-enabled service optimizations in supply chain digital twin. IEEE Trans. Serv. Comput. (2022)
13. Li, Y., Gai, K., et al.: Intercrossed access controls for secure financial services on multimedia big data in cloud systems. ACM Trans. MCCA (2016)
14. Qiu, M., Qiu, H., et al.: Secure Data Sharing Through Untrusted Clouds with Blockchain-enabled Key Management. The 3rd SmartBlock, pp. 11–16, China (2020)
15. Qiu, H., Zheng, Q., et al.: Topological graph convolutional network-based urban traffic flow and density prediction. IEEE Trans. ITS (2020)
16. Qiu, M., Qiu, H.: Review on image processing based adversarial example defenses in computer vision. In: IEEE 6th International Conference BigDataSecurity, pp. 94–99, USA (2020)
17. Qiu, H., Dong, T., et al.: Adversarial attacks against network intrusion detection in IoT systems. IEEE Internet Things J. **8**(13), 10327–10335 (2020)
18. Gao, X., Qiu, M.: Energy-Based Learning for Preventing Backdoor Attack. KSEM , vol. 3, pp. 706–721 (2022)

19. Zhang, L., Qiu, M., Tseng, W., Sha, E.: Variable partitioning and scheduling for MPSoC with virtually shared scratch pad memory. J. Sign. Process. Syst. **58**(2), 247–265 (2010)
20. Hu, F., Lakdawala, S., et al.: Low-power, intelligent sensor hardware interface for medical data preprocessing. IEEE Trans. Infor. Tech. Biomed. **13**(4), 656–663 (2009)
21. Caleb, H.M., et al.: On partial identification of the pure direct effect. arXiv preprint arXiv: 1509.01652 (2015)
22. Tyler, J.V.W.: A unification of mediation and interaction: a four-way decomposition. Epidemiology, vol. 25.5, p. 749. Cambridge, Mass (2014)
23. Judea, P.: Direct and indirect effects. Probabilistic and Causal Inference: The Works of Judea Pearl, pp. 373–392 (2022)
24. Mittinty, M.N., Vansteelandt, S.: Longitudinal mediation analysis using natural effect models. Am. J. Epidemiol. **189**(11), 1427–1435 (2020)
25. Clare, P.J., Dobbins, T.A., Mattick, R.P.: Causal models adjusting for time-varying confounding—a systematic review of the literature. Int. J. Epidemiol. **48**(1), 254–265 (2019)
26. Baldwin, S.A., Larson, M.J.: An introduction to using Bayesian linear regression with clinical data. Behav. Res. Ther. **98**, 58–75 (2017)
27. Fan, Y., et al.: Applications of structural equation modeling (SEM) in ecological studies: an updated review. Ecol. Process. **5**(1), 1–12 (2016). https://doi.org/10.1186/s13717-016-0063-3
28. Huang, J., Yuan, Y.: Bayesian dynamic mediation analysis. Psychol. Methods **22**(4), 667 (2017)
29. Robins, J.M., Hernan, M.A., Brumback, B.: Marginal structural models and causal inference in epidemiology. Epidemiology **11**(5), 550–560 (2000)
30. Tyler, J.V.W., Tchetgen, E.J.: Mediation analysis with time varying exposures and mediators. J. Royal Statist. Soc. Ser. B (Statistical Methodology) **79.3**, 917–938 (2017)

Track Obstacle Real-Time Detection of Underground Electric Locomotive Based on Improved YOLOX

Caiwu Lu[1], Fan Ji[2](✉), Naixue Xiong[3], Song Jiang[1,2], Di Liu[1,2], and Sai Zhang[1,2]

[1] Xi'an University of Architecture and Technology, Xi'an 710055, China
jiangsong@live.xauat.edu.cn

[2] Xi'an Key Laboratory of Intelligent Industrial Sensing Computing and Decision Making, Xi'an 710055, China
1090375697@qq.com

[3] Sul Ross State University, Alpine, TX 79830, USA

Abstract. Based on the influence of dark obstacles caused by insufficient light in an underground mine on the driving safety of an electric locomotive. This paper proposes an improved YOLOX target detection algorithm to effectively identify and classify the track obstacles of the unmanned electric mine locomotive. On the basis of the YOLOX target detection network, the CBAM attention module is added to the CSPDarket and the FPN part of the feature pyramid, and the loss function of YOLO head part is replaced by SIOU. The collected image data of track obstacles of electric locomotive under different lighting conditions are used as the training set. The Pytorch deep learning framework is used to construct an object detection model for training and verification. Experiments show that the average accuracy and recall rate of the improved YOLOX underground electric locomotive track obstacle detection model can reach 93.05% and 88.29%, and the speed is improved to 45.3 fps. Compared with other target detection models, this model can better realize the accuracy and real-time detection of underground electric locomotive track obstacles. It provides the basis for the intelligence of underground mine transportation equipment.

Keywords: The underground mine · Underground electric locomotive · Object detection · CBAM · SIOU

1 Introduction

In recent years, with the rapid development of computer [1–4], network [5–7], big data [8–10], and unmanned driving technology [11, 12], underground mine transportation equipment has been gradually transformed into intelligent [13, 14]. Compared with the traditional road, the mine has insufficient illumination [15] and low visibility, which leads to the low definition of the image captured by the camera mounted on the rail motor vehicle, which is difficult to effectively detect [16]. Therefore, it is of great significance to study the fast and accurate detection [17–19] of track obstacles in front of the underground electric locomotive [20].

M. Qiu et al. (Eds.): SmartCom 2022, LNCS 13828, pp. 236–246, 2023.
https://doi.org/10.1007/978-3-031-28124-2_22

Some scholars at home and abroad have studied the detection methods of track obstacles. Yinping Zhang et al. [21] proposed a method based on information fusion of machine vision and millimeter wave radar. This method introduced the idea of information fusion [22] to obtain the final detection results of the environmental perception system of coal mine ground rail transportation. Ren Wang et al. [23] proposed a track edge extraction algorithm based on the combination of gray scale and gradient amplitude. Deqiang He et al. [24] proposed a detection network based on Mask R-CNN [25]. Although the Mask-RCNN model with ResNet[26] as the backbone feature extraction network can achieve high accuracy, it is difficult to meet the requirements of fast detection in the driving of unmanned vehicles. H. Hamadi et al. [27–29] proposed a color segmentation method based on decision support.

In view of the above analysis and the safety threats brought by traditional mining methods to production personnel in the process of mineral resources mining [30, 31], the underground unmanned electric locomotive is more suitable for mines than the manual driving electric locomotive. This paper proposes a real-time detection method of electric locomotive track obstacles based on improved YOLOx. It provides real-time and reliable obstacle warning for underground unmanned locomotives [32, 33].

The arrangement of this paper is as follows. The first part introduces the algorithm and the specific improvement method of the algorithm. In the second part, the experiment is carried out on the actual problem of electric locomotive in underground mine. The third part analyzes the performance of the improved algorithm according to the experimental results. Finally, conclusions and future research directions are given.

2 Underground Obstacle Detection Model

2.1 The Object Detection Network – YOLOX

The baselines of YOLOX are YOLOv3-SPP [34, 35] and Darknet53, and the Focus network [36, 37] structure is added to the Darknet53 backbone. In the FPN [38] feature pyramid part, the three effective feature layers were fused. For the prediction head part, in order to reduce the adverse effects of classification and regression [39, 40], classification and regression are implemented separately and finally integrated together. However, the loss function used in the whole network training is still the traditional border center point and height and width IOU [41, 42] loss, which is consistent with YOLO3.

2.2 Improved YOLOX Underground Obstacle Detection Model

Although the performance of YOLOX detection network has been significantly improved compared with the previous YOLO series, the deep image features will still be lost with the increase of network depth in the feature extraction process. At the same time, the underground rail locomotive needs to have high real-time performance in obstacle detection in order to make a quick subsequent response [43, 44], so it is imperative to improve the model to improve the network performance. This paper mainly adds CBAM module to improve semantic information in CSPDarket and FPN part of feature pyramid [45], and uses SIoU function in Bbox regression in YOLOHead to improve the convergence speed of model training. The improved overall structure diagram is shown in Fig. 1.

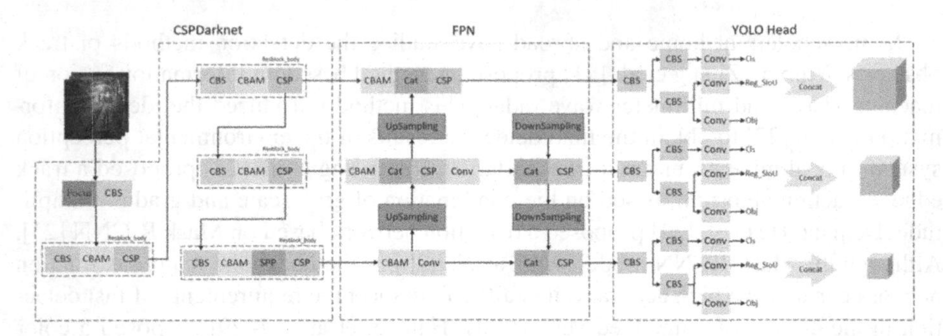

Fig. 1. Improved YOLOX network structure

2.3 Dual-Channel Attention Mechanism: CBAM

In this paper, we add attention to the CSPDarknet backbone network to improve the sensitivity of the model and help the network focus more on important features [46, 47] and suppress unnecessary features. Therefore, this paper uses the lightweight attention mechanism CBAM to improve the performance and accuracy of YOLOX network. The attention mechanism CBAM [48] structure is shown in Fig. 2.

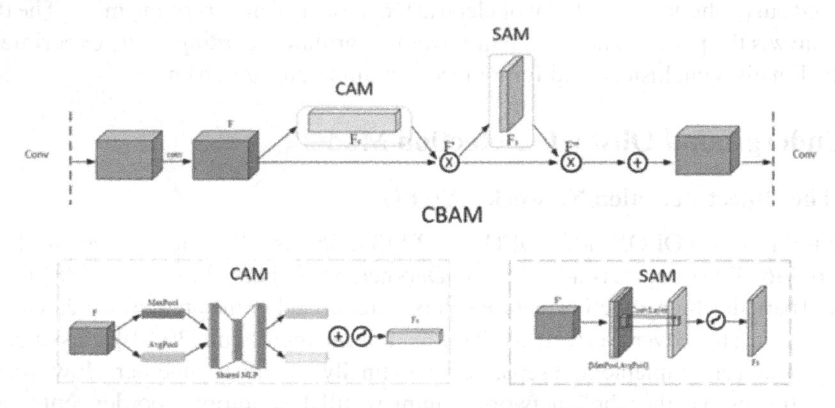

Fig. 2. The attention mechanism CBAM

He first step of the CBAM attention mechanism is that the channel attention mechanism takes the input feature map F through global maximum pooling and global average pooling based on width and height, respectively, to obtain a 1 × 1 × C feature map. Then, the feature map was input into the shared MLP neural network. The obtained feature map is activated by the sigmoid function to generate the output feature map F_c, which is formulated as following:

$$F_c = \sigma(MLP(MaxPool(F)) + MLP(AvgPool(f))) \tag{1}$$

F' is obtained by excellence multiplication of the output feature map F_c and the input feature map F. The formula is as following:

$$F' = \sigma\left(F\left(F_0\left(F^c_{max}\right)\right) + FF_0\left(F^c_{avg}\right)\right) = F_c \otimes F \tag{2}$$

F′ was used as the input feature of the spatial attention mechanism. The spatial attention mechanism carries out global maximum pooling and global average pooling of the input feature map F′ based on the channel, and the obtained feature map is concentrated by the channel, and a 7 × 7 convolution is used to reduce the dimension. Finally, the Sigmoid function is used to activate the operation to generate the output feature map F_S, and the formula is as following:

$$F_s = \sigma\left(f^{7\times7}(MaxPool(F_c)) + f^{7\times7}(AvgPool(F_c))\right) \tag{3}$$

The obtained F_S is then multiplied with the input feature F' of the spatial attention mechanism to obtain the final generated feature map F″, whose formula is as following:

$$F'' = F\otimes_s F' \tag{4}$$

Therefore, CBAM attention mechanism can be used to mix channel information and spatial information to extract image features without occupying a lot of computing power, in order to improve the accuracy and efficiency of object detection tasks.

2.4 Loss Function – SIOU

The prediction of the YOLO Head for each feature layer can be expressed as following:

$$Out_{pre} = (H, W, 4 + 1 + num\_classes) \tag{5}$$

H and W represent the width and height of the output feature map, respectively. 4 represents the Reg prediction result, which can be divided into 2 + 2. The first 2 represents the offset of the center point of the prediction box compared with the feature point, and the second 2 represents the width and height of the prediction box compared with the logarithmic exponent parameter. 1 represents the prediction result of Obj, which reflects the probability that each feature point prediction box contains an object. Num_classes represents the probability that each feature point corresponds to a certain class of objects.

Aiming at the insensitivity of IOU to the distance and Angle of the actual box and the predicted box, this paper uses the just released SIOU [49], which improves the Bbox regression in the network. Compared with IOU, SIOU improves the inference speed and accuracy during model training. Since SIOU considers the angle between the center point of the real box and the prediction box, the number of variables that are not related to the distance is reduced to the greatest extent, and the convergence speed of the prediction box is improved.

Based on the SIOU loss, the loss function in the entire YOLOHead includes the Bbox regression loss, the category loss, and the confidence loss. Therefore, the total loss function can be expressed as following:

$$\begin{aligned} Loss = Loss_{siou} + Loss_{obj} + Loss_{cls} = 1 - IoU + \frac{\Delta + \Omega}{2} \\ - \sum_{i=0}^{s^2}\sum_{j=0}^{B} I_{ij}^{obj}\left[C' \log C + (1 - C')\log(1 - C)\right] \end{aligned}$$

$$- \gamma_{noobj} \sum_{i=0}^{s^2} \sum_{j=0}^{B} I_i^{noobj} [C' \log C + (1 - C') \log(1 - C)]$$

$$- \sum\nolimits_{i=0}^{s^2} I_{ij}^{obj} \sum\nolimits_{c \in classes} [P'(logP) + (1 - p')log(1 - P)] \tag{6}$$

3 Underground Obstacle Detection Test

The data acquisition equipment uses Canon camera. The hardware platform of the test equipment is: operating system Windows10, graphics card model is NVIDIA RTX3050Ti; Software configuration: Python 3.6, Pytorch version 1.7.0.

In this paper, Zero_DCE [50, 51] module is added to the data preprocessing part to obtain enhanced images. Figure 3 shows some original images with typical characteristics and the corresponding effect after Zero_DCE processing.

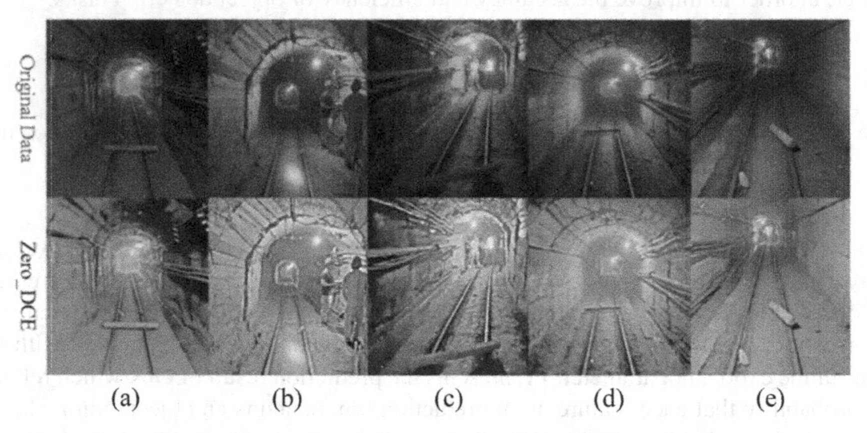

Fig. 3. Comparison before and after Zero_DCE treatment

4 Result and Analysis

4.1 Effectiveness Analysis of Improved YOLOx Model

In order to verify the effect of the improved model checking, the YOLOx model before and after the improvement and the common type models YOLOv3, YOLOv4 and Faster-RCNN are trained with the same training parameters and data sets. The loss change curve and related data are generated according to the training log recorded by the improved YOLOx, as showed in Fig. 4, Fig. 5 and Table 1. Figure 4 and 5 shows that the improved Yolox model has a smoother curve than other commonly used models, and the convergence occurs earlier.

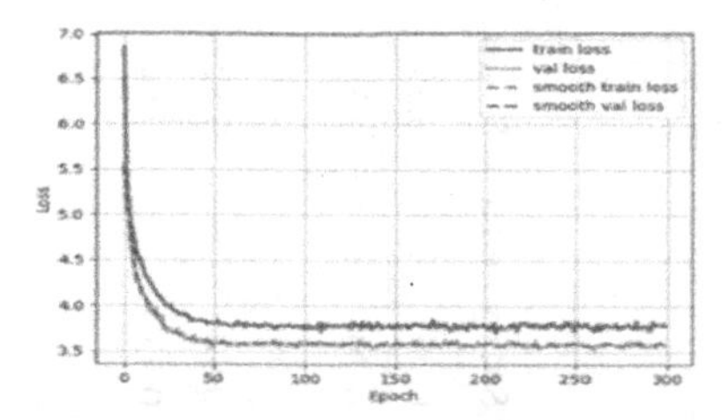

Fig. 4. Change of YOLOx Loss curve after improvement

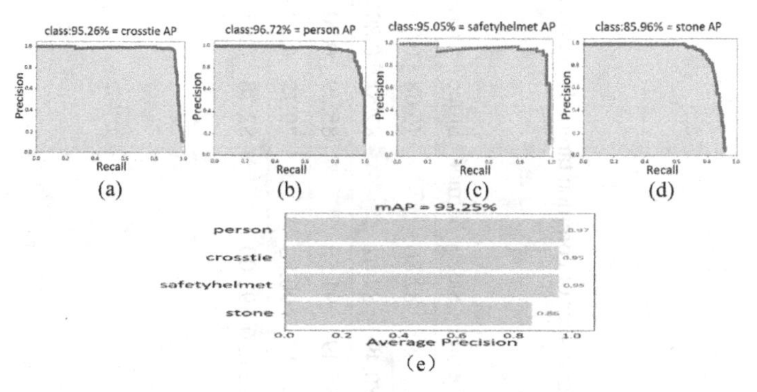

Fig. 5. Improved YOLOx all types of AP and mAP

Table 1 shows the results according to the statistics. According to the results in the table, when the improved YOLOx model performs underground track obstacle detection, the speed is 28.6, 16, 13.6 and 2.9fps higher than Fast-RCNN, YOLOv4, YOLOv5 and the original YOLOx. In terms of accuracy, it is 4.65, 2.65, 2.19 and 1.35 percentage points higher than Fast-RCNN, YOLOv4, YOLOv5 and original YOLOx, respectively. In terms of recall rate, it is increased by 9.39, 4.36, 0.82, 0.76 percentage points respectively. At the same time, the F_1 of the improved YOLOx model is 0.905, which is 0.1925 higher than Faster-RCNN, 0.095 higher than YOLOv4, 0.06 higher than YOLOv5, and 0.02 higher than the original YOLOx. It shows that the improved YOLOx has better performance.

Table 1. Improved YOLOx evaluation index parameter of common target detection model

Model	FPS	P /%				R/%				F_1			
		Crosstie	Person	Safety helmet	Stone	Crosstie	Person	Safety helmet	Stone	Crosstie	Person	Safety helmet	Stone
Faster_RCNN	16.7	92.56	87.42	84.62	88.95	75.58	80.49	82.86	76.68	0.73	0.68	0.65	0.79
YOLO v4	29.3	93.27	84.62	92.19	91.52	83.10	80.49	89.62	82.51	0.89	0.81	0.79	0.75
YOLO v5	31.7	95.54	82.50	93.06	92.34	87.56	85.49	91.56	85.29	0.90	0.86	0.84	0.78
YOLO x	42.4	96.58	85.56	92.87	91.78	89.37	90.15	90.83	79.78	0.93	0.89	0.90	0.82
Improved YOLOx	45.3	97.72	88.00	93.88	92.60	91.85	92.630	92.00	82.78	0.95	0.90	0.93	0.84

4.2 Underground Obstacle Detection Ablation Experiment

In order to verify the influence of the dual-channel attention mechanism on the performance of the YOLOx model, an ablation experiment is carried out in this paper, and the results are shown in Table 2.

Table 2. Table of ablation parameters of YOLOx

	CBAM	SA	SE	Mosaic	P/%	FPS
1	—	—	—	—	90.12	44.3
2	√	—	—	—	92.41	44.8
3	√	—	—	√	93.05	45.3
4	—	√	—	—	91.23	44.2
5	—	√	—	√	92.21	44.9
6	—	—	√	—	91.14	44.1
7	—	—	√	√	91.76	45.1

As showed in the above table, compared with the addition of SA, SA + Mosaic, SE and SE + Mosaic, the model is increased by 1.82%, 0.84%, 1.91% and 1.29%, respectively. The model meets the requirements of deploying in the underground unmanned rail motor vehicle.

5 Conclusion

In order to solve the problem of the blending of background and target in the underground scene, this paper adds the Zero_DCE module for dealing with dim background in the data preprocessing part, adds the CBAM attention mechanism in the detection model to focus more on the important information of the object target and suppress irrelevant information, and finally replaces the loss function with SIOU. The convergence speed and accuracy of the algorithm are improved. Experiments show that. The speed, accuracy and recall rate of the improved model reach 45.3 fps, 93.05%and 88.29%respectively. In the future work, we will try more effective image enhancement algorithms and preprocess the experimental data to reduce the limitation of sample defects. In addition, feature pyramids with different depths were added to eliminate redundant information to achieve more efficient detection in complex environments.

References

1. Qiu, M., Li, H., Sha, E.: Heterogeneous real-time embedded software optimization considering hardware platform. ACM Symposium on Applied Computing, pp. 1637–1641 (2009)
2. Qiu, M., Xue, C., Shao, Z., Sha, E.: Energy minimization with soft real-time and DVS for uniprocessor and multiprocessor embedded systems. IEEE DATE Conference, pp. 1–6 (2007)
3. Qiu, M., Jia, Z., et al.: Voltage assignment with guaranteed probability satisfying timing constraint for real-time multiproceesor DSP. J. Sign. Process. Syst. (2007)
4. Qiu, M., Yang, L., et al.: Dynamic and leakage energy minimization with soft real-time loop scheduling and voltage assignment. IEEE Trans. VLSI **18**(3), 501–504 (2009)
5. Niu, J., Gao, Y., et al.: Selecting proper wireless network interfaces for user experience enhancement with guaranteed probability. JPDC **72**(12), 1565–1575 (2012)
6. Qiu, M., Xue, C., Shao, Z., et al.: Efficient algorithm of energy minimization for heterogeneous wireless sensor network. IEEE EUC, pp. 25–34 (2006)
7. Qiu, M., Chen, Z., Ming, Z., Qin, X., Niu, J.: Energy-aware data allocation with hybrid memory for mobile cloud systems. IEEE Syst. J. **11**(2), 813–822 (2014)
8. Gai, K., Qiu, M., Elnagdy, S.: A novel secure big data cyber incident analytics framework for cloud-based cybersecurity insurance. IEEE BigDataSecurity (2016)
9. Li, J., Ming, Z., et al.: Resource allocation robustness in multi-core embedded systems with inaccurate information. J. Syst. Archi. **57**(9), 840–849 (2011)
10. Li, Y., Gai, K., et al.: Intercrossed access controls for secure financial services on multimedia big data in cloud systems. ACM Trans. Multimedia Comp. Comm. App. (2016)
11. Lipson, H., Kurman, M.: Driverless: Intelligent Cars & the Road Ahead (2016)
12. Zablocki, É., Ben-Younes, H., Pérez, P., et al.: Explainability of deep vision-based autonomous driving systems: review and challenges. Int'l J. Comput. Vis. 1–28 (2022)
13. Wu, C., Ju, B., et al.: UAV autonomous target search based on deep reinforcement learning in complex disaster scene. IEEE Access **7**, 117227–117245 (2019)
14. Huang, S., Zeng, Z., Ota, K., Dong, M., Wang, T., Xiong, N.N.: An intelligent collaboration trust interconnections system for mobile information control in ubiquitous 5G networks. IEEE Trans. Network Sci. Eng. **8**(1), 347–365 (2020)
15. Ning, Z., Mao, S.: Mei Li Mine non-uniform illumination video image enhancement algorithm based on illumination adjustment. J. China Coal Soc. **42**(8), 8 (2017)
16. Chen, S., Cheng, Z., Zhang, L., et al.: SnipeDet: attention-guided pyramidal prediction kernels for generic object detection. Pattern Recogn. Lett. **152**, 302–310 (2021)
17. Hu, F., Lakdawala, S., et al.: Low-power, intelligent sensor hardware interface for medical data preprocessing. IEEE Trans. Info. Tech. Bio. **13**(4), 656–663 (2009)
18. Qiu, H., Dong, T., Zhang, T., Lu, J., Memmi, G., Qiu, M.: Adversarial attacks against network intrusion detection in IoT systems. IEEE IoT J. **8**(13), 10327–10335 (2020)
19. Qiu, H., Zheng, Q., et al.: Topological graph convolutional network-based urban traffic flow and density prediction. IEEE Trans. ITS (2020)
20. Chen, Y., Li, Z.: An effective approach of vehicle detection using deep learning. Comput. Intell. Neurosci. (2022)
21. Zhang, Y.:esearch on Environmental Perception System of coal mine Surface Rail Transportation. China University of Mining and Technology (2020)
22. You, H., Wang, G.: Multi-sensor information fusion and its application. Publishing House of Electronics Industry (2007)
23. Wang, R.: Research on anti-collision system of underground locomotive based on machine vision. Chongqing University (2012)

24. He, D., Li, K., Chen, Y., et al.: Obstacle detection in dangerous railway track areas by a convolutional neural network. Meas. Sci. Technol. **32**(10), 105401 (2021)
25. He, K., Gkioxari, G., Dollár, P., et al.: Mask r-cnn. In: Proceedings of the IEEE International Conference on Computer Vision, pp. 2961–2969 (2017)
26. He, K., Zhang, X., Ren, S., et al.: Identity mappings in deep residual networks. In: European Conference on Computer Vision, pp. 630–645. Springer, Cham (2016)
27. Hamadi, H., Supriyono, Riansah, D.: Detection and measurement of obstacles on a track using color segmentation with background subtraction and morphological operation. J. Phys. Conf. Ser. **1436**(1), 012026 (2020)
28. Xia, F., Hao, R., et al.: Adaptive GTS allocation in IEEE 802.15. 4 for real-time wireless sensor networks. J. Syst. Architect. **59**(10), 1231–1242 (2013)
29. Wu, C., Luo, C., et al.: A greedy deep learning method for medical disease analysis. IEEE Access **6**, 20021–20030 (2018)
30. Gao, Y., Xiang, X., et al.: Human action monitoring for healthcare based on deep learning. IEEE Access **6**, 52277–52285 (2018)
31. Yao, Y., Xiong, N., et al.: Privacy-preserving max/min query in two-tiered wireless sensor networks. Comput. Math. Appl. **65**(9), 1318–1325 (2013)
32. Wang, G., Ding, H., Yang, Z., et al.: TRC-YOLO: a real-time detection method for lightweight targets based on mobile devices. IET Comput. Vision **16**(2), 126–142 (2022)
33. Ruan, Z., Cao, J., Wang, H., et al.: Adaptive feedback connection with a single-level feature for object detection. IET Comput. Vis. (2022)
34. Ge, Z., Liu, S., Wang, F., et al.: Yolox: exceeding yolo series in 2021. arXiv preprint arXiv: 2107.08430 (2021)
35. Huang, J., Huang, Y., Huang, H., et al.: An improved YOLOX algorithm for forest insect pest detection. Comput. Intell. Neurosci. (2022)
36. Redmon, J., Farhadi, A.: Yolov3: An incremental improvement. arXiv preprint arXiv:1804. 02767 (2018)
37. Zhu, X., Lyu, S., Wang, X., et al.: TPH-YOLOv5: Improved YOLOv5 based on transformer prediction head for object detection on drone-captured scenarios. In: Proceedings of the IEEE/CVF International Conference on Computer Vision, pp. 2778–2788 (2021)
38. Zhao, T., Wei, X., Yang, X.: Improved YOLO v5 for railway PCCS tiny defect detection. In: 14th International Conference on Advanced Computational Intelligence (ICACI), pp. 85–90. IEEE (2022)
39. Lin, T.Y., Dollar, P., Girshick, R., et al.: Feature pyramid networks for object detection. In: IEEE Conference on Computer Vision and Pattern Recognition (CVPR) (2017)
40. Yang, X., Yan, J.: On the arbitrary-oriented object detection: classification based approaches revisited. Int. J. Comput. Vis. **130**(5), 1340–1365 (2022)
41. Cui, Y., et al.: Joint classification and regression for visual tracking with fully convolutional siamese networks. Int. J. Comput. Vis. 1–17 (2022). https://doi.org/10.1007/s11263-021-01559-4
42. Wu, S., Yang, J., Wang, X., et al.: Iou-balanced loss functions for single-stage object detection. Pattern Recogn. Lett. **156**, 96–103 (2022)
43. Zhang, W., Zhu, S., Tang, J., Xiong, N.: A novel trust management scheme based on Dempster-Shafer evidence theory for malicious nodes detection in wireless sensor networks. J. Supercomput. **74**(4), 1779–1801 (2018)
44. Zhao, J., Huang, J., et al.: An effective exponential-based trust and reputation evaluation system in wireless sensor networks. IEEE Access **7**, 33859–33869 (2019)
45. Cheng, H., Xie, Z., et al.: Multi-step data prediction in wireless sensor networks based on one-dimensional CNN and bidirectional LSTM. IEEE Access **7**, 117883–117896 (2019)
46. Zhang, T., Jin, B., Jia, W.: An anchor-free object detector based on soften optimized bi-directional FPN. Comput. Vis. Image Underst. **218**, 103410 (2022)

47. Liu, Y., Zhang, Y., Liu, S., et al.: Salient object detection by aggregating contextual information. Pattern Recogn. Lett. **153**, 190–199 (2022)
48. Wang, Y., Li, Y., Guo, X., et al.: CDANet: common-and-differential attention network for object detection and instance segmentation. Pattern Recog. Lett. **158**, 48–54 (2022)
49. Gevorgyan, Z.: SIoU Loss: More Powerful Learning for Bounding Box Regression. arXiv preprint arXiv:2205.12740 (2022)
50. Woo, S., Park, J., Lee, J.-Y., Kweon, I.S.: CBAM: convolutional block attention module. In: Ferrari, V., Hebert, M., Sminchisescu, C., Weiss, Y. (eds.) ECCV 2018. LNCS, vol. 11211, pp. 3–19. Springer, Cham (2018). https://doi.org/10.1007/978-3-030-01234-2_1
51. Guo, C., Li, C., Guo, J., et al.: Zero-reference deep curve estimation for low-light image enhancement. In: IEEE/CVF Conference on Computer Vision and Pattern Recognition, pp. 1780–1789 (2020)

Research on Sharding Strategy of Blockchain Based on TOPSIS

Jun Liu, Xu Shen(✉), Mingyue Xie, and Qi Zhang

Chongqing University of Posts and Telecommunications, Chongqing, China
junliu@cqupt.edu.cn, xushen0417@163.com,
s211201034@stu.cqupt.edu.cn

Abstract. Cryptocurrency applications with blockchain technology as the underlying architecture have gradually developed into a new means of payment, and are expanding to all walks of life with the support of cryptography and consensus algorithms. Due to the disadvantages of low throughput and high latency, the blockchain has seriously hindered the widespread use of upper-layer applications and cannot meet the growing demand for users and transaction volumes. Drawing on the sharding idea of traditional databases, blockchain sharding, as a representative of on-chain scaling solutions, greatly improves the throughput of the blockchain system. At present, most of the network sharding schemes in the sharded blockchain adopt a strategy based on random sharding. This strategy does not take into account the performance of the node itself, resulting in large performance differences between different shards, further reducing the throughput of the entire system. In addition, the aggregation behavior of malicious nodes may also occur, reducing the security of the system. Aiming at the performance of each node, this paper proposes a sharding strategy based on the approximate ideal solution model (TOPSIS). Through the TOPSIS model, the nodes are scored according to the hardware performance of the node, the response time to the transaction and the results, etc., and the nodes are allocated to the corresponding shards according to the scoring results. The sharding strategy based on this model balances the performance differences among shards and improves the throughput of the entire system.

Keywords: Blockchain · Sharding · Distributed Systems

1 Introduction

With the rapid development of computer capability [1–3] and cloud computing [4–6], data security [7–9] and privacy protection [10–12] become a critical issue. Currently, blockchain techniques emerged as a promising approach to ensure security and privacy for various applications. With the continuous development of blockchain technology [13, 14], the computing power of the entire network continues to expand due to the increase in the number of nodes participating in the consensus protocol, and the resulting transaction volume is also increasing. The blockchain will face serious challenges [15,

M. Qiu et al. (Eds.): SmartCom 2022, LNCS 13828, pp. 247–257, 2023.
https://doi.org/10.1007/978-3-031-28124-2_23

16]. Specifically, the Bitcoin network consumes a lot of computing power due to the consensus process. The throughput of current system is maintained at a low level.

While other centralized payment processing systems can achieve a throughput of 1,200 to 56,000 [17]. If you blindly increase the resources of participating consensus nodes, it will in turn affect the decentralization and security of the entire system [18, 19]. How to achieve scalability in system performance [20–22] on the premise of ensuring decentralization and security is a problem that requires in-depth research [23, 24]. The current solutions to the blockchain scalability problem include on-chain and off-chain scaling, among which sharding technology is the representative of on-chain scaling, as shown in Table 1.

Sharding was first proposed by Luul et al. in Elastico [25]. The main idea is to divide the network into smaller committees, each of which handles a set of disjoint transactions. By performing consensus and confirming transactions in parallel [6, 26–28], the throughput of the entire network increases approximately linearly with the increase of nodes participating in the consensus protocol in the entire network [29–31]. However, the previous sharding protocols were all based on the strategy of random sharding. By randomly allocating nodes in the entire network to different shards, each shard processed transactions in parallel to improve throughput [32–34]. However, in the sharded blockchain, the computing power is allocated to each shard, and the attack cost of the attacker is further reduced, and then a 1% attack on the sharded blockchain occurs.

We propose a network sharding strategy based on TOPSIS. The TOPSIS model can help us conduct a comprehensive evaluation of nodes the performance of each node is comprehensively evaluated through the evaluation information set composed of the computing power of the node, the number of blocks packaged, the time to process transactions, and the failure rate. Through multiple rounds of implementation testing, the overall time complexity of the proposed model is controlled within a very small range. The key contributions are summarized as follows:

Table 1. Sharding Blockchain Technology Comparison

Sharding	Elastico	OmniLedger	Zilliqa	Ethereum2.0
Model	UTXO	UTXO	Account	Account
Consensus	PBFT	POW	PBFT	BFT
Throughput	40 tx/s^2	3500 tx/s^2	N/A	N/A
Delay	800	800	N/A	N/A
Network syn	Partial sync	Partial sync	Asynchronous	Partial sync
Smart contract	×	×	√	√

We apply the TOPSIS evaluation model to the block chain network node sharding strategy [35, 36], replacing the random sharding strategy in the previous sharding scheme. By fully considering the comprehensive performance advantages of each node through the proposed performance indicators [37], the blockchain can break through the performance bottleneck and achieve greater performance improvement [38].

The rest of the paper is organized as follows. Section 2 reviews the related work on the blockchain technology. Section 3 presents an introduction of the considered sharding mechanism. Section 4 reports experimental evaluation. Finally, conclusions are presented in Sect. 5.

2 Related Work

2.1 Committee-Based Consensus

PeerCensus is the first committee-based consensus protocol, which uses the PBFT protocol internally, as shown in Fig. 1. PeerCensus does not give a solution on how to avoid malicious nodes gathering in a single committee, so the protocol cannot ensure the security of transactions [39]. The scheme proposed in ByzCoin adds multi-signature on the basis of the committee, and improves the throughput of the entire blockchain system through the introduction of this technology [40, 41]. Also, like PeerCensus, this scheme does not achieve a relatively high level of security and is subject to Byzantine error attacks [42, 43].

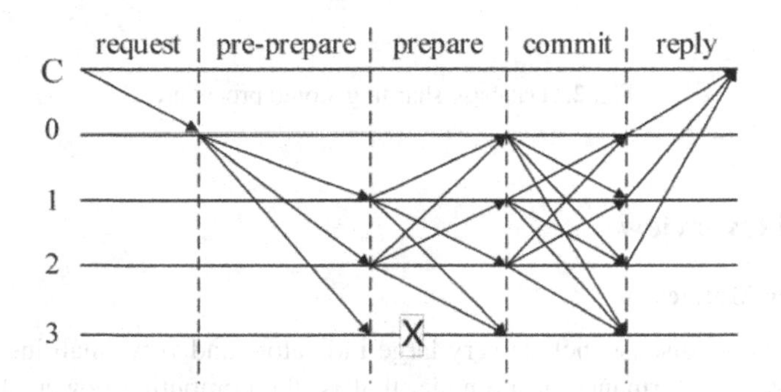

Fig. 1. PBFT algorithm process

2.2 Shard-Based Consensus

Drawing on the idea of database sharding, as shown in Fig. 2, Luul et al. proposed the first public chain sharding protocol Elastico. Its main job is to divide the network into small committees [44], and these small independent committees handle their own affairs independently, thereby improving the throughput of the entire network. That is, the number of these divided committees grows linearly with the degree of division, and each individual committee has members belonging to that committee. Decide which set of transactions to agree to in parallel by running a classic Byzantine consensus protocol. In Elastico's design, the following five steps are performed for each epoch: (1) identity establishment and committee formation [45]; (2) committee coverage setting; (3) consensus within the committee; (4) final Consensus broadcasting; (5) Generate random values for epoch.

Kokoris-Kogias et al. proposed Omniledger [36], which relieves the storage pressure of nodes through data shard storage, and realizes the processing of cross-shard transactions through a two-phase commit protocol. OmniLedger combines the ideas of ByzCoin and Hybird Consensus, and proposes miners that solve the pow puzzle as its validator set. In addition, it uses the RoundHound protocol, which guarantees long-term security and scalability as each epoch runs. Finally, OmniLedger proposes Atomix in order to ensure the atomicity of transactions, that is, when transactions are located in different shards, the atomic state can be maintained. Through a two-phase "lock/unlock" protocol, both parties to a transaction in the network can submit a cross-shard transaction completely correctly, or obtain a "rejection proof" to abort and unlock the state affected by the partially completed transaction [46, 47].

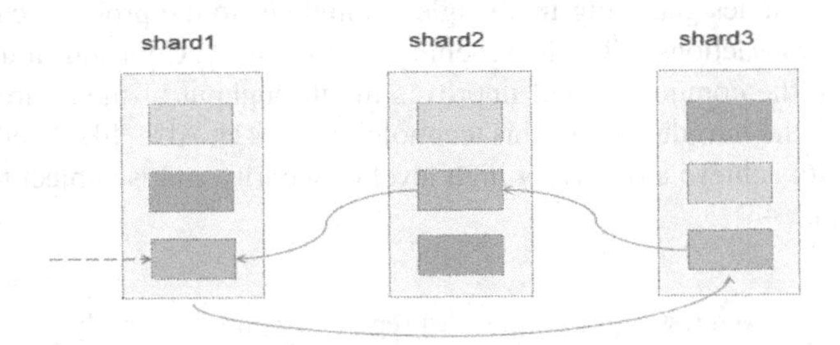

Fig. 2. Database sharding stored procedure

3 Model Overview

3.1 Scoring Metrics

The indicators we discuss include very large indicators and very small indicators: one is the hardware performance of the node, that is, the computing power of the node's CPU (Central Processing Unit) and GPU (Graphics Processing Unit). The higher the computing power of the node, the higher its score. One is the number of blocks the node had packaged, which indirectly reflect the activity level of a node [48, 49].

The larger the number of blocks packaged by the node, the higher the participation and activity in the network [50], and the higher the score accordingly. Another is the time a node takes to process a transaction [51]. The smaller the time to confirm the transaction, the higher the score. The last indicator is the failure rate of the node, which is an important factor to measure a node to maintain the stability of the entire network. The lower the failure rate of a node, the higher the node's score.

3.2 Mathematical Model

We set the multi-attribute decision object set as $A = \{a_1, a_2, \ldots, a_n\}$. The set of indicators to measure the value of object attributes is $I = \{x_1, x_2, \ldots, x_m\}$. The attribute vector composes of m attribute values of each object $a_i (i = 1, 2, \ldots, n)$ in the object set is $[a_{i1}, a_{i2}, \ldots, a_{im}] (i = 1, 2, \ldots, n)$. Set up a positive ideal solution C^+ and a negative ideal solution C^-. Specific steps are as follows:

(1) Import the corresponding data through the proposed evaluation index, construct the original decision matrix A, and perform data preprocessing, A is expressed as follows:

$$A = \left(a_{ij}\right)_{n*m} \quad (1)$$

The decision matrix A can be further expressed in detail as:

$$A = \begin{bmatrix} ComPower(1) & BlockNum(1) & FaultRate(1) & ConfirmRate(1) \\ ComPower(2) & BlockNum(2) & FaultRate(2) & ConfirmRate(2) \\ \vdots & \vdots & \vdots & \vdots \\ ComPower(n) & BlockNum(n) & FaultRate(n) & ConfirmRate(n) \end{bmatrix}$$

The extremely large indicator $ComPower(n)$ represent the computing power of the nth node in the network; the largest indicator $BlockNum(n)$ represent the number of blocks that had been packaged by the nth node in the network; The very small indicator $FaultRate(n)$ represent the failure rate of the nth node in the network; the very small indicator $ConfirmRate(n)$ represent the confirmation time of the nth node in the network for a transaction.

We first deal with very small attribute indicators using the linear variation method:

$$A^* = \begin{bmatrix} ComPower(1)^* & BlockNum(1)^* & FaultRate(1)^* & ConfirmRate(1)^* \\ ComPower(2)^* & BlockNum(2)^* & FaultRate(2)^* & ConfirmRate(2)^* \\ \vdots & \vdots & \vdots & \vdots \\ ComPower(n)^* & BlockNum(n)^* & FaultRate(n)^* & ConfirmRate(n)^* \end{bmatrix}$$

Then used the vector normalization method to normalize the transformed decision matrix A^*, and get the normalized matrix $B = (b_{ij})_{n*m}$, where:

$$b_{ij} = \frac{a_{ij}^*}{\sqrt{\sum_{i=1}^{n} a_{ij}^{*2}}} i = 1, 2, \ldots, n \quad j = 1, 2, \ldots, m \quad (2)$$

(2) Calculate the weighted normalization matrix by the obtained weight values:

$$C = \left(c_{ij}\right)_{n*m} = B \cdot W$$

(3) Determine the positive ideal solution C^+ and the negative ideal solution C^-:

$$C^+ = \left[C_1^+, C_2^+, \ldots, C_m^+\right] \quad (4)$$

$$C^- = \left[C_1^-, C_2^-, \ldots, C_m^-\right] \quad (5)$$

(4) Calculate the distance d^+ from each object to be evaluated to the positive ideal solution C^+, and the distance d^- to the negative ideal solution C^-:

$$d_i^+ = \sqrt{\sum_{j=1}^{m} \left(c_{ij} - c_j^+\right)^2} \quad \mathrm{i} = 1, 2, \ldots, \mathrm{n} \quad (6)$$

$$d_i^- = \sqrt{\sum_{j=1}^{m} \left(c_{ij} - c_j^-\right)^2} \quad \text{i} = 1, 2, \ldots, \text{n} \tag{7}$$

(5) Calculate the relative closeness of each object to be evaluated:

$$f_i = \frac{d_i^-}{d_i^+ + d_i^-} \quad \text{i} = 1, 2, \ldots, \text{n} \tag{8}$$

Finally, each object to be evaluated can be sorted from large to small according to its relative proximate degree, and the object with the greater relative proximate degree is ranked higher.

In general, the sharding scheme based on the TOPSIS model is determined by the node's scoring mechanism, allocation strategy, and generation of transaction blocks. The functions of each part are:

Scoring Mechanism. After leader collects the verification results of the transaction by the member nodes in the shard, it will score the nodes according to the four evaluation indicators set previously.

Node Allocation Strategy. Committee tries to balance the proportion of nodes with different scores in each shard as much as possible according to the scoring results of the nodes. Then within each shard, a node with a score higher than the average is randomly selected as the leader node.

Generate Transaction Block. The leader collects the transaction verification results of the member nodes in the shard, packs the results into blocks and sends them to the committee for processing. After the committee receives it, it is packaged into a transaction block and added to the blockchain.

4 Experiment Analysis

4.1 Experiment Description

Due to the advantages of Go language compared to other programming languages in implementing distributed systems, we used Go language as the implementation language of blockchain systems.

We implemented a blockchain system based on random sharding strategy and a blockchain system based on TOPSIS sharding scheme respectively. The blockchain system based on the random sharding strategy used the Elastico protocol as the underlying framework, and the blockchain system based on the TOPSIS model modified the sharding process in the above system to achieve the function of scoring each node in the network. There were three types of nodes in the system: committee nodes, intra-shard leader nodes, and intra-shard member nodes.

The shard leader node processed the transactions broadcasted in the network after receiving them, and the committee nodes collected the processing results in each shard and package the results into blocks and added them to the blockchain. In our system, the committee nodes would score the leader nodes in each shard, and the score would be used as the basis for reassigning nodes in the next epoch.

We deployed 200 machines with similar performance in the local area network to form a blockchain network environment, and simulated 1,000 blockchain nodes, which were evenly distributed in each machine, and each node was independent from the rest. Since we needed the performance of each node to be different, network bandwidth and processor performance were not required. We mainly compared whether the throughput of the two systems to be evaluated had improved significantly.

4.2 Experimental Design and Analysis

In order to better simulate the performance difference between nodes, the experiment was carried out by means of a large number of repeated verification transactions, which can exclude the influence of network communication delay between nodes on the experimental results.

In the experiment, 2000 nodes were set up and divided into several groups. The performance of each group of nodes was different, and the method of repeated verification transactions was used to make the performance difference of the nodes better affect the scoring results, as shown in Table 2. In addition, we also simulated the effect of different shard size on the experimental results, the experiments were carried out under different shard numbers, and the experimental results were compared.

Table 2. Comparison of average throughput of two blockchain systems

	TOPSIS	Random
Throughput	3066.3	2630.2

Under the conditions of different shard sizes in the blockchain system, we obtained the following results through experiments, and the results are shown in Fig. 3 and Fig. 4.

The experimental data were all throughput data obtained after running 50 epochs. However, through a horizontal comparison, it was found that the throughput of the blockchain system based on the TOPSIS sharding strategy had increased by nearly 40% compared with the traditional blockchain system based on the random sharding strategy. The throughput of our proposed blockchain system can be stably higher than the traditional system based on the random sharding strategy.

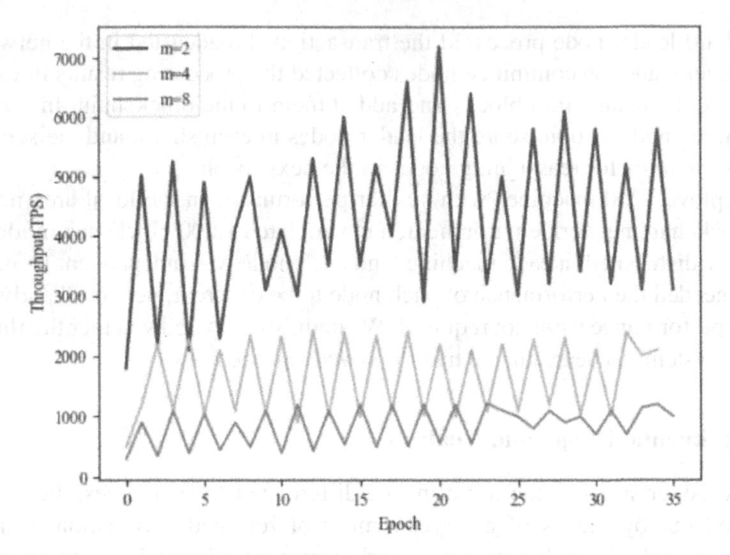

Fig. 3. Throughput Comparison Results

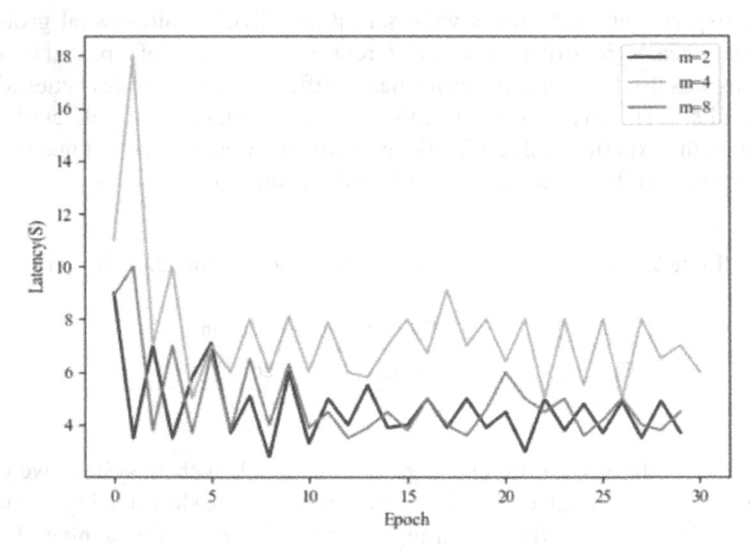

Fig. 4. Delay comparison results

5 Conclusion

This paper proposed a model based on the TOPSIS allocation strategy. The TOPSIS model could score the nodes in the blockchain according to the proposed indicators, and shard according to the scoring results, so that the scores of the nodes in different shards were balanced, thereby greatly improving the transaction throughput of the blockchain. Through comparative experiments, it was verified that the sharding scheme based on the TOPSIS model had higher transaction throughput than Elastico when there were performance differences between nodes in the network. The results of comprehensive experiments show that the sharding scheme proposed in this paper can cope with complex network environments and can better adapt to actual needs.

References

1. Qiu, M., Li, H., Sha, E.: Heterogeneous real-time embedded software optimization considering hardware platform. In: ACM Symposium on Applied Computing, pp. 1637–1641 (2009)
2. Qiu, M., Jia, Z., et al.: Voltage assignment with guaranteed probability satisfying timing constraint for real-time multiproceesor DSP. J. Signal Proc. Syst. (2007)
3. Qiu, M., Yang, L., et al.: Dynamic and leakage energy minimization with soft real-time loop scheduling and voltage assignment. IEEE Trans. VLSI **18**(3), 501–504 (2009)
4. Niu, J., Gao, Y., et al.: Selecting proper wireless network interfaces for user experience enhancement with guaranteed probability. JPDC **72**(12), 1565–1575 (2012)
5. Qiu, M., Xue, C., Shao, Z., et al.: Efficient algorithm of energy minimization for heterogeneous wireless sensor network. IEEE EUC, pp. 25–34 (2006)
6. Qiu, M., Chen, Z., Ming, Z., Qin, X., Niu, J.: Energy-aware data allocation with hybrid memory for mobile cloud systems. IEEE Syst. J. **11**(2), 813–822 (2014)
7. Gai, K., Qiu, M., Elnagdy, S.: A novel secure big data cyber incident analytics framework for cloud-based cybersecurity insurance. IEEE BigDataSecurity (2016)
8. Li, Y., Gai, K., et al.: Intercrossed access controls for secure financial services on multimedia big data in cloud systems. ACM Trans. MCCA (2016)
9. Gao, X., Qiu, M.: Energy-based learning for preventing backdoor attack. KSEM, vol. 3, pp. 706–721 (2022)
10. Gai, K., Qiu, M., Chen, L., Liu, M.: Electronic health record error prevention approach using ontology in big data. IEEE 17th HPCC (2015)
11. Qiu, M., Qiu, H.: Review on image processing based adversarial example defenses in computer vision. In: IEEE 6th International Conference BigDataSecurity, pp. 94–99. USA (2020)
12. Qiu, H., Dong, T., et al.: Adversarial attacks against network intrusion detection in IoT systems. IEEE Internet Things J. **8**(13), 10327–10335 (2020)
13. Nakamoto, S.: Bitcoin: a peer-to-peer electronic cash system [EB], bitcoin.org (2009)
14. Liu, J., Xie, M., Chen, S., Ma, C., Gong, Q.: An improved DPoS consensus mechanism in blockchain based on PLTS for the smart autonomous multi-robot system. Inform Sci. **575**, 528–554 (2021)
15. Gai, K., Zhang, Y., et al.: Blockchain-enabled service optimizations in supply chain digital twin. IEEE Trans. Service Comput. (2022)
16. Qiu, M., Qiu, H., et al.: Secure Data Sharing Through Untrusted Clouds with Blockchain-enabled Key Management. The 3rd SmartBlock, pp. 11–16, China (2020)
17. Chen, S., Xie, M., Liu, J.: Improvement of the DPoS Consensus Mechanism in Blockchain Based on PLTS. IEEE BigDataSecurity (2021)
18. Liu, J., Zhao, J., Huang, H., Xu, G.: A novel logistics data privacy protection method based on blockchain. Multimedia Tools Appl. 1–21 (2022). https://doi.org/10.1007/s11042-022-12836-w
19. Xu, G., Dong, J., Ma, C.: A certificateless encryption scheme based on blockchain. Peer-to-Peer Network. Appl. **14**(5), 2952–2960 (2021). https://doi.org/10.1007/s12083-021-01147-w
20. Qiu, M., Xue, C., Shao, Z., Sha, E.: Energy minimization with soft real-time and DVS for uniprocessor and multiprocessor embedded systems. In: IEEE DATE Conference, pp. 1–6 (2007)

21. Li, J., Ming, Z., et al.: Resource allocation robustness in multi-core embedded systems with inaccurate information. J. Syst. Architect. **57**(9), 840–849 (2011)
22. Zhang, L., Qiu, M., et al.: Variable partitioning and scheduling for MPSoC with virtually shared scratch pad memory. J. Signal Process. Syst. **58**(2), 247–265 (2010)
23. Qiu, H., Qiu, M., Lu, R.: Secure V2X communication network based on intelligent PKI and edge computing. IEEE Network **34**(2), 172–178 (2019)
24. Qiu, H., Zeng, Y., Guo, S., Zhang, T., Qiu, M., Thuraisingham, B.: Deepsweep: an evaluation framework for mitigating DNN backdoor attacks using data augmentation. In: ACM Asia Conference on Computer and Communications (2021)
25. Xu, G., Zhang, J., Wang, L.: An Edge Computing Data Privacy-Preserving Scheme Based on Blockchain and Homomorphic Encryption. ICBCTIS'21 (2021)
26. Qiu, M., Guo, M., et al.: Loop scheduling and bank type assignment for heterogeneous multi-bank memory. JPDC **69**(6), 546–558 (2009)
27. Qiu, M., Sha, E., et al.: Energy minimization with loop fusion and multi-functional-unit scheduling for multidimensional DSP. JPDC **68**(4), 443–455 (2008)
28. Qiu, M., Ming, Z., Li, J., et al.: Three-phase time-aware energy minimization with DVFS and unrolling for chip multiprocessors. JSA **58**(10), 439–445 (2012)
29. Hu, F., Lakdawala, S., et al.: Low-power, intelligent sensor hardware interface for medical data preprocessing. IEEE Trans. Infor. Tech. Biomed. **13**(4), 656–663 (2009)
30. Qiu, H., Zheng, Q., et al.: Topological graph convolutional network-based urban traffic flow and density prediction. IEEE Trans. ITS (2020)
31. Qiu, M., Khisamutdinov, E., et al.: RNA nanotechnology for computer design and in vivo computation. Philosoph. Trans. Royal Soc. A (2013)
32. Zhang, K., Kong, J., et al.: Multimedia layout adaptation through grammatical specifications. Multimedia Syst. **10**(3), 245–260 (2005)
33. Qiu, M., Zhang, K., Huang, M.: Usability in mobile interface browsing. Web Intell. Agent Syst. J. **4**(1), 43–59 (2006)
34. Qi, D., Liu, M., et al.: Exponential synchronization of general discrete-time chaotic neural networks with or without time delays. IEEE TNN **21**(8), 1358–1365 (2010)
35. Luu, L., Narayanan, V., et al.: A secure sharding protocol for open blockchains. 2016 ACM SIGSAC, pp. 17–30 (2016)
36. Xie, M., Liu, J., et al.: Primary node election based on probabilistic linguistic term set with confidence interval in the PBFT consensus mechanism for blockchain. Complex Intell Syst. https://doi.org/10.1007/s40747-022-00857-9
37. Gan, C., Saini, A., et al.: Blockchain-based access control scheme with incentive mechanism for eHealth systems: patient as supervisor. Multimed Tools Appl. 1–17 (2020)
38. Zhu, Q., Loke, S., et al.: Applications of distributed ledger technologies to the internet of things: a survey. ACM Comput. Surv. **52**(6), 1–34 (2020)
39. Kokoris-Kogias, E., Jovanovic, P., et al.: Omniledger: a secure, scale-out, decentralized ledger via sharding. IEEE SP, pp. 583–598 (2018)
40. Zhou, Y., Liu, T., Tang, F.: A privacy-preserving authentication and key agreement scheme with deniability for IoT. Electronics **8**(4) (2019)
41. Zhang, X., Yijun, Y.: Research on digital copyright management system based on blockchain technology. In: IEEE 3rd ITNEC (2019)
42. Bin, C., Shouming, H., Daquan, F.: Impact of Network Load on Direct Acyclic Graph Based Blockchain for Internet of Things. 2019 CyberC (2019)
43. Ning, Z., et al.: Intelligent resource allocation in mobile blockchain for privacy and security transactions: a deep reinforcement learning based approach. Sci. China Inf. Sci. **64**(6), 1–16 (2021). https://doi.org/10.1007/s11432-020-3125-y

44. Tang, F., Pang, J., Cheng, K., Gong, Q.: Multiauthority traceable ring signature scheme for smart grid based on blockchain. Wireless Commun. Mobile Comput. **2021**, 1–9 (2021). https://doi.org/10.1155/2021/5566430
45. Fu, W., Wei, X., Tong, S.: An improved blockchain consensus algorithm based on raft. Arab. J. Sci. Eng. **46**(9), 8137–8149 (2021). https://doi.org/10.1007/s13369-021-05427-8
46. Zhu, Z., Qi, G., Zheng, M.: Blockchain based consensus checking in decentralized cloud storage. Simul. Model Pract. Tt. **102** (2020)
47. Chauhan, A., Malviya, O.P., et al.: Blockchain and scalability. In: IEEE QRS-C, pp. 122–128 (2018)
48. Kim, S., Kwon, Y., Cho, S.: A survey of scalability solutions on blockchain. In: IEEE ICTC, pp. 1204–1207 (2018)
49. Al-Bassam, M., Sonnino, A., et al.: Chainspace: a sharded smart contracts platform. arXiv: 1708.03778 (2017)
50. Gao, Y., Nobuhara, H.: A proof of stake sharding protocol for scalable blockchains. 2017 APAN, vol. 44, pp. 13–16 (2017)
51. Wang, W., Hoang, D.T., et al.: A survey on consensus mechanisms and mining strategy management in blockchain networks. IEEE Access **7**, 22328–22370 (2019)

Parallel Pileup Correction for Nuclear Spectrometric Data on Many-Core Accelerators

Zikang Chen[1,2], Xiangcong Kong[1,2], Xiaoying Zheng[1,2](✉), Yongxin Zhu[1,2](✉), and Tom Trigano[3]

[1] Shanghai Advanced Research Institute, Chinese Academy of Sciences, Shanghai, China
{chenzk,kongxiangcong2019}@sari.ac.cn
[2] University of Chinese Academy of Sciences, Beijing, China
{zhengxy,zhuyongxin}@sari.ac.cn
[3] Shamoon College of Engineering, Ashdod, Israel
thomast@sce.ac.il

Abstract. Spectroscopy devices suffer from the pulse pileup phenomenon, caused by overlapping of the signals. The energy-domain based pileup correction algorithm estimates the pulse energy distribution by measuring the duration and energies of pileups directly and does not need to identify each individual pulses. The correction algorithm can efficiently recovers the energy spectrum even under a very high photon arrival rate. However, the correction algorithm is sequential in nature and is slow when the energy resolution is high. A fast parallel implementation of the original correction algorithm is proposed in this paper. The parallel counterpart leverages state-of-the-art many-core system technology and achieves a nearly linear acceleration when the problem size scales. The speedup ratio exceeds 1,000 when the energy spectrum is split into 2,048 bins.

Keywords: Pileup correction · GPU · Parallel · X-ray spectroscopy

1 Introduction

Gamma spectroscopy is the study of the energy spectra of gamma ray sources, present in applications such as in the nuclear industry, geochemical investigation, and astrophysics. These sources emit particles which are recorded with a spectroscope, designed to measure the spectral power distribution of a radioactive source. The incident radiation generates a signal that allows to determine the energy of the incident particle. When these signal emissions are detected and analyzed with a spectroscopy system, the histogram of the recorded energies (known as energy spectrum in the field) can be produced. Many spectroscopy systems such as the ones based on Germanium detectors suffer from pulses pileup characterized by overlapping of the signals. This phenomenon deteriorates the energy spectrum and causes count losses due to random coincidences.

Various approaches have been developed over time to address the issue of pules pileups. These approaches can be categorized into time-domain based and

M. Qiu et al. (Eds.): SmartCom 2022, LNCS 13828, pp. 258–267, 2023.
https://doi.org/10.1007/978-3-031-28124-2_24

energy-domain based methods [1]. Time-domain based strategies attempt to detect the occurrence of piles in the time domain, the identified pileups being afterwards either discarded or compensated by numerical approaches such as maximum likelihood estimation [2]. Tough relevant for moderate radioactive activities, these approaches are eventually limited when the photon arrival rate increases [3]. On the other hand, energy-domain based strategies do not seek to identify and characterize individual pulses, but rather the pulses energy spectrum by means of statistical methods [1,4–6]. Energy-domain based strategies circumvent the need to identify individual pulses, but still require considerable computations and do not always satisfy the requirement of real-time processing.

We consider in this paper the acceleration of pileup correction algorithm developed in [5]. This recursive correction algorithm estimates the pulse energy histogram from measurements of the duration and energies of overlapping pulses. Compared with other time-domain algorithms, this approach can efficiently eliminate the pileup effect even under a high photon arrival rate. Furthermore, its lower algorithmic complexity makes it more fitting for real-time implementation compared with the method presented in [4]. However, due to its sequential nature, the algorithm is still slow when the number of energy histogram bins is large and the energy resolution is high.

With the development of computer hardware [7–9] and new algorithms [10–12], parallel data processing techniques [13–15] can be applied in many different areas, such as health care [16], autonomous drive [17], and traditional industries [18]. We propose to deploy the NVIDIA GPU *(Graphics Processing Unit)* technology to accelerate the pileup correction algorithm. GPUs are one of several available coprocessors that feature a large number of cores, and are specialized for compute-intensive and massively parallel computations; they became the *de facto* solution for parallel acceleration [19,20]. Custom acceleration of algorithms using GPU have become prevalent in high performance computing. For example, Kong et al. [21] performed custom acceleration of astronomical coherent dispersion algorithm on GPU to achieve real-time processing for pulsar searching, while Huang et al. [22] presented a CPU-GPU collaborative framework to accelerate the Bulletproofs protocol for blockchain applications. Qiu et al. [23] also explored multiprocessor algorithms for embedded systems. We illustrate in this paper the challenges encountered when applying GPU technology to perform pileup correction, and how to circumvent them. The numerical results show that the parallel implementation on GPU efficiently corrects the pileups with a significant speedup ratio exceeding 1000. The acceleration ratio remains nearly linear as the problem size scales up.

The rest of the paper is organized as follows. In Sect. 2, the pulses pileup problem is described and the original sequential algorithm is presented. In Sect. 3, a parallel implementation is elaborated. Section 4 presents the numerical results and the conclusion is drawn in Sect. 5.

2 Problem Statement

The response of an X-ray or gramma-ray detector to incident photons can be modeled as

$$s[n] = \sum_{k=1}^{\infty} \Phi_k[n - \lceil T_k \rceil], \tag{1}$$

where the arrival times $\{T_k, k \geq 1\}$ are unknown and form a Poisson process with know intensity λ, and Φ_k is the k-th recorded electrical pulse with duration X_k and energy Y_k. We assume that $\{(X_k, Y_k), k \geq 1\}$ is a sequence of independent and identically distributed integer-valued discrete random variables, independent of $\{T_k, k \geq 1\}$, with finite expectation and common probability mass function $p_{X,Y}$. In practice, both X_k and Y_k cannot be observed in experiments. Instead, the pileup's duration X'_k and energy Y'_k can be inferred from the experiments (as described in Fig. 1), and the common probability mass function is denoted by $p_{X',Y'}$.

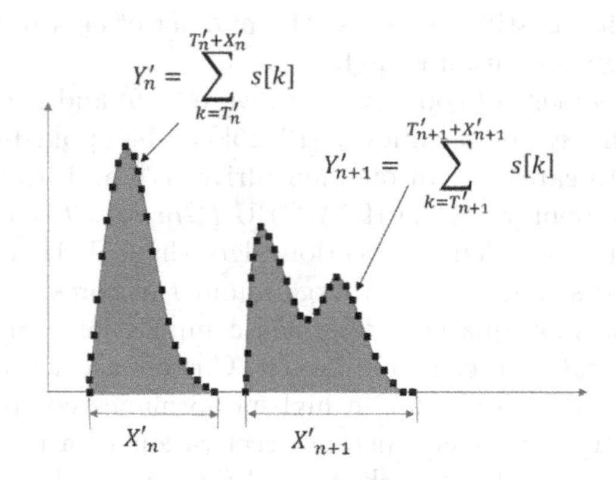

Fig. 1. A pileup example. The first electrical signal is not a pileup, so $X_k = X'_k$ and $Y_k = Y'_k$, whereas the second signal is a pileup, which yields $Y'_{k+1} > Y_{k+1}$ and $X'_{k+1} > X_{k+1}$.

The pileup correction algorithm described in [5] estimates the marginal distribution of the pulse's energy p_Y given the distribution of the observed durations and energies of pileups $\{(X'_k, Y'_k)\}$. We recall it for convenience in Algorithm 1. In this algorithm, the joint probability mass function $p_{X',Y'}$ is estimated by a bi-dimensional histogram on $[\![0, N]\!] \times [\![0, M]\!]$, where N and M are the number of histogram bins for duration and energy, respectively. This algorithm efficiently recovers the pulses' energy distribution from the histogram of the duration and energies of overlapping pulses, i.e., the pileups. However, Algorithm 1 is sequential in nature and it recursively corrects the probability of each energy bin, which is slow when the number of energy bins M is large and the energy resolution is high.

Algorithm 1. Algorithm proposed by [5].

1: INIT: $y[0][0] \leftarrow 1, y[n][m] \leftarrow 0$ for other n, m
2: **for** $n = 0, 1, \ldots N$ **do**
3: **for** $m = 0, 1, \ldots M$ **do**

$$y[n,m] \leftarrow e^{-\lambda} y[n-1,m] + (1-e^{-\lambda}) \sum_{k=0}^{n-1} \sum_{l=0}^{m} y[n-1-k,l] p_{X',Y'}[k,m-l] \quad (2)$$

4: **end for**
5: **end for**
6: **for** $n = 0, 1, \ldots N$ **do**
7: **for** $m = 0, 1, \ldots M$ **do**

$$k[n,m] \leftarrow \frac{y[n,m]}{\lambda y[n,0]} - \frac{1}{m} \sum_{k=1}^{m-1} (m-k) \frac{y[n,k]}{y[n,0]} k[n,m-k] \quad (3)$$

8: **end for**
9: **end for**
10: **for** $m = 0, 1, \ldots M$ **do**

$$p_Y(m) \leftarrow k[N,m] - k[N-1,m]$$

11: **end for**

3 Parallel Implementation

As discussed in Sect. 2, Algorithm 1 is not suited for real-time implementation when the number of energy bins M is large. The vigorous development and application of multi-core technology such as GPU has brought new opportunities for the acceleration of pileup correction. In this section, we introduce a parallel implementation of Algorithm 1 on GPU.

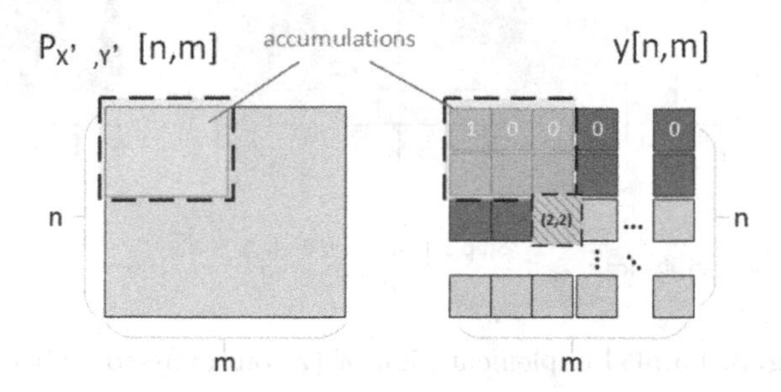

Fig. 2. CPU-based implementation of Eq. 2 computation.

It is noticed that the computation of (2) is the most time consuming step of Algorithm 1. As shown in Fig. 2, the computation of $y[n][m]$ involves all $y[k][l]$ for $k < n, l \leq m$. In the sequential implementation of Algorithm 1, each element of the matrix y is calculated one by one. Therefore the algorithm complexity depends on the size of matrix y, and as the size grows, the overall algorithm is slow. Fortunately, the accumulation term $\sum_{k=0}^{n-1}\sum_{l=0}^{m} y[n-1-k,l]p_{X',Y'}[k,m-l]$ can be parallelized in a many-core GPU environment and the execution speed is expected to increase.

We describe the parallel implementation of pileup correction in Algorithm 2. There are two challenges faced by the migration of pileup correction to a GPU environment. First, better parallelism can be achieved by element-wise scheduling for matrix y. The idea is to assign an individual GPU core to each element of matrix y so that the accumulation can be computed simultaneously. Figure 3 illustrates how parallelization is achieved by element-wise scheduling. For an $N \times M$ matrix y, NM GPU cores are required and N steps are executed in order to compute matrix y. At step i, the GPU cores that assigned to elements with row index $n, \cdots, N-1$ calculate its accumulative as in (2) with partial information obtained from step $i-1$. Also, the GPU cores assigned to elements with row index n will finish the computation of (2) and their results will be accessed at step $i+1$. The second challenge is that the iterative accumulation requires frequent memory access, which makes the memory bandwidth a bottleneck and leads to a high memory latency. Shared memory of GPU is generally faster than global memory. Therefore, shared memory is used to cache the i-th row of matrix y at step i. At step $i+1$, the program can access the shared memory to load the previous result with a low memory latency.

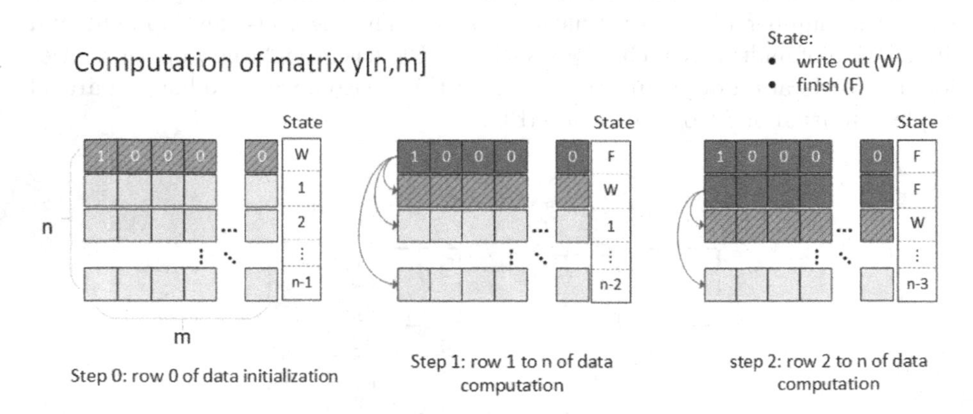

Fig. 3. Parallel implementation of (2) on many-core GPU.

Algorithm 2. Parallel implementation.

1: **for** each GPU core $P_{n,m}$ **do**
2: **if** $n == 0, m == 0$ **then**

$$y[0][0] \leftarrow 1$$

3: **else**

$$y[n][m] \leftarrow 0$$

4: **end if**
5: **end for**
6: ##Compute y[][]
7: **for** $i = 1, \ldots N$ **do**
8: **for** each GPU core $P_{n,m}$ **do**
9: **if** $n \geq i$ **then**

$$y[n,m] \leftarrow y[n,m] + (1 - e^{-\lambda}) \sum_{l=0}^{m} y[i-1,l] p_{X',Y'}[n-i, m-l] \quad (4)$$

10: **end if**
11: **if** $n == i$ **then**

$$y[n,m] \leftarrow y[n,m] + e^{-\lambda} y[i-1,m]$$

12: **end if**
13: **end for**
14: **end for**
15: ##Compute k[][]
16: **for** each GPU core $P_{n,m}$ **do**

$$k[n,m] \leftarrow \frac{y[n,m]}{\lambda y[n,0]}$$

17: **end for**
18: **for** $j = 1, \ldots M$ **do**
19: **for** each GPU core $P_{n,m}$ **do**
20: **if** $m > j$ **then**

$$k[n,m] \leftarrow k[n,m] - \frac{j}{m} k[n,j] \frac{y[n,m-j]}{y[n,0]}$$

21: **end if**
22: **if** $n == N$ **then**

$$p_Y(m) \leftarrow k[n,m] - k[n-1,m]$$

23: **end if**
24: **end for**
25: **end for**

3.1 Time Complexity

For the original implementation of Algorithm 1, the nested loop for calculating the matrix y (Line 2 to Line 5) dominates the time complexity. In the loop, the calculation of $y[n, m]$ as in (2) involves an accumulation operation of time complexity $O(NM)$, and thus the complexity of the nested loop is $O(N^2M^2)$. Therefore, the overall time complexity of Algorithm 1 is $O(N^2M^2)$.

Regarding the parallel implementation of Algorithm 2, for each GPU core, it first goes through a loop of N steps (Line 6 to Line 13). At each step, each core runs an accumulation operation of complexity $O(M)$. Secondly, each GPU core goes through a loop of M steps (Line 16 to Line 23), and each step involves a constant number of operations. Therefore, there are NM GPU cores, and for each GPU core, its time complexity is $O(NM)$.

4 Numerical Results

The ideal density used in [5] was used for generating simulated pileup durations and energies. It consists of a mixture of six Gaussian and one gamma distribution to simulate the Compton background. The probability density of the mixture f_Y is proportional to

$$0.5g + 10N(40, 1) + 10N(112, 1) + N(50, 2) + N(63, 1) + 2N(140, 1) + N(200, 1), \tag{2}$$

where $N(\mu, \sigma^2)$ is the probability density of a normal distribution with mean μ and variance σ^2. The probability density of the gamma distribution is given by $g(x) = (0.5 + x/200)e^{-(0.5+x/200)}$ for $x \geq 0$.

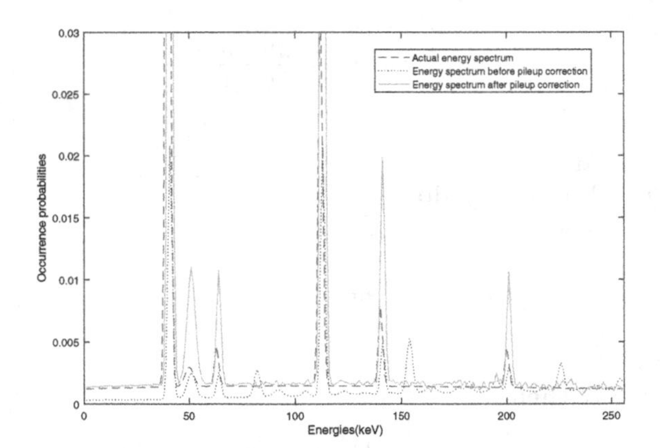

Fig. 4. Pileup correction by Algorithm 2. $\lambda = 2,000,000$ photons per second. $N = 16, M = 256$.

Algorithm 1 is experimented with an Intel(R) Xeon(R) Gold 6230 CPU, and Algorithm 2 is experimented on the same platform with an additional NVIDIA Tesla V100S PCIe 32 GB card. Figure 4 shows that the energy spectrum is efficiently corrected by parallel Algorithm 2: all the fake spikes (at 80 KeV, 152 KeV and 224 KeV) caused by pulse overlapping are discarded, and the obtained estimation of the energy spectrum almost perfectly matches the objective density even with a very high photon arrival rate of $\lambda = 2,000,000$.

Next, we demonstrate the performance of Algorithm 2 in terms of execution speed. The speedup ratio ρ is defined in (3), where t_1 and t_2 are the run time of Algorithm 1 and Algorithm 2, respectively.

$$\rho = t_1/t_2. \tag{3}$$

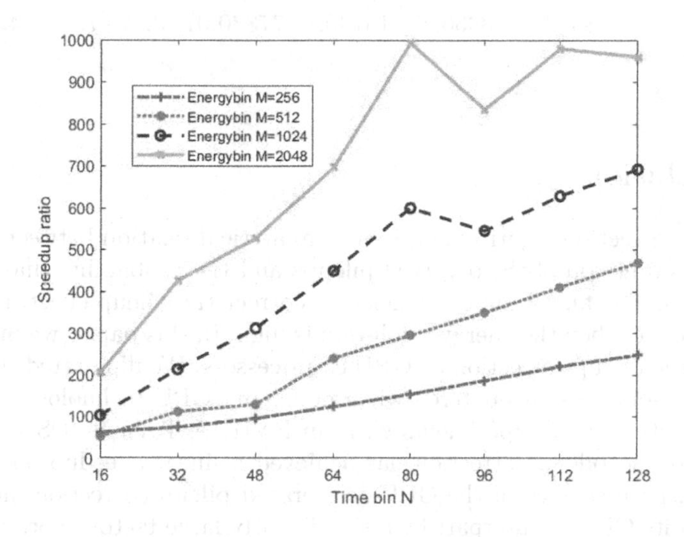

Fig. 5. Speedup ratio ρ under different combinations of N and M.

Table 1 presents the run time of Algorithm 1 and Algorithm 2 with different combination of energy bins and duration bins. With duration bin $N = 128$ and energy bin $M = 2048$, the speedup ratio ρ exceeds $1000x$. In Fig. 5, the speedup ratio ρ is illustrated. The results in Fig. 5 demonstrates a linear speedup as the numbers of duration and energy bins N, M increase. It is obvious that at the current problem scale, the NVidia V100 has sufficient cores to maximize the parallelism and the execution speed of the pileup correction has been greatly improved.

Table 1. Running time (ms)

	Before parallelization				After parallelization			
#duration bins	#energy bins							
	256	512	1024	2048	256	512	1024	2048
16	32.346	58.043	227.769	921.014	0.52	1.11	2.22	4.5
32	69.8	225.831	879.442	3844.21	0.94	2.02	4.13	9.04
48	133.541	379.837	1994.05	9704.03	1.4	2.94	6.4	18.41
64	224.961	927.954	3709.74	17404.6	1.82	3.87	8.26	24.97
80	342.999	1406	6104.29	33034	2.23	4.77	10.17	33.27
96	490.585	2017.26	9609.35	41016.6	2.64	5.78	17.61	49.13
112	669.725	2733.88	13035.6	58235.5	3.04	6.66	20.72	59.47
128	859.955	3630.45	16730.1	77330.01	3.47	7.75	24.16	80.59

5 Conclusion

The pileup correction algorithm derives an analytical relation between the probability mass function of the observed pileups and the probability mass function of the pulses. The algorithm can efficiently correct the pileup effect, but is very slow in practice when the energy resolution is high. In this paper, we proposed to accelerate the pileup correction by GPU coprocessors. We illustrated the difficulties and the solutions encountered when deploying GPU technology to perform pileup correction. Our experiments with an NVIDIA Tesla V100S PCIe 32 GB showed that the pileup correction has achieved a linear speedup on the GPU. The overall performance of the GPU-accelerated pileup correction method can outperform its CPU counterpart by a significantly large factor, more than 1000, which is important in real-time processing nuclear spectrometric data in practice.

Acknowledgment. This work was supported partially by the National SKA Program of China (Grant No. 2020SKA0120202), the National Natural Science Foundation of China (Grant No. U2032125), the Science and Technology Commission of Shanghai Municipality (Grant No. 21511101400), and Shanghai Talent Development Fund (Grant No. E1322E1). Sincere gratitude to all the people who helped me during this period.

References

1. Mclean, C., Pauley, M., Manton, J.H.: Non-parametric decompounding of pulse pile-up under gaussian noise with finite data sets. IEEE Trans Signal Process **68**, 2114–2127 (2020)
2. Bolic, M., Drndarevic, V., Gueaieb, W.: Pileup correction algorithms for very-high-count-rate gamma-ray spectrometry with Nai(Tl) detectors. IEEE Trans. Instru. Meas. **59**(1), 122–130 (2010)

3. McLean, C., Pauley, M., Manton, J.H,: Limitations of decision based pile-up correction algorithms. In: IEEE Workshop SSP, 2018, pp. 693–697 (2018)
4. Trigano, T., Dautremer, T., et al.: Pile-up correction algorithms for nuclear spectrometry. In: IEEE ICASSP, vol. 4, pp. iv/441-iv/444 Vol. 4 (2005)
5. Trigano, T., Barat, E., et al.: Fast digital filtering of spectrometric data for pile-up correction. IEEE Signal Proc. Lett. **22**(7), 973–977 (2015)
6. Trigano, T., Souloumiac, A., Montagu, T., et al.: Statistical pileup correction method for HPGe detectors. IEEE Trans. Signal Process. **55**(10), 4871–4881 (2007)
7. Qiu, M., Li, H., Sha, E.: Heterogeneous real-time embedded software optimization considering hardware platform. In: ACM SAC, pp. 1637–1641 (2009)
8. Li, J., Ming, Z., et al.: Resource allocation robustness in multi-core embedded systems with inaccurate information. J. Syst. Architect. **57**(9), 840–849 (2011)
9. Qiu, M., Jia, Z., et al..: Voltage assignment with guaranteed probability satisfying timing constraint for real-time multiproceesor DSP. J. VLSI Signal Process. Syst. Signal Image Video Technol. **46**, 55–73 (2007)
10. Qiu, M., Yang, L., et al.: Dynamic and leakage energy minimization with soft real-time loop scheduling and voltage assignment. IEEE Trans. Very Large Scale Integr. **18**(3), 501–504 (2009)
11. Qiu, M., Xue, C., et al..: Energy minimization with soft real-time and DVS for uniprocessor and multiprocessor embedded systems. In: IEEE DATE, 2007, pp. 1–6 (2007)
12. Zhang, L., Qiu, M., et al.: Variable partitioning and scheduling for MPSoC with virtually shared scratch pad memory. J. Signal Process. Stys. **58**(2), 247–265 (2018)
13. Hu, F., Lakdawala, S., et al.: Low-power, intelligent sensor hardware interface for medical data preprocessing. IEEE Trans. Info. Tech. Bio. **13**(4), 656–663 (2009)
14. Qiu, M., Sha, E., et al.: Energy minimization with loop fusion and multi-functional-unit scheduling for multidimensional DSP. J. Paralell Distrib. Comput. **68**(4), 443–455 (2008)
15. Shao, Z., Wang, M., et al.: Real-time dynamic voltage loop scheduling for multi-core embedded systems. IEEE Trans. Cir. Sys. II **54**(5), 445–449 (2007)
16. Gai, K., et al.: Electronic health record error prevention approach using ontology in big data. In: IEEE 17th HPCC (2015)
17. Qiu, H., Zheng, Q., et al.: Topological graph convolutional network-based urban traffic flow and density prediction. In: IEEE Trans. Intell. Transp. Syst. **22**. 4560–4569 (2020)
18. Gai, K., Qiu, M., et al.: In-memory big data analytics under space constraints using dynamic programming. Fut. Gen. Ccomput. Syst. **83**, 219–227 (2018)
19. Wu, G., Zhang, H., et al.: A decentralized approach for mining event correlations in distributed system monitoring. J. Paralell Distrib. Comput. **73**(3), 330–340 (2013)
20. Qiu, M., Khisamutdinov, E., et al.: RNA nanotechnology for computer design and in vivo computation. Philoso. Trans. Ser. A **371**(2000), 20120310 (2013)
21. Kong, X., Zheng, X., Zhu, Y., et al.: Custom computing design and implementation for multiple dedispersion with GPU. In: CSCloud/EdgeCom, pp. 103–108 (2021)
22. Huang, Y., Zheng, X., Zhu, Y., et al..: CPU-GPU collaborative acceleration of bulletproofs-a zero-knowledge proof algorithm. In: IEEE ISPA/BDCloud/SocialCom/SustainCom, 2021, pp. 674–680 (2021)
23. Qiu, M., et al.: Energy minimization with soft real-time and DVS for uniprocessor and multiprocessor embedded systems. In: IEEE DATE, 2007, pp. 1–6 (2007)

Component Extraction for Deep Learning Through Progressive Method

Xiangyu Gao[1], Meikang Qiu[2(✉)], and Hui Zhao[3]

[1] New York University, New York City, NY, USA
xg673@nyu.edu
[2] Dakota State University, Madison, SD 57042, USA
Meikang.Qiu@dsu.edu
[3] Educational Information Technology Lab., Henan University, Kaifeng 475000, China
zhh@henu.edu.cn

Abstract. Machine learning has shown great impact in a lot of applications. Within all types of tools, deep learning should be one of the most important techniques thanks to its ability to capture the correlation between the input features and output results. However, the relatively long training time and high computation complexity remain a big problem in deep learning. In addition, the impossibility to explain the model makes it harder for us to look for alternatives to fix the bad fitting results. Therefore, this paper aims at improving the deep learning model training result by proposing a principal component extraction algorithm. Compared with the previous *Principal Component Analysis* (PCA) methods, this algorithm creatively consider not only the original input components but also the computed variables in the first hidden layer in neural network so as to capture more representative components. The experiment shows that compared to previous PCA method, this can better capture the principal components from all input variables.

Keywords: Deep learning · Machine learning · Principal component analysis · Deep neural network · Computation complexity · Back propagation

1 Introduction

Machine learning has become quite popular across different areas including data-driven trading [12], cyber security [10] and social media [11]. Specifically, all of these papers propose algorithms that purely rely on the data rather than people's professional knowledge to make better decision because the results are objective without any influence from human's bias. For example, Gao and Qiu [12] set a momentum element to measure the S&P 500 index [7] market. Whenever the momentum element reaches a threshold, this trading strategy will execute the corresponding buy/sell actions. With the energy [34,37,38] and security [30,31] requirement in various application, how to processing data more efficient is a hot research area.

M. Qiu et al. (Eds.): SmartCom 2022, LNCS 13828, pp. 268–279, 2023.
https://doi.org/10.1007/978-3-031-28124-2_25

Among all machine learning techniques, deep learning [20] should the most widely used one given its good performance among a broad spectrum of applications. The well-known deep learning case is AlphaGo [39], which trained deep neural network to achieve 99.8% winning rate against other Go programs and human being Go players. To be specific, in deep learning [29,32], multiple processing layers are used to learn representations of data [16,27,36], which increase the possibility that such model can precisely find the correlation between the input data and output result. Therefore, as for several complex problems, deep learning might be a good fit for them.

However, deep learning still have several drawbacks [24]. On the one hand, it takes long time for us to train such model because of complicated structure and a lot of parameters. As for this problem, even after Hecht-Nielsen developed back propagation [15] idea and Kung [19] proposed effective neural architecture search algorithm, sometimes the training process is still not short. On the other hand, given the fact that it is usually hard to explain why deep learning can/cannot work, it is hard for people to attribute the bad model training result to either no correlation between input and output or unsuitable model selection. To make things worse, people's blind eyesight towards deep learning prevent them from investing more money or time to such area.

In order to tackle all these concerns above, in this paper, we aim at presenting a novel *Principal Component Analysis* (PCA) method especially for *Deep Neural Network* (DNN). Previous PCA [25] method takes all features as an input and extract some principal components from eigenvalues and eigenvectors of the feature matrix. The goal for this method is to reduce the dimension of the input variable space so that it can make the model training problem easier. However, such method sometimes should not be sufficient in the situation of DNN. Generically speaking, the model training complexity and computation complexity are proportional to the size of input features dimension. Therefore, it would not help much if the PCA fails to significantly reduce the size of the input data. If we only focus on reducing the dimension of input features but relax a little bit on the component extraction accuracy by lowering the PCA score threshold, it might affect the final fitting result in DNN. With respect to such issue, in the paper written by Kung [19], it considers doing PCA over the space spanned by input features and complement subspace spanned by the neuron data in hidden layer for effective PCA analysis. In this paper, our approach ignores the complement subspace from the hidden layer's neuron data matrix and takes both into consideration and hence include all calculated features in the first hidden layer into the feature pool for PCA. We believe this could be a more effective way to finally get the components. There are three main contributions of this paper:

- An overview of several drawbacks of deep learning framework together with the explanation of why traditional PCA might not be suitable for DNN.
- A novel algorithm to modify the current PCA methods for the DNN.
- The implementation of the proposed algorithm to compare its performance against the present ones.

The remainder of this paper is organized as follows. Section 2 summarizes the background of deep learning features and the PCA techniques. Then, Sect. 3

presents the motivation of our paper by listing several drawbacks of current PCA methods so as to highlight the necessity to develop a new method in this background. Furthermore, Sect. 4 subsequently gives a detailed description of our algorithm, followed by several experiment results in Sect. 5. Finally, conclusions are shown in Sect. 6.

2 Background

In this section, we are going to overview two methods: deep learning and *Principal Component Analysis* (PCA). Both of them are popularly used in the machine learning framework. Several important techniques of deep learning and PCA will be highlighted as well.

2.1 Deep Learning

With the rapid development of computer capability [21,35], software [8,41,43] and cloud computing [9,22,33], large amounts of data can be generated and machine learning becomes a hot area. Deep learning is a subfield of machine learning. It tries to simulate the functionality of our brain so as to build the connection between the input space and the output space. When we take a closer look at the deep learning framework shown in Fig. 1, it is easy for us to build the analogy between nodes in Fig. 1 and cells within our brain [45]. When an input data comes into the *Deep Neural Network* (DNN), it will be fed into the input layer. All nodes in the previous layer will be used to compute the output that shows up in the next layer. This process will last layer by layer until the final results are generated from the output layer.

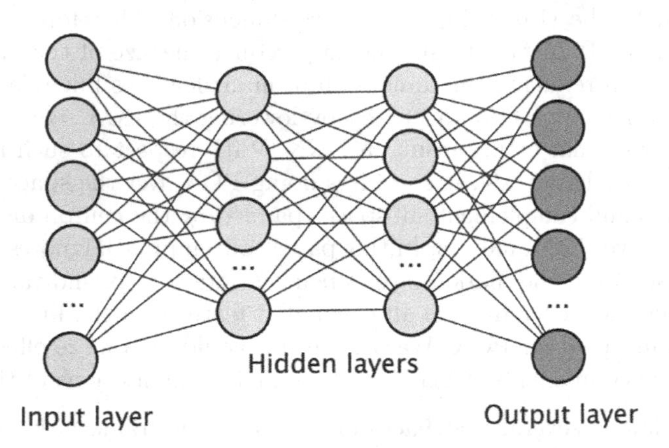

Fig. 1. The visualization of deep neural network.

Given its similarity between human brain and DNN [3], deep learning shows significant impact on many real-world applications. In other words, in practice the input/output relationship is hard to describe by a simple function, therefore the DNN might be a reasonably good solution to depict such phenomenon. By building computational models that are composed of multiple processing layers, the networks can present complex input/output mapping functionalities by dividing them into small but simple functions, each of which is represented by one layer-to-layer connection. Then, the combination of these multiple layers can formulate a complex function.

With in the DNN platform, several techniques have been used to improve its performance in terms of training speed and model fitting effect. On the one hand, since there are many layers together with a lot of parameters requiring training, back propagation [15] method has been proposed to speed up the training process. To be specific, it goes back and forth in the neural network structure and update the parameters' value accordingly by calculating the derivatives of each node. Main advantages of back propagation include simplicity to program, flexibility and fast implementation. On the other hand, people are also eager to find suitable functions between layers to easily capture the relationship between the input and the output. We call these functions as activation functions or transfer functions. Popularly used activation functions contain linear combination, sigmoid, tanh and relu. People are still interested in the way to select the suitable activation functions [14] in different scenarios.

2.2 Principal Component Analysis

The developed data storage technology enables it easy to record large amount of data. In a lot of situations, the dimension of the data might be too high to train quickly. In reality, high-dimensional are common in biology [17] and finance [4] field, when multiple features are measured for each sample. There are usually several problems with the high-dimension data. An obvious one is that the model training time will increase exponentially as the dimension increases, which makes the parameters' value hard to obtain. In addition, usually the error increases with the increase in the number of features, mainly because of some redundant information stored. One of the most important ones might be related to the case that high-dimension features require relatively larger amount of data to feed. Otherwise, it may cause over-fitting [23] to occur.

When it comes to the high-dimension data as the input, people prefer doing some data cleaning before training them into the specific model. *Principal component analysis* (PCA) and *linear discriminant analysis* (LDA) [42] are two popular options, which are visualized in Fig. 2. At a high level, these methods look for linear combination of features which can best explain the data. Then, some threshold will be defined to pick from these generated features to reach a balance between the number of selected features and how good these features can represent the data. LDA aims at finding a feature space that can maximize the separability between groups while PCA focuses on finding the direction of maximum variation in the data set. There are many other dimension reduction techniques [5] as well but all of them share the same ultimate goal that they

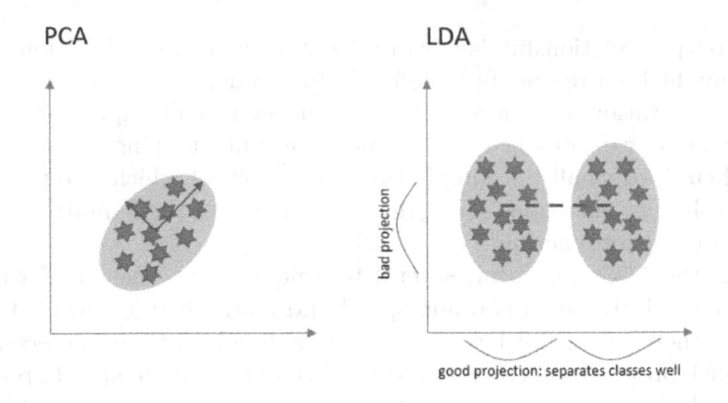

Fig. 2. An example to show how backdoor attacks work.

want to represent the data as much as possible with the minimum number of extracted features.

3 Motivation

We will present the motivation of our paper in this section. To be specific, after overviewing the DNN model in deep learning field and some dimension reduction methods (e.g., PCA and LDA), we will present several existing drawbacks for these techniques. These potential problems will motivate us to consider new approaches to achieving better model training performance.

3.1 Information Loss in PCA

General steps of PCA include data standardization, covariance matrix computation, eigenvectors and eigenvalues calculation, feature vectors selection and data recasting to the principal components. After going through all these steps, we can easily generate the principal components whose dimension is usually smaller than original features. Therefore, the PCA method contributes a lot to machine learning techniques because it can remove correlated features; then, if all the remaining components are independent to each other, it could reduce the overfitting and speed up the computation process because the requirement of the data size is proportional to the dimension of the input features.

Although all these benefits of PCA might be helpful in several situations, there still exist a lot of problems [44], one of which is the information loss. Note that feature vectors selection is one of the important steps in PCA and it must be the case that after selecting feature vectors, we have to lose some information hidden in the input feature vectors. Sometimes, if the original input features are relatively independent across each other and we still force the PCA to reduce the input dimension, it might cause to the fact that quite much useful information is eliminated, which will have side effect on the model training. Moreover, this might hurt the model training because compared with the original input features the principal components are hard to interpret.

3.2 Lack of Interpretability in DNN

Although the relative complicate DNN structure makes it possible for deep learning to gain good model fitting result, it becomes harder for people to explain the fitting result reasonably. In other words, no matter how good or bad the data fits into such model, we cannot tell why it is the case, let alone coming up convincing ideas to make further improvement. Most of the trials are related to change the model structure to expect better result.

Typically, DNN is regarded as the black-box function [6] which maps a given input to a corresponding output. Multiple layers, many nodes per layer and complex layer-to-layer connections make the model space relatively large. Therefore, it increases the possibility that some part of the model space can precisely describe the input/output relationship. We can refer to such phenomenon to the fact that a lot of continuous function can be written as the Taylor formula. As the number of terms within the Taylor formula increases, the value difference between such formula and the target function will become negligible.

However, such invisibility of the DNN might cause some problems in reality. If people cannot have a full eyesight of the inner structure within the model, it is hard to explain why it works well, which sometimes make it risky to rely fully on it. For instance, in financial area, even DNN can obtain good fitting result in the existing training data, the limitation [26] to interpret such result leads to the reality that most companies still hesitate to get that involved in automated trading. Different from models such as linear regression, where if the training result is good it means the input and output relationship is linear, people dare not bet on DNN no matter how good performance it has in the training data.

4 Our Approach

In this section, we want to leverage the pseudo code to concretely show each step of our algorithm design, followed by the description of the intuition behind. Generically speaking, through expanding the scope of candidate features used for PCA by adding the calculated variables in the first hidden layer, we believe this can help extract representative components for better DNN model training.

4.1 Algorithm Design

We are now presenting the algorithm concretely below. It offers a new approach to doing the PCA for DNN.

Compared with directly doing PCA over the input features, our proposed algorithm runs model training over a three-layer neural network at first. After the model training, we include all nodes in the hidden layer into input features space because we think this will be a valuable addition to the existing features

Algorithm 1. ***Progressive PCA*** **Algorithm**

Require: M input features, size of the data set N, activation function of DNN F, the maximum number of layers of DNN T, and number of eigenvalues K for selection.
Ensure: The updated set of features that originate from the modified PCA methods on input features.

1: Set up a Deep Neural Network structure with only three layers (one input layer, one output layer and one hidden layer).
2: Train the three-layer DNN on the input data set.
3: Record both the input features value and the values from all nodes in the hidden layer into a set S.
4: Implement PCA on the collected set S in order to extract K principal components.

5: Take the extracted K components as the input and train deep neural network model over the data.
6: Output the DNN training result.
Output results: The final results include all parameters of the trained DNN framework together with the fitting score.

pool. Then, we would do a PCA over the extended feature set to extract K principal components that would be finally used for DNN training later. More details are shown in the algorithm description part.

4.2 Intuition Analysis

We believe our algorithm is a good fix to the existing PCA algorithm and want to mention its potential benefits intuitively. First of all, general PCA method tries to get the minimum number of principal components that can represent the maximum information within the input features. Usually, these components would be some linear combination of existing components. However, if most of the features are already relatively orthogonal to each other, PCA cannot have a good effect in terms of the dimension reduction. Therefore, we want to add more derivative features from DNN framework into the feature space for PCA.

In addition, our proposed algorithm can strike a balance between short running time and good component extraction effect. As is known to all, training a DNN might be time-consuming so we do not want to train a neural network with a lot of hidden layers at first. Then, we decide the number of hidden layer to be only one. This decision can help us not only generate new features quickly but also increase the size of feature set for component extraction later. Then, after the PCA, we can consider DNN with multiple hidden layers so as to achieve good model training result.

5 Experiments

This section is mainly about the experiment results of our proposed algorithm and its comparison against others. In the traditional DNN training process, people would directly feed the data into the neural network pipeline and train such model using feedforward and backpropagation methodology. We want to see how much improvement our algorithm can provide. We are concerned of two main metrics: accuracy [18] and training time [2]. Accuracy is an important determinant since it directly decides whether the correlation between input and output could be represented by DNN. Besides, we also want to get the reasonably good result the sooner the better; hence, the training time is another point worth paying attention to. In our experiment setup, we regard the number of epochs as the concrete index to measure the length of training time.

5.1 Data Description

In our experiment, we leverage our model to fit the open source data of house price [1]. This is because usually the price of a house depends on many features. Within the data set, the output Y variable is binary, where 1 means the price is above the median and 0 means the price is below the median. Ten features (e.g., overall quality, overall condition, distance to surrounding utilities) are considered as important determinants to the final price. There are 1,460 data points in total within the data set.

We utilize related packages in python (e.g., sklearn [28], keras [13]) to implement our algorithm. Concretely speaking, house price data is stored in the csv format. We read the data from csv file and store it into the pandas data frame. Then, we implement several preprocessing techniques into the data with inserted packages such as standardization. The standardized data is split into training set and testing set. In our setting, we put the proportion of training data to be 70% of the whole data set while the remaining 30% are used as the testing set. The DNN model from keras package will be trained in the training data and we test its performance over testing data set.

5.2 Results

When it comes to the results comparison, we compare our method against pure DNN model because we want to see how much improvement such proposed algorithm can bring. As for the metrics, we take into consideration accuracy [18] and loss [40] value.

The comparison results are illustrated in Fig. 3 and Fig. 4. In general, given the high accuracy and low loss value, we can see that DNN should be a suitable candidate model for house price prediction. First of all, it is easy to witness that the big trend is that the fitting results get better and better as the number of epochs increases, even if the marginal improvement decrements. If we directly feed the data into DNN without any PCA, the performance gap between training data and testing data is larger, which means that the model has a trend to overfit the training data set but does not work well on the testing set. In comparison, the difference of the performance between the training set and testing set is much smaller if we implement our proposed PCA into the data set beforehand.

In addition, we also compare the convergence speed of both DNN with and without our proposed PCA techniques in Fig. 5. In Fig. 3 and Fig. 4, we set

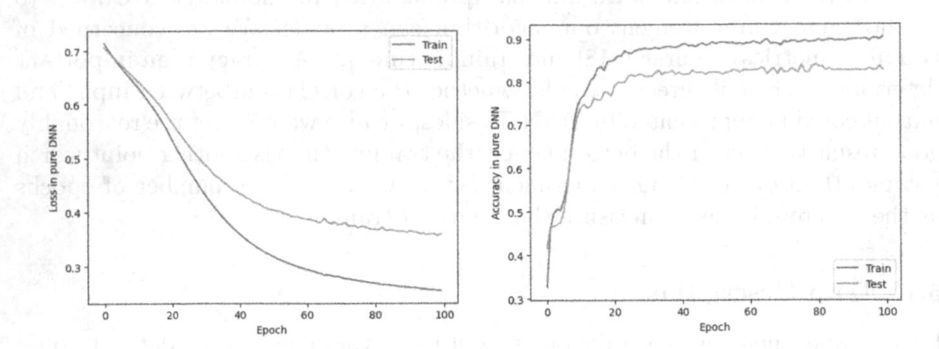

Fig. 3. The loss and accuracy measurement in DNN framework.

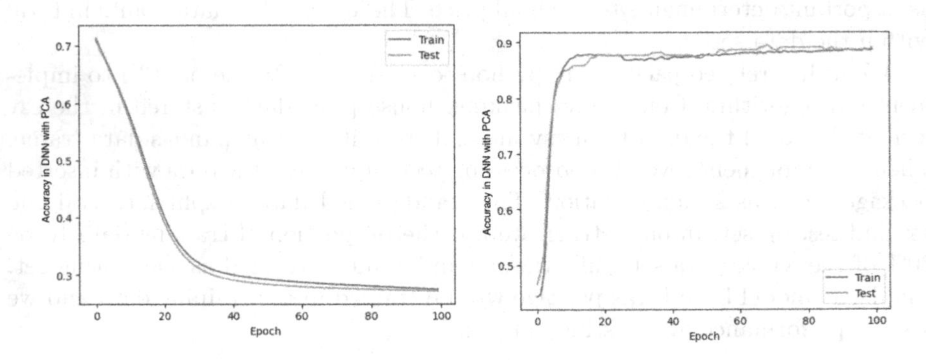

Fig. 4. The loss and accuracy measurement with proposed PCA implemented.

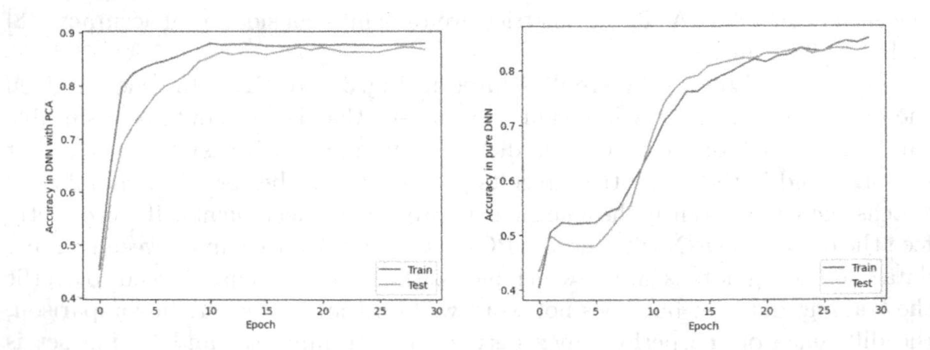

Fig. 5. Accuracy comparison when the number of epoch is 30.

the number of epochs to be 100 because it might be necessary for us to train the model longer for better result. In Fig. 5, we make an assumption that the computation power is relatively insufficient which forces the user to train the model within a short period of time. Then we reduce the epoch number from 100 down to 30.

Then, we can find that if we implement our proposed PCA before DNN model training, it takes less than five epochs to grow up to a platform whose accuracy is over 80%, while this is not the case for direct DNN model training. The figure tells us that after the epoch number is greater than twenty, the accuracy level starts to be larger than 80%. A reasonable explanation to such phenomenon is that our proposed PCA has already extracted the important information from the data set successfully, and then it make the DNN model training process easier; However, pure DNN model does not have any preprocessing work, which let the model training task more difficult to build the correlation between input and output features.

6 Conclusions

In this paper, we introduced a novel component extraction method for deep neural network, an idea that puts both hidden layer nodes and input features together for *Principal Component Analysis* (PCA). Our approach is an extension of existing PCA tool, which decides to include more relative elements for better component generation. The experiment results shown that in terms of house price prediction, compared with not using our techniques, our model's fitting result improves and the gap between training data set and testing data set is smaller. We hope this proposed algorithm can motivate more deep learning users to implement in more application fields.

References

1. Zillow's home value prediction Kaggle competition data. https://www.kaggle.com/c/zillow-prize-1/data
2. Bennett, K., Mangasarian, O.: Neural network training via linear programming. Technical report, University of Wisconsin-Madison Department of Computer Sciences (1990)
3. Brahma, P., Wu, D., She, Y.: Why deep learning works: a manifold disentanglement perspective. IEEE Trans. Neural Netw. Learn. Syst. **27**(10), 1997–2008 (2015)
4. Breymann, W., Dias, A., Embrechts, P.: Dependence structures for multivariate high-frequency data in finance. Quant. Financ. **3**(1), 1 (2003)
5. Carreira-Perpinán. M.: A review of dimension reduction techniques. Department of Computer Science. University of Sheffield. Tech. Rep. CS-96-09, 9:1–69 (1997)
6. Chakraborty, S., et al.: Interpretability of deep learning models: a survey of results. In: 2017 IEEE smartworld, ubiquitous intelligence, pp. 1–6. IEEE (2017)
7. Frino, A., Gallagher, D.: Tracking s&p 500 index funds. J. Portf. Manag. **28**(1), 44–55 (2001)

8. Gai, K., Qiu, M., Chen, L., Liu, M.: Electronic health record error prevention approach using ontology in big data. In: IEEE 17th HPCC (2015)
9. Gai, K., Qiu, M., Elnagdy. S.: A novel secure big data cyber incident analytics framework for cloud-based cybersecurity insurance. In: 2016 IEEE 2nd International Conference on Big Data Security on Cloud (BigDataSecurity) (2016)
10. Gao, X., Qiu, M.: Energy-based learning for preventing backdoor attack. In International Conference on Knowledge Science, Engineering and Management, pp. 706–721 (2022)
11. Gao, X., Qiu, M.: Recommendation system design for social media using reinforcement learning. In: 2022 IEEE 9th International Conference on Cyber Security and Cloud Computing, pp. 6–11 (2022)
12. Gao, X., Qiu, M., He, Z.: Big data analysis with momentum strategy on data-driven trading. In: IEEE ISPA, pp. 1328–1335 (2021)
13. Gulli, A., Pal, S.: Deep Learning with Keras. Packt Publishing Ltd., Birmingham (2017)
14. Hayou, S., Doucet, A., Rousseau, J.: On the selection of initialization and activation function for deep neural networks. arXiv preprint arXiv:1805.08266 (2018)
15. Hecht-Nielsen, R.: Theory of the backpropagation neural network. In: Neural Networks for Perception, pp. 65–93. Elsevier, Academic Press (1992)
16. Hu, F., Lakdawala, S., et al.: Low-power, intelligent sensor hardware interface for medical data preprocessing. IEEE Trans. Infor. Tech. Biomed. **13**(4), 656–663 (2009)
17. Hua, J., Tembe, W., Dougherty. E.: Feature selection in the classification of high-dimension data. In: 2008 IEEE International Workshop on Genomic Signal Processing and Statistics, pp. 1–2, IEEE (2008)
18. Johnson, V., Rogers, L.: Accuracy of neural network approximators in simulation-optimization. J. Water Resour. Plan. Manag. **126**(2), 48–56 (2000)
19. Kung. S.: XNAS: a regressive/progressive NAS for deep learning. ACM Trans. Sensor Netw. (TOSN) (2022)
20. LeCun, Y., Bengio, Y., Hinton, G.: Deep learning. nature **521**(7553), 436–444 (2015)
21. Li, J., Ming, Z., et al.: Resource allocation robustness in multi-core embedded systems with inaccurate information. J. Syst. Arch. **57**(9), 840–849 (2011)
22. Li, Y., Gai, K., et al.: Intercrossed access controls for secure financial services on multimedia big data in cloud systems. ACM Trans. Multim. Compu., Commu., Appli, **12**(4) (2016)
23. Liu, B., Wei, Y., Zhang, X., Yang, Q.: Deep neural networks for high dimension, low sample size data. In: IJCAI, pp. 2287–2293 (2017)
24. Marcus, G.: Deep learning: a critical appraisal. arXiv preprint arXiv:1801.00631 (2018)
25. Martinez, A., Kak, A.: PCA versus LDA. IEEE Trans. Pattern Anal. Mach. Intell. **23**(2), 228–233 (2001)
26. Ng, C.: The future of AI in finance. The AI Book: The Artificial Intelligence Handbook for Investors, Entrepreneurs and FinTech Visionaries, pp. 6–8 (2020)
27. Niu, J., Gao, Y., Qiu, M., Ming, Z.: Selecting proper wireless network interfaces for user experience enhancement with guaranteed probability. J. Parallell Distrib. Comput. **72**(12), 1565–1575 (2012)
28. Pedregosa, F., Varoquaux, G., Gramfort, A., Michel, V., Thirion, B., Grisel, O., et al.: Scikit-learn: machine learning in python. J. Mach. Learn. Res. **12**, 2825–2830 (2011)

29. Qiu, H., Dong, T., et al.: Adversarial attacks against network intrusion detection in IoT systems. IEEE Internet of Things J. **8**(13), 10327–10335 (2020)
30. Qiu, H., Qiu, M., Lu, R.: Secure V2X communication network based on intelligent PKI and edge computing. IEEE Netw. **34**(2), 172–178 (2019)
31. Qiu, H., Zeng, Y., Guo, S., et al.: Deepsweep: an evaluation framework for mitigating DNN backdoor attacks using data augmentation. In: ACM ASIA Conference on Computer and Communications Security (2021)
32. Qiu, H., Zheng, Q., et al.: Topological graph convolutional network-based urban traffic flow and density prediction. IEEE Trans. Intell. Trans. Syst. **22**, 4560–4569 (2020)
33. Qiu, M., Chen, Z., Ming, Z., Qin, X., Niu, J.: Energy-aware data allocation with hybrid memory for mobile cloud systems. IEEE Syst. J. **11**(2), 813–822 (2014)
34. Qiu, M., Jia, Z., Xue, C., Shao, Z., Sha, E.: Voltage assignment with guaranteed probability satisfying timing constraint for real-time multiproceesor DSP. J. VLSI Signal Proc. Sys. **46**, 55–73 (2007)
35. Qiu, M., Li, H., Sha, E.: Heterogeneous real-time embedded software optimization considering hardware platform. In Proceedings of the 2009 ACM Symposium on Applied Computing (SAC), Honolulu, Hawaii, USA, 9-12 March 2009, pp. 1637–1641 (2009)
36. Qiu, M., Xue, C., Shao, Z., Zhuge, Q., Liu, M., Sha, E.H.-M.: Efficent algorithm of energy minimization for heterogeneous wireless sensor network. In: Sha, E., Han, S.-K., Xu, C.-Z., Kim, M.-H., Yang, L.T., Xiao, B. (eds.) EUC 2006. LNCS, vol. 4096, pp. 25–34. Springer, Heidelberg (2006). https://doi.org/10.1007/11802167_5
37. Qiu, M., Xue, C., Shao, Z., Sha. E.: Energy minimization with soft real-time and DVS for uniprocessor and multiprocessor embedded systems. In: 2007 Design, Automation Test in Europe Conference Exhibition, pp. 1–6 (2007)
38. Qiu, M., Yang, L., Shao, Z., Sha, E.: Dynamic and leakage energy minimization with soft real-time loop scheduling and voltage assignment. IEEE Trans. Very Large Scale Integr. **18**(3), 501–504 (2009)
39. Silver, D., et al.: Mastering the game of go without human knowledge. Nature **550**(7676), 354–359 (2017)
40. Takase, T., Oyama, S., Kurihara, M.: Effective neural network training with adaptive learning rate based on training loss. Neural Netw. **101**, 68–78 (2018)
41. Tao, L., Golikov, S., et al.: A reusable software component for integrated syntax and semantic validation for services computing. In: IEEE Symposium on Service-Oriented System Eng., pp. 127–132 (2015)
42. Yu, H., Yang, J.: A direct LDA algorithm for high-dimensional data-with application to face recognition. Pattern Recogn. **34**(10), 2067–2070 (2001)
43. Zhang, K., Kong, J., et al.: Multimedia layout adaptation through grammatical specifications. Multimedia Syst. **10**(3), 245–260 (2005)
44. Zhao, W., Chellappa, R., Nandhakumar, N.: Empirical performance analysis of linear discriminant classifiers. In: Proceedings. 1998 IEEE Computer Society Conference on Computer Vision and Pattern Recognition (Cat. No. 98CB36231), pp. 164–169. IEEE (1998)
45. Zheng, X., Chen, W., Li, M., Zhang, T., You, Y., Jiang, Y.: Decoding human brain activity with deep learning. Biomed. Signal Process. Control **56**, 101730 (2020)

LowFreqAttack: An Frequency Attack Method in Time Series Prediction

Neal N. Xiong[1], Wenyong He[2(✉)], and Mingming Lu[2]

[1] National Engineering Laboratory for Educational Big Data, Central China Normal University, Wuhan, China

[2] School of Computer Science and Engineering, Central South University, ChangSha 410083, HuNan, China

hewenyong@csu.edu.cn

Abstract. Time series prediction has become an important research direction in data mining because of the time-varying pattern of data in various fields. However, time series prediction suffers from the problem of vulnerability to adversarial example attack, which leads to models making wrong decisions in critical application scenarios and causing great losses to people's lives and properties. In addition, there is relatively little attacks research on time series prediction, and the existing attack methods simply migrate classical-attack methods in the image to time series prediction. On the one hand, it not only without fully considering the characteristics of temporal data but also without comparing and analyzing the effects of those classical-attack methods on time series prediction models. On the other hand, there is no comparative analysis of the effectiveness of these classical attack methods in different time series prediction methods. To address the above problems, this paper firstly compares the effectiveness of the attack methods on some time series prediction models and analyzes the inner mechanism of these time series prediction models. In addition, this paper finds that the defense ability of those models is related to their ability to portray the overall trend of time series data. Therefore, this paper further propose the new attack method, LowFreqAttack. The experimental results show that LowFreqAttack can attack the three existing time series prediction models better than the existing attach methods.

Keywords: Time series prediction · Adversarial attack · Frequency

1 Introduction

Time series prediction is an important research area in data mining and machine learning. On the one hand, time series prediction is widely used in various fields such as healthcare [1–6], traffic flow prediction [7–10]. On the other hand, time series prediction can reflect the intrinsic patterns of time-series data over time. In recent years, given the successful application of deep learning in fields such as

M. Qiu et al. (Eds.): SmartCom 2022, LNCS 13828, pp. 280–289, 2023.
https://doi.org/10.1007/978-3-031-28124-2_26

computer vision and natural language processing. Some researchers have recently started to use it for time series prediction tasks as well [12–17].

However, deep learning is extremely vulnerable in adversarial settings. Attackers generate adversarial examples by finding perturbations that are not easily detectable by humans, and these examples can cause prediction models to produce erroneous outputs with high confidence, which can severely affect the decision-making of some critical applications. Recently, it has been shown that deep learning models are equally vulnerable to attacks on time series prediction tasks [20,21]. For example, in a user health monitoring system, a fraudster can make the patient miss the best treatment time for the onset of the disease by faking the illusion that all health indicators of the onset patient are not abnormal, thus seriously endangering the patient's life [24–29]. However, there is relatively little research on time series prediction attacks. Therefore, this paper explores the time series-based counterattack. First, CNN, LSTM, and transformers [30] are used as the base models for time series prediction. The robustness is compared under the gradient-based attack approach. Experiments show that the ability of the three models to defend against attacks decreases in the order of transformer, CNN, and LSTM. Subsequently, to analyze the reasons for the different robustness of each model, this paper finds that the defense ability of these three models is related to their ability to portray the overall trend of the time series data by comparing and analyzing the inner mechanism of these time series prediction models.

To deeply analyze the reason for the robustness of the transformer, this paper finds inspiration from the application of the transformer in images. As early as in the field of vision, researchers points out that the transformer makes more use of the global information of the image for prediction [31], where the global information can be understood as the overall contour features in the image. Therefore, this paper proposes a more low-frequency component attack method Low-Frequency Attack (LowFreqAttack) based on this feature. And the attack performance of LowFreqAttack is compared with the existing attack methods on some existing defense methods. The experiments show that LowFreqAttack still guarantees better attack capability, thus verifying the effectiveness of the LowFreqAttack.

The remainder of this article is organized as follows. In Sect. 2, we review the related literature on time series forecasting and transformer research in images. Section 3 elaborates some attack methods and the design method. We analysis the results in Sect. 4, and this article is concluded in Sect. 5 with a summary of potential future studies.

2 Related Work

2.1 Research on Time Series Forecasting

Time series prediction has become an important research direction in data mining and is now widely used in various areas of life [32–34]. However, studies have shown that deep learning models are vulnerable to attacks [18]. While there have been many significant studies on adversarial attacks on data such as images, text, and graph structures, the opposite is true for time series, for which there is a paucity of research. Forestier [20] pioneered the study of the vulnerability of deep learning models on time series adversarial examples by applying some attack methods on images [35,36] to time sequence classification task, which reduces the accuracy of the classification model, thus pointing out the same vulnerability to attacks on the application of time series data. Mode studied the adversarial attacks on multivariate time series prediction and verified the vulnerability of deep learning models to adversarial attacks by conducting experiments on datasets in multiple scenarios. Recently, M. Qiu et al. [13,22] proposed privacy-preserving wireless communications using bipartite matching in social big data and a novel topological graph for convolutional network-based urban traffic flow and density prediction [23]. Wu [21] performed adversarial attacks on some recent time series prediction models and showed that even very small perturbations can equally easily cause significant degradation in the performance of the models. As for the defense methods for time series prediction, the existing defense methods are mainly developed from the perspective of data preprocessing. Juan [37] proposed an efficient data preprocessing method that takes advantage of the correlation of continuous time data and uses mean substitution for data noise reduction. Fahad Algarni [38] used the idea that input discretization can mitigate adversarial attack ideas, and then proposed a defense method based on thermometer coding, which effectively improves the robustness of the model.

2.2 Transformer Research in Images

Recently, the transformer, which has shown great potential in the field of natural language processing [30,39,40], can reach the same level of Convolutional Neural Network (CNN) in various tasks of computer vision [41,42], and the transformer is usually able to show better robustness. Alexey et al [41] first proposed the transformer architecture for image recognition, thus pioneering Vision Transformer (Vit), which not only has the same high accuracy as CNN but also shows better model robustness. Subsequently, Naseer [43] compared the robustness of three different architectures of the transformer, as well as some better performing CNN models considering the presence of severe occlusion, region shifting, spatial alignment, adversarial and natural perturbations in images, and again, found that the transformer has better robustness.

3 Method

3.1 Overview

Time Series. The set $X = [x_1, x_2, ..., x_T], x_T \in R^n$ formed by the values corresponding to T consecutive unit times can be called a time series.

Time Series Prediction. Input a time series $X = [x_1, x_2, ..., x_T]$, and predict the value $x_(T + h)$ of $(T + h)$, where h denotes the horizontal offset relative to the current moment, which is used to control the number of units of time between the corresponding moment of the predicted value of the prediction task and the current moment, and different offsets are set according to different tasks.

Time Series Adversarial Example. On a given time series X $= [x_1, x_2, ..., x_T]$, the attacker finds a small perturbation η based on the characteristics of the input time series, thus generating a new time series $X' = [x_1', x_2', ..., x_T']$, where $X' = X + \eta$, also called the adversarial sample. When the adversarial sample makes the model prediction accuracy decrease significantly, it means the attack is successful.

3.2 Gradient-Based Attack Method

FGSM Goodfellow et al. [35] proposed the Fast Gradient Sign Method (FGSM) for generating image adversarial examples, which successfully makes GoogleNet models predict errors. FGSM generates perturbations by computing the gradient of the input. In order to generate perturbations faster, the method computes the gradient of only one step. The computational procedure of FGSM for generating perturbations is shown Eq. 1.

$$X' = X + \epsilon \cdot sign(\nabla_x J(X, \hat{Y})), \tag{1}$$

BIM Kurakin et al. [36] proposed an iterative version (Basic Iterative Method) based on FGSM. The idea of BIM is to generate adversarial examples by moving smaller steps instead. Thus BIM iterates more times compared to the single-step perturbation generation method of FGSM. After each gradient-based generation of adversarial examples, cropping is performed, thus ensuring that the range obtained from each calculation lies in [X-α, X+α].

3.3 The Analysis of Attack Effect

The robustness of CNN, LSTM, and Transformer under the attack methods of FGSM and BIM were compared, and some of the results are shown in Fig. 1.

As can be seen from Fig. 1, the prediction performance of the transformer performs the best in both cases with the same perturbation factor ϵ. In particular, the performance of those is almost equal in the case of no perturbation. As the perturbation gradually increases, the performance of CNN and LSTM decreases dramatically, but the performance of the transformer decreases more slowly, which indicates that the transformer has better robustness under the same level of attack. In this paper, we argue that Transformer is better able to resist the noise of high-frequency signals because the window of its input data tends to be large, which is conducive to preserving the sequence correlation for a longer period. In addition, the transformer learns the attention weights among

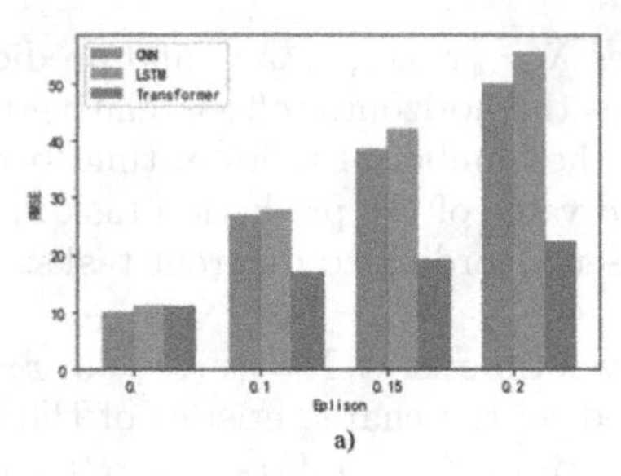

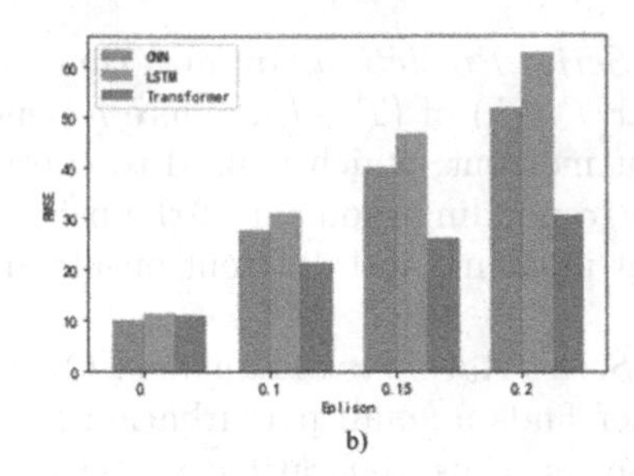

Fig. 1. a) The Comparison of results under FGSM attack [35]. b) The Comparison of results under BIM attack [36].

the corresponding data at each time point through the self-attention mechanism, which makes it more dependent on the overall sequence to complete the prediction. However, noise tends to manifest itself locally in the few samples that are disturbed and have less impact on the global features on which the transformer relies to make decisions. Therefore, the transformer is still able to ensure a certain level of robustness even under large disturbances. On the contrary, CNN and LSTM focus more on the time series inside a segment window and are therefore much more affected than the transformer.

Based on the above phenomenon, this paper argues that transformer has more affinity for low-frequency features. Therefore, this paper then proposes a frequency domain-based perturbation attack method LowFreqAttack.

3.4 LowFreqAttack

The process of LowFreqAttack is divided into four processes. (1) Generating perturbation by attack methods FGSM or BIM. (2) Transformation of the perturbation into multiple components in the frequency domain. (3) Sieving the high-frequency components in the frequency domain. (4) Transforming the sieved perturbation back to the time domain to obtain a more low-frequency perturbation.

4 Result Analysis

4.1 The Overview of Experimental Data

The data used in this paper comes from Caltrans' Performance Measurement System (PeMS). This system collects traffic flow in real-time through detectors on the freeway. In this paper, we use two columns of this dataset (Time, Traffic). The time column represents the instantaneous moment of the detector feedback data, with 5 min as the feedback interval. The traffic flow column, on the other hand, indicates the number of vehicles that passed through this detector during the interval from the previous moment to the current moment. Among them, the training set (2016/01/04–2016/02/29) includes 7776 data and the test set (2016/03/04–2016/03/31) contains a total of 4320 data.

4.2 The Effectiveness Comparison Under the Defense Method

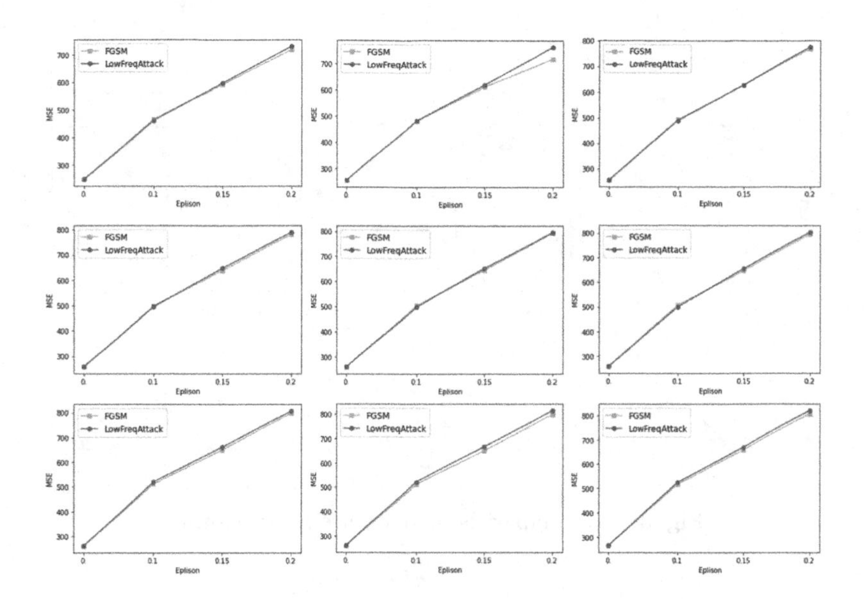

Fig. 2. The comparison under Kalman filtering.

To better show the performance between the lowfreqattack and other attack methods, we compare the attack effect by performing the attack under two defense methods. We use the two defense methods: Kalman filter and wavelet transform. We obtain the experimental results under wavelet transform as shown in Fig. 2. The nine plots in the figure represent the results of LowFreqAttack compared with the FGSM attack method under the wavelet transform defense method. The wavelet transform defense control factor is a quantile factor, ranging from 0.1 to 0.9, and the closer the value is to 1, the stronger the wavelet defense is. The horizontal coordinate in Fig. 2 indicates the perturbation factor ϵ, with values ranging from 0.1, 0.15, 0.2, which is used to control the size of the perturbation, the larger the value the larger the perturbation the vertical coordinate in Fig. 2 is MSE, which is used to indicate the deviation of the prediction result the larger the value, the better the attack effect, then the better the attack method. When the perturbation factor is 0.1, the LowFreqAttack attack effect exceeds the FGSM by about 0.63% on average when the perturbation factor is 0.15, the LowFreqAttack attack effect exceeds the FGSM by about 1.52% on average when the perturbation factor is 0.2, the LowFreqAttack attack effect exceeds the FGSM by about 1.07% on average. Therefore, based on the above experimental results, shows that LowFreqAttack is more aggressive than the FGSM method for the same wavelet defense strength.

Figure 3 shows the comparison of the experimental results under the Kalman filter defense method. The nine plots in Fig. 3 represent the results of LowFreqAttack compared with the FGSM attack method under the Kalman filter defense

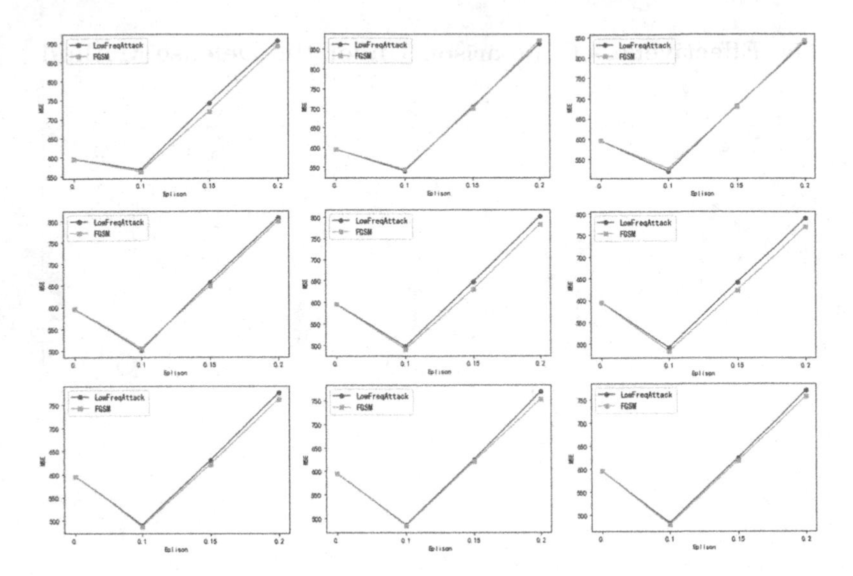

Fig. 3. The comparison under wavelet transform.

method. The Kalman filtering defense control factor, Karman-level, ranges from 0.1 to 0.9, and the closer the value is to 0.1, the stronger the Kalman filtering defense is. In Fig. 3, the horizontal coordinate is the perturbation factor ϵ, the larger the value the larger the perturbation. When the perturbation factor is 0.1, the LowFreqAttack attack effect exceeds the FGSM by about 0.73% on average. When the perturbation factor is 0.15, the LowFreqAttack attack effect exceeds the FGSM by about 1.25% on average. When the perturbation factor is 0.2, the LowFreqAttack attack effect exceeds the FGSM by about 1.20%. Therefore, based on the above experimental results, it shows that LowFreqAttack is more aggressive than the FGSM method with the same Kalman filtering defense strength.

5 Conclusion

In this paper, we compared and analysed the robustness of CNN, LSTM, and transformer on time series prediction tasks under the attack gradient-based approach. We found that the transformer is more robust. In addition, this paper analysed the reasons why the transformers are more robust in the frequency domain and also points out that the transformer is vulnerable to low-frequency perturbation attacks. Based on the vulnerability of transformers, this paper proposed a more low-frequency attack method, LowFreqAttack. Comparing LowFreqAttack with other attack methods under the same defense method, it is found that LowFreqAttack can still maintain better attack capability, thus verifying its effectiveness.

Acknowledgements. This work was partially supported by the National Natural Science Foundation of China under Grant No. U20A20182 and 62177019.

References

1. Yadav, P., Steinbach, M., Kumar, V., Simon, G.: Mining electronic health records (EHRS) a survey. ACM Compu. Surv. **50**(6), 1–40 (2018)
2. Kaushik, S., et al.: Ai in healthcare: time-series forecasting using statistical, neural, and ensemble architectures. Front. Big Data **3**, 4 (2020)
3. Xia,F., Hao, R., Li, J., Xiong, N., Yang, L.T., Zhang. Y.: Adaptive GTS allocation in IEEE 802.15. 4 for real-time wireless sensor networks. J. Syst. Archit. **59**(10), 1231–1242 (2013)
4. Kumar, P., et al.: PPSF: a privacy-preserving and secure framework using blockchain-based machine-learning for IOT-driven smart cities. IEEE Trans. Netw. Sci. Eng. **8**(3), 2326–2341 (2021)
5. Chunxue, W., Luo, C., Xiong, N., Zhang, W., Kim, T.-H.: A greedy deep learning method for medical disease analysis. IEEE Access **6**, 20021–20030 (2018)
6. Cheng, H., Xie, Z., Shi, Y., Xiong, N.: Multi-step data prediction in wireless sensor networks based on one-dimensional CNN and bidirectional LSTM. IEEE Access **7**, 117883–117896 (2019)
7. Zhang, L., et al.: SATP-GAN: self-attention based generative adversarial network for traffic flow prediction. Transportmetrica B Transp. Dyn. **9**(1), 552–568 (2021)
8. Yao, Y., Xiong, N., Park, J.H., Ma, L., Liu, J.: Privacy-preserving max/min query in two-tiered wireless sensor networks. Comput. Math. Appl. **65**(9), 1318–1325 (2013)
9. Chunxue, W., et al.: UAV autonomous target search based on deep reinforcement learning in complex disaster scene. IEEE Access **7**, 117227–117245 (2019)
10. Fu, A., Zhang, X., Xiong, N., Gao, Y., Wang, H., Zhang. J.: VFL: a verifiable federated learning with privacy-preserving for big data in industrial IOT. IEEE Trans. Indust. Inform. (2020)
11. Shafiq, M., Tian, Z., Bashir, A.K., Du, X., Guizani., M.: Corrauc: a malicious bot-IoT traffic detection method in IOT network using machine-learning techniques. IEEE Internet of Things J. **8**(5), 3242–3254 (2020)
12. Shastri, S., Singh, K., Kumar, S., Kour, P., Mansotra, V.: Time series forecasting of Covid-19 using deep learning models: India-USA comparative case study. Chaos, Solit, Fract **140**, 110227 (2020)
13. Qiu, M., Gai, K., Xiong, Z.: Privacy-preserving wireless communications using bipartite matching in social big data. Futur. Gener. Comput. Syst. **87**, 772–781 (2018)
14. Zhou, Y., Zhang, Y., Hao, L., Xiong, N., Vasilakos, A.V.: A bare-metal and asymmetric partitioning approach to client virtualization. IEEE Trans. Serv. Comput. **7**(1), 40–53 (2014)
15. Lin, C., He, Y., Xiong. N.: An energy-efficient dynamic power management in wireless sensor networks. In: Fifth International Symposium on Parallel Distributed Computing (2006)
16. Qiu, H., Qiu, M., Zhihui, L.: Selective encryption on ECG data in body sensor network based on supervised machine learning. Inf. Fusion **55**, 59–67 (2020)
17. Qiu, M., Zhang, L., Ming, Z., Chen, Z., Qin, X., Yang, L.T.: Security-aware optimization for ubiquitous computing systems with seat graph approach. J. Comput. Syst. Sci. **79**(5), 518–529 (2013)

18. Chakraborty, A., Alam, M., Dey, V., Chattopadhyay, A., Mukhopadhyay, D.: A survey on adversarial attacks and defences. CAAI Trans. Intell. Technol. **6**(1), 25–45 (2021)
19. Karim, F., Majumdar, S., Darabi, H.: Adversarial attacks on time series. IEEE Trans. Pattern Anal. Mach. Intell. **43**(10), 3309–3320 (2020)
20. Fawaz, H.I., Forestier, G., Weber, J., Idoumghar, L., Muller, P.-A. Adversarial attacks on deep neural networks for time series classification. In: 2019 International Joint Conference on Neural Networks (IJCNN), pp. 1–8. IEEE (2019)
21. Tao, W., Wang, X., Qiao, S., Xian, X., Liu, Y., Zhang, L.: Small perturbations are enough: adversarial attacks on time series prediction. Inf. Sci. **587**, 794–812 (2022)
22. Gai, K., Qiu, M., Elnagdy, S.A.: A novel secure big data cyber incident analytics framework for cloud-based cybersecurity insurance. In: 2016 IEEE 2nd International Conference on Big Data Security on Cloud (BigDataSecurity), IEEE International Conference on High Performance and Smart Computing (HPSC), and IEEE International Conference on Intelligent Data and Security (IDS), pp. 171–176. IEEE (2016)
23. Qiu, H., et al.: Topological graph convolutional network-based urban traffic flow and density prediction. IEEE Trans. Intell. Transp. Syst. **22**(7), 4560–4569 (2020)
24. Ma, T., Xiao, C., Wang. H., Health-ATM: A deep architecture for multifaceted patient health record representation and risk prediction. In: Proceedings of the 2018 SIAM International Conference on Data Mining, pp. 261–269. SIAM (2018)
25. Chen, Y., Zhou, L., Pei, S., Zhiwen, Yu., Chen, Y., Liu, X., Jixiang, D., Xiong, N.: KNN-block dbscan: fast clustering for large-scale data. IEEE Trans. Syst. Cybernet. Syst **51**(6), 3939–3953 (2019)
26. Zhao, J., Huang, J., Xiong, N.: An effective exponential-based trust and reputation evaluation system in wireless sensor networks. IEEE Access **7**, 33859–33869 (2019)
27. Gao, Y, et al.: Human action monitoring for healthcare based on deep learning. IEEE Access **6**, 52277–52285 (2018)
28. Zhang, W., Zhu, S., Tang, J., Xiong, N.: A novel trust management scheme based on Dempster-Shafer evidence theory for malicious nodes detection in wireless sensor networks. J. Supercomput. **74**(4), 1779–1801 (2018)
29. Huang, S., Zeng, Z., Ota, K., Dong, M., Wang, T., Xiong. N.N.: An intelligent collaboration trust interconnections system for mobile information control in ubiquitous 5g networks. IEEE Trans. Netw. Sci. Eng. **8**(1):347–365 (2020)
30. Vaswani, A., et al.: Attention is all you need. In: 30th Proceedings on Advances in Neural Information Processing Systems (2017)
31. Tuli, S., Dasgupta, J., Grant, E., Griffiths, T.J.: Are convolutional neural networks or transformers more like human vision? arXiv preprint arXiv:2105.07197 (2021)
32. Zhang, K., Gençay, R., Ege Yazgan, M.: Application of wavelet decomposition in time-series forecasting. Econ. Lett. **158**, 41–46 (2017)
33. Ismail Fawaz, Hassan, Forestier, Germain, Weber, Jonathan, Idoumghar, Lhassane, Muller, Pierre-Alain.: Evaluating surgical skills from kinematic data using convolutional neural networks. In: Frangi, Alejandro F.., Schnabel, Julia A.., Davatzikos, Christos, Alberola-López, Carlos, Fichtinger, Gabor (eds.) MICCAI 2018. LNCS, vol. 11073, pp. 214–221. Springer, Cham (2018). https://doi.org/10.1007/978-3-030-00937-3_25
34. Qiu, M., Chen, Z., Ming, Z., Qin, X., Niu, J.: Energy-aware data allocation with hybrid memory for mobile cloud systems. IEEE Syst. J. **11**(2), 813–822 (2014)
35. Goodfellow, I.J., Shlens, J., Szegedy. C.: Explaining and harnessing adversarial examples. arXiv preprint arXiv:1412.6572 (2014)

36. Kurakin, A., Goodfellow, J.J., Bengio, S.: Adversarial examples in the physical world. In: Artificial Intelligence Safety and Security, pp. 99–112. Chapman and Hall/CRC (2018)
37. Cortés-Ibáñez, J.A., et al.; An industrial world application case study: preprocessing methodology for time series. Inf. Sci. **514**, 385–401 (2020)
38. Yang, Z., Abbasi, I.A., Algarni, F., Ali, S., Zhang. M.: An IoT time series data security model for adversarial attack based on thermometer encoding. Secur. Commun. Netw. **2021** (2021)
39. Devlin, J., Chang, M., Lee, K., Toutanova. K.: Pre-training of deep bidirectional transformers for language understanding. arXiv preprint arXiv:1810.04805 (2018)
40. Brown, T., et al. : Language models are few-shot learners. In: 33rd Proceedings Conference on Advances in Neural Information Processing Systems, pp. 1877–1901 (2020)
41. Dosovitskiy, A., et al.: An image is worth 16×16 words: transformers for image recognition at scale. arXiv preprint arXiv:2010.11929 (2020)
42. Liu, Z., et al,.: Swin transformer: hierarchical vision transformer using shifted windows. In: Proceedings of the IEEE/CVF International Conference on Computer Vision, pp. 10012–10022 (2021)
43. Naseer, M.M., et al.: Intriguing properties of vision transformers. In: 34th Proceedings of Advances in Neural Information Processing Systems, pp. 23296–23308 (2021)

A Novel Machine Learning-Based Model for Reentrant Vulnerabilities Detection

Hui Zhao[1], Peng Su[2], and Meikang Qiu[3](✉)

[1] Educational Information Technology Laboratory, Henan University, Kaifeng, China
zhh@henu.edu.cn
[2] Software School, Henan University, Kaifeng, China
supeng@henu.edu.cn
[3] Dakota State University, Midison, SD, USA
qiumeikang@ieee.org

Abstract. Machine learning-based models are one of the main methods for detecting reentrant vulnerabilities. However, these models extract smart contract features only from a single form, resulting in incompleteness and inaccuracy of features. To address this problem, we propose a novel machine learning-based model for reentrant vulnerabilities detection. We extract and fuse features from abstract syntax trees, opcodes, control flow graph basic blocks, and combine machine learning algorithms for reentrant vulnerabilities detection. Additionally, to address the time-consuming problem of manual labeling, we also propose an approach for automatically adding dataset labels. We perform experiments on Smartbugs and SolidiFi-benchmark datasets and results show that our model outperforms existing models.

Keywords: Smart contracts · Machine learning · Reentrant vulnerabilities · Dataset labels · Detection

1 Introduction

In recent years, blockchain has become a hotspot of research and application. We begin to focus on information security and information privacy [7,13,21]. Privacy stealing and privacy protection are a long-term confrontation process [18]. Blockchain technology [2,6,8] has many applications in modern information technology and computing environments. This technology uses key features, such as distributed ledgers and tamper resistance, to use distributed computing resources [11,14,20]. As one of the key components of blockchain [5], smart contracts play a vital role in achieving automation functions. The concept of smart contracts was proposed by Nick Szabo [23]. When the condition is triggered, the smart

M. Qiu et al. (Eds.): SmartCom 2022, LNCS 13828, pp. 290–299, 2023.
https://doi.org/10.1007/978-3-031-28124-2_27

contract will execute the corresponding transaction. Smart contracts cannot find suitable application scenarios at that time. Therefore, the smart contract cannot develop quickly. Until the advent of blockchain technology, smart contracts begin to attract people's attention and gradually became a research hotspot. However, security problems of smart contracts are beginning to be exposed. One of the most destructive attacks is the reentrant attack.

Reentrant attacks are one of the most destructive attacks in blockchain systems [9,22] caused by smart contract vulnerabilities [3]. When an attacker recursively calls the **draw(.)** of the target contract to withdraw funds from the target contract, a reentry attack will occur. When the contract cannot update its balance after sending the funds, the attacker can call the withdrawal function continuously to exhaust the contract funds. A well-known reentrant attack is "**TheDAO**" attack which caused a loss of 60 million US dollars [16].

Machine learning-based vulnerabilities detection has been proven effective [10,19]. However, there are two drawbacks to this technique. First, the problem of noise code interference of similar contracts. At the current stage, smart contracts are mostly deployed on blockchains by programmers. To save time, programmers often use copy and paste while writing smart contracts. Therefore, this situation has produced a lot of similar contracts [26]. Vulnerabilities are usually caused by a few statements, but similar contracts contain many statements that are not related to the vulnerability. We call it noise codes [12]. Second, the time-consuming issue of manual marking. At the current stage, researchers usually use traditional methods such as Mythril and Slither to detect data sets. Then, they manually add tags to the data set according to the detection results [17].

In order to detect whether smart contracts have reentrant vulnerabilities, we propose a novel machine learning-based model for reentrant vulnerabilities' detection. Our proposed model parses smart contracts into abstract syntax trees and opcodes. By sharing node and opcode categorization techniques, we extract feature vectors from abstract syntax trees and opcodes. Our model parses smart contracts into control flow graphs and extracts basic blocks containing transaction features from them. We concatenate the basic blocks and perform word embeddings on them through the n-grams algorithm. Our model concatenates these three types of vectors to obtain the full feature vector for the smart contract. Our proposed model automatically labels vectors through code embedding and similarity detection techniques. Under six machine learning models, our method is compared with existing vulnerability detection methods. Experiments show that our proposed model improve the accuracy of reentrant vulnerabilities' detection. Moreover, our accuracy rate is much higher than some common smart contract detection tools [1]. Our proposed model automatically add labels to vectors, which solves the time-consuming problem of manually labeling data.

The main contributions of this paper are as follows. (1) We extract and fuse features from smart contract abstract syntax trees, opcodes, and control flow graph basic blocks, which enhances feature completeness and accuracy. (2) We automatically add labels to smart contracts through code embedding technology and similarity comparison technology, which solves the time-consuming manual labeling. (3) We perform vulnerability detection on the extracted smart contract features under the Smartbugs and SolidiFi-benchmark datasets and 6 machine learning algorithms, and the results show that our model outperforms existing models.

The remainder of this paper is organized in the following order. Section 2 describes the design of our proposed model, which is applied to reentrant vulnerabilities detection. Section 3 describes our proposed algorithms. We analyze the experimental results in Sect. 4 and conclude the paper in Sect. 5.

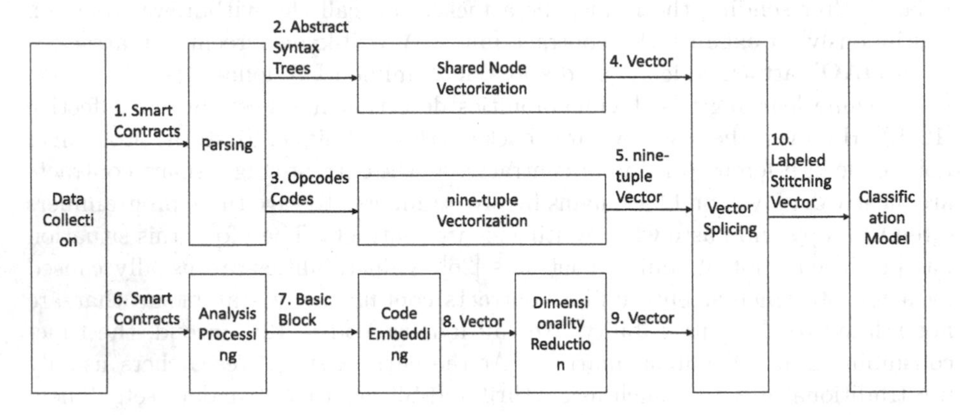

(a) Collecting smart contracts and extracting smart contract features

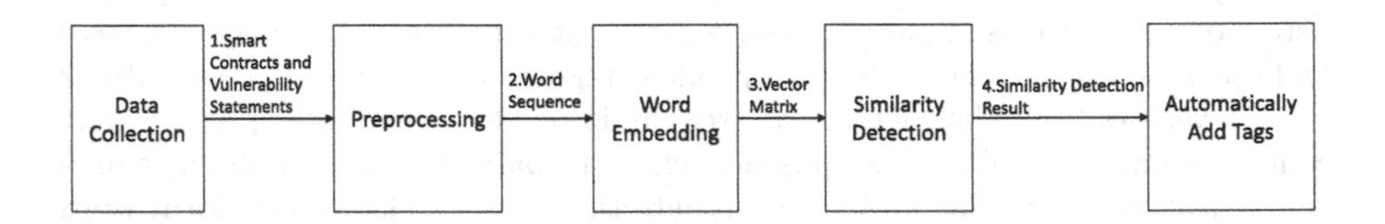

(b) Collecting vulnerability statements and adding smart contract tags

Fig. 1. The overall workflow of our proposed model.

2 Proposed Model

2.1 Overview

Figure 1(a) shows the step of collecting smart contracts and extracting smart contract features. The system parses smart contracts into abstract syntax trees and extracts features from smart contracts by solving shared nodes. The system parses the smart contract into opcodes and obtains the nine-tuple vector of the smart contract by traversing the opcodes. The system parses the smart contract into a control flow graph, from which it extracts basic blocks containing transaction characteristics. These basic blocks are converted into vectors by word embedding technology and dimensionality reduction is performed on them. The system concatenates three types of vectors. Our proposed system uses six machine learning algorithms as classification models for experiments.

Figure 1(b) shows the step of collecting vulnerability statements and adding smart contract tags. The system relies on code embedding and space vector comparison for automatic labeling. Through data preprocessing, smart contract statements and vulnerability statements are converted into word sequences. The system uses the **FastText** tool to train it into word vectors and sums to get the sentence vector matrix. Through the results of similarity detection, labels are automatically added to the smart contracts.

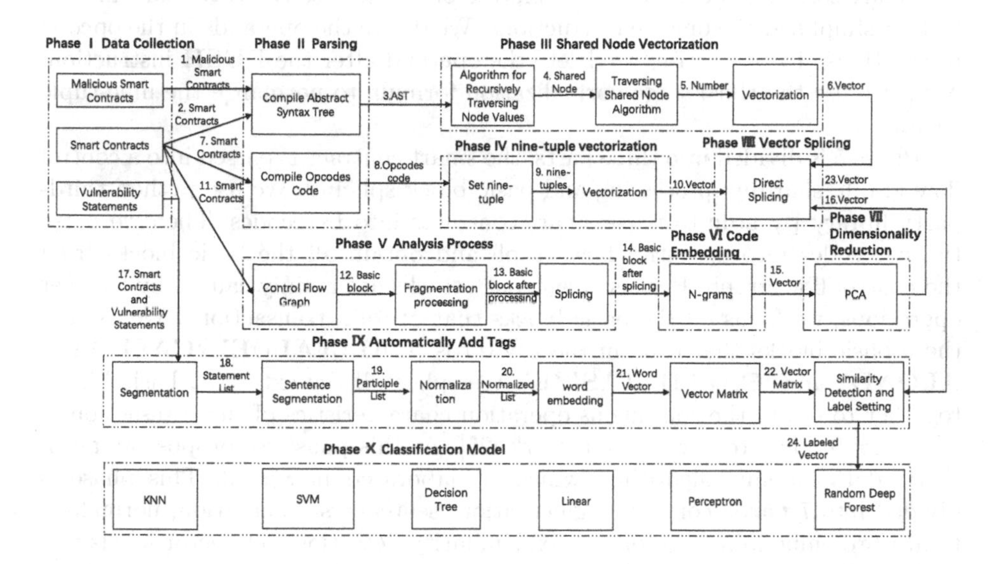

Fig. 2. The phase diagram of our proposed model.

2.2 Model Design

The design scheme of the model mainly includes 10 phases. Data collection will be completed in the first phase. The second phase is responsible for the parse of the smart contract. The third phase realizes the vectorization of shared nodes. The nine-tuple vectorization will be completed in the fourth phase. The fifth phase is to analyze and process the smart contract. The sixth phase is to perform code embeddings on basic blocks. The seventh phase is to reduce the dimensionality of the vector. The eighth phase is responsible for vector splicing. The ninth phase realizes automatically add tags. The tenth phase is responsible for the classification model. The schematic diagram of the model construction method is shown in Fig. 2. Next, we will describe the core phases within our model.

In the third phase, due to our goal is to obtain the shared nodes between the malicious smart contract and the smart contract set, we first need to obtain the values of all nodes in the smart contract abstract syntax tree. In order to facilitate data manipulation, we convert the abstract syntax tree in JSON format to dictionary format. Each abstract syntax tree contains three types of nodes. The three types are **list**, **dictionary** and **string**. Our goal is to get the node value of all string types. Through recursive traversal list and dictionary, we get all the node values of the abstract syntax tree. Through traversal of shared node number, we obtain the number of shared nodes of the two abstract syntax trees.

In the fourth phase, we divide opcode instructions into 9 categories, namely PUSH, DUP, SWAP, LOG, OPERATION, LOGIC, COMPARISON, VALUE1, VALUE2. By counting the number of each category in the opcode instructions of the smart contract, we get a nine-tuple vector. To reduce the traversal time, we further simplified the opcode instructions. We delete the operands in the opcode instructions. To be precise, we delete the operand after the **PUSH** instruction. Next, we use the third-stage normalization formula to normalize the nine-tuple vector.

Phase 5 is divided into three parts: the smart contract is parsed into a control flow graph, sharding processing and basic block splicing. We use python third-party library **py-solc** to parse smart contracts into bytecodes. Then we parse the bytecodes into a control flow graph and obtain all the basic blocks from the control flow graph. Because reentrant attacks are mainly caused by transfer operations, we focus on the basic blocks that contain transactions. We extract these basic blocks that contain the keywords **'CALLVALUE'**, **'CALLDATALOAD'**, and **'CALLDATASIZE'**. We splice all the extracted basic blocks together to retain the continuous operation characteristics of the transaction.

Phase 9 refers to our previous work [25]. At this phase we propose an automatic data labeling algorithm, which is elaborated in Sect. 3. This phase is divided into 7 parts: contract segmentation, sentence segmentation, normalization, word embedding, vector matrix, similarity detection, set vector labels.

In order to prove that the features we extracted are feasible, we perform experimental evaluation under six machine learning models. Through experimentation, we can observe whether our method works and which model is better for our method.

3 Algorithm

In this section, we will describe our proposed *Automatic Data Labeling* (ADL) Algorithm. The goal of the algorithm is to automatically add labels to datasets, thereby reducing the time-consuming manual labeling process.

The input to the Automatic Data Labeling Algorithm is a set of smart contracts, which we denote by **SCS**. The smart contract set includes training set, test set and malicious statement set. The output of this algorithm is the label set. This set includes labels for all smart contracts in the training and testing sets. We denote the output of this algorithm by **Label_set**.

Algorithm 1. Automatic Data Labeling Algorithm

```
Input: SCS
Output: Label_set
 1: function LABEL(SCS)
 2:     op_1 ← Sentence segmentation and word segmentation are performed on SCS.
 3:     op_2 ← Perform word embeddings on op_1.
 4:     train_set,test_set,eyi_statement_set ← Divide the op_2.
 5:     for smart_contract in train_set do
 6:         for statement in smart_contract do
 7:             for statement_eyi in eyi_statement_set do
 8:                 if The similarity between statement and statement_eyi is greater than
    the threshold then
 9:                     smart_contract's label is set to 1.
10:                 else
11:                     smart_contract's label is set to 0.
12:                 end if
13:             end for
14:         end for
15:     end for
16:     Label_set ← labels for train_set and test_set. return Label_set
17: end function
```

The Automatic Data Labeling Algorithm is applied in the ninth phase of our proposed model. The goal of the algorithm is to obtain the labels of all smart contracts in the training and test sets. When the label is 1, it means that the smart contract has a vulnerability risk. When the label is 0, it means that the smart contract is a security contract.

The workflow of the Automatic Data Labeling Algorithm is: (1) Input smart contracts into the algorithm. (2) Sentence segmentation and word segmentation for smart contracts. (3) Perform word embedding on the segmented smart contracts. (4) Divide smart contracts into training set, test set and malicious statement set. (5) All the sentences of each smart contract in the training set are checked for similarity with the malicious sentence set. According to the comparison between the detection result and the threshold, the label of the contract is set. (6) The test set and the training set are treated the same way. (7) Save all labels in **Label_set**.

Time Complexity. The algorithm uses a three-level **for** loop. The first **for** loop is used to traverse all smart contracts in the training set. The second level of **for** loop is used to traverse all statement vectors in the smart contract. The third layer of **for** loop is used to traverse all the sentence vectors in the malicious sentence set. The test set and training set operate in the same way. Therefore, the time complexity of this algorithm is $O(n^3)$.

Table 1. Experiment settings

Contract sources	reference [4]
Abstract syntax tree parser tool	py-solc
Opcode parsing tool	py-solc
Control flow graph parsing	EVM-CFG-BUILDER
Divider	newline ('\n')
Word segmentation tool	Jieba
Word embedding tool	FastText,n-grams
Training vector parameters	MinCount =1 lr = 0.05 dim = 20 loss = ns
Threshold	0.92

4 Experiments and Results

4.1 Experiment Configuration

We ran the assessment in the following environment. We used Python's third-party library **jieba** as a word segmentation tool. We also used Python (v3.7) to implement our model in the Pychar IDE. The hardware platform was Lenovo ideapad 300-15ISK laptop, equipped with Windows 10 operating system, 2.3 GHz CPU, i5 version kernel and 8 GB 2691 MHz LPDDR3 memory [25]. The experiment settings were shown in Table 1.

Our proposed model adopted the n-grams algorithm for word embedding in the sixth phase. We conducted experiments when n was equal to 1, 2 and 3 respectively. We defined three different algorithms as 1-grams, 2-grams, 3-grams. We used these three names to denote the overall model with different algorithms. Selected existing methods included Xu et al.'s method [24], Wang et al.'s method [17], and NeuCheck [15]. NeuCheck [15] used three different n-grams algorithms to detect vulnerabilities. We defined as NeuCheck_1, NeuCheck_2, NeuCheck_3.

4.2 Experimental Results

Figure 3 showed the experimental comparison between our method and existing methods under the KNN model. Among the existing methods, Xu et al.'s method had the highest precision rate, reaching 86%. Our method achieved 84% precision rate, which was 2% lower than existing methods. The precision rate of existing methods was stable at 70% to 86%. Among the existing methods, Wang et al.'s method had the highest macro average, reaching 87%. The macro average of our method was the highest, reaching 89%.

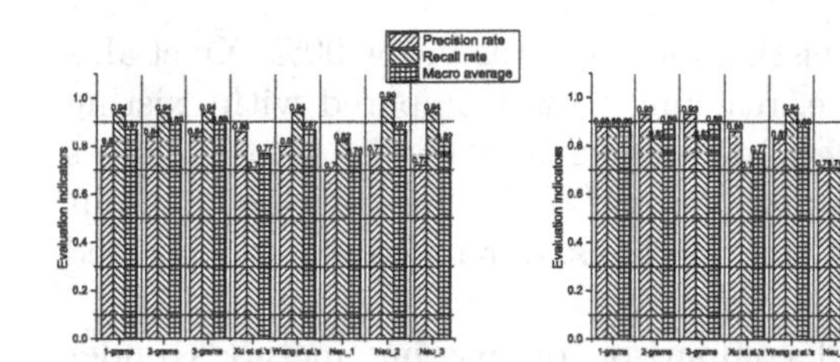

Fig. 3. Comparison of existing methods under KNN model.

Fig. 4. Comparison of existing methods under SVM model.

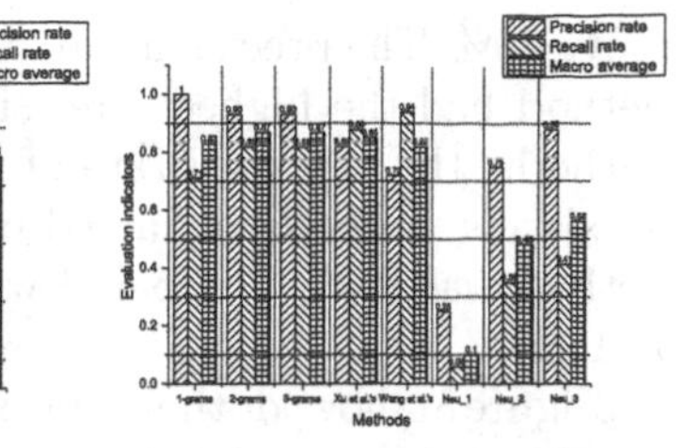

Fig. 5. Comparison of existing methods under Decision Tree model.

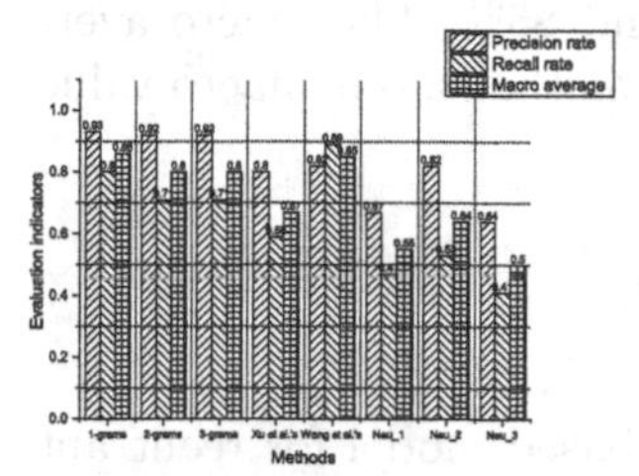

Fig. 6. Comparison of existing methods under Random Deep Forest model.

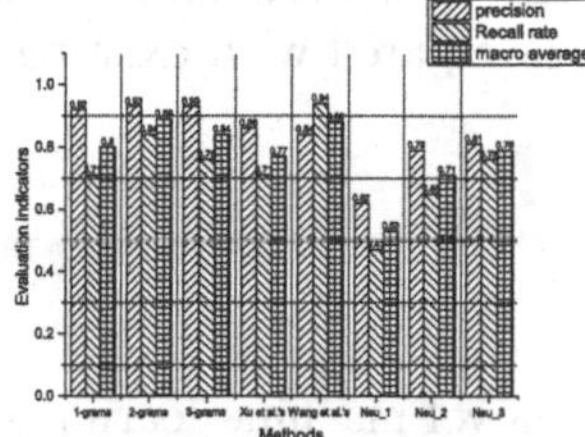

Fig. 7. Comparison of existing methods under Linear model.

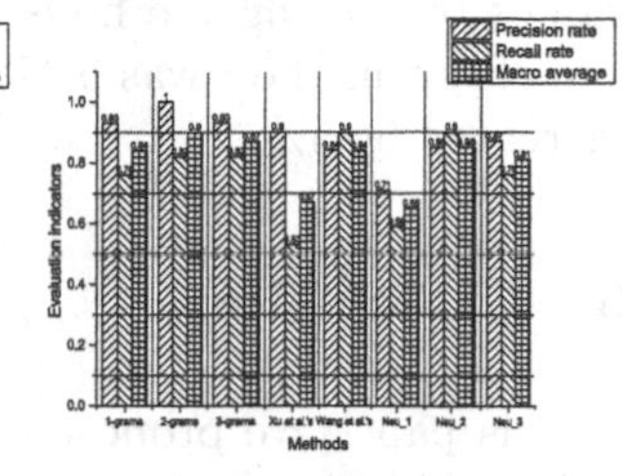

Fig. 8. Comparison of existing methods under Perceptron model.

Figure 4 showed the experimental comparison between our method and existing methods under the SVM model. Among the existing methods, Xu et al.'s method had the highest precision rate, reaching 86%. Our method achieved 93% precision rate, which was much larger than existing methods. The precision rate of existing methods was stable at 71% to 93%. Among the existing methods, Wang et al.'s method had the highest macro average, reaching 88%. The macro average of our method was the highest, reaching 89%.

Figure 5 illustrated an experimental comparison of existing methods under the Decision Tree model. The vulnerability detection precision rate of existing methods remained between 70% and 88%. The vulnerability detection precision rate of our methods were all maintained above 93%, with a maximum of 100%. Among the existing methods, the method of Xu et al. had the highest macro average, reaching 85%. Our method achieved an average of 87% macros, a 2% increase over existing methods.

Figure 6 illustrated an experimental comparison of existing methods under the Random Deep Forest model. The vulnerability detection precision rate of existing methods remained between 64% and 82%. The vulnerability detection precision rate of our methods were all maintained above 92%, with a maximum of 93%. Among the existing methods, the method of Wang et al. had the highest macro average, reaching 85%. Our method achieved an average of 86% macros, a 1% increase over existing methods.

Figure 7 showed the experimental comparison of existing methods under the Linear model. In vulnerability detection precision rate, our method remained above 90%. The precision rate of the existing methods was below 90%. Xu et al.'s method had the highest precision rate, reaching 86%. Compared with existing methods, the precision rate of our method was increased by 7%. Macro values of existing methods remained between 53% and 88%. The macro average of our method was 89%. Compared with existing methods, our macro value increased by 1%.

Figure 8 showed the experimental comparison of existing methods under the Perceptron model. In vulnerability detection precision rate, our method remained above 93%. The precision rate of the existing methods was below 90%. Xu et al.'s method had the highest precision rate, reaching 90%. Compared with existing methods, the precision rate of our method was increased by 10%. Macro values of existing methods remained between 65% and 85%. The macro average of our method was 90%. Compared with existing methods, our macro value increased by 5%.

5 Conclusions

In this paper, we proposed a novel machine learning-based model for reentrant vulnerabilities detection. Our model parsed smart contracts into abstract syntax trees, control flow graphs and opcodes. We extracted features from abstract syntax trees, control flow graphs and opcodes and concatenated the extracted features. To solve the time-consuming problem of manual labeling, our model implemented automatic labeling. With code embedding and similarity detection technology, our model automatically added tags to the smart contract set. We performed experiments on Smartbugs and SolidiFi-benchmark datasets and results showed that our model outperformed existing models.

Acknowledgment. Natural Science Foundation of Shandong Province (Grant No. ZR2020ZD01).

References

1. Badruddoja, S., Dantu, R., et al.: Making smart contracts smarter. In: IEEE International Conference on Blockchain and Cryptocurrency (2021)
2. Claudia, P., Tudor, C., Marcel, A., et al.: Blockchain based decentralized management of demand response programs in smart energy grids. Sensors **18**, 162 (2018)
3. Dong, C., Li, Y., Tan, L.: A new approach to prevent reentrant attack in solidity smart contracts. In: CCF China Blockchain Conference (2019)
4. Durieux, T., Ferreira, J., Abreu, R., Cruz, P.: Empirical review of automated analysis tools on 47,587 ethereum smart contracts. In: ACM/IEEE 42nd International Conference on Software Engineering(2019)
5. Gai, K., Guo, J., Zhu, L., Yu, S.: Blockchain meets cloud computing: a survey. IEEE Comm. Surv. Tutor. (99), 1-1(2020)
6. Gai, K., Qiu, M., Zhao, H., Tao, L., Zong, Z.: Dynamic energy-aware cloudlet-based mobile cloud computing model for green computing. J. Netw. Comput. Appl. 59, 46–54 (2016)

7. Gai, K., Wu, Y., et al.: Privacy-preserving energy trading using consortium blockchain in smart grid. IEEE Trans. Indust. Inform. **15**,3548–3558 (2019)
8. Gai, K., Wu, Y., et al.: Differential privacy-based blockchain for industrial internet-of-things. IEEE Trans. Indust. Inform. (99), 1-1 (2020)
9. Gai, K., Zhang, Y., et al.: Blockchain-enabled service optimizations in supply chain digital twin. IEEE TSC (2022)
10. Gao, X., Qiu, M.: Energy-based learning for preventing backdoor attack. In: KSEM (3). pp. 706–721 (2022)
11. Hu, F., Lakdawala, S., et al.: Low-power, intelligent sensor hardware interface for medical data preprocessing. IEEE TITB **13**(4), 656–663 (2009)
12. Jianjun, H., Songming, H., et al.: Hunting vulnerable smart contracts via graph embedding based bytecode matching. IEEE Trans. Inf. Forens. Secur. **16**, 2144–2156 (2021)
13. Li, Y., Gai, K., et al.: Intercrossed access controls for secure financial services on multimedia big data in cloud systems. In: ACM Trans. Multim. Cmput. Commun. Appl.**12** (2016)
14. Li, Y., Song, Y., et al.: Intelligent fault diagnosis by fusing domain adversarial training and maximum mean discrepancy via ensemble learning. IEEE Trans. Inform. J. **17**(4), 2833–2841 (2020)
15. Lu, N., Wang, B., et al.: Neucheck: a more practical ethereum smart contract security analysis tool. Softw. Pract. Exp. **51**(7) (2021)
16. Marta, M., Norberto, M.: Consecuencias penales y tributarias a la modificación fraudulenta de los smart contracts. especial referencia al caso the dao. CEFLegal: revista práctica de derecho. Comentarios y casos prácticos (2020)
17. Pouyan, M., Yu, W., Reza, S.: Machine learning model for smart contracts security analysis. In: 17th International Conference on Privacy, Security and Trust (2019)
18. Qiu, H., Kapusta, K., et al.: All-or-nothing data protection for ubiquitous communication: Challenges and perspectives. Inf. Sci. **502**, 434–445 (2019)
19. Qiu, H., Qiu, M., Lu, R.: Secure V2X communication network based on intelligent PKI and edge computing. IEEE Netw. **34**(2), 172–178 (2019)
20. Qiu, H., Zheng, Q., et al.: Topological graph convolutional network-based urban traffic flow and density prediction. IEEE Trans. Intell. Transp. Syst. **22**(7), 4560–4569 (2020)
21. Qiu, M., Qiu, H.: Review on image processing based adversarial example defenses in computer vision. In: IEEE 6th International Conference on BigDataSecurity. pp. 94–99 (2020)
22. Qiu, M., Qiu, H., et al.: Secure data sharing through untrusted clouds with blockchain-enabled key management. In: 2020 3rd International Conference on Smart BlockChain (SmartBlock), pp. 11–16 (2020)
23. Szabo, N.: Formalizing and securing relationships on public networks. First Monday (1997)
24. Xu, Y., Hu, G., You, L., Cao, C.: A novel machine learning-based analysis model for smart contract vulnerability. Secur. Commu. Netw. **2021**, 5798033 (2021)
25. Zhao, H., Su, P., et al.: Gan-enabled code embedding for reentrant vulnerabilities detection. In: 23rd International Conference on Knowledge Engineering and Knowledge Management (2021)
26. Zhipeng, G., Vinoj, J., Lingxiao, J., et al.: Smartembed: a tool for clone and bug detection in smart contracts through structural code embedding. In: IEEE International Conference on Software Maintenance and Evolution (2019)

Deep Learning Object Detection

Jingnian Liu[1,2], Weihong Huang[1,2](✉), Lijun Xiao[1,2], Yingzi Huo[1,2,3], Huixuan Xiong[1,2], Xiong Li[4], and Weidong Xiao[5]

[1] School of Computer Science and Engineering, Hunan University of Science and Technology, Xiangtan 411201, China
{whhuang,ljxiao}@hnust.edu.cn
[2] Hunan Key Laboratory for Service Computing and Novel Software Technology, Xiangtan 411201, China
[3] Guangdong Financial High-Tech Zone "Blockchain +" Fintech Research Institute, Foshan 528253, China
[4] School of Computer Science and Engineering, University of Electronic Science and Technology of China, Xiangtan 411201, China
[5] School of Software Engineering, Xiamen University of Technology, Xiamen 361024, China
xiaoweidong@xmut.edu.cn

Abstract. Object detection techniques are a major part of computer vision research, with large-scale applications in industrial, scientific and other scenarios. Technologies such as face detection, medical image detection, autonomous driving, and traffic detection have played a significant role in people's lives. With the rapid development of deep learning, many application areas, such as image classification, text classification, machine translation, etc., have achieved breakthrough success in combination with deep learning. R-CNN brings object detection into the era of deep learning, and its advantage compared with traditional methods is that the former requires personnel to extract features manually, while the latter uses deep learning to extract features automatically, which greatly improves efficiency, simplifies operation, and opens a new era of object detection research. First, this paper provides an overview of deep learning-based object detection backbone networks, reviews and analyzes milestone object detection algorithms, compares commonly used datasets, summarizes applications, and finally concludes the paper.

Keywords: Computer Vision · Deep Learning · Neural Network · Object-Detection · R-CNN

1 Introduction

The key task of object detection is to correctly identify objects (e.g., humans, animals, vehicles, and logo text) in a picture and to determine the location of the object [1]. By means of a rectangular edge box, to locate the detected object and to distinguish between classes of objects. Object detection has an important role in industrial scenes, scientific research, etc. And similarly, other tasks such as classification, segmentation, motion

M. Qiu et al. (Eds.): SmartCom 2022, LNCS 13828, pp. 300–309, 2023.
https://doi.org/10.1007/978-3-031-28124-2_28

estimation, and scene understanding are also fundamental problems in computer vision [2].

Most traditional object detection algorithms are constructed based on artificially constructed features [3], similar to the Viola-Jones detector [4], *Histogram of Oriented Gradients* (HOG) [5] etc. The structure of these early traditional algorithms is generally divided into three steps: informative region selection, feature extraction and classification [6], such models have obvious drawbacks and shortcomings, such as not fast convergence and poor migration ability on new datasets.

The development of deep learning [7–10] and storage technology [11, 12] has promoted the breakthrough of target detection. DCNN has excellent feature extraction and data migration capabilities, and its emergence has changed the field of object detection. The DCNN network AlexNet [13] was introduced in 2012, which has since opened the era of deep learning research boom.

In this paper, deep learning-based object detection techniques are reviewed and sorted out, and the main part will be organized as described below. In Sect. 2, the main deep convolutional neural network models are reviewed, and their architectures and performances are concisely described. Section 3 summarizes important object detection algorithms from the past to the present, and their structures are carefully analyzed and compared. Section 4 reviews the commonly used datasets and evaluation criteria in object detection. Section 5 summarizes the main current application. Section 6 concludes the paper.

2 Backbone Network for Object Detection

The rise of deep learning [14–16], big data [17, 18], computer capability [19–21], and cloud computing [22, 23], has also led to the development of object detection. In the object detection task, we usually use convolutional neural networks to extract features from images for subsequent recognition and localization of objects, which is a very important part of the object detection field. In the following, we will focus on reviewing landmark networks in deep learning.

2.1 AlexNet

AlexNet is one of the seminal works in the field of deep learning, which opened a new era of modern deep learning. The earliest convolutional neural network was LeNet [24], which perfectly solved the handwritten digit recognition task and achieved an average accuracy of 98% on the MNIST dataset. Alexet uses a deeper and wider network compared to LeNet, consisting of five convolutional layers, three maximum pooling layers, and three fully connected layers. Each convolution layer uses multiple channels to enhance information processing capability. The activation function between the intermediate layers is performed by relu to speed up the model convergence, while a new regularization technique dropout is used on the first two fully connected layers in order to cope with the overfitting problem. The last fully connected layer is passed through softmax, which produces a vector of size 1000 to represent the distribution of categories. AlexNet achieved the best result on the ImageNet LSVRC-2010 dataset at that time, with an accuracy rate of 62.5% and 83% on top-1 and top-5 classifications (Fig. 1).

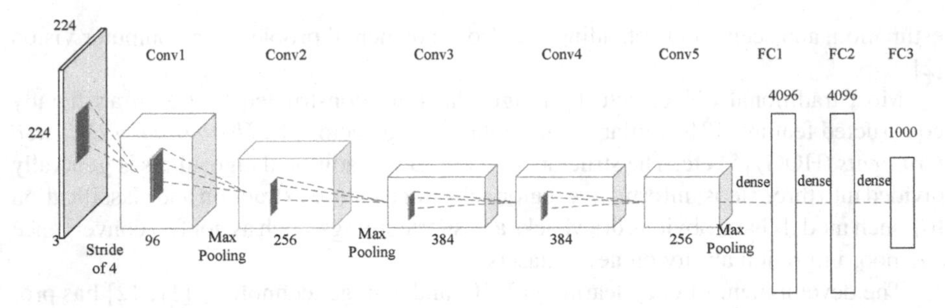

Fig. 1. Schematic diagram of AlexNet structure.

2.2 VGG

VGG has deeper layers and more parameters than AlexNet. VGG uses a convolutional kernel of 3 × 3 with a step size of 1 to get more information and details about the object from the picture, and uses the ReLu Layer after each convolutional layer. The dominant structures are VGG16 and VGG19, the former consisting of 13 convolutional layers and 3 fully connected layers, and the latter consisting of 16 convolutional layers and 3 fully connected layers [25]. Combining multi-crop and dense evaluation, using scale jittering to resize the images, we achieved second place in the classification task of the ILSVRC-2014 challenge with an error rate of 7.3% and won first place in the localization task. VGG proved at that time that a deeper and wider network would lead to higher accuracy.

2.3 GoogLeNet

Since the start of the deep learning boom, convolutional networks have moved in a deeper and broader direction with more parameters. But larger models also require higher computational costs and are more likely to cause overfitting of data, so how to achieve high accuracy with lower computational costs is the core goal of GoogLeNet [26]. Inception block is an important concept in GoogLeNet. The structure of the Inception model consists of four parallel paths with different sizes of convolutional kernels (5 × 5, 3 × 3, 1 × 1) to extract information simultaneously, 1 × 1 convolutional kernel can change the dimension and reduce the parameters while achieving the purpose of deepening the network and interacting with information across channels [27]. GoogLeNet won the ILSVRC-2014 challenge, outperforming other networks of the same period with an error rate of 6% in the classification task [28].

2.4 Resnet

The deeper layers will cause gradient disappearance, gradient explosion, and degradation problems will occur, and 56 layers will not even perform as well as 26 layers. Resnet introduced Batch Normalization [29] to solve the gradient disappearance and gradient explosion, and proposed residual to solve the degradation problem. The residual module notates the mapping stacked by convolutional layers as H(x), and the first input of these layers is noted as x. By way of a shortcut connection, we can choose to skip some layers.

Rather than expect stacked layers to approximate H(x), we explicitly let these layers approximate a residual function F(x) := H(x) − x [30]. The advantage of this is that at least it does not make the parameters worse, it is able to train deeper models, effectively solving the degradation problem, and the depth of the model can even reach more than 1000 layers. Resnet won first place in all kinds of tasks in the ILSVRC 2015 competition [30]. Some subsequent networks, similar to Densenet [31], and ShuffLeNet [32], were also inspired by the ideas in Resnet and were born.

3 The Architectures of Object Detection

Object detection is not limited to the identification of a particular object, but requires the detection and localization of many objects within an image. Traditional object detection algorithms usually require manual efforts to extract features, which has inherent drawbacks such as poor performance on new datasets and inefficient operation. DCNNs employ deep convolutional networks to automatically extract object features, greatly improving efficiency and speed. In the following, we review the important algorithms and models based on DCNN in the field of object detection (Table 1).

Table 1. Comparison of mainstream algorithms for target detection.

Methods	Backbone	Highlights	Year
R-CNN	AlexNet	The first application of deep neural networks to object detection	2014
Fast R-CNN	VGG16	Put the whole image and its bounding boxes into the neural network	2015
Faster R-CNN	VGG16	Use the neural network to generate proposals to improve efficiency	2016
Yolo	GoogLeNet(Modified)	A neural network is used to complete the work of bounding box generation and feature extraction	2015
SSD	VGG-16	The accuracy is guaranteed while the speed is guaranteed	2016

3.1 Two Stage

R-CNN. R-CNN [33] is the first model that successfully applies deep learning to object detection, and the module is designed as described below. 2000 region proposals are obtained by Selective search, fixing all region proposals to the same size, then applying Alext to each region proposal for feature extraction and outputting a vector of 4096 sizes, and finally classifying them by a trained SVM to remove the candidate bounding boxes with IOU values larger than a threshold by NMS. Finally, a trained regression model is used to predict the correction of its bounding box by training four parameters, centroid, height and width.

R-CNN obtained the best results in the then VOC2007, VOC2010 and other object detection challenge competitions. However, R-CNN has to perform feature extraction for all 2000 region proposals, and there are many crossovers between the region proposals and redundant feature extraction operations, resulting in slow speed, large space occupation, and the need to train AlexNet, SVM and regressor individually.

Fast R-CNN. In response to a series of problems with R-CNN, Fast R-CNN was born in 2015. Fast R-CNN uses VGG16 as the Backbone of the network, which not only makes a breakthrough in speed but also improves the accuracy rate than before.

The specific operation is region selection first (this step is consistent with R-CNN), unlike R-CNN which puts each region proposal for feature extraction, Fast R-CNN extracts features by putting the whole image and the region proposals on the image into VGG16 at once. The Rol pooling layer [34] is used for selecting the region of interest and feeding the resulting feature vectors into two fully connected layers. One of which is responsible for discriminating the category and one is responsible for locating the anchor box to the correct position. The Fast RCNN uses multi-task loss jointly train classification and bounding-box regression so that two tasks share convolution features.

Faster R-CNN. Although Fast R-CNN has been effective, traditional methods of generating candidate bounding boxes such as selective search still have the problem of taking a long time. Faster R-CNN generates detection boxes directly using RPN [35] based on Fast R-CNN (RPN is a fully connected neural network), which further improves the running speed. This is done as follows: (1) generating a large number of anchors (2) RPN determines whether all the anchors contain objects, but not their categories, and (3) adjusting the positions of the anchors to get more reasonable proposals. The ROI pooling layer is used to adjust the vectors to a uniform size, and then output to the fully connected layer. The Faster R-CNNN is different from the Compared with previous R-CNN series, the region selection task also uses a deep learning approach, which greatly improves the operational efficiency.

3.2 One Stage

Yolo. Traditional two-stage models need two steps to generate a bounding-box and predict object types. In contrast, Yolo is an end-to-end model in which the predicted bounding boxes and object classes are obtained by only one network.

Yolo's Backbone framework is inspired by the GoogLeNet model [36] and consists of 24 convolutional layers, and 2 fully connected layers, but instead of using Inception blocks, a 1×1 convolution is used behind a 3×3 convolution. The full connected layer vector in Yolo is Reshaped into a three-dimensional tensor with a size of $7 \times 7 \times 30$ [36], which is responsible for predicting the object if its center falls in a cell. 7×7 represents the division of the image into 7×7 cells, and 30 represents the generation of two bounding boxes per cell, each predicting five values (c, x, y, w, h), and 20 categories.

Yolo is inferior to Faster R-CNN in terms of accuracy, but faster than Faster R-CNN. The Yolo positioning accuracy is not enough, multiple targets are close together, the target is too small, and the detection effect is not good. J. Redmon et al. subsequently proposed yolov2 [37], yolov3 [38] (Fig. 2).

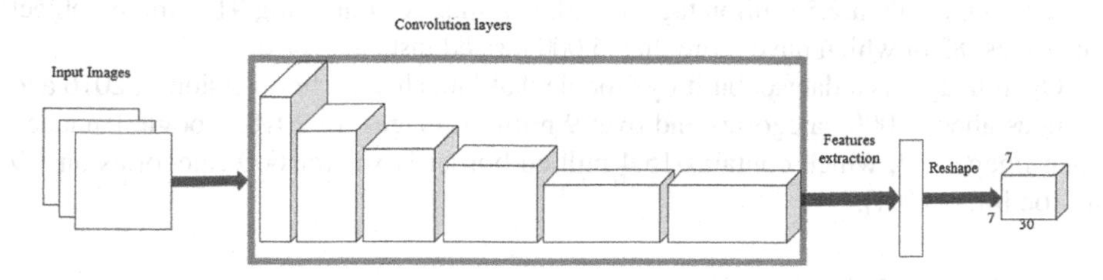

Fig. 2. Schematic diagram of Yolo structure.

SSD. In order to ensure the speed and accuracy at the same time, Liu W et al. proposed SSD, which is the same as the popular detection model nowadays, SSD combines the whole detection process into a single deep neural network, combining the advantages of Faster R-CNN and Yolo, and its speed is faster than the one stage model of the same period, yolo v1 is faster and has higher accuracy, 59 FPS with MAP 74.3% on VOC2007 test, vs Faster R-CNN 7 FPS with MAP 73.2% or YOLO 45 FPS with MAP 63.4% [39].

The Backbone of SSD is VGG16, which uses a base network to extract features, generating multiple anchor frames at each pixel on a feature map at different scales, predicting bounding boxes and categories for each anchor frame. The convolution layer halves the height and width of the input image to arrive at fitting small objects with the bottom layer and large objects with the top layer.

4 Datasets and Evaluation Criteria

4.1 Datasets

Datasets are an important part of the object detection task to train parameters and evaluate models. This subsection will systematically review the classic datasets that have made outstanding contributions and advanced research in the field of object detection.

The creation of the VOC (2005–2012) challenge made an important contribution to the development of the computer vision field. The most commonly used ones are VOC07 and VOC12, which have been extended to 20 classes of objects compared to VOC05. The training set size of VOC07 [40] has been increased to 5k and has more than 12k labeled objects. In contrast, the training set size of VOC12 reached 11k and had 16k labeled objects [41].

The ImageNet Large Scale Visual Recognition Challenge (ILSVRC) (2010–2017) [42], which had made irreplaceable contributions to the development of image classification, target detection and other fields. The ImageNet dataset, created under the auspices of Stanford professor Feifei Li, contains over 14 million tagged images, and 1000 classes of objects, including over 500k images for the target detection class and 200 classes.

MS-COCO is the most challenging dataset at present, with a huge scale and a high status in the industry, mainly used for object detection, instance segmentation and other scenarios.MS-COCO has fewer categories than ImageNet, but more instances per category, with more than 2.5 million tags in 320,000 images, containing 91 common object categories, 82 of which have more than 5,000 tagged instances [43].

Open images is a dataset built by Google that launched its first version in 2016 and includes about 6,000 categories and over 9 million images. In 2018, Google launched Open Images V4, which contains 15.4 million border boxes for 600 categories on 1.9 million images [44].

4.2 Evaluation Criteria

Evaluation criteria are used to measure how good the network is on the dataset. There are many different kinds of evaluation criteria in the object detection task, such as recall, accuracy, mean average precision (MAP), FPS, etc. The following is an analysis of the evaluation criteria in object detection data.

Intersection over union (IOU) is the ratio of intersection and union of predicted and true bounding-box. If the IOU is greater than the threshold value, the prediction is considered as True Positive(TP), and if the IOU is less than the threshold value, the prediction is considered as False Positive(FP). If the object in the bounding box is not detected by the model, it is recorded as False Negative (FN). Precision measures the percentage of correct predictions while recall measures the correct predictions with respect to the ground truth 2.

$$Precision = TP/(TP + FP)\#(1) \tag{1}$$

$$Recall = TP/(TP + FN)\#(2) \tag{2}$$

Based on the above equation, Average Precision is calculated for each class separately. Average Precision of all classes is averaged to obtain *mean Average Precision* (mAP) and MAP is used to compare the performance between detectors.

5 Applications

5.1 Face Detection

Face detection has been an important application scenario in the field of object detection, where the task goal is to find out the face in the image and determine its location, and the traditional face detection is mainly done by manually extracting the face features and then using a sliding window to match out the face in the image, where the representative algorithm is VJ detector [4]. Object detection has achieved great success since it entered the era of deep learning, and face detection algorithms are inextricably linked with general-purpose object detection algorithms such as the RCNN family. A cascaded CNN containing multiple cascaded DCNN classifiers was proposed [45], which improves the speed of face detection and solves the problems caused by illumination and angle in some realistic applications. To improve the problem of multi-pose and face occlusion recognition, [46] was proposed subsequently.

5.2 Text Detection

Text is one of the most important information carriers in human society and is a necessary part of people's lives. Text detection in images has important applications in many aspects, such as intelligent traffic, recognizing road signs and slogans. Used for information extraction, automatic recognition of text in natural scenes can save a lot of resources and protect customer privacy.

In the early days, text detection was usually extracted manually, but in the era of deep learning, features are usually extracted automatically using neural networks. This has greatly improved efficiency and simplified the workflow.

Mainstream text detection is divided into two ways, one is to first detect the text with generic object detection and then identify the text content, including image pre-processing, feature representation, sequence modeling (or character segmentation recognition), and prediction. Among the representative algorithms are [47] and others. And one is the end-to-end recognition approach, in which text detection and text recognition were previously divided into two separate problems, while end-to-end systems unite them into one, and recently, building real-time and efficient end-to-end systems has become a new trend in the community [48].

6 Conclusions

This paper reviewed the evolution of object detection, focusing on the contribution of deep learning-based object detection to the industry and research development, as well as comparing its advantages and where it has advanced compared to traditional approaches. The architecture of landmark backbone networks for target detection, such as AlexNet, VGG, GoogleNet, and ResNet was analyzed. Deep learning-based target detection algorithms such as R-CNN, Fast R-CNN, and Faster R-CNN were summarized and reviewed, and their differences from traditional object detection algorithms are analyzed and their characteristics were compared. The architectures of One Stage algorithms YOLO and SSD were concisely outlined, and their advantages and disadvantages were compared with Two Stage algorithms, and their operational effects were illustrated. Four major datasets in the field of object detection were introduced and the evaluation criteria of the models were parsed. Finally, a summary of the classic applications in target detection.

References

1. Xiao, Y., Tian, Z., Yu, J., et al.: A review of object detection based on deep learning. Multimedia Tools Appl. **79**(33), 23729–23791 (2020)
2. Zaidi, S.S.A, Ansari, M.S., Aslam, A., et al.: A survey of modern deep learning based object detection models. Digital Signal Processing, p. 103514 (2022)
3. Zou, Z., Shi, Z., Guo, Y., et al.: Object detection in 20 years: a survey. arXiv preprint arXiv: 1905.05055 (2019)
4. Viola, P., Jones, M.: Rapid object detection using a boosted cascade of simple features, IEEE CVPR 2001, vol. 1, pp. I-I (2001)
5. Dalal, N., Triggs, B.: Histograms of oriented gradients for human detection. In: 2005 IEEE CVPR, vol. 1, pp. 886–893.3 (2005)

6. Zhao, Z.Q., Zheng, P., Xu, S., et al.: Object detection with deep learning: a review. IEEE Trans. Neural Networks Learn. Syst. **30**(11), 3212–3232 (2019)
7. Liang, W., Long, J., Li, K.C., et al.: A fast defogging image recognition algorithm based on bilateral hybrid filtering. ACM TOMM **17**(2), 1–16 (2021)
8. Diao, C., Zhang, D., Liang, W., et al.: A novel spatial-temporal multi-scale alignment graph neural network security model for vehicles prediction. In: IEEE TITS (2022)
9. Peng, L., Peng, M., Liao, B., et al.: Improved low-rank matrix recovery method for predicting miRNA-disease association. Sci. Rep. **7**(1), 1–10 (2017)
10. Xiao, W., Tang, Z., Yang, C., et al.: ASM-VoFDehaze: a real-time defogging method of zinc froth image. Connect. Sci. **34**(1), 709–731 (2022)
11. Wang, J., Luo, W., Liang, W., et al.: Locally minimum storage regenerating codes in distributed cloud storage systems. China Commun. **14**(11), 82–91 (2017)
12. Liang, W., Huang, Y., Xu, J., et al.: A distributed data secure transmission scheme in wireless sensor network. Int'l J. Distrib. Sensor Netw. **13**(4), 1550147717705552 (2017)
13. Krizhevsky, A., Sutskever, I., Hinton, G.E.: Imagenet classifification with deep convolutional neural networks. In: NIPS (2012)
14. Qiu, H., Dong, T., et al.: Adversarial attacks against network intrusion detection in IoT systems. IEEE Internet Things J. **8**(13), 10327–10335 (2020)
15. Qiu, H., Zheng, Q., et al.: Topological graph convolutional network-based urban traffic flow and density prediction. IEEE Trans. ITS (2020)
16. Hu, F., Lakdawala, S., et al.: Low-power, intelligent sensor hardware interface for medical data preprocessing. IEEE Trans. Info. Tech. Biome. **13**(4), 656–663 (2009)
17. Gai, K., Qiu, M., Elnagdy, S.: A novel secure big data cyber incident analytics framework for cloud-based cybersecurity insurance. IEEE BigDataSecurity (2016)
18. Li, Y., Gai, K., et al.: Intercrossed access controls for secure financial services on multimedia big data in cloud systems. ACM TMCCA (2016)
19. Qiu, M., Chen, Z., Ming, Z., Qin, X., Niu, J.: Energy-aware data allocation with hybrid memory for mobile cloud systems. IEEE Syst. J. **11**(2), 813–822 (2014)
20. Qiu, M., Xue, C., Shao, Z., Sha, E.: Energy minimization with soft real-time and DVS for uniprocessor and multiprocessor embedded systems. In: IEEE DATE Conference, pp. 1–6 (2007)
21. Qiu, M., Jia, Z., et al.: Voltage assignment with guaranteed probability satisfying timing constraint for real-time multiproceesor DSP, JSPS. Springer (2007)
22. Niu, J., Gao, Y., Qiu, M., Ming, Z.: Selecting proper wireless network interfaces for user experience enhancement with guaranteed probability. JPDC **72**(12), 1565–1575 (2012)
23. Li, J., Ming, Z., et al.: Resource allocation robustness in multi-embedded systems with inaccurate information. J. Syst. Architect. **57**(9), ore e840–849 (2011)
24. LeCun, Y., Bottou, L., Bengio, Y., et al.: Gradient-based learning applied to document recognition. Proc. IEEE **86**(11), 2278–2324 (1998)
25. Simonyan, K., Zisserman, A.: Very deep convolutional networks for large-scale image recognition. arXiv preprint arXiv:1409.1556 (2014)
26. Dhillon, A., Verma, G.K.: Convolutional neural network: a review of models, methodologies and applications to object detection. Progress Artific. Intell. **9**(2), 85–112 (2019). https://doi.org/10.1007/s13748-019-00203-0
27. Liu, L., Ouyang, W., Wang, X., et al.: Deep learning for generic object detection: a survey. Int. J. Comput. Vis. **128**(2), 261–318 (2020)
28. Szegedy, C., Liu, W., Jia, Y., et al.: Going deeper with convolutions. In: Proceedings of the IEEE Conference on Computer Vision and Pattern Recognition, pp. 1–9 (2015)
29. Ioffe, S., Szegedy, C.: Batch normalization: Accelerating deep network training by reducing internal covariate shift. International Conference on Machine Learning. PMLR, pp. 448–456 (2015)

30. He, K., Zhang, X., Ren, S., et al.: Deep residual learning for image recognition, IEEE CVPR, pp. 770–778 (2016)
31. Huang, G., Liu, Z., Van Der Maaten, L., et al.: Densely connected convolutional networks, IEEE CVPR, pp. 4700–4708 (2017)
32. Zhang, X., Zhou, X., Lin, M., et al.: Shufflenet: an extremely efficient convolutional neural network for mobile devices. In: IEEE CVPR, pp. 6848–6856 (2018)
33. Girshick, R., Donahue, J., Darrell, T., et al.: Rich feature hierarchies for accurate object detection and semantic segmentation. In: IEEE CVPR, pp. 580–587 (2014)
34. Girshick, R.: Fast r-cnn. In: Proceedings of the IEEE International Conference on Computer Vision, pp. 1440–1448 (2015)
35. Ren, S., He, K., Girshick, R., et al.: Faster r-cnn: towards real-time object detection with region proposal networks. Adv. Neural Info. Process. Syst. **28** (2015)
36. Redmon, J., Divvala, S., Girshick, R., et al.: You only look once: Unified, real-time object detection. In: IEEE Conference on Computer Vision and Pattern Recognition, pp. 779–788 (2016)
37. Redmon, J., Farhadi, A.: YOLO9000: better, faster, stronger. In: Proceedings of the IEEE Conference on Computer Vision and Pattern Recognition, pp. 7263–7271 (2017)
38. Redmon, J., Farhadi, A.: Yolov3: an incremental improvement. arXiv preprint arXiv:1804. 02767 (2018)
39. Liu, W., Anguelov, D., Erhan, D., et al.: SSD: single shot multibox detector. In: European Conference on Computer Vision, pp. 21–37. Springer, Cham (2016)
40. Everingham, M., Winn, J.: The PASCAL visual object classes challenge 2007 (VOC2007) development kit. Int. J. Comput. Vis **88**(2), 303–338 (2010)
41. Everingham, M., Winn, J.: The PASCAL visual object classes challenge 2012 (VOC2012) development kit. Pattern Analysis Statistical Modelling and Computational Learning, Technical Report 2012, pp. 1–45 (2007)
42. Russakovsky, O., Deng, J., Su, H., et al.: Imagenet large scale visual recognition challenge. Int. J. Comput. Vis. **115**(3), 211–252 (2015)
43. Lin, T.-Y., Maire, M., Belongie, S., et al.: Microsoft coco: Common objects in context. In: Fleet, David, Pajdla, Tomas, Schiele, Bernt, Tuytelaars, Tinne (eds.) Computer Vision – ECCV 2014: 13th European Conference, Zurich, Switzerland, September 6-12, 2014, Proceedings, Part V, pp. 740–755. Springer, Cham (2014). https://doi.org/10.1007/978-3-319-10602-1_48
44. Kuznetsova, A., Rom, H., Alldrin, N., et al.: The open images dataset v4. Int. J. Comput. Vis. **128**(7), 1956–1981 (2020)
45. Li, H., Lin, Z., Shen, X., et al.: A convolutional neural network cascade for face detection. In: IEEE CVP, pp. 5325–5334 (2015)
46. Shi, X., Shan, S., Kan, M., et al.: Real-time rotation-invariant face detection with progressive calibration networks. IEEE CVPR, pp. 2295–2303 (2018)
47. Wang, T., Wu, D.J., Coates, A., et al.: End-to-end text recognition with convolutional neural networks. In: 21st IEEE ICPR, pp. 3304–3308 (2012)
48. Chen, X., Jin, L., Zhu, Y., et al.: Text recognition in the wild: a survey. ACM Comput. Surv. **54**(2), 1–35 (2021)

A Decision-Making Method for Blockchain Platforms Using Axiomatic Design

Jun Liu, Qi Zhang(✉), Ming-Yue Xie, and Ming-Peng Chen

Chongqing University of Posts and Telecommunications, Chongqing 400065, China
s211201034@stu.cqupt.edu.cn

Abstract. For companies using blockchain technology, it is critical to select the most suitable blockchain platform to develop enterprise applications. However, it is still a challenge for enterprises. As an important part of modern decision science, multi-criteria decision-making can well solve the problem of blockchain platform selection. Blockchain platforms integrate various blockchain technologies, and enterprises also need to consider multiple different criteria in the decision-making process. Therefore, this paper will use a heterogeneous multi-criteria decision-making method to solve the blockchain platform selection problem. First, the blockchain platform alternatives and evaluation criteria used for decision-making are identified. Second, blockchain platform alternatives are evaluated with appropriate fuzzy numbers based on defined evaluation criteria. Then, the original evaluations are consistency and normalized to obtain a normalized evaluation. Next, the improved information content formulas of the axiomatic design is proposed to obtain the information content of each normalized evaluation. Then, the weights of all evaluated criteria are obtained using the entropy weight method. Finally, the total weighted information content of each blockchain platform alternative is obtained. With validation, the decision-making model of blockchain platform proposed in this paper has a strong reference value.

Keywords: Blockchain platform · Multi-criteria decision-making · Heterogeneous information · Axiomatic design · Information content · Entropy weight method

1 Introduction

As a novel technology combination integrating existing technologies such as computing power [1–3], peer-to-peer communication [4–6], cryptography [7–9], and distributed data storage [10–12], blockchain [13, 14] is widely used by numerous enterprises because of its characteristics of anonymity, openness, security, information immutability, and decentralization [15–17]. Blockchain has gone through three generations since 2008. The latest generation of blockchain is not limited to the financial industry [18], but has been widely used in various industries, such as the Internet of Things, logistics [19], public services, digital rights, and insurance [20–22]. In these contexts, developing blockchain applications for various industries is the current focus [23–25]. As a result,

M. Qiu et al. (Eds.): SmartCom 2022, LNCS 13828, pp. 310–318, 2023.
https://doi.org/10.1007/978-3-031-28124-2_29

industries are increasingly inseparable from blockchain technology [26, 27]. Blockchain platforms integrate various blockchain technologies and can shorten the development and deployment cycle of blockchain applications in various industries. Therefore, it is more beneficial to use existing blockchain platforms to develop blockchain applications.

With the development of blockchain technology, there are a large number of blockchain platforms on the market and are used in various businesses of enterprises. However, with the rapid increase in the number of blockchain platforms, it has become a challenge for enterprises to select the suitable blockchain platform. Different blockchain platforms provide different functions, and the functions required by enterprises in different industries are also different. As a decision-making method, multi-criteria decision-making can well solve the decision-making problem of blockchain platforms. Supporting this argument, paper [28] proposed a decision-making model for blockchain platforms, [29] proposed a decision-making framework for business blockchain platforms, and [30] proposed a decision-making model for public blockchain platforms.

In order to solve the problem of blockchain platform selection, this paper proposes an axiomatic design-based blockchain platform decision-making method with the following main contributions:

- Transforming the blockchain platform selection problem into a heterogeneous multi-criteria decision-making.
- Considering that one type of fuzzy number [31–36] is not sufficient to express multiple criteria, this paper provides interval number, real number, and triangular fuzzy number to express the evaluation information based on different criteria.
- The formula for the information content of the axiomatic design is improved, and its value falls between 0 and 1, and the larger the better. It can avoid the situation that the weights of the evaluated criteria cannot be calculated and the ranking of the solved alternatives is not reasonable when the information content is ∞ [36–38].

The rest of the paper is organized as follows. Section 2 introduces related concepts, including entropy weight method, and axiomatic design. Section 3 describes the decision-making process of blockchain platform based on axiomatic design, and Sect. 4 summarizes the whole paper.

2 Preliminaries

In this section, we will briefly introduce some necessary basics related to this thesis. Firstly, the concept of entropy weight method will be introduced. Finally, the concepts related to axiomatic design will be introduced.

2.1 The Concept of Entropy Weight Method

The entropy weight method is an objective method to determine the weight of criteria. In the entropy weight method, according to the definition of information entropy, for a certain index, the entropy value can be used to determine the dispersion degree of a certain index, the smaller the information entropy value, the greater the dispersion degree of the index, the greater the influence (i.e. weight) of the index on the comprehensive evaluation, if all the values of a certain index are equal, the index does not work in the comprehensive evaluation. The calculation steps of the entropy weight method are as follows.

Step 1: Obtain the normalized matrix with n rows and m columns.
Step 2: Calculate the entropy value e_j of criterion *j*.

$$e_j = -l\sum_{i=1}^{n} p_{ij}\ln p_{ij}, p_{ij} = \frac{x_{ij}}{\sum_{i=1}^{n} x_{ij}}, l = \frac{1}{\ln n}, 0 \le e_j \le 1 \tag{1}$$

Step 3: Calculate the degree of divergence (d_j) of average intrinsic information contained by criterion *j*.

$$d_j = 1 - e_j \tag{2}$$

Step 4: Calculate the weight W_j for criterion *j*.

$$W_j = \frac{d_j}{\sum_{j=1}^{m} d_j} \tag{3}$$

2.2 The Concept of Axiomatic Design

Axiomatic design is proposed by Suh, which can provide a theoretical basis for designers to choose the most reasonable design solutions and establish scientific and systematic standards for design activities.

The most important concept in axiomatic design is the existence of design axioms, the first one being the axiom of independence and the second one being the axiom of information. These axioms are expressed as follows:

1. Independence Axiom: Maintain the independence of functional requirements (FRs).
2. Information axiom: Minimize information content.

The independence axiom states that the independence of functional requirements (FRs) must always be maintained, and that FRs are defined as the smallest set of independent requirements that describe the design goals.

The information axiom states that the design with the least information content is the best design among the designs that satisfy the independent axiom. The information content is the probability of satisfying a given functional requirement. For a functional requirement FR_m, its information content I_m is defined as follows:

$$I_m = \log_2(\frac{1}{P_m}) \tag{4}$$

P_m is the probability of achieving the functional requirement FR_m, the notation "log" means that the logarithm is base 2 logarithm. The probability of achieving a functional requirement is determined by both the degree to which the designer wants to achieve it (the design range) and the degree to which the design can achieve it (the system range). The common part of the design range and system range is the area where an acceptable solution exists. As shown in Fig. 1. Therefore, under the uniform probability density function:

$$P_m = \frac{common\ \ range}{system\ \ range} \tag{5}$$

Then the information content I_m can be defined as:

$$I_m = \log_2(\frac{system\ \ range}{common\ \ range}) \tag{6}$$

Based on the axiomatic design, we can get that in a multi-criteria decision-making, the design range of benefit criteria is the maximum evaluated value of the alternative under the corresponding criteria. The cost criteria can be transformed into benefit criteria. The independence axiom of axiomatic design requires that functional requirements are independent of each other, and therefore the criteria in this research are independent of each other.

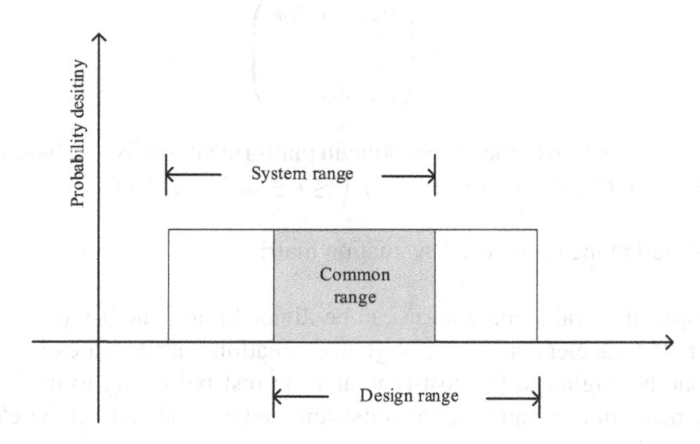

Fig. 1. Design range, system range, common range and probability density function of FR

3 Decision-Making Process for Blockchain Platforms

In this section, we will introduce the decision-making process for blockchain platforms. The steps of decision-making include: determining the blockchain platform alternatives and evaluation criteria, evaluating all blockchain platforms based on each evaluation criterion to get the original evaluation matrix, consistency and normalized the original evaluation matrix to get the normalized evaluation matrix, calculating the information content of each element in the normalized evaluation matrix, using the entropy weight method to get the weight of each evaluation criterion, calculate the total weighted information content of each blockchain platform alternative, and rank the blockchain platform alternatives. The decision-making process is described in Fig. 2, and the detailed decision-making process is described below.

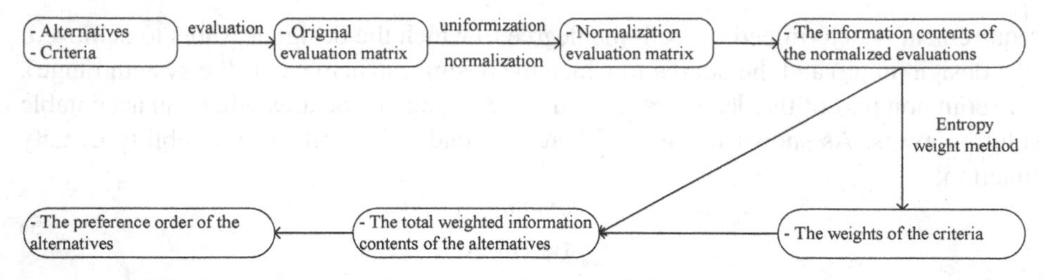

Fig. 2. Decision-making Process for Blockchain Platform Alternatives

Step 1: Let the n blockchain platform alternatives be $\{A_1, A_2, ..., A_n\}$ and the m evaluation criteria be $\{C_1, C_2, ..., C_m\}$.
Step 2: Constructing the original evaluation matrix.

The evaluation of n blockchain platform alternatives based on m evaluation criteria leads to the following decision matrix.

$$A = \begin{pmatrix} a_{11} & \cdots & a_{1m} \\ \vdots & \ddots & \vdots \\ a_{n1} & \cdots & a_{nm} \end{pmatrix}$$

a_{nm} denotes the evaluation of the nth blockchain platform alternative A_n based on the mth evaluation criterion C_m, where $0 \le a_{ij} \le 1,\ 1 \le i \le n,\ 1 \le j \le m$.

Step 3: Constructing the normalized evaluation matrix.

In this paper, the evaluation criteria can be divided into benefit-type and cost-type. It can be seen that the elements in the original evaluation matrix A are divided into two categories, one belonging to the cost-type and the rest belonging to the benefit-type. Next, the elements in the matrix A are consistent and normalized, all the elements are transformed into benefit-type elements.

For the interval number $a_{ij} = [k_{ij}, l_{ij}]$ in the original evaluation matrix, where $0 \le k_{ij} \le l_{ij} \le 1$, the following equation can be obtained, where b_{ij} is the normalized evaluation of $a_{ij} = [k_{ij}, l_{ij}]$.

$$b_{ij} = \begin{cases} [\frac{k_{ij}-k_j^{\min}}{l_j^{\max}-k_j^{\min}}, \frac{l_{ij}-k_j^{\min}}{l_j^{\max}-k_j^{\min}}]\ (benefit\ \ criterion) \\ [\frac{l_j^{\max}-l_{ij}}{l_j^{\max}-k_j^{\min}}, \frac{l_j^{\max}-k_{ij}}{l_j^{\max}-k_j^{\min}}]\ \ (\text{cost}\ \ criterion) \end{cases} \tag{7}$$

For the real number $a_{ij} = r_{ij}$ in the original evaluation matrix, where $0 \le r_{ij} \le 1$, the following equation can be obtained, where b_{ij} is the normalized evaluation of $a_{ij} = r_{ij}$.

$$b_{ij} = \begin{cases} \frac{r_{ij}-r_j^{\min}}{r_j^{\max}-r_j^{\min}} & (benefit\ \ criterion) \\ \frac{r_j^{\max}-r_{ij}}{r_j^{\max}-r_j^{\min}} & (\text{cost}\ \ criterion) \end{cases} \tag{8}$$

For the triangular fuzzy number $a_{ij} = (m_{ij}, n_{ij}, p_{ij})$ in the original evaluation matrix, where $0 \le m_{ij} \le n_{ij} \le p_{ij} \le 1$, the following equation can be obtained, where b_{ij} is the normalized evaluation of $a_{ij} = (m_{ij}, n_{ij}, p_{ij})$.

$$b_{ij} = \begin{cases} \left(\frac{m_{ij}-m_j^{\min}}{p_j^{\max}-m_j^{\min}}, \frac{n_{ij}-m_j^{\min}}{p_j^{\max}-m_j^{\min}}, \frac{p_{ij}-m_j^{\min}}{p_j^{\max}-m_j^{\min}}\right) & (\textit{benefit criterion}) \\ \left(\frac{p_j^{\max}-p_{ij}}{p_j^{\max}-m_j^{\min}}, \frac{p_j^{\max}-n_{ij}}{p_j^{\max}-m_j^{\min}}, \frac{p_j^{\max}-m_{ij}}{p_j^{\max}-m_j^{\min}}\right) & (\text{cost } \textit{criterion}) \end{cases} \tag{9}$$

Finally, the following decision matrix can be obtained.

$$B = \begin{pmatrix} b_{11} & \dots & b_{1m} \\ \vdots & \ddots & \vdots \\ b_{n1} & \cdots & b_{nm} \end{pmatrix}$$

where b_{nm} denotes the normalized evaluation obtained after consistent and normalized processing of the a_{nm} in the original evaluation matrix, where $0 \le b_{ij} \le 1$, $1 \le i \le n$, $1 \le j \le m$.

Step 4: Calculate the information content in each element of the normalized evaluation matrix.

For the interval numbers $b_{ij} = [\overline{k_{ij}}, \overline{l_{ij}}]$ and $b_j^* = [\overline{k_{ij}}_{\max}, \overline{l_{ij}}_{\max}]$, where $0 \le \overline{k_{ij}} \le \overline{l_{ij}} \le 1, 0 \le \overline{k_{ij}}_{\max} \le \overline{l_{ij}}_{\max} \le 1$, the following equation can be obtained, where $I'(b_{ij}, b_j^*)$ is the information content of $b_{ij} = [\overline{k_{ij}}, \overline{l_{ij}}]$.

$$I'(b_{ij}, b_j^*) = \begin{cases} \log_2\left(1 + \frac{\overline{l_{ij}}-\overline{k_{ij}}_{\max}}{\overline{l_{ij}}-\overline{k_{ij}}}\right) & (\overline{k_{ij}}_{\max} < \overline{l_{ij}}) \\ 0 \quad (\overline{l_{ij}} \le \overline{k_{ij}}_{\max}) \end{cases} \tag{10}$$

For the real numbers $b_{ij} = \overline{r_{ij}}$ and $b_j^* = \overline{r_{ij}}_{\max}$, where $0 \le \overline{r_{ij}} \le 1, 0 \le \overline{r_{ij}}_{\max} \le 1$, the following equation can be obtained, where $I'(b_{ij}, b_j^*)$ is the information content of $b_{ij} = \overline{r_{ij}}$.

$$I'(b_{ij}, b_j^*) = \log_2\left(1 + \frac{\overline{r_{ij}}}{\overline{r_{ij}}_{\max}}\right) \quad (\overline{r_{ij}} \le \overline{r_{ij}}_{\max}) \tag{11}$$

For the triangular fuzzy numbers $b_{ij} = (\overline{m_{ij}}, \overline{n_{ij}}, \overline{p_{ij}})$ and $b_j^* = (\overline{m_{ij}}_{\max}, \overline{n_{ij}}_{\max}, \overline{p_{ij}}_{\max})$, where $0 \le \overline{m_{ij}} \le \overline{n_{ij}} \le \overline{p_{ij}} \le 1$, $0 \le \overline{m_{ij}}_{\max} \le \overline{n_{ij}}_{\max} \le \overline{p_{ij}}_{\max} \le 1$, the following equation can be obtained, where $I'(b_{ij}, b_j^*)$ is the information content of $b_{ij} = (\overline{m_{ij}}, \overline{n_{ij}}, \overline{p_{ij}})$.

$$I'(b_{ij}, b_j^*) = \begin{cases} \log_2\left(1 + \frac{(\overline{p_{ij}}-\overline{m_{ij}}_{\max})^2}{(\overline{p_{ij}}-\overline{m_{ij}})(\overline{p_{ij}}-\overline{n_{ij}}+\overline{n_{ij}}_{\max}-\overline{m_{ij}}_{\max})}\right) & (\overline{p_{ij}} > \overline{m_{ij}}_{\max}) \\ 0 \quad (\overline{p_{ij}} \le \overline{m_{ij}}_{\max}) \end{cases} \tag{12}$$

Step 5: Calculate the weight of each evaluation criterion using the entropy weight method.

Step 5.1: Constructing the matrix for the calculation, the matrix can be obtained from Step 4 as follows:

$$I' = \begin{pmatrix} I'_{11} & \cdots & I'_{1m} \\ \vdots & \ddots & \vdots \\ I'_{n1} & \cdots & I'_{nm} \end{pmatrix}$$

where I'_{nm} denotes the information content of b_{nm} in the normalized evaluation matrix, where $0 \leq I'_{ij} \leq 1,\ 1 \leq i \leq n,\ 1 \leq j \leq m$

Step 5.2: Calculate the entropy of each evaluation criterion.

$$e_j = -h \sum\nolimits_{i=1}^{n} p_{ij} \ln p_{ij}, p_{ij} = \frac{I'_{ij}}{\sum_{i=1}^{n} I'_{ij}}, h = \frac{1}{\ln n}, 0 \leq e_j \leq 1 \tag{13}$$

Step 5.3: Calculate the degree of divergence of average intrinsic information contained for each evaluation criterion.

$$d_j = 1 - e_j \tag{14}$$

Step 5.4: Calculate the weights for each evaluation criterion.

$$W_j = \frac{d_j}{\sum_{j=1}^{m} d_j} \tag{15}$$

Step 6: Calculate the total weighted information content for each blockchain platform alternative, where $I'(A_i)$ is the total weighted information content of the alternative A_i.

$$I'(A_i) = \sum\nolimits_{j=1}^{m} I'_{ij} W_j \tag{16}$$

Step 7: Ranking of all blockchain platform alternatives.

Each blockchain platform alternative A_i is ranked according to $I'(A_i)$, and the larger the $I'(A_i)$, the better the blockchain platform alternative A_i.

4 Conclusion

With the continuous development of blockchain technology, the number of blockchain platforms in the market is increasing. However, selecting the suitable blockchain platform is still a problem. To address the above background, this paper proposed a heterogeneous multi-criteria decision-making method based on axiomatic design to solve the blockchain platform selection problem. The blockchain platform decision-making method proposed in this paper can help relevant practitioners to select the suitable blockchain platform more quickly and efficiently.

References

1. Qiu, M., Jia, Z., et al.: Voltage assignment with guaranteed probability satisfying timing constraint for real-time multiproceesor DSP. J. Signal Process. Syst. (2007)
2. Qiu, M., Yang, L., Shao, Z., Sha, E.: Dynamic and leakage energy minimization with soft real-time loop scheduling and voltage assignment. IEEE TVLSI **18**(3), 501–504 (2009)
3. Qiu, M., Xue, C., Shao, Z., Sha, E.: Energy minimization with soft real-time and DVS for uniprocessor and multiprocessor embedded systems. In: IEEE DATE Conference, pp. 1–6 (2007)
4. Niu, J., Gao, Y., et al.: Selecting proper wireless network interfaces for user experience enhancement with guaranteed probability. JPDC **72**(12), 1565–1575 (2012)
5. Qiu, M., Xue, C., Shao, Z., et al.: Efficient algorithm of energy minimization for heterogeneous wireless sensor network. In: IEEE EUC, pp. 25–34 (2006)
6. Qiu, M., Li, H., Sha, E.: Heterogeneous real-time embedded software optimization considering hardware platform. In: ACM Symposium on Applied Computing, pp. 1637–1641 (2009)
7. Qiu, H., Dong, T., Zhang, T., et al.: Adversarial attacks against network intrusion detection in IoT systems. IEEE Internet Things J. **8**(13), 10327–10335 (2020)
8. Gai, K., Qiu, M., Elnagdy, S.: A novel secure big data cyber incident analytics framework for cloud-based cybersecurity insurance. In: IEEE BigDataSecurity (2016)
9. Gao, X., Qiu, M.: Energy-based learning for preventing backdoor attack. In: KSEM, no. 3, pp. 706–721 (2022)
10. Qiu, M., Chen, Z., Ming, Z., Qin, X., Niu, J.: Energy-aware data allocation with hybrid memory for mobile cloud systems. IEEE Syst. J. **11**(2), 813–822 (2014)
11. Qiu, M., Qiu, H., et al.: Secure data sharing through untrusted clouds with blockchain-enabled key management. In: SmartBlock 2020, China, pp. 11–16 (2020)
12. Li, J., Ming, Z., et al.: Resource allocation robustness in multi-core embedded systems with inaccurate information. J. Syst. Arch. **57**(9), 840–849 (2011)
13. Qiu, M., Qiu, H.: Review on image processing based adversarial example defenses in computer vision. In: IEEE BigDataSecurity, Baltimore, USA, pp. 94–99 (2020)
14. Gai, K., Zhang, Y., et al.: Blockchain-enabled service optimizations in supply chain digital twin. IEEE Trans. Serv. Comput. (2022)
15. Xie, M.-Y., Liu, J.: A survey on blockchain consensus mechanism: research overview, current advances, and future directions. Int. J. Intell. Comput. Cybern., 1–27 (2022)
16. Xie, M.-Y., Liu, J., Chen, S.-Y., Xu, G.-X., Lin, M.-W.: Primary node election based on probabilistic linguistic term set with confidence interval in the PBFT consensus mechanism for blockchain. Complex Intell. Syst. (2022). https://doi.org/10.1007/s40747-022-00857-9
17. Liu, J., Xie, M.-Y., Chen, S.-Y., Ma, C., Gong, Q.-H.: An improved DPoS consensus mechanism in blockchain based on PLTS for the smart autonomous multi-robot system. Inf. Sci. **575**, 528–541 (2021)
18. Li, Y., Gai, K., et al.: Intercrossed access controls for secure financial services on multimedia big data in cloud systems. ACM Trans. Multimedia Comput. Commun. Appl. (2016)
19. Liu, J., Zhao, J., Huang, H., Xu, G.: A novel logistics data privacy protection method based on blockchain. Multimedia Tools Appl. **81**(17), 23867–23877 (2022)
20. Qiu, H., Zheng, Q, et al.: Topological graph convolutional network-based urban traffic flow and density prediction. IEEE Trans. Intell. Transp. Syst. **99** (2020)
21. Li, Y.-B., Song, Y., Jia, L., et al.: Intelligent fault diagnosis by fusing domain adversarial training and maximum mean discrepancy via ensemble learning. IEEE Trans. Ind. Inf. **17**(4), 2833–2841 (2021)

22. Hu, F., Lakdawala, S., Hao, Q., et al.: Low-power, intelligent sensor hardware interface for medical data preprocessing. IEEE Trans. Inf Technol. Biomed. **13**(4), 656–663 (2009)
23. Gai, K.-K., Wu, Y.-L., Zhu, L.-H., et al.: Permissioned blockchain and edge computing empowered privacy-preserving smart grid networks. IEEE Internet Things J. **6**(5), 7992–8004 (2019)
24. Qiu, M.-K., Gai, K.-K., Xiong, Z.-G.: Privacy-preserving wireless communications using bipartite matching in social big data. Future Gener. Comput. Syst. Int. J. eScience. **87**, 772–781 (2018)
25. Gai, K.-K., Fang, Z.-K., Wang, R.-L., et al.: Edge computing and lightning network empowered secure food supply management. IEEE Internet Things J. **9**(16), 14247–14259 (2022)
26. Hijazi, A.A., Perera, S., Alashwal, A.M., et al.: Enabling a single source of truth through BIM and blockchain integration. In: International Conference on Innovation, Technology, Enterprise, and Entrepreneurship 2019, pp. 24–25 (2019)
27. Perera, S., Nanayakkara, S., Rodrigo, M., et al.: Blockchain technology: Is it hype or real in the construction industry? J. Ind. Inf. Integr. **17** (2020)
28. Farshidi, S., et al.: Decision support for blockchain platform selection: three industry case studies. IEEE Trans. Eng. Manag. **67**(4), 1109–1128 (2020)
29. Büyüközkan, G., Tüfekçi, G.: A decision-making framework for evaluating appropriate business blockchain platforms using multiple preference formats and VIKOR. Inf. Sci. **571**, 337–357 (2021)
30. Tang, H.-M., Shi, Y., Dong, P.-W.: Public blockchain evaluation using entropy and TOPSIS. Expert Syst. Appl. **117**, 204–210 (2019)
31. Chen, C.-H.: A novel multi-criteria decision-making model for building material supplier selection based on entropy-AHP weighted TOPSIS. Entropy **22**(2) (2020)
32. Kumar, R., Bilga, P.-S., Singh, S.: Multi objective optimization using different methods of assigning weights to energy consumption responses, surface roughness and material removal rate during rough turning operation. J. Clean. Prod. **164**, 45–57 (2017)
33. Chen, P.-Y.: Effects of the entropy weight on TOPSIS. Expert Syst. Appl. **168** (2021)
34. Khan, M.J., et al.: The renewable energy source selection by remoteness index-based VIKOR method for generalized intuitionistic fuzzy soft sets. Symmetry **12**(6) (2020)
35. Ghadikolaei, A.S., Madhoushi, M., Divsalar, M.: Extension of the VIKOR method for group decision making with extended hesitant fuzzy linguistic information. Neural Comput. Appl. **30**(12), 3589–3602 (2017). https://doi.org/10.1007/s00521-017-2944-5
36. Feng, J.-H., Xu, S.-X., Li, M.: A novel multi-criteria decision-making method for selecting the site of an electric-vehicle charging station from a sustainable perspective. Sustain. Cities Soc. **65** (2021)
37. Chen, X., et al.: Matching demanders and suppliers in knowledge service: a method based on fuzzy axiomatic design. Inf. Sci. **346**, 130–145 (2016)
38. Büyüközkan, G., Karabulut, Y., Arsenyan, J.: RFID service provider selection: an integrated fuzzy MCDM approach. Measurement **112**, 88–98 (2017)

An Auxiliary Classifier GAN-Based DDoS Defense Solution in Blockchain-Based Software Defined Industrial Network

Yue Zhang[1], Keke Gai[2](✉), Liehuang Zhu[2], and Meikang Qiu[3]

[1] School of Computer Science and Technology, Beijing Institute of Technology, Beijing 100081, China
3220201017@bit.edu.cn

[2] School of Cyberspace Science and Technology, Beijing Institute of Technology, Beijing 100081, China
{gaikeke,liehuangz}@bit.edu.cn

[3] Beacom College of Computer and Cyber Sciences, Dakota State University, Madison, SD 57042, USA
qiumeikang@ieee.org

Abstract. As an emerging technology, *Software-Defined Industrial Networks* (SDIN) appears to be a vital technical approach for powering up new manufacturing modes due to its higher-level scalability and controllability. However, a few threats are still restricting the implementation of SDIN and *Distributed Denial of Service* (DDoS) is one of the common attacks. In this paper, we focus on the DDoS issue in SDIN, and propose a blockchain-empower SDIN scheme and an *Auxiliary Classifier Generative Adversarial Networks* (AC-GAN)-based DDoS attack detection model. Our experiment evaluations have demonstrated the effectiveness and performance of our proposed approach.

Keywords: Blockchain · Software defined industrial networks · Distributed denial of service · Auxiliary classifier generative adversarial networks · Artificial intelligence · Security

1 Introduction

Software Defined Industrial Network (SDIN) is a novel technical paradigm that provides flexible and manageable network control capabilities for complex industrial network systems, especially heterogeneous networks deployed in cross-organizations [9]. SDIN separates the network control and data plane, overcoming limitations such as unscalability and low efficiency caused by static configuration in traditional industrial networks [2].

Despite multiple observable advantages offered by SDIN, vulnerabilities still exist in the industrial network context. Prior study [11] has argued that *Distributed Denial of Service* (DDoS) attack is one of critical threats for SDIN. First, considering the application plane in SDIN, the programmable interface is

M. Qiu et al. (Eds.): SmartCom 2022, LNCS 13828, pp. 319–328, 2023.
https://doi.org/10.1007/978-3-031-28124-2_30

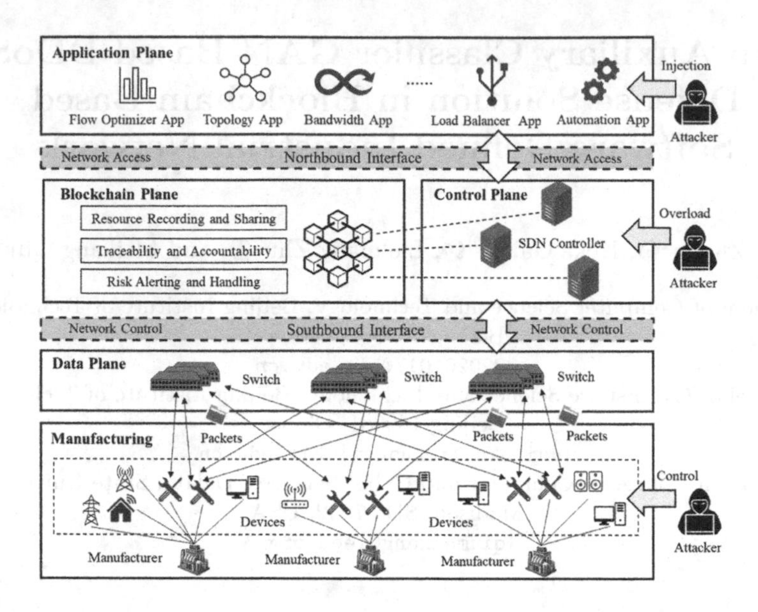

Fig. 1. The architecture of the blockchain-empowered SDIN.

often exploited by attackers. Second, the centralized network controller is easy to become the attack target. Finally, the threat is further enlarged along with the large amount of *Internet of Things* (IoT) devices deploying in the industrial network, as IoT devices generally offer limited defense capability [14] and are easily controlled by attackers.

To address above security issues, a variety of DDoS attack detection methods have emerged [3,12]. Among them, deep learning has gained widespread attention due to its autonomy and accuracy. However, deep learning models tend to become sensitive and inaccurate when detecting adversarial DDoS attacks[1]. Since the distributed and traceable features of blockchain [5,6,8] provide novel ideas for improving SDIN security, blockchain-based security solutions have also gained popularity [1,7,13]. However, there is currently no specific defense method against DDoS attacks in the blockchain-enabled SDIN architecture.

In this work, we focus on adversarial DDoS attack detection and mitigation in the blockchain-empowered SDIN system. As illustrated in Fig. 1, we propose the architecture of blockchain-empowered SDIN, in which we also mark locations where adversaries may launch DDoS attacks. Multiple controllers act as participants in the blockchain to record application flows and network behaviors as transactions into the distributed ledger. After distributing the control plane to multiple controllers, it is necessary to ensure resource consensus among all controllers. In addition, deriving prior achievements, we propose an *Auxiliary Classifier Generative Adversarial Networks* (AC-GAN) scheme to detect

[1] Adversarial attacks refer to that the attacker deliberately adds certain imperceptible interference to input samples, causing the misjudgment of the prediction model.

adversarial DDoS attacks and adopt varied data flow strategies via the controller to lower down the attack opportunities.

Main contributions of this work are threefold, as given in the followings:

- This paper proposes a scheme using deep learning to defend against DDoS attacks in blockchain-empowered SDIN.
- We present a blockchain-based SDIN architecture that decentralizes the control plane to multiple controllers, thus preventing single points of failure. The behavior of application flows can be tracked and audited via the blockchain, helping to monitor and replay network state for debugging and recovery.
- We design an AC-GAN algorithm to detect three types of DDoS attacks in blockchain-based SDIN environment, including TCP, UDP and ICMP flood attacks. In particular, AC-GAN generates adversarial samples for data enhancement, which can effectively improve the detection accuracy.

The rest of this work are organized in the following order. Sections 2 and 3 describe the proposed method and core algorithms, respectively. Then, Sect. 4 provide our experiment evaluations and results. Finally, conclusions of this work are drawn in Sect. 5.

2 Model Design

2.1 Blockchain-Based SDIN Model

Figure 1 illustrates the higher-level architecture of the blockchain-based SDIN model, which consists of five parts, including application plane, control plane, blockchain plane, data plane and physical plane.

Since a single controller in SDIN is vulnerable to DDoS attacks and causes a single point of failure, we decentralize the control plane to multiple controllers based on blockchain technology [15]. Specifically, multiple controllers in SDIN serve as clients of the blockchain platform for resource sharing and transaction recording. Resource sharing means that the network resources managed by each controller can be shared after reaching a consensus among all controllers. Transaction recording refers to the storage of network traffic, behavior and state into the blockchain in the form of transactions.

In the proposed model, we design the following three types of transactions.

- **Entity registration transaction.** Each entity participating in SDIN, including applications, controllers, and switches, needs to register identity information before joining the blockchain. Through the authentication access mechanism, the attack cost of the adversary is increased.
- **Application flow transaction.** Transactions are generated when an application sends flow information to a controller, which is mainly used to record and monitor the behavior of the application.
- **Network event transaction.** Network events provided by switches also generate transactions uploaded to the blockchain, which are mainly dynamic

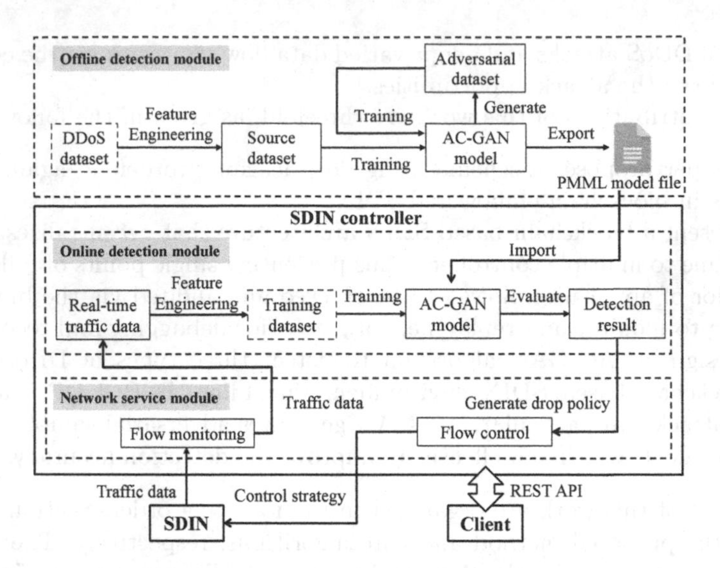

Fig. 2. Processing flow of DDoS attack prevention model based on AC-GAN.

states triggered by application events and *Packet_IN* messages. This kind of transaction records the state of network in time, so as to facilitate the timely detection of network issues.

Through above three transactions, entities, information flows and network events in SDIN are all recorded in blockchain, which facilitates tracking and auditing of network status. Besides, smart contracts can be designed at blockchain plane to further enhance the security of SDIN system. For example, when a large number of abnormal flows are detected, smart contracts can issue policies to relieve the burden on the controller or alert the controller to failure in time.

2.2 AC-GAN-Based Anti-DDoS Model

Figure 2 illustrates the process flow of AC-GAN-based anti-DDoS model in the blockchain-empowered SDIN system, including data pre-processing, feature engineering, traffic acquisition, DDoS detection and attack mitigation.

Data preprocessing is responsible for preliminary processing of the original DDoS dataset, including data cleaning, interpolation and other operations.

Feature engineering is mainly responsible for feature construction and selection. Traffic characteristics used in the proposed model include attribute (categorical) features [4] and statistical features. As for attribute features, we use the *Random Forest* (RF) method to rank the importance of original categorical features, so as to select the top effective characteristics. Each analysis interval T needs to group and count flow attributes. The statistical features constructed by the proposed model are as follows.

- **Avg_Packets** means the average packets number of flow in T.
- **Avg_Bytes** means avetage packets bytes of flow in T.
- **E(srcIP)** represents the entropy of the source IP address during T, which can be calculated in Eq. 1. $srcIP_i$ is the number of occurrences of the source IP address within the given time interval T. Hence, N is the total number of occurrences of all source IP addresses, defined in Eq. 2. In the same way, we can obtain the **E(dstIP)**, **E(srcPort)**, and **E(dstPort)**.

$$E(srcIP) = -\sum_{i=1}^{n}(\frac{srcIP_i}{N})log_2(\frac{srcIP_i}{N}). \tag{1}$$

$$N = -\sum_{i=1}^{n} srcIP_i. \tag{2}$$

- **PGR** represents the port growth rate, which can be defined as $PGR = \Delta Port/T$, where the $\Delta Port$ is the change number of ports during T.
- **IPGR** represents IP address growth rate, which is defined as $IPGR = \Delta IP/T$, where the ΔIP is the change number of IP addresses during T.

Model training refers to building the deep learning classifier to detect adversarial DDoS attacks. AC-GAN consits a generator and a discriminator. The input of generator is composed of the one-hot encoding tensor of the training data classification information **c** and the tensor of the noise data **z**. Equation 3 represents the false data output by the generator.

$$X_{fake} = G(\boldsymbol{c}, \boldsymbol{z}). \tag{3}$$

Equations 4 to 6 formalizes the optimization objective of generator loss function, i.e. maximizing the difference between classification loss L_C and authenticity judgment loss L_S.

$$\boldsymbol{Max}\ L_G = \boldsymbol{Max}\ (L_C - L_S). \tag{4}$$

$$L_C = E[logP(C = c|X_{fake})]. \tag{5}$$

$$L_S = E[logP(S = fake|X_{fake})]. \tag{6}$$

As shown in Eqs. 7 to 9, the optimization goal of discriminator is to distinguish the real data from the generated data as much as possible, while being able to effectively classify data. The input of discriminator includes real data X_{real}, fake data X_{fake} (generated by the generator) and data label c. The discriminator outputs P(C|X) and P(S|X), where P(C|X) is the probability distribution of input data with respect to classification labels, and P(S|X) is the probability distribution of source data authenticity judgment (whether the data is real or fake). The last layer of the discriminator is two parallel dense (fully connected) layers to obtain the results of P(C|X) and P(S|X), respectively.

$$\boldsymbol{Max}\ L_D = \boldsymbol{Max}\ (L'_C + L'_S). \tag{7}$$

$$L'_C = E[logP(C = c|X_{real})] + E[logP(C = c|X_{fake})]. \tag{8}$$

$$L'_S = E[logP(S = real|X_{real})] + E[logP(S = fake|X_{fake})]. \tag{9}$$

L'_C is loss function of data classification accuracy, which is defined as logarithmic probability of correct classification. L'_S is loss function for judging the authenticity of data, which is defined as log-likelihood of the correct source data.

Traffic acquisition is mainly responsible for obtaining the key features required for traffic classification through the SDIN controller.

DDoS attack detection is completed by calling the trained machine learning model. These obtained traffic characteristics are used as the input of the classification prediction module, and the prediction result of whether the traffic is a DDoS attack flow is output.

DDoS attack mitigation means that the controller takes corresponding measures according to the detection results. When a traffic is detected as normal, it will be transmitted safely according to the original forwarding rules. Once the detection module detects the attack traffic, the proposed model automatically activates the DDoS attack mitigation module.

3 AC-GAN Training and Classification Algorithm

The *AC-GAN Training and Classification* (ATC) algorithm conducts model training for offline DDoS attack detection. Main phases of Algorithm 1 are described as follows:

1. The generator model and discriminator model of AC-GAN are defined in Steps 1 to 11. The main function of generator is using noise data and label information to generate adversarial samples. The discriminator is mainly responsible for judging the authenticity of the data and giving the classification result of the DDoS attack flow.
2. Steps 12 to 23 define loss functions of the generator and discriminator.
3. Steps 24 to 32 complete data processing and feature selection process. Read in offline DDoS dataset and perform feature processing.
4. The training process of the AC-GAN model, which calls the G, D, G_Loss and D_Loss functions defined above, is given in Steps 33 to 44. In step 45, we evaluate the performance of proposed model on testing dataset.
5. Save training parameters to PMML file to improve the detection efficiency when calling the model across platforms.

FLOP is usually used to measure the time complexity of a deep learning model, which determines the time for model training and prediction. Dense layers and convolutional layers are used in our model. The *FLOP* of dense layers is $\sum_{i=1}^{D}(2 \times In_i - 1) \times Out_i$, where D represents the number of dense layers, In represents the input dimension of the hidden layer, and Out represents the output dimension of the hidden layer. The *FLOP* of convolutional layers is $\sum_{j=1}^{L} M_j^2 \times K_j^2 \times C_{j-1} \times C_j$, where L represents the number of convolutional layers, M represents the side length of feature map, K represents the side length of kernel, and C_j represents the number of convolution kernels in the j_{th} layer. Total time complexity of training model is approximately the sum of the above.

Algorithm 1. AC-GAN-enabled Training and Classification Algorithm

Require: Generator model, Discriminator model, DDoS training dataset, Data classification label
Ensure: Trained model in PMML format

```
    /*Define the generator model of AC-GAN*/
 1: function G(Classified traffic c, Noise information z)
 2:     Define network structure of generator model G
 3:     Use c and z to generate adversarial samples X_fake
 4:     return X_fake
 5: end function
    /*Define the discriminator model of AC-GAN*/
 6: function D(X_train, c)
 7:     Define network structure of discriminator model D
 8:     Determine whether the data is real or fake
 9:     Determine the classification of traffic data
10:     return isReal_out, DDoS_class_out
11: end function
    /*Define the loss function of generator*/
12: function G_Loss(fake_out, fake_class_out, c)
13:     Calculate fake losss L_S by Eq. 6
14:     Calculate classification loss L_C by Eq. 5
15:     L_G ← L_C - L_S
16:     return L_G
17: end function
    /*Define the loss function of discriminator*/
18: function D_Loss(real_out, real_class_out, fake_out, c)
19:     Calculate authenticity loss L'_S by Eq. 9
20:     Calculate classification loss L'_C by Eq. 8
21:     L_D ← L'_C + L'_S
22:     return L_D
23: end function
    /*Data reading and processing*/
24: Read the DDoS dataset with classification label c
25: Data preprocessing
26: for ∀ characteristic columns in the dataset do
27:     for ∀ characteristic X do
28:         X ← (X − X_min) ÷ (X_max − X_min)
29:     end for
30: end for
31: Feature engineering for feature selection and extraction
32: Divide the dataset to X_train, Y_train, X_test, Y_test
    /*AC-GAN model training process*/
33: Generate noise z based on the distribution of X_train
34: while i ≤ preset iterations & stop condition not met do
35:     for each epoch do
36:         X_fake = G(Y_train,z)
37:         fake_out, fake_class_out = D(X_fake, Y_train)
38:         real_out, real_class_out = D(X_real, Y_train)
39:         g_loss = G_Loss(fake_out, fake_class_out, Y_train)
40:         d_loss = D_Loss(real_out, real_class_out, fake_out, Y_train)
41:         Calculate the gradient of g_loss and d_loss
42:     end for
43:     i ← i + 1
44: end while
45: Validate model performance on X_test, Y_test
46: Export AC-GAN model to PMML file return PMML
```

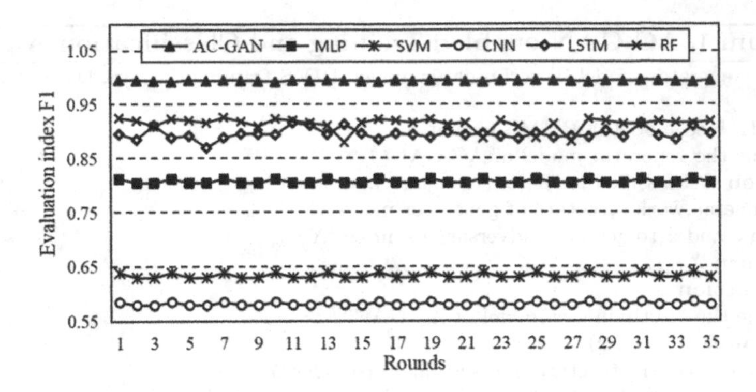

Fig. 3. F1 results of algorithms under multiple rounds of experiments.

4 Experiments and the Results

4.1 Experiment Environment and Configuration

When simulating the setting up of blockchain-based SDIN scenario, we use Hyperledger Fabric 1.4 blockchain platform and Mininet 2.3 emulator to customize network topology with Python 3.7 on the VirtualBox virtual machine with the Ubuntu 18.04 operating system. In order to improve development efficiency, the virtual network is connected to the Floodlight 1.2 controller.

The original dataset used for offline model training are TCP SYN flood flow, UDP flood flow and $TCMP$ flood flow in $CICDDoS$2019, as well as *Benign* normal flow in $CICDS$2017. The adversarial data samples generated by the AN-GAN network are also used for training.

4.2 Experiment Results

We compare the performance of *Multi-layer Perceptron* (MLP) [10], *Convolutional Neural Networks* (CNN) [3], *Long-Short Term Memory* (LSTM) [3], *Support Vector Machines* (SVM), basic *Random Forest* (RF) with the designed AC-GAN algorithm on the adversarial DDoS attack dataset from four perspectives, namely *Precision*, *Recall*, *Accuracy* and *F1*.

We repeat 35 rounds of model trainging as well as verification, and the $F1$ results of AC-GAN and other algorithms are shown in Fig. 3. It can be concluded that the performance of the AC-GAN algorithm is stable and performs well on $F1$. We give the performance of above algorithms in specific percentage in Table 1. It can be found that AC-GAN has achieved better performance in $Precision$, $Accuracy$ and $F1$. The $Recall$ rate of CNN can reach 100%, and its $Accuracy$ performance is poor on adversarial attack dataset. In addition, the performance of MLP and SVM in DDoS detection is not satisfactory. LSTM and RF methods show high $Recall$, and both have poor performance in $Accuracy$.

After launching the attack, the SDIN controller detects the attack and issues mitigation strategies to the flow table in time. Figure 4 compares the mitigation

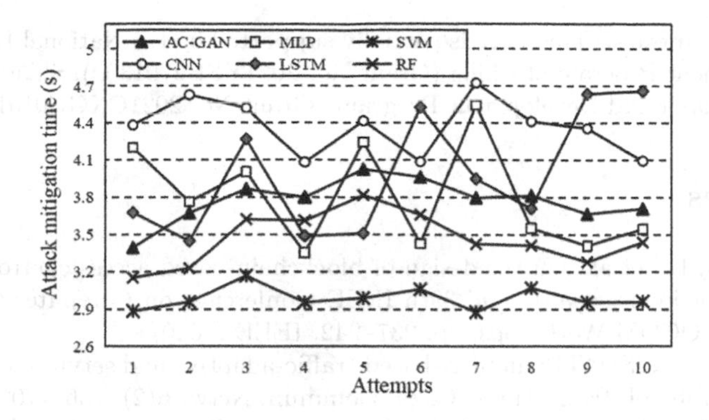

Fig. 4. Comparison of mitigation time under TCP flood attack.

Table 1. Comparison of different models on four detection indicators

Algorithm	Precision	Recall	Accuracy	F1 Score
AC-GAN	**99.658%**	99.318%	**99.402%**	**99.488%**
MLP	89.549%	79.557%	82.610%	84.258%
CNN	58.4952%	**100.000%**	58.495%	73.813%
LSTM	86.464%	99.574%	90.633%	92.557%
SVM	77.588%	54.258%	64.075%	63.859%
RF	86.642%	**100.000%**	90.981%	92.843%

time of adopting different algorithms when launching 10 TCP Flood attacks. Although the mitigation time of SVM is slightly faster than other methods, its detection accuracy is flawed. Moreover, there is little difference in mitigation time between AC-GAN and other algorithms.

In summary, experiment evaluations depict that the AC-GAN approach can effectively detect and mitigate DDoS attacks in SDIN environment. The performance of the scheme have been evidenced and adoptable results are received.

5 Conclusions

Aiming at security issues in SDIN system, this paper proposes a blockchain-empowered SDIN scheme and a deep learning-based attack detection model to defend adversarial DDoS attacks. The blockchain records all entities, information flows and network events in SDIN, monitoring the state of devices and networks to improve security. We also develop an AC-GAN method to generate adversarial attack samples, which improves the detection accuracy of DDoS attacks by increasing the model sensitivity. The performance of our proposed scheme had been evaluated and demonstrated to be an adoptable approach in SDIN.

Acknowledgements. This work is partially supported by the National Key Research and Development Program of China (Grant No. 2021YFB2701300), Shandong Provincial Key Research and Development Program (Grant No. 2021CXGC010106).

References

1. Chattaraj, D., et al.: On the design of blockchain-based access control scheme for software defined networks. In: 39th IEEE Conference on Computer Communications, INFOCOM Workshops, pp. 237–242. IEEE (2020)
2. Chen, J., et al.: SDATP: an SDN-based traffic-adaptive and service-oriented transmission protocol. IEEE Trans. Cogn. Commun. Netw. **6**(2), 756–770 (2019)
3. Gadze, J.D., Bamfo-Asante, A.A., Agyemang, J.O., Nunoo-Mensah, H., Opare, K.A.: An investigation into the application of deep learning in the detection and mitigation of DDoS attack on SDN controllers. Technologies **9**(1), 14 (2021)
4. Gai, K., Qiu, M., Thuraisingham, B.M., Tao, L.: Proactive attribute-based secure data schema for mobile cloud in financial industry. In: 17th IEEE HPCC, pp. 1332–1337. IEEE, New York, USA (2015)
5. Gai, K., Wu, Y., et al.: Differential privacy-based blockchain for industrial Internet-of-Things. IEEE Trans. Ind. Inform. **16**(6), 4156–4165 (2020)
6. Gai, K., Zhang, Y., Qiu, M., Thuraisingham, B.: Blockchain-enabled service optimizations in supply chain digital twin. IEEE Trans. Serv. Comput. (99), 1–12 (2022)
7. Gao, Y., Chen, Y., Lin, H., Rodrigues, J.J.P.C.: Blockchain based secure IoT data sharing framework for SDN-enabled smart communities. In: 39th IEEE Conference on Computer Communications, INFOCOM Workshop, pp. 514–519. IEEE (2020)
8. Guo, Y., et al.: A privacy-preserving auditable approach using threshold tag-based encryption in consortium blockchain. In: International Conference on Smart Computing and Communication, pp. 265–275. Springer, Cham (2021). https://doi.org/10.1007/978-3-030-97774-0_24
9. Liu, K., Xiao, K., Dai, P., Lee, V., Guo, S., Cao, J.: Fog computing empowered data dissemination in software defined heterogeneous VANETs. IEEE Trans. Mob. Comput. (99), 1 (2020)
10. Nugraha, B., et al.: Deep learning-based slow DDoS attack detection in sdn-based networks. In: 2020 IEEE NFV-SDN, .pp. 51–56. IEEE, Madrid, Spain (2020)
11. Tayfour, O., Marsono, M.: Collaborative detection and mitigation of distributed Denial-of-Service attacks on Software-Defined-Network. Mob. Netw. Appl. **25**, 1338–1347 (2020)
12. Wang, R., et al.: An entropy-based distributed DDoS detection mechanism in Software-Defined-Networking. In: 2015 IEEE Trustcom/BigDataSE/ISPA, vol. 1, pp. 310–317. IEEE (2015)
13. Weng, J., Weng, J., Liu, J., Zhang, Y.: Secure Software-Defined Networking based on blockchain.arxiv.org/abs/1906.04342 (2019)
14. Xu, Y., Liu, Y.: DDoS attack detection under SDN context. In: IEEE INFOCOM 2016-the 35th annual IEEE ICCC, pp. 1–9. IEEE, San Francisco, CA, USA (2016)
15. Zhang, Y., Gai, K., Xiao, J., Zhu, L., Choo, K.R.: Blockchain-empowered efficient data sharing in Internet of Things settings. IEEE J. Select. Areas Commun. (99), 1–1 (2022)

Blockchain-Based Fairness-Enhanced Federated Learning Scheme Against Data Poisoning Attack

Shan Jin[1], Yong Li[1(✉)], Xi Chen[2], Ruxian Li[2], and Zhibin Shen[2]

[1] School of Electronic and Information Engineering, Beijing Jiaotong University, Beijing 100081, China
liyong@bjtu.edu.cn
[2] Linklogis, Shenzhen 518063, China

Abstract. The federated learning technology provides a new method for data integration, which realizes sharing of a global model and prevent the leakage of user's original data information. In order to resist data poisoning attack from some participants, ensure reliability and accuracy of the global model, and ensure fairness of the aggregation process in federated learning, we propose a blockchain-based fairness enhanced federated learning scheme. The *accuracy* of global model and *fairness* of the aggregation process is guaranteed by an adaptive aggregation algorithm which can defense data poisoning attack. The *reliability* of federated learning process is ensured by recording the entire process of the model training on the blockchain and using digital signatures. The *privacy* of each participant of federated learning is protected by public key encryption combined with the use of random numbers. Theoretical analysis and experiments show that the scheme can protect privacy of each participant, mitigate data poisoning attack and ensure the reliability and fairness of the entire federated learning process.

Keywords: Blockchain · Data Poisoning Attack · Federated Learning · Privacy Protection

1 Introduction

Nowadays, the existence of data silos has seriously affected the integration of data. The existence of data silos and the strengthening of data privacy regulation presents challenges to artificial intelligence. In this conflict between data availability and privacy protection [1–3], "*privacy-preserving computing*" technology can promote the process of data consolidation. Federated learning (FL) [4, 5], one privacy-preserving computing technology, supports secure multi-party machine learning: local data is maintained on the participant's device and the updates to the model are aggregated through a secure aggregation algorithm. Federated learning generally includes a central server and various participants. Participants upload their local models or gradients to the central server, afterwards the server aggregates the global model and then sends it to each participant to continue the next iteration.

However, traditional federated learning assumes a trusted central server for coordination. Due to the centralized form of federated learning, it faces some problems such as

M. Qiu et al. (Eds.): SmartCom 2022, LNCS 13828, pp. 329–339, 2023.
https://doi.org/10.1007/978-3-031-28124-2_31

the single point of failure problem, the lack of verifiability of the aggregation results, etc. In addition, some malicious participants can harm the performance of the global model by carrying out a poisoning attack [6, 7]. Considering above problems, some studies [9–12] have tried to combine blockchain [8] with federated learning, and complete federated learning tasks by storing model updates on the blockchain.

Most schemes which considered poisoning attacks only discuss the scheme's performance in the presence of poisoning attacks. However, there is another situation in practice, that is, all participants are honest, and no one carries out poisoning attacks. In this case, most schemes may misjudge an honest participant as a poisoner, making the scheme unfair. It's necessary to design a scheme that can ensure the fairness of federated learning as well as resist poisoning attack.

In this paper, we propose a blockchain-based federated learning framework and an A-BFL (*Adaptive Blockchain based Federated Learning*) algorithm which can resist data poisoning attack and ensure the fairness of the aggregation process. In the end, we perform experiments on the MNIST [13] and Fashion-MNIST [14] datasets to test performance of our scheme. The contributions of our work are as follows:

- A blockchain-based federated learning framework is proposed, which combines the traditional federated learning with blockchain to ensure the training process and results are traceable and cannot be tampered with.
- We design an A-BFL aggregation algorithm to resist data poisoning attack, which can identify most of the poisoners, reduce the probability of misjudgment and ensure the fairness of aggregation.
- We use the deposit mechanism in the blockchain to ensure the task requester's reliability. If the task requester performs malicious actions, such as uploading a wrong accuracy on the test dataset, etc., its deposit will be deducted as a penalty.
- Public key encryption and digital signature are used to ensure that no one except the task requester can obtain the local models and to prevent local models from being stolen.

The rest of this paper is organized as follows. In Sect. 2, we review related works. Then, we discuss the design of the scheme in Sect. 3 and test the performance of our scheme through experiments on the actual datasets in Sect. 4. Finally, we summarize the paper in Sect. 5.

2 Related Works

In the process of FL, some participants may carry out data poisoning attack to drop the accuracy of the global model. There is no proper method for resisting poisoning attack in schemes [9, 10, 15] which combined blockchain with FL. Schemes [16, 17] focused on the quality of the global model, which can prevent poisoning attacks to a certain extent.

Algorithms used in traditional machine learning such as Krum/Multi-Krum [18], Zeno [19], etc. can also be used to resist data poisoning attack in FL. For example, the first P2P machine learning system [20] proposed by Shayan et al. filtered malicious model updates through multi-Krum algorithm. These traditional algorithms require to input the number of poisoners in advance, however it's difficult to do this in reality. Unlike the traditional algorithms, schemes [21, 22] filtered out poisoned models by setting the threshold. However, in the absence of poisoning attacks, some normal models will also be filter out, making these schemes unfair.

In addition to filter out models, schemes [23, 24] adjusted the weights of models during the aggregation process according to the quality of models. If a local model is of low quality, it will be weighted less when aggregating. However, homomorphic encryption and the calculation under ciphertext make them non-universal.

In addition, some algorithms used in traditional machine learning such as Median [25] (take the median value of the local models or gradients submitted by the participants) are also suitable for FL.

3 Blockchain Based Federated Learning

Figure 1 shows the pipeline of the proposed framework. The framework includes task requester, task participants, blockchain, and cloud platform.

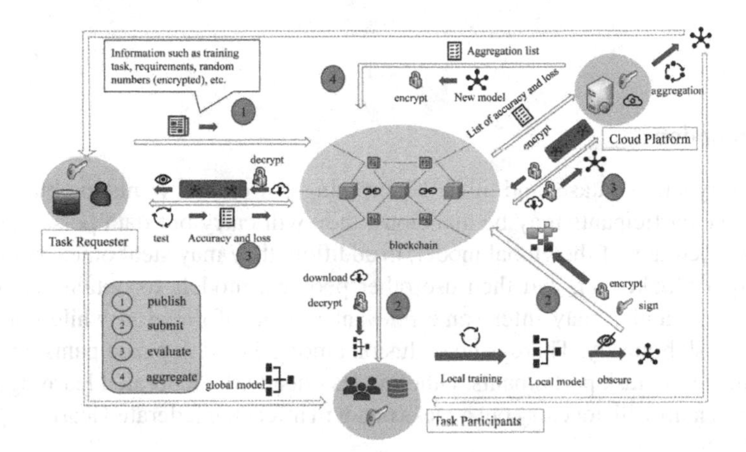

Fig. 1. An overview of the proposed blockchain based federated learning framework.

- Task Requester (TR for short): The requester of the federated learning task, responsible for publishing the task to the blockchain and collaborating with the cloud platform to complete the aggregation of global model. TR can obtain all the local models submitted by task participants. To reward task participants and cloud platform, TR should pay for them.
- Cloud Platform (CP): It is responsible for calculating the distance between models submitted by participants and cooperating with TR to complete model aggregation.
- Task Participants (P): They are responsible for training local models, who can be from different institutions in the same industry with similar datasets. Task participants should own eligible local datasets. (Assuming that the local datasets of all participants are independent and identically distributed).

- Blockchain: It records the process of model submission and aggregation.

For ease of reference, the symbols that appeared in this paper and corresponding descriptions are listed in Table 1.

Table 1. Symbols and their descriptions

Symbol	Description	Symbol	Description
SGD	stochastic gradient descent	m	dimension of the model
N	number of participants	r	m random numbers
T	max iteration of FL	G	gradient
L	local training epochs	R	obscured model
η	learning rate	sk	private key
x	training data vector	pk	public key
y	training data labels	pop	percentage of poisoners
B	small batches of training data	iop	intensity of poisoning attack
ω	model	I	number of iterations

3.1 Assumption

TR has a very small dataset and limited computing power, and wants to obtain a better model. Some participants may be malicious, they will carry out data poisoning attack to drop the accuracy of the global model; in addition, they may steal other participants' models from blockchain, and then use other people's models to obtain rewards. CP is semi-honest, and it may infer some relevant model information while completing computing task honestly. There is no collusion among TR, task participants, and CP, as well as among the task participants. Other nodes outside the federated learning process on the blockchain will not carry out attacks that will affect the federated learning process.

3.2 Blockchain-Based Federated Learning Framework

In this paper, we mainly focus on representative poisoning attacks: *label-flipping attack*. For example, malicious participants incorrectly mark the samples with the true label to the wrong label for misclassification. Our main ideas are as follows:

The federated learning process can proceed only if the data is successfully uploaded to the blockchain. However, uploading the data to blockchain also takes some time. In order to achieve high accuracy with fewer iterations and reduce waiting time for participants, task participants need to complete several training epochs locally before uploading local models to the blockchain. Inspired by the previous work [23], we combine asymmetric encryption with random numbers to ensure that CP and anyone in blockchain who do not participant in the federated learning task cannot obtain the original model data.

To resist data poisoning attack, we proposed an A-BFL aggregation algorithm that can not only identify the poisoners as much as possible, but also guarantee fairness when there are no poisoners. The aggregation algorithm mainly considers the following factors: the distance (use the same method as in [18]) between the local models, and the performance (accuracy and loss) of the local models on the test dataset. Each step is described in detail as follows, and the A-BFL algorithm is shown in Fig. 2.

3.3 Details of Our Scheme

1. *Publishing the task*

TR determines the training task and training parameters, uploads the following content to the blockchain.

(1) Training task, the target accuracy of global model or the maximum training iteration T; the participant's local training epochs L, the waiting time for the submission of the local model for each iteration, etc.
(2) The hash value of the test dataset D_{test} (each participant should get the original test dataset after the overall training process to verify that the TR has uploaded the correct results when testing the accuracy and loss of the local models).
(3) Initial global model $[\![\boldsymbol{\omega}^0]\!]_{pk_n}(n \in [1, N])$(encrypted with each participant's public key). The initial m random numbers $\mathbf{r}^0 = \{r, i \in [1, m]\}$ are selected according to the dimension m of the initial global model, encrypted with the public keys of each participant.

In addition, TR needs to pay a deposit to the blockchain, and the deposit will be redeemed only when more than half of the participants agree with the training results at the end of the training.

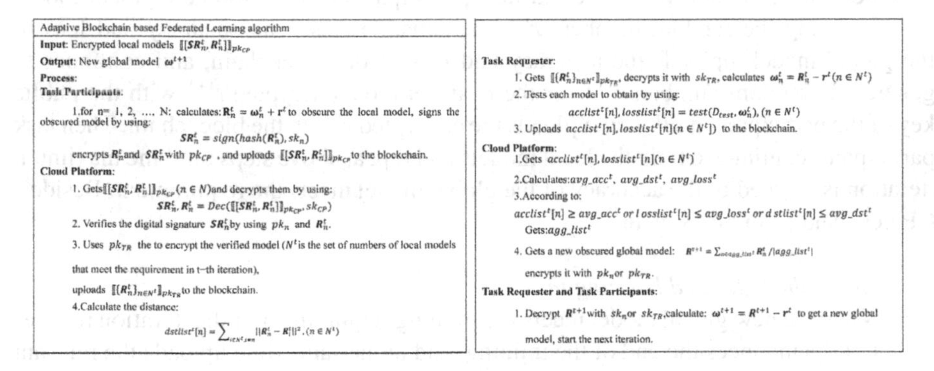

Adaptive Blockchain based Federated Learning algorithm

Input: Encrypted local models $[\![[SR_n^t, R_n^t]]\!]_{pk_{CP}}$

Output: New global model $\boldsymbol{\omega}^{t+1}$

Process:

Task Participants:

1.for n= 1, 2, …, N: calculates: $\mathbf{R}_n^t = \boldsymbol{\omega}_n^t + \mathbf{r}^t$ to obscure the local model, signs the obscured model by using:

$$SR_n^t = sign(hash(R_n^t), sk_n)$$

encrypts R_n^t and SR_n^t with pk_{CP} , and uploads $[\![[SR_n^t, R_n^t]]\!]_{pk_{CP}}$ to the blockchain.

Cloud Platform:

1. Gets $[\![[SR_n^t, R_n^t]]\!]_{pk_{CP}}(n \in N)$ and decrypts them by using:

$$SR_n^t, R_n^t = Dec([\![[SR_n^t, R_n^t]]\!]_{pk_{CP}}, sk_{CP})$$

2. Verifies the digital signature SR_n^t by using pk_n and R_n^t.

3. Uses pk_{TR} the to encrypt the verified model (N^t is the set of numbers of local models that meet the requirement in t−th iteration),

uploads $[\![\{R_n^t\}_{n \in N^t}]\!]_{pk_{TR}}$ to the blockchain.

4.Calculate the distance:

$$dstlist^t[n] = \sum_{i \in N^t, i \neq n} \|R_n^t - R_i^t\|^2, (n \in N^t)$$

Task Requester:

1. Gets $[\![\{R_n^t\}_{n \in N^t}]\!]_{pk_{TR}}$, decrypts it with sk_{TR}, calculates $\boldsymbol{\omega}_n^t = R_n^t - r^t (n \in N^t)$

2. Tests each model to obtain by using:

$$acclist^t[n], losslist^t[n] = test(D_{test}, \boldsymbol{\omega}_n^t), (n \in N^t)$$

3. Uploads $acclist^t[n], losslist^t[n](n \in N^t])$ to the blockchain.

Cloud Platform:

1.Gets $acclist^t[n], losslist^t[n](n \in N^t)$

2.Calculates: avg_acc^t, avg_dst^t, avg_loss^t

3.According to:

$acclist^t[n] \geq avg\_acc^t$ or $losslist^t[n] \leq avg\_loss^t$ or $dstlist^t[n] \leq avg\_dst^t$

Gets: agg_list^t

4. Gets a new obscured global model: $R^{t+1} = \sum_{n \in agg\_list^t} R_n^t / |agg\_list^t|$

encrypts it with pk_n or pk_{TR}.

Task Requester and Task Participants:

1. Decrypt R^{t+1} with sk_n or sk_{TR}, calculate: $\boldsymbol{\omega}^{t+1} = R^{t+1} - r^t$ to get a new global model, start the next iteration.

Fig. 2. A-BFL algorithm

2. *Submitting the model*

Take the first iteration as an example:

Participants randomly select the local data of small batch B and calculate the loss (cross entropy loss function): $L_f(B, \omega) = \frac{1}{|B|}\sum_{(\mathbf{X}_i, y_i)\in B} L_f(\omega, \mathbf{X}_i, y_i)$. Then each participant calculates the gradient of the loss function: $\boldsymbol{G} = \nabla_\omega L_f(B, \omega)$. Finally, the new local model is calculated: $\omega^l = \omega^{l-1} - \eta \cdot \mathbf{G}$.

Each participant repeats the above training process until the number of training epochs reaches L. The local model at this time is recorded as $\boldsymbol{\omega}_n^0 = \boldsymbol{\omega}^L$. Then each participant obscures the local model with m random numbers: $\mathbf{R}_n^0 = \omega_n^0 + \mathbf{r}^0$, hashes the obscured model and sign it with its private key: $\boldsymbol{SR}_n^0 = Sign(hash(\boldsymbol{R}_n^0), sk_n)$.

Then participants encrypt their respective models by using the public key of CP: $[\![[\boldsymbol{SR}_n^0, \boldsymbol{R}_n^0]]\!]_{\boldsymbol{pk_{CP}}} = \boldsymbol{Enc}([\boldsymbol{SR}_n^0, \boldsymbol{R}_n^0], \boldsymbol{pk_{CP}})$, upload $[\![[\boldsymbol{SR}_n^0, \boldsymbol{R}_n^0]]\!]_{\boldsymbol{pk_{CP}}}$ to the blockchain.

3. *Model evaluation and aggregation*
Taking the t-th iteration as an example, within the specified waiting time, CP gets $\boldsymbol{SR}_n^t, \boldsymbol{R}_n^t = \boldsymbol{Dec}([\![[\boldsymbol{SR}_n^t, \boldsymbol{R}_n^t]]\!]_{\boldsymbol{pk_{CP}}}, \boldsymbol{sk_{CP}})$

TR, task participants and CP cooperate with each other to evaluate and aggregate the local models by executing the A-BFL algorithm shown in Fig. 2. In the A-BFL algorithm, CP can calculate: $avg\_acc^t = (\max(acclist^t[n]) + \min(acclist^t[n]))/2$, $avg\_loss^t = (\max(losslist^t[n]) + \min(losslist^t[n]))/2$ and $avg\_dst^t = (\sum_{n\in N^t} dstlist^t[n])/|N^t|$. Then CP uploads the public key list agg_list^t (participants whose local model meet (1)) and the public key list of all participants all_list^t (all participants who submit model in the t-th iteration) to the blockchain. The corresponding task participants in the agg_list^t can get paid from blockchain.

$$acclist^t[n] \geq avg\_acc^t \text{or } losslist^t[n] \leq avg\_loss^t \text{or } dstlist^t[n] \leq avg\_dst^t \quad (1)$$

Next, CP can aggregate the models that meet the requirement (1) to obtain a new obscured global model: $\mathbf{R}^{t+1}$. TR and task participants can get the new global model ω^{t+1} by using the random numbers $\boldsymbol{r}^t$. After this, TR tests the accuracy and loss of the global model, uploads the accuracy and loss to the blockchain, and continues to generate m random numbers $\boldsymbol{r}^{t+1}$ for the next iteration, encrypted $\boldsymbol{r}^{t+1}$ with the public key of the participant in all_list^t, uploads the encrypted $\boldsymbol{r}^{t+1}$ to the blockchain. Then task participants continue to train the local models and repeat above steps until the maximum iteration is reached or the accuracy of the global model meets the requirement. Besides, CP gets paid from blockchain.

4. *End of the federated learning task*
(1) When the new global model meets the training requirement or the iteration reaches T, TR announces the end of the training, and at the same time uploads the relevant information of the test dataset to the blockchain (for example, if it is stored in the cloud, the link of the test dataset needs to be uploaded to the blockchain). After the task participants obtain the dataset, they can verify whether the hash value is the same as the hash value uploaded by the TR in the beginning.
(2) Each participant can use the historical submitted models on the blockchain to verify the accuracy and loss on the test dataset, and compares them with the results uploaded by TR on the blockchain. If more than half of the participants believe that

there is no problem with the aggregation process, then the deposit will be redeemed to TR; otherwise, as a penalty, TR will not be able to get the deposit. At this point, the training process is finished, the TR and eligible task participants have obtained the global model.

4 Experiments Analysis and Comparison

The experiments were run on a PC equipped with a 4-core I5-6300H processor at 2.30 GHz and 12 GB RAM. We implemented label flipping attack and defense algorithms in Python, using the TensorFlow and sklearn packages. In our experiments, participants train the local model using the Multilayer Perceptron (MLP) network in TensorFlow. We compare the performance of schemes [21, 25] with our A-BFL, the evaluation criteria include the accuracy in test dataset and the misjudgment rate. Regarding the misjudgment rate, the description is as follows: It mainly consider two aspects, one is the probability that the local models submitted by poisoners are regarded as the normal models: *WTC* (Wrong to Correct), on the other hand is the probability of excluding global models submitted by honest participants when the global model is aggregated: *CTW* (Correct to Wrong).

Suppose the set of all participants is expressed as $\boldsymbol{P} = \{\boldsymbol{P}_{\mathrm{h}}, \boldsymbol{P}_{\mathrm{p}}\}$, the total amount $N = |\boldsymbol{P}|$, the number of honest participants is $N_{\mathrm{h}} = |\boldsymbol{P}_{\mathrm{h}}|$, and the number of poisoners is $N_{\mathrm{p}} = \left|\boldsymbol{P}_{\mathrm{p}}\right|$. The set of the participants selected to participate in the aggregation in aggregation process is donated as $\boldsymbol{P}_{\mathrm{chose}}$. Then, $WTC = (\boldsymbol{P}_{\mathrm{chose}} \cap \boldsymbol{P}_{\mathrm{p}})/N_{\mathrm{p}}$, $CTW = ((\mathbf{P} - \mathbf{P}_{\mathrm{chosen}}) \cap \mathbf{P}_{\mathrm{h}})/N_{\mathrm{h}}$. Total misjudgment rate can be calculated: $err\_rate = WTC + CTW$.

The federated average algorithm is denoted as: "Fed-Avg", the aggregation algorithm proposed in [21] is denoted as "SFL", and the method of taking the median [25] value of the model (or gradient) is denoted as "Median". We use the percentage of the modified labels to indicate the intensity of data poisoning attack. The symbols used in the following description are described in Sect. 3.

1) Performance under data poisoning attack
   It is assumed that $N = 20$, $L = 5$, $pop = 0.3$, $iop = 1$. "Fed-Avg" is set as the baseline (the dotted line in the figure). As shown in Fig. 3, when some participants carry out data poisoning attack, compared with the baseline, the accuracy of Fed-Avg dropped by nearly 10% in the first few iterations. The Median algorithm requires more parameter interaction rounds to achieve better accuracy. SFL and A-BFL algorithm can achieve an accuracy comparable to the baseline. The total misjudgment rate of SFL and A-BFL is almost 0, so that A-BFL can identify most of the poisoners and exclude them to get a better global model (Because the total misjudgment rate is always 0 in most time when there exist poisoners, we do not show the experiment results here).
   When setting $N = 20, pop = 0.3, I = 10$, the impact of the poisoning attack intensity (*iop*) on the accuracy of the global model is shown in Fig. 3. Compared with the baseline, the A-BFL algorithm can always maintain a high accuracy, which ensures that TR can obtain the desired training results.

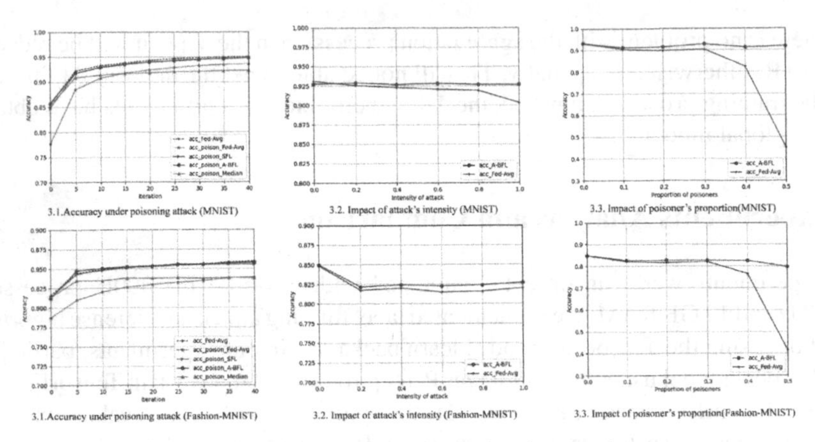

Fig. 3. Experiments under poisoning attack

When setting $N = 20$, $iop = 1$, $I = 10$, the impact of the proportion of the poisoners (*pop*) on the accuracy of the global model is shown in Fig. 3. When the proportion of poisoners is less than 50%, with the increase of the proportion of poisoners, the accuracy of Fed-Avg algorithm drops sharply. In contrast, A-BFL algorithm can still maintain a high accuracy. In the case of more than half of the participants become the poisoners, the accuracy of A-BFL algorithm will be seriously reduced. Because in the model evaluation stage, the local models submitted by honest participants are very different from the poisoning local models, which will make the honest participants have a high probability of being judged as poisoners, resulting in a decrease in the accuracy (It's not shown here).

2) Performance without data poisoning attack
In addition to considering how to identify and exclude the poisoned models, we also consider how to prevent local models submitted by honest participants from being regarded as the poisoned models when all the participants are honest. Figure 4 shows that without the poisoners, the performance of the proposed A-BFL algorithm is close to the baseline. In addition, we consider how to ensure the fairness of the algorithm when there is no poisoning attack. Because the honest participant may be identified as poisoners, which leads to normal local models cannot participate in the aggregation, we take the probability of misjudgment into consideration. Figure 4 shows that compared to SFL algorithm, A-BFL algorithm can greatly reduce the probability of misjudgment. Besides, from Fig. 4 we can see that the misjudge rate of SFL varies greatly, that is because SFL algorithm uses the accuracy of the model as the only criterion for identifying the poisoned models.
In addition, TR can choose to set a tolerable maximum number of nonparticipation aggregation times MT, and participants who exceed this number of times will not be eligible to continue participating in the subsequent training process. CP makes judgment based on the aggregation history recorded on the blockchain. If the local model submitted by a participant does not meet the requirement in consecutive MT iterations of training, then the public key of that participant will be uploaded to the blockchain. And after this iteration, the local model uploaded by that participant

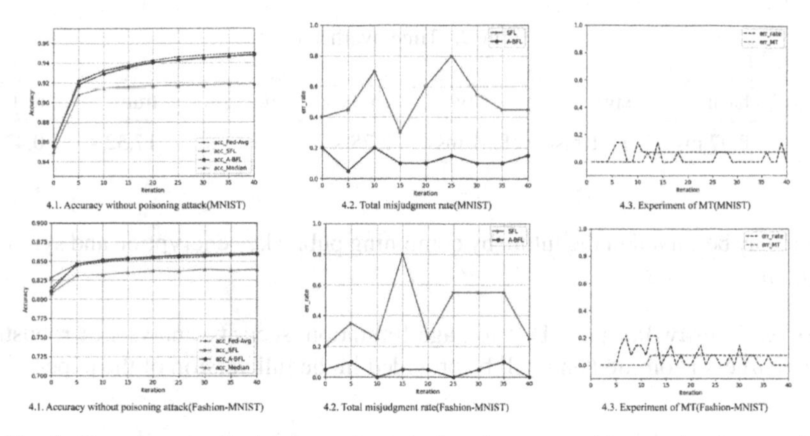

Fig. 4. Experiments of misjudgment rate and performance without poisoning attack

will no longer be aggregated, and TR also will no longer send m random numbers for obscuring to that participant.

3) Experiments of MT
Under the setting of MT, if one participant whose local model does not meet the requirements in consecutive MT iterations, then that participant will be excluded from the federated learning task. If one participant whose local model does not meet the requirements in a single iteration, it will not be excluded.
In addition to the total misjudgment rate mentioned above, we use a parameter $err\_MT = (\boldsymbol{P}_{\text{ex}} \cap \boldsymbol{P}_h)/N_h$ to represent the probability of honest participants being excluded from the federated learning task ($\boldsymbol{P}_{\text{ex}}$ is the set of participants who have excluded from the federated learning task).
It is assumed that $N = 20$, $L = 5$, $pop = 0.3$, $iop = 1$ and MT is 4. As shown in Fig. 4, the poisoners are excluded in the first MT iterations. After that, even if some honest participants are regarded as the poisoner in some iteration, they will not be excluded from the federated learning task, unless they are regarded as the poisoner in a consecutive MT iteration. The setting of MT can make our scheme fairer, prevent misjudge the honest participants as the poisoners as much as possible.
4) Time overhead
Some cryptographic algorithms such as hash function are used in our scheme. We perform experiments to illustrate the time spent by our scheme. RSA cryptosystem is used in the experiment to encrypt, decrypt, sign, and verify. The length of the private key is set to 1024. We do ten times of experiment and take the average as the results.
The final results are shown in Table 2. The decryption of the model takes the longest time. Other processes take less time compared to the process of training one local model which is necessary in traditional federated learning.

Remark on Decryption Time Overhead. TR only need to perform one decryption operation in an iteration to obtain all local models. However, when the number of task participants is too large, TR will spend too much time in decrypting the model set. This

Table 2. Time overhead

Process	hash	sign	verify	encrypt	decrypt	train	test
Time	8.07 ms	1.71 ms	8.51 ms	2.28 s	48.11 s	17.32 s	0.47 s

problem may be solved in the future by combining public key encryption and symmetric encryption.

Remark on Security Analysis. Due to page limitation, security analysis of robustness, privacy, fairness is omitted and will be provided in the full version of the paper.

5 Conclusion

In this paper, we proposed a blockchain-based federated learning scheme, focusing on the defense against data poisoning attack and the privacy protection of each participant. In practice, TR can be specific agency, such as government, who need a lot of data to train the model but don't have sufficient data. And the proposed scheme can be applied in multiple scenarios such as Internet of Things (IoT) and Mobile Edge Computing (MEC) [26, 27].

Through experiments on public datasets, it can be seen that the proposed scheme can effectively resist data poisoning attack, and in the absence of data poisoning attack, the scheme can also ensure a low probability of misjudgment. In addition, our scheme can protect the privacy of each participant, guarantee the fairness of federated learning and reliability of the TR, ensure high accuracy of the global model.

References

1. Qiu, M., Gai, K., Xiong, Z.: Privacy-preserving wireless communications using bipartite matching in social big data. Future Gener. Comput. Syst. **87**, 772–781 (2018)
2. Gai, K., Qiu, M.: Blend arithmetic operations on tensor-based fully homomorphic encryption over real numbers. IEEE Trans. Ind. Inf. **14**(8), 3590–3598 (2017)
3. Qiu, H., Qiu, M., Lu, Z.: Selective encryption on ECG data in body sensor network based on supervised machine learning. Inf. Fus. **55**, 59–67 (2020)
4. McMahan, B., Moore, E., Ramage, D., Hampson, S., y Arcas, B.A.: Communication-efficient learning of deep networks from decentralized data. In: AISTATS, pp. 1273–1282 (2017)
5. Li, T., Sahu, A.K., Talwalkar, A., Smith, V.: Federated learning: challenges, methods, and future directions. IEEE Signal Process. Mag. **37**(3), 50–60 (2020)
6. Goldblum, M., et al.: Dataset security for machine learning: data poisoning, backdoor attacks, and defenses. IEEE Trans. Pattern Anal. Mach. Intell. (2022)
7. Biggio, B., Nelson, B., Laskov, P.: Poisoning attacks against support vector machines. arXiv preprint arXiv:1206.6389 (2012)
8. Nakamoto, S.: Bitcoin: a peer-to-peer electronic cash system. Decent. Bus. Rev., 21260 (2008)
9. Zhang, Z., Yang, T., Liu, Y.: SABlockFL: a blockchain-based smart agent system architecture and its application in federated learning. Int. J. Crowd Sci. **4**(2), 133–147 (2020)

10. Arachchige, P.C.M., Bertok, P.,. Khalil, I, Liu, D., Camtepe, S., Atiquzzaman, M.: A trustworthy privacy preserving framework for machine learning in industrial IoT systems. IEEE Trans. Ind. Inf. **16**(9), 6092–6102 (2020)
11. Wang, Z., Hu, Q.: Blockchain-based federated learning: a comprehensive survey. arXiv preprint arXiv:2110.02182 (2021)
12. Qu, Y., et al.: Decentralized privacy using blockchain-enabled federated learning in fog computing. IEEE Internet Things J. **7**(6), 5171–5183 (2020)
13. LeCun, Y., Bottou, L., Bengio, Y., et al.: Gradient-based learning applied to document recognition. Proc. IEEE **86**(11), 2278–2324 (1998)
14. Xiao, H., Rasul, K., Vollgraf, R.: Fashion-mnist: a novel image dataset for benchmarking machine learning algorithms. arXiv preprint arXiv:1708.07747 (2017)
15. Mugunthan, V., Rahman, R., Kagal, L.: BlockFLow: an accountable and privacy-preserving solution for federated learning. arXiv preprint arXiv:2007.03856 (2020)
16. Lu, Y., Huang, X., Dai, Y., Maharjan, S., Zhang, Y.: Blockchain and federated learning for privacy-preserved data sharing in industrial IoT. IEEE Trans. Ind. Inform. **16**(6), 4177–4186 (2020)
17. Li, Z., Liu, J., Hao, J., Wang, H., Xian, M.: CrowdSFL: a secure crowd computing framework based on blockchain and federated learning. Electronics **9**(773), 1– 21 (2020)
18. Blanchard, P., et al.: Machine learning with adversaries: Byzantine tolerant gradient descent. In: Proceedings of the Neural Information Processing Systems (NeurIPS), pp. 119–129 (2017)
19. Xie, C., Koyejo, S., Gupta, I.: Zeno: distributed stochastic gradient descent with suspicion-based fault-tolerance. In: 36th International Conference on Machine Learning, PMLR 97, pp. 6893–6901 (2019)
20. Shayan, M., Fung, C., Yoon, C.J.M., Beschastnikh, I.: Biscotti: a blockchain system for private and secure federated learning. IEEE Trans. Parallel Distrib. Syst. **32**(7), 1513–1525 (2021)
21. Liu, Y., Peng, J., Kang, J., et al.: A secure federated learning framework for 5G networks. IEEE Wirel. Commun. **27**(4), 24–31 (2020)
22. Qi, Y., Shamim Hossain, M., Nie, J., Li, X.: Privacy-preserving blockchain-based federated learning for traffic flow prediction. Future Gener. Comput. Syst. **117**, 328–337 (2021)
23. Liu, X., Li, H., Xu, G., Chen, Z., Huang, X., Lu, R.: Privacy-enhanced federated learning against poisoning adversaries. IEEE Trans. Inf. Forensics Secur.**16**, 4574–4588 (2021)
24. Ma, Z., Ma, J., Miao, Y., Li, Y., Deng, R.H.: ShieldFL: mitigating model poisoning attacks in privacy-preserving federated learning. IEEE Trans. Inf. Forensics Secur. **17**, 1639–1654 (2022)
25. Yin, D., Chen, Y., Ramchandran, K., Bartlett, P.: Byzantine-robust distributed learning: towards optimal statistical rates. In: 35th International Conference on Machine Learning, PMLR 80, pp. 5650–5659 (2018)
26. Gai, K., Wu, Y., Zhu, L., et al.: Differential privacy-based blockchain for industrial Internet-of-Things. IEEE Trans. Ind. Inf. **16**(6), 4156–4165 (2019)
27. Xu, Y., Lu, Z., Gai, K., et al.: BESIFL: blockchain empowered secure and incentive federated learning paradigm in IoT. IEEE Internet Things J., 2327–4662 (2021)

Cryptography of Blockchain

Ying Long[1,2], Yinyan Gong[1,2](✉), Weihong Huang[1,2], Jiahong Cai[1,2], Nengxiang Xu[1,2], and Kuan-ching Li[1,2]

[1] School of Computer Science and Engineering, Hunan University of Science and Technology, Xiangtan 411201, China
G18873530267@163.com, whhuang@hnust.edu.cn, jiahongcai@mail.hnust.edu.cn

[2] Hunan Key Laboratory for Service Computing and Novel Software Technology, Xiangtan 411201, China

Abstract. With the development of digital currencies and 5G technology, blockchain has gained widespread attention and is being used in areas such as healthcare, industry and smart vehicles. Many security issues have also been exposed in the course of blockchain applications. Cryptography can ensure the security of data on the blockchain, the integrity and validity of data as well as the ability to authenticate users and anonymize them. This article therefore examines the cryptography underlying blockchain security issues, providing an overview of cryptographic homomorphic encryption, zero-knowledge proofs and secure multi-party computation commonly used in blockchains. At the same time, the development of quantum computing is bound to affect existing cryptographic systems, and blockchains applying these cryptographic systems are bound to be hit hard, so this article discusses four of the most promising post-quantum cryptography techniques available: hash-based public key cryptography, code-based public key cryptography, multivariate public key cryptography, and lattice-based public key cryptography.

Keywords: Blockchain · Homomorphic encryption · Post quantum cryptography · Secure multi-party computation · Zero-knowledge proof

1 Introduction

In 2008, Satoshi Nakamoto introduced the concept of Bitcoin, a decentralized virtual currency, in his published paper "Bitcoin: A Peer-to-Peer Electronic Money System" [1]. Blockchain is a data structure that organizes blocks of data in a chain in chronological order and is capable of verifying, tracing, and reliably storing data on the chain through cryptography to ensure that the data on the blockchain is not tampered with and cannot be forged. The consistency of data on the blockchain is ensured through transaction signatures, consensus mechanisms and cross-chain technologies [2]. Blockchain technology is decentralized, traceable, tamper-proof, complete and open and transparent [3], which have attracted the attention of academia and industry. Moreover, blockchain technology will be a revolutionary technology to solve the trust crisis in the future society [4].

M. Qiu et al. (Eds.): SmartCom 2022, LNCS 13828, pp. 340–349, 2023.
https://doi.org/10.1007/978-3-031-28124-2_32

As blockchain continues to develop, there is a greater demand for data protection, anonymity and untraceability [5] in many fields. Blockchain is no longer only used for virtual currencies [6] but is also being extended to various fields such as healthcare, copyright protection and finance. For example, blockchain is currently the most effective solution for personal privacy protection and sharing [7]. As part of the big data trend [8], the growing scale of the Internet of Things (IoT) [9] and the sharing of its data requires the use of blockchain. Many insider attacks are caused by trust issues, and blockchain can solve the problem of trust between untrustworthy users [10]. However, the rise of blockchain technology in various fields has brought about many security issues. For example, privacy protection and transaction protection. Also, with the development of quantum computing, which may break many cryptographic systems [11], the cryptography of blockchain will be severely challenged. With the emergence of various attacks, various cryptography-based blockchain security protection techniques are gradually developed [12]. Therefore, this paper will study the cryptography techniques in blockchain.

2 Blockchain Overview

2.1 The Data Structure of Bitcoin

In blockchain technology, a great deal of cryptographic knowledge is used to ensure the system's security. Cryptography greatly protects the privacy of data on the blockchain [13]. The blockchain is actually made up of blocks connected one by one, each block generating a hash value from the previous block [14]. The user can then verify the correctness of the data on the chain. Furthermore, if an attacker wants to tamper with the data on the blockchain, he must change all the blocks after that one (Fig. 1).

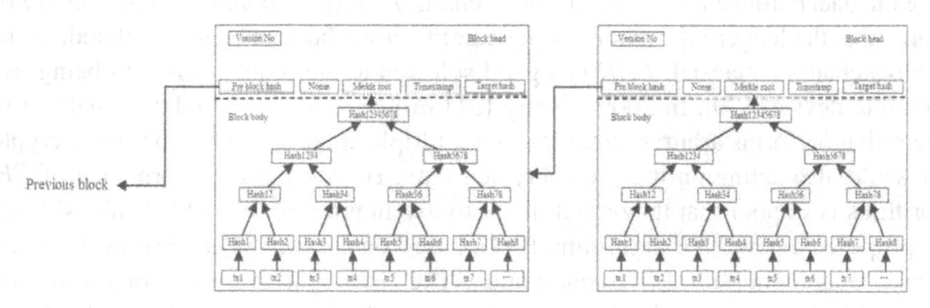

Fig. 1. Bitcoin structure diagram.

A block contains a block header and a block body. The block header holds the previous block's hash value, version number, random number Nonce, timestamp, Merkle root and the target hash value. In the block body is a Merkle tree, a typical binary tree. Its root is formed by the hashes of all the transactions in the block; the leaf nodes of the Merkle tree are the hashes generated by the transactions packed into the block, and the values of the non-leaf nodes are generated by concatenating the hashes of their children into a string and then hashing them, in this way working from the bottom up to generate the hash of the Merkle root. This structure allows a quick look at whether the transactions packed in the block have been altered and, if found to have been altered, a quick way to locate the altered transaction.

2.2 Security Challenges Facing Blockchain

Blockchain is a promising and growing technology but also faces many challenges. These challenges arise from the existing computer system [15–17] and network architecture [18–20], the consensus mechanisms used in the blockchain and the need for data protection [21–23]. With the development of blockchain and the development and promotion of the application of 5G technology, blockchain is gradually applied to various industries such as healthcare [24], industry [25], and finance [26]. While blockchain is widely used, it also raises a series of security and privacy issues. The digital currencies used in blockchain have also suffered many security threats, with attacks on trading platforms, theft of currencies and crimes committed by hackers and criminals using blockchain's anonymous transactions occurring frequently. At the same time, privacy breaches [27, 28] in blockchain can also make the skeptical public of blockchain. These challenges are very detrimental to the development and innovation of blockchain.

3 Typical Cryptography

3.1 Homomorphic Encryption

The idea of homomorphic encryption was first introduced by Rivest, Adleman and Detouzos [29] (the R and A in "RSA") in 1978. Homomorphic data encryption allows direct manipulation of the encrypted data without the need for preliminary decryption of the operands. The effect of manipulating the encrypted data is the same as manipulating the data before encryption. In a blockchain, *FHE* (*fully homomorphic encryption*) ensures that the ledger information is not compromised but can be manipulated, even if the blockchain is attacked. *FHE* is a good solution to the problem of data being used on remote devices [30]. In 2009, Gentry [31] proposed a secure and reasonable *FHE* system that performs arbitrary addition and multiplication operations on the encrypted data while also acting on the pre-encrypted data. However, the performance of *FHE* algorithms is so poor that they are difficult to use in practice. In 2011, Brakerski et al. [32] proposed a new *FHE* algorithm, BVG, based on *Learning With Errors* (LWE), an alternative assumption to lattice encryption. The BGV system uses a somewhat more practical LWE assumption than the system proposed by Gentry in 2009. In 2013, Gentry [33] et al. proposed a simpler, *FHE* algorithm GSW based on LWE. In their scheme, they proposed a way to construct *FHE* of a new technique known as the approximate eigenvector method.

3.2 Zero-Knowledge Proofs

Zero-knowledge proof means that the prover does not need to reveal anything about the verification to the verifier who can also do the verification. With the use of zero-knowledge proofs in blockchain, other nodes can verify the legitimacy and correctness of a transaction even if both parties do not reveal any information about the transaction. Zero-knowledge proofs are divided into interactive zero-knowledge proofs and non-interactive zero-knowledge proofs. Interactive zero-knowledge proofs require multiple interactions between the verifier and the prover, and the verifier improves the trustworthiness of the prover by performing multiple verifications to the prover. Non-interactive zero-knowledge proofs, on the other hand, allow the verifier and the prover to interact once with the aid of a machine. An overview of the development of zero-knowledge proofs is given next (Figs. 2 and 3).

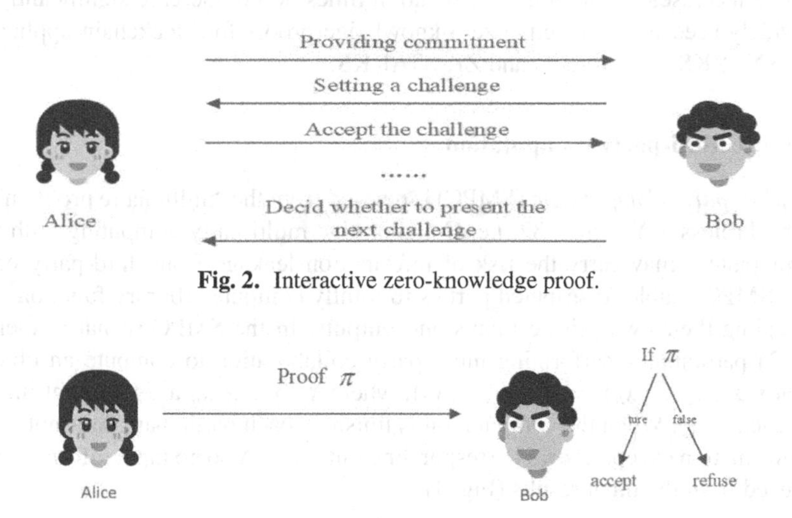

Fig. 2. Interactive zero-knowledge proof.

Fig. 3. Non-interactive zero-knowledge proof.

The concept of zero-knowledge proofs was introduced by S. Goldwasser, S. Micali and C. Rackoff [34] in 1988. After introducing this concept, zero-knowledge proofs have also been present in the overview of the theory. The emergence of blockchains and the need for data confidentiality has facilitated scholarly research on zero-knowledge proofs, which can address the difficulty of aligning blockchain privacy protection with data transparency [35]. In 2010, Groth proposed the key theory of zero-knowledge proofs ZK-SNARK [36] (zero-knowledge succinct non-interactive knowledge proofs). The provers can prove the correctness of their provided proofs mathematically to the verifiers without providing information about the proofs as the verifiers do. Subsequent scholars have worked on ZK-SNARK to reduce verification time and improve efficiency. The Pinocchio [37] protocol, proposed in 2013, is an improved version of ZK-SNARK, and in 2015 the blockchain application Zcash was used to build ZK-SNARK [38], a widespread application of zero-knowledge proofs. In 2016, Groth [39] proposed Groth16, which is also based on an improved version of ZK-SNARK with asymmetric pairing, and the proof will be more efficient.

ZK-SNARKS requires public reference strings for provers and verifiers for trustworthy settings when performing zero-knowledge proofs, but these public strings are again provided by a small group of people, and thus are vulnerable to attack by some malicious nodes [39]. As a result, research now prefers to discard trusted settings. In 2018, the

Bulletproofs algorithm was introduced to eliminate the need for trusted settings. It is a more efficient algorithm that produces proofs of logarithmically transformed size, which would be very beneficial for storing proofs in the blockchain. Moreover, Bulletproofs can also merge and compress proofs of the same scope, reducing the size of the space occupied by the blockchain. In 2018, a new zero-knowledge proof scheme ZK-STARKS was also proposed [40], a zero-knowledge proof scheme that does not require trustworthy settings. ZK-STARKS has better scalability, and its proof and verification times are linearly and logarithmically related to the initial computation time, respectively. As the initial size increases, its proof and verification times do not increase significantly. The more widely used non-interactive zero-knowledge proofs for blockchain applications are ZK-SNARKS, Bulletproofs and ZK-STARKS.

3.3 Secure Muti-party Computation

Secure Muti-party Computation (SMPC) is derived from the "millionaire problem" proposed by Professor Yao in 1982, i.e. Collaborative multi-party computing with third-party guarantees may carry the risk of information leakage from third-party organizations. SMPC enables distributed parties to jointly compute arbitrary functions without revealing their own private inputs and outputs. In the SMPC scenario, there are $n(n \geq 2)$ participants performing multi-party collaboration to compute an objective function $f(x_1, x_2, ..., x_n) = (y_1, y_2, ..., y_n)$, where $x_1, x_2, ..., x_n$ are the input information of each party. When the computation is finished, each participant does not get any other information except its own corresponding output y_i, Also no input information can be deduced from the input results (Fig. 4).

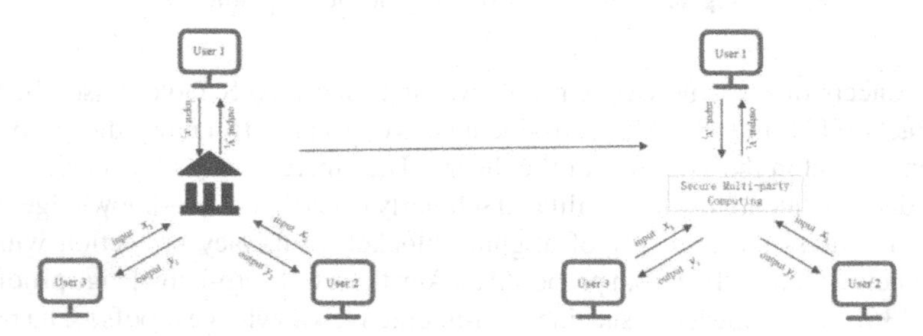

Fig. 4. Secure multi-party calculation.

SMPC provides input correctness, computational correctness and output independence to analyze and capture the value in the data while protecting the privacy of the data. Similar to blockchain, secure multi-party computing supports collaborative computing by uniting untrusted users without a trusted third party. SMPC can collaborate with untrusted users in the blockchain without compromising their privacy to perform analytical calculations, analytical modelling, etc., on some sensitive data. It is also beneficial for blockchain applications to industries that require data analysis and storage, such as healthcare and industry. SMPC can be used in multi-signature, secret sharing, and random number generation, and the wallet ZenGo [41], a wallet released by Kzen, does not require the use of mnemonic word and keys, but rather uses a gated signature method that combines the advantages of multi-signature and secret sharing. However, there are still difficulties that need to be addressed in the application of SMPC in blockchain, such as the fact that SMPC requires the participation of multiple honest nodes, malicious nodes may collude in the computation [42], and the efficiency of SMPC is low when the network transmission rate is low.

4 Post-quantum Cryptography

If quantum computing develops as expected, it will inevitably disrupt existing cryptographic systems. For example, Bernstein and Daniel J. [43] point out that quantum computing's well-known Shor [44] algorithm and Grover [45] algorithm will have an impact on existing cryptographic systems. The Shor algorithm can theoretically solve the underlying mathematical problems on which the security of public key cryptographic algorithms depends, such as discrete logarithms and large integer decomposition problems. So these public-key algorithms, such as RSA and DSA, should not be secure. And Grover's algorithm will halve the security effect of some symmetric cryptographic algorithms and hash algorithms, requiring an increase in key length. Although it is only for theoretical threats, to take a long-term view, it is necessary to research early cryptography that can resist quantum attacks.

Post-Quantum Cryptography (PQC) resists an attacker even if the attacker has a quantum computer, also known as anti-quantum cryptography. The current mainstream schemes for PQC are hash-based public key cryptography, code-based public key cryptography, multivariate public key cryptography and lattice-based public key cryptography.

4.1 Lattice Based Cryptography

A lattice is a set of points in a high-dimensional space. Let $b_1, b_2, ..., b_n$ be a linearly independent set of bases ($n \leq m$) in R^m and the lattice be the set of all linear combinations of integer coefficients of this set of bases. That is.

$$L(B) = \left\{ \sum_{i=1}^{n} x_i b_i, x_i \in Z, i = 1, 2, ..., n \right\} \tag{1}$$

The security of cryptographic algorithms relies on the underlying mathematical problem. Furthermore, the two main difficulties in lattice problems: are the difficulty of solving the shortest vector and the difficulty of solving the nearest vector. These problems have worst-case difficulty [46]. Many scholars have conducted many studies on lattice problems. The most famous algorithm is the LLL proposed by H. Lenstra, A. Lenstra and Lovasz in 1982 [47], however it can only solve the shortest vector in polynomial time with an approximation factor $(1+\varepsilon)\sqrt{4/\sqrt{3}}^{(n-1)/2}$ of (where it is a constant). Thus lattice-based cryptography is quantum resistant.

In the "Third Status Report on the Post-NIST Quantum Cryptographic Standardization Process" - NISTIR 8413 published in July 2022, four algorithms to be standardized were announced. And three of these are all lattice-based cryptographic schemes. The lattice-based cryptography algorithms have a better balance of security, public and private key size, and computational speed and are considered one of the most promising post-quantum cryptographic algorithms [48].

4.2 Hash-Based Signature Algorithm

The hash-based signature algorithm was proposed by Leslie Lamport in 1979, but compared to other signature schemes, it did not to be widely used because it could produce relatively long signatures. With the arrival of the threat of quantum computing, it is gradually gaining attention again because hash-based signature counting has quantum-resistant properties, such as being resistant to attacks by Shor algorithms. It is one of the algorithms that have the potential to replace the traditional signature algorithm [49]. The one proposed by Leslie Lamport is a single hash signature, which cannot sign multiple messages, and was later improved by Ralph Merkle to form a multiple signature algorithm based on the Merkle tree. The public key is the root of the Merkle, and the key is each leaf node in the Merkle tree. The quantum resistance of Hash-based signature algorithm is based on the collision resistance of the Hash function because the current quantum algorithms cannot find the collision of the Hash. Swati Kumari [50] proposed an enhanced hash-based post-quantum cipher (PQC) architecture called signature-based Merkle hash multiplication (SMHM) algorithm. The hash Merkle signature-based algorithm is enhanced by using the Bernoulli-Karatsuba multiplication algorithm. Konstantinos Chalkias [51] proposed a scalable post-quantum cryptography scheme based on Merkle tree signatures suitable for blockchains and distributed ledgers, which can utilize dedicated chains or image structures to reduce the cost of key generation, signing, verification, and the size of signatures.

4.3 Code Based Cryptography

The code-based cryptosystem is derived from McEliece [52]. The algorithm is based on the integrable binary Goppa code called classical McEliece. The encryption and decryption of the McEliece cryptosystem are fast and secure. However, it is rarely used in practice because of the large size of the key, so one of the subsequent directions of research on code-based cryptography is to reduce the size of its key. The general linear decoding hard problem on which McEliece cryptographic algorithm is based is the NP-hard problem [53], so coding-based cryptography is very promising in quantum-resistant cryptography. Moreover, the NIST post-quantum cryptographic algorithm standard collection has coding-based cryptography second only to lattice-based cryptography. It is mainly used in public key encryption algorithms and only two for signature algorithms.

4.4 Multivariate-Based Cryptography Regime

The security of the multivariate-based cryptography regime relies on solving the mathematical problem of solving a system of random multivariate quadratic polynomial equations over a finite field, which is nondeterministic polynomial time-hard. There is no finite algorithm for solving this problem. The *multivariate quadratic* polynomial problem is to find a solution in a system of quadratic polynomial equations in a given finite field. Since multivariate based cryptographic systems emerged late, they still need a lot of research and experiments to prove their security [54]. Although earlier multivariable-based signature systems have been breached and are no longer secure, multivariable-based signature algorithms are small in signature size and fast in inflammation. Therefore, multivariate based signature schemes are still very promising, and multivariate based signature algorithms are the most numerous in the NIST post-quantum cryptographic algorithm standards collection.

5 Conclusion

With the application of the blockchain, the blockchain needs to meet various different needs for data protection, multi-party participation and collaboration, and identity authentication in the face of different scenarios, and cryptography is crucial to the development of blockchain applications. In this paper, some classical cryptography and post-quantum cryptography in blockchain were studied. First, the origin of blockchain and its concepts were introduced, and the structure of Bitcoin and the security challenges it faces were presented. Subsequently, some classical cryptographic homomorphic encryption, zero-knowledge proofs and secure multi-party computation used in blockchains were investigated. Finally, four more promising post-quantum cryptograms were introduced for quantum computing attacks.

References

1. Nakamoto, S.: Bitcoin: a peer-to-peer electronic cash system. Decent. Bus. Rev., 21260 (2008)
2. Liang, W., Xiao, L., Zhang, K., et al.: Data fusion approach for collaborative anomaly intrusion detection in blockchain-based systems. IEEE Internet Things J. (2021)
3. Kumar, P., Kumar, R., et al.: PPSF: a privacy-preserving and secure framework using blockchain-based machine-learning for IoT-driven smart cities. IEEE Trans. Netw. Sci. Eng. **8**(3), 2326–2341 (2021)
4. He, W., Zheng, H.: Literature review on block chain: technology, principle and development. J. Phys. Conf. Ser. **1848**(1), 012166 (2021)
5. Xu, Z., Liang, W., Li, K.C., et al.: A time-sensitive token-based anonymous authentication and dynamic group key agreement scheme for industry 5.0. IEEE TII (2021)
6. Gorkhali, A., Li, L., Shrestha, A.: Blockchain: a literature review. J. Manag. Anal. **7**(3), 321–343 (2020)
7. Liang, W., et al.: PDPChain: a consortium blockchain-based privacy protection scheme for personal data. IEEE Trans. Reliab., 1–13 (2022). https://doi.org/10.1109/TR.2022.3190932
8. Long, J., Liang, W., Li, K.C., et al.: A regularized cross-layer ladder network for intrusion detection in industrial Internet-of-Things. IEEE Trans. Ind. Inform. (2022)
9. Liang, W., Xie, S., Cai, J., et al.: Novel private data access control scheme suitable for mobile edge computing. China Commun. **18**(11), 92–103 (2021)
10. Zhao, J., Huang, J., et al.: An effective exponential-based trust and reputation evaluation system in wireless sensor networks. IEEE Access **7**, 33859–33869 (2019)
11. Nejatollahi, H., Dutt, N., Ray, S., et al.: Post-quantum lattice-based cryptography implementations: a survey. ACM Comput. Surv. (CSUR) **51**(6), 1–41 (2019)

12. Li, X., Liao, J., Kumari, S., Liang, W., Wu, F., Khan, M.K.: A new dynamic id-based user authentication scheme using mobile device: cryptanalysis, the principles and design. Wirel. Pers. Commun. **85**(1), 263–288 (2015). https://doi.org/10.1007/s11277-015-2737-z
13. Liang, W., Xie, S., Cai, J., et al.: Deep neural network security collaborative filtering scheme for service recommendation in intelligent cyber-physical systems. IEEE IoT J. (2021)
14. Liang, W., Ning, Z., Xie, S., et al.: Secure fusion approach for the internet of things in smart autonomous multi-robot systems. Inf. Sci. **579**, 468–482 (2021)
15. Qiu, M., Jia, Z., et al.: Voltage assignment with guaranteed probability satisfying timing constraint for real-time multiproceesor DSP. J. Signal Proc. Syst. (2007)
16. Qiu, M., Yang, L., et al.: Dynamic and leakage energy minimization with soft real-time loop scheduling and voltage assignment. IEEE TVLSI **18**(3), 501–504 (2009)
17. Qiu, M., Xue, C., et al.: Energy minimization with soft real-time and DVS for uniprocessor and multiprocessor embedded systems. In: IEEE DATE Conference, pp. 1–6 (2007)
18. Qiu, M., Chen, Z., et al.: Energy-aware data allocation with hybrid memory for mobile cloud systems. IEEE Syst. J. **11**(2), 813–822 (2014)
19. Qiu, M., Xue, C., Shao, Z., et al.: Efficient algorithm of energy minimization for heterogeneous wireless sensor network. In: IEEE EUC, pp. 25–34 (2006)
20. Li, J., Ming, Z., et al.: Resource allocation robustness in multi-core embedded systems with inaccurate information. J. Syst. Architect. **57**(9), 840–849 (2011)
21. Raikwar, M., Gligoroski, D., Kralevska, K.: SoK of used cryptography in blockchain. IEEE Access **7**, 148550–148575 (2019)
22. Qiu, H., Dong, T., et al.: Adversarial attacks against network intrusion detection in IoT systems. IEEE Internet Things J. **8**(13), 10327–10335 (2020)
23. Gai, K., Qiu, M., Elnagdy, S.: A novel secure big data cyber incident analytics framework for cloud-based cybersecurity insurance. In: IEEE BigDataSecurity (2016)
24. Hu, F., Lakdawala, S., et al.: Low-power, intelligent sensor hardware interface for medical data preprocessing. IEEE Trans. Inf. Tech. Biomed. **13**(4), 656–663 (2009)
25. Qiu, H., Zheng, Q., et al.: Topological graph convolutional network-based urban traffic flow and density prediction. IEEE Trans. ITS (2020)
26. Li, Y., Gai, K., et al.: Intercrossed access controls for secure financial services on multimedia big data in cloud systems. ACM Trans. Multimedia Comput. Commun. Appl. (2016)
27. Qiu, M., Qiu, H., et al.: Secure data sharing through untrusted clouds with blockchain-enabled key management. In: The 3rd SmartBlock, Zhengzhou, China, October 2020, pp. 11–16 (2020)
28. Gai, K., Zhang, Y., et al.: Blockchain-enabled service optimizations in supply chain digital twin. IEEE Trans. Serv. Comput. (2022)
29. Rivest, R.L., Adleman, L., Dertouzos, M.L.: On data banks and privacy homomorphisms. Found. Secure Comput. **4**(11), 169–180 (1978)
30. Gai, K., Qiu, M.: Blend arithmetic operations on tensor-based fully homomorphic encryption over real numbers. IEEE Trans. Ind. Inf. **14**(8), 3590–3598 (2017)
31. Gentry, C.: Fully homomorphic encryption using ideal lattices. In: Proceedings of the Forty-First Annual ACM Symposium on Theory of Computing, pp. 169–178 (2009)
32. Brakerski, Z., Vaikuntanathan, V.: Fully homomorphic encryption from ring-LWE and security for key dependent messages. In: Rogaway, P. (eds.) CRYPTO 2011. LNCS, vol. 6841, pp. 505–524. Springer, Heidelberg (2011). https://doi.org/10.1007/978-3-642-22792-9_29
33. Gentry, C., Sahai, A., Waters, B.: Homomorphic encryption from learning with errors: conceptually-simpler, asymptotically-faster, attribute-based. In: Canetti, R., Garay, J.A. (eds.) CRYPTO 2013. LNCS, vol. 8042, pp. 75–92. Springer, Heidelberg (2013). https://doi.org/10.1007/978-3-642-40041-4_5
34. Goldwasser, S., Micali, S., Rackoff, C.: The knowledge complexity of interactive proof-systems. In: Providing Sound Foundations for Cryptography: On the Work of Shafi Goldwasser and Silvio Micali, pp. 203–225 (2019)

35. Chor, B., Goldwasser, S., Micali, S., et al.: Verifiable secret sharing and achieving simultaneity in the presence of faults. In: 26th IEEE Symposium on Foundations of Computer Science (SFCS), pp. 383–395 (1985)
36. Groth, J.: Short pairing-based non-interactive zero-knowledge arguments. In: Abe, M. (eds) ASIACRYPT 2010. LNCS, vol. 6477, pp. 321–340. Springer, Heidelberg (2010). https://doi.org/10.1007/978-3-642-17373-8_19
37. Parno, B., Howell, J., Gentry, C., et al.: Pinocchio: nearly practical verifiable computation. Commun. ACM **59**(2), 103–112 (2016)
38. Banerjee, A., Clear, M., Tewari, H.: Demystifying the role of zk-SNARKs in Zcash. In: IEEE Conference on Application, Information and NETWORK SECURITY (AINS), pp. 12–19 (2020)
39. Groth, J.: On the size of pairing-based non-interactive arguments. In: Fischlin, M., Coron, J.S. (eds.) EUROCRYPT 2016. LNCS, vol. 9666, pp. 305–326. Springer, Heidelberg (2016). https://doi.org/10.1007/978-3-662-49896-5_11
40. Sasson, E.B., Chiesa, A., Garman, C., et al.: Zerocash: decentralized anonymous payments from bitcoin. In: 2014 IEEE Symposium on Security and Privacy, pp. 459–474. IEEE (2014)
41. Lindell, Y.: Fast secure two-party ECDSA signing. In: Katz, J., Shacham, H. (eds.) CRYPTO 2017. LNCS, vol. 10402, pp. 613–644. Springer, Cham (2017). https://doi.org/10.1007/978-3-319-63715-0_21
42. Wang, Z., Cheung, S.C.S., Luo, Y.: Information-theoretic secure multi-party computation with collusion deterrence. IEEE Trans. Inf. Forensics Secur. **12**(4), 980–995 (2016)
43. Bernstein, D.J., Lange, T.: Post-quantum cryptography. Nature **549**(7671), 188–194 (2017)
44. Shor, P.W.: Algorithms for quantum computation: discrete logarithms and factoring. In: 35th Annual Symposium on Foundations of Computer Science, pp. 124–134. IEEE (1994)
45. Grover, L.K.: A fast quantum mechanical algorithm for database search. In: Proceedings of the Twenty-Eighth Annual ACM Symposium on Theory of Computing, pp. 212–219 (1996)
46. Esgin, M.F., Steinfeld, R., Sakzad, A., Liu, J.K., Liu, D.: Short lattice-based one-out-of-many proofs and applications to ring signatures. In: Deng, R., Gauthier-Umaña, V., Ochoa, M., Yung, M. (eds.) ACNS 2019. LNCS, vol. 11464, pp. 67–88. Springer, Cham (2019). https://doi.org/10.1007/978-3-030-21568-2_4
47. Lenstra, A.K., Lenstra, H.W., Lovász, L.: Factoring polynomials with rational coefficients. Mathematische Annalen **261**, 515–534 (1982)
48. Micciancio, D., Regev, O.: Lattice-based cryptography. In: Bernstein, D.J., Buchmann, J., Dahmen, E. (eds.) Post-Quantum Cryptography, pp. 147–191. Springer, Heidelberg (2009). https://doi.org/10.1007/978-3-540-88702-7_5
49. Merkle, R.C.: Secrecy, authentication, and public key systems. Stanford University (1979)
50. Kumari, S., Singh, M., Singh, R., et al.: Signature based Merkle Hash Multiplication algorithm to secure the communication in IoT devices. Knowl. Based Syst. **253**, 109543 (2022)
51. Chalkias, K., Brown, J., Hearn, M., et al.: Blockchained post-quantum signatures. In: IEEE iThings/GreenCom/CPSCom/SmartData, pp. 1196–1203 (2018)
52. McEliece, R.J.: A public-key cryptosystem based on algebraic. Coding Thv **4244**, 114–116 (1978)
53. Chaulet, J., Sendrier, N.: Worst case QC-MDPC decoder for McEliece cryptosystem. In: 2016 IEEE International Symposium on Information Theory (ISIT). IEEE, pp. 1366–1370 (2016)
54. Ding, J., Yang, B.Y.: Multivariate public key cryptography. In: Bernstein, D.J., Buchmann, J., Dahmen, E. (eds.) Post-Quantum Cryptography, pp. 193–241. Springer, Heidelberg (2009). https://doi.org/10.1007/978-3-540-88702-7_6

An Efficient Detection Model for Smart Contract Reentrancy Vulnerabilities

Yuan Li[1], Ran Guo[2(✉)], Guopeng Wang[3(✉)], Lejun Zhang[1,2(✉)], Jing Qiu[2], Shen Su[2], Yuan Liu[2], Guangxia Xu[2], and Huiling Chen[4]

[1] College of Information Engineering, Yangzhou University, Yangzhou 225127, China
{MX120200526,zhanglejun}@yzu.edu.cn

[2] Cyberspace Institute Advanced Technology, Guangzhou University, Guangzhou 510006, China
{guoran,qiujing,yuanliu,xugx}@gzhu.edu.cn

[3] Engineering Research Center of Integration and Application of Digital Learning Technology, Ministry of Education, Beijing 100039, China
wangguopeng@sohu.com

[4] Department of Computer Science and Artificial Intelligence, Wenzhou University, Wenzhou 325035, China

Abstract. We propose a novel smart contract re-entry vulnerability detection model based on BiGAS. The model combines a BiGRU neural network that introduces an attention mechanism with an SVM. We start from the data features of smart contracts, learn the model layer by layer to achieve feature extraction and vulnerability identification, introduce batch normalization, Dropout processing and use improved model classifiers to improve the vulnerability identification accuracy, model convergence speed and generalization capability of smart contracts. We had conducted numerous experiments, and the experimental results showed that BiGAS Detection Model has a strong vulnerability detection ability. The accuracy of vulnerability detection reached 93.24%, and the F1-score was 93.17%. We compared our approach with advanced automated audit tools and other deep learning-based vulnerability detection methods. The conclusion was that our method is significantly better than the existing advanced methods in detecting smart contract reentrancy vulnerabilities.

Keywords: Smart contract · Reentrancy vulnerability · Bidirectional gated recurrent neural network · SVM

1 Introduction

With the development of blockchain technology, smart contracts have attracted a lot of attention in recent years. They are widely used because they can reduce the cost of trust compared with traditional contracts. Smart contracts are made up of computer code written by humans, which inevitably carries the risk of flaws and vulnerabilities. If the vulnerabilities are exploited by hackers, they will lead to the theft of assets that cannot

M. Qiu et al. (Eds.): SmartCom 2022, LNCS 13828, pp. 350–359, 2023.
https://doi.org/10.1007/978-3-031-28124-2_33

be recovered, ultimately causing huge losses to users. Therefore, the current research on smart contract vulnerability detection is particularly important.

A smart contract represents the transaction requirements of two parties to a contract in the form of a digital rule, which usually contains a set of predefined states. When a user on the blockchain makes a transaction (calling the address of a contract), the contract can perform a transfer of state driven by the transaction [1]. However, there is no effective way to guarantee the security of smart contract code [2]. Smart contracts can simplify and accelerate the development of various applications, but they also bring some problems [3]. In 2016, hackers attacked decentralized autonomous organization (DAO) using a reentrancy vulnerability, resulting in the theft of over 3.6 million ETH. Since then, Ether split into two chains, ETH and ETC [4]. In May 2021, the flash.sx smart contract suffered a reentrancy attack in which approximately 1.2 million EOS and 462,000 USDT were stolen one after another. Although it passed a security audit before then, the reentrancy attack was not detected in time. We propose a new approach that goes beyond the rule-based framework. The contributions of the proposed model are listed below.

1. The BiGAS Detection Model is proposed to solve the problem that the use of Softmax as a recurrent neural network classifier leads to the lack of generalization ability of the smart contract vulnerability identification model and cannot be better applied to vulnerability detection and classification.
2. BiGAS Detection Model combines bidirectional gated recurrent neural network (BiGRU), attention mechanism and SVM. The loss function of SVM is squared hinge loss (or maximum edge loss). For the problem of judging whether there is a vulnerability, using the hinge loss does have good performance, and using the squared hinge loss will make the model more robust on this basis.
3. BiGAS Detection Model is used for smart contract vulnerability detection. The performance is compared with existing traditional vulnerability detection tools and vulnerability detection tools incorporating deep learning. The experimental results show that our deep learning-based approach outperforms state-of-the-art smart contract vulnerability detection tools and our method has high practicality and application value in smart contract reentrancy vulnerability detection.

The article is divided into five chapters, which are structured as follows. The Introduction section introduces the security issues of smart contract reentrancy vulnerabilities in recent years and the significance and contribution of our research. The Related Work section presents existing practical research combining SVM with neural networks. The construction and principles of the model are described in Materials and Methods. The Results and Discussion section describes the experimental procedure and analysis of the results. We compared the proposed method with other currently available methods in the experimental section. The Conclusion section summarizes the strengths, weaknesses and future research directions of our study.

2 Related Work

There are many existing studies that combine SVM with neural networks for various applications. Han Qiu et al. propose a machine learning-based selective encryption design that combines SVM with BSNs [5]. A novel approach to component assembly inspection based on mask R-CNN and Support Vector Machines was proposed by Huang et al. [6]. The method combines CNN and SVM to construct an SVM classification model to identify assembly defects. Pedro F. et al. proposed an object detection method based on discriminatively trained Part-Base V models [7]. They combine a margin-sensitive approach for data-mining hard negative examples with a formalism they call latent SVM. Agarap A F combines convolutional neural network and SVM to construct image classification architecture [8]. Ross Girshick et al. pro-posed rich feature hierarchies for accurate object detection and semantic segmentation [9]. Abdulrahman proposed a time series classification method combining SVM with an echo state network (ESN), replacing the linear readout function of the output layer with the radial basis function kernel of the SVM [10]. Yichuan Tang proposed Deep Learning using Linear Support Vector Machines demonstrate a small but consistent advantage of replacing the softmax layer with a linear support vector machine [11]. Agarap et al. proposed a neural network architecture combining gated recurrent unit and support vector machine for intrusion detection in network traffic data [12].

3 Materials and Methods

3.1 Vulnerability Detection Processing

The process of vulnerability detection for smart contracts by the model we designed (BiGAS Detection Model) is as follows. Figure 1 shows the diagram of the model we designed.

(1) Perform data clean up on the original smart contract. Remove blank lines, spaces, comments, non-ASCII values, and other information that is not relevant to the reentrancy vulnerability. Then, the code statements related to specific operations in the program code are extracted, and these semantically and functionally related code fragments are combined into small code fragments for analysis.
(2) Convert the smart contract into a smart contract fragment after data cleaning has been performed. First, the user-defined variables and functions are mapped to symbolic names, respectively. Information such as keywords and III operators of the smart contract are to be preserved in this process. After that, the contract fragments represented by the symbols are divided into a series of token lists preserving the semantic order by word partition analysis and converted to vector representation by word2vec.
(3) The data are randomly initialized and the training and test sets are divided in a ratio of 8:2. The data from the training set is fed into a BiGRU model with a self-attentive mechanism for feature extraction learning. BiGRU can extract deep-level features from the input vectors. The BiGRU model can be considered as consisting of two parts, the forward GRU and the reverse GRU, and the unit state of the GRU is calculated from the model input and its learning parameter values.

(4) The loss function of the SVM is used in the classification phase to compare the input data (i.e., predicted values) with the target. The decision function of SVM is used to calculate the prediction labels, and the calculated loss values are used to measure how well the network predictions match the expected results.
(5) We use an optimization algorithm to minimize the losses, and we use the Adam optimizer, which has significant advantages [13]. It is simple to implement and computationally efficient. The parameters are updated without the scaling transformation of the gradient, the hyperparameters are well interpreted and the learning rate is usually adjusted automatically without adjustment or with only little fine-tuning. It is well suited for scenarios with large-scale data and parameters for unstable objective functions. The optimizer uses the loss values to update the weights of the network.
(6) The above steps are repeated in the experiment by multiple epochs for the purpose of training the model and outputting the optimized BiGAS detection model after the training.
(7) The test set is fed into the trained model to predict the labels of contract fragments to determine the presence of re-entry vulnerabilities, and the predictions are compared with the target data to obtain the evaluation metrics for the model.

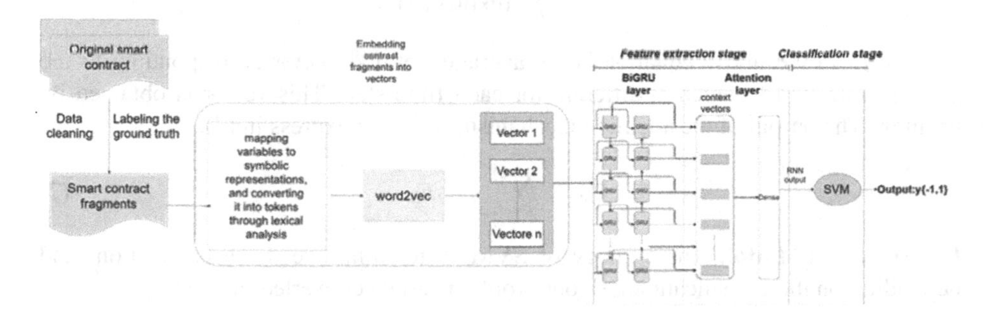

Fig. 1. Vulnerability detection processing.

3.2 BiGAS Detection Model

BiGRU Neural Network. The following procedure calculates the hidden state of the GRU at the time step t. We use $h_t = GRU(x_t, h_{t-1})$ to symbolize the following process. Our number of samples is n, the number of inputs is d, the number of hidden cells is h, the input vector $x_t \in R^{n*d}$, $h_{t-1} \in R^{n*h}$ at the tth time step, the reset gate r_t and the update gate z_t are expressed as below.

$$r_t = \sigma(x_t W_{xr} + h_{t-1} W_{hr} + b_r) \tag{1}$$

$$z_t = \sigma(x_t W_{xz} + h_{t-1} W_{hz} + b_z) \tag{2}$$

$$\tilde{h_t} = \tanh(x_t W_{xh} + (r_t \odot h_{t-1}) W_{hh} + b_h) \tag{3}$$

$$h_t = z_t \odot h_{t-1} + (1 - z_t) \odot \tilde{h_t} \tag{4}$$

Self-attention Mechanism. To highlight the importance of the impact of different keywords on the reentrancy vulnerability, a self-attentive mechanism is introduced after the BiGRU layer of the model to assign weights to different words. The input of the Attention mechanism layer is h_t, which is the hidden vector of the output processed by the BiGRU neural network layer in the previous layer.

First, we input the hidden layer vector to the fully connected layer (with weight W and bias b) to obtain u_t, , which is stated in (5).

$$u_t = \tanh(Wh_t + b) \tag{5}$$

Then, we use this vector to calculate the calibration vector α_t (weights normalized by the attention mechanism), which is stated in (6).

$$\alpha_t = \frac{\exp(u^T u)}{\sum_t (\exp(u^t u))} \tag{6}$$

The u^t in the above equation is the average optimal vector corresponding to the current time step t, which is different for each time step. This vector is obtained by training. The output of the attention mechanism layer we express in (7).

$$s_t = \sum\nolimits_{i=1}^{t} \alpha_t h_i \tag{7}$$

Classification Modul. The strategy of SVM is to minimize the loss function, and depending on the classification task our problem can be converted into (8).

$$L = \min_{w} \frac{1}{2} \|w\|^2 + C \sum\nolimits_{i=1}^{n} \max\left(0, 1 - y_i\left(w^T x_i + b_i\right)\right) \tag{8}$$

Briefly, when SVM is introduced as the last layer of a neural network model, the parameters will be learned by optimizing the objective function of the SVM. In addition, the process uses the loss function of SVM instead of the cross-entropy function to calculate the loss values. In the above equation, L is the loss value, w is the weight of the output layer, n denotes the length of the output vector, i.e., the number of labels, y_i denotes the factual label of the unique thermal encoding, x_i denotes the input, b_i denotes the bias, C is the penalty level of the error sample, and T is the transposition character.The double-regularized SVM is denoted as L2-SVM, one of the improved algorithms of SVM, converts soft intervals into hard intervals, i.e., linearly indistinguishable into linearly distinguishable. In this case, the objective function is simply a constant factor 1/C added to the diagonal of the kernel function, which can be treated as a minor modification of the kernel function. It is differentiable and more stable compared to L1-SVM, so L2-SVM is used for the network structure in this paper, and its mathematical representation is as (9).

$$L = \min_{w} \frac{1}{2}\|w\|_2^2 + C\sum\nolimits_{i=1}^{n} \max\left(0, 1 - y_i\left(w^T x_i + b_i\right)\right)^2 \tag{9}$$

The decision function of the support vector machine generates a score vector for each classification, in order to obtain a mathematical representation of the predicted label of the data as (10).

$$predicted_{class} = argmax(sign(wx + b)) \tag{10}$$

4 Results and Discussion

4.1 Evaluation Metrics

The whole experiment is divided into a training phase and a testing phase. Our job in the training phase is to optimize the model parameters by learning the loopholes to get a trained model. The testing phase of our work is to use the test data as input to the trained model and output the prediction results of vulnerability detection. The prediction results are compared with the real tags to measure the performance of our model. In this paper, we used widely used metrics, including accuracy (ACC), true positive rate or recall (TPR), false positive rate (FPR), precision (PRE), and F1-score (F1).

The whole experiment is divided into a training phase and a testing phase. Our job in the training phase is to optimize the model parameters by learning the loopholes to get a trained model. The testing phase of our work is to use the test data as input to the trained model and output the prediction results of vulnerability detection. The prediction results are compared with the real tags to measure the performance of our model. In this paper, we used widely used metrics, including accuracy (ACC), true positive rate or recall (TPR), false positive rate (FPR), precision (PRE), and F1-score (F1). The experimental model is built on a computer with Intel Core (TM) i7-10875H CPU, NVIDIA GeForce GTX 2060 GPU, and 16 GB RAM.As for the parameter settings, for each model, we use a 10-fold cross-validation method to select and train the best parameter values for training, corresponding to the effectiveness of the reentrancy vulnerability detection. The settings of all parameters we give in Table 1.

4.2 Experimental Results and Performance Comparison

We compare the performance of our model with an advanced automatic safety analysis. Table 4 shows the comparison of experimental results data. The experimental results show that: (1) The vulnerability detection performance of existing automated security analysis needs to be improved. The Oyente [14] with the highest accuracy reached only 71.50%, the Oyente with the highest recall (TPR) reached only 50.84%, and the Securify [15] with the highest F1 value reached only 52.79%. (2) Our model (BiGAS Detection Model) has a high accuracy compared to the state-of-the-art tools. The accuracy of our model is 93.75%, which is 22.25% higher than that of the state-of-the-art Oyente. In

Table 1. Setting of hyperparameters

Hyper-parameters	BIGSA
Batch size	64
Cell size	300
Dropout rate	0.3
Recurrent dropout rate	0.5
Epochs	120
Learning rate	0.001
SVM C	0.5

addition, F1-score of our model is 93.44%, which is 40.65% higher than the F1-score of Securify. (3) Our model achieves a recall of 98.27%, which is much higher than that of advanced automatic safety analysis. This indicates that our model has a lower false positive rate. (4) All the metrics of our model outperform advanced automated security analysis. The experimental data shows that our approach has great potential for smart contract reentrancy vulnerability detection. We plot the results in Fig. 2.

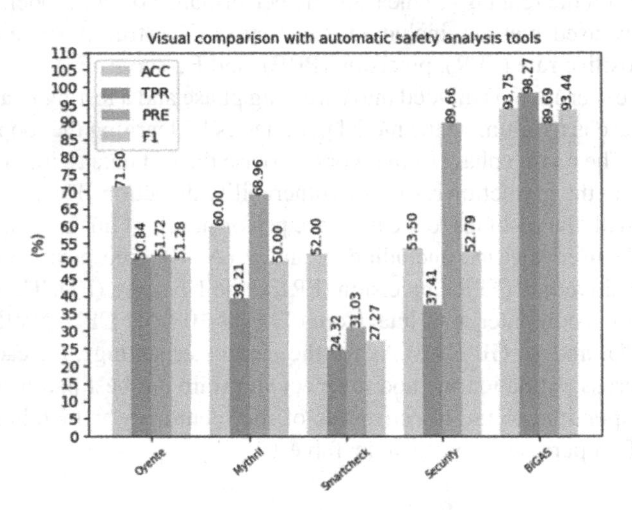

Fig. 2. Visual comparison with existing automatic safety analysis tools

To verify the effectiveness of our model, we also compared it with other neural network-based vulnerability detection methods, and the experimental results are shown in Table 2. Based on the data in Table 2, the following conclusions can be drawn. (1) According to models 1–5 in the table (sequential model): for our dataset, GRU performs slightly better than LSTM; bidirectional recurrent neural network is slightly better than unidirectional; introducing attention mechanism in the model will improve the model performance. (2) All metric values of our method are higher than existing neural network-based vulnerability detection methods, indicating the research significance and development potential of our method in combining deep learning with smart contract vulnerability detection. (3) Comparing our model with the model replaced with Softmax classifier, the accuracy of our model (the model combined with SVM) is 4.47% higher than that combined with Softmax, the recall of our model is 8.86% higher than that combined with Softmax, the precision of our model is 3.27% higher than that combined with Softmax, and the F1-score of our model is 5.29% higher than that combined with Softmax. From the experimental data of these two models, we can conclude that our choice of SVM as the classifier of the model greatly improves the performance of the model. From these two sets of values, SVM as a classifier gives our BIGAS Dection Model a higher vulnerability detection capability. And even without SVM (i.e., BiGRU-ATT-Softmax model) shows that it is not inferior to the vulnerability detection capability of existing more advanced methods, which indicates that the flexible framework we build is more suitable for the task of detecting reentrancy vulnerabilities in smart contracts.

Table 2. Performance comparison with Neural Network Based Methods.

Models	Metrics			
	ACC (%)	TPR (%)	PRE (%)	F1 (%)
RNN	78.84	75.65	82.10	79.56
GRU	83.33	73.28	81.66	81.79
LSTM	82.78	76.41	85.26	80.06
BLSTM	85.36	85.57	85.23	84.56
BLSTM-ATT [16]	87.27	88.48	86.45	87.14
DR-GCN [17]	81.47	81.80	72.36	76.39
TMP [17]	84.48	82.63	74.06	78.11
CGE [18]	89.15	87.62	85.24	86.41
BiGRU-ATT-Softmax	89.28	89.41	86.42	88.15
BiGAS	93.75	98.27	89.69	93.44

As shown in Fig. 3, we plot the ROC curves of different methods under the same threshold value. We use it to compare the classification effect of different methods for vulnerability detection. The area under the ROC curve is also called the AUC value, and the larger the value, the better the classification effect of the model. From the figure, our method has the largest AUC value of 0.9474. This indicates that our method has the best classification of vulnerable and non-vulnerable, i.e., the best vulnerability detection performance.

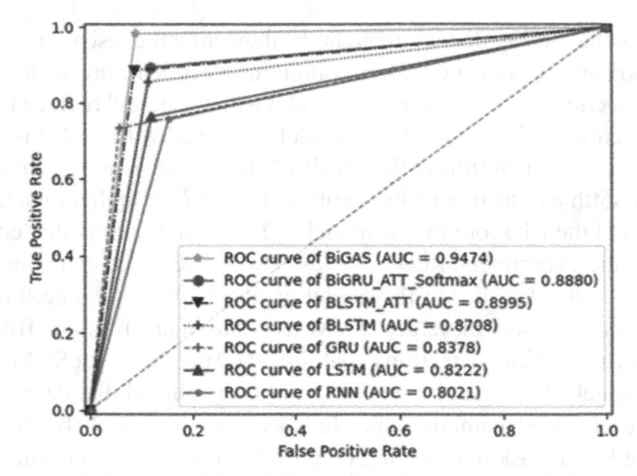

Fig. 3. ROC curves of Recurrent Neural Networks, this figure symbolizes the performance visualization results of different method

5 Conclusions

In this paper, we proposed a novel smart contract reentrancy vulnerability detection model based on BiGAS (which combines BiGRU-ATT and SVM). Starting from the original data features, the model learns layer by layer to achieve feature extraction and vulnerability identification, introducing batch normalization, Dropout processing and using SVM to improve the model classifier to improve the vulnerability identification accuracy, model convergence speed and generalization capability of smart contracts. In addition, we compared with existing security analysis and deep learning-based vulnerability detection methods. The results of the experiments showed that BiGAS Detection Model has better performance in terms of prediction accuracy, F1-score and classification effect. All the experimental data confirmed the effectiveness and practicality of our approach in dealing with the smart contract reentrancy vulnerability detection problem, which has certain advantages over the existing models.

The shortcoming of our research is that the model was only used for smart contract reentrancy vulnerability detection. Our future work will consider combining the expert knowledge model to detect more types of smart contract vulnerabilities such as timestamp dependencies, integer overflows, and other vulnerabilities on top of that. Our goal is to achieve a feasible and efficient model with wider applications.

Funding Statement. This work is sponsored by the National Natural Science Foundation of China under grant number No. 62172353. And Innovation Fund Program of the Engineering Research Center for Integration and Application of Digital Learning Technology of Ministry of Education under grant number No. 1221045.

References

1. Zhang, K.F., Zhang, S.L., Jin, S.: The security research of blockchain smart contract. J. Inf. Secur. Res. **5**(3), 192–206 (2019)

2. Zou, W.Q., Lo, D., Kochhar, P.S.: Smart contract development: challenges and opportunities. IEEE Trans. Softw. Eng. **47**, 2084–2106 (2019). https://doi.org/10.1109/TSE.2019.2942301
3. Hu, T., Liu, X., Chen, T.: Transaction-based classification and detection approach for Ethereum smart contract. Inf. Process. Manag. **58**(2), 102462 (2021). https://doi.org/10.1016/j.ipm.2020.102462
4. Amiet, N.: Blockchain vulnerabilities in practice. ACM Digit. Libr. **2**(2), Article no. 8 (2021)
5. Qiu, H., Qiu, M., Lu, Z.: Selective encryption on ECG data in body sensor network based on supervised machine learning. Inf. Fus. **55**, 59–67 (2020)
6. Huang, H., Wei, Z., Yao, L.: A novel approach to component assembly inspection based on mask R-CNN and support vector machines. Information **10**, 282 (2019)
7. Felzenszwalb, P.F., Girshick, R.B., McAllester, D., Ramanan, D.: Object detection with discriminatively trained part-based models. IEEE Trans. Pattern Anal. Mach. Intell. **32**(9), 1627–1645 (2010). https://doi.org/10.3390/info10090282
8. Agarap, A.F.: An architecture combining convolutional neural network (CNN) and support vector machine (SVM) for image classification. Comput. Sci. (2017)
9. Girshick, R., Donahue, J., Darrell, T., Malik, J.: Rich feature hierarchies for accurate object detection and semantic segmentation. In: 2014 IEEE Conference on Computer Vision and Pattern Recognition, pp. 580–587 (2014)
10. Alalshekmubarak, A., Smith, L.S.: A novel approach combining recurrent neural network and support vector machines for time series classification. In: 2013 9th International Conference. Proceedings: Innovations in Information Technology (IIT), Al Ain, United Arab Emirates, pp. 42–47 (2013)
11. Tang, Y.: Deep learning using linear support vector machines (2013)
12. Agarap, A.F.M.: A neural network architecture combining gated recurrent unit (GRU) and support vector machine (SVM) for intrusion detection in network traffic data. In: Proceedings: the 2018 10th International Conference on Machine Learning and Computing (ICMLC), 26–30 (2018)
13. Kingma, D.P., Ba, J.: Adam: a method for stochastic optimization (2014)
14. Luu, L., Chu, D.H., Olickel, H.: Making smart contracts smarter. In: The 2016 ACM SIGSAC Conference. Proceedings: Computer and Communications Security (CCS), New York City, NY, USA, pp. 254–269 (2016)
15. Tsankov, P., Dan, A., Drachsler-Cohen, D.: Securify: practical security analysis of smart contracts. In: The 2018 ACM SIGSAC Conference. Proceedings: Computer and Communications Security, Toronto, Canada, pp. 67–82 (2018)
16. Qian, P., Liu, Z., He, Q.: Towards automated reentrancy detection for smart contracts based on sequential models. IEEE Access **8**, 19685–19695 (2020)
17. Zhuang, Y., Liu, Z., Qian, P.: Smart contract vulnerability detection using graph neural network. In: The Twenty-Ninth International Joint Conference on Artificial Intelligence, IJCAI, pp. 3283–3290 (2020)
18. Liu, Z., Qian, P., Wang, X.: Combining graph neural networks with expert knowledge for smart contract vulnerability detection. IEEE Trans. Knowl. Data Eng. (2021)

Using Convolutional Neural Network to Redress Outliers in Clustering Based Side-Channel Analysis on Cryptosystem

An Wang[1,2], Shulin He[3], Congming Wei[1(✉)], Shaofei Sun[1], Yaoling Ding[1,4], and Jiayao Wang[5]

[1] School of Cyberspace Science and Technology, Beijing Institute of Technology, Beijing 100081, China
{wangan1,7520220100,dyl19}@bit.edu.cn

[2] State Key Laboratory of Cryptology, P.O. Box 5159, Beijing 100878, China

[3] School of Computer Science and Technology, Beijing Institute of Technology, Beijing 100081, China
hsl21@bit.edu.cn

[4] State Key Laboratory of Information Security (Institute of Information Engineering, Chinese Academy of Sciences), Beijing 100093, China

[5] The 31461 Unit of the Chinese People's Liberation Army, Shenyang 110001, China

Abstract. Blockchain, designed with cryptographic technology, is widely used in the financial area, such as digital billing and cross-border payments. Digital signature is the core technology in it. However, digital signatures in public key cryptosystems face the threat of simple power analysis in Side-Channel Analysis (SCA). The state-of-the-art simple power analysis based on clustering mostly will appear outliers in the process of analysis, which will reduce success rate of key recover. In this paper, we propose a new SCA method with clustering algorithm Density-Based Spatial Clustering of Applications with Noise (DBSCAN) and deep learning technology Convolutional Neural Network (CNN), called DBSCAN-CNN, to analyze public key cryptosystems. We cluster data with DBSCAN firstly. Then we train a CNN model based on the trusted clustering results. Finally, we classify the outliers of clustering results by the trained model. We mount the proposed method to analyze an FPGA-based elliptic curve scalar multiplication power trace which is desynchronized by simulating random delay. The experimental results show that the error rate of the proposed method is at least 69.23% lower than that of the classical clustering method in SCA.

Keywords: Side-Channel Analysis · Outlier detection · DBSCAN · Convolutional Neural Network · Public-key cryptosystems

1 Introduction

Blockchain is widely used in various fields [3,4], which digital signature algorithm is one of the core technologies in it. Side-Channel Analysis (SCA) [6] is a method

M. Qiu et al. (Eds.): SmartCom 2022, LNCS 13828, pp. 360–370, 2023.
https://doi.org/10.1007/978-3-031-28124-2_34

in which an analyst uses physical signal changes to recover secret information, such as secret key used in digital signature algorithm. One of the analysis scenarios in SCA is non-profiled. It's consistent with the actual situation, because it's that the analyst can only obtain side-channel information on the target device, e.g., Simple Power Analysis (SPA) [6].

In recent years, machine learning has a great impact on some well-known countermeasures after being introduced into SCA [8,13]. For example, clustering-based SCA can relieve countermeasures against SPA, which clusters data in an unsupervised scenario according to "Samples of the same clusters are like each other while samples of different clusters are significantly different".

The most used clustering algorithm in SCA is K-Means [13,14], which requires the number of clusters to be specified. K-Means is a hard clustering algorithm based on the distance between cluster centers and sample points. So, it only deals with convex clusters data and cannot find outliers. These problems increase the computational complexity of cryptographic operations classification. So, we propose to use Density-Based Spatial Clustering of Applications with Noise (DBSCAN) [12] instead of K-Means. Because DBSCAN can well avoid the above problems.

Clustering generally requires Principal Components Analysis (PCA) to perform a projection into a lower dimensional space for computational reasons. In this process, the loss of information may cause outliers which possibly have error labels after clustering. These outliers still contain feature information which can be used to distinguish between clusters. In addition, side-channel countermeasures, such as random delay, may cause the time samples corresponding to the same operation on the power trace to be misaligned, and bring some outliers [1], which will have a significant impact on the accuracy of clustering in SCA.

In order to eliminate the influence of outliers in clustering, outlier detection and processing are usually taken to replace outliers in the preprocessing of SCA [9,10]. However, if there are a lot of outliers, it will lead to error classification of the cryptographic operations. We choose deep learning to mine effective features in outliers and improve the accuracy of clustering. Applying deep learning to execute non-profiled SCA is a new way [7].

Among them, Convolutional Neural Network (CNN) is a suited model in SCA because of its ability to desynchronize. In 2019, Carbone et al. [2] used CNN models to attack the hardware RSA implementation and succeeded. Therefore, we use CNN to redress the error labels of outliers, which will improve the accuracy of clustering-based SCA.

Nascimento et al. [9] improved the unsupervised RSA horizontal clustering attack framework. They discussed the problem of outlier processing. In order to ensure high efficiency, they used the median of normal data to replace outliers that detected by distance from mean or Turkey's range test. Perin et al. [10] proposed to use deep learning to correct the error bits generated after K-Means clustering horizontal attack. For outlier detection and handling, they used the Turkey test method to detect and replace the outliers with the median value. But the approach of median replacement makes the analyst lose some information.

Because they are normal data perturbed by unconventional behaviors, outliers should contain some characteristics of normal data to a certain extent.

Contributions. In this paper we propose a new SCA method DBSCAN-CNN. It only needs to collect a power trace on the target device, and then performs analysis using unsupervised clustering algorithm and deep learning.

- Combining clustering and deep learning, we propose DBSCAN-CNN SCA. Our method is unsupervised overall. Firstly, we use DBSCAN to cluster the lower-dimensional power segments after PCA projection. Then the original power segments is divided into training set and testing set according to the *clustered* labels and *outlier* labels. Finally, we train CNN with training set and test it with testing set. Through repeatability experiments, we prove that the accuracy of key recovery through our method is about 99%.
- Deep learning is used to extract effective features in outliers. Different from outlier processing methods such as median replacement, elimination, and increase samples to reduce noise impact in previous work, we use the effective feature mining method "deep learning" to help classifying outliers.
- We compare our proposed method with the classical clustering methods (i.e. K-Means and DBSCAN) in SCA. The results show that the error rate of our method is at least 69.23% lower than that of the classical clustering methods.

2 Preliminaries

2.1 Public-Key Cryptosystems and Simple Power Analysis

In public-key cryptosystem, the communication parties use different keys: public key and private key. The sender encrypts the message with the receiver's public key. The receiver decrypts the message with its own private key. The security of existing public key cryptosystems depends on difficult mathematical problems, which makes it impossible to obtain the corresponding private key by calculation when the public key is known. Representative algorithms include: RSA, ElGamal, and ECC.

The core of ECC is scalar multiplication (Algorithm 1). Among them, R and P represent points on the elliptic curve and k is an integer operating parameter.

Algorithm 1. Scalar Multiplication

Input: P, $k = (k_{n-1}k_{n-2} \dots k_1k_0)$
Output: $R = kP$

```
1 R = P
2 for i = n − 1 down to 0 do
3     R = 2R
4     if k_i = 1 then
5         R = R + P
6     end
7 end
8 return R
```

In Algorithm 1, there are two different operations: Double and Add. When k_i is 0, one times Double operation. When k_i is 1, one times Double operation and Add operation. Therefore, we can have following statements: (1) There are more Double operations than Add operations; (2) There is no continuous Add operations; (3) There is continuous Double operations; (4) An add operation always follows behind a Double operation.

SPA differentiates different cryptographic operations based on the difference in amplitude of the power trace. It's often used to analyze the square-and-multiply algorithm of RSA and the double-and-add algorithm of ECC. Next, we take an ECC power trace as an example to introduce how to perform SPA.

Figure 1 is a fragment of power trace during scalar multiplication calculation in ECC algorithm. There are obviously two patterns, one for Double operation "D" and one for Add operation "A". Double operation corresponds to a higher energy value while Add operation corresponds to a lower energy value. According to the above statements about two kinds of operations, we can uniquely obtain the binary scalar sequence corresponding to the operation sequence, to restore the scalar value used in the scalar multiplication process.

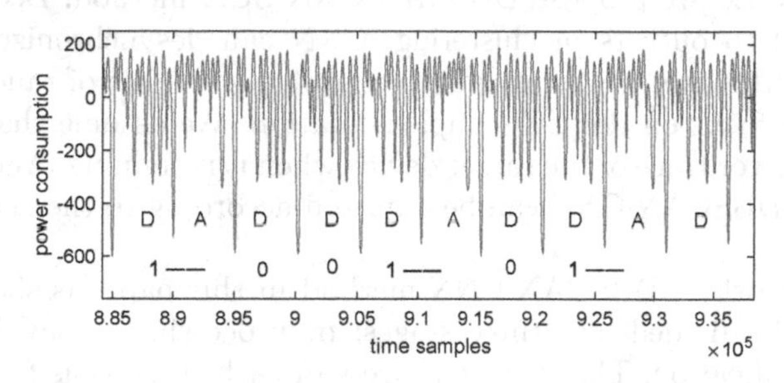

Fig. 1. SPA on An ECC Power Trace

When cryptanalysts want to obtain the private key in public key cryptosystem through SPA, they need to have certain cryptographic theory and analysis experience. But the keys for public-key cryptosystem are usually very long, SPA is time-consuming and laborious. In practice, cryptographic products will adopt some countermeasures against SPA and protect core chip security. Therefore, artificial intelligence methods such as machine learning and deep learning are introduced to improve state-of-the-art SCA techniques for the analysis of public-key cryptosystem.

2.2 Outlier Detection and Clustering

Outlier detection is an active research field in data processing, whose goal is to distinguish between normal and abnormal data [5]. Clustering is a common unsupervised machine learning method in outlier detection and SCA.

In this paper, we chose DBSCAN as clustering method to analyze ECC. DBSCAN is a density clustering algorithm [12]. This method doesn't need to specify the number of clusters. And it can pick out abnormal data points. The parameters ϵ and θ are used to control the size of determination range of the search data points in clustering. The parameter ϵ represents the domain search radius around the data point. The parameter θ on behalf of the minimum number of data points within the search radius. According to this group of parameters, DBSCAN find core points, border points and noise from all points. Border points have the same clustered labels as their own core points.

Obviously, DBSCAN can make the points in the high-density area cluster into the same classes, and marks points that stray far outside the clusters as outliers. Therefore, by debugging the parameters ϵ and θ, we can make the number of correct labels of clustered data as much as possible and ensure that the cryptographic operations with the obtained clustered labels is not ambiguous. Then we can get reliable training set and further analyze outlier data points picked out, to reduce the computational complexity.

3 DBSCAN-CNN Side-Channel Analysis Method

In view of the shortcomings of the median replacement and rejection of outliers in the previous work, we propose DBSCAN-CNN SCA method. Because random delay will lead to outliers in clustering, CNN can desynchronize and extract features. So, DBSCAN-CNN can effectively solve the effect of random delay.

This paper focus on the ECC implementation. We assume that the analyst can collect a power trace of the target device when it is running. And the network depth used in DBSCAN-CNN can be adjusted according to the actual situation of the dataset.

The framework of DBSCAN-CNN method in this paper is shown in Fig. 2, which is mainly divided into three stages: preprocessing, DBSCAN clustering and CNN classification. The specific process of each stage is as follows:

Preprocessing Step 1.1 Collect a power trace T from the target device. The trace contains L time samples.

Step 1.2 Cut the trace T into several power segments $\{t_0, ..., t_u\}$, according to previous knowledge in Sect. 2 about the number of operations in ECC.

DBSCAN Clustering

Step 2.1 Reduce the high-dimensional segments to three-dimensional by PCA, get the three-dimensional data points $\{p_0, ..., p_u\}$, the coordinate of p_u is represented by (PC_1, PC_2, PC_3).

Step 2.2 Cluster data points $\{p_0, ..., p_u\}$ through DBSCAN. Adjust parameters ϵ and θ and get clustering results: normal clusters and outliers. There is $K (K \geq 2)$ classes in normal clusters.

Step 2.3 Divide the data into training set and testing set according to clustering results. The segments corresponding to normal clusters $\{t_0, ..., t_r\}$ is training data and their labels is $\{y_0, ..., y_r\}$, here $y_i \in [1, ..., K]$ for $i \in [0, new]$. The segments corresponding to outliers $\{t_0, ..., t_s\}$ is testing data. Also, $r + s + 1 = u$.

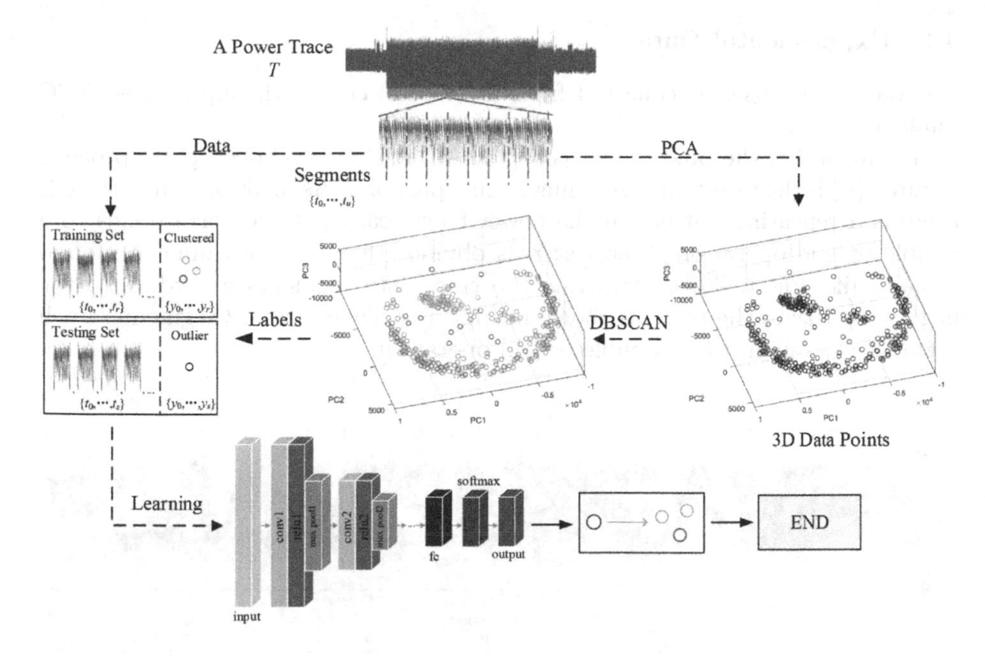

Fig. 2. DBSCAN-CNN Framework

CNN Classifying

Step 3.1 Extend the training set by data augmentation i.e., data $\{t_0, ..., t_{new}\}$ and labels $\{y_0, ..., y_{new}\}$, here $y_i \in [1, ..., K]$ for $i \in [0, new]$.

Step 3.2 Build CNN model with depth D. Its conventional kernel size is 3*3 and padding is 1.

Step 3.3 Train the CNN model on training set obtained in Step 3.1.

Step 3.4 Test the trained model with testing set. The output is K classification labels.

In [5], Järvinen et al. have given the method about how to recover key when they have K kinds of labels. In this case, there are $K!$ possible mapping π between labels and cryptographic operations. If K is small, the attacker can try all mappings. Then they find the correct best mapping, and finally recover the key bits.

4 Experimental Results

In this section, we take an FPGA-based ECC implementation as example to verify the performance of DBSCAN-CNN. All experiments were performed on MATLAB R2022a, and the CNN we used is designed using the Deep Network Designer App. Our experimental device is a laptop with 16 GB memory and 2.4 GHz CPU.

4.1 Experimental Data

The dataset we used is collected from a terminal chip with unprotected ECC implementation.

Figure 3(a) is the power trace collected during the ECC decryption process. Figure 3(b) is the result after zooming in on a part of it. As mentioned in Sect. 2.1, there is a repeating pattern in the power trace, each pattern corresponds to a double or add operation. However, it is obvious that leakage cannot be found visually like Fig. 1. Then we divide the trace into segments according to the negative peaks of the trace as shown in Fig. 3(c). There are 372 segments. Each segment corresponds to a double or add operation.

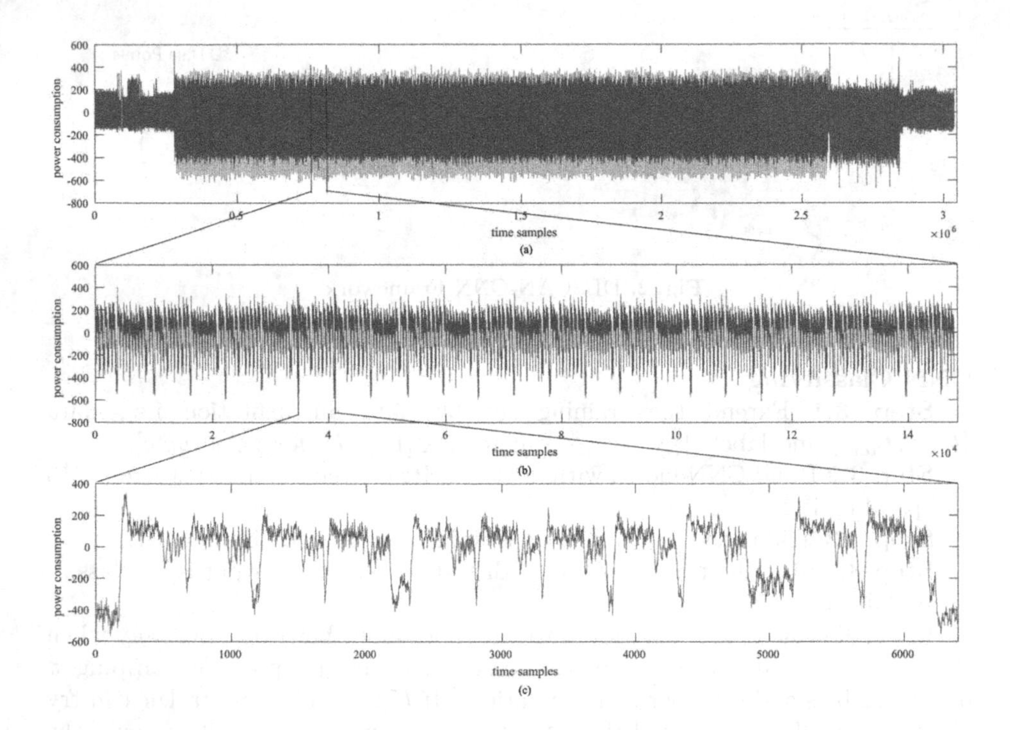

Fig. 3. ECC Power Trace

It is common to simulate the effect of countermeasures by modifying the dataset [11]. In order to show a random delay scenario, we desynchronized the power trace to simulate random delay countermeasure by inserting redundant traces into segments. Here's how we simulate:

We randomly select 20 segments from the segment set and insert square waves with different signal-to-noise ratio (SNR), duty cycle and sample length at random position. Some of the square waves we used are shown in Fig. 4.

The simulation results are shown in Fig. 5. Red stars indicate the insertion position.

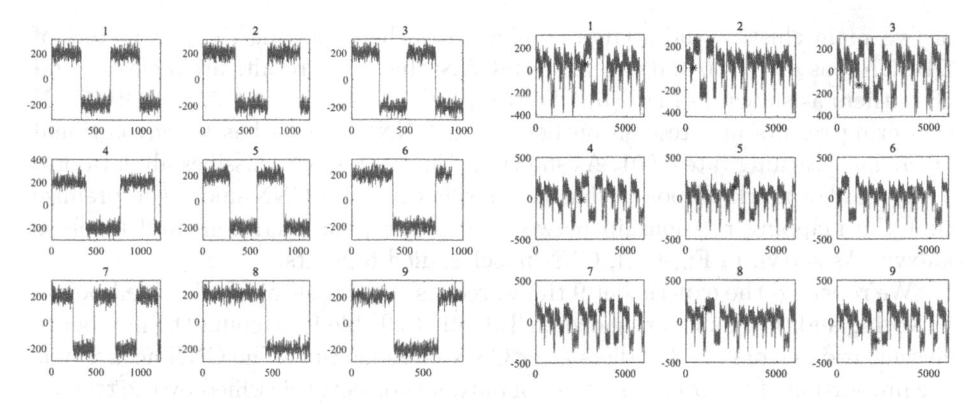

Fig. 4. The Redundant Square Waves

Fig. 5. Power Segments After Random Delay Simulation

4.2 Experiment Results

In subsequent chapters, we use "D" representing Double operations and "A" representing Add operations.

The results of DBSCAN-CNN are shown in Fig. 6. As Figure 6(a) shows, after reducing the data to 3 dimensions using PCA, the data presents a distribution

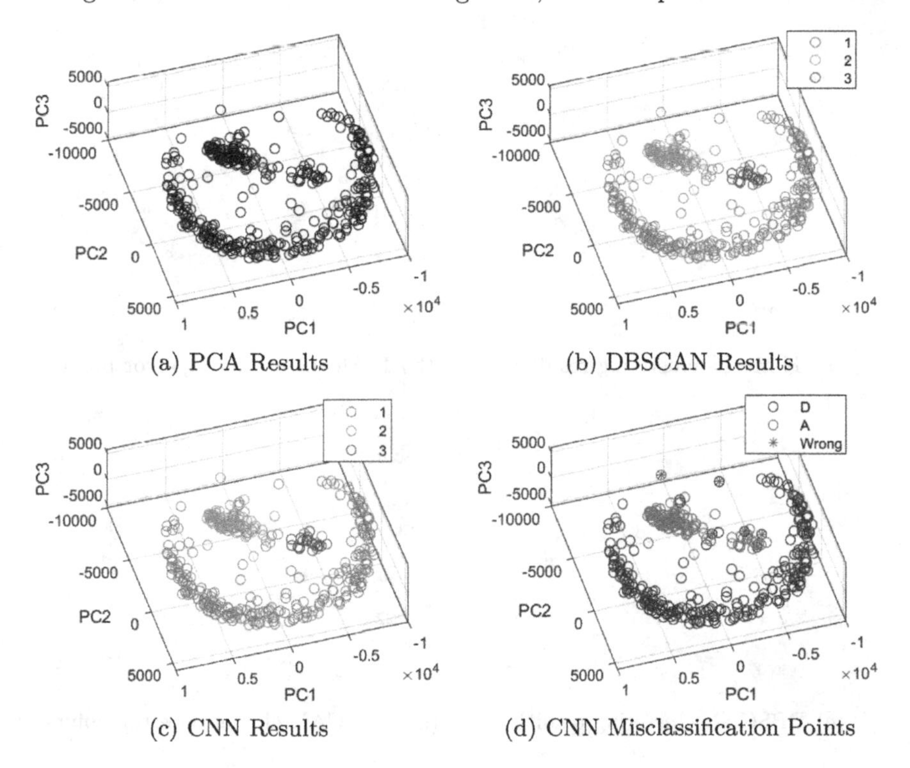

(a) PCA Results

(b) DBSCAN Results

(c) CNN Results

(d) CNN Misclassification Points

Fig. 6. DBSCAN-CNN Experiment Results

of two main clusters and a large number of outliers. Setting the parameters of DBSCAN as $\varepsilon = 1900$ and $\theta = 15$, DBSCAN clustering results are 3 classes and 33 outliers as shown in Fig. 6(b). We train a CNN on classes 1, 2, 3 of DBSCAN clustering results and test on outliers. The CNN we used has 9 convolutional layers and learning rate 0.001. As shown in Fig. 6(c), CNN classifies all data into 3 classes. We mark all points that are misclassified by CNN under the premise that the mapping relationship between cryptographic operations and labels is known. As shown in Fig. 6(d), CNN misclassified 6 points.

We repeated the experiment 9 times, redress the labels of misclassified points by voting and the results are shown in Table 1. In Table 1, we count the number of classification errors of 33 outliers after CNN classification. The CNN accuracy is the proportion of the overall number of outliers correctly classified by the trained model. Here, the DBSCAN-CNN classification accuracy obtained by voting is improved to 87.9%, and the number of misclassification are reduced to 4.

Table 1. Repeated Experimental Results

Experiment index	1	2	3	4	5	6	7	8	9	Vote
CNN misclassification count	3	7	5	5	8	6	5	6	4	4
CNN accuracy	0.909	0.788	0.848	0.848	0.758	0.818	0.848	0.818	0.879	**0.879**

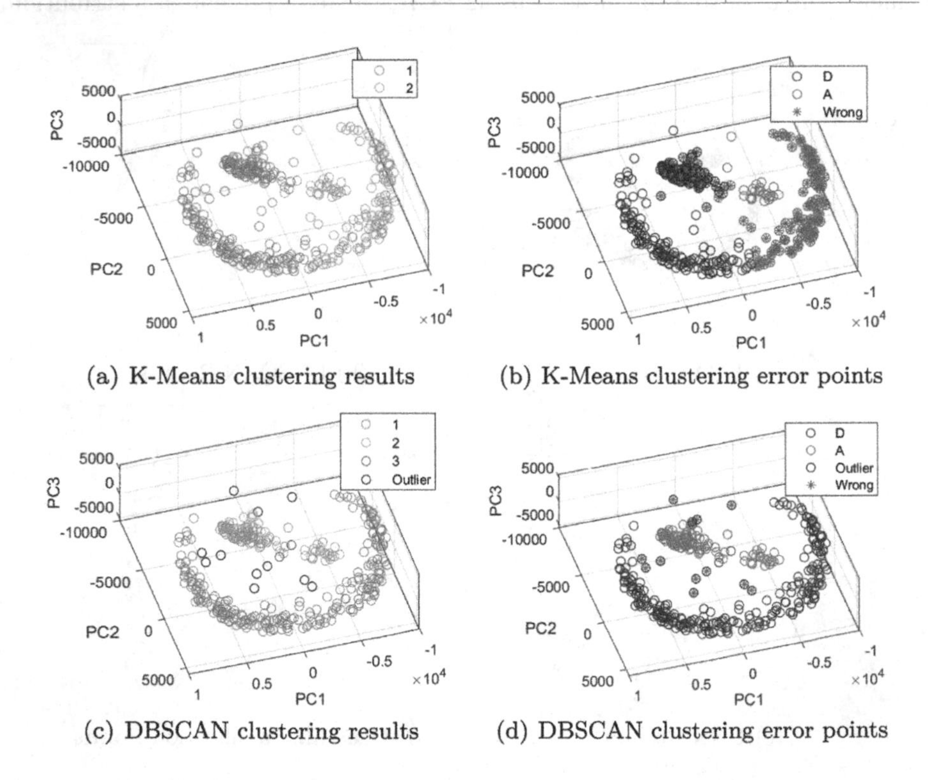

(a) K-Means clustering results
(b) K-Means clustering error points
(c) DBSCAN clustering results
(d) DBSCAN clustering error points

Fig. 7. Comparative Experiment

4.3 Comparisons

Finally, we also compare DBSCAN-CNN with classical clustering SCA methods K-Means and DBSCAN.

K-Means result is shown in Fig. 7(a), points are directly divided into two clusters from the middle.

As shown in Fig. 7(b), there are many misclassified points. DBSCAN result is shown in Fig. 7(c) with parameters as $\varepsilon = 2200$ and $\theta = 2$. As shown in Figure 7(d), there are few misclassified points. We repeated the experiment 8 times: the average number of K-Means errors is 165 and the average number of DBSCAN errors is 13. It can be found from the comparative experiments that under the same dataset, the accuracy of K-Means is 55.65%, that of DBSCAN is 96.51%, and that of DBSCAN-CNN is 98.92%. The method proposed in this paper is better than single clustering SCA method when there are outliers during clustering.

In Table 2, we list error numbers, error rate and time consumption for three methods. Compared with classical clustering SCA, DBSCAN-CNN always has fewer misclassified points number, and the misclassification rate is reduced by at least 69.23%, which confirms that our proposed method is effective for clustering with outliers. Our method takes more time than classical methods because this time includes the training duration of the CNN model. But it's still within an acceptable range. Overall, if an attacker use our proposed method, he can complete the correction of errors and eventually recover secret information with less brute force complexity while spending some time.

Table 2. Average Error Nums, Error Rate and Time Consumption

Method	Error nums	Error rate	Time consumption
K-Means	165	0.4435	0.451695 s
DBSCAN	13	0.0349	0.403635 s
DBSCAN-CNN	4	0.0101	303.97 s

5 Conclusion

This paper proposed a new SCA method for public-key cryptosystems in the scenario, which used deep learning to recover outliers in clustering results. As mentioned before, the accuracy of clustering results was limited by abnormal (but correct) data, so attackers needed to process these abnormal data. The method we proposed can extract correct features contained in abnormal data and label these data correctly. Among classical clustering SCA methods, K-Means can't distinguish outliers, DBSCAN can only distinguish outliers but can't process them. Our proposed DBSCAN-CNN trained CNN on normal data and tested on abnormal data to improve the classification accuracy of abnormal data. The experimental results showed that compared with classical clustering SCA, our method reduced error rate of clustering results by at least 69.23%.

Acknowledgement. This work is supported by National Key R&D Program of China (Nos. 2022YFB310 3800, 2021YFB3101500), National Natural Science Foundation of China (Nos. 62272047, 62002021), Beijing Institute of Technology Research Fund Program for Young Scholars, and Cryptographic Application Industry Chain Supply and Demand Docking Platform of New Energy and Intelligent Connected Vehicle Industry (2021-0181-1-1).

References

1. Basora, L., Olive, X., Dubot, T.: Recent advances in anomaly detection methods applied to aviation. Aerospace **6**(11), 117 (2019)
2. Carbone, M., et al.: Deep learning to evaluate secure RSA implementations. IACR Trans. Cryptograph. Hardw. Embed. Syst. **2019**(2), 132–161 (2019)
3. Gai, K., Guo, J., Zhu, L., Yu, S.: Blockchain meets cloud computing: a survey. IEEE Commun. Surv. Tutor. **22**(3), 2009–2030 (2020)
4. Gai, K., Wu, Y., Zhu, L., Zhang, Z., Qiu, M.: Differential privacy-based blockchain for industrial internet-of-things. IEEE Trans. Industr. Inf. **16**(6), 4156–4165 (2019)
5. Roche, T., Imbert, L., Lomné, V.: Side-channel attacks on blinded scalar multiplications revisited. In: Belaïd, S., Güneysu, T. (eds.) CARDIS 2019. LNCS, vol. 11833, pp. 95–108. Springer, Cham (2020). https://doi.org/10.1007/978-3-030-42068-0_6
6. Kocher, P., Jaffe, J., Jun, B.: Differential power analysis. In: Wiener, M. (ed.) CRYPTO 1999. LNCS, vol. 1666, pp. 388–397. Springer, Heidelberg (1999). https://doi.org/10.1007/3-540-48405-1_25
7. Kwon, D., Kim, H., Hong, S.: Non-profiled deep learning-based side-channel preprocessing with autoencoders. IEEE Access **9**, 57692–57703 (2021)
8. Lerman, L., Poussier, R., Markowitch, O., Standaert, F.X.: Template attacks versus machine learning revisited and the curse of dimensionality in side-channel analysis: extended version. J. Cryptogr. Eng. **8**(4), 301–313 (2018)
9. Nascimento, E., Chmielewski, Ł: Applying horizontal clustering side-channel attacks on embedded ECC implementations. In: Eisenbarth, T., Teglia, Y. (eds.) CARDIS 2017. LNCS, vol. 10728, pp. 213–231. Springer, Cham (2018). https://doi.org/10.1007/978-3-319-75208-2_13
10. Perin, G., Chmielewski, Ł., Batina, L., Picek, S.: Keep it unsupervised: horizontal attacks meet deep learning. IACR Trans. Cryptogr. Hardw. Embed. Syst, **2021**(1), 343–372 (2021)
11. Picek, S., Perin, G., Mariot, L., Wu, L., Batina, L.: Sok: Deep learning-based physical side-channel analysis. Cryptology ePrint Archive (2021)
12. Ram, A., Jalal, S., Jalal, A.S., Kumar, M.: A density based algorithm for discovering density varied clusters in large spatial databases. Int. J. Compu. Appl. **3**(6), 1–4 (2010)
13. Ravi, P., Jungk, B., Bhasin, S.: Single trace electromagnetic side-channel attacks on fpga implementation of elliptic curve cryptography. In: 2019 Joint International Symposium on Electromagnetic Compatibility, Sapporo and Asia-Pacific International Symposium on Electromagnetic Compatibility (EMC Sapporo/APEMC), pp. 1–4. IEEE (2019)
14. Tang, M., et al.: Side-channel attacks in a real scenario. Tsinghua Sci. Technol. **23**(5), 586–598 (2018)

Secure File Outsourcing Method Based on Consortium Blockchain

Xuan You[1], Changsong Zhou[1], Zeng Chen[2], Yu Gu[1], Rui Han[1], and Guozi Sun[1](✉)

[1] School of Computer Science, Nanjing University of Posts and Telecommunications, Nanjing 210023, Jiangsu, China
sun@njupt.edu.cn

[2] Jiangsu Cancer Hospital, Jiangsu, China

Abstract. With the development of the big data era, the amount of data has entered an explosive growth phase. Limited by the constraints of cost, efficiency, and security of self-built storage systems, enterprises are forced to outsource files to cloud storage systems. However, the lack of file security and auditability in cloud storage systems continues to threaten the security of outsourced files. This paper designs and implements BFSOut, which is a secure file outsourcing method based on consortium blockchain. It uses Hyperledger Fabric and Interplanetary File System (IPFS) as the underlying storage engine, which solves the problem of cloud storage security issues. In BFSOut, in order to ensure the security of outsource files, the client-side offline block encryption is used. Furthermore, a dynamic hybrid encryption scheme is adopted, making the overall encryption effect more Efficient. Experimental performance analysis show that the system has good performance.

Keywords: Secure file outsourcing · Hybrid encryption · Hyperledger Fabric

1 Introduction

With the rapid development of big data and cloud computing, more and more enterprises are using cloud storage services. Therefore, a large amount of data in the enterprise is stored in the cloud platform, such as employee record sheets, company financial data, and so on. When sharing files, users only need to share the link through the client software provided by the cloud platform, such as Dropbox and Google Drive. However, this way of outsourcing data faces great security threats and privacy leaks [1,2]. The user loses control of a file when it is uploaded to the cloud storage platform in plaintext. Although the platform provides the function of deleting the file, this deletion is executed in a soft method,

This work was supported by the National Natural Science Foundation of China (No. 61906099), the Open Fund of Key Laboratory of Urban Land Resources Monitoring and Simulation, Ministry of Natural Resources (No. KF-2019-04-065).

M. Qiu et al. (Eds.): SmartCom 2022, LNCS 13828, pp. 371–380, 2023.
https://doi.org/10.1007/978-3-031-28124-2_35

as the automatic backup mechanism of the cloud platform will automatically retain a copy of the file. Which is worse, cloud service providers may leak user data to unauthorized entities to obtain illegal profits. In addition, third-party cloud platforms also have the problem of opaque auditing and the inability to guarantee integrity [3].

For the problems existing in a cloud platform, blockchain technology is an ideal solution [4,5]. Blockchain is a sharable database, and participants in different networks, sharing data by synchronizing ledgers. The ledger has the characteristics of traceability, decentralization, and immutable, which promotes credible interaction in an untrusted environment [6,7]. The blockchain uses smart contracts to execute the transaction process. Smart contracts are automatically executable codes that control operations according to contracts or agreements. It has transaction logic that can control the entire business life cycle [8]. Through the use of smart contracts, transactions on the blockchain can be fully automated and run without any human intervention. The use of blockchain technology [9] in the file-sharing system can make the file-sharing process more transparent, and its immutable feature ensures the integrity and traceability of the data. There are many methods to establish a blockchain network. Public blockchains require a lot of computing power to ensure the fairness of transactions. The privacy of transactions is extremely low or there is even no privacy at all [10], which are important considerations for enterprises. The consortium blockchain is controlled by multiple organizations, and these organizations can share the responsibility of maintaining the blockchain. When all participants need to obtain permission and are responsible for the blockchain, the consortium blockchain is the ideal choice [11]. In addition, the consortium blockchain has higher throughput and lower transaction delay, which can improve the overall performance of the system.

From the perspective of file itself, a feasible way to ensure the security of the sensitive information is to use some encryption scheme for encrypting before uploading the file to an untrusted server, while owner of the file is responsible for it. The security strength of the file depends on the effectiveness of the encryption key and the complexity of the encryption algorithm [12]. Common systems use a dynamic encryption method to encrypt all files in a unified manner as file is unstructured data.

Our Contributions. The main contributions of this paper are demonstrated as follows:

1. This paper designs a file security outsourcing method based on consortium blockchain, to ensure the balance of file security and efficiency in the process of file outsourcing. It also explains the process of file outsourcing.
2. In the process of file outsourcing, a dynamic hybrid encryption scheme based on the combination of symmetric and asymmetric encryption algorithms to provides fine-grained protection for a single file.
3. For each operation step of the BFSOut system, its performance and consumption are tested, and the block generation of blockchain is tested.

Paper Organization. The remainder of this paper is structured as follows. The second part reviews some work related to the method. The third part explains the specific details. The fourth part tests the method from multiple angles. Finally, the conclusion is in the fifth part.

2 Related Work

2.1 Distributed Blockchain Storage Systems

Azaria et al. developed a system called MedRec [13], which effectively manages and stores medical records based on blockchain. Using unique blockchain properties, MedRec can also be used for identity verification, accountability, and data sharing. Li Z et al. designed a distributed peer-to-peer network architecture based on the blockchain protocol [14] to ensure the security and stability of cloud data transmission and sharing. Moses et al. aiming at the problem that fingerprint templates are vulnerable to security attacks due to their asymmetry [15], proposed to protect encrypted fingerprint templates through the symmetric peer-to-peer network and symmetric encryption.

2.2 Data Outsoucing and Sharing Based on Blockchain

Li et al. designed a medical data fusion distributed privacy management system based on Fabric and IPFS [16] to overcome the disadvantages of traditional hospital data separation and solve the problems of electronic medical record data sharing and tracking difficulties. Tang et al. designed a data storage model based on Ethereum [17]. The purpose is to use smart contracts to implement some back-end service functions, ensuring the transparency, openness, and traceability of services while effectively resisting DDoS attacks. Shen et al. proposed a secure access control scheme based on CP-ABE [18] to solve the problems of cloud storage decentralization and data sharing.

3 Secure File Outsourcing Method

In this section, we will first introduce BFSOut's secure file outsourcing model in detail. Then we will demonstrate the dynamic hybrid encryption model.

3.1 Secure File Outsourcing Model

This model designs a file outsourcing method with decentralized features based on the consortium blockchain, which allows data owners to upload or share files to any address that exists on the blockchain, while the file receiver can safely download the files in their address. Its model is shown in Fig. 1.

In the file outsourcing model, it is mainly divided into the sender, receiver, dynamic hybrid encryption and decryption algorithm, smart contract and Fabric and IPFS network. The main process of this model is listed as follows:

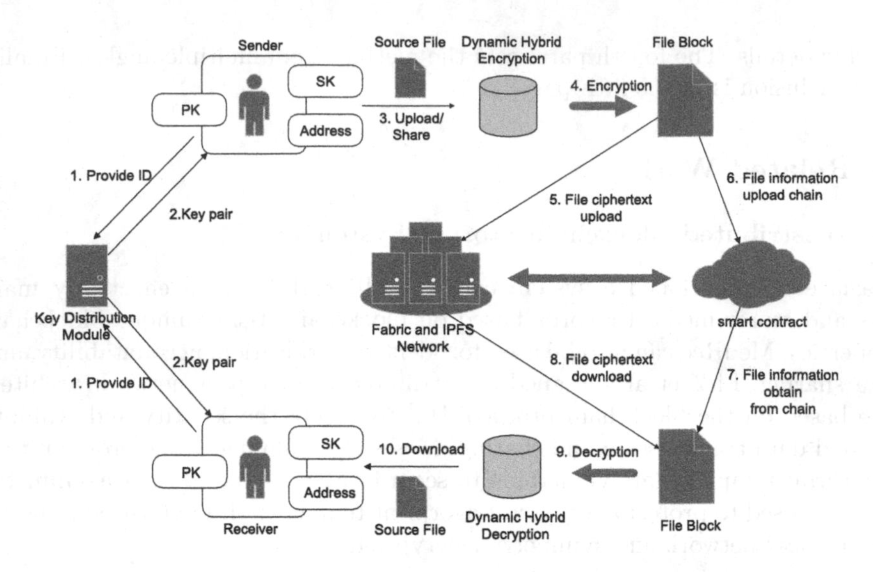

Fig. 1. Secure file outsourcing model

1. **System initialization.** The sender and receiver use the BFSOut client to generate a unique ID (step 1) and obtain the user's public and private key pair through the built-in key distribution module of the client (step 2). Then the public key is converted into a 272-bit address by the address conversion algorithm. At the same time, the public key and address are stored in the Fabric platform through a smart contract.
2. **File upload.** In the file upload process, the sender first initiates a file upload and sharing request (step 3). The client generates file blocks using the dynamic hybrid encryption algorithm (step 4). The file block is divided into file stream ciphertext(f_{sc}) and file digest ciphertext(f_{dc}). The f_{sc} is stored in the IPFS network in the form of slices (step 5). The f_{dc} should be written to the receiver's address through a smart contract and is permanently stored as a file record (step 6).
3. **File sharing.** The smart contract automatically determines whether the address of the file upload is the receiver's address. If the address is confirmed to be correct, the smart contract will write the f_{dc} to the corresponding receiver's address.
4. **File download.** The receiver confirms that the sender shared the file by checking its address. The smart contract will be called first to obtain the information of this file (step 7). Then download the f_{sc} through the file Hash to the IPFS network (step 8). Next, the original file is decrypted through the decryption module (step 9). Finally, the system will check the validity and correctness of the file, download and save the file in the local disk (step 10).

3.2 Dynamic Hybrid Encryption Model

In the hybrid encryption scheme, the symmetric encryption algorithm is AES, and the asymmetric encryption uses the ECC. Symmetric encryption is usually faster than asymmetric encryption, but its security is lower, so they are often used in combination. The method uses AES encrypt real data, and ECC to encrypt parameters such as the key to the symmetric encryption algorithm. This method is more efficient in encryption, and the key is easier to manage. To enhance the file security level, we take the block-based encryption method. The file is divided into several blocks of the same size, while each block is responsible for different encryption processing.

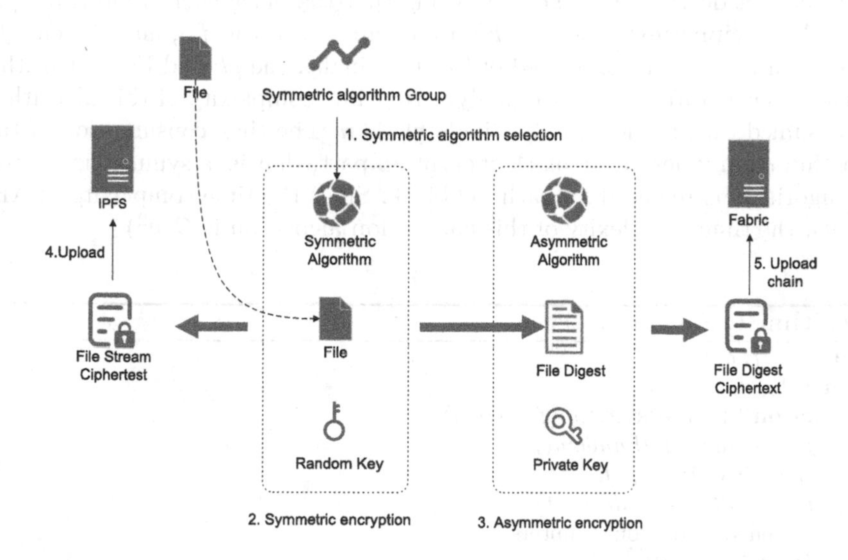

Fig. 2. Dynamic hybrid encryption model

When a file is uploaded through the client program, the client's encryption module will be triggered. To prevent file transmission from being stolen, the file encryption and decryption process are completely offline. The entire encryption process is shown in Fig. 2. When file encryption is performed, it is first choose symmetric algorithm from Symmetric algorithm group (step 1). Then the system will generate a random symmetric key, use the key to perform symmetric encryption on the source file to generate the file stream ciphertext f_{sc} (step 2), and then the model will generate the file digest ciphertext f_{dc} by asymmetric encryption with some parameters, including symmetric encryption algorithm, symmetric key and random vector in block encryption (step 3). Finally, the model will perform a persistence operation, upload thef_{sc} to the IPFS server for persistent storage (step 4), and store the f_{dc} in the Fabric through a smart contract (step 5). The algorithm description is as follows.

Algorithm 1 describes the encryption process of this method. First, an original encryption key f_{s_k} is randomly generated when the encryption module is invoked after the file (f) is read. Then,f_{s_k} and f_{id} are combined into file encryption key f_k by Hash-based Message Authentication Code ($HMAC$). To prevent the key from being tampered with, the algorithm used by $HMAC$ is $HMAC - MD5$. In the block encryption process, each file block (B) will firstly generate a unique counter (ctr). Then f_{id} and ctr_i generate a unique IV_i vector associated with the file block through the XOR function. Then it uses the encryption algorithm (em) to encrypt the file block to get the file ciphertext block (CB_i). The purpose is to prevent the same or similar file blocks from being encrypted into the same ciphertext output, so that if an attacker decrypts one file, he cannot decrypt the others, ensuring that the block encryption is unique. Next, all file ciphertext blocks (CB) are composed of the f_{sc}, and the em, f_k, and ctr parameters are composed of the f_d. Finally, the pk and ECC algorithm are used to generate the f_{dc}. To analyze the time complexity of this algorithm, it is assumed that the length of a single file is n. The time consumption of this algorithm mainly lies in the block encryption part, that is, a symmetric encryption algorithm is required for each file block. Since the time complexity of AES is $O(n)$, the time complexity of this encryption algorithm is $O(n^2)$.

Algorithm 1. File Encryption

input: f_{id}, f, em
output: f_{dc}, f_{sc}
1: **function** FILEENCRYPTION(f_{id}, f, em)
2: $f_{s_k} \leftarrow GenerateRandom()$
3: $f_k \leftarrow HMAC(f_{s_k}, f_{id})$
4: $B \leftarrow Call\ CreatBlocks(f)$
5: //Generate random counters
6: **for** $\forall B_i \in B$ **do**
7: $ctr_i \leftarrow GenerateRandom()$
8: **end for**
9: //Blocks encryption using encryption method
10: **for** $\forall B_i \in B$ **do**
11: $IV_i \leftarrow f_{id} \bigoplus ctr_i$
12: $CB_i \leftarrow Encrypt(em(B_i, f_k, IV_i))$
13: **end for**
14: $f_{sc} := (CB_1, ..., CB_n)$
15: $f_d := (em, f_k, ctr)$ // generate file digest
16: $f_{dc} \leftarrow Encrypt(ECC(fd, pk))$ // Keys encrytion using ECC
17: **return** f_{dc}, f_{sc}
18: **end function**

The decryption process is reversed. Algorithm 2 describes the decryption steps of the file. The user takes the private key (sk) to decrypt f_{dc} into related parameters, and then uses these parameters to decrypt the file block CB and

Algorithm 2. File Decryption

```
input: f_id, f_sc, f_dc
output: f
 1: function FileDecryption(f_id, f_sc, f_dc)
 2:     //Keys decryption using ECC
 3:     em, f_k, ctr ← Decrypt(ECC(f_dc, sk))
 4:     //Blocks decryption using encryption method
 5:     for ∀B_i ∈ B do
 6:         IV_i ← f_id ⊕ ctr_i
 7:         B_i ← Decrypt(em(CB_i, f_k, IV_i))
 8:     end for
 9:     f := (B_1, ..., B_n)
10:     return f
11: end function
```

compose the file. Again, since decryption requires one decryption for each file block, all time complexity is $O(n^2)$.

4 Experimental Evaluation

In this section, we analyze and test BFSOut's block generation and the system performance is tested.

4.1 Blockchain Block Analysis

The paper uses Fabric 2.0, the time for a single block generation is adjusted to 0.1 s, the maximum number of transactions in a block is 10, and the smart contract is written in the Go language. In the BFSOut system, there are system initialization phases, file upload phases, and file sharing phases that will generate blocks. The public key and address will be uploaded during the system initialization phase, and a block will be generated in this phase. The file upload stage will generate the f_{dc} and Hash, and at the same time generate the encryption related parameters. The information of the file itself will generate a block, and the storage of other parameters will generate a block. So the stage will generate 2 blocks. During the file-sharing phase, the parameters are rewritten into the receiver address. So the file-sharing process will generate a blocks.

The blockchain system in BFSOut builds a pseudo-cluster composed of two organizations and a single channel as a test environment. Each organization has two Peer nodes. The blockchain system is initialized into 7 blocks, including 1 genesis block, 2 update anchor node blocks, and 4 instantiation chain code blocks.

4.2 System Performance Test

The performance test of BFSOut is to test the upload and download performance of various files of different sizes through experiments. The experiment is carried out in an experimental environment composed of 3 computers. One of the machines is used as the server of IPFS and Fabric, and the other two machines take the responsibility of the file sender and receiver respectively. The server machine is running on the Centos 7.9 system and the client machine is running on Windows 10 system. The server machine is configured with Intel Xeon E-2225G processor, the IPFS uses version 0.4.23, and all components of the client are written using Go1.14.4.

The execution time is measured by considering the entire process of uploading and sharing on BFSOut. The test file uses a data set made of text and compressed files, and the data size ranges from 1 M to 1000 M. The result of each experiment is the average of 10 runs to reduce the randomness of a single test. The size of the page cache is one page, usually 4 K. When linux reads and writes a file, it is used to cache the logical content of the file, thus speeding up access to the disk's images and data. In order to ensure the accuracy of the results obtained, the page cache is refreshed between each test.

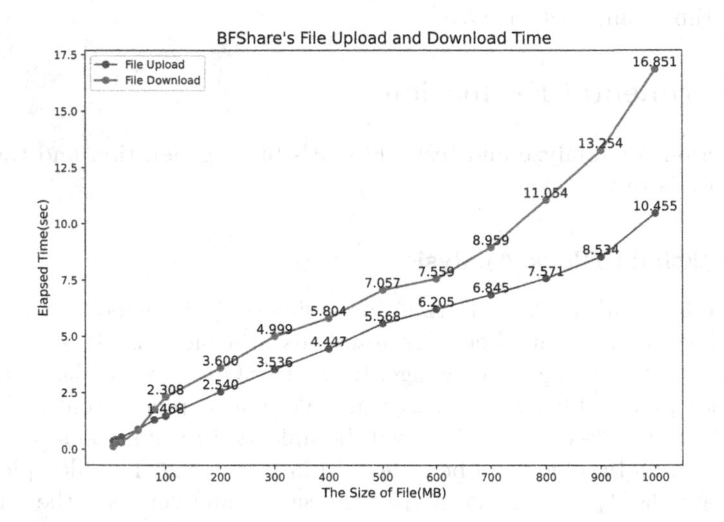

Fig. 3. BFSOut file upload and download time consumption

First, we evaluate the performance of BFSOut in uploading and downloading files of different sizes. Figure 3 shows a comparison of the time taken to write and read files to the server using the BFSOut. This performance data includes the entire file operation process, that is, file hybrid encryption and decryption, file storage and download, and related information insertion and query on the blockchain during file reading and writing. In the whole process, the time-consuming file upload and download increase linearly with the file size. When the file is less than 100M, the uploading time is slightly longer than the downloading time. This is mainly because the file uploading process will generate new blocks and increase the time consumption. In general, the download time

Table 1. BFSOut file upload and download time consumption (sec)

Method	File upload				File download			
	100 MB	200 MB	500 MB	1000 MB	100 MB	200 MB	500 MB	1000 MB
Read/Write	0.063	0.212	0.222	0.839	0.002	0.148	0.325	0.558
File Enc/Dec	0.189	0.365	0.912	1.804	0.161	0.326	0.782	1.799
Keys Enc/Dec	0.046	0.085	0.167	0.316	0.039	0.072	0.153	0.307
Query/Insert	0.276	0.261	0.274	0.216	0.069	0.062	0.069	0.074
IPFS Up/Down	0.894	1.617	3.993	7.280	2.037	2.992	5.728	14.113
Sum	1.468	2.540	5.568	10.455	2.308	3.600	7.057	16.851

is higher than the upload process, and the gap becomes more obvious as the file size increases.

Table 1 shows the measured time of each process during uploading, downloading, and file sharing using BFSOut. Among them, the file encryption algorithm uses AES, and the asymmetric encryption uses the ECC-512. From the table, it can be concluded that most of the time taken for file upload and download from IPFS and file encryption, the proxy re-encryption process and the asymmetric encryption process also linearly increase with the size of the file.

5 Conclusion

This paper introduced BFSOut, a secure file outsourcing system based on the blockchain Hyperledger Fabric. It was mainly to guaranteed the security and efficiency of outsourcing files to third-party untrusted servers and file sharing. On the premise of ensuring file security, BFSOut proposed a dynamic hybrid encryption scheme based on symmetric and asymmetric encryption algorithms to made the overall encryption effect more efficient in the process of file outsourcing.

In future work, this paper intended to use machine learning to model the dynamic encryption algorithm selection part to made the algorithm selection more accurate [20]. And explored the applicability of parallel encryption or file stream encryption in BFShare to further improved the encryption efficiency and resource consumption of the system.

References

1. Sanchez-Gomez, A., Diaz, J., Hernandez-Encinas, L., Arroyo, D.: Review of the main security threats and challenges in free-access public cloud storage servers. In: Daimi, K. (ed.) Computer and Network Security Essentials, pp. 263–281. Springer, Cham (2018). https://doi.org/10.1007/978-3-319-58424-9_15
2. Qiu, M., Gai, K., Xiong, Z.: Privacy-preserving wireless communications using bipartite matching in social big data. Futur. Gener. Comput. Syst. **87**, 772–781 (2018)
3. Gan, Q., Wang, X., Li, J., Yan, J.: Enabling online/offline remote data auditing for secure cloud storage, Cluster Comput. **24**, 3027-3-041 (2021)

4. Stodt, J., Schönle, D., Reich, C., Ghovanlooy Ghajar, F., Welte, D., Sikora, A.: Security audit of a blockchain-based industrial application platform. Algorithms **14**(4),121 (2021)
5. Gai, K., Wu, Y., Zhu, L., Zhang, Z., Qiu, M.: Differential privacy-based blockchain for industrial internet-of-things. IEEE Trans. Industr. Inf. **16**(6), 4156–4165 (2019)
6. Aitzhan, N.Z., Svetinovic, D.: Security and privacy in decentralized energy trading through multi-signatures, blockchain and anonymous messaging streams. IEEE Trans. Dependable Secure Comput. **15**(5), 840–852 (2016)
7. Sun, C., Huang, C., Zhang, H., Chen, B., An, F., Wang, L., Yun, T.: Individual tree crown segmentation and crown width extraction from a heightmap derived from aerial laser scanning data using a deep learning framework. Front. Plant Sci. **13** (2022)
8. Nasir, M.H., Arshad, J., Khan, M.M., Fatima, M., Salah, K., Jayaraman, R.: Scalable blockchains-a systematic review. Futur. Gener. Comput. Syst. **126**, 136–162 (2022)
9. Gai, K., Guo, J., Zhu, L., Yu, S.: Blockchain meets cloud computing: a survey. IEEE Commun. Surv. Tutor **22**(3), 2009–2030 (2020)
10. Ghesmati, S., Fdhila, W., Weippl, E.: Studying bitcoin privacy attacks and their impact on bitcoin-based identity methods. In: González Enríquez, J., Debois, S., Fettke, P., Plebani, P., van de Weerd, I., Weber, I. (eds.) BPM 2021. LNBIP, vol. 428, pp. 85–101. Springer, Cham (2021). https://doi.org/10.1007/978-3-030-85867-4_7
11. Aggarwal, S., Kumar, N.: Chapter sixteen-hyperledger. Adv. Comput. **121**, 323–343 (2021)
12. Sinha, K., Priya, A., Paul, P.: K-RSA: secure data storage technique for multimedia in cloud data server. J. Intell. Fuzzy Syst. **39**(3), 3297–3314 (2020)
13. Azaria, A., Ekblaw, A., Vieira, T., Lippman, A.: Medrec: using blockchain for medical data access and permission management. In: 2016 2nd International Conference on Open and Big Data (OBD). IEEE, pp. 25–30 (2020)
14. Li, Z., Barenji, A.V., Huang, G.Q.: Toward a blockchain cloud manufacturing system as a peer to peer distributed network platform. Rob. Comput. Integr. Manuf. **54**, 133–144 (2018)
15. Acquah, M.A., Chen, N., Pan, J.-S., Yang, H.-M., Yan, B.: Securing fingerprint template using blockchain and distributed storage system. Symmetry **12**(6), 951 (2020)
16. Li, L., Yue, Z., Wu, G.: Electronic medical record sharing system based on hyperledger fabric and interplanetary file system. In: 2021 the 5th International Conference on Compute and Data Analysis, pp. 149–154 (2021)
17. Tang, X., Guo, H., Li, H., Yuan, Y., Wang, J., Cheng, J.: A DAPP business data storage model based on blockchain and IPFS. In: Sun, X., Zhang, X., Xia, Z., Bertino, E. (eds.) ICAIS 2021. LNCS, vol. 12737, pp. 219–230. Springer, Cham (2021). https://doi.org/10.1007/978-3-030-78612-0_18
18. Lai, J., Deng, R.H., Li, Y.: Fully secure cipertext-policy hiding CP-ABE. In: Bao, F., Weng, J. (eds.) ISPEC 2011. LNCS, vol. 6672, pp. 24–39. Springer, Heidelberg (2011). https://doi.org/10.1007/978-3-642-21031-0_3
19. Shao, J., Cao, Z.: CCA-secure proxy re-encryption without pairings. In: Jarecki, S., Tsudik, G. (eds.) PKC 2009. LNCS, vol. 5443, pp. 357–376. Springer, Heidelberg (2009). https://doi.org/10.1007/978-3-642-00468-1_20
20. Qiu, H., Qiu, M., Lu, Z.: Selective encryption on ECG data in body sensor network based on supervised machine learning. Inf. Fusion **55**, 59–67 (2020)

Fabric Smart Contract Read-After-Write Risk Detection Method Based on Key Methods and Call Chains

Feixiang Ren and Sujuan Qin(✉)

State Key Laboratory of Networking and Switching Technology, Beijing University of Posts and Telecommunications, Beijing 100876, China
qsujuan@bupt.edu.cn

Abstract. Fabric is currently the most popular consortium chain platform with a modular architecture that provides high security, elasticity, flexibility and scalability. Smart contracts realize the automatic execution of transactions and the operation of reconciliation data. The Fabric platform supports general programming languages to write smart contracts. However, in the development process of smart contracts, due to insufficient understanding of the underlying operating logic of smart contracts, developers are prone to introduce some risky operations, resulting in a mismatch between the execution logic of smart contracts and business logic, resulting in a lot of losses. The read-after-write risk is a relatively complex and common security risk in smart contracts. Currently, many detection tools cannot detect this risk. There is an urgent need for a solution that can quickly and accurately detect the read-after-write risk in smart contracts. This paper proposes a static analysis smart contract read-after-write risk detection method based on key methods and call chains. The scheme extracts key method patterns on the abstract syntax tree, identifies and locates key methods with risks, greatly reduces the interference of useless nodes on detection, and realizes rapid detection. By constructing the key method call chain, the real call scene is restored according to the call type and attribute of the key method. After experimental verification, compared with the current popular smart contract risk detection tool Revive^CC, the tool proposed in this paper has higher detection accuracy and can more accurately locate the read-after-write risk in smart contracts.

Keywords: Fabric Blockchain · Smart Contract · Static Analysis · Read-after-Write Risk

1 Introduction

Fabric is an open-source permissioned blockchain platform founded by the Linux Foundation. Since its inception, the Fabric platform has dominated enterprise-level blockchain deployments, at least 13 companies with over 1 billion dollars in revenue have built their strategic net-works on Fabric platform, far more than any other technology or platform in this type of deployment [1]. A wide range of applications in the Internet of Things

M. Qiu et al. (Eds.): SmartCom 2022, LNCS 13828, pp. 381–392, 2023.
https://doi.org/10.1007/978-3-031-28124-2_36

[2], supply chain [3], privacy data [4] and other fields have been deployed on the Fabric platform.

Smart contract is a digital agreement that uses algorithms and programs to formulate contract terms, which is deployed on the blockchain, and can be automatically executed according to the rules [5]. Smart contracts allow for trusted transactions without third parties that are traceable and irreversible. Smart contracts are generally used to manage digital assets, have extremely high economic value [6], and are also the most vulnerable to attack. A research report pointed out [7] that the smart contract layer is the weakest link in the blockchain system and one of the important incentives for blockchain security incidents in recent years. The most representative case is TheDao Attack in 2016 [8], where attackers used a reentrancy vulnerability [9] on smart contracts to steal 3.64 million ETH, causing a loss of about $60 million.

Fabric provides a trusted execution environment for smart contracts that can be written in general-purpose programming languages [10]. The digital assets in Fabric are stored in the state database, and are mainly updated in an appending manner through the read-write interface of the smart contract [11], so there are a large number of read-write operations in the smart contract. However, as a distributed system, Fabric does not meet read and write consistency [12], so there is a risk of read-after-write. The read-after-write risk specifically refers to the fact that the data can-not be read immediately after a data is successfully written during the operation of the smart contract, and the value observed at this time is still the value that was not updated before writing. The read-after-write risk brings severe challenges to the business on the Fabric blockchain. In the process of processing assets, the operations performed may not match the actual assets, resulting in serious errors in business logic. At present, the tools for the read-after-write risk detection of Fabric smart contracts have the shortcomings of low detection accuracy and single detection scene, and cannot cover a large number of read and write logic in actual projects. Therefore, it is urgent to find a solution that can detect the risk of write-after-write more comprehensively. This paper uses static analysis technology to implement a read-after-write risk detection method based on key methods and key method call chains, and through experimental verification, the detection method proposed in this paper makes up for the shortcomings of current tools for read-after-write risk detection.

The rest of the paper is organized as follows. Section 2 introduces the related work of smart contract security risk detection. Section 3 analyzes the execution process of the Fabric smart contract and the principle of the read-after-write risk. Section 4 presents a read-after-write risk detection scheme based on key methods and method call chains. Section 5 conducts relevant experiments on the scheme of this paper, and provides a comparison between the scheme of this paper and the Revive^CC [13] tool. Finally, Sect. 6 draws our conclusions.

2 Related Work

At present, the static detection work for Fabric smart contracts mainly focuses on feature matching. By analyzing the smart contract source code or intermediate code form, fast matching is realized on the basis of a predefined feature library to detect whether there is a corresponding smart contract. Security risks. In 2019, K. Yamashita et al. [14]

proposed to form a signature library through static analysis of smart contracts to detect security risks in smart contracts. This scheme converts the source code of the smart contract into an abstract syntax tree, and preliminarily determines whether there is a security risk by detecting whether the package and the defined variables introduced by the smart contract are related to the defined risk. It is a coarse-grained detection scheme. P. Lv, Y. Wang, et al. [15] followed the idea of K. Yamashita et al., and refined the process of feature matching into three parts: packet detection, instruction detection, and logic detection. The package detection module detects whether there is a package that can introduce random sources through package dependency analysis. The instruction detection module mainly detects the existence of global variables and function calls that can cause security risks based on the form of intermediate code. The logic detection module detects by analyzing the function call relationship. Whether there is logic for read after write risk. Then, a signature database is formed according to the static code features of the three security risks, and the type and location of the risks are determined by matching the signature database. Since this scheme loads the entire abstract syntax tree during static analysis, the overall detection efficiency is low. Revive^CC is an open source detection tool for the security risks of Fabric smart contracts, which is widely used in the community. Revive^CC can support multiple security risk detection, and has independent detection logic for each smart contract security risk. Unfortunately, Revive^CC is designed with a single function in a smart contract as the detection target, which leads to certain limitations in the detection of smart contract security risks spread across functions.

Based on the above analysis, it can be seen that the existing smart contract security risk detection schemes and tools cannot detect the read-after-write risk well, and have the problems of low detection accuracy and single detection scenario.

3 Read-After-Write Risk on Fabric

3.1 Read-After-Write Risk Root Cause Analysis

The read-after-write risk refers to the fact that the data cannot be read immediately after a data is successfully written during the operation of the smart contract, and the value observed at this time is still the value that was not updated before writing.

The generation of read-after-write risk is closely related to the Fabric transaction process. In Fabric, transactions follow the execute-order-validate phases. Any operation that writes transaction data on the blockchain to the ledger database is performed after the transaction is completed and the result of the transaction is verified. To maintain data consistency, Fabric does not support read-after-write operations which means that in the same transaction of Fabric, the data updated recently cannot be read.

As shown in Fig. 1, assuming that the initial state of the ledger database is {"key":"0"}. In line 6 of the code, the function reads after writing to update the value of the key to "1" by calling the write method, and then in line 8 of the code attempts to obtain this updated value by calling the read method. However, because the two actions of reading and writing are in the execution stage of the same transaction, the update operation of the ledger database will not occur during the period, and the value of the key will become "1" only after the transaction is committed, so the return value obtained at this time is the original state. Therefore, there is a risk of inconsistency in read and write, resulting in errors in business logic.

```
1   func ReadYourWrite(){
2       //Value of key in database befor update is 0
3       key:="key"
4       data:="1"
5
6       err:=stub.PutState(key,[]byte(data))
7       //through PutState method setting value of key to 1
8       response,_:=GetState(key)
9       //through GetState method getting value of key,cause result has not
10      //been submitted,value of key is still 0
11  }
```

Fig. 1. Read-after-write risk code sample

3.2 Read-After-Write Risk Detection Analysis

Although the existing Fabric smart contract security risk detection has achieved good results, there are still some problems. These problems mainly focus on the detection of read-after-write risks, and there are a large number of false negatives in the detection of read-after-write risks in existing solutions. Unlike other risks on Fabric smart contracts, read-after-write risks have two notable characteristics:

1. Read-after-write risk is a risk caused by both the write method and the read method. Other risks on smart contracts such as range query risk are only associated with one method.
2. Read-after-write risk has strict requirements on the execution order of the write method and the read method, that is, executing the write method first and then the read method following.

After researching the current smart contract security risk detection tools and solutions, this paper analyzes the two main factors that cause the poor performance of read-after-write risk detection:

1. Combination of multiple functions leads to read-after-write risks.
   In actual projects, it is very common to decouple the read and write operations of the database. In the process of writing business code, developers call the read and write interfaces in different functions, which makes the risk of read-after-write not easy to be detected. In Fig. 2, the function Cross calls function A and function B respectively, resulting in indirect calls to the write method and the read method, which together lead to the risk of read-after-write.
2. Special statements affect the execution order of multiple methods.
   The read-after-write risk needs to satisfy that the write method has priority over the read method in the execution order. Existing detection tools do this by comparing where the two methods are located in the code. As shown in the Fig. 3, the physical

location of the write method is before the read method. However, due to the particularity of the delayed execution keyword, the delayed execution of the code block will add a method call to the function call list. This method is always called after the function returns, which means that the write method will actually be called after the read method, so the write-after-reading risks will not arise.

```
1   func A(key,data){
2       stub.PutState(key,data)
3   } //function A through PutState method set value of key
4   func B(key){
5       stub.GetState(key)
6   } //function B through GetState method get value of key
7   func Cross(){
8       key:="key"
9       data:="data"
10      A(key,[]byte(data))
11      B(key)
12  } //function Cross indirectly call function A and function B
```

Fig. 2. Example of read-after-write risk caused by multi-functions

```
1 v func DeferFunc(){
2       key:="key"
3       data:="data"
4       defer stub,PutState(key,[]byte(data))
5       //Affected by defer keyword, PutState method execute later than GetState method
6       stub.GetState(key)
7   }
```

Fig. 3. An example of the effect of a particular statement on the read and write order

None of the existing detection schemes can well support these two scenarios, so a detection scheme is needed that can cover these two scenarios and improve the detection accuracy of read-after-write risks.

4 Detection Scheme Based on Key Method and Key Method Call Chain

Aiming at the current security risk problem of read after write in Fabric smart contracts, this paper uses the technology of static analysis to implement a detection method based on key methods and key method call chains.

4.1 Overview

The overall detection scheme consists of three parts: preprocess, identify key method, and construct key method call chain, as shown in Fig. 4. Our scheme first defines key methods, and uses pattern matching to extract features related to read-after-write risks from the abstract syntax tree form of smart contracts. Further, by summarizing the different trigger scenarios of read-after-write risk, and constructing the key method invocation chain according to the multiple attributes of the key method, to reflect the real execution sequence of the key method in the process of invoking the smart contract.

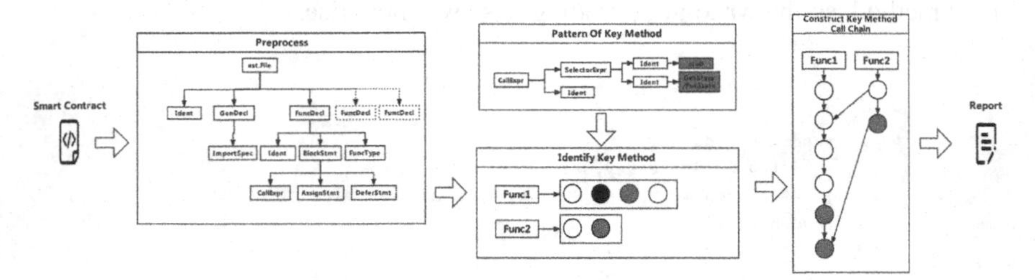

Fig. 4. Overall process of read-after-write risk detection

4.2 Preprocessing

This paper obtains rich semantic and grammatical information of smart contracts through preprocessing. Preprocessing includes two parts:

1. Convert the smart contract source code into an abstract syntax tree
   Use Golang official toolkits Scanner and Parser [16] to perform lexical and semantic analysis on smart contracts written in Golang and generate abstract syntax trees, as Fig. 5 shows.
2. Prune the abstract syntax tree
   Prune the converted abstract syntax tree including externally pruning nodes that are not related to analysis outside the body of functions such as GenDecl and its sub-nodes and internally pruning unrelated nodes such as Literal nodes inside the function body.

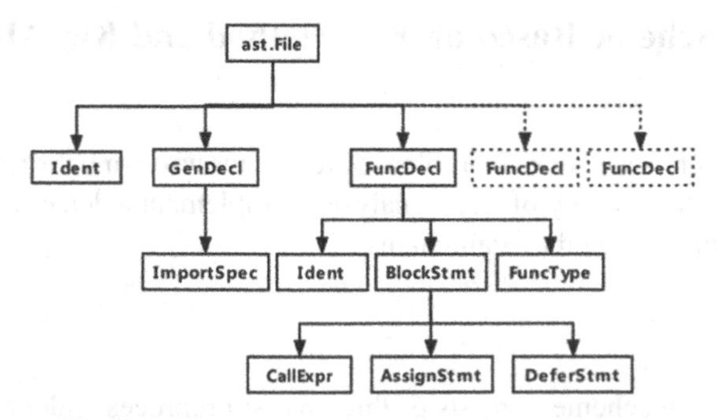

Fig. 5. Abstract syntax tree form of smart contract

4.3 Identify Key Methods

The key method is used to define the method invocation information related to the read-after-write risk on the Fabric smart contract. As shown in Table 1, the key methods defined in this paper mainly include the following attributes:

(1) Name represents the name of the key method
(2) Type indicates the key method whether under defer statement.
(3) Pos represents the position in the code where the key method is located.
(4) Param represents the parameters of the key method call.
(5) Function indicates whether the key method is a function call.

Table 1. Attributes of key methods

Attribute	Value Type	Content
Name	String	PutState, GetState or function name
Type	Int	1 represents ordinary method or 2 in special method
Pos	token.Pos	Indicates the location of key method in the source code
Param	[]byte{}	Parameters representing key method calls
Function	Bool	True means it is a function call, False is the opposite

The purpose of defining key methods is to describe the read-after-write risk in smart contracts more clearly. Identifying key methods is the first step in detecting read-after-write risks. After the abstract syntax tree converted from the smart contract source code is preprocessed, there are still a large number of redundant nodes. This paper uses pattern matching to quickly identify key methods in the abstract syntax tree. According to the different calling methods, the key methods can be divided into two categories:

1. Direct call
Direct call refers to the read and write methods that are directly called in the function. For the key method of directly calling the class, there is always a three-level sub-tree structure on the abstract syntax tree as shown in Fig. 6, which is called the key method pattern in this paper. The three-layer sub-tree structure is embodied as follows: the root node is the CallExpr node, the successor node is the SelectorExpr node, and the successor nodes of the SelectorExpr node are two Ident nodes, the specific content is stub and PutState or GetState, which means PutState or GetState two read and write methods. It can be seen that the type and structure of the abstract syntax tree nodes that can represent the read and write methods are very fixed. Key methods can be quickly identified by matching such structures in an abstract syntax tree. It should be noted that when a structure of key method is found, in order to determine whether the key method exists in the special statement, it is necessary to trace back to the parent node of the CallExpr node. If the parent node of the CallExpr node is

the DeferStmt node, it means that the key method exists in the special statement, and the Type value is set to 2.

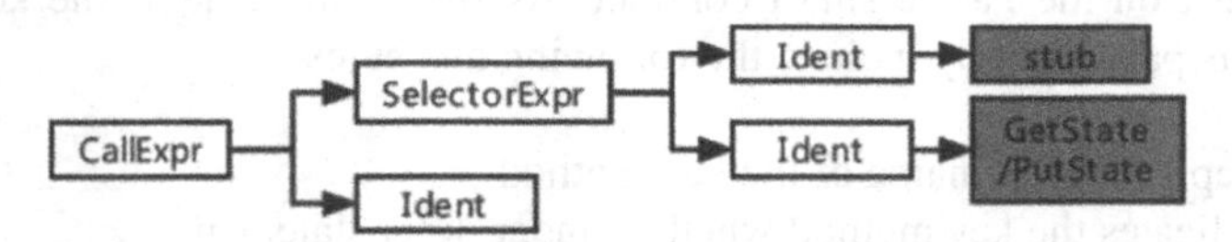

Fig. 6. Pattern of key method

2. Indirect call
   Indirect call refers to functional call. Different from the direct call, the indirect call implements the call to the read and write methods by calling the function. To identify indirect call, traverse all FuncDecl child nodes in turn starting from the root node of the abstract syntax tree, that is, traverse the first layer nodes of the abstract syntax tree (the root node is layer 0). Extract the function name from the FuncDecl function child node structure, and maintain a global list of function names. Starting from the node declared with the FuncDecl function, traverse the subtree with the FuncDecl node as the root node. During the traversal process, if the node type meets the CallExpr call expression, extract the Ident identifier in the CallExpr node. If the Ident identifier exists in the global function name list, indicating that there is a calling relationship between functions.

4.4 Construct Key Method Call Chains

The key method call chain is a one-way linked list composed of a collection of key methods, used to represent the actual calling sequence of key methods in the smart contract.

Through the identification of key methods, all the key methods in each function body are collected, including the position attribute of the key method, but there may be problems if the position attribute is simply used to construct the call chain of the key method.

Figure 7 shows the call chain constructed using only the position value. In the function Func1, according to the context of the position, a logical sequence of read-after-write is formed. However, if the writing method exists in the defer statement, due to the characteristics of the defer statement, the delayed execution will delay the execution of the writing method, so that the sequence of read-after-write is not formed. The actual execution sequence is shown in Fig. 8.

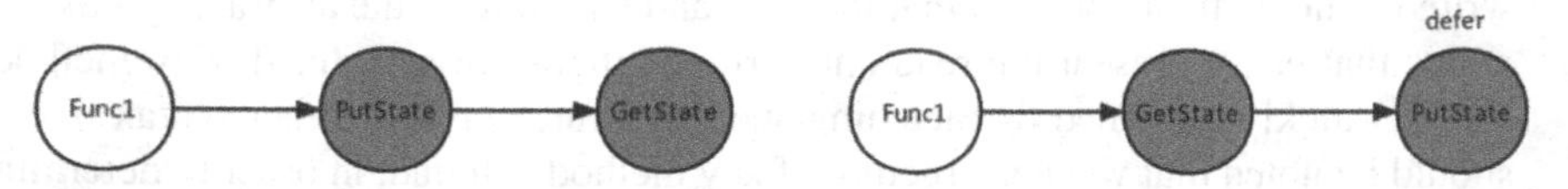

Fig. 7. Constructed by position attribute **Fig. 8.** Under defer statement

Another problem that needs to be solved is the impact of invocation within functions on the calling sequence of key methods. There may be a mutual calling relationship between functions. As shown in Fig. 9, the function body of Func1 executes the write method first, and then calls another function, Func2. In the Func2, the read method is executed. In this way, the logic of writing first and then reading will be formed through this indirect call, so it is necessary to consider the call between functions.

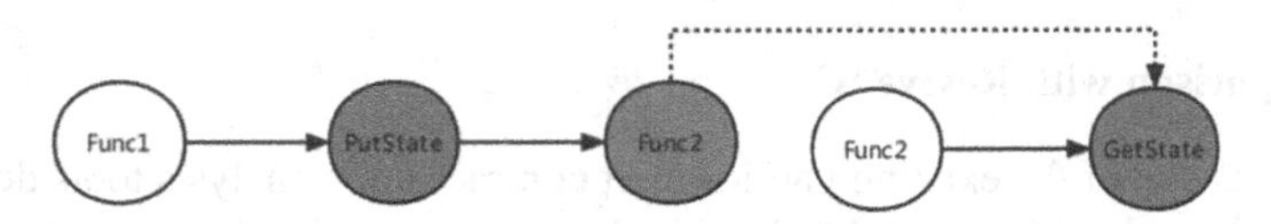

Fig. 9. Read-after-write risk caused by indirect call

Considering the execution characteristics of the defer statement, the defer statement will be executed when the function exits, that is to say, the defer statement will be later than ordinary statement. Another point worth noting is that when there are multiple defer statements, they will be executed according to the order of execution of the stack, that is, the order of first-in-last-out, that is to say, the first defer statement defined by defer will be executed last. In order to represent the actual call sequence of key methods in the function body, the method call chain cannot be constructed simply based on the position attribute, but should be considered in combination with the position attribute, type attribute and function attribute. The process of constructing the key method call chain is as follows:

1. Scan key methods
   According to the types of key methods, the set of key methods is divided into two sub-sets. If the key method type is ordinary, add the key method to set 1; if the key method type is special, add the key method to set 2.
2. Construct sub-chain
   For the elements in set 1, according to the position attribute of the key method, the sub-chain S1 is formed from small to large. For the elements in set 2, the sub-chain S2 is formed from large to small according to the position attribute of the key method.
3. Union sub-chains
   Connect the obtained sub-chain S2 to the sub-chain S1, and get the key method call chain S0.
4. Expand call chain
   Node is expanded into the key method call chain corresponding to the function attribute until there is no key method marked as true in the key method call chain, and the key method call chain S is returned.

5 Evaluation

This section compares the proposed scheme with the current static detection scheme Revive^CC to demonstrate the advantages of the scheme.

5.1 Dataset

This experiment collects 120 smart contracts for the actual application of Fabric from the opensource platform GitHub. Among them, the read-after-write risk samples accounted for 24 samples. Specifically, the number of read-after-write risk samples in the multi-function scenario was 5, and the read-after-write risk caused by special statements accounted for 7 samples.

5.2 Comparison with Revive^CC

Revive^CC is one of the existing Fabric smart contract static analysis tools designed to detect security vulnerabilities related to blockchain to help developers write clean and secure smart contracts. This paper sets up two groups of experiments, one group adopts the detection scheme based on the key method and the key method call chain in this paper, and the other group adopts Revive^CC for control experiments. By performing read-after-write risk detection on the collected 120 smart contracts, and counting the number of detected risk samples, the number of false negatives and false positives in the samples is manually screened, and finally the overall accuracy rate is calculated.

Table 2 shows the difference in read-after-write risk between the two detection schemes. The number of risk samples, false negatives, false positives, and accuracy detected by Revive^CC are 14, 5, 2, and 85.71%, respectively, while the number of risk samples, false negatives, false positives, and accuracy rates are 25, 2, 3, and 88%, respectively. The number of read-after-write risks detected by the scheme proposed in this paper is 1.8 times that of Revive^CC, and the accuracy rate is improved by 2.3%, which is significantly better than Revive^CC. Both in the special sentence scenario and multi-function scenario, the false negative rate and accuracy rate of the tool proposed in this paper are better than Revive^CC.

Table 2. Comparison of our tool and Revive^CC on read-after-write risk detection

Type	Tool	Risk samples	False negatives	False positives	Accuracy rate
Read-after-write Risk	Revive^CC	14	5	2	85.71%
	Our Tool	**25**	**2**	**3**	**88.00%**
Special Sentence	Revive^CC	2	4	1	50.00%
	Our Tool	**6**	**0**	**1**	**83.33%**
Multi-function	Revive^CC	0	7	0	0.00%
	Our Tool	**7**	**1**	**1**	**85.71%**

The accuracy and false negative rate of our scheme are better than Revive^CC. But the overall false positive rate is slightly higher than Revive^CC. The main reason is that, by observing the distribution of read-after-write risk samples, we can see that one false positive sample was introduced when detecting the read-after-write risk caused by multi-functions, resulting in the overall number of false positives changing from 2 to 3. Through research, this paper finds that the false positive samples are generated because when the risk of read after write is detected, the parameters of the read and write methods may have aliasing problems, resulting in that the read and write methods do not operate on the same object. One problem is that the alias relationship of the method parameters cannot be identified and analyzed, which leads to the misunderstanding that the behavior of read after write occurs.

6 Conclusion

In this paper, we proposed a static analysis-based detection scheme for read-after-write risk in Fabric smart contracts. By identifying key methods, it quickly located nodes which were closely linked that may cause read-after-write risk. This paper also proposed to construct a chain of key method calls to reflect the actual execution sequence between key methods, which covered more read-after-write risk scenarios. Experiments showed that the solution proposed in this paper made up for the shortcomings of the existing mainstream smart contract detection tools when detecting the risk of read-after-write and showed certain application value.

Acknowledgments. This work is supported by National Key R&D Program of China under Grant 2021YFB2700400.

References

1. Web3.0 Prospective Research Report [R]TBI (2022)
2. Qiu, H., Qiu, M., Memmi, G., Ming, Z., Liu, M.: A dynamic scalable blockchain based communication architecture for IoT. In: Qiu, M. (eds.) Smart Blockchain. SmartBlock 2018 (2018).
3. Li, H., Gai, K., Zhu, L., Jiang, P., Qiu, M.: Reputation-based trustworthy supply chain management using smart contract. In: Qiu, M. (eds) ICA3PP 2020. LNCS, vol. 12454, pp. 35–49. Springer, Cham (2020). https://doi.org/10.1007/978-3-030-60248-2_3
4. Gai, K., Wu, Y., Zhu, L., Zhang, Z., Qiu, M.: Differential privacy-based blockchain for industrial Internet-of-Things. IEEE Trans. Ind. Inf. **16**(6), 4156–4165 (2020). https://doi.org/10.1109/TII.2019.2948094
5. Shao, Q., Jin, C., Zhang, Z., Qian, W., Zhou, A.: Blockchain: architecture and research progess. Chin. J. Comput. **41**(05), 969–988 (2018)
6. Wu, X., Qiu, H., Zhang, S., et al.: ChainIDE 2.0: facilitating smart contract development for consortium blockchain. In: IEEE INFOCOM 2020 - IEEE Conference on Computer Communications Workshops (INFOCOM WKSHPS), pp. 388–393 (2020)
7. Blockchain security capability evaluation and analysis report. CAICT (2021)
8. Chinen, Y., Yanai, N., Cruz, J.P., Okamura, S.: RA: Hunting for re-entrancy attacks in ethereum smart contracts via static analysis. In: 2020 IEEE International Conference on Blockchain (Blockchain), pp. 327–336 (2020). https://doi.org/10.1109/Blockchain50366.2020.00048
9. Yan, X., Wang, S., Gai, K.: A semantic analysis-based method for smart contract vulnerability. In: 2022 IEEE 8th International Conference on Big Data Security on Cloud (BigDataSecurity), IEEE International Conference on High Performance and Smart Computing, (HPSC) and IEEE International Conference on Intelligent Data and Security (IDS), pp. 23–28 (2022)

10. Androulaki, E., Barger, A., Bortnikov, V., et al.: Hyperledger fabric: a distributed operating system for permissioned blockchains. In: Proceedings of the Thirteenth EuroSys Conference (EuroSys 2018), Article no. 0, pp. 1–15. Association for Computing Machinery, New York (2018). https://doi.org/10.1145/3190508.3190538
11. FabricLedger[EB/OL] (2022). https://hyperledger-fabric.readthedocs.io/en/latest/ledger/ledger.html
12. Vukolic, M.: Rethinking permissioned blockchains. In: Proceedings of the ACM Workshop on Blockchain, Cryptocurrencies and Contracts (2017)
13. Revive^CC [EB/OL] (2019). https://github.com/sivachokkapu/revive-cc
14. Yamashita, K., Nomura, Y., Zhou, E., Pi, B., Jun, S.: Potential risks of hyperledger fabric smart contracts. In: 2019 IEEE International Workshop on Blockchain Oriented Software Engineering (IWBOSE), pp. 1–10 (2019). https://doi.org/10.1109/IWBOSE.2019.8666486
15. Lv, P., Wang, Y., Wang, Y., Zhou, Q.: Potential risk detection system of hyperledger fabric smart contract based on static analysis. In: 2021 IEEE Symposium on Computers and Communications (ISCC), pp. 1–7 (2021). https://doi.org/10.1109/ISCC53001.2021.9631249
16. Go-parser [EB/OL] (2022). https://pkg.go.dev/go/parser

A Critical-Path-Based Vulnerability Detection Method for tx.origin Dependency of Smart Contract

Hui Zhao[1](✉) and Jiacheng Tan[2]

[1] Educational Information Technology Laboratory, Henan University, Kaifeng, China
zhh@henu.edu.cn

[2] Software School, Henan University, Kaifeng, China
15515439053@163.com

Abstract. Smart contracts are one of the most successfully applied technologies on the blockchain, which are decentralized and immutable. Smart contracts cannot be modified once deployed. Therefore, security detection of smart contracts before deployment is essential. Some smart contracts may have tx.origin dependency vulnerabilities. In this paper, we propose a critical path vulnerability detection method for detecting tx.origin dependency vulnerabilities in smart contracts. Then, in order to solve the problem that the traditional search algorithm cannot determine the critical path, we propose a path determination method based on path priority. Our method determines the critical path in the control flow graph, which enables us to detect the vulnerabilities existing in smart contracts more quickly. The experimental results show that our method is more efficient than the existing technology and the false positive rate is lower.

Keywords: smart contract · critical path · vulnerability detection · control flow graph

1 Introduction

Encrypted currency is widely considered one of the most disruptive technologies of the past few years [1]. With the rapid development of computer hardware [2–4], parallel and distributed computing [5–7], and networks [8–10], blockchain has emerging as a promising technique. Smart contracts are programs running on the blockchain [11,12] that store code and state on the ledger and can send and receive virtual currency [13]. Smart contracts can be viewed as code-based contracts that execute transactions without third-party supervision. These transaction records are trustworthy and traceable, but contract transactions are irreversible once completed [14]. The node user interacts with the smart contract by

M. Qiu et al. (Eds.): SmartCom 2022, LNCS 13828, pp. 393–402, 2023.
https://doi.org/10.1007/978-3-031-28124-2_37

calling the internal functions of the smart contract [15]. The security of smart contracts is a hot research topic today [16]. Due to the irreversible and immutable nature of smart contracts, making it critical to check their correctness before deployment [17]. Existing vulnerability detection methods include formal verification, symbolic execution, intermediate relation representation, fuzzing, and deep learning. However, most smart contract detection tools still have a high false positive rate and a long audit time. This paper proposes a method to determine the critical path through dangerous instructions to detect tx.origin dependency vulnerabilities, which can effectively improve the detection efficiency and reduce the false positive rate.
We summarize our core contributions as follows:

- We propose a method to identify critical paths by dangerous instructions to discover tx.origin dependency vulnerabilities in smart contracts.
- We propose an algorithm to judge the critical path to identify tx.origin dependency vulnerabilities in smart contracts.

The structure of this paper is organized as follows: Sect. 2 provides an introduction to related work. Section 3 analyzes our focus on the issues. Section 4 introduces our proposed method in detail. Section 5 gives a detailed description of the core algorithms used in our method. Section 6 conducts experiments comparing our method with other existing methods. Section 7 summarizes this paper.

2 Related Work

Since 2016, due to the rapid growing of big data [18–20] and cybersecurity [21,22] techniques, frequent attacks on smart contract vulnerabilities and resulting economic losses. The security [23,24] of smart contracts has received extensive attention [16]. Checking based on symbolic execution. Oyente is the first symbolic execution tool to perform vulnerability analysis [25]. Muller et al. proposed a symbolic execution-based tool Mythril, which is mainly used to detect common smart contract security problems. Securify analyzes the dependency graph of smart contracts through symbolic analysis [26]. It extracts precise contract semantic information from code.

Based on fuzzy testing. ContractFuzzer is the first fuzzification framework based on Ethereum platform. Regurad is a fuzzing analysis tool mainly for smart contract reentrancy vulnerabilities [27]. sFuzz combines the strategies in AFL blurrs with adaptive strategies for branches that are hard to find [28]. Based on the intermediate method representation. Slither transforms the Solidity source code of smart contracts into an intermediate representation of SlithIR [29]. Vandal performs abstract interpretation and transformation of bytecodes in a logical manner through a decompiler. [30]. Ethir is also an analysis tool based on the EVM bytecode level [31]. SmartCheck detects smart contract vulnerabilities by transforming the Smart Contract Solidity source code into an XML intermediate representation relationship and exploiting XPath patterns.

3 Problem Definition

In this section, we describe the research problem. Analyze how the problem was identified. Describes how our critical paths are determined.

3.1 Problem Description

There is a global variable **tx.origin** in Ethereum smart contracts. It can backtrack the entire call stack to return the address of the contract that originally initiated the call. When the smart contract uses this variable for user authentication or authorization, the attacker can use the characteristics of **tx.origin** to create a corresponding attack contract to steal ether. For example, the attacker calls the withdrawal function of the victim contract in his own Fallback function, and induces the victim contract to transfer ether to the attacker contract. Undetectable exception due to authentication via **tx.origin**. Thus, all the ether in the victim's contract is transferred to the attacker's contract account.

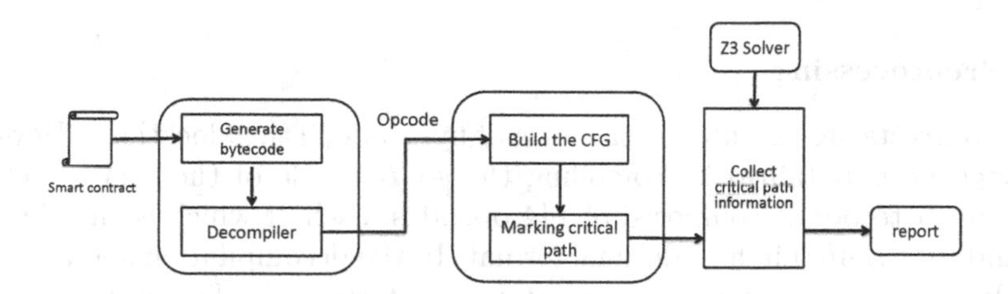

Fig. 1. GLP Smart Contract Vulnerability Check

3.2 Problem Analysis

We analyze the contract paths related to the receipt and sending of money. For example, if the contract never sends any Ether. To determine if the contract can receive ether, we check to see if the contract has a payment function. Starting with Solidity version 0.4.x, contracts are allowed to receive ether only if the public functions in the contract are declared with the keyword payable.

When the Solidity compiler compiles a non-payble function. It inserts a sequence of opcode instructions before the function body to indicate that ether cannot be accepted. Therefore, we check whether the contract allows to receive ether. We check that each path contains the above sequence of instructions.

3.3 Critical Path Determination

To allow for more effective critical path analysis, we prioritize paths based on their likelihood of revealing vulnerabilities.For each path, we compute a criticality score and determine the path based on the score. The criticality score is calculated as follows: Let Pa be a path and V be the set of properties that Pa violates.

$$criticalpath(Pa) = \frac{\sum_{s \in V} \cdot \alpha_s}{\epsilon \cdot depth(Pa)} \tag{1}$$

where α_s is a constant representing the severity of the violation of property s. *depth(Pa)* is the depth bound of the path Pa, the number of times the function is called. ϵ is a positive constant. Intuitively, the critical score is designed such that the more critical the property violated by the path, the higher the score. And the more properties it violates, the higher the score.

4 Method

In response to the above problems, we propose a method to detect tx.origin dependency vulnerabilities based on critical paths. In this section, we introduce our method based on control flow graph critical path detection (GLP). Figure 1 is the flow of our method GLP.

4.1 Preprocessing

Smart contracts are executed in the form of bytecode in the blockchain. However, the bytecode is obtained by compiling the source code of the smart contract. Ethereum bytecode is composed of 144 opcodes, each of which is encoded as a byte and represented in hexadecimal format. In the decompilation operation, use the solc compiler to compile the smart contract source code .sol file to obtain the EVM virtual machine bytecode of the contract. The generated bytecode is divided into deployment code, runtime bytecode and auxdata. Then decompile the runtime bytecode. Starting from the first byte of the runtime bytecode, compare each byte in the runtime bytecode in turn to obtain the instruction value corresponding to each bytecode.

4.2 Build Control Flow Graph

A control flow graph is a directed graph structure. Each of the vertices corresponds to the basic block of the program. There are no branching jump instructions in this basic block, and the basic block starts with a branch purpose and ends with a branch.

The Ethereum Virtual Machine currently supports 144 instructions. Among them, the two instructions JUMP and JUMPI represent unconditional jump and conditional jump respectively. They can generate branches of execution paths. The instructions JUMP and JUMPI will be used as the end of the basic block. They indicate a jump to the next basic block. The instruction JUMPDEST identifies the destination offset for JUMP and JUMPI jumps. And the jump destination offset of the instructions JUMP and JUMPI can only be the offset where the JUMPDEST instruction is located. Therefore, JUMPDEST needs to be the beginning of the basic block. REVERT, RETURN and INVALID in EVM indicate termination. They are also the end of the basic block. We prioritize the basic blocks in the control flow graph according to JUMPDEST. Then perform more detailed block partitioning according to JUMP and JUMPI instructions.

4.3 Critical Path Generation

In determining the critical path, we mainly mark the paths that may have vulnerabilities, and then detect the paths that may have vulnerabilities. We do not need to instrument all paths in the generated control flow graph. Through this method, we can improve the detection efficiency and reduce the average detection time of a single contract.

When using the global variable tx.orgin for authentication and authorization operations, the smart contract is at risk of being attacked. We get the bytecode of the contract by preprocessing the contract. The control flow graph of the contract is created through bytecode to simulate the running process of the smart contract. When generating the contract control flow graph, we judge the information in the generated block in each path. We mainly focus on dangerous instructions such as contract calls and reading and modification of balances, such as CALL, CALLVALUE, CALLDATALOAD, SLOAD, SSTORE instructions, etc. We label the paths with this information and get the critical path. We label the paths with this information and get the critical path. Then we judge the instruction information of the block in the critical path, whether there is the use of ORIGIN for authentication and authorization. When it exists, the contract may be vulnerable. The z3 solver is called on the path with dangerous information to judge the reachability of the path. Finally output the detection result.

5 Algorithms

This section mainly describes the algorithms for critical path determination. As shown in Algorithm 1. The algorithm for this purpose is applied to the collection of critical path sections. The algorithm mainly introduces the process of determining the critical path. The algorithm takes the generated control flow graph as input, and then judges the block information generated in the control flow graph. When the current block contains dangerous instructions, such as CALL, CALLVALUE, CALLDATALOAD, SLOAD, SSTORE, etc. Then mark the current path, and then traverse the marked path. When the marked path has an instruction to use ORIGIN for authentication and authorization, the current path is added as a critical path. Finally output the critical path.

In Algorithm 2. The algorithm is applied to the tx.origin dependency vulnerability detection section. The algorithm mainly introduces the use of the generated critical path information to judge whether the contract has vulnerabilities. The algorithm takes the generated critical path as input. First, we traverse all the generated critical paths. And make judgments on all critical paths. If the current path contains dangerous instructions that meet the requirements, collect

Algorithm 1. Select critical path.

```
Input CFG.
Onput path.
 1: for blocks[i] in CFG.blocks do
 2:    if blocks[i].instruction in CFG.blocks.instruction then
 3:       a ← blocks[i]
 4:    end if
 5:    if boleanjudge(a) then
 6:       newpath ← a
 7:    end if
 8: end for
 9: for newpath in CFG.blocks do
10:    if boleanMark(newpath) then
11:       pathAdd(newpath)
12:    end if
13: end for
return path
```

the current path constraints. Continue to traverse the information of the current path. Determine whether the current path contains the opcode feature of the tx.origin dependency vulnerability. If the current path contains the opcode characteristics of tx.origin dependency vulnerability. Then call the solver to solve the constraint of the current path to judge the accessibility of the current path. Finally, return the detection result.

boleanjudge(n) is A bool value. When the current command contains dangerous commands such as CALL, CALLVALUE, CALLDATALOAD, SLOAD, SSTORE, etc., the bool value is true. *boleanMark(p)* is A bool value. When the marked path basic block contains the ORIGIN directive, the bool value is true. *Collect(p)* represents a collection. When the judgment conditions are met, the constraints of the current path are collected. *boleanFinal(p)* is A bool value. Determine whether the current path contains the opcode feature of the tx.origin dependency vulnerability. *Solver(p)* is A bool value. Call the solver to judge the reachability of the current path.

6 Evaluation

This section includes the environment in which we conduct experiments, data sources, comparative experiments with other detection tools, and analysis of results. We first introduce the data sources and experimental configuration, and then evaluate the effectiveness of our method GLP in detecting tx.origin dependency vulnerabilities in real smart contracts. We conducted experiments on the accuracy of Mythril, SmartCheck, Vandal, and ContractGuard in detecting this vulnerability, and compared GLP with these four tools in terms of accuracy.

Algorithm 2. Vulnerability detection.

```
Input path.
Onput result
 1: for CurrentPath in path do
 2:    if current opcode is dangerous instruction. then
 3:       Code ← current opcode
 4:       Code.Path ← CurrentPath
 5:    end if
 6: end for
 7: for opcode in code do
 8:    if opcode is a characteristic of tx.origin . then
 9:       Collect(Code.Path)
10:    end if
11:    if boleanFinal(Code.Path) then
12:       result ← Solver(CurrentPath)
13:    end if
14: end for
return result
```

6.1 Experiment Configuration

The dataset we use for evaluation is 6500 real smart contracts collected from Ethereum. We divided the collected smart contracts into three datasets. The number of contracts in the first dataset is 500, including 100 vulnerable contracts. The number of contracts in the second dataset is 1,000, of which 150 contain vulnerable contracts. The number of contracts in the third dataset is 5000, including 200 vulnerable contracts. We also added smart contracts that correctly use the global variable tx.origin to the dataset. This experiment mainly compares the three aspects of true positive, false positive and false negative of each tool. Finally, the average time to detect vulnerabilities by each tool is compared. We use these three datasets to make an empirical evaluation of GLP. We compare the performance of GLP with that of Mythril, SmartCheck, Vandal, and ContractGuard and show that GLP outperforms all systems in terms of accuracy and runtime. The experiment is mainly implemented in the Linux system.

6.2 Experiments and Results

We compared the smart contract detection accuracy of GLP and four detection tools, Mythril, SmartCheck, Vandal and ContractGuard, under the same experimental environment. We conducted three sets of experiments for verification. As shown in Fig. 2. In the figure, we use TP to represent the number of true positives, FN to represent the number of false negatives, and FP to represent the number of false positives. We use the first dataset for experiments. We observed that Mythril, SmartCheck, Vandal and ContractGuard have different degrees of false positives and false negatives, and Mythril and SmartCheck have the most false positives and false negatives. Our method outperforms all tools in terms of contract detection accuracy.

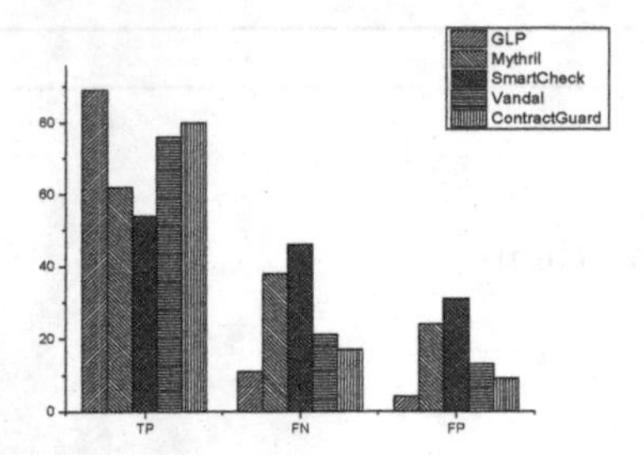

Fig. 2. The comparison of TP FN and FP on setting one.

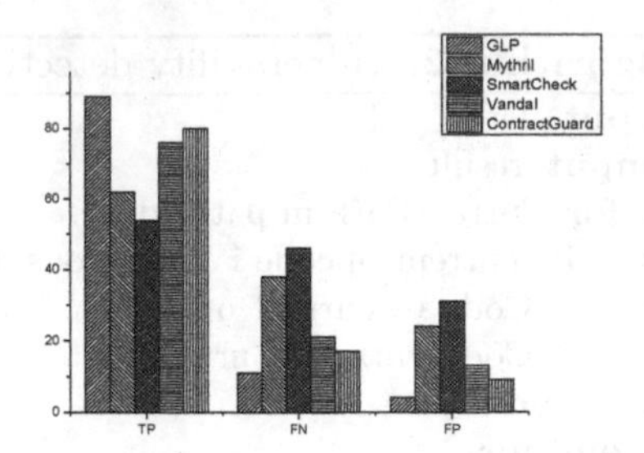

Fig. 3. The comparison of TP FN and FP on setting two.

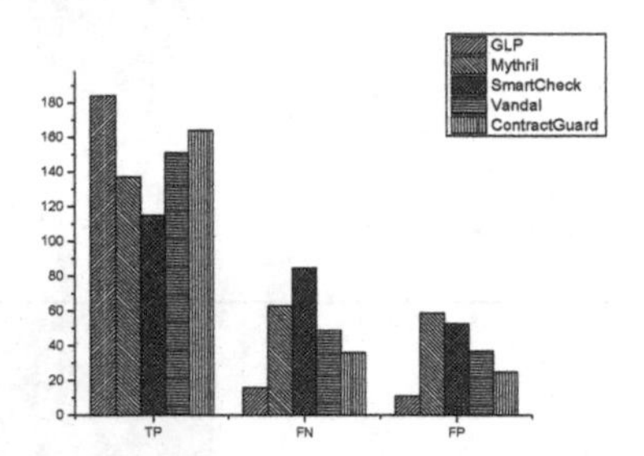

Fig. 4. The comparison of TP FN and FP on setting three.

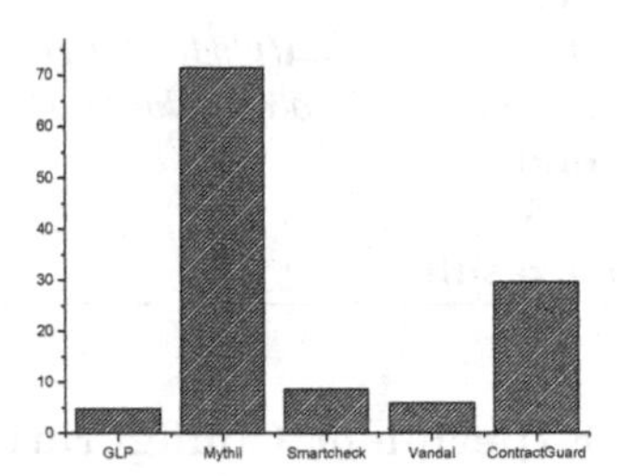

Fig. 5. Average detection time (seconds).

Figure 3 shows the results of the second group of experiments were compared. As the number of smart contracts increases. Mythril, SmartCheck, Vandal and ContractGuard experienced a significant increase in false positives and false negatives. We see that GLP has the highest accuracy rate and lowest false positive rate compared to all other tools.Fig. 4 shows the results of the second group of experiments were compared. In the last set of experiments, we added 200 contracts with vulnerabilities to 5000 smart contracts. We see that SmartCheck detected the least number of vulnerable contracts, followed by Mythril. Out of these tools, we observed that GLP flagged more vulnerable contracts and had a small number of false positives and false negatives. As the number of smart contracts increases. Accuracy rates of Vandal and ContractGuard detections decreased, while false positive rates increased. The experimental results show that Vandal and ContractGuard are superior to Mythril and SmartCheck in detection accuracy. Experiments show that our method outperforms other tools in accuracy.

Finally, we also compared the time taken for detection. The average detection time comparison results are shown in Fig. 5. Mythril was used the longest in average detection time. The average detection time of Mythril was 71.4s. The second is ContractGuard, which has an average detection time of 29.6s. The average detection times of SmartCheck and Vandal are 8.6s and 5.9s, respectively. The average detection time of GLP is the shortest compared to other tools.

The average detection time of GLP is 4.7s. In terms of total detection time, Mythril and ContractGuard are the two longest-used tools. The total detection time of Mythril was 129.1 h. The total detection time of ContractGuard is 53.5h. GLP is the least used in total detection time compared to other tools. The total detection time of GLP is 8.4h. Comparing processing time, GLP uses the least time. Experiments show that compared with other tools, GLP can effectively reduce the false positive rate and false positive rate, and can effectively reduce the detection time.

Overall, our experimental results show that GLP is effective in discovering tx.origin dependency vulnerabilities and determining smart contract correctness. Compared with existing detection methods, GLP is overall better than existing detection methods in terms of accuracy and detection time.

7 Conclusion

We proposed a method to detect **tx.origin** dependency vulnerabilities existing in smart contracts by determining the critical path. We show examples of vulnerable contracts and how to attack them, and propose a critical path method to detect vulnerabilities. Finally, we experimented with real smart contracts on Ethereum. Compared with existing methods, our method performs better overall in detection efficiency and false positive rate, indicating that our method is both efficient and effective.

Acknowledgement. Natural Science Foundation of Shandong Province (Grant No. ZR2020ZD01).

References

1. Zheng, Z., Xie, S., et al.: Blockchain challenges and opportunities. Int. J. Web Grid Serv. **14**(4), 352–375 (2018)
2. Qiu, M., Li, H., Sha, E.: Heterogeneous real-time embedded software optimization considering hardware platform. In: ACM SAC, pp. 1637–1641 (2009)
3. Qiu, M., Jia, Z., et al.: Voltage assignment with guaranteed probability satisfying timing constraint for real-time multiproceesor DSP. J. Sign. Proc. Sys. (2007)
4. Qiu, M., Yang, L., et al.: Dynamic and leakage energy minimization with soft real-time loop scheduling and voltage assignment. IEEE TVLSI **18**(3), 501–504 (2009)
5. Qiu, M., Xue, C., Shao, Z., Zhuge, Q., Liu, M., Sha, E.H.-M.: Efficent algorithm of energy minimization for heterogeneous wireless sensor network. In: Sha, E., Han, S.-K., Xu, C.-Z., Kim, M.-H., Yang, L.T., Xiao, B. (eds.) EUC 2006. LNCS, vol. 4096, pp. 25–34. Springer, Heidelberg (2006). https://doi.org/10.1007/11802167_5
6. Qiu, M., Xue, C., et al.: Energy minimization with soft real-time and DVS for uniprocessor and multiprocessor embedded systems. In: IEEE DATE, pp. 1–6 (2007)
7. Zhang, L., Qiu, M., et al.: Variable partitioning and scheduling for MPSoC with virtually shared scratch pad memory. JSPS **58**(2), 247–265 (2018)

8. Niu, J., Gao, Y., et al.: Selecting proper wireless network interfaces for user experience enhancement with guaranteed probability. JPDC **72**(12), 1565–1575 (2012)
9. Qiu, M., Chen, Z., et al.: Energy-aware data allocation with hybrid memory for mobile cloud systems. IEEE Syst. J. **11**(2), 813–822 (2014)
10. Li, Y., Gai, K., et al.: Intercrossed access controls for secure financial services on multimedia big data in cloud systems. ACM TMCCA (2016)
11. Qiu, M., Qiu, H., et al.: Secure data sharing through untrusted clouds with blockchain-enabled key management. In: 3rd SmartBlock conference, pp. 11–16 (2020)
12. Gai, K., Zhang, Y., et al.: Blockchain-enabled service optimizations in supply chain digital twin. IEEE TSC (2022)
13. Ouyang, L., Wang, S., Yuan, Y., Ni, X., Wang, F.Y.: Smart contracts: architecture and research progresses. Acta Automatica Sinica **45**, 445–457 (2019)
14. Qiu, M., Gai, K., Xiong, Z.: Privacy-preserving wireless communications using bipartite matching in social big data. FGCS **87**, 772–781 (2018)
15. Qiu, M., Zhang, L., et al.: Security-aware optimization for ubiquitous computing systems with seat graph approach. J. Comput. Syst. Sci. **79**(5), 518–529 (2013)
16. Fu, M., Wu, L., Hong, Z., Feng, W.: Research on vulnerability mining technique for smart contracts. J. Comput. Appl. **39**, 1959 (2019)
17. Qiu, H., Qiu, M., Lu, Z.: Selective encryption on ECG data in body sensor network based on supervised machine learning. Inf. Fusion **55**, 59–67 (2020)
18. Li, J., Ming, Z., et al.: Resource allocation robustness in multi-core embedded systems with inaccurate information. J. Syst. Arch. **57**(9), 840–849 (2011)
19. Hu, F., Lakdawala, S., et al.: Low-power, intelligent sensor hardware interface for medical data preprocessing. IEEE Trans. Inf. Tech. Biomed. **13**(4), 656–663 (2009)
20. Qiu, H., Zheng, Q., et al.: Topological graph convolutional network-based urban traffic flow and density prediction. IEEE Trans. ITS **22**, 4560–4569 (2020)
21. Qiu, H., Dong, T., et al.: Adversarial attacks against network intrusion detection in IoT systems. IEEE IoT J. **8**(13), 10327–10335 (2020)
22. Gai, K., Qiu, M., Elnagdy, S.: A novel secure big data cyber incident analytics framework for cloud-based cybersecurity insurance. In: IEEE BigDataSecurity conference (2016)
23. Gao, X., Qiu, M.: Energy-based learning for preventing backdoor attack. KSEM **3**, 706–721 (2022)
24. Qiu, M., Qiu, H.: Review on image processing based adversarial example defenses in computer vision. In: IEEE 6th BigDataSecurity, pp. 94–99 (2020)
25. Luu, L., Chu, D.H., Olickel, H., Saxena, P., Hobor, A.: Making smart contracts smarter. In: the 2016 ACM SIGSAC Conference (2016)
26. Tsankov, P., Dan, A., Drachsler-Cohen, D., Gervais, A., Buenzli, F., Vechev, M.: Securify: practical security analysis of smart contracts. In: ACM (2018)
27. Liu, C., Liu, H., et al.: Reguard: finding reentrancy bugs in smart contracts. In: IEEE/ACM International Conference on Software Engineering: Companion (2018)
28. Tai, D.N., Long, H.P., et al.: sfuzz: an efficient adaptive fuzzer for solidity smart contracts. In: 42nd ICSE (2020)
29. Feist, J., Grieco, G., Groce, A.: Slither: a static analysis framework for smart contracts. In: IEEE/ACM 2nd WETSEB (2019)
30. Brent, L., Jurisevic, A., Kong, M., Liu, E., Scholz, B.: Vandal: a scalable security analysis framework for smart contracts (2018)
31. Albert, E., Gordillo, P., Livshits, B., Rubio, A., Sergey, I.: EthIR: a framework for high-level analysis of Ethereum bytecode (2018)

A Dynamic Taint Analysis-Based Smart Contract Testing Approach

Hui Zhao[1], Xing Li[2], and Keke Gai[3(✉)]

[1] Educational Information Technology Laboratory, Henan University, Kaifeng, China
zhh@henu.edu.cn

[2] Software School, Henan University, Kaifeng, China
lx@henu.edu.cn

[3] School of Cyberspace Science and Technology, Beijing Institute of Technology, Beijing, China
gaikeke@bit.edu.cn

Abstract. Due to the unique global state and transaction sequence characteristics of smart contracts, the detection method based on a single test case cannot improve the vulnerability detection rate during contract detection. The current contract testing methods based on genetic algorithms have not yet solved the problems caused by these characteristics. Therefore, we propose an adaptive fuzzing method based on dynamic taint analysis and genetic algorithm, SDTGfuzzer. SDTGfuzzer focuses on dynamic taint analysis to collect runtime information as feedback, and focuses on solving the challenges brought by global variables and transaction sequences for contract testing. Genetic Algorithms work well in test case generation for fuzzing. Therefore, SDTGfuzzer optimizes the genetic algorithm based on an efficient and lightweight multi-objective adaptive strategy, focusing on solving the problem that the contract constraints cannot be covered due to the global state. Experimental results show that our method has a higher vulnerability detection rate than other tools for detecting contract vulnerabilities.

Keywords: Smart Contracts · Vulnerability Detection · Fuzzing · Genetic Algorithms · Taint Analysis

1 Introduction

With the development of computer capability [14,16] and distributed cloud [6,13], blockchain [1,15] is emerging as very promising techniques for various applications. Fuzz testing, as a common vulnerability detection method, is widely used in program detection. Smart contracts are different from traditional C language and other languages in terms of operating environment and program characteristics. Therefore, when designing a fuzzing method for smart contracts, it is necessary to consider the properties specific to smart contracts.

M. Qiu et al. (Eds.): SmartCom 2022, LNCS 13828, pp. 403–413, 2023.
https://doi.org/10.1007/978-3-031-28124-2_38

There are three main challenges when fuzzing contracts. 1) Influence of transaction sequence on global variables 2) Influence of global variables on constraints. 3) The impact of transaction-related variables on constraints. Although various fuzzy testing tools have been proposed to detect contract vulnerabilities, these problems are still not well solved.

This paper mainly solves the problems brought by the above three challenges for contract testing. We propose SDTGfuzzer, a fuzzing method combining dynamic taint analysis and genetic algorithm. SDTGfuzzer uses dynamic taint analysis to obtain the pendencies of variables in the contract, and improves the efficient and lightweight multi-objective adaptive strategy based on sfuzz [11].

Our main contributions are as follows:

- For the impact of transaction sequences on contract testing, we divide transaction sequence generation into two steps, and use feedback to generate transaction sequences that are easier to find potential vulnerabilities.
- For the challenges brought by global variables for fuzzing, we modify the genetic algorithm based on an efficient and lightweight multi-objective adaptive strategy. It can better solve the problem of contract testing under the condition that the value of global variables is uncertain.
- We propose a smart contract dynamic taint analysis and genetic algorithm fuzzing method, SDTGfuzzer. The model centers on dynamic taint analysis, captures runtime feedback, and optimizes for features such as smart contract special transaction variables.

The remainder of the paper is organized as follows: We introduced the relevant work in Sect. 2 . We introduces the system architecture of SDTGfuzzer in Sect. 3. Section 4 Algorithms for Methods In Sect. 5, we conduct experiments to compare our method with other methods. Concludes is in Sect. 6.

2 Related Work

Various tools are proposed to detect contract vulnerabilities. teEther [5], SGUARD [10], Sereum [17] ,Osiris [18], Mythril [9] use symbolic execution methods for contract testing. But the symbol execution has the problem of facing the path explosion. Although machine learning [3,7,12] is used for contract detection, it is itself less interpretable. Static analysis produces many false positives

Fuzzing is widely used for contract detection due to low false positives. ContractFuzzer [4] uses black-box testing to detect contracts. Reguard [8] converts specific smart contracts into the traditional language C++. Ethploit [21] constructs pollution graphs to generate target sequences. sfuzz [11] combines policies in AFL fuzzers and lightweight multi-objective adaptive policies. ADF-GA [20] uses the exception stop statement as the basis to generate test cases. In addition, Harvey [19] uses gray box testing for contract fuzzing, and ILF [2] uses a combination of deep learning and fuzz testing for vulnerability detection.

Many fuzzing methods are used for fuzzing of smart contracts, but the research on the problems caused by the transaction sequence and global variables faced by fuzzing is still insufficient. In addition, we found that although the genetic algorithm is widely used in contract fuzzing, it does not solve the problems caused by the global state and transaction sequence. Therefore, we introduce dynamic taint analysis and improve the genetic algorithm of lightweight multi-objective adaptive strategy. Use dynamic taint analysis to obtain feedback, and solve the problem of global state and transaction sequence faced by genetic algorithm applied to fuzzing through program feedback.

3 SDTGfuzzer Method

The main task of this section is to construct transactions. A smart contract transaction consists of four parts: FROM, TO, VALUE, and DATA. Figure 1 shows the logic for test parameter generation. In order to solve the FROM, VALUE, DATA required for transaction generation.

3.1 Feedback Stage

Since storage is stored in the form of *key−value* pairs, it is possible to obtain read and write operations on those storage variables in the function. Dependencies between constraints and storage variables are available through constraints and storage variables. Due to the influence of the storage variable, individual transactions directly affect each other. When the global variable directly or indirectly affects the branch constraints, the coverage of the branch cannot be completely dependent on the unit test of the function, and the calling order between functions needs to be considered. Take SLOAD as source and JUMPI opcode as sink. Record the *key* value in the storage variable to SLOAD. The storage variable flowing into the branch will have an effect on the operands of the opcode, which will affect the test.

3.2 Transaction Sequence Generation

The first step in generating a transaction sequence is to look for a danger transaction function. The second step generates a sequence of transactions based on the selected function.

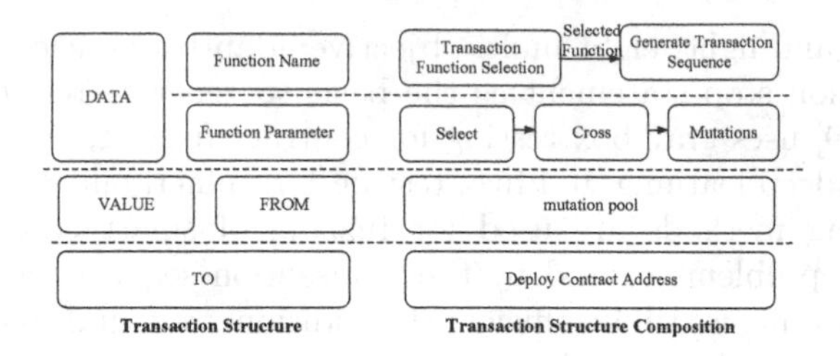

Fig. 1. Overview of invocation model parameter generation.

Transaction Function Selection. The first step in transaction sequence generation is transaction function selection. The method selects functions by taking coverage, dangerous opcodes and abnormally stopped branch statements as important influencing factors. We first combine the dangerous opcode and block coverage as the first set of parameters. The more dangerous opcodes in uncovered blocks, the greater the probability of a vulnerability. We combine the Abnormally stopped branch and branch coverage as the second set of parameters. The more Abnormally stopped branch that are not covered, the harder it is to code overwrite the contract We uses the roulette algorithm to pick the probabilities.

Generate Transaction Sequence. The second step in transaction sequence generation is generate transaction sequence. Smart transaction may modify global variables, thereby affecting the coverage of contract branches. It is for the above reasons that some contracts can only execute the logic in a specific order. The storage variable is a variable stored globally, which greatly affects the transaction sequence of the contract. To this end, it is necessary to obtain the situation of the storage variables associated with the uncovered branches, and then obtain the dependencies on the storage variables.

At this stage, the functions obtained in the Transaction Function Selection stage are first analyzed. There may be multiple storage variables in the function, which will have different effects on the Constraint branch. Different storage variables affect different Constraint branches. In order to reduce the impact of irrelevant storage variables on JUMPI constraints, and generate transaction sequences more accurately. When performing dynamic taint analysis, use SLOAD as the taint source and JUMPI as the taint sink to record the *key* of the storage variable that affects JUMPI. That is, construct < constraints, Storage variables >. Only these storage variables directly or indirectly affect the constraints of the function Constraint, and they are regarded as global variables that affect the transaction sequence. The storage variable is composed of *key* and value, and a storage variable can be located by *key*. According to the collected < Storage variables, function name>, that is, the SLOAD and SSTORE information in each function. In addition, due to the principle of assigning first and then using.

The SSTORE of the *key* flowing into JUMPI will have a greater impact on the contract. Collection of functions that write to SSTORE variables, which help to generate transaction sequences.

Let the transaction sequence be $Tra = \langle fun_1 ... fun_n \rangle$. Where Fun_n is the danger function chosen from transaction function selection. Analyze Fun_n to get the JUMPI constraint that SLOAD flows into. Obtain a JUMPI whose storage quantity is greater than 0 from Fun_n's uncovered JUMPI, and then obtain the set of *keys* that affect the storage variable of this branch. Search for the function that writes the *key* set in $< key, SLOAD, SSTORE >$. And then make the function part of the transaction sequence $\langle fun_1 ... fun_{n-1} \rangle$. At the same time, the sequences with increased coverage are stored.

3.3 Genetic Algorithm-Based Test Cases

A complete transaction call should include functions and their corresponding parameters. The second step of the invocation module faces the problem of test case generation. Efficient and lightweight multi-objective adaptive strategies in sFuzz had been shown to be very effective for generating test cases. However, the algorithm used in sFuzz does not consider the influence of the uncertainty of the value of global variables, which makes some vulnerabilities undetectable. Factors that affect branching in a contract may be parameters of functions or due to global variables. Therefore, on the basis of considering the transaction sequence, it is necessary to consider the case where branch constraints depend on global variables. For the case where the branch constraint does not depend on the global variable in the function. We take the method used by sfuzz, combined with the factor of coverage, to generate test cases for uncovered branches.

Genetic algorithm first considers generative fitness. Two kinds of fitness are adopted, one for the branch constraints that do not depend on global variables, and one for considering branch constraints that depend on global variables. The first method does not consider global variables. The fitness algorithm has better coverage for hard constraints. According to the branch constraints, the branch distance is determined by using the comparison opcode in the EVM. The smaller the branch distance, the more suitable the test case is to survive, and it is taken as the selected test case t. In this method, the branch distance is no longer obtained for the covered branch, so that the algorithm is more focused on covering the uncovered code.

Second, consider the impact of global variables, as shown in Algorithm 1. During each execution of the genetic algorithm, for the branch constraints that are not covered and have global variables flowing in, the branch distance is obtained in the same way as the first fitness. However, the test cases with small distance from the branch do not necessarily contribute to the coverage of the branch. Rather, other transactions indirectly affect the branch by affecting global variables. Therefore, for the fitness of this branch, functions with global variable

Algorithm 1. getTestCase

```
Require: fun
Ensure: returnParameter, relatedFunTestCase
 1: if There is an uncovered branch in fun then
 2:   if storageDependencyNum[fun] > 0 then
 3:     if η then
 4:       relationFuns ← [ ]
 5:       for key in compareStorage [comparepc] do
 6:         relationFuns.append (writefun(key))
 7:       end for
 8:       totalLength ← λ
 9:       for totalLength>0 do
10:         if iscfun(relationFuns) then
11:           relatedFunTestCase[cfun] ← (cfun related test case, fun current branch
              distance)
12:         end if
13:       end for
14:     end if
15:   else
16:     GeneticAlgorithm ()
17:     return returnParameter,relatedFunTestCase
18:   end if
19: else
20:   outFunctionGeneticAlgorithm()
21:   return returnParameter
22: end if
```

dependencies need to be considered. In order to get the test cases that depend on the function, it is necessary to traverse the historical transactions. Get the position of the current test case in the whole testing process, look for the dependent function from this position forward, and record the test case of the dependent function that affects it.

After obtaining the fitness, the selection operation needs to be performed according to the fitness. There are two options. The first option: a function whose branching constraints do not depend on global variables uses branching distance as an important factor to select test cases suitable for survival. For each branch in the function, a test case with the smallest branch distance is selected as the fit for survival test case. The second option: For the sequence $< f_a, f_b >$, the current function f_a indirectly affects the function f_b by affecting the global variable. While using the first method to select test cases suitable for covering branches in f_a, it is also necessary to select test cases suitable for covering branches in f_b. In this method, the test case of f_a which makes the branch distance of f_b small is regarded as the test case suitable for survival.

3.4 Build Mutation Pool

In order to test the contract more quickly, we built a mutation pool and used it in the mutation phase of the genetic algorithm. We build mutation pools for hard-coded and transaction-related global variables. Hardcoding helps in some cases to generate test cases. The system uses dynamic taint analysis to obtain hard codes. Take CALLDATALOAD and CALLDATACOPY as the source to obtain the direct dependencies between transaction parameters and constraints. With comparison opcodes as tainted sinks, comparison opcodes usually exist as constraints. The operands to get the comparison opcode are used to generate hardcoding. The transaction-related global variables of the blockchain have a greater impact on the coverage of fuzzing. The system uses dynamic taint analysis to build a special global variable mutation pool to quickly cover constraints that are dependent on special global variables.

4 Algorithm

In order to deal with the impact of global variables on fuzzing, the impact of global variables on branching needs to be considered in the genetic algorithm. To do this, the fitness between functions needs to be calculated, and the algorithm is shown in Algorithm 1. This algorithm is used for the second fitness generation. The algorithm mainly introduces how to generate survival-fit test cases in relationFuns that affect fun coverage. The algorithm takes *fun* that reaches the genetic algorithm condition as input, outputs the test case returned by *returnParameter*, and *relatedFunTestCase*, which is the test case suitable for survival in the correlation function. *callFunHistory* is the call status of transactions in history, and *storageDependencyNum* is the number of dependencies between global variables and branches. The read-write relationship of the storage variable in the *funStorage* is in the form of { *key*: {'SLOAD': function name, 'SSTORE': the function name}}, and the direct relationship between storage and the comparison operator in the *compareStorage* is in the form of {functionName: {comparePC:{*key*}}}. *comparePC* is the pc for comparing opcodes.

η is a bool value that is true when a branch *comparepc* is selected from the branch, and the branch contains a global variable dependency . *writefun(key)* is the name of the function in *funStorage* that performs the SSTORE operation on the *key*. *writefun(key)* is the relationship between the storage variable and the function name .λ is the subscript of the position where the test case is found from back to front in *callFunHistory*. *callFunHistory* is the historical transaction information *.iscfun(relationFuns)* is a bool value. Traverse *callFunHistory* from λ forward, when the function cfun in relationFuns is found, the bool value is true.

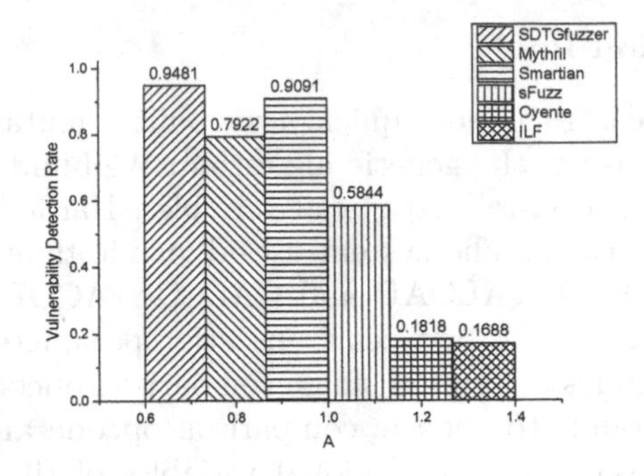

Fig. 2. Contract detection rate for dataset.

5 Experiment Evaluation

In this section, we introduces the content of the experiment, and verifies the effectiveness of this method by comparing it with advanced fuzzing tools.

5.1 Experiment Configuration

We used dataset tests the accuracy of our method for a given labeled contract. For the design of the dataset, We adopted smartbugs' SB Curated dataset. And we used 4 types of vulnerabilities to test for unchecked return value exceptions, block-dependent and timestamp-dependent vulnerabilities, and Arithmetic. For contracts with inheritance relationship, we only tested the last contract. For datasets marked by smartbugs, Arithmetic, Bad Randomness and Time Manipulation each have a contract that does not directly cause the vulnerability.

We writing system in python and runs on the win10 system. The system uses the solc version of 0.4.25+commit.59dbf8f1.Windows.msvc,and use Ganache as a private chain platform. The CPU is Intel(R) Core(TM) i5-9500 CPU @ 3.00 GHz and the memory is DDR4, 8 G. We ran each method for up to 10 min

5.2 Results and Analysis

In this section, We test each tool based on the number of vulnerable contracts detected for each contract.

Testing on the Dataset. We tested each method in the database. Compared with sFuzz, SDTGfuzzer has been greatly improved in contract Vulnerability. The results are shown in Fig. 2. Since Smartian, sFuzz, and Mythril simultaneously support vulnerabilities in all datasets, we calculate their accuracy under

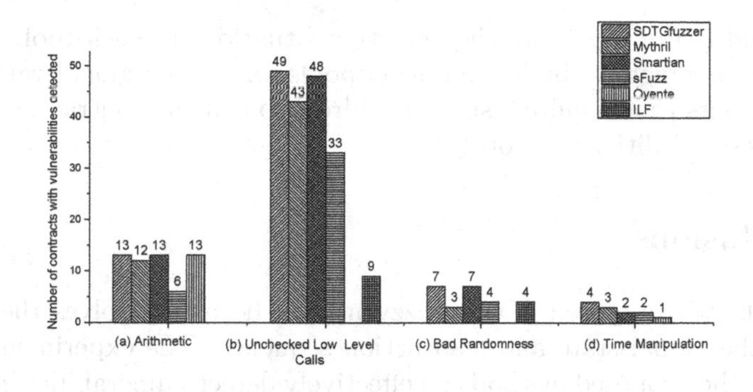

Fig. 3. Detection results of each method in dataset.

the dataset. Compared to smatian, we can detect 3.9% more vulnerabilities in contracts. The reason for the low detection rates of Oyente and ILF is that they do not fully support these vulnerabilities. Oyente does not support Unchecked Low Level Calls, and ILF does not support Arithmetic. Figure 2 shows the detection rates of these methods in the smartbugs dataset.

For the test in Arithmetic. The detection result is shown in part (a) of Fig. 3. Since sFuzz does not perform vulnerability analysis for multiplication, the multiplication related vulnerabilities cannot be detected. The vulnerability detection method proposed by DTGfuzzer for Arithmetic refers to the method of Osiris, but the method of Osiris does not carry out detailed analysis on integer overflow between functions, so it cannot detect cross-contract vulnerabilities. Our method can further analyze cross-contract vulnerabilities by designing taint analysis tools.

Because SDTGfuzzer adopts dynamic taint analysis, it can avoid the problem that sFuzz must use interactive contracts to cause exceptions to detect unchecked return value vulnerability. Compared with Smartian, SDTGfuzzer can still obtain a better vulnerability detection rate within a certain period of time. We found that smartian is not good at detecting contracts that require a specific address to call. SDTGfuzzer can easily find the correct address of the calling contract by using the mutation pool, and unlock the contract address on the private chain. It is possible to override contracts that need to be called from a specific address. The detection result of unchecked return value vulnerability is shown in part (b) of Fig. 3.

Both Bad Randomness and Time Manipulation are essentially dependent on the blockchain. For these dependencies, sFuzz may have false positives. Since sFuzz does not analyze the data flow, it cannot correctly judge the flow of variables, and false positives may be generated. There may be situations where an environment variable is not used by a dangerous operation, but sFuzz reports a vulnerability. Since our tool uses dynamic taint analysis, it can track the opcode and determine whether it flows into JUMPI or CALL to determine vulnerability, thereby avoiding false positives. Meanwhile, through the mutation pool, our method can quickly cover the branches constrained by environmental variables.

Parts (c) and (d) of Fig. 3 are the detection situations of each tool. Compared to smartian, our method builds mutation pools for special global variables. For some constraints that require a specific address to call the contract, our method can pass these conditions smoothly.

6 Conclusions

In this work, We proposed SDTG fuzzy model. It mainly solves the problems caused by the global state and transaction sequence. The experimental results showed that the proposed method can effectively detect vulnerability in contract.

Acknowledgement. Natural Science Foundation of Shandong Province (Grant No. ZR2020ZD01).

References

1. Gai, K., Zhang, Y., et al.: Blockchain-enabled service optimizations in supply chain digital twin. IEEE TSC (2022)
2. He, J., Balunović, M., et al.: Learning to fuzz from symbolic execution with application to smart contracts. In: ACM CCS, pp. 531–548 (2019)
3. Hu, F., Lakdawala, S., et al.: Low-power, intelligent sensor hardware interface for medical data preprocessing. IEEE TITB **13**(4), 656–663 (2009)
4. Jiang, B., Liu, Y., Chan, W.: Contractfuzzer: Fuzzing smart contracts for vulnerability detection. In: 33rd IEEE/ACM International Conference ASE, pp. 259–269 (2018)
5. Krupp, J., Rossow, C.: {teEther}: Gnawing at ethereum to automatically exploit smart contracts. In: 27th USENIX Security Symposium (USENIX Security 18), pp. 1317–1333 (2018)
6. Li, Y., Gai, K., et al.: Intercrossed access controls for secure financial services on multimedia big data in cloud systems. ACM TMCCA (2016)
7. Li, Y., Song, Y., et al.: Intelligent fault diagnosis by fusing domain adversarial training and maximum mean discrepancy via ensemble learning. IEEE TII **17**(4), 2833–2841 (2020)
8. Liu, C., Liu, H., et al.: Reguard: finding reentrancy bugs in smart contracts. In: 2IEEE/ACM 40th International Conference ICSE-Companion, pp. 65–68 (2018)
9. Mueller, B.: A framework for bug hunting on the Ethereum blockchain (2017)
10. Nguyen, T.D., Pham, L.H., Sun, J.: SGUARD: towards fixing vulnerable smart contracts automatically. In: IEEE Symposium on Security and Privacy (SP), pp. 1215–1229 (2021)
11. Nguyen, T.D., Pham, L.H., Sun, J., Lin, Y., Minh, Q.T.: sFuzz: an efficient adaptive Fuzzer for solidity smart contracts. In: Proceedings of the ACM/IEEE 42nd International Conference on Software Engineering, pp. 778–788 (2020)
12. Qiu, H., Zheng, Q., et al.: Topological graph convolutional network-based urban traffic flow and density prediction. IEEE TITS **22**(7), 4560–4569 (2020)
13. Qiu, M., Chen, Z., et al.: Energy-aware data allocation with hybrid memory for mobile cloud systems. IEEE Syst. J. **11**(2), 813–822 (2014)

14. Qiu, M., Jia, Z., et al.: Voltage assignment with guaranteed probability satisfying timing constraint for real-time multiproceesor DSP. J. Signal Proc. Syst. **46**, 55–73 (2007)
15. Qiu, M., Qiu, H., et al.: Secure data sharing through untrusted clouds with blockchain-enabled key management. In: 3rd SmartBlock Conference, pp. 11–16 (2020)
16. Qiu, M., Yang, L., et al.: Dynamic and leakage energy minimization with soft real-time loop scheduling and voltage assignment. IEEE TVLSI **18**(3), 501–504 (2009)
17. Rodler, M., Li, W., Karame, G.O., Davi, L.: Sereum: Protecting existing smart contracts against re-entrancy attacks. arXiv preprint: arXiv:1812.05934 (2018)
18. Torres, C.F., Schütte, J., State, R.: Osiris: hunting for integer bugs in Ethereum smart contracts. In: Proceedings of the 34th Annual Computer Security Applications Conference, pp. 664–676 (2018)
19. Wüstholz, V., Christakis, M.: Harvey: A Greybox Fuzzer for smart contracts. In: 28th ACM European Software Engineering Conference and Symposium on the Foundations of Software Engineering, pp. 1398–1409 (2020)
20. Zhang, P., Yu, J., Ji, S.: ADF-GA: data flow criterion based test case generation for Ethereum smart contracts. In: Proceedings of the IEEE/ACM 42nd International Conference on Software Engineering Workshops, pp. 754–761 (2020)
21. Zhang, Q., Wang, Y., et al.: ETHPLOIT: from fuzzing to efficient exploit generation against smart contracts. In: IEEE 27th Int'l Conf. on Software Analysis, Evolution and Reengineering (SANER), pp. 116–126 (2020)

Construction Practice of Cloud Billing Message Based on Stream Native

Xiaoli Huang[1], Andi Liu[2], Yizhong Liu[2](✉), Li Li[1], Zhenglin Lv[1], and Fan Wang[1]

[1] China Mobile Information Technology Co., Ltd., Shenzhen, China
{huangxiaoliit,liliit,lvzhenglin,wangfanit}@chinamobile.com
[2] School of Cyber Science and Technology, Beihang University, Beijing, China
{liuandi,liuyizhong}@buaa.edu.cn

Abstract. It is necessary to accelerate digital development to make digital economy, society and government shine brightly in the future. Digital technology and the real economy will be deeply integrated, and a large number of new industries and new models will emerge. Cloud computing is an important industry to create further advantages in the digital economy. As the key to the development of enterprise business, cloud computing is the decisive factor. The ability to support billing for the entire cloud is a vital link, and it is the core capability of the whole system. It requires high accuracy and performance. Based on cloud billing, this paper analyzes the challenges cloud computing service support faces. It studies the architecture upgrade, builds the cloud service billing support capability based on stream native technology, meets the flexible billing needs of cloud services, and realizes the rapid and efficient operation support of cloud services. This system has been put into production and played a significant role strongly supporting the high-quality development needs of cloud business billing.

Keywords: Stream Native · Cloud Computing Billing · Billing Message

1 Introduction

With the requirement of digital economy, society and government, a large number of new industries and new models appear and develop quickly. Cloud computing [1,2] has recently emerged as a buzzword in the distributed computing community and one of the scenarios of the digital outline. Cloud computing is a model for enabling convenient, on-demand network access to a shared pool of configurable computing resources (e.g., networks, servers, storage, applications, and services). It can be rapidly provisioned and released with minimal management effort or service provider interaction [3].

There are five essential elements of cloud computing: on-demand self-service, broad network access, resource pooling, rapid elasticity and measured service

M. Qiu et al. (Eds.): SmartCom 2022, LNCS 13828, pp. 414–427, 2023.
https://doi.org/10.1007/978-3-031-28124-2_40

[4]. Depending on the different types of cloud service, cloud community has three service models: *Software as a service* (SaaS), *Platform as a Service* (PaaS) and *Infrastructure as a Service* (IaaS). Furthermore, cloud community can also be classified as private cloud, community cloud, public cloud and hybrid cloud according to the deployment.

However, the development of cloud computing technology also has many bottlenecks [5,6]. The existing cloud computing market is highly centralized. A few technology giants dominate the market share. Relying on their highly centralized server resources, they monopolize the entire cloud computing market and enjoy high profits, leading to high computing service prices. Blockchain-based distributed cloud computing infrastructure will allow on-demand, secure, low-cost access to the most competitive computing infrastructure. Blockchain technology achieves the ledger's consistency among distributed nodes through a specific consensus [7,8]. Distributed cloud computing based on blockchains can arouse transactions between participants triggered by off-chain behaviors, such as providing data sets in real-time, transferring files, performing calculations, and providing professional services [9]. Besides, sharding blockchains [10–12] are prominent solving tools to realize scalability for cloud computing. Also, blockchain-based access control methods [13] could be adopted to keep data safe in cloud environments.

Current researches focus on accelerating the iterative upgrade of cloud operating systems, promoting technological innovations such as ultra-large-scale distributed storage [14,15], elastic computing, data virtual isolation and improving cloud security [16,17]. They are the basis of hybrid cloud industry solutions, system integration and maintenance management. Cloud computing is an important industry to create new advantages of digital economy. By giving full play to the benefits of cloud computing's massive data, artificial intelligence and rich application scenarios, it can promote the deep integration of digital technology and the real economy, enable the transformation and upgrading of traditional industries, spawn new industries, new formats and new models, and strengthen new engines for economic development.

Our Contributions. This paper builds a cloud service support system to support stream native billing capabilities. It surpasses the application of cloud computing technology and meets the diverse needs of different vertical industries. In Sect. 2, we introduce the structure and each part property of our cloud support system. The system charges cloud service with microservice and docker technologies. In Sect. 3, we list the challenges to build a cloud service billing support and build our support system in Sect. 4. The system can carry out multi-dimensional and differentiated billing designs for users. Finally, we summarize our system capabilities and show its combination with state-of-art blockchain technologies in Sect. 5.

2 Cloud Service Support System

2.1 Cloud Support Technology Architecture Based on Cloud-Native

The system adopts a "cloud-native" architecture with microservices + containerization as the core technical support. It can flexibly scale the support capability according to the system's real-time business volume, effectively improving each business's processing efficiency and performance. Figure 1 is the technical architecture of cloud support system.

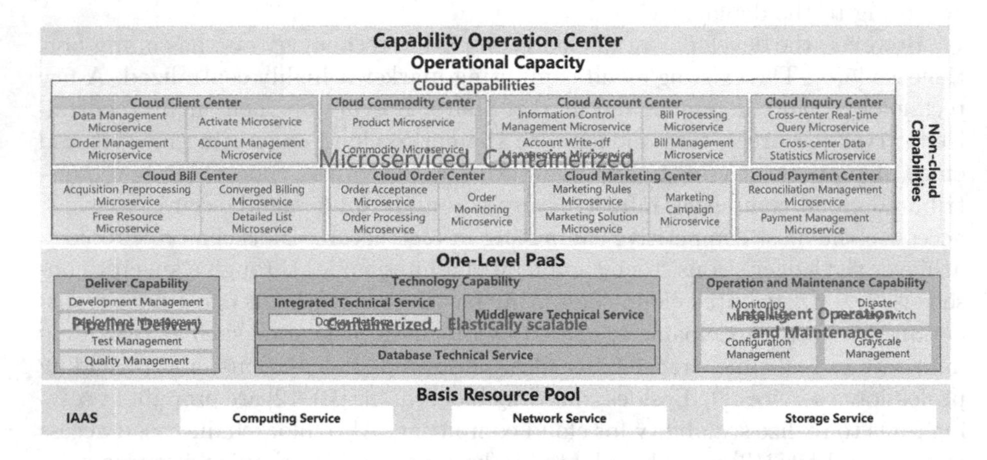

Fig. 1. Technical Architecture of Cloud Support

Cloud-Native Definition. Cloud-native includes a set of applied patterns that help enterprises deliver business software quickly, continuously, reliably, and at scale. Cloud-native consists of microservice architecture, DevOps, and agile infrastructure represented by containers.
Cloud-native applications changed the application architecture, development method, deployment and maintenance technology. It achieves cloud elasticity, dynamic scheduling, automatic scaling and other capabilities that traditional IT does not have. The most significant feature of cloud-native applications is that new services can rapidly deploy.

Cloud-Native Features. There are three classic cloud-native features: microserviced, containerized and DevOps.

(1) Microserviced
Microservice is a style of software architecture that composes large, complex applications in a modular fashion based on small functional blocks focused on a single responsibility and function. It follows the principle of a single operation, independent deployment of services, decoupling of functions, high scalability, simplified development testing and deployment, and friendly to development teams. Improve the overall agility and maintainability of the application through loose coupling. An application container is a lightweight runtime environment that provides applications with files,

variables, and information bases required for their operation, thereby maximizing portability, supporting distributed deployment, elastic scaling, and massive business access.

(2) Containerized
Containers provide applications with an isolated running space: each container contains an entire, complete user environment space, and changes in one container will not affect the running environment of other containers. Container technology uses namespaces for space isolation, in which the container can access files through the mount point of the file system and how many resources each container can use through groups.

(3) DevOps
DevOps (the combination of Development and Operations) is a set of best practice methodologies that provides agile collaboration, code management, build engine, automated testing, automated deployment and other capabilities. Facilitates collaboration and communication among IT stakeholders (including development, operations, and testing) throughout the application and service lifecycle, resulting in:

- Continuous Integration: Easily switch from development to testing and operations.
- Continuous Deployment: Release continuously or as often as possible.
- Continuous Feedback: Seek rapid feedback at all application and service lifecycle stages (Fig. 2).

2.2 Key Technologies and Implementation Schemes

Microservice Technology Framework and Implementation. The system microservice framework follows the principle of "high cohesion and loose coupling". It divides the cloud business support services into microservices and microservices transformation. It builds a microservice governance system to improve the governance capabilities of services. A critical problem that needs to be solved is the communication and invocation problem between microservices (service discovery and service routing). The cloud support system adopts the ServiceMesh microservice framework. It is a new generation of microservice architecture currently promoted by CNCF (*Cloud-Native Computing Foundation*). ServiceMesh is a complex infrastructure layer for handling communication between services. It is essentially a service network composed of network agents. These agents are deployed next to the user's application, and the application's code is unaware of their presence. Through the sidecar of ServiceMesh, the service application itself and the microservice governance framework are decoupled entirely. Business developers can focus on the business itself. Microservice framework engineers focus on microservice orchestration and governance to solve cloud-native applications. The complexity of the microservice architecture ensures reliable and fast delivery of applications. ServiceMesh is an intelligent service network, and the sidecar is a critical component in this intelligent network - a smart router. Relying on Sidecar, ServiceMesh provides a complete set

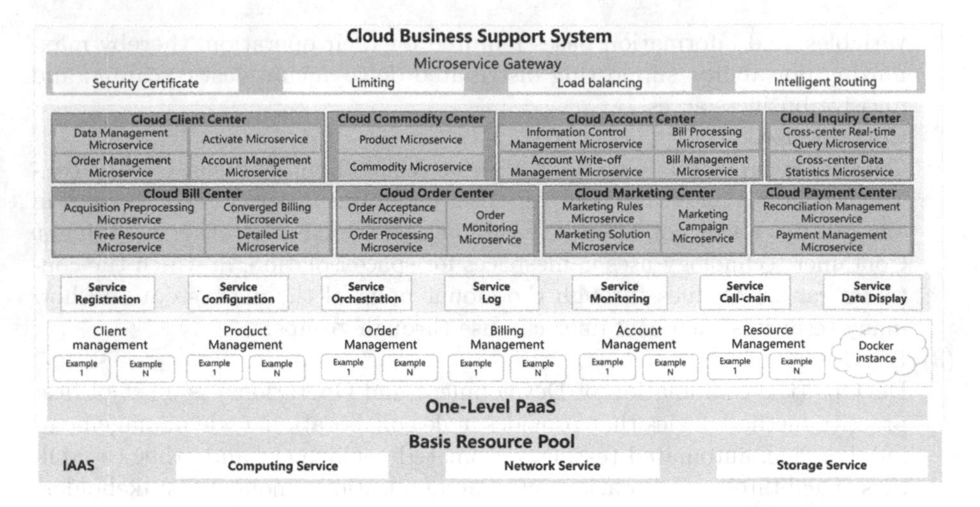

Fig. 2. Cloud Supports Microservice Business Architecture

of microservice architecture solutions, including advanced functions such as service routing, load balancing, traffic control, and circuit breaker. The ServiceMesh components are more streamlined, the interaction efficiency between services is improved, and the business response is faster. The microservice architecture for the cloud support system is shown in Fig. 3. It includes application service layer, microservice PaaS layer and IaaS layer.

Application service layer. First, we introduce the application service layer.

(1) CRM microservice division
Our system divided CRM into 11 microservices:
1. Account management microservice: Realize cloud customer account opening, customer data synchronization, customer manager change and other customer-related management services.
2. Order management microservice: Realize the generation of cloud business orders, order instance management, and order attribute management services.
3. Activation-type microservice: When ordering cloud basic password products, it synchronizes EC enterprise information to the cloud platform and sends basic password products to the cloud platform for activation services.
4. Marketing rule microservice: Implement management services for marketing rule elements, templates and relationships.
5. Marketing activity microservice: Realize management services for marketing activity information, marketing target groups, and marketing scenarios.
6. Order acceptance microservice: Realize order creation, order modification, order cancellation, order closing, order feedback and other services. They are the core services supporting channel business sales acceptance.

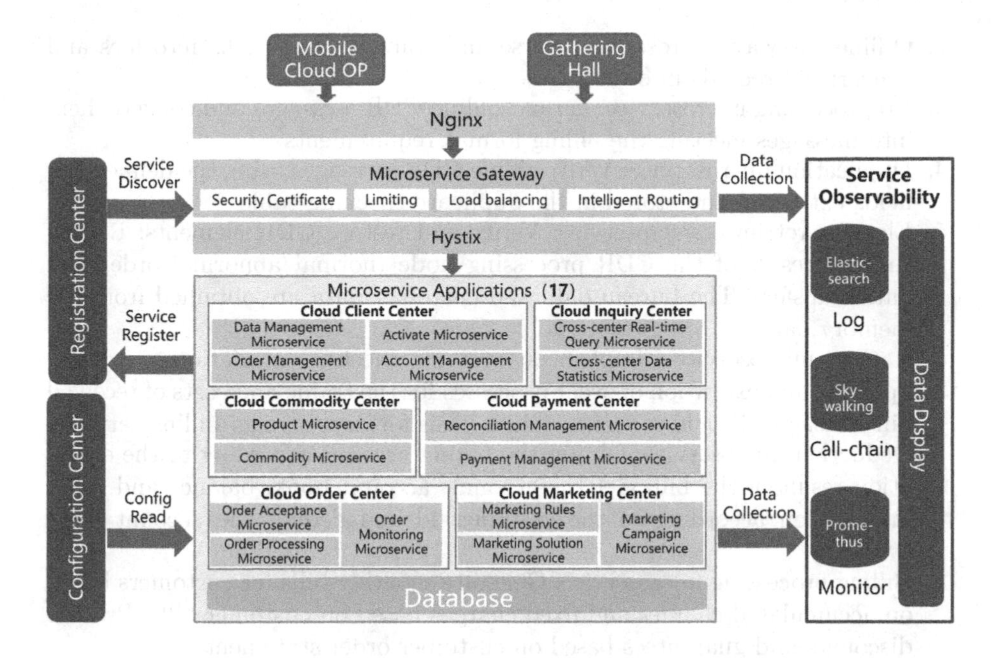

Fig. 3. Cloud Supports Microservice Technology Architecture

7. Order management microservice: realize the functions of order decomposition, scheduling, merging, completion, work order dispatching, feedback, and filling, and complete the execution of external or workflow according to the business, and realize the core scheduling service of sales performance.
8. Product microservice: Realize configuration management of related attributes that constitute products, including configuration services for product creation, product catalogs, product attributes, and product changes.
9. Commodity microservice: Realize total life cycle management of commodities and commodity-related configuration services.
10. Cross-center query microservice: Realize data cross-center common query to solve the situation after the system is microserviced. Each center is independent of the other and cannot be directly related to query access.
11. Payment management microservice: When cloud Internet customers make payments, the cloud business supports the service of invoking unified payment capabilities.

(2) Microservice division of billing and accounting system

The billing and accounting system of the cloud support system is divided into 12 microservices:

1. Collection microservice: Realize the acquisition of original offline bills from the cloud platform and bill reconciliation auditing.

2. Offline gateway microservice: Parse and convert offline bills into files and convert bill records into messages.
3. Preprocessing microservice: Format offline CDR messages and convert them into messages meeting the billing format requirements.
4. Duplication microservice: Verify offline CDR messages using the defined key fields extracted and stored in the duplicate checking table.
5. Element retrieval microservice: Verify and restore CDR elements. Return the call result of the CDR processing node (normal/abnormal order and data transfer). The bureau data and customer data are obtained from the memory bank.
6. Rating microservice: Implement various checks, calculations, records, updates, and execution of event processes for the traffic and costs of received bill messages. Provide necessary data basis for downstream billing services.
7. Deduction microservice: Obtain the billing account according to the evaluation result of the bill. Get the account, account book, balance and other information according to the customer id and deduct the account book cyclically.
8. Billing processing microservice: Generate monthly bills for customers based on accumulated charges and fixed charges based on customer bills. Provide discounts and guarantees based on customer order statements.
9. Billing management microservice: Provide external billing queries, billing data statistics and other services according to the bills generated by the billing processing.
10. Account write-off management microservice: Obtain customer payment records from the centralized ERP system and match them against unwritten bills. Provide the matching results to the front page for display. According to the displayed results, the account manager updates billing record including unpaid amount, bill status, etc., and triggers the signal control startup service after confirming the write-off.
11. Credit control management microservice: Remind customers who exceed the payment cycle and operate shutdown according to the customer's credit control level. Receive the startup request of account write-off management microservice and perform the startup operation.
12. Balance management microservice: Receive customer balance recharge requests and change the balance. Receive the deduction request sent by the batching microservice and change the balance.

Microservice PaaS layer. Next, we introduce the microservice PasS layer.

The service registration center is realized through the PaaS platform as well as the circuit breaker, current limit, and downgrade of services. PaaS platform also provides functions such as service deployment, port, domain name management, and automatic capacity expansion.

IaaS layer. Finally, we describe the IaaS layer.

IaaS layer is provided by IT Ningbo resource pool, including physical equipment, power security, equipment access control, equipment security configuration optimization, and regular security inspections of physical equipment.

Containerization Technology Framework and Implementation. The system adopts a containerized architecture based on k8s+docker, packages applications into containers, and releases them to run on the PaaS platform, realizing the transformation from traditional resource-centric to application-centric. Applications do not need to care about the process running on the platform. On which host only resource requirements need to be raised. Kubernetes is used to manage Docker containers, containers are isolated from each other, and microservice applications form clusters through Docker to provide high-availability services. Kubernetes can manage each container effectively. It has the following characteristics.

1. Separation of application and configuration
   Applications use port numbers, IPs, etc. in the service configuration center. The service configuration center provides unified storage, change, and version maintenance for all configuration files and management. In the subsequent application startup and deployment, the operation and maintenance personnel can associate the application configuration with the application to complete the deployment to avoid repackaging the image every time when the configuration is modified.
2. Application image asset construction
   The basis for container operation is the image, and each application writes a docker file to generate a container image.
3. Ability Arrangement
   The atomic service interfaces are combined according to the expected output capabilities, and the service orchestration layer realizes the control of the calling sequence and service status between the atomic service interfaces to ensure the integrity and consistency of the output capabilities. Service routing can realize multiple instances of the original aggregation capability. It ensures each atomic service request and parameter of the orchestration call correctly passed to the appropriate service instance and returns the result and status.
4. The application provides a health check interface
   The application provides a survival and service availability check interface for the K8S to call. Once the application is unavailable, the K8S will kill the original container and restart the new container to improve the reliability of the application.
5. Application log
   Stream collection method transmits the logs generated by the computing nodes to ELK or big data platform in real-time, which realizes the real-time display of business report data.

Cloud Service Charging Framework and Implementation. In the model of cloud service billing 1.0, each module interacts thru files and processes offline CDR in the traditional way shown in Fig. 4.

Traditional offline CDR file processing must go through multiple steps, such as collection, preprocessing, deduplication and rating. It relies on a high-performance distributed shared file system with large demand storage and high

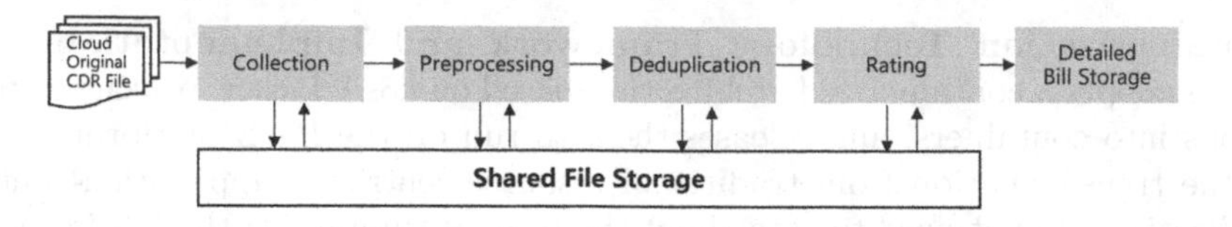

Fig. 4. Traditional Offline CDR File Processing

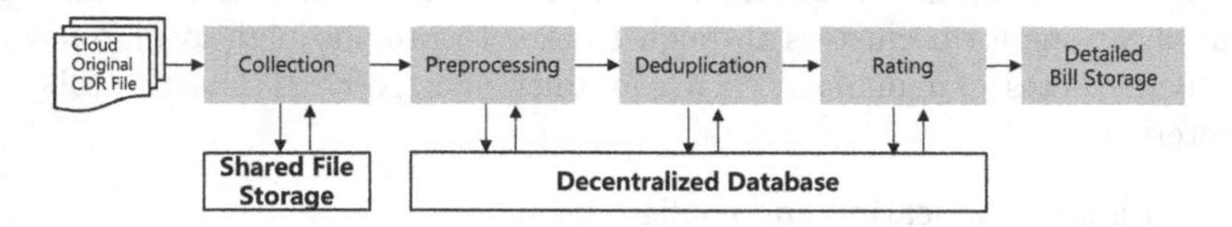

Fig. 5. Cloud-based and Distributed Billing

cost. Dealing with traditional offline file processing flaws, cloud service billing 2.0 in Fig. 5 meets cloud-based and distributed processing. It generally realized X86 and elastic expansion under a service-oriented, centralized capacity building, in line with the development trend of the industry. The bottom layer adopts distributed database technology to support the storage of many bills, achieving high reliability, maintainability and online capacity expansion.

3 Challenges Faced by Cloud Service Billing Support

In business development, the cloud is the key to the success of government and enterprise affairs in the cloud. The billing capability is an essential part of the cloud business and the core capability of the entire system, which requires high accuracy and performance. At the same time, the number of cloud business customers, subscriptions and bills continue to grow rapidly, placing higher requirements on billing efficiency and system resource utilization. The cloud business has thrived with various products, increasing business complexity, and complex and diverse billing methods. Cloud services have higher and higher requirements on billing scenarios and complexity and require billing systems to have higher support efficiency and more substantial support capabilities (Fig. 6).

4 Cloud Service Billing Support

4.1 Stream Native Architecture

The new generation of stream native (Pulsar) technology architecture is based on the interactive data flow mode and adopts a flexible and unified message processing model. It realizes the message nation of the whole system process and solves the high I/O brought by the existing file processing method, which achieves high throughput and low latency CDR processing capability.

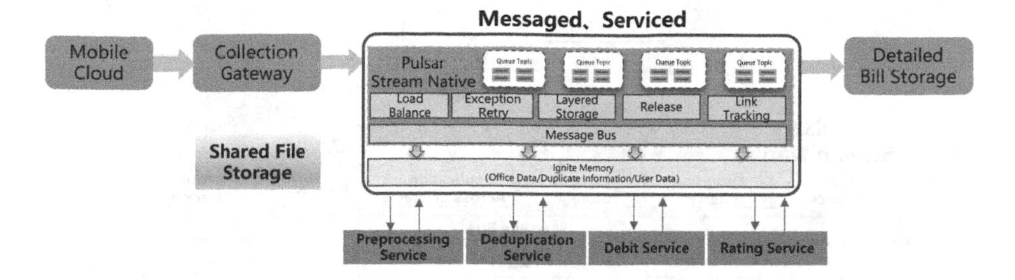

Fig. 6. Stream Native Application Framework Diagram in Billing System

1. *Acceleration of Message Processing.* We build a distributed stream billing system based on the new stream native queue component combined with the stream processing concept. It realizes the full message bearing of the billing process and reduces the number of file landings to improve the overall execution efficiency. 2. *Consolidation and streamlining of business links.* By re-abstracting the existing billing business process, dividing the functions and merging the links, we encapsulate the business implementation with a functional and service-oriented architecture that supports multi-mode business carrying. 3. *Visualized process standard* By using the call chain tracking and APM components to track the link call relationship and performance in units of billing "request packets". The standardized and general RPC protocol is used between business components, which is convenient for upgrades and function expansion. 4. *Balanced and hierarchical resource storage.* The system introduces distributed message storage components to strengthen the utilization of local disk resources. With existing memory libraries, distributed file storage, and shared storage, they formed a five-layer storage architecture that decouples high-performance shared storage dependencies and reduces storage expansion costs. 5. *Center-level cluster takeover.* The cross-regional replication and synchronization of message data support the switching between message processing clusters, allowing quick taking over other centers with billing center-level exceptions. 6. *Non-interruption online business.* With the release capability of pulsar components, the billing process will not be interrupted to realize the ability of lightweight and fast support for billing when the version is upgraded. A set of billing logic can realize various supports for service, batch and real-time processing (Fig. 7). The billing module of the cloud support system connects to the Pulsar stream to input and output data streams, and implements functions such as message adaptation, process engine, service management, and configuration management through the service control CN node. The business carries modules such as preprocessing, duplicate checking, pricing, deduction, and auditing. The Pulsar streaming native technology implements streaming processing, load balancing, and exception retry for billing. The offline gateway microservice converts the bill file into a message and puts it in the Pulsar stream native message queue for business control acquisition and processing. The business control performs message adaptation according to the type of the CDR message, and the process execution engine

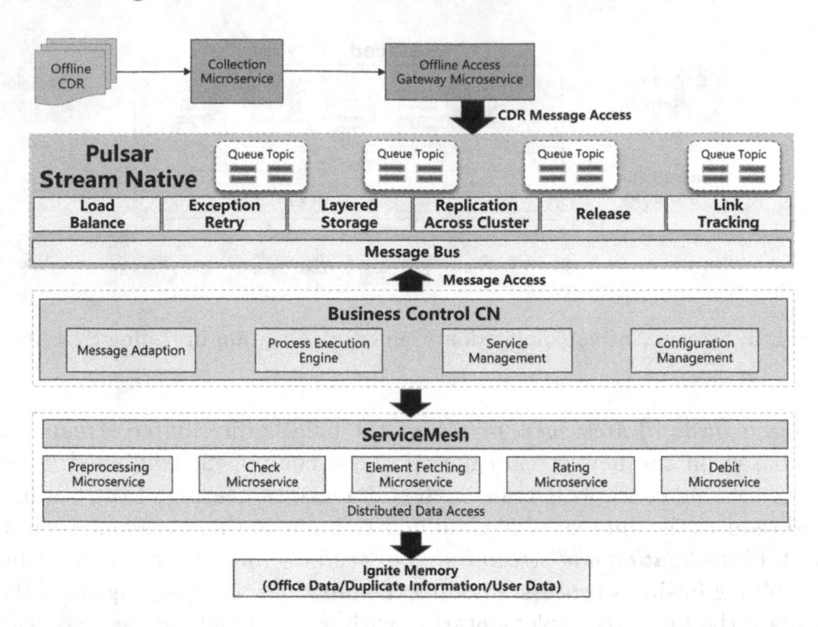

Fig. 7. Stream Native Application Implementation Diagram in Billing System

calls different microservices for message processing. After the message processing is completed, the message is put into the Pulsar stream native message queue for business control calls for further processing. After the billing message is processed, the Pulsar stream natively outputs the processed message as a file for subsequent detailed bill storage and other applications. The ability to support lightweight and fast billing is realized through the application of the streaming native architecture in the billing system. A set of billing logic can realize multiple support modes of service, batch, and real-time processing. At the same time, the service-oriented business access capability can better support the development of vertical industries and provides various external service capabilities, including authority billing, pay-per-view, and differentiated rate billing.

4.2 Security Policy of Data Transmission Based on Stream Native Billing Capability

In the process of data transmission using pulsar, based on the security needs of the project itself, it is necessary to encrypt the transmitted data. There are usually two types of encryption algorithms: symmetric and asymmetric. In symmetric encryption, the same key is used for encryption and decryption; in asymmetric encryption, two keys are used. Generally, the public key is used for encryption and the private key is used for decryption. Symmetric encryption and decryption are relatively fast, while asymmetric encryption and decryption take a long time and the speed is relatively slow. Because stream native has extremely high real-time requirements for data processing, asymmetric encryption algorithms cannot meet the needs in terms of processing efficiency. Symmetric encryption is used here. The stream native in the EBOSS system includes a large amount of data information, such as the current user's order data, the message body itself is large, and the message length is further expanded after AES encrypts the data. The encryption takes a long time, which affects the timeliness of message delivery and increases. It increases the network load pressure of the streaming native system. To solve this problem, we construct Huffman coding trees to improve the efficiency of data encryption and decryption. It reduces the stress of data network transmission. In the traditional Huffman algorithm, data compression is based on character statistics of the current data to be compressed and then encoded. However, in combination with stream native application scenarios, if each stream data is constructed with a separate Huffman coding tree, the cost of processing massive stream data will be very high, and the transmission efficiency will also be affected. Considering that in our business system, different single stream data have high similarity and repetition, we use pre-timed to construct a Huffman coding tree based on global stream data and load it into each stream native application node. The compression method that only encodes and decodes based on the existing code tree maximizes the data compression ratio on the premise of ensuring compression performance.

In the future, cloud business support will continue to study system implementation based on technologies such as artificial intelligence, neural networks, and blockchain systems, continuously optimizing system capabilities in terms of intelligence. With blockchain consensuses, such as PBFT [18], Hotstuff [19] and SSHC [20], different clouds can communicate as nodes in blockchain system and achieve agreement on the computing results. As one of the most promising and appealing notions in blockchain technology, the self-enforcing and event-driven features of smart contracts is also adapted to cloud computing application deployment [21]. Some digital rights management [22], cross-chain technology [23], and cloud storage [24] could be combined with cloud computing [25] to realize a better application. At the same time, we will continuously improve the professionalism and flexibility of cloud business support based on in-depth support and understanding of cloud business.

5 Conclusion

This paper systematically studied the architecture upgrade and built the cloud service billing support capability based on stream native technology. The upgraded cloud service billing capability can meet the multi-scenario billing needs of the cloud business and realize the intelligent operation support of the cloud business. The system had been put into production and had achieved remarkable results, which strongly supported the high-quality development of cloud business.

Acknowledgment. This paper is supported by the National Natural Science Foundation of China (U21B2021, 62202027, 61972018, 61932014), Yunnan Key Laboratory of Blockchain Application Technology (202105AG070005-YNB202206).

References

1. Zou, J., He, D., et al.: Integrated blockchain and cloud computing systems: a systematic survey, solutions, and challenges. ACM Comput. Surv. **54**(8), 1–36 (2022)
2. Okegbile, S.D., Cai, J., Alfa, A.S.: Performance analysis of blockchain-enabled data-sharing scheme in cloud-edge computing-based IoT networks. IEEE IoT J. **9**(21), 21520–21536 (2022)
3. Mell, P., Grance, T.: Draft NIST working definition of cloud computing, Referenced on June. 3rd **15**(32), 2 (2009)
4. Dillon, T., Wu, C., Chang, E.: Cloud computing: issues and challenges. In: 24th IEEE Conference on Advanced Information Networking and Applications, pp. 27–33 (2010)
5. Qiu, M., Chen, Z., et al.: Energy-aware data allocation with hybrid memory for mobile cloud systems. IEEE Syst. J. **11**(2), 813–822 (2014)
6. Gai, K., Qiu, M., Elnagdy, S.: A novel secure big data cyber incident analytics framework for cloud-based cybersecurity insurance. In: IEEE BigDataSecurity (2016)
7. Liu, Y., Liu, J., Zhang, Z., Xu, T., Yu, H.: Overview on consensus mechanism of blockchain technology. J. Cryptol. Res. **6**(4), 395–432 (2019)
8. Liu, Y., Liu, J., Zhang, Z., Yu, H.: A fair selection protocol for committee-based permissionless blockchains. Comput. Secur. **91**, 101718 (2020)
9. Liu, Y., Liu, J., Hei, Y., Xia, Yu., Wu, Q.: A secure cross-shard view-change protocol for sharding blockchains. In: Baek, J., Ruj, S. (eds.) ACISP 2021. LNCS, vol. 13083, pp. 372–390. Springer, Cham (2021). https://doi.org/10.1007/978-3-030-90567-5_19
10. Liu, Y., Liu, J., et al.: Building blocks of sharding blockchain systems: Concepts, approaches, and open problems. Comput. Sci. Rev. **46**, 100513 (2022)
11. Liu, Y., Liu, J., Yin, J., Li, G., Yu, H., Wu, Q.: Cross-shard transaction processing in sharding blockchains. In: Qiu, M. (ed.) ICA3PP 2020. LNCS, vol. 12454, pp. 324–339. Springer, Cham (2020). https://doi.org/10.1007/978-3-030-60248-2_22
12. Kokoris-Kogias, E., Jovanovic, P., et al.: Omniledger: a secure, scale-out, decentralized ledger via sharding. In: IEEE SP, pp. 583–598 (2018)
13. Liu, Y., Qiu, M., Liu, J., Liu, M.: Blockchain-based access control approaches. In: IEEE CSCloud, pp. 127-132 (2021)
14. Li, J., Ming, Z., et al.: Resource allocation robustness in multi-core embedded systems with inaccurate information. J. Syst. Arch. **57**(9), 840–849 (2011)
15. Hu, F., Lakdawala, S., et al.: Low-power, intelligent sensor hardware interface for medical data preprocessing. IEEE TITB **13**(4), 656–663 (2009)
16. Li, Y., Gai, K., et al.: Intercrossed access controls for secure financial services on multimedia big data in cloud systems. ACM TMCCA **12**, 1–18 (2016)
17. Qiu, M., Qiu, H., et al.: Secure data sharing through untrusted clouds with blockchain-enabled key management. In: 3rd SmartBlock Conference, pp. 11–16 (2020)
18. Castro, M., Liskov, B., et al.: Practical byzantine fault tolerance. In: OsDI, vol. 99, pp. 173–186 (1999)
19. Yin, M., Malkhi, D., et al.: HotStuff: BFT consensus with linearity and responsiveness. In: ACM Symposium on Principles of Distributed Computing, pp. 347–356 (2019)
20. Liu, Y., Liu, J., et al.: SSHC: a secure and scalable hybrid consensus protocol for sharding blockchains with a formal security framework. IEEE Trans. Dependable Secur. Comput. **19**(3), 2070–2088 (2022)

21. Hu, B., Zhang, Z., et al.: A comprehensive survey on smart contract construction and execution: paradigms, tools, and systems. Patterns **2**(2), 100179 (2021)
22. Hei, Y., Liu, J., et al.: Making MA-ABE fully accountable: a blockchain-based approach for secure digital right management. Comput. Netw. **191**, 108029 (2021)
23. Hei, Y., Li, D., Zhang, C., Liu, J., Liu, Y., Wu, Q.: Practical agentchain: a compatible cross-chain exchange system. Future Gener. Comput. Syst. **130**, 207–218 (2022)
24. Hei, Y., Liu, Y., Li, D., Liu, J., Wu, Q.: Themis: an accountable blockchain-based P2P cloud storage scheme. Peer-to-Peer Netw. Appl. **14**(1), 225–239 (2021)
25. Qiu, C., Yao, H., Jiang, C., Guo, S., Xu, F.: Cloud computing assisted blockchain-enabled internet of things. IEEE Trans. Cloud Comput. **10**(1), 247–257 (2022)

A Fine-Grained Access Control Framework for Data Sharing in IoT Based on IPFS and Cross-Blockchain Technology

Jiasheng Cui, Li Duan(✉), Mengchen Li, and Wei Wang

Beijing Key Laboratory of Security and Privacy in Intelligent Transportation, Beijing Jiaotong University, Beijing, China
{21120469,duanli,mengchen.li,wangwei1}@bjtu.edu.cn

Abstract. *Internet of Things* (IoT) data from different trust domains is usually shared to assist in providing more services, where privacy sensitive information of shared data will be leaked or accessed without authorization. The traditional centralized access control method is difficult to adapt to the current dynamic and distributed large-scale IoT environment, and there is a risk of the single point of failure. To address these challenges, we propose a fine-grained access control framework for shared data based on cross-blockchain technology and *Interplanetary File System* (IPFS). In this framework, we firstly introduce a cross-blockchain module to realize cross-domain data sharing and solve the problem of data isolation between different data domains in IoT. Then IPFS is used to store the shared data, avoiding the risk of centralized storage. Combining symmetric encryption algorithm with ciphertext policy attribute based encryption (CP-ABE) algorithm, the fine-grained access control of shared data is guaranteed. In addition, the blockchain is applied to store the decryption key and the storage address of the original data, which records the authorization operation of access transactions and audits the access behavior of users. Experimental results show that the proposed scheme can provide higher performance compared to centralized access control methods.

Keywords: cross-blockchain · IPFS · data sharing · CP-ABE · fine-grained access control

1 Introduction

With the development and implementation of 5G, *Internet of things* (IoT), big data, artificial intelligence and other technologies, especially with the continuous increase of IoT device [1], a large amount of data has also been generated, which needs to be stored, processed and analyzed to create value [2,3]. The traditional cloud based access control schemes [4] cannot prevent malicious cloud servers from disclosing users' data, the privacy of the stored data is damaged [5,6].

At present, the blockchain-based cloud data management schemes [7] that have been proposed still have problems such as the risk of cloud server centralization and the low storage capacity of blockchain network nodes. What is more,

M. Qiu et al. (Eds.): SmartCom 2022, LNCS 13828, pp. 428–438, 2023.
https://doi.org/10.1007/978-3-031-28124-2_41

the data domains of the IoT [8] are not interconnected, which also results in different degrees of data isolation. Cross-blockchain technology [9] is an important method to realize the data interaction between different blockchains, and can be used to construct the communication infrastructure between different blockchain networks composed of IoT devices in different trust domains.

At the same time, secure data sharing has high requirements for the privacy [10], confidentiality and security [11]. Therefore, fine-grained access control has become an important means to protect user data privacy and share data efficiently and securely.

Interplanetary File System (IPFS) [12] is a network transmission protocol aimed at creating persistent and distributed storage and sharing files. As a distributed storage solution, IPFS has the characteristics of low cost, high efficiency and high security. Compared with cloud server, IPFS is a distributed storage solution, which solves the single point of failure problem caused by cloud storage.

As an open, transparent, tamper proof and traceable emerging decentralized network, blockchain technology is widely used in various fields. Distributed blockchain network can well solve the defects of centralized access control scheme. The exit of any accounting node in the blockchain will not affect the stability of the whole system. The blockchain programmable smart contract [13] can realize data sharing in the state of encryption, while automatically processing user data access requests, effectively avoid the disclosure of sensitive information caused by the openness and transparency of the blockchain.

In this paper, we use the cross-blockchain technology and the IPFS to design a novel data sharing scheme in the IoT scenario. In this method, IPFS is used to realize distributed data storage, blockchain is used to realize shared data storage, cross-blockchain module is introduced to realize cross-domain data sharing. Since CP-ABE technology [14] can make specific access policies, only the data users who meet the attributes can access the data successfully, which can flexibly manage user's access rights. The main contributions of this paper are as follows:

- This scheme adopts the IPFS to realize the distributed storage of shared data, avoiding the risk of centralized storage.
- This scheme uses blockchain to realize the on-chain storage of shared data. The introduced cross-blockchain module can realize cross-blockchain data sharing, which solves the problem of data isolation between different trust domains in traditional IoT.
- This scheme uses CP-ABE technology to encrypt the key, which can flexibly specify the access authority of a single user. Only when the attribute meets the access policy can the access be successful. It solves the problem of fine-grained access control of shared data.

The rest of this paper is organized as follows. In Sect. 2, some related work about the existing data access control methods is presented. The proposed model, data access process and security analysis are proposed in Sect. 3. To demonstrate the validity of our method, experiment analysis is presented in Sects. 4. The application scenario of the proposed scheme is introduced in Sects. 5. Finally, the conclusion is shown in Sect. 6.

2 Related Work

In order to achieve data secure sharing and privacy protection [15], a personal data management system based on blockchain technology [16] was proposed. This system combines blockchain and non-blockchain storage to construct a privacy-focused personal data management platform, which can better protect users' data privacy [17]. Unfortunately, this scheme still uses a centralized cloud to store data. In order to solve the confidentiality and privacy protection of data in the IoT system, a blockchain architecture system for IoT [18] was proposed, which used *Attribute Based Encryption* (ABE) technology to achieve data access control to solve the privacy and confidentiality of data shared in the blockchain IoT ecosystem. Among them, the data recorded by the sensors of the IoT needed to be transmitted to the cluster head for centralized processing and encryption, which still has the risk of centralization.

In addition, in order to solve the problem that traditional access control methods cannot support the security of private data access control process in the current IoT, an auditable and attribute-based access control system based on blockchain [19] was proposed. [20] proposed a privacy-protecting medical data sharing scheme based on blockchain, which is supplemented by proxy re-encryption and zero-knowledge proof technology. At the same time, a decentralized access control mechanism based on blockchain and ABE [21] was proposed. This scheme redesigns blockchain transaction, token encryption, token initialization and token update schemes to achieve cross-domain, fine-grained and flexible permission management.

In [22], a fine-grained access control scheme for attribute revocation in the blockchain IoT system was proposed. This scheme combines chameleon hash and ABE technology on the multi-layer blockchain scheme to realize the dynamic update of attributes. However, the key generation and distribution of this scheme heavily relies on a centralized authority. [23] proposed an intelligent usage-based insurance system premium competition scheme with privacy protection based on cross-chain. This scheme uses cross-chain technology to connect multiple blockchains to form an open multi-chain premium competition ecosystem. In order to solve the interconnection of the system and the need of data sharing among multiple organizations in cross-organizational collaboration, a blockchain-based access control scheme [24] was proposed. This method uses the consortium blockchain to establish a trusted environment. Then the *Role-Based Access Control* (RBAC) model is deployed in this environment using multi-signature protocol and smart contract approach.

3 The Proposed Scheme

3.1 System Model

In this section, we describe a data sharing scheme based on IPFS and cross-blockchain. The system model is shown in Fig. 1. The model includes four entities: source blockchain, target blockchain, cross-blockchain module, and IPFS. Among them, the source blockchain contains the data owner, who owns the data can be shared. The target blockchain contains the data user, who needs to use the shared data and can interact with the cross-blockchain module. Besides, IPFS can generate Content Identifier (CID), which is the storage address of the original data. In our scheme, assume that IPFS is not trusted, the execution process of the designed system is as follows:

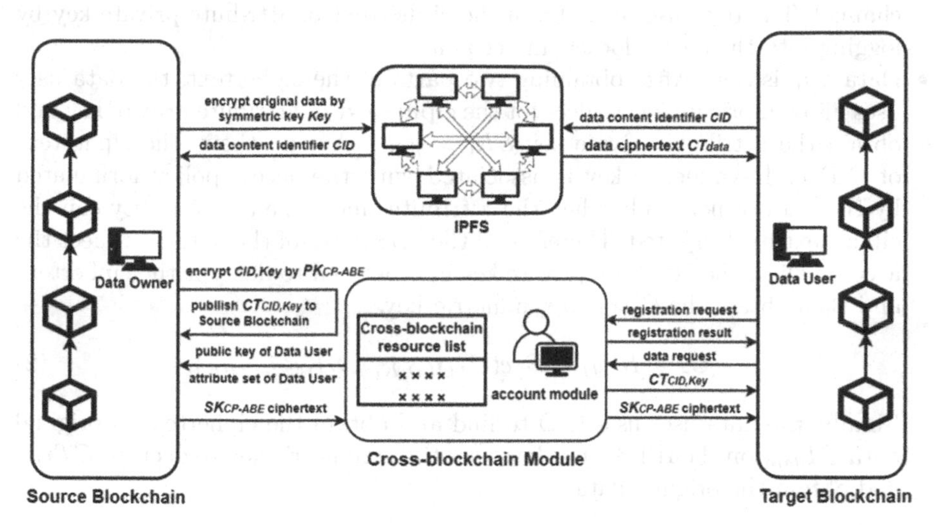

Fig. 1. The model of cross-blockchain data sharing access control.

- System initialization: The data user sends a registration request to the cross-blockchain account module. After the cross-clockchain account module verifies the registration request, it helps the user complete the registration process and return the registration result. Then, the data owner generates the system public key PK_{CP-ABE} and system master key MK_{CP-ABE}.
- Data publishing: The data owner encrypts the original data using a symmetric key Key to obtain the ciphertext CT_{data}. The data owner uploads the ciphertext CT_{data} to IPFS and receives the unique content identifier CID corresponding to the data. Then, the data owner encrypts CID and the symmetric key Key using the system public key PK_{CP-ABE} in conjunction with the access policy

$$CT_{CID,Key} = Enc(PK_{CP-ABE}, Policy, CID, Key) \tag{1}$$

  The ciphertext is associated with the access policy, and can be decrypted only if the attributes meet the access policy in the ciphertext. Then the data owner publishes the ciphertext of data $CT_{CID,Key}$ and access control policy to the source blockchain.
- Data request: The data user on the target blockchain logs in the cross-blockchain account module, checks the cross-blockchain resource list, and sends a data request to data user by calling the relevant contract. If the contract is called successfully and the attribute set of the data user meets the access policy, then the smart contract will send the ciphertext $CT_{CID,Key}$ of the CID and symmetric key to the data user, and send the public key

and attribute set of the data user to the data owner. The data owner uses the attribute set to generate the attribute private key SK_{CP-ABE} and uses public key of the data user to encrypt the attribute private key to generate ciphertext of attribute private key. The data owner sends the cpabe-key ciphertext to the cross-blockchain account of the data user through the secure channel. The data user can obtain the ciphertext of attribute private key by logging into the cross-blockchain account.

- Data acquisition: After obtaining two parts of the ciphertext, the data user uses his own private key to decrypt the ciphertext of attribute private key and obtain the attribute private key SK_{CP-ABE}. Among them, the ciphertext of CID and symmetric key is associated with the access policy formulated by the data owner. Only when the attributes meet the access policy can the ciphertext be decrypted. Therefore, if the attributes of the data user meet the access policy, the attribute private key can be used to decrypt the ciphertext and then obtain the CID and symmetric key.

$$(CID, Key) = Dec(CT_{CID,Key}, SK_{CP-ABE}) \tag{2}$$

Finally, the data user uses CID to find and obtain the ciphertext of original data CT_{data} on the IPFS, and then uses the symmetric key to decrypt CT_{data} and obtain the original data.

3.2 Data Access Process

In our proposed scheme, the data access process is realised by the following five algorithms, including *Setup*(), *KeyGen*(), *Encrypt*(), *Match*() and *Decrypt*(), the detailed process is as follows:

***Setup*() :** The input of the initialization algorithm is set as the secure parameter λ. Firstly, the algorithm randomly selects the bilinear group G_0 and the bilinear map $e : G_0 \times G_1 \rightarrow G_T$ whose generators are g and order p. The whole set of attributes $\Omega = \{a_1, a_2, \cdots, a_n\}$ and random elements $t_1, t_2, \cdots, t_n, t_{n+1}, t_{n+m} \in Z_p^*$ are then generated. The algorithm randomly selects parameters $\alpha_1, \beta_1 \in Z_p^*$ to generate the main private key $MK_{CP-ABE} = (\beta_1, g^{\alpha_1})$ and the main public key $PK_{CP-ABE} = \{G_0, g, h_1 = g^{\beta_1}, e(g, g)^{\alpha_1}\}$.

***KeyGen*() :** The key generation algorithm first takes the user's attribute set U as input. The algorithm first selects a random number $r_1 \in Z_p^*$, figures out $d_1 = g^{a_1 - r_1}$, then randomly select $r_j \in Z_p^*$ for each property in attribute set U, figures out $d_j = g^{a_j - r_j}$ and generates the attribute private key $SK_{CP-ABE} = (d_1 = g^{a_1 - r_1}, \forall j \in U : d_j = g^{a_j - r_j})$.

***Encrypt*() :** The encryption algorithm is run by the data owner and consists of the following two sub-algorithms:

- *DataEncrypt()*. The data owner randomly selects a string from the key space as the symmetric key Key and figures out $CT_{data} = Enc_{Key}(data)$, which indicates that *data* is encrypted by AES algorithm, and the encryption key is Key. The data owner uploads the ciphertext to IPFS and records the CID returned by IPFS.

- *KeyEncrypt()*. In order to encrypt the Key and CID under the access policy T, the algorithm first selects a random number $s_1 \in Z_p$ and assigns it as the root node of tree T, then it assigns a shared secret to each non-leaf node in T in a recursive way. Given that the value of this node is s_1, if this node is an AND operation and its leaf node is not marked, mark this node and assign a random value $s_i(1 \leq s_i \leq p-1)$ to each of its child nodes. Set the last child node to $s_t = s_1 - \sum_{i=1}^{t-1} s_i \ mod\ p$. If the node is an OR operation, mark the node and set the value of all its children to s_1. For every leaf node $a_{j,i} \in T$, it calculates $c_{j,i} = g_{t_j s_i}$, where i is the serial number of the attribute in the tree. The data plaintext that needs to be encrypted is $M = (CID, Key)$, and then it calculates $c^* = Me(g,g)^{a_1 s_1}$, $c_1 = g^{s_1}$ respectively. Finally, it can output $CT_{CID,Key} = (T, c^*, c_1, \forall a_{j,i} \in T)$ as ciphertext.

Match() **:** The smart contract first determines whether the attributes of the user U meet the access control policy T. If the attributes do not meet the access control policy, the access fails. Otherwise, $result = CT_{CID,Key}$, the $result$ is returned to the user.

Decrypt() **:** The decryption algorithm is run by the data user and consists of the following two sub-algorithms:

- *KeyDecrypt()*. If the user attribute set U does not satisfy T, decryption fails. Otherwise, a minimum subset of U satisfying T is selected for calculation $\prod e(c_{j,i}, d_j) = \prod e(g^{t_j s_i}, g^{rt_j - 1}) = e(g,g)^{r_1 s_1}, e(c_1, d_1) \cdot e(g,g)^{r_1 s_1} = e(g^{s_1}, g^{a_1})$, and calculate the plaintext $\frac{c^*}{e(g^{s_1}, g^{a_1})} = M = (CID, Key)$.
- *DataDecrypt()*. Data user downloads the ciphertext of original data CT_{data} from IPFS according to CID and figures out $data = Dec_{Key}(CT_{data})$, which indicates that CT_{data} is decrypted by a symmetric key.

In order to ensure the correctness of the model, our original data is encrypted and stored on IPFS, and the storage address of the data, namely CID, uniquely corresponds to the ciphertext of the original data. Therefore, as long as the user obtains the correct CID, he can accurately search and obtain the required ciphertext of original data on IPFS. At the same time, in order to achieve the requirement of fine-grained access control, our access control strategy adopts an access structure tree composed of many attributes. In the access structure tree, each leaf node is an attribute set by the data owner, and the attribute value is the secret value passed to the node by the parent node. Only when the attributes of the user accurately meet the access structure tree, the ciphertext of CID and the symmetric key can be obtained and decrypted, and then the ciphertext of the original data can be decrypted to obtain the original data.

3.3 Security Analysis

In this section, we will describe the security analysis of our scheme from the aspects of confidentiality and privacy.

1) Confidentiality. To ensure that no adversary can decrypt the published data, we use the symmetric encryption algorithm *DataEncrypt*() and CP-ABE algorithm to ensure confidentiality. In the model, the data owner first encrypts the data by a symmetric key and then uploads it to IPFS. The symmetric key is stored on the blockchain. Even if IPFS is malicious, it cannot obtain the clear text of the data. The reason is that its attributes cannot match the access policy, the ciphertext of symmetric key cannot be obtained and decrypted. In addition, the data owner uses CP-ABE algorithm *KeyEncrypt*() to encrypt the symmetric key, and then uploads the ciphertext to the blockchain. Only data user with attribute private key can decrypt the ciphertext, which makes the confidentiality of the entire system more complete.
2) Privacy. In order to ensure that the privacy of the data is well satisfied, the data owner encrypts the data before uploading the data to IPFS, and then uploads the ciphertext. The key information is stored on the blockchain, and only the users who meet the access policy can successfully obtain it. Even if the IPFS can obtain the ciphertext data uploaded by the data owner, it cannot obtain any useful information. In order to achieve secure access control, the CP-ABE attribute private key is generated by *KeyGen*() algorithm. Only the data owner know it. This part avoids the problem of exposing access policies and realizes the privacy protection of attributes in the access policy. In addition, the symmetric key and CID are encrypted by *KeyEncrypt*() before being uploaded to the blockchain to ensure their privacy.

4 Experiment Analysis

In terms of experiments, we implemented a prototype to analyze the feasibility and performance of the scheme. The specific configuration of the experimental platform and environment are as follows: the operating system is Ubuntu 20.04.2 LTS, the processor is AMD Ryzen 7 5800U with Radeon Graphics, 4GB memory, and the programming languages are C++ and Solidity. The cross-blockchain module is mainly implemented through WeCross [25], which enables interoperability between heterogeneous blockchains.

Table 1. Comparison of upload speed between cloud server and IPFS

Size of data (KB)	1000	5000	10000	15000	20000
The upload speed of cloud (M/s)	1.01	0.95	0.99	1.01	1.03
The upload speed of IPFS (M/s)	10.95	37.25	52.51	64.28	70.57

Table 2. Comparison of download speed between cloud server and IPFS

Size of data (KB)	1000	5000	10000	15000	20000
The download speed of cloud (M/s)	31.26	35.84	26.14	30.56	35.26
The download speed of IPFS (M/s)	10.98	47.54	74.67	97.06	122.75

Firstly, we generated a group of test data on Linux to compare the upload and download speeds of IPFS and cloud servers. The experiment had five groups of tests in total. Table 1 and Table 2 show the size of test data, upload speed and download speed of the two schemes respectively. Each group of test was carried out 100 experiments on IPFS and cloud server respectively. Finally, we calculated the average value of the 100 results. The test results are shown in Fig. 2. Since IPFS uses content-based addresses instead of domain-based addresses and just needs to simply validate the hash of the content, so it can get faster transfer speeds than cloud server.

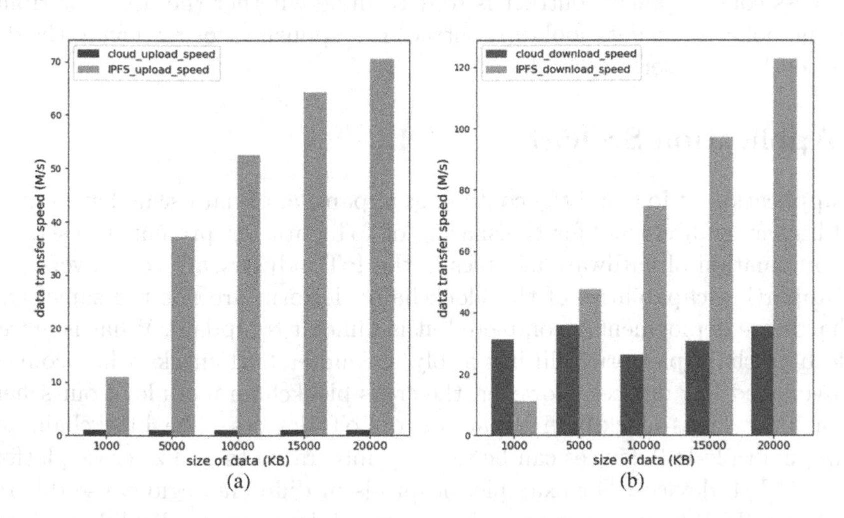

Fig. 2. (a) The comparison of upload speed between cloud and IPFS, and (b) the comparison of download speed between cloud server and IPFS.

From the experimental results, we can see that the upload speed of data on IPFS is much faster than that on cloud server. And when the data size is about 1000 KB, the download speed of data on IPFS is slower than that of the cloud server. But with the data size increasing, the download speed of data on IPFS is much faster than that of the cloud server. In other words, when the data is very small, the cloud server may be faster than IPFS download speed, and when the data is larger, the IPFS download speed is more advantageous. Therefore, the data transmission speed of IPFS can well meet the needs of users.

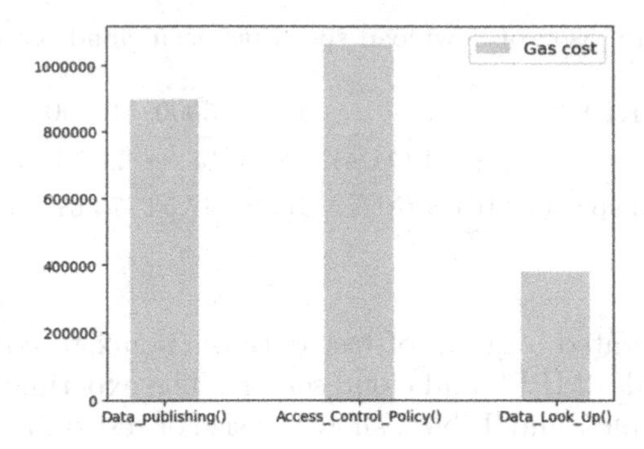

Fig. 3. The gas cost of deploying smart contracts.

In addition, we also test the gas cost of three smart contracts on Remix. The test results are shown in Fig. 3. The data publishing contract is responsible for storing data that needs to be published to the blockchain. The function of the access control policy contract is to determine whether the user's attributes meet the policy. The data look-up contract is responsible for retrieving the data requested by the user.

5 Application Scenario

The application of IoT and the continuous expansion of data scale have put forward higher requirements for the sharing of IoT data. At present, the selection and combination of hardware modules in the IoT industry are very diverse, and the supporting capabilities of the blockchain platform are not the same. Once the hardware deployment is completed, it is difficult to update. What is more, a single blockchain platform will inevitably encounter bottlenecks when connecting diversified IoT devices. However, the cross-blockchain module of our scheme supports the cross-blockchain expansion of IoT devices. The blockchain connecting multiple IoT devices can be securely integrated to realize cross-platform linkage of IoT devices. For example, hospitals in different regions use different underlying blockchain systems to record medical data. For medical data sharing in different regions, our scheme uses CP-ABE access control strategy, which can provide effective privacy and security protection for medical data and realize secure sharing of patient medical data between hospitals in different regions. In addition, when the IoT devices in one region need to obtain and use the data of IoT devices in another region, the cross-blockchain data access control function of this scheme can realize the secure data sharing between IoT devices in different regions.

6 Conclusion

In this paper, we proposed a data sharing access control model based on IPFS and cross-blockchain, which realizes the secure sharing of cross-blockchain data between users in different data domains. Secondly, we designed an access control scheme based on CP-ABE, and control the access rights of data users through the permission of smart contracts. The IPFS-based data storage method reduces the storage overhead on the blockchain and avoids the risk of centralized storage. Finally, the experiment shows that the scheme can achieve the autonomous access control of data on the blockchain, and can realize the secure data sharing between heterogeneous IoT domain.

Acknowledgments. This work was supported by the National Key R&D Program of China under Grant 2020YFB1005604, the National Natural Science Foundation of China under Grant No.61902021 and No.62272031, the Beijing Natural Science Foundation under Grant No.4212008, the Open Foundation of Information Security Evaluation Center of Civil Aviation, Civil Aviation University of China under Grant No. ISECCA-202101.

References

1. Gai, K., Choo, K.K.R., et al.: Privacy-preserving content-oriented wireless communication in internet-of-things. IEEE IoT J. **5**(4), 3059–3067 (2018)
2. Qiu, M., Chen, Z., et al.: Energy-aware data allocation with hybrid memory for mobile cloud systems. IEEE Syst. J. **11**(2), 813–822 (2014)
3. Li, J., Ming, Z., et al.: Resource allocation robustness in multi-core embedded systems with inaccurate information. J. Syst. Arch. **57**(9), 840–849 (2011)
4. Fan, K., Tian, Q., Wang, J., et al.: Privacy protection based access control scheme in cloud-based services. China Commun. **14**(1), 61–71 (2017)
5. Li, Y., Gai, K., et al.: Intercrossed access controls for secure financial services on multimedia big data in cloud systems. ACM Trans. Multimedia Comput. Commun. Appl. (2016)
6. Qiu, H., Dong, T., et al.: Adversarial attacks against network intrusion detection in IoT systems. IEEE IoT J. **8**(13), 10327–10335 (2020)
7. Zhu, L., Wu, Y., Gai, K., et al.: Controllable and trustworthy blockchain-based cloud data management. Future Gener. Comput. Syst. **91**, 527–535 (2019)
8. Gai, K., Wu, Y., Zhu, L., et al.: Differential privacy-based blockchain for industrial internet-of-things. IEEE TII **16**(6), 4156–4165 (2019)
9. Borkowski, M., Frauenthaler, P., Sigwart, M., et al.: Cross-blockchain technologies: review, state of the art, and outlook. White paper (2019)
10. Liu, X., Liu, J., Zhu, S., et al.: Privacy risk analysis and mitigation of analytics libraries in the android ecosystem. IEEE TMC **19**(5), 1184–1199 (2019)
11. Li, L., Liu, J., Cheng, L., et al.: CreditCoin: a privacy-preserving blockchain-based incentive announcement network for communications of smart vehicles. IEEE TITS **19**(7), 2204–2220 (2018)
12. IPFS Homepage. http://ipfs.tech/. Accessed 29 Oct 2022
13. Wang, W., Song, J., Xu, G., et al.: ContractWard: automated vulnerability detection models for Ethereum smart contracts. IEEE TNSE **8**(2), 1133–1144 (2020)
14. Bethencourt, J., Sahai, A., Waters, B.: Ciphertext-policy attribute-based encryption. In: IEEE Symposium on Security and Privacy (SP 2007), pp. 321–334 (2007)
15. Qiu, M., Gai, K., Xiong, Z.: Privacy-preserving wireless communications using bipartite matching in social big data. Future Gener. Comput. Syst. **87**, 772–781 (2018)

16. Zyskind, G., Nathan, O.: Decentralizing privacy: using blockchain to protect personal data. In: IEEE Security and Privacy Workshops, pp. 180–184 (2015)
17. Wang, W., Shang, Y., He, Y., et al.: BotMark: automated botnet detection with hybrid analysis of flow-based and graph-based traffic behaviors. Inf. Sci. **511**, 284–296 (2020)
18. Rahulamathavan, Y., Phan, R.C.W., Rajarajan, M., et al.: Privacy-preserving blockchain based IoT ecosystem using attribute-based encryption. In: IEEE International Conference on Advanced Networks and Telecommunications Systems (ANTS), pp. 1–6 (2017)
19. Han, D., Zhu, Y., Li, D., et al.: A blockchain-based auditable access control system for private data in service-centric IoT environments. IEEE TII **18**(5), 3530–3540 (2021)
20. Li, T., Wang, H., He, D., et al.: Blockchain-based privacy-preserving and rewarding private data sharing for IoT. IEEE IoT J **9**, 15138–15149 (2022)
21. Ren, W., Sun, Y., Luo, H., et al.: SILedger: a blockchain and ABE-based access control for applications in SDN-IoT networks. IEEE TNSM **18**(4), 4406–4419 (2021)
22. Yu, G., Zha, X., Wang, X., et al.: Enabling attribute revocation for fine-grained access control in blockchain-IoT systems. IEEE T. Eng. Manag. **67**(4), 1213–1230 (2020)
23. Yi, L., Sun, Y., et al.: CCUBI: a cross-chain based premium competition scheme with privacy preservation for usage-based insurance. Int. J. Intell. Syst. **37**, 11522–11546 (2022)
24. Gai, K., She, Y., Zhu, L., et al.: A blockchain-based access control scheme for zero trust cross-organizational data sharing. ACM TOIT (2022)
25. WeCross Homepage. https://wecross.readthedocs.io/zh_CN/latest/. Accessed 29 Oct 2022

Research on Diabetes Disease Development Prediction Algorithm Based on Model Fusion

Wenyu Shao[1(✉)], Xueyang Liu[2], Wenhui Hu[2], Xiankui Zhang[3], and Xiaodong Zeng[3]

[1] School of Software and Microelectronics, Peking University, Beijing, China
swy1798@stu.pku.edu.cn

[2] National Engineering Research Center for Software Engineering, Peking University, Beijing, China
{liuxueyang,huwenhui}@pku.edu.cn

[3] School of Software and Microelectronics, Peking University, Beijing, China
{zxkui1999,zengxiaodong}@stu.pku.edu.cn

Abstract. In today's world, with the deepening of population aging, chronic diseases have become the main diseases which affecting human health. Diabetes is a common chronic disease. Its incidence rate is high and rising year by year. For patients with diabetes, it is very important to predict the development of the disease and the possible complications for their follow-up treatment and recovery. However, the existing prediction of diabetes is mostly limited to the prediction of the incidence rate of patients, and only uses collaborative filtering or features for prediction, and rarely uses the patient's condition information to construct sequences and predict the development of the disease. Our aim is to make a accurate prediction on the development of diabetes patients and the possible complications. We need to use the historical development information of patients. Therefore, we propose a sequence based model fusion prediction algorithm, which effectively fuses the sequence and feature information. We use the high-order Markov Chains with attention mechanism as the basic learner for learning sequence information, and we also use XGBoost and CatBoost as the basic learner for learning feature information. Finally, LightGBM is used as a meta learner to fuse the output of the base learner. Experiments on the data of diabetes patients show that our method achieves better results than the original sub learner.

Keywords: Diabetes Patients · Development Prediction · Markov Chains · CatBoost · Sequence Prediction · Model Fusion

1 Introduction

Diabetes is a chronic disease with a large number of patients and a high incidence rate. It is of positive significance in disease control and complication prevention to predict the disease development and possible complications of diabetes patients. With the development of medical informatization, various medical institutions have also accumulated a large number of patient's medical electronic data in the process of patient diagnosis and treatment. Making full use of these data and mining valuable laws from them will help

M. Qiu et al. (Eds.): SmartCom 2022, LNCS 13828, pp. 439–449, 2023.
https://doi.org/10.1007/978-3-031-28124-2_42

doctors adjust treatment plans in time according to the development of patients' conditions, improve treatment efficiency, and help patients better control their own disease development and conduct follow-up recovery and treatment more scientifically [1].

In recent years, researchers have made many explorations in the research and prediction of diabetes. Peijie Du used statistical methods to collect the statistical data of diabetes incidence rate, prevalence and risk factors of community residents over 40 years old in Zhengzhou, China, and summarized the main factors leading to the disease; Pal et al. used naive Bayes, decision tree, SVM and other machine learning algorithms to establish a prediction model for retinopathy, and analyzed its prediction performance and so on. However, most of the existing diabetes prediction is only used to predict the incidence rate. They mainly use collaborative filtering or human characteristics to predict the incidence rate, rather than using the diagnosis and treatment sequence of diabetes patients to predict the disease development [2].

Sequence prediction is to learn the change rule of user' behavior or preference from the user' behavior sequence, so as to predict the subsequent behavior of the user [3]. In the prediction of the development of diabetes, namely learning the rule of the change of the patient's condition from the historical diagnosis and treatment sequence of diabetes patients, so as to predict the development degree of the patient's condition. Sequence prediction can make good use of rich historical data and mine corresponding rules. In this paper, the high-order Markov Chains with attention mechanism is used to learn the sequence information of the disease development of diabetes patients, which helps to achieve a more accurate prediction of the disease development of diabetes [4].

Model fusion is a method to complete learning tasks by building multiple learners and combining them. A learner is a model. Different models have their own advantages. By combining multiple models, model fusion gives play to the advantages of each model, and obtains a learner with stronger effects. Stacking is a relatively advanced model fusion method. Its idea is to train multiple base learners based on the original data, and then combine the prediction results of the base learners into a new training set to train a new learner. The new learner is also called a meta learner, so as to reduce the dependence on a single learner and improve the performance of the model [5]. In this paper, high-order Markov Chains with attention mechanism, XGBoost and CatBoost are selected as the base learners of model fusion, and LightGBM is selected as the meta learner for the experiment. In feature engineering, XGBoost is used to obtain the highest accuracy, CatBoost is used to process category features, and LightGBM is used to achieve the highest accuracy as quickly as possible. From the analysis of the experimental results, it can be seen that the effect of the fused model is better than any single model.

The content organization structure of this article is as follows:

Section 1: Introduction. First, the status quo of diabetes disease prediction is introduced. Secondly, the applicability of sequence prediction in diabetes prediction is analyzed, and the method of model fusion in this paper are introduced. Finally, the main work of this paper is introduced.

Section 2: Related work. Firstly, the disadvantages of traditional recommendation algorithms are introduced. Secondly, the advantages of sequence based recommendation algorithm are introduced. Finally, we introduce the high-order Markov algorithm combined with the attention mechanism and the model fusion technology.

Section 3: Our Methods. First, we introduce the high-order Markov algorithm combined with attention mechanism. Secondly, the basic learner of the feature-based prediction algorithm for the development of diabetes is introduced. Finally, the specific scheme design of model fusion is introduced.
Section 4: Experiment and result analysis. The experimental environment is introduced, and the experimental results are displayed and analyzed.
Section 5: Conclusion and future work. First, the work of this paper is summarized, and then the future research direction is proposed.

2 Related Work

In recent years, with the rapid development of the Internet and the popularity of mobile devices, the amount of information that people are exposed to has reached an explosive level. The recommendation system can help people find the information they need accurately and quickly from the redundant information. The recommendation algorithm is the core factor that affects the accuracy of the recommendation system. There is a large amount of historical preference information of users in such a large amount of data. Although traditional recommendation algorithms, such as content-based recommendation algorithm and collaborative filtering algorithm, have achieved good results on the whole, they have the problem of insufficient model expression ability. In particular, it can only simulate the interaction between users and projects statically, and can only capture the general preferences of users, and can not make full use of the existing historical preference information of users. While the sequence based recommendation algorithm generally considers that the user's behavior at a certain time is determined by the user's historical behavior. It regards the user's historical behavior or preference as a dynamic and sequential sequence, and through such serialization, it understands and learns the user's historical behavior, so as to predict the user's subsequent behavior or preference more accurately [6]. Sequence recommendation algorithms include Markov chain based recommendation algorithm and deep learning based recommendation algorithm. Both of them make prediction and recommendation based on user historical behavior sequence.

2.1 High Order Markov Chains with Attention Mechanism

Markov chain is a kind of random time series with no aftereffect. It considers that the sequence state of a random sequence process at a certain time is only related to the state of the previous time, and not related to the state at an earlier time. It is a mainstream sequence modeling method. This method is widely used in the research of population quality of life, disease prediction and prevention, and has good effect. The factorized Markov Chains (FPMC) model was proposed in 2010 as a method for sequence prediction by combining matrix decomposition and Markov Chains by Rendle et al. The model learns a corresponding transfer matrix for each user, and then uses maximum likelihood estimation to solve the parameters. However, it has sparsity and long tail distribution on many data sets [7]. Fusing similarity models with Markov Chains (Fossil) model is a high-order Markov Chains model proposed by Ruining He et al. in 2016 [8]. The model smoothly combines similarity based methods with Markov Chains, uses similarity based methods to model users' long-term preferences, and uses high-order Markov Chains to model users' short-term preferences, Then, the combination of long-term and short-term preferences realizes personalized sequence prediction for sparse data and long tail data sets. The meaning of high-order is to extend the traditional view that a state at a certain time is only related to the state at the previous time to the state at the previous multiple times. The number of related states is reflected as the order. This parameter can be set by yourself. Therefore, we can make full use of the user's historical behavior sequence, learn more information through the high-order Markov Chains, and high-order Markov Chains have stronger practical significance. On this basis, we think that we should consider that there may be differences in the degree of influence of the states at the previous times on the current state. Therefore, we add the attention mechanism to the algorithm, which can help adaptively learn the weight of the influence of the past states on the current state, so as to achieve stronger practical significance and more accurate recommendation effect [9].

2.2 Model Fusion

Model fusion is to train multiple learners according to a certain method, that is, to train multiple models. It has been proved mathematically that with the increase of the number of individual classifiers, the error rate of the integration will decrease exponentially, and eventually it will be zero. Model fusion can make different models learn from each other and achieve the effect of model optimization. Stacking is a relatively advanced model fusion method, which uses raw data to train multiple base learners, and then combines the prediction results of the base learners into a new training set to train a new learner. The new learner is also called a meta learner. The most common stacking structure is that only one layer of meta learners is stacked on the base learner, and the single-layer stacking can flexibly select their favorite models on the base learner and the meta learner. This structure improves the effect of the model at the cost of increasing relatively small complexity [10].

3 Our Method

In order to predict the development of diabetes patients, the historical diagnosis and treatment information of patients is very important. We use the high-order Markov Chains with attention mechanism to learn the historical diagnosis and treatment sequence information of diabetes patients. The characteristics of patients are also essential for the prediction of disease development. Therefore, we selected several classical machine learning algorithms to learn the characteristics of patients and then predict the disease development [11]. Finally, we use model fusion to fuse the above sub models to obtain the final model.

3.1 High Order Markov Chains with Attention Mechanism

In this paper, the high-order Markov Chains with attention mechanism is selected as one of the basic learners. The operation steps of the model are shown in Fig. 1.

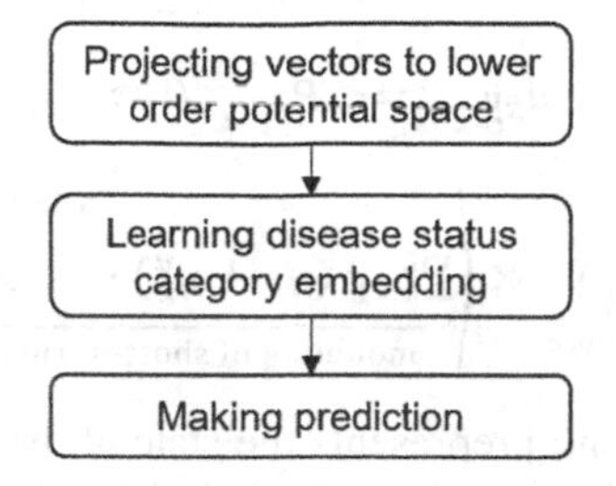

Fig. 1. Running steps of the model

First, the high-order Markov Chains with attention mechanism will project the patient's historical state of illness category vector into the low-order potential space, and then the patient's long-term state of illness information is modeled.

$$p_u(j \mid i) \propto \underbrace{\sum_{j' \in \mathcal{I}_u^+ \setminus \{j\}} < \vec{P}_i, \vec{Q}_j >}_{\text{modeling of long-term condition information}} \tag{1}$$

$\vec{Q}_j$ represents the state of disease at the final time to be predicted in each disease sequence, $\vec{P}_i$ represents the patient's condition at each historical moment before the final moment. Then, the high-order Markov Chains is used to model the short-term condition state information of the patient, and the probability that the condition state j becomes the state of the patient's next condition under the given condition state sequence is calculated.

$$p_u(j \mid \underbrace{S_{t-1}^u, S_{t-2}^u, \ldots, S_{t-L}^u}_{\text{patient short-term state sequence}}) \propto \left(\underbrace{\sum_{k=1}^{L} \overbrace{(\eta_k + \eta_k^u)}^{\text{Personalized weighting factor}} \cdot \vec{P}_{S_{t-k}^u}, \vec{Q}_j}_{\text{modeling of short-term condition information}} \right) \tag{2}$$

$S_{t-1}^u, S_{t-2}^u, \ldots, S_{t-L}^u$ represents the patient's short-term state sequence, and each patient has a global deviation vector $\eta_1^u, \eta_2^u, \ldots, \eta_L^u$. The basic idea is that each patient's previous state of illness has different weights for the high-order smoothness.

The attention mechanism is realized by calculating the similarity between the historical state of illness and the target state. The specific method is to calculate the implicit vector of the historical state of illness category in the learning sequence and the implicit vector inner product of the state to be predicted, and replace the original vector with the newly generated implicit vector of the historical state of illness category. Thus, different weights are assigned to the state of illness at different times in the history of diabetes patients, It reflects the influence degree of the historical state of illness at different times on the target state.

$$\vec{a}_{i,j} = < \vec{P}_i, \vec{Q}_j > \tag{3}$$

$$p_u(j \mid i) \propto \underbrace{\sum_{j' \in \mathcal{I}_u^+ \setminus \{j\}} \left(\overbrace{\vec{a}_{i,j}}^{\text{attention weight}} \cdot \vec{P}_i, \vec{Q}_j \right)}_{\text{modeling of long-term condition information}} \tag{4}$$

$$\vec{a}_{S^u_{t-k},j} = < \vec{P}_{S^u_{t-k}}, \vec{Q}_j > \tag{5}$$

$$p_u(j|\underbrace{S^u_{t-1}, S^u_{t-2}, \ldots, S^u_{t-L}}_{\text{patient short-term state sequence}}) \propto \left(\underbrace{\sum_{k=1}^{L} (\eta_k + \eta^u_k) \cdot \overbrace{\vec{a}_{S^u_{t-k},j}}^{\text{attention weight}} \cdot \vec{P}_{S^u_{t-k}}, \vec{Q}_j}_{\text{modeling of short-term condition information}} \right) \tag{6}$$

u represents the current patient, j represents the state of the patient at the target time, i and S^u_trepresents the patient's state of illness at different historical times, $\mathcal{I}^+_u$ indicates the patient's condition state set, $\vec{P}_i$ represents the hidden vector of the patient's state of illness at the historical time, $\vec{Q}_j$ represents the hidden vector of the patient's disease state at the target time, $\vec{a}_{i,j}$ denotes attention weight, η Represents a global parameter shared by all patients, η^u represents a personalized weighted scalar for a specific patient, which is used to control the relative weight of the long-term and short-term patient's state of illness.

Finally, the maximum a posteriori probability (MAP) estimation and stochastic gradient descent (SGD) method are used to solve the model parameters, and the algorithm training and model optimization are carried out to obtain the final model. By calculating the conditional transition probability of the target disease state, the matching probability of the candidate disease state to be recommended is calculated using the softmax function and the obtained prediction probability to generate the K disease states with the highest probability as the final prediction result. The final prediction result formula of the high-order Markov model combined with the attention mechanism is as follows, β_j is the offset term used to normalize the long-term dynamic component.

$$\hat{p}_{u,t,j} = \underbrace{\beta_j}_{\text{offset term}} + \underbrace{\sum_{j' \in \mathcal{I}^+_u \setminus \{j\}} \left(\overbrace{\vec{a}_{i,j}}^{\text{attention weight}} \cdot \vec{P}_i, \vec{Q}_j \right)}_{\text{modeling of long-term condition information}} + \left(\underbrace{\sum_{k=1}^{L} (\eta_k + \eta^u_k) \cdot \overbrace{\vec{a}_{S^u_{t-k},j}}^{\text{attention weight}} \cdot \vec{P}_{S^u_{t-k}}, \vec{Q}_j}_{\text{modeling of short-term condition information}} \right) \tag{7}$$

3.2 Feature-Based Basic Learners for Predicting the Development of Diabetes

We use feature engineering technology for feature selection firstly, and then use two classical machine learning models as the basic learners for predicting the development of diabetes.

3.2.1 Feature Engineering

Feature engineering is particularly important in machine learning. In this experiment, the characteristics of diabetes patients have a very important impact on the development of diabetes. Through the observation and analysis of the data set, we can obtain the personal information of diabetes patients, including basic information, previous history, family history, personal history, marriage and childbearing history, etc., the basic information including the patient's age, height, weight, occupation, etc. For the needs of the later experiments, we standardized continuous digital features, and coded discrete classification features using one-hot code [12].

After preliminary screening, we selected 81 dimensions of features. In order to ensure the effectiveness of features, we use logistic regression and XGBoost to process features of patients. XGBoost is an integrated learning method. Its corresponding model is multiple CART trees with dependency. By adding CART trees that can improve the overall effect, it uses features to split, fit the residuals of the last prediction, and continuously reduce losses. At the same time, in order to reduce the risk of over fitting, it limits the number of leaf nodes by adding penalties in the objective function. The algorithm finally adds the predicted value of each tree together as the final predicted value [13]. CatBoost is a GBDT framework with fewer parameters and high accuracy that supports category variables based on symmetric decision tree based learners. It uses combined category features and can take advantage of the relationship between features, which greatly enriches feature dimensions and can efficiently and reasonably handle category features [14].

3.2.2 Learning and Training

By inputting the personal information of patients with diabetes, we can output the probability corresponding to the development status of diabetes patients in the next stage. About the objective function, we choose the cross entropy loss function as the objective function, and the calculation formula is as follows.

$$L(\hat{y}) = -\sum\nolimits_{i=1}^{m} y_i \log(\hat{y_i}) + (1 - y_i)\log(1 - \hat{y_i}) \tag{8}$$

$\hat{y}$ is the matching probability of the disease development state of the diabetes patient in the next stage predicted by the base learner, and m is the number of candidate states.

3.3 Model Fusion

The method framework of this paper is shown in Fig. 2. Our method is a model fusion algorithm based on stacking, and the method steps are as follows.

- Firstly, preprocess the original data to obtain the characteristics of diabetes patients and the development sequence of their state of illness.
- Secondly, we choose the high-order Markov Chains with attention mechanism, XGBoost and CatBoost as the base learners; Among them, the inputs of XGBoost and CatBoost are the characteristics of diabetes patients, and the inputs of the high-order Markov Chains with attention mechanism are the state sequence of diabetes patients. The output result of these basic learners is the matching probability of the development state of diabetes patients in the next stage.
- Then, we choose LightGBM as the meta learner to obtain the highest accuracy in the shortest possible model training time; We take the output of the base learner as the

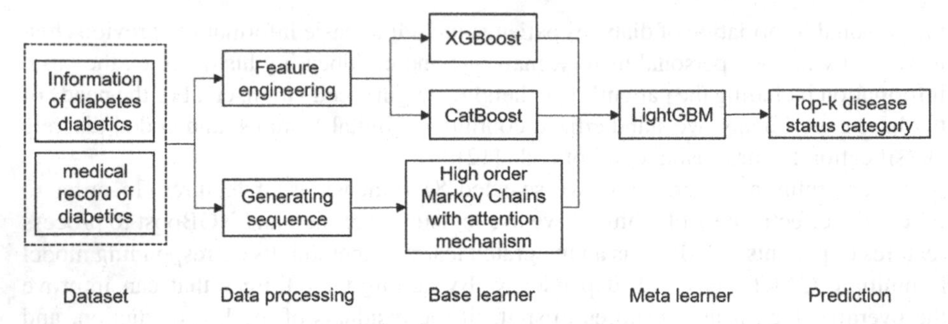

Fig. 2. Framework diagram of our method

input of the meta learner and learn through the meta learner LightGBM; The output result of the meta learner is the final matching probability of the development state of diabetes patients in the next stage.

- Finally, according to the final matching probability, predicting the development status of the Top-k next stage of diabetes patients.

4 Experiment and Result Analyzing

4.1 Experimental Setting

The data set in this paper comes from the patient visit data of some hospitals in a certain area. The final experimental data set is obtained by manual labeling, rule screening and personal information data masking. Our experimental data set includes two types of tables: diabetes patient information table and diabetes patient diagnosis and treatment record table. The information table of diabetes patients records the relevant information of diabetes patients, including the patient's age, height, weight, occupation, previous history, family history, personal history, marriage and childbirth history, etc.; The diagnosis and treatment record of diabetes patients is the diagnosis and treatment information of the diabetes patients from admission to discharge. We obtain the historical disease development sequence of each diabetes patient from the diagnosis and treatment record table of diabetes patients, and select the influencing factors of disease development from the information table of diabetes patients as the characteristics of the algorithm.

4.1.1 Classification of Disease State

According to the development stage of diabetes, the patient's condition was classified. The state of the disease is mapped into a two bit numeric code. Among them, the first one represents the major category of diabetes to indicate whether the patient has diabetes complications and the types of complications. The second one represents the severity of the disease corresponding to the current major category of diabetes. The specific method is to normalize each index examined by the patient, convert it into a numerical value, and then carry out a weighted average. Finally, sixty disease status labels are obtained.

4.1.2 Generating Disease State Sequence

Based on the diagnosis and treatment record table of diabetes patients and the classification results of the disease status, the corresponding codes of the current disease status of the patients are calculated according to the classification of the disease status and the current detection indicators of the patients, thus generating the sequence of the patient's disease development. Then for sequence s, assuming its total length is n, s [k] is the prediction target (k ∈ [0, n]) of sequence s [0: k]. Therefore, a sequence with a total length of n can generate n − 1 pieces of training data. Among all the data, 60% of the data of diabetes patients were used for training, 20% for validation, and the remaining 20% for testing.

4.2 Results and Analysis

We chose Recall and NDCG (normalized discounted cumulative gain) to evaluate the effect of the experiment.

Recall is also called recall rate, which means the probability of being predicted as positive samples in the actual positive samples, so as to find out how many of the actual positive samples are predicted as positive.

$$Recall = \frac{TP}{TP + FN} \tag{9}$$

TP represents the number of positive samples predicted to be positive, and $TP + FN$ represents the total number of positive samples actually.

NDCG (normalized cumulative gain) uses the ratio of DCG and IDCG of each patient as the normalized score, and DCG (Discounted cumulative gain) can evaluate the recommended list. Its core idea is that when there is a relatively high correlation result in the lower position of the search result list, punishment should be imposed on the evaluation score, and the punishment proportion is related to the logarithmic value of the corresponding result location. Therefore, it takes into account the factor of sorting order. IDCG is the maximum DCG value under ideal conditions.

$$DCG_p = \sum\nolimits_{i=1}^{p} \frac{2^{rel_i} - 1}{\log_2(i+1)} \tag{10}$$

$$IDCG_p = \sum\nolimits_{i=1}^{|REL|} \frac{2^{rel_i} - 1}{\log_2(i+1)} \tag{11}$$

$$nDCG_p = \frac{DCG_p}{IDCG_p} \tag{12}$$

In these formulas, i indicates different sorting positions in the recommendation results, and rel_i represents the correlation at i, p represents the recommended number of results, and $|REL|$ indicates the optimal ranking of recommendation results.

In order to evaluate the performance of our method, we conducted experiments on data.The experimental results are shown in Table 1. It can be seen from the results that our method is superior to the other base learner in terms of Recall and NDCG, and we have achieved better prediction performance. Model fusion combines Markov's good at processing sequence data, XGBoost and CatBoost's use of rich feature information, makes each sub model give full play to its advantages, uses other sub models to make up for its shortcomings, and finally achieves better results than a single model [15]. Affected by the relatively detailed classification of disease state, the overall accuracy of the model is not particularly ideal. However, in general, the model has a high accuracy in predicting the development of diabetes patients and has good practical significance.

Table 1. Experimental Results Of Our Method And Baseline

Method	Top-5		Top-10		Top-20	
	Recall	NDCG	Recall	NDCG	Recall	NDCG
XGBoost	0.7529	0.5483	0.8064	0.5681	0.8237	0.5735
CatBoost	0.7697	0.5452	0.8023	0.5729	0.8317	0.5798
High order Marcov Chains with attention mechanism	0.7876	0.5782	0.8325	0.5977	0.8754	0.6201
Our method	0.8254	0.5949	0.8751	0.6198	0.9192	0.6317

5 Conclusion and Future Work

In this experiment, we quantified the examination indicators of diabetes patients and obtained the disease degree value of the patients. Then we used the high-order Markov Chains with attention mechanism to learn the disease development sequence of diabetes patients, and used XGBoost and CatBoost to learn the influence of the characteristics of the patient's personal information on the patient's disease development. We choose these three algorithms as the base learners, and we chose LightGBM as the meta learner for model fusion. We took the output of the base learner as the input of the meta learner and output the final prediction result through the meta learner. The experimental results show that compared with using a single model, the model fusion method has obvious advantages. Moreover, Markov model and model fusion technology are widely used in the research of population quality of life, disease prediction and prevention, and have good effects. This experiment proves that Markov model and model fusion technology also have good effects in using the patient's disease progression sequence to predict the patient's disease progression. In future work, we will try the combination of other algorithms in model fusion, including changing the number and types of learners, so as to explore whether we can further enhance the effect of the model and explore the application of this model fusion in other scenes.

References

1. Guomin, J., Liu, Y.: Research on development opportunities and countermeasures of "Internet plus chronic disease management." China Med. Insur. **07**, 46–52 (2022)

2. He, B.: Prediction of Diabetes Based on Convolutional Neural Network. Southwest University (2019)
3. Wangrui, J., Peixi, K., Zhuju, Y., et al.: Multivariable time series forecasting using model fusion. Inf. Sci. **585** (2022)
4. Wangjia, M., Liuqiu, P., Zhangming, L., et al.: Effectiveness of different screening strategies for type 2 diabete on preventing cardiovascular diseases in a community-based Chinese population using a decision-analytic Markov model. J. Peking Univ. (Med. Ed.) **54**(03), 450–457 (2022)
5. Liu, W., Zouwei, H., Luyan, J., et al.: Research on image recognition of Chinese medicinal materials based on transfer learning and model fusion. J. Hunan Univ. Tradit. Chin. Med. **42**(05), 809–814 (2022)
6. Puhong, F., Shaojian, F., Zhangxiao, W., et al.: Sequence recommendation of fusing dynamic interest preference and feature information. J. Yunnan Univ. (Nat. Sci. Ed.) **44**(04), 708–717 (2022)
7. Rendle, S., Freudenthaler, C., Schmidt-Thieme, L.: Factorizing personalized Markov chains for next-basket recommendation. World Wide Web (2010)
8. He, R., McAuley, J.: Fusing similarity models with Markov chains for sparse sequential recommendation. CoRR (2016)
9. Wujian, F.: Research of Attentional Mechanisms Involved in the Recognition of Diabetes Retinopathy. Guangdong Normal University of Technology (2022)
10. Duanqian, W.: Personal Credit Risk Assessment Based on Stacking Fusion Model. Shandong University (2021)
11. Chensi, H., Zhangyun, Q.: Machine learning-based prediction model of type 2 diabetes mellitus complications. Chin. J. Med. Libr. Inf. **29**(11), 31–38 (2020)
12. Dengxiu, Q., Xiewei, H., Liufu, C., et al.: A prediction model for advertising click conversion rate based on feature engineering. Data Collect. Process. **35**(05), 842–849 (2020)
13. Yang, J., Guan, J.: A heart disease prediction model based on feature optimization and smote-Xgboost algorithm. Information **13**(10) (2022)
14. Câlburean, P.-A., Grebenişan, P., Vacariu, V., et al.: Prediction of 3-year all-cause death in a percutaneous coronary intervention registry using machine learning: a comparison between random forest and CatBoost algorithms. Appl. Med. Inform. **43** (2021)
15. Ma, S., Cuijian, F., Xiaowei, D., et al.: Deep learning-based data augmentation and model fusion for automatic arrhythmia identification and classification algorithms. Comput. Intell. Neurosci. **2022** (2022)

Smart-Contract Vulnerability Detection Method Based on Deep Learning

Zimu Hu[1](✉), Wei-Tek Tsai[1,2], and Li Zhang[1]

[1] Digital Society and Blockchain Laboratory, Beihang University, Beijing, China
{huzmu049,leezhang}@buaa.edu.cn
[2] Beijing Tiande Technologies, Beijing, China
tsai@tiandetech.com

Abstract. With the rapid development of blockchain technology, smart contracts (SCs) applied in digital currency transactions have been widely used. However, SCs often have vulnerability in their code that allow criminals to exploit them to steal associated digital assets. Benefiting from the development of machine learning technology and the improvement of hardware performance, one can use deep learning techniques to analyze code and detect vulnerabilities. This paper proposes an innovative combination of opcode sequences and abstract syntax trees for source code parsing. And a method based on the combination of self-attention mechanism and bidirectional long-short term memory neural network is proposed to detect the vulnerability of SCs after word embedding. Experimentation results show that the two parsing methods can complement each other and effectively improve the accuracy of vulnerability detection.

Keywords: Blockchain · Smart Contract · Deep Learning · Vulnerability Detection

1 Introduction

With the development of blockchain (BC), smart contracts (SCs) running on top of a BC have received significant attention. Since 2009, BC has gone through several significant innovations with different architectures, consensus protocols, and transaction processes. For example, instead of single BC systems completing all the steps of financial transactions, new BC systems often perform only one step of a financial transaction, such as KYC (Know Your Customers), trading, or settlement. As a BC is responsible for only one step of a financial transaction, the complexity of the involved BC is significantly reduced. Thus, the overall system can have much better scalability, optimization, and performance. Passed through the first generation of BC system Bitcoin, the second generation of BC system Ethereum, and then to the third generation of BC with multi-chain structure ChainNet [1]. Nowadays, with the emergence of concepts such as the metaverse, BC is undergoing innovation, and there will be a more powerful and well-regulated BC system architecture. SC, as an electronic agreement that can execute contract terms

M. Qiu et al. (Eds.): SmartCom 2022, LNCS 13828, pp. 450–460, 2023.
https://doi.org/10.1007/978-3-031-28124-2_43

without mutual trusts, is of great significance in both legal theory and the development of computer technology.

SC code is stored and run on a BC. Due to the BC characteristics, it is difficult to change, and the content of BC can be seen by all participants. If the code with vulnerability is submitted to the chain, it will be easy to be maliciously exploited. Sometimes it can even lead to serious losses. The DAO incident, an attack against SC [2], appeared in the Ethereum, and caused huge economic losses. Attackers used a reentrancy vulnerability to steal more than 3.6 million ETH, causing losses of up to $70 million. Because of this, vulnerability detection of SC has become an important research topic.

Some studies have used methods of symbolic execution [3, 4] and fuzzing [8, 9]. However, the efficiency of detection is slow, and it relies too much on expert knowledge. There are also some studies that use machine learning-based methods [10, 12, 13]. Some of them use source code as inputs while others use opcode. There are 145 different opcodes designed in the Ethereum Yellow Book. Each opcode corresponds to an operation.

This paper proposes a deep-learning-based vulnerability-detection method for SCs. This approach parses the SC source code into opcode and abstract syntax tree (AST). And the two vectors are represented respectively, and the neural network model is used for feature learning. Then, the vectors are fused, and the obtained fusion features are classified to obtain whether the code has vulnerabilities. In the deep learning method, we use RNN, GRU, and LSTM (Long Short-Term Memory) to learn the features. Meanwhile, we add the attention mechanism.

This paper is organized as follows: Sect. 2 introduces related work; in Sect. 3 proposes a vulnerability detection method based on the combination of self-attention mechanism and BiLSTM neural network; Sect. 4 discusses experimental results; Sect. 5 finally concludes this paper.

2 Related Work

The existing research mainly focuses on three general directions, which are symbolic execution, fuzzing, and machine learning. Symbolic execution can use symbols to analyze unknown variables, and it can use the information stored on the BC outside the SC to deduce it as symbols. Slither [3] proposed by Feist, S-gram [4] proposed by Liu, Manticore [5] proposed by Mossberg and Osiris [6] proposed by Torres all use symbolic execution to detect SC vulnerability. Although symbolic execution can detect vulnerabilities in SCs, it may not be able to solve with the constraints of the program are too complex. At the same time, it cannot get rid of the heavy dependence on the prior knowledge of experts. The time required to detect a SC often ranges from tens to hundreds of seconds, depending on the complexity of the code.

Fuzzing is another detection method. For example, Torres's study [7], Sereum [8], and Jiang's Contractfuzzer [9]. These dynamic analysis tools can execute the code. However, with dynamic analysis, analysts need to implement and execute attack patterns to prevent vulnerabilities in advance, so analysts need to be fully aware of attacks on SCs. Therefore, these dynamic analysis tools rely heavily on the rule setting, and on the prior knowledge of experts. At the same time, dynamic analysis takes a long time to execute, because the probability of detecting a contract vulnerability can be higher only if more inputs are executed. Such detection methods often need to run for hundreds of seconds to detect vulnerabilities.

With the improvement of machine learning technology, SC vulnerability detection methods based on machine learning have been proposed. It uses machine learning to learn the features of SCs, to classify the code into two cases: vulnerability and no vulnerability. Ashizawa proposes Eth2Vec [10], a neural network training tool based on the PV-DM model [11]. There are also some SC analysis tools that cannot automatically extract features and require human input. For example, the studies of Momeni [12] and Wang [13] use classical machine learning algorithms for analysis, but they may output wrong results when the code to be analyzed is rewritten. Qian proposed VulDeeSmartContract [14], an automatic feature extraction method based on Word2Vec and LSTM [16], but it can target re-entrancy vulnerability only.

3 Methodology

3.1 Overview

This paper proposes SC vulnerability detection method based on deep learning is mainly divided into the following contents: First, the SC code is preprocessed into opcode sequences and abstract syntax tree sequences. Next, the Word2Vec [15] model is used to process the input into vectors. Then, a method based on the combination of self-attention mechanism [17] and BiLSTM [16] is used for feature learning, and finally a fully connected network is used to classify vulnerabilities.

There are many types vulnerabilities of SCs, and this paper mainly analyzes four of them: integer overflow, integer underflow, timestamp dependency, and reentrancy vulnerability. Integer vulnerability is divided into two cases: integer overflow and underflow. Overflow or underflow occurs if the stored integer is larger or smaller than the maximum or minimum value of the data type. If an attacker exploits the overflow or underflow, it may cause problems such as infinite loops or failure of execution conditions [18]. Timestamp dependency vulnerability is the use of timestamp related variables in the SC [2]. Some malicious miners will set the timestamp within the allowed range to benefit the SC and even steal the digital asset. The reentrancy is exploited in the DAO incident [2]. A reentrancy can be achieved if the attacker designs the callback function so that some function of the contract, such as a money transfer function, is called again.

The opcode sequences generated are generated from source code. Firstly, the SC source code.sol file should be compiled to generate the opcode file. After that, according to the division of each contract in the source code, the opcode to each contract is stored separately. In SCs written in Solidity, each.sol file can contain multiple contracts. This can be compared to the structure of multiple classes in a source file in other object-oriented programming languages. The structure of the SC.sol file is shown in Fig. 1.

Instead of storing all opcodes in a source file together as many existing studies, this paper chooses to divide according to contracts. This can provide a finer granularity for code vulnerability detection, and can detect which contract is vulnerable in a source code file. The SC vulnerability detection method proposed in this paper is based on the BiLSTM neural network model, that can consider the connections between contexts in

```
1  // Sample.sol
2  contract A {
3      function a() public {}
4      function b() public {}
5  }
6
7  contract B {
8      uint32 i;
9  }
```

Fig. 1. The structure of.sol file

the SC. It can detect vulnerabilities with high accuracy in a short time, and combine the self-attention mechanism to further improve the effect of the model. Figure 2 illustrates the vulnerability detection algorithm.

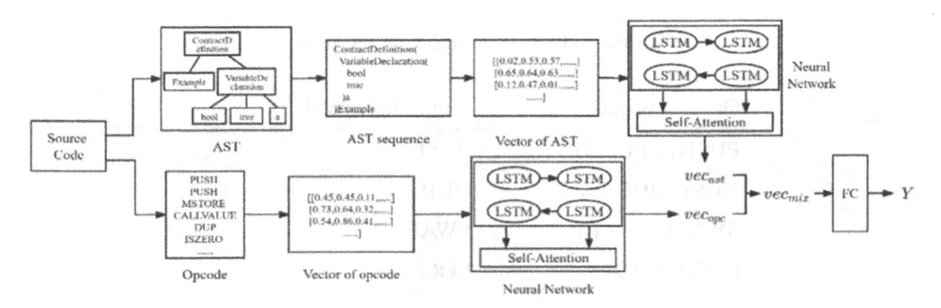

Fig. 2. Vulnerability Detection Algorithm

The vulnerability-detection algorithm consists of four steps: Firstly, the source code of the SC is preprocessed, and the abstract syntax tree sequence and opcode sequence are generated; Secondly, the word embedding operation is performed on the two sequences to generate the vectorized representations; Thirdly, the vectorized representation matrix put into the neural network for feature learning, and obtain vec_{opc} and vec_{ast}; Finally, binary classification detection was performed. To make full use of the extracted feature vectors, the algorithm concatenates vec_{opc} and vec_{ast} to obtain vec_{mix}. Then the vec_{mix} is input into the fully connected layer, and the classification result shows whether there is a vulnerability.

3.2 Smart Contract Preprocessing

SC code is written in Solidity, that contains a lot of redundant content, including comments and identifiers. Therefore, it is not appropriate to use the source code of SCs for vulnerability detection. This paper uses two different methods to preprocess the SC, compiling it into opcode and extracting the AST from it.

Since there are many kinds of opcodes in the Ethereum virtual machine, it needs to be simplified. By analyzing the function and structure of opcodes, this paper finds some opcodes that can be reduced. The simplification of these opcodes with similar semantics can reduce the features for vectorization, and reduce the complexity of feature learning. The opcodes are simplified according to the reduction rules, shown in Table 1.

By analyzing the characteristics of Solidity syntax, this paper designs a rule for structured traversal of Solidity AST. An AST does not represent every detail that in the code, but it preserves important parts of the source code. In Solidity, the structure represented by AST can be mainly divided into Definition, Expression, Statement and TypeName structure. For example, the Definition structure mainly includes Event, Function, Structure and Contract Definition. To better reflect the syntax structure information contained in Solidity, a rule for structured traversal sequence is formulated. First, each structure in the AST is represented by the structure type name and the node name, with the structure name placed first. Then use parentheses to indicate the contents of the subtree, followed by the struct type name or node name. For the Definition structure, after the closing parenthesis, there is a definition identifier. In Expression structures, structure name is used at the end again.

Table 1. Opcode simplification rules

Original opcode	Simplified opcode
PUSH1~ PUSH32	PUSH
DUP1~DUP16	DUP
SWAP1~SWAP16	SWAP
LOG0~LOG4	LOG

3.3 Embedding

After processing the SC code into opcode and AST, it is not possible to directly input this information into a neural network. The input must be in the form of vectors. Therefore, opcode and AST must be vectorized. This paper uses word embedding Word2Vec model. First, all the input will be decomposed to extract all the words and use one-hot encoding, but the dimension of one-hot encoding is large, and is not convenient to use. Therefore, the CBOW (Continuous Bag-of-Words) [15] model will be used to learn the word embedding matrix. In this way, all the words can be mapped to a short vector through the word embedding matrix, and the dimension of the vector can be reduced while the relationship between the identifiers is preserved. Using this method, all opcodes and AST can be vectorized.

3.4 Feature Learning

Once the code is processed into a word vector, it can be used for feature learning. This step uses the LSTM model. There are RNN, GRU, and LSTM in recurrent neural networks. GRU and LSTM, are both optimized models based on RNN. In terms of performance, both are able to capture long distance dependencies, unlike RNN, which can only pay attention to the previous word. And code vulnerabilities are not only related to the previous word, but may be related to several words in the context. Through analysis, it can be found that the number of parameters in the GRU is less than that of the LSTM. When the features to be learned are relatively simple, this difference is not obvious, and may even produce gains. But when the features are more complex, such as the features of the SC vulnerability, this difference will show up. At the same time, some studies have also found that when the amount of data to be learned is large, the effect of LSTM is also better than that of GRU, and this is also confirmed by our experimentation. As the number of parameters of GRU is small, the time cost required for training is also low. However, in the task of vulnerability detection, the effect of the model is prioritized over the training cost, so this paper finally chooses to use the LSTM.

LSTM is an improved recurrent neural network. There are gradient vanishing and gradient explosion problems in ordinary recurrent neural networks, which is also the reason why ordinary recurrent neural networks cannot deal with the problem of long-distance dependence. The idea of LSTM is to add a state C to preserve the long-term state based on the original recurrent neural network. The hidden layer h_t can be calculated by the following formula.

$$\begin{cases} C_t = f_t \times C_{t-1} + i_t \times \tilde{C}_t \\ f_t = \sigma\left(W_f\left[h_{t-1}, x_t\right] + b_f\right) \\ \tilde{C}_t = tanh\left(W_C\left[h_{t-1}, x_t\right] + b_C\right) \\ i_t = \sigma\left(W_i\left[h_{t-1}, x_t\right] + b_i\right) \\ o_t = \sigma\left(W_o\left[h_{t-1}, x_t\right] + b_o\right) \\ h_t = o_t \times tanh(C_t) \end{cases} \tag{1}$$

Meanwhile, this method adds the attention mechanism, and chooses the Self-Attention mechanism. It is a variant of the attention mechanism [17], that can better find the correlation within the data features and deal with long-distance dependencies. Through this method, the relationship between the input sequences can be further analyzed. By adding the self-attention layer, the model can focus on the key words, so that the effect of the whole model is enhanced. This paper inputs all the output of the BiLSTM layer into the self-attention mechanism layer, so that the attention mechanism can fully learn. In the self-attention mechanism, the three parameters W, b and u need to be trained. W and b parameters are used to obtain u_t, and the calculation formula of vec is as follows.

$$u_t = tanh(W[h_{Lt}, h_{Rt}] + b) \tag{2}$$

$$a_t = softmax\left(u_t^T \cdot u\right) \tag{3}$$

$$vec = \sum_{t=1}^{T} a_t[h_{Lt}, h_{Rt}]$$

4 Experimental Verification

4.1 Evaluation Metrics

As a classification model, the SC vulnerability detection model needs to use certain metrics to evaluate its classification effect. The evaluation metric needs to be able to show the effect of the model accurately, and can compare the pros and cons of different models. There are four types of SC vulnerabilities studied in this paper, which are Integer overflow (IOF), Integer underflow (IUF), timestamp dependency (TD), and reentrancy (REN). Detecting each vulnerability is a binary classification. For the binary classification problem, the evaluation metrics used in this paper are accuracy (ACC), Recall and F1 score (F1). We calculate the metrics using the following formula: $ACC = \frac{TP+TN}{TP+TN+FP+FN}$, $Precision = \frac{TP}{TP+FP}$, $Recall = \frac{TP}{TP+FN}$ and $F1 = \frac{2*Precision*Recall}{Precision+Recall}$.

4.2 Data Sets

According to the method of selecting data sets in existing research, such as the work of Momeni [12] and Wang [13], this paper collects source code, makes labels, and builds its own data set Contract5000 in the same way. At the same time, the datasets in the study of Eth2Vec [10] are used for experimentation. The base case statistics of the two datasets are shown in Table 2. By collecting a large number of SC source code from Etherscan, the Contract5000 dataset retains 5321 SC codes that are different from each other at the opcode level by compiling to opcode and removing duplication. Among them, there are 20327 contracts, 8560 contracts have IOF, 3396 contracts have IUF, 452 contracts have TD, and 96 contracts have REN. We label each of these contracts separately, with the presence of vulnerabilities marked as 1 and the absence of vulnerabilities marked as 0. The Eth2Vec dataset contains 5000 SC source code, which contains a total of 17237 contracts. There are 8730 IOF, 3594 IUF, 323 TD, and 58 REN.

Table 2. Dataset statistics

Data Set	Source Code	Contracts	IOF	IUF	TD	REN
Eth2Vec	5000	17237	8730	3594	323	58
Contract5000	5321	20327	8560	3396	452	96

4.3 Training Environment and Parameter Settings

We use Keras and Tensorflow to build the model, and the language we use is Python. The hardware environment used for all the experiments is CPU: Intel Xeon at 2.50 GHz, GPU at RTX 2080 Ti and 43 GB Memory. In terms of parameter setting, for each dataset, we use 7:3 to divide the training set and test set, and choose binary cross entropy as the Loss function. The optimal gradient descent algorithm Adam was used in all the experiments. For the learning rate lr, we experimented with multiple values, and to prevent the model from overfitting, we added a dropout layer and set the parameter dr. We used [5, 10, 20] for the word vector dimension vop of opcodes, and [50, 80, 100] for the word vector dimension vast of abstract syntax trees. In the experimental results, we use the following parameters: lr = 0.001, dr = 0.2, batch = 128, vop = 10, vast = 50.

4.4 Results and Comparison

In the experiment, we use the Eth2Vec and Contract5000 dataset to train and test the model. During the training process of Eth2Vec and Contract5000, the changes of Acc, Loss and F1-score of the training set are shown in Fig. 3 and Fig. 4.

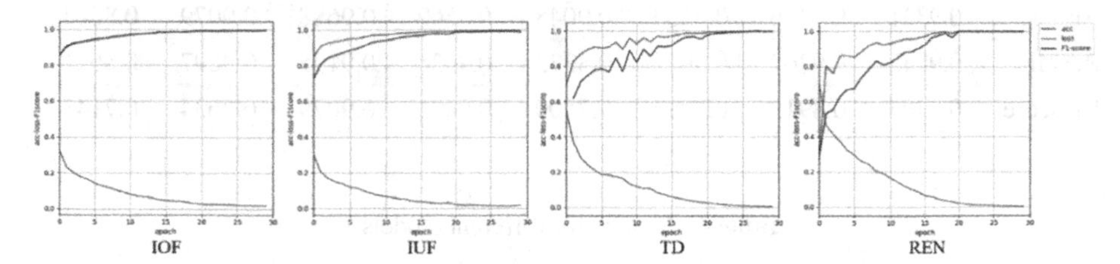

Fig. 3. Eth2Vec training set ACC, Loss, and F1 curves

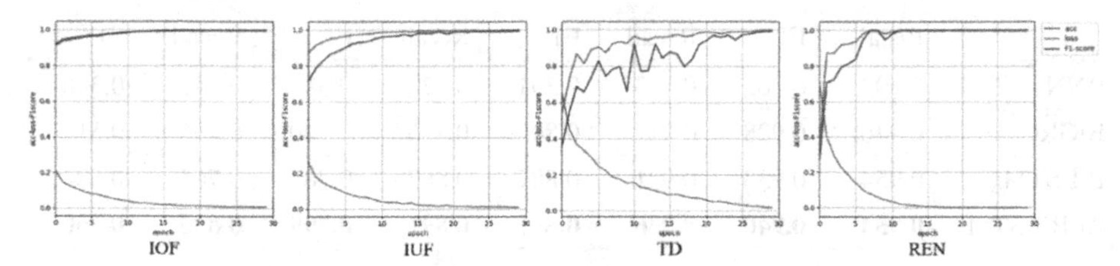

Fig. 4. Contract5000 training set ACC, Loss, and F1 curves

The final effect in the test set of the two datasets is shown in Table 3. In the experiments, we respectively experiment the effects of RNN, BiGRU, BiLSTM and AttBiLSTM models on various vulnerabilities in this part, each model will input the opcode sequence and the abstract syntax tree sequence for learning at the same time, and the recurrent neural network hidden layer hyperparameter Settings of each model are the same. In the self-attention mechanism layer, we pass the output sequence of the hidden layer of BiLSTM at different moments for learning, so that the self-attention layer learns the whole sequence. The other parameters of the model remain the same, and the Eth2Vec dataset is uniformly used for the corresponding experiments. The division of training and test set is 7:3, and the specific experimental effects are shown in Table 4.

From the data in the table, one can see that when simple RNN is used for vulnerability detection, except for the integer overflow vulnerability, the detection performance of the others of vulnerabilities is relatively poor, and the features of vulnerabilities cannot be well identified. However, the model using BiGRU and BiLSTM can achieve good detection effects on the four kinds of vulnerabilities, which also shows that the characteristics of vulnerabilities are indeed context-related. Only by taking full account of the context, can we achieve more accurate detection of vulnerabilities. When comparing the effect of BiGRU and BiLSTM, we find that the effect of using BiLSTM model is slightly higher than that of BiGRU. Although GRU is improved based on LSTM, it has fewer parameters than LSTM, so it is normal that the actual effect is slightly worse than LSTM. Considering the time of model training, GRU takes less time than LSTM.

Table 3. Test set effect of Eth2Vec and Contract5000

Metrics	Dataset							
	Eth2Vec				Contract5000			
	IOF	IUF	TD	REN	IOF	IUF	TD	REN
Acc	0.9331	0.9476	0.9233	0.9048	0.9569	0.9688	0.9079	0.8844
Recall	0.9542	0.9365	0.8912	0.8333	0.9626	0.9195	0.8897	0.8621
F1-score	0.9404	0.8913	0.7994	0.7500	0.9495	0.9079	0.7634	0.7143

Table 4. Results for different models

Models	Type							
	IOF		IUF		TD		REN	
	Recall	F1	Recall	F1	Recall	F1	Recall	F1
RNN	0.793	0.762	0.493	0.391	0.320	0.383	0.647	0.386
BiGRU	0.946	0.928	0.912	0.848	0.866	0.740	0.706	0.511
BiLSTM	0.954	0.933	0.921	0.882	0.887	0.785	0.765	0.684
AttBiLSTM	**0.954**	**0.940**	**0.936**	**0.891**	**0.891**	**0.799**	**0.833**	**0.750**

Since the problem of this paper is the vulnerability detection of SCs, the final effect of the model is more important, and the requirement for the time of model training is not very high. Therefore, the LSTM neural network can better meet the requirements of this paper.

AttBiLSTM represents the BiLSTM network using self-attention mechanism. AttBiLSTM is the best one among the four network models in the table. In the IOF and IUF, the advantage of AttBiLSTM model is not obvious. However, on the REN, the effect of the network model with the self-attention mechanism has a more prominent performance, which indicates that the self-attention mechanism layer has learned more information, which improves the performance of the model. Combined with the whole comparison test, we can see that adding the self-attention mechanism can indeed improve the classification effect of the model, especially for the REN, that proves that the method proposed in this paper is effective. To verify the effectiveness of the vulnerability detection method proposed in this paper, this paper selects five related works for comparison, and the comparison results with related works are shown in Table 5.

Through the experimental data, the effect of this paper is better than that of SVM and Eth2Vec. The method proposed in this paper can effectively detect vulnerabilities. Within comparison with ABCNN, we find that the method in this paper is superior to the comparison work in the detection of IOF and IUF, which indicates that under the condition of sufficient learning, BiLSTM has better ability to extract sequence features than TextCNN. However, it is slightly inferior to the comparison work on the REN. This is since the number of positive samples of the REN category in the dataset is small,

Table 5. Comparison result with related work

Models	Type							
	IOF		IUF		TD		REN	
	Recall	F1	Recall	F1	Recall	F1	Recall	F1
SVM [12]	0.731	0.783	0.392	0.500	0.028	0.053	0.078	0.123
Eth2Vec [10]	0.576	0.701	0.560	0.637	0.170	0.273	0.548	0.615
ABCNN [18]	0.830	0.858	0.903	0.886	0.832	0.870	0.862	0.820
DR-GCN [19]	—	—	—	—	0.789	0.749	0.809	0.764
CGE [20]	—	—	—	—	0.881	**0.878**	**0.876**	**0.864**
AttBiLSTM	**0.954**	**0.940**	**0.936**	**0.891**	**0.891**	0.799	0.833	0.750

and it is difficult to learn the features of the REN, which affects the final effect of the model. Within comparison with DR-GCN and CGE, since these two works both detect TD and REN, only the effects of these two vulnerabilities can be compared. The method in this paper is better than DR-GCN, but it is slightly inferior to CGE. Under the same conditions, the expert mechanism is used in CGE, which theoretically improves the effect of the model. In addition, CGE uses the callback point and edge specifically for the REN, and further optimizes the detection of the REN. However, the work of DR-GCN and CGE uses graph neural network, which cannot detect long code and needs to use specific methods to intercept fragments in the code, so it is not more universal than the method in this paper. Through the comparison of related work, the method proposed in this paper has a good performance in IOF and IUF, and can also show good results in TD and REN. In comparison with other methods, the method in this paper also has certain advantages, which can well complete the task of SC vulnerability detection.

5 Conclusion

To detect Ethereum SC vulnerabilities efficiently, this paper proposes a SC vulnerability detection method based on self-attention mechanism and BiLSTM neural network that integrates multiple code extraction techniques. Experiments show that the SC vulnerability detection method proposed in this paper can detect vulnerability with great coverage and efficiency. While this paper analyzes four kinds of vulnerability, the same method can be used with a different dataset to detect other kinds of vulnerability.

Acknowledgment. This work is supported by Chinese Ministry of Science and Technology (Grant No. 2018YFB1402700).

References

1. Tsai, W.-T., et al.: ChainNet. Oriental Publishing House, Beijing (2020)

2. Atzei, N., Bartoletti, M., Cimoli, T.: A survey of attacks on ethereum smart contracts (SoK). In: Maffei, M., Ryan, M. (eds.) POST 2017. LNCS, vol. 10204, pp. 164–186. Springer, Heidelberg (2017). https://doi.org/10.1007/978-3-662-54455-6_8
3. Feist, J., Greico, G., Groce, A.: Slither: a static analysis framework for smart contracts. In: 2019 IEEE/ACM 2nd International Workshop on Emerging Trends in Software Engineering for Blockchain (WETSEB). IEEE (2019)
4. Liu, H., Liu, C., Zhao, W., et al.: Smashing smart: towards semantic-aware security auditing for Ethereum smart contracts. In: 2018 33rd IEEE/ACM International Conference on Automated Software Engineering (ASE), pp. 814–819. IEEE (2018)
5. Mossberg, M., Manzano, F., Hennenfent, E., et al.: Manticore: a user-friendly symbolic execution framework for binaries and smart contracts. In: 2019 34th IEEE/ACM International Conference on Automated Software Engineering (ASE), pp. 1186–1189. IEEE (2019)
6. Torres, C.F., Schütte, J., State, R.: Osiris: hunting for integer bugs in Ethereum smart contracts. In: Proceedings of the 34th Annual Computer Security Applications Conference, pp. 664–676 (2018)
7. Torres, C.F., Baden, M., Norvill, R., et al.: AEGIS: shielding vulnerable smart contracts against attacks (2020)
8. Rodler, M., Li, W., Karame, G.O., et al.: Sereum: protecting existing smart contracts against re-entrancy attacks. arXiv preprint arXiv:1812.05934 (2018)
9. Jiang, B., Liu, Y., Chan, W.K.: Contractfuzzer: fuzzing smart contracts for vulnerability detection. In: 2018 33rd IEEE/ACM International Conference on Automated Software Engineering (ASE), pp. 259–269. IEEE (2018)
10. Ashizawa, N., Yanai, N., Cruz, J.P., et al.: Eth2Vec: learning contract-wide code representations for vulnerability detection on Ethereum smart contracts. In: Proceedings of the 3rd ACM International Symposium on Blockchain and Secure Critical Infrastructure, pp. 47–59 (2021)
11. Le, Q., Mikolov, T.: Distributed representations of sentences and documents. In: International Conference on Machine Learning. PMLR, pp. 1188–1196 (2014)
12. Momeni, P., Wang, Y., Samavi, R.: Machine learning model for smart contracts security analysis. In: 2019 17th International Conference on Privacy, Security and Trust (PST), pp. 1–6. IEEE (2019)
13. Wang, W., Song, J., Xu, G., et al.: Contractward: automated vulnerability detection models for ethereum smart contracts. IEEE Trans. Netw. Sci. Eng. (2020)
14. Qian, P., Liu, Z., He, Q., et al.: Towards automated reentrancy detection for smart contracts based on sequential models. IEEE Access **8**, 19685–19695 (2020)
15. Mikolov, T., Chen, K., Corrado, G., et al.: Efficient estimation of word representations in vector space. Comput. Sci. (2013)
16. Hochreiter, S., Schmidhuber, J.: Long short-term memory. Neural Comput. **9**(8), 1735–1780 (1997)
17. Vaswani, A., Shazeer, N., Parmar, N., et al.: Attention is all you need. Adv. Neural Inf. Process. Syst. **30** (2017)
18. Sun, Y., Gu, L.: Attention-based machine learning model for smart contract vulnerability detection. J. Phys. Conf. Ser. **1820**(1), 012004 (2021)
19. Zhuang, Y., Liu, Z., Qian, P., et al.: Smart contract vulnerability detection using graph neural network. In: IJCAI, pp. 3283–3290 (2020)
20. Liu, Z., Qian, P., Wang, X., et al.: Combining graph neural networks with expert knowledge for smart contract vulnerability detection. IEEE Trans. Knowl. Data Eng. (2021)

Automatic Smart Contract Generation with Knowledge Extraction and Unified Modeling Language

Peiyun Ran[1], Mingsheng Liu[2(✉)], Jianwu Zheng[3,4(✉)], Zakirul Alam Bhuiyan[5], Jianhua Li[6], Gang Li[7], Shiyuan Yu[8], Lifeng Wang[9], Song Tang[10], and Peng Zhao[11]

[1] School of Cyberspace Security of Beihang University, Beijing, China
rpy233@buaa.edu.cn
[2] Shijiazhuang Institute of Railway Technology, Shijiazhuang 050041, Hebei, China
liums601001@sina.com
[3] Key Laboratory of Traffic Safety and Control of Hebei Province, Shijiazhuang, China
[4] School of Transportation, Shijiazhuang Tiedao University, Shaoxing, Hebei, China
zhengjw@stdu.edu.cn
[5] Department of Computer and Information Sciences Fordham University JMH 334, E Fordham Road, Bronx, NY 10458, USA
mbhuiyan3@fordham.edu
[6] School of Information Science and Technology, Shijiazhuang TieDao University, No. 17 East North Second Ring Road, Changan District, Shijiazhuang, China
[7] Zhongke Zidong Information Technology (Beijing) Co., Ltd., Beijing, China
ligang@people-ai.cn
[8] Beijing Public Security Burea, Beijing, China
[9] Hebei Huaye Jike Information Technology Co., Ltd., Shijiazhuang, China
[10] Institute of Applied Mathematics of Hebei Academy of Sciences, Shijiazhuang, China
[11] The First Hospital of Hebei Medical University, Shijiazhuang, China
zhaopeng@jyyy.com.cn

Abstract. Since the launch of Ethereum in 2013, the smart contract has been a momentous part of the blockchain systems due to its character of automatic execution. The generation of smart contracts has also attracted extensive attention from the academic community. However, the preparation and generation of smart contracts are still mainly manual so far, which limits the scalability of the smart contracts. In this paper, we put forward a new method to generate smart contracts automatically based on knowledge extraction and Unified Modeling Language (UML), which can significantly accelerate the generation of smart contracts. We will describe this method in more detail based on the logistics supply chain.

Keywords: blockchain · smart contract · automatic code generation · knowledge extraction · unified modeling language

M. Qiu et al. (Eds.): SmartCom 2022, LNCS 13828, pp. 461–474, 2023.
https://doi.org/10.1007/978-3-031-28124-2_44

1 Introduction

Blockchain has attracted broad academic attention since its inception. With decentralized and tamper-proof characteristics, blockchain can establish a distributed ledger and help solve the trust issue. Cryptography technology is primarily used to ensure data security in a decentralized scenario, including, but not limited to, encryption, digital signatures, and hash functions. Nowadays blockchain technology has been widely applied in many scenarios such as finance, agriculture, the medical industry, logistics, and so on. The focus of research is often on improving its scalability [1].

In a blockchain system, smart contracts are common. Smart contract was first proposed in the 1990s. It is essentially a type of computer program that can be executed automatically when certain conditions are met [2]. Due to this attribute of smart contracts, they can be observed and executed by multiple untrusted nodes in a distributed system. The characteristics of blockchain are quite consistent with those of smart contracts. The smart contracts, combined with cryptography technology which ensures the difficulty of tampering or intervening in the content and execution of contracts, help to establish a set of guidelines for distrustful nodes to follow together.

Smart contracts have been successfully implemented in multiple blockchain systems such as Ethereum, Hyperledger Fabric, and so on. While making a splash in the blockchain, smart contracts are also faced with a series of problems, the most crucial among which is its generation. There are already some Turing complete languages, such as Go and Solidity that can be used for its generation. However, when it comes to smart contract coding, since the absence of a unified standard for developing decentralized solutions for business processes, the development of smart contracts, especially their automatic generation is challenging [3]. In addition, the tamper resistance of smart contracts brings difficulties to error correction. All these issues limit the scalability of smart contracts and the overall efficiency of the blockchain system. Therefore, researches on the automatic generation and security check of smart contracts have drawn more attention in recent years [4].

To fundamentally solve the aforementioned problem, we still need to start with the natural language (NL) contract. In this paper, we introduced natural language processing (NLP) technology to preprocess the original NL contracts and extract knowledge from them. Based on the result of knowledge extraction, we can further understand the various state transition processes experienced during the operation of smart contracts. In addition, the entities involved in the contract and the relationship between them will also be extracted. Using the extracted knowledge, we can first construct the smart contract template of the involved fields. Since the execution of smart contracts is essentially the running process of a finite state machine, the template will be presented in the form of a state diagram. Based on this template, the template will be further filled according to the specific contract content. When the state diagram template is populated, we can further convert the state diagram into a formal smart contract code. In [3], Smart contract code is proven to be generated from UML state diagrams. In this paper, we use the UML modeling method to model and populate the smart contract template. Specifically, we first define all possible states of the contract, use these states to create a state diagram template, and then supplement the events that may occur in each state, especially those that may cause state transition, according to the various actions that the subject may take in each state.

We will take the application of blockchain in the logistics field as an example to show the process of automatically generating smart contracts according to our method. In the logistics industry, goods generally pass through the process of shipment, transportation and unloading inspection. Due to the relative lack of mandatory supervision, there may be many violations in this process. The enforceability of smart contracts helps to better monitor both parties and reduces the occurrence of defaults.

2 Related Work

Automatic code generation is considered an effective method to improve the degree of automation and the final quality of software development and is widely used across academia and industry. At present, the technology of automatic code generation is mainly used in fields such as automatic code completion, and its purpose is to help programmers write programs.

In this paper, we aim to automatically generate smart contract codes according to NL contracts. This is automatic code generation based on function description, which is more challenging. In [5], a computer algorithm named Deepcoder is proposed, which can be used for automatic code writing. To generate codes, deep learning was used along with a graphical user interface (GUI). [6] takes advantage of UML and abstract syntax tree(AST) to generate JAVA codes. When it comes to smart contracts, [7] provides a method based on domain-specific ontologies and semantic rules to generate templates of smart contracts first, then insert constraints of the contracts by manipulating an AST. Moreover, [8] introduces the concept of controlled NL and uses it to rewrite the traditional NL contract, so that the computer can understand the contract content more accurately. In addition, this paper also proposes using a state diagram to describe the execution process of smart contracts. In [9], a code development platform is introduced to auto-generate and deploy the smart contract, which simplifies the deployment of a supply chain management (SCM) solution on the blockchain. The creator (the user who creates a supply chain) only needs to enter personal details and company or organization details through the user interface, and then submit product details, including its name and personalized attributes.

3 Main Technologies

3.1 Knowledge Extraction

Knowledge extraction, whose target is to extract knowledge from unstructured data from different sources and store it in the knowledge map, is an important technology to realize the automatic construction of large-scale knowledge

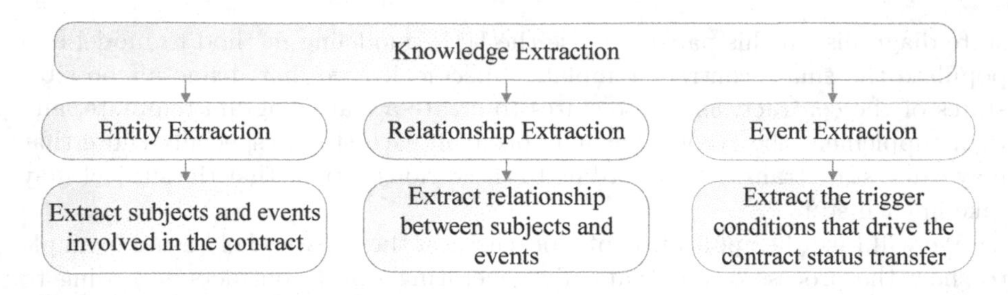

Fig. 1. Knowledge extraction classification and its role in processing NL contracts

atlases in NLP. Knowledge extraction mainly includes three sub-tasks: named entity recognition, relationship extraction, and event extraction. The goals of the three are to determine the entities in the text, determine the relationship between entities [10], detect the occurrence of specific types of events, and extract parameters respectively [11]. At present, there are many methods for knowledge extraction, especially event extraction. The methods based on are mainly graph convolutional networks [12], heterogeneous information networks [13], reinforcement learning [14], and so on. There are similar methods for entity and relationship extraction using graph neural networks [15] and heterogeneous information networks [16].

As shown in Fig. 1, in our method, we mainly use entity extraction and event extraction to analyze contract participants and events that trigger contract terms. We also use relationship extraction to analyze the relationship between multiple entities that may be involved in the contract. After this step, the contract written in NL will be simplified and structured.

Contracts written in NL generally exist in an unstructured form, and contract terms will be converted into structured data after knowledge extraction. This has laid the foundation for the establishment and filling of the smart contract template. With these structured data, it will be easier to analyze the execution process of smart contracts. For example, the contract specifies the possible breach of contract and the punishment method. After event extraction, the breach of contract and the punishment method can be extracted completely. In addition, unnecessary and redundant words can be deleted, which will make the computer more accurate in understanding and implementing the contract terms. In addition, the probability of errors in the process of generating the contract state diagram and even the code will be reduced accordingly.

3.2 UML Class Diagrams Generation

UML class diagram is a powerful modeling tool. As a result, it is closely related to object-oriented (OO) programming and can help developers fully understand relevant modeling for businesses. UML class diagrams show a set of classes, interfaces, collaborations, and their relationships. Before modeling with the UML class diagram, we must specify the business requirements. In the context of smart contracts, pre-written NL contracts are essential. We can start with NL contracts, first describe the requirements in NL, and then define the objects, classes, and methods involved in the contracts through object-oriented programming. Then we can model the implementation process of smart contracts on this basis.

Though mapping from user requirements to UML diagrams provides great convenience, it is difficult to complete the direct conversion from contract requirements written in NL to UML class diagrams, since almost all NLs contain redundancy, fuzziness, and imprecision. In addition, artificially prepared contracts may sometimes even have omissions, resulting in more serious consequences. Therefore, this research has always been a hot spot in the academic circle, especially in the field of NLP [17]. Some methods have been proposed to solve this difficulty. In [17], an approach which can maps user's requirements to UML class diagram based on the MDA is proposed. With the rule-based NLP approach, the processed text will be mapped into an XMI file first, then the XMI file will be converted to UML diagram by a CASE tool such as ArgoUMLFurther, [18] proposed a kind of heuristic rule based on NLP. They developed a set of pre-defined rules to extract OO concepts such as classes, attributes, methods, and relationships to generate a UML class diagram from the given requirements. The paper also referred to the reconstruction and normalization of NL text, whicprovided inspiration for our preprocessing.

There is one thing in common between [17] and [18]. Both of them reconstruct the NL contract based on certain semantic or grammatical rules and then construct the classes and objects in the UML class diagram. We do not subscribe to this idea. Instead, we use knowledge extraction to directly structure the contract content, which saves the trouble of defining semantic rules.

3.3 UML Class Diagrams to Smart Contract

There have been previous studies on knowledge graphs embedding [19]. As mentioned earlier, since the smart contract will execute itself after meeting the trigger conditions, we can regard it as a finite state machine, and the process of executing a smart contract as a state transition process when a finite state machine runs. Coincidentally, the UML class diagram is also a powerful tool to describe the running process of finite state machines. Therefore, it is not difficult to imagine that transforming UML class diagrams into smart contracts will be an effective way to automatically generate smart contracts.

In [3], the algorithm for model transformation and generation of executable smart contracts is outlined. This paper also compares their generated smart contract with the original contract, and the results are highly similar, which further illustrates the feasibility of transforming UML class diagram into smart contract. Besides, the modeling method for smart contract state diagram is not unique. In addition to UML class diagram, [9] used another modeling language named State Chart XML (SCXML). Although the modeling languages used are

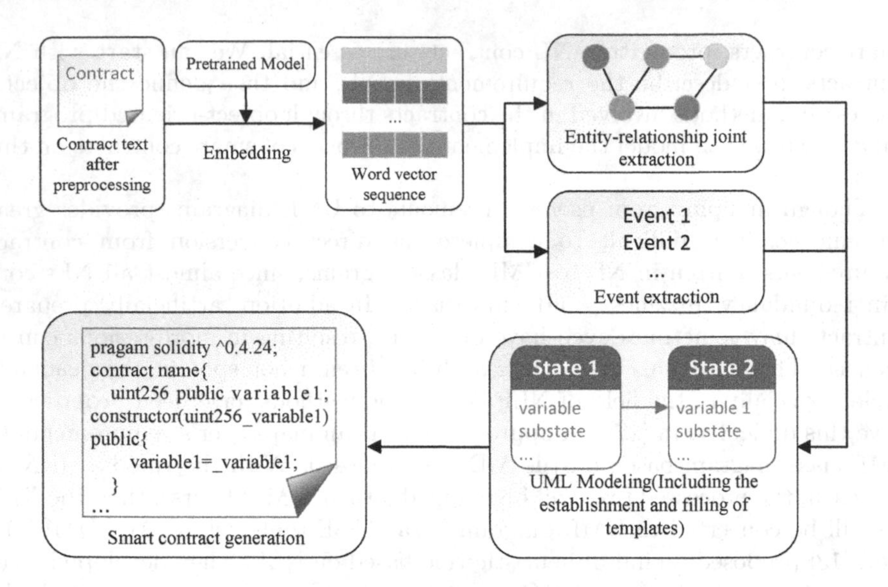

Fig. 2. Flow chart of the framework

different, these methods are based on the state diagram to realize the automatic generation of smart contracts. The method proposed in this paper is also based on this idea, using the state diagram constructed by UML to generate smart contracts.

4 Methods

In this section, we will introduce the whole process of automatically generating smart contracts from natural language contracts under our method. Figure 2 gives a flow chart of the framework.

4.1 Preparation and Preprocessing of NL Contracts

Firstly, we need to Clarify the rights and obligations of all parties involved in the contract to avoid possible mistakes in natural language contracts. As previously stated, almost all the NLs have more or less redundancy, fuzziness, and imprecision. Therefore, it is necessary for us to preprocess the natural language contract after its content is determined. For example, we can observe the following two contract terms. One term is 'Party B will be fined if it releases false information and other acts that damage Party A's reputation'. The other term is 'Party A shall test the goods (iron ore) and compare them with the specified indicators. If the iron grade difference is greater than 1.5 and not greater than 2.5, 500 yuan/vehicle will be deducted each time. If the difference is greater than 2.5 and not greater than 3.5, all the vehicle freight will be deducted'. It is straightforward that the former term's description is hard for computers to understand because it does not clearly define the so-called "false information" and "damage Party A's reputation". So, this kind of terms are not suitable for smart contract execution. In contrast, the latter term has a much clearer definition of the default of substandard goods, which is also much easier for computers to understand.

The above example illustrates the importance of the accurate description of contract terms to computer understanding of contracts. Therefore, in the process of preprocessing NL contracts, we need to focus on the regulations that are not fully and clearly described in the contract and modify or delete them so that the regulations can be fully understood by the computer.

4.2 Data Structured Representation

After preprocessing the NL contracts, we can further analyze the contract with the method of knowledge extraction. Specifically, we need to use entity extraction and relationship extraction to extract the parties involved in the contract (such as contract participants) and their relationships, and use event extraction to extract all events that may occur in the process of contract execution.

In practice, according to [10] and [20], we can jointly extract entities and their relationships. We will introduce a multi-round question and answer mechanism to "poll" the contract content, so as to obtain the parties involved in the contract and their relationships. Given that the smart contract application scenario is dominated by blockchain, which is a distributed ledger, the subjects involved can also be roughly considered as contract participants and their property or employees. Before multiple rounds of Q&A, we can set the question template in advance, and then fine-tune this template according to the changes in the application scenarios. For example, in the scenario of the logistics industry, in addition to the contract between Party A and Party B, the subjects involved also involve a series of entities such as the goods transported, the employees involved in the transportation, and the assets of the entrusting party, and their relationship networks need to be clarified without any omission. If we change scenarios, such as car leasing, then the entities we involve include not only the lessee and the leasing company, but also the information about related vehicles, the qualifications of the leasing company, the driver's license of the lessee, etc. Although the entities involved in the above two scenarios are different, the common parts can be summarized, such as the rights and obligations of Party A and Party B, both parties, certain items to be traded, and the qualifications or characteristics.

Similarly, inspired by [11] and [21] Event extraction can also take the form of similar multi-round question and answer. The only difference is that it focuses more on the process of contract execution. It needs to take into account every event in the contract execution process, and further consider the contract state transition when an event occurs. Therefore, the accuracy of NL contracts description directly determines the quality of event extraction results. In addition, when designing the template for multiple rounds of question and answer, we also need to focus on the state changes during the implementation of the smart contract. For each state, we need to ask about the possible behavior of the subject in

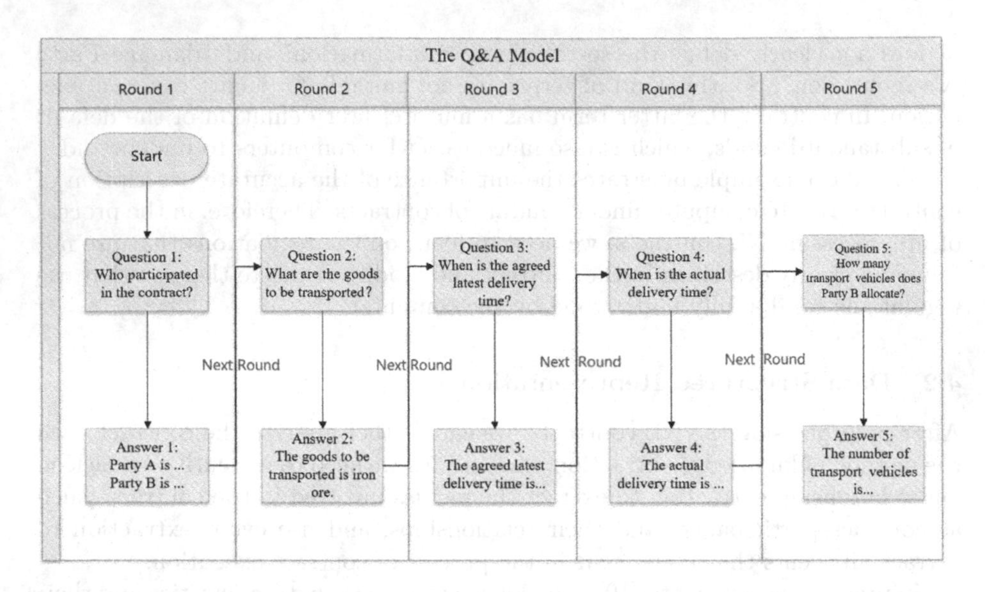

Fig. 3. QA for extracting the event in the loading stage of the logistics contract

the state and the state changes that may be caused by it. Finally, the results of multiple rounds of inquiry are output in a structured form of triple groups of subject, behavior, and state changes. Figure 3 is an example of some questions and answers involved in the loading phase of the logistics contract for event extraction.

As for the generation of dialogue, we learned from the mode in [11], which is divided into two modules. One is to generate a question and answer based on the current contract statement, and the other is to generate the predictive answer based on the existing statement. In different scenarios, Module 1 will generate a question and answer dialogue template according to the characteristics of the scenario. Correspondingly, module 2 extracts parameters that may be used as responses on the basis of existing contract statements to fill in the template. Specifically, the filling process needs to use the statement embedding vector obtained in the previous stage to locate the parameters to be selected (such as the keywords used for answering). We can use the following probability formula to calculate the probability of the k-th word (P_k) being selected: [11]

$$P_k = \frac{exp(H \cdot WE_k)}{\sum_{i=1}^{N} exp(H \cdot WE_i)} \tag{1}$$

In the formula, WE refers to the final representation of the i-th word embedded in the sentence, H refers to the vector that maps a word vector to a scalar. H will change due to different scenarios.

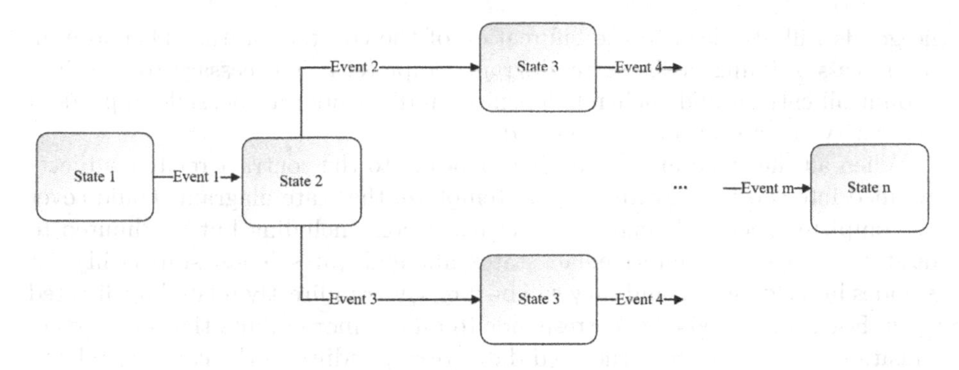

Fig. 4. A template of state diagram

4.3 Construction of Smart Contract State Diagram

With the help of the structured contract information obtained from the previous knowledge extraction, we have clarified the subject and state information involved in the smart contract. Next, we will build the state diagram of the smart contract. We will use a UML class diagram to describe the change of contract state. Therefore, in the process of UML modeling, the established objects will also focus on different states of the contract. The involved entity objects can perform certain activities in each state and cause corresponding state changes.

The construction of a smart contract state diagram based on structured data is mainly divided into two steps. First, it is necessary to define all the state changes experienced in the whole process of smart contract execution and the sequence of states, and then draw the general framework of the UML state diagram. The state graph drawn at this stage only contains the states experienced by the contract. Different nodes of the graph represent different states, and the edges point to the order of the surface states. This completes the construction of the state diagram template. Figure 4 shows a template of a state diagram. The specific contents need to be further supplemented.

Then we will further fill in the state diagram template according to the entities involved in the contract and their possible behaviors in each phase. Specifically, for a certain subject involved in the contract, he may play different roles in different stages of contract execution and lead the execution of the contract to different directions. For example, for the goods that need to be transported in the logistics contract, they are in the state of being unloaded for transportation in the preparation stage before transportation. Whether they are delivered on time or not will lead the contract execution into two different states: "normal delivery, normal contract execution" and "default of the carrier, overdue delivery". In the transportation stage, whether the goods are in good condition during transportation also determines the trend of the contract status. If the goods are in good condition, the contract will be executed normally. If the goods are damaged or lost, it will be deemed that the carrier is in breach of contract. Similarly, in the unloading inspection stage, the quality inspection results of the goods will also lead to the bifurcation of the contract status. Therefore, in the process of filling in the state diagram template, it is necessary to take into account all entities and their relationships, and consider all possible impacts of each entity on the contract state trend.

When all the structured data corresponding to the contract related subjects are filled into the UML state diagram template, the state diagram should cover the complete process of smart contract execution, including but not limited to the start state, various intermediate states, and end states. These states will exist as nodes in the diagram, and they will be directly or indirectly related by directed edges. Each connected edge corresponds to one or more events that cause state transitions. When all the structured data corresponding to the contract related subjects are filled into the UML state diagram template, the state diagram should cover the complete process of smart contract execution, including but not limited to the start state, various intermediate states, and end states. These states will exist as nodes in the diagram, and they will be directly or indirectly linked by directed edges. Each connected edge corresponds to one or more events that cause state transitions. It is worth mentioning that the starting state of each state diagram is unique, but the ending state may not be unique. This indicates that the execution of the contract may have different trends.

4.4 Transformation from State Diagram to Code

After the UML state diagram describing the smart contract is constructed, we can further convert it into smart contract code. Since UML class diagrams are closely related to object-oriented programming, UML class diagrams were used to generate JAVA code at an early stage [22,23], based on object-oriented methods. Their ideas are much the same, almost all of them use the state changes depicted by UML class diagrams to build the hierarchical state, concurrent state, and historical state of the corresponding code [24].

For smart contracts, the mainstream method is based on Model Driven Architecture(MDA). The MDA-based approach makes the development of smart contracts more systematic and helps to change the dilemma that there is no unified approach to smart contract development [25].

In addition, there are tools that can directly translate UML state diagrams into the Solidity codes. For example, Eclipse has introduced an extension package called UML to Solidity. It contains a UML profile and a set of Acceleo code generators to model smart contracts in UML and generate Solidity code. Usable with the Papyrus UML modeler for Eclipse. In addition, the UML state diagram here is drawn using Papyrus, which is a UML 2 modeling tool based on the Eclipse platform and supports the UML 2 standard specified by the Object Management Group (OMG) and the second-generation Diagram Interchange (DI2) standard.

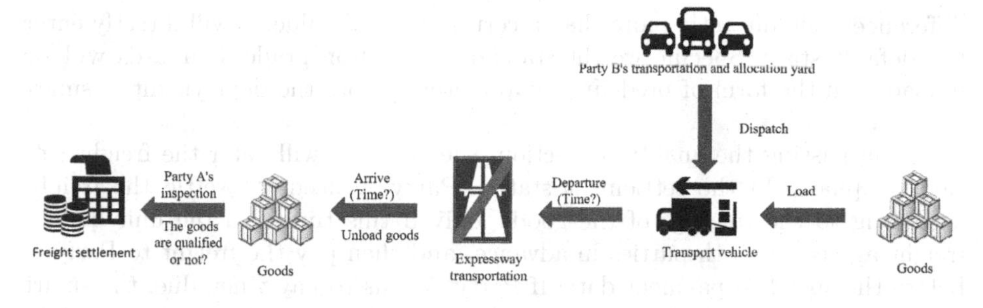

Fig. 5. A general process of logistics transportation

5 Example

In this section, we will give an example of automatic generation of smart contracts for logistics and transportation to explain our method in more detail. Figure 5 illustrates the general process of logistics and transportation.

In a logistics contract, Party A and Party B are generally the shipper and carrier respectively. The contract will first include the basic information of both parties, such as company name, legal representative, address, and then the period of validity of the contract. Both Party A and Party B are participants, which can be considered two different objects of the same category. The validity period of the contract indicates the starting and ending time of the code.

When a logistics contract is formally implemented, it is generally considered that it can be divided into preparation stage, transportation stage, arrival inspection stage and freight settlement stage. In addition, a special default status should be set to deal with the default of both parties. Each stage can be considered as a state that the contract needs to pass through.

In the preparation stage, the goods (assumed to be iron ore) have not been loaded. In this state, goods and trucks are the main players. In addition, the latest departure time and the destination of goods is also involved, which can be used as a variable. If the goods are delivered on time, it will enter the next stage (transportation stage); otherwise, it will be treated as Party B's breach of contract and enter the default state.

During the transport stage, the main body involved is typically the goods being transported. It is the state of the goods during the transportation process that determines whether they enter the default state. In addition, weather, road conditions, and road fees need to be considered. These will all exist in the form of variables.

If the goods are delivered to the place specified by Party A on time, it will enter the goods inspection stage. At this stage, Party A will first weigh the received goods (iron ore) to check whether the weight is up to the standard. If it is up to standard, samples will be taken for testing, focusing on the difference between the weight of iron ore and the iron grade value and the standard value, and determine whether freight is deducted according to the difference. If the difference is significantly more than a certain critical value, it will directly enter the default state. Specific weight standards and iron grade standards will be uploaded in the form of predefined parameters before the deployment of smart contracts.

Upon passing the quality inspection, the contract will enter the freight settlement phase. In the settlement status, Party A needs to settle the freight according to the tonnage of the goods arrived this time and the unit quality freight agreed by both parties in advance, and then pay the freight to Party B before the specified payment date. If Party A fails to pay when due, the smart contract will automatically execute the payment process and transfer the freight payable by Party A to Party B's account. The variables to be set at this stage are mainly freight per ton of goods, payment period of Party A, etc.

The above is the normal state experienced during the execution of the contract, while the abnormal state is the default state mentioned previously. The abnormal state will be divided into several sub states, each of which corresponds to the possible breach of contract at a certain stage of contract execution. For example, in the preparation stage, if the goods are not delivered on time, it will enter default sub state 1, which will be regarded as inheriting the sub class of default status, and the other sub states are the same. In each sub class, the handling methods for violating the corresponding provisions of the contract will be clearly specified. This will include but not limited to a certain amount of liquidated damages, fines, and even termination of the contract.

To date, the state refinement of the logistics contract has been completed. This process will be completed by knowledge extraction based on multiple rounds of questions and answers. It should be emphasized that data preprocessing before refining is essential. The entities and variables involved in the contract mentioned above must be explicitly written into the NL contract after data preprocessing. According to the structured data extracted from knowledge, we can generate the state diagram as shown in the figure. After the smart contract state diagram is determined, we can directly use the Eclipse plug-in to convert it into Solidity code. So far, the source code of the smart contract has been generated.

6 Conclusion

In this paper, we propose an automatic generation method of smart contracts based on NLP technology and UML. Based on the knowledge extraction technology in NLP, we used multi-round dialogue to extract the entity relationship joint extraction and event extraction of the preprocessed NL contract, used the obtained structured contract data for UML state diagram modeling, and finally transformed the UML state diagram into a smart contract.

The method of knowledge extraction can eliminate the relatively complex definition of semantic rules in the process of transformation from natural language to UML class diagram, and the use of UML, a unified language, to model smart contract state diagram has a similar effect, without the need to define additional domain-specific languages for each application scenario. Therefore, we have reason to believe that the automatic generation of smart contracts based on these two technologies will further shorten the development cycle of smart contracts. This will improve the scalability of smart contracts based on existing methods.

Acknowledgements. This work was supported by the S&T Program of Hebei through grant 20310101D.

References

1. Liu, L., et al.: A recursive reinforced blockchain performance evaluation and improvement architecture: maximising diversity to improve scalability. Int. J. Commun. Syst. e5315 (2022)
2. Zheng, Z., Xie, S., Dai, H.-N., Chen, X., Wang, H.: Blockchain challenges and opportunities: a survey. Int. J. Web Grid Serv. **14**(4), 352–375 (2018)
3. Jurgelaitis, M., Čeponienė, L., Butkienė, R.: Solidity code generation from UML state machines in model-driven smart contract development. IEEE Access **10**, 33465–33481 (2022)
4. Liu, L., Wei-Tek Tsai, Md., Bhuiyan, Z.A., Peng, H., Liu, M.: Blockchain-enabled fraud discovery through abnormal smart contract detection on ethereum. Futur. Gener. Comput. Syst. **128**, 158–166 (2022)
5. Balog, M., Gaunt, A.L., Brockschmidt, M., Nowozin, S., Tarlow, D.: Deepcoder: learning to write programs. arXiv preprint arXiv:1611.01989 (2016)
6. Veeramani, A., Venkatesan, K., Nalinadevi, K.: Abstract syntax tree based unified modeling language to object oriented code conversion. In: Proceedings of the 2014 International Conference on Interdisciplinary Advances in Applied Computing, pp. 1–8 (2014)
7. Choudhury, O., Rudolph, N., Sylla, I., Fairoza, N., Das, A.: Auto-generation of smart contracts from domain-specific ontologies and semantic rules. In: 2018 IEEE International Conference on Internet of Things (iThings) and IEEE Green Computing and Communications (GreenCom) and IEEE Cyber, Physical and Social Computing (CPSCom) and IEEE Smart Data (SmartData), pp. 963–970. IEEE (2018)
8. Tateishi, T., Yoshihama, S., Sato, N., Saito, S.: Automatic smart contract generation using controlled natural language and template. IBM J. Res. Dev. **63**(2/3), 6–1 (2019)
9. Asawa, K., Kukreja, S., Gondkar, R.: An NCDP for developing a blockchain based dynamic supply chain management with auto-generation of smart contract. In: 2021 26th International Conference on Automation and Computing (ICAC), pp. 1–6. IEEE (2021)
10. Li, X., et al.: Entity-relation extraction as multi-turn question answering. arXiv preprint arXiv:1905.05529 (2019)
11. Li, Q., et al.: Reinforcement learning-based dialogue guided event extraction to exploit argument relations. IEEE/ACM Trans. Audio Speech Lang. Process. **30**, 520–533 (2021)
12. Peng, H., et al.: Fine-grained event categorization with heterogeneous graph convolutional networks. arXiv preprint arXiv:1906.04580 (2019)
13. Peng, H., et al.: Streaming social event detection and evolution discovery in heterogeneous information networks. ACM Trans. Knowl. Discov. Data **15**(5), 1–33 (2021)

14. Peng, H., Zhang, R., Li, S., Cao, Y., Pan, S., Philip, Yu.: Reinforced, incremental and cross-lingual event detection from social messages. IEEE Trans. Pattern Anal. Mach. Intell. **45**(1), 980–998 (2022)
15. Peng, H., Zhang, R., Dou, Y., Yang, R., Zhang, J., Yu, P.S.: Reinforced neighborhood selection guided multi-relational graph neural networks. ACM Trans. Inf. Syst. (TOIS) **40**(4), 1–46 (2021)
16. Peng, H., et al.: Lime: low-cost and incremental learning for dynamic heterogeneous information networks. IEEE Trans. Comput. **71**(3), 628–642 (2021)
17. Ben Abdessalem Karaa, W., Ben Azzouz, Z., Singh, A., Dey, N., Ashour, A.S., Ben Ghazala, H.: Automatic builder of class diagram (ABCD): an application of UML generation from functional requirements. Softw. Pract. Exp. **46**(11), 1443–1458 (2016)
18. Abdelnabi, E.A., Maatuk, A.M., Abdelaziz, T.M., Elakeili, S.M.: Generating UML class diagram using NLP techniques and heuristic rules. In: 2020 20th International Conference on Sciences and Techniques of Automatic Control and Computer Engineering (STA), pp. 277–282 (2020)
19. Peng, H., Li, H., Song, Y., Zheng, V., Li, J.: Differentially private federated knowledge graphs embedding. In: Proceedings of the 30th ACM International Conference on Information & Knowledge Management, pp. 1416–1425 (2021)
20. Eberts, M., Ulges, A.: Span-based joint entity and relation extraction with transformer pre-training. arXiv preprint arXiv:1909.07755 (2019)
21. Du, X., Cardie, C.: Event extraction by answering (almost) natural questions. arXiv preprint arXiv:2004.13625 (2020)
22. Niaz, I.A., Tanaka, J., et al.: Mapping UML statecharts to java code. In: IASTED Conference on Software Engineering, pp. 111–116 (2004)
23. Usman, M., Nadeem, A.: Automatic generation of java code from UML diagrams using UJECTOR. Int. J. Softw. Eng. Appl. **3**(2), 21–37 (2009)
24. Sunitha, E.V., Samuel, P.: Automatic code generation from UML state chart diagrams. IEEE Access **7**, 8591–8608 (2019)
25. Kai, H., Zhu, J., Ding, Y., Bai, X., Huang, J.: Smart contract engineering. Electronics **9**(12), 2042 (2020)

Blockchain Scalability Technologies

Nengxiang Xu[1,2], Jiahong Cai[1,2], Yinyan Gong[1,2], Huan Zhang[1,2], Weihong Huang[1,2(✉)], and Kuan-ching Li[1,2]

[1] School of Computer Science and Engineering, Hunan University of Science and Technology, Xiangtan 411201, China
jiahongcai@mail.hnust.edu.cn, whhuang@hnust.edu.cn
[2] Hunan Key Laboratory for Service Computing and Novel Software Technology, Xiangtan 411201, China

Abstract. As the underlying implementation technology of the current main-stream digital currency, blockchain can establish a trusted distributed system without relying on third-party trusted institutions or a privacy-protect system. The decentralized characteristics of blockchain have broad application scenarios, such as the Internet of Things, financial technology and other fields. However, the scalability of the current blockchain is seriously insufficient, such as limited throughput in performance, small storage capacity, and difficulty in scaling functions. This paper introduces the definition and technical classification of scalability, analyzes the current problems of scalability and briefly introduces the current main-stream expandable technologies such as sharding, on-chain and off-chain storage, off-chain payment channel and cross-chain technology from two aspects of performance and function, as well as the principles and ideas of these technologies. Finally, the research progress of current blockchain extension technology is summarized, and the problems faced by the current extension scheme are pointed out, which provides a direction for future research work.

Keywords: Blockchain · Cross-chain · Off-chain · Scalability · Sharding Introduction

1 Introduction

In 2008, Nakamoto published a technical white paper [1], Bitcoin: "a point-to-point electronic cash payment system". The Bitcoin system first proposed and implemented a decentralized digital payment system that uses blockchain technology as the underlying support technology and does not rely on trusted third parties in an open network [2]. The security trust model in Bitcoin system is greatly different from the existing commercial digital payment system [3]. In a Bitcoin system, trust between users comes from trust in the entire payment system [4]. The continuous development of blockchain has brought more possibilities in many areas. However, the development of blockchain technology is still subject to the trilemma [5]. The throughput of Bitcoin is 7 transactions per second [6], and Ethereum is 10–20 transactions per second. Scalability has become the biggest bottleneck for blockchain applications [7], as shown in Fig. 1. Aiming at the problems

M. Qiu et al. (Eds.): SmartCom 2022, LNCS 13828, pp. 475–484, 2023.
https://doi.org/10.1007/978-3-031-28124-2_45

faced by blockchain scalability, this paper introduces the current mainstream scalability technologies in Sect. 3, analyzes their characteristics and compares them. In Sect. 4, the technical bottlenecks of these problems are analyzed, and the future research directions are pointed out.

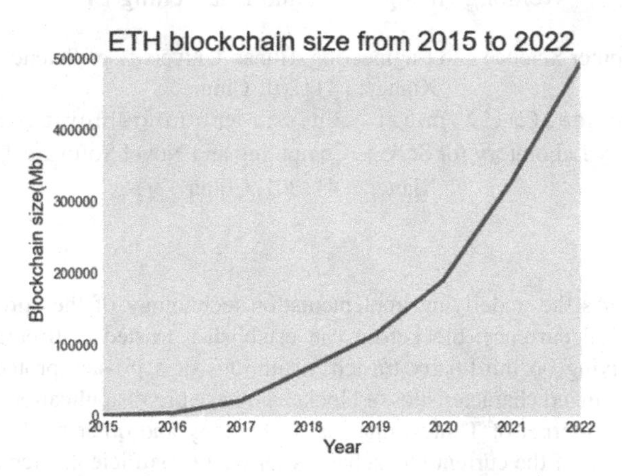

Fig. 1. ETH block size from 2015 to 2022 in Mb.

2 Correlation Technique

With the development of computer hardware [8–10], networks [11–13], and new algorithms [14–16], large amounts of data [17–19] can be generated in a short period. Data mining and machine learning [20–22] techniques had been applied in various datasets. However, the concerns about privacy and security [23–25] become a big challenge. Blockchain is a promising tool for solving this problem [26–28]. The current mainstream methods for improving blockchain scalability contains two aspects.

2.1 Sharding Technology

In the blockchain system based on sharding technology, the formation of a committee for transaction verification and consensus has reduced the storage burden of nodes to a certain extent and achieved the effect of scalability.

2.2 Payment Channel off Chain

Spilman proposed the micro-payment channel protocol [29], which establishes one-way channel cumulative transactions for small high-frequency transactions in the blockchain without the need for blockchain consensus.

3 Research Statue

This chapter will begin to introduce several blockchain extension methods mentioned above.

3.1 Scalable Performance

Performance-scalable methods include sharding and off-chain payment channels. The sharding mechanism was first applied to traditional distributed databases, and Elastico [30] and Zilliqa innovatively applied this technology to the consensus mechanism of the blockchain. Several typical sharding methods and offline payment channels will be introduced below.

Elastico. The core idea of the Elastico sharding protocol is to divide the nodes in the blockchain network into committees. Each committee has the same number of members. The disjoint transaction subsets in the blockchain are assigned to the committee and the PBFT consensus protocol is independently performed internally. Elastico defines an intra-fragment consensus process as an Epoch, in which the agreement is carried out in five steps, as shown in Fig. 2. There are several key points in the Elastico protocol. First, in the authentication and confirmation committee grouping phase, each processor now locally selects their own identity information group (IP, PK) [31], representing their IP address and public key, respectively [7]. In order for the network to accept their identity, each processor must also find a PoW solution (nonce), which must correspond to their own chosen identity, and everyone in the system can verify the nonce. Using this method can effectively avoid sybil attack. But this is not enough; to prevent the node from calculating the nonce value in advance and then submitting it in the current era, the system also introduces an epoch Randomness variable as a random source for an era. This random source was published after the last step of the previous era was generated. The Nonce value calculated by the node needs to meet the following conditions [30].

$$H(epochRandomness|PK||IP|nonce) \leq 2^{(r-D)} \tag{1}$$

Among them, D is a pre-set workload proof difficulty, r refers to the number of output bits generated by calculating the hash value H, and the final calculated hash value will also be used as the ID of the node. The last s bits of the node ID will determine the committee grouping of the node. When assigning the committee, if each member independently confirms the nodes in the committee through broadcast, the allocated message overhead $O(n^2)$, Elastico establishes a directory committee to determine the committee grouping results, and broadcasts them to each node, so that the message overhead is controlled at $O(n^c)$, where c is the number of committee nodes. The directory committee consists of the first c nodes that first calculate the nonce value.

In the consensus phase, the committee group conducts partial consensus through the PBFT protocol [32]. The final consensus is carried out by the final committee established by Elastico. The final committee is randomly selected and responsible for broadcasting the consensus results to the blockchain network nodes and calculating the next random source to generate new nonce values.

As the first sharding scheme, Elastico provides many guiding ideas for the next research. In terms of scalability, the Elastico scheme greatly improves the blockchain throughput, which is almost linear growth. However, Elastico also has some defects. For example, the PBFT protocol used by the consensus within the committee has too many communication rounds. It is a protocol with extremely high communication complexity, so the protocol delay is as high as 100 s. At the same time every time after an era of redistribution committee also brought a lot of time consumption.

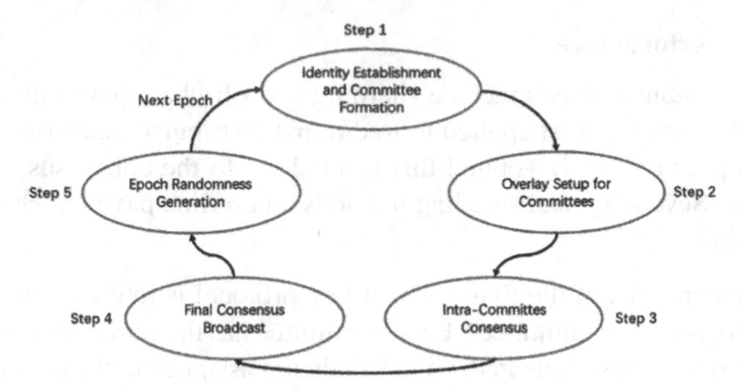

Fig. 2. Elastico workflow.

OmniLedger. OmniLedger [33] was proposed by Philipp et al. The OmniLedger protocol improves the sharding tendency of the Elastico protocol and the lack of cross-shard transaction guarantees, while increasing transaction throughput. First, the Elastico sharding mechanism has a nonce value calculation to determine the grouping. Even if the random source is announced after the end of an era, it still cannot avoid the aggregation of malicious nodes. OmniLedger determines the location of the slice based on the randomly generated rndn, and rndn uses the unbiased random number generation scheme RandHound, which avoids the centralization problem caused by third-party participation, and also solves the tendency sharding problem caused by calculating the nonce value, as shown in Fig. 3. RandHound requires a Leader to generate unbiased random numbers, which is determined by the VRF-based leader election algorithm. In the consensus algorithm [34], OmniLedger and Elastico protocol use the same PBFT consensus algorithm. Not the same with Elastico, OmniLedger divides the fragment into a sub-chain, and then uses the PBFT consensus on the fragment sub-chain. OmniLedger named this consensus algorithm ByzCoinX.

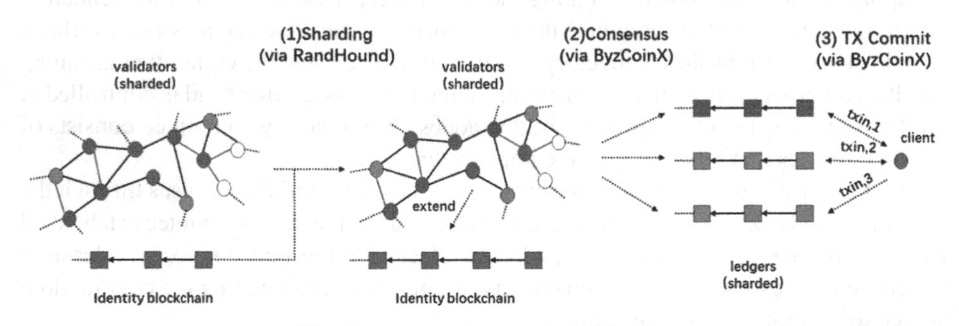

Fig. 3. OmniLeder sharding structure.

In dealing with cross-shard transactions, OmniLedger is completed by the user (client). If it is necessary to transfer the unspent transaction output (UTXO) of different fragments to other fragments, the input fragments (ISs) only need to provide proof, and the user locks these fragments. If all the input fragments provide proof, the user will send the proof to the output fragments (OSs). As long as one of the fragments rejects the transaction, unlock all the fragments and return these unspent transaction outputs. OmniLedger's cross-slice transaction processing method does not require communication between slices. The disadvantage is that users may be lazy in some application scenarios, resulting in the infeasibility of cross-slice transactions, as shown in Fig. 4.

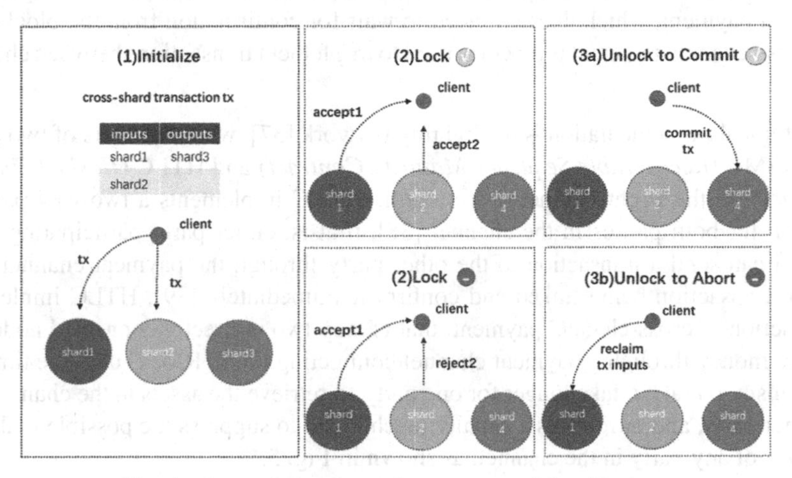

Fig. 4. Cross-chain transaction progress of OmniLedger.

For small transactions in slices, OmniLedger adopts the 'trust-but-verify' authentication mechanism. When small transactions accumulate to a certain amount, a small number of nodes perform the first verification, and then the other part of the node performs the second verification. The two-layer verification ensures that malicious behavior can be detected in a timely manner, and users can choose to perform only one verification. The verification method of cumulative confirmation improves the transaction throughput to a certain extent. Table 1. Compares Elastico with OmniLedger from three different aspects.

Table 1. Compare between Elastico and OmniLedger.

Approach	Sharding method	Consensus protocol	Efficiency
Elastico	Pow solution	PBFT	Depend on PBFT and duration of an epoch
OmniLedger	RandHound	ByzCoinX	Depend on ByzCoinx

Off-Chain Payment Network. The payment channel network deals with high-frequency small transactions through off-chain processing, which indirectly improves the transaction throughput of the blockchain system [34]. Among them, the blockchain plays the role of an arbitration platform. In the process of channel payment, if the two parties disagree on the status of the channel, the perpetrator can be punished by triggering a penalty transaction (the penalty transaction will be confirmed on the blockchain). The payment network under the chain mainly includes two protocols, payment channel protocol and cross channel payment protocol [35]. The payment channel protocol implements off-chain payment, which does not need to wait for confirmation from the blockchain. The cross-channel payment protocol is used to implement transactions between channels [36].

A typical implementation is the lighting network [37], which consists of two protocols: RSMC (*Recoverable Sequence Maturity Contract*) and HTLC (*Hashed Timelock Contract*). In the Lightning network system, RSMC implements a two-way payment function for both parties in the channel [38], that is, either party participating in the channel can send a transaction to the other party through the payment channel without the transaction being linked and confirm it immediately [39]. HTLC implements the function of cross-channel payment, that is, any two indirectly connected nodes can transfer money through a payment channel connecting them. RMSC uses the timelock mechanism to make it take longer for one party to retrieve the assets in the channel than the other party, and introduces a penalty mechanism to suppress the possible malicious behavior of any party in the channel, as shown in Fig. 5.

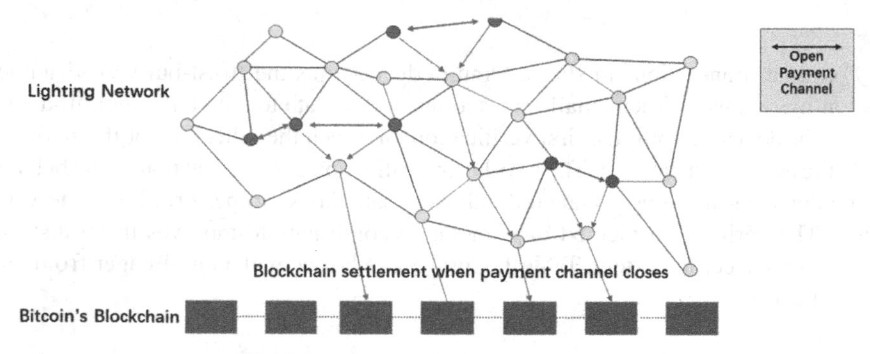

Fig. 5. Payment path of lighting network.

3.2 Function-Expandable

At present, blockchain also has shortcomings in function scalability. At present, there is a lack of credible data exchange mechanism between different blockchain systems, which makes it impossible to share information between different blockchain systems, so as to realize more complex commercial applications. This paper introduces several mainstream cross-chain technologies.

Notary Model. Notary Schemes is a cross-chain technology that is easy to maintain and scale. It mainly realizes data collection, transaction confirmation and verification by electing notaries between blockchains. The main representative program is Ripple [40].

Sidechain Technology. In 2014, BlockStream proposed sidechain technology [41] to solve the scalability problem of Bitcoin system. It is a blockchain system independent of the main chain, using bidirectional anchoring technology to make the side chain attached to the main chain. The bidirectional anchoring performs the internal exchange of assets between the main chain and the side chain at a preset rate. The side chain technology uses the managed mode or the SPV mode to act as the connecting party of the main chain and the side chain. The managed mode is divided into two types: single managed and alliance managed.

3.3 Other Storage Solutions

The literature [42] proposes a blockchain light node verification protocol FlyClient. To reduce the number of blocks downloaded by light-node clients during a validation transaction, FlyClient uses a probabilistic validation mechanism to download the blockhead validation chain for log(n) (n is the number of blocks). If the blockchain is validated as a valid chain, the light-node client only needs to store one block information to verify that the blockchain has the transactions it looks for.

The literature [43] proposes a lightweight node ESPV. Because the request for new blocks in the blockchain system is large and the request for old blocks is small, ESPV adopts a fully redundant storage strategy to store new blocks. The new block is updated using the sliding window mechanism [44]. When a new block is generated, the last block in the window is deleted and the new block is added to the window.

The literature [45] proposes a layered edge cloud blockchain structure LayerChain. LayerChain stores blockchain data from IIoT (*Industry Internet of things*) [46] devices in a three-tier architecture of cloud blockchain layer [47], edge blockchain layer, and IIoT device layer [48].

4 Summarized and Prospected

This section will point out the challenges faced by these methods and provide some possible directions for future research work.

4.1 Challenges for Scalable Programs

Efficiency. Under the chain payment network, the side chain technology performs transaction processing and data storage outside the blockchain. When data exchange with the blockchain is needed, the efficiency is not high [49].

Security. The Elastico sharding protocol uses the method of calculating the nonce value to confirm the committee grouping in the formation committee stage, so it is very likely that malicious nodes will gather [50]. Light nodes still rely on transaction data provided by all nodes when conducting transaction verification, which may also cause security problems.

4.2 Future Research Direction

Research on Efficient Data Query Model. In the side chain technology, the data sharing between the side chain and the main chain will cause a relatively large burden on the system. The transactions submission in the payment channel under the chain also needs to transmit data with the block chain. Therefore, the next step can be to study efficient data query mode.

Research on Secure Data Protection Mechanism. In the current blockchain technology research, security is still a very important direction. For example, in the Elastico sharding protocol, how the formation of the committee avoids the aggregation of malicious nodes. Therefore, the next step is to study secure data protection mechanisms [51].

5 Conclusion

This paper introduced some current technologies to improve the scalability of blockchain from two aspects: performance and function. These technologies have different emphases on scalability, security and decentralization. Finally, the challenges faced by the current blockchain scalability technology were analyzed, and the future direction was prospected.

References

1. Nakamoto, S.: Bitcoin: a peer-to-peer electronic cash system (2008). https://bitcoin.org/bitcoin.pdf
2. Huang, Y., et al.: A novel identity authentication for FPGA based IP designs. In: 17th IEEE TrustCom/BigDataSE, pp. 1531–1536 (2018)
3. Xia, F., Hao, R., et al.: Adaptive GTS allocation in IEEE 802.15. 4 for real-time wireless sensor networks. J. Syst. Archit. **59**(10), 1231–1242 (2013)
4. Liang, W., Yang, Y., Yang, C., et al.: A consortium blockchain-based privacy protection scheme for personal data. IEEE Trans. Reliab., 1–13 (2022)
5. Gai, K., Hu, Z., Zhu, L., et al.: Blockchain meets dag: a blockdag consensus mechanism. In: Conference on Algorithms and. Architecture for Parallel Processing, pp. 110–125 (2020)
6. Antonopoulos, A.M.: Mastering Bitcoin: Unlocking Digital Cryptocurrencies. O'Reilly Media, Inc. (2014)
7. Xu, Z., et al.: A time-sensitive token-based anonymous authentication and dynamic group key agreement scheme for industry 5.0. IEEE TII **18**(10), 7118–7127 (2022)
8. Qiu, M., Jia, Z., et al.: Voltage assignment with guaranteed probability satisfying timing constraint for real-time multiproceesor DSP. JSPS (2007)
9. Qiu, M., Li, H., Sha, E.: Heterogeneous real-time embedded software optimization considering hardware platform. In: ACM Symposium on Applied Computing, pp. 1637–1641 (2009)
10. Qiu, M., Xue, C., Shao, Z., Sha, E.: Energy minimization with soft real-time and DVS for uniprocessor and multiprocessor embedded systems. In: IEEE DATE Conference, pp. 1–6 (2007)

11. Niu, J., Gao, Y., et al.: Selecting proper wireless network interfaces for user experience enhancement with guaranteed probability. JPDC **72**(12), 1565–1575 (2012)
12. Qiu, M., Chen, Z., Ming, Z., Qin, X., Niu, J.: Energy-aware data allocation with hybrid memory for mobile cloud systems. IEEE Syst. J. **11**(2), 813–822 (2014)
13. Qiu, M., Liu, J., Li, J., et al.: A novel energy-aware fault tolerance mechanism for wireless sensor networks. In: IEEE/ACM International Conference on GCC (2011)
14. Shao, Z., Wang, M., et al.: Real-time dynamic voltage loop scheduling for multi-core embedded systems. IEEE Trans. Circuits Syst. II **54**(5), 445–449 (2007)
15. Qiu, M., et al.: Efficent algorithm of energy minimization for heterogeneous wireless sensor network. In: Sha, E., Han, S.K., Xu, C.Z., Kim, M.H., Yang, L.T., Xiao, B. (eds.) EUC 2006. LNCS, vol. 4096, pp. 25–34. Springer, Heidelberg (2006). https://doi.org/10.1007/11802167_5
16. Qiu, M., Yang, L., et al.: Dynamic and leakage energy minimization with soft real-time loop scheduling and voltage assignment. IEEE TVLSI **18**(3), 501–504 (2009)
17. Li, J., Ming, Z., et al.: Resource allocation robustness in multi-core embedded systems with inaccurate information. J. Syst. Architect. **57**(9), 840–849 (2011)
18. Qiu, M., Khisamutdinov, E., et al.: RNA nanotechnology for computer design and in vivo computation. Philos. Trans. R. Soc. A (2013)
19. Qiu, M., Sha, E., Liu, M., et al.: Energy minimization with loop fusion and multi-functional-unit scheduling for multidimensional DSP. JPDC **68**(4), 443–455 (2008)
20. Qiu, H., Zheng, Q., et al.: Topological graph convolutional network-based urban traffic flow and density prediction. IEEE Trans. ITS (2020)
21. Qiu, M., Qiu, H.: Review on image processing based adversarial example defenses in computer vision. In: IEEE 6th BigDataSecurity, Baltimore, MD, USA, pp. 94–99 (2020)
22. Hu, F., Lakdawala, S., et al.: Low-power, intelligent sensor hardware interface for medical data preprocessing. IEEE Trans. Info. Technol. Biomed. **13**(4), 656–663 (2009)
23. Qiu, H., Dong, T., et al.: Adversarial attacks against network intrusion detection in IoT systems. IEEE Internet Things J. **8**(13), 10327–10335 (2020)
24. Li, Y., Gai, K., et al.: Intercrossed access controls for secure financial services on multimedia big data in cloud systems. ACM Trans. Multimedia Comput. Commun. Appl. (2016)
25. Gai, K., Qiu, M., Elnagdy, S.: A novel secure big data cyber incident analytics framework for cloud-based cybersecurity insurance. In: IEEE BigDataSecurity (2016)
26. Qiu, M., Qiu, H., et al.: Secure data sharing through untrusted clouds with blockchain-enabled key management. In: The 3rd SmartBlock 2020, China, pp. 11–16, October 2020
27. Gai, K., Zhang, Y., et al.: Blockchain-enabled service optimizations in supply chain digital twin. IEEE Trans. Serv. Comput. (2022)
28. Gao, X., Qiu, M.: Energy-based learning for preventing backdoor attack. KSEM (3), 706–721 (2022)
29. Chan, E., Lesani, M.: Brief announcement: brokering with hashed timelock contracts is NP-hard. In: ACM PODC, Virtual Event, Italy, pp. 199–202. ACM (2021)
30. Luu, L., Narayanan, V., Zheng, C., Baweja, K., Gilbert, S., Saxena, P.: A secure sharding protocol for open blockchains. In: ACM SIGSAC Conference on Computer and Communications Security, Vienna, Austria, pp. 17–30. ACM (2016)
31. Liang, W., Xie, S., Li, X., Long, J., Xie, Y., Li, K.-C.: A novel lightweight PUF-based RFID mutual authentication protocol (2017)
32. Xia, F., et al.: Adaptive GTS allocation in IEEE 802.15.4 for real-time wireless sensor networks. J. Syst. Archit. **59**(10, Part D), 1231–1242 (2013)
33. Kokoris-Kogias, E., et al.: Omniledger: a secure, scale-out, decentralized ledger via sharding. In: IEEE Symposium on Security and Privacy, SP 2018, pp. 583–598 (2018)
34. Cheng, H., Xie, Z., et al.: Multi-step data prediction in wireless sensor networks based on one-dimensional CNN and bidirectional LSTM. IEEE Access **7**, 117883–117896 (2019)

35. Chen, Y., Zhou, L., et al.: KNN-BLOCK DBSCAN: fast clustering for large-scale data. IEEE Trans. Syst. Man Cybern. Syst. **51**(6), 3939–3953 (2019)
36. Pan, C.: Research on blockchain scalability technology. Master's thesis, Shanghaijiaotong University (2019)
37. Fang, Z., Gai, K., Zhu, L., Xu, L.: LNBFSM: a food safety management system using blockchain and lightning network. In: Qiu, M. (eds.) ICA3PP 2020. LNCS, vol. 12454, pp. 19–34. Springer, Cham (2020). https://doi.org/10.1007/978-3-030-60248-2_2
38. Zhao, J., Huang, J., et al.: An effective exponential-based trust and reputation evaluation system in wireless sensor networks. IEEE Access **7**, 33859–33869 (2019)
39. Yao, Y., Xiong, N., et al.: Privacy-preserving max/min query in two-tiered wireless sensor networks. Comput. Math. Appl. **65**(9), 1318–1325 (2013)
40. Armknecht, F., Karame, G.O., Mandal, A., Youssef, F., Zenner, E.: Ripple: overview and outlook. In: Conti, M., Schunter, M., Askoxylakis, I. (eds.) Trust 2015. LNCS, vol. 9229, pp. 163–180. Springer, Cham (2015). https://doi.org/10.1007/978-3-319-22846-4_10
41. Back, A., Corallo, M., Dashjr, L., et al.: Enabling blockchain innovations with pegged sidechains (2014)
42. Bünz, B., Kiffer, L., et al.: Super-light clients for cryptocurrencies. In: 2020 IEEE Symposium on Security and Privacy, SP 2020, pp. 928–946 (2020)
43. Zhao, Y., Niu, B., Li, P., Fan, X.: A novel enhanced lightweight node for blockchain. In: Zheng, Z., Dai, HN., Tang, M., Chen, X. (eds.) BlockSys 2019. CCIS, vol. 1156, pp. 137–149. Springer, Singapore (2020). https://doi.org/10.1007/978-981-15-2777-7_12
44. Fu, A., Zhang, X., et al.: VFL: a verifiable federated learning with privacy-preserving for big data in industrial IoT. IEEE Trans. Ind. Inform. (2020)
45. Yao, Y., Liu, S., et al.: A hierarchical edge-cloud blockchain for large-scale low-delay industrial internet of things applications. IEEE TII **17**(7), 5077–5086 (2021)
46. Zhang, Y., Gai, K., et al.: Blockchain-empowered efficient data sharing in Internet of Things settings. IEEE J. Sel. Areas Commun. 1 (2022)
47. Gai, K., Guo, J., Zhu, L., Shui, Y.: Blockchain meets cloud computing: a survey. IEEE Commun. Surv. Tutor. **22**, 2009–2030 (2020)
48. Wang, J., Luo, W., Liang, W., et al.: Locally minimum storage regenerating codes in distributed cloud storage systems. China Commun. **14**(11), 82–91 (2017)
49. Hafid, A., et al.: Scaling blockchains: a comprehensive survey. IEEE Access **8**, 125244–125262 (2020)
50. Kumar, P., Kumar, R., et al.: PPSF: a privacy-preserving and secure framework using blockchain-based machine-learning for IoT-driven smart cities. IEEE Trans. Netw. Sci. Eng. **8**(3), 2326–2341 (2021)
51. Gao, F., Zhu, L., Gai, K., Zhang, C., Liu, S.: Achieving a covert channel over an open blockchain network. IEEE Netw. **34**(2), 6–13 (2020)

Blockchain Applications in Smart City: A Survey

Shuo Wang[1,2], Zhiqi Lei[1,2], Zijun Wang[1], Dongjue Wang[3], Mohan Wang[4], Gangqiang Yang[5], and Keke Gai[1,2](✉)

[1] School of Cyberspace Science and Technology, Beijing Institute of Technology, Beijing 100081, China
{3220215214,3220211023,3120221290,gaikeke}@bit.edu.cn
[2] Yangtze Delta Region Academy of Beijing Institute of Technology, Jiaxing, Zhejiang, China
[3] School of Cyber Engineering, Xidian University, Xi'an 710071, China
[4] Faculty of Information Science and Engineering, Ocean University of China, Qingdao 266400, China
[5] School of Information Science and Engineering, Shandong University, Qingdao, Shandong 266237, China
g37yang@sdu.edu.cn

Abstract. Smart cities bring new ideas to solve the social, economic, and environmental problems existing in traditional cities. However, the services of smart cities suffer from centralized data storage and untrustworthiness. As a decentralized distributed database, blockchain provides distributed and trusted infrastructure technology for the construction of smart cities. In this survey, we conduct a brief survey of the literature on blockchain applications in smart cities. We review the application of blockchain in smart cities from the four main application scenarios of energy, transportation, medical care, and manufacturing. Then, we discuss the realization of blockchain-based smart cities from the perspectives of privacy and storage. We believe this survey provides new ideas for the development of smart cities.

Keywords: Blockchain · Smart City · Smart Contract · Distributed Storage

1 Introduction

With the continuous increase of urban population, the medical level, convenient transportation, education level and living environment have been dramatically improved in cities. A large number of citizens are mitigating to bigger-sized cities for seeking a better living. However, a large growth of urban population also has brought many issues, for example, the limited energy resource, traffic congestion, air pollution, lack of medical resource, and manufacturing shortage. Constructing a digital transformation towards smart city, i.e., integrating Information Technology (IT)-based solutions with city infrastructure, is an alternative solution to solving some known city issues. The goal of the smart city is to build

M. Qiu et al. (Eds.): SmartCom 2022, LNCS 13828, pp. 485–494, 2023.
https://doi.org/10.1007/978-3-031-28124-2_46

Table 1. Summary of the application of blockchain in specific fields

Ref.s	Fields	Description
[6]	Smart grid	Blockchain-based data protection model; strengthening the defense of smart grid against external network attacks
[7]	Medical	Hybrid storage model; improving the scalability of blockchain for the medical field
[8]	Transportation	Blockchain-based Internet of Vehicles trust system; realizing transportation detection management
[9]	Supply sys	Ethereum-based supply chain system; ensuring the security and traceability of the information

a sustainable city through new technologies. New technologies (e.g., blockchain [1,2]) provide technical supports for the construction of smart cities. This survey focuses on new adoptions of blockchain in smart city.

Blockchain [3–5] is known as a distributed data ledger that jointly maintains system tamper-proof, immutability, and traceability in a decentralized and trustless manner. Blockchain-based smart cities can solve some existing city issues. For example, decentralization and trustless mechanisms secure the execution of distributed environments without a trusted third party; immutability ensures security of data transmissions; traceability is conducive to the realization of the full life cycle management of block storage data; the deployment of smart contracts, moreover, realizes automated executions of transactions and expand applications in various fields in smart city.

Many prior researches have explored blockchain applications from various dimensions, for instance, as shown in Table 1. However, investigations targeting at blockchain adoptions in smart city are rare. To make up for the application gap in this field, this survey mainly synthesizes blockchain-based smart city solutions from perspectives of transportation, medical, energy [10], and manufacturing. In addition, we further discuss major factors of blockchain-based smart city in order to provide a fundamental view of the blockchain adoptions in smart city.

The rest of this survey is organized in the following order. Section 2, 3, 4, 5 elaborate on the four main application scenarios of blockchain-based smart cities, namely medical, transportation, energy and manufacturing, respectively. The discussions of the smart city are presented in Sect. 6. Finally, we concluded the work in Sect. 7.

2 Blockchain-Based Smart Energy

Energy is a crucial foundation for urban economic development. The traditional centralized energy system is no longer appropriate for the current situation, and the energy system is gradually evolving from centralized to dispersed [11]. Blockchain includes attributes like anonymity, decentralization, and traceability that can be utilized to change the way the energy system operates.

The smart grid [12] is a new grid infrastructure that supports the two-way delivery of energy and information. A smart grid must effectively manage the data related to distributed energy. To ensure the secure sharing of distributed energy data, a safe and auditable data exchange system [13] for smart grids based on blockchain was proposed, which establishes a framework for data exchange that includes fine-grained access control, computation of sensitive data, and data traceability. Furthermore, to ensure the security of private information, an out-of-chain smart contract based on the trusted execution environment is designed in this scheme.

The smart grid is vulnerable to cyberattacks because network communication technology based on TCP/IP and Ethernet protocols exposes it to the public network. To strengthen the smart grid's defenses against outside cyberattacks, a distributed data protection model [6] based on blockchain was presented. In this model, smart meters are used as nodes in the blockchain to capture and store power data to ensure data security. Aiming at the security problems of smart grids in an open environment, Bera et al. [14] proposed a blockchain-based access control for the smart grid, which provides anonymity and non-traceability.

The decentralization of blockchain is consistent with distributed energy trading. There are many designs that combine blockchain and energy trading [15–18]. To ensure privacy in smart grid energy transactions, Gai et al. [15] presented a consortium blockchain-based energy trading system. To defend against linking attacks and malicious data mining algorithm attacks, this scheme protects private data such as energy trading trends by building an account mapping mechanism on the smart contract.

In addition, some scholars have been interested in the efficiency of the energy trading process. Ali et al. [16] proposed an adaptive network model for P2P energy trading. In this model, the smart contract is used to create and manage prosumer groups, which improves the scalability of the prosumer grouping mechanism and eventually improves the overall system performance. For distributed energy, the use of P2P energy trading can reduce energy consumption costs. Khalid et al. [17] designed a hybrid P2P energy trading model based on blockchain. This scheme makes use of three smart contracts to realize a complex energy trading market. Among them, the master smart contract manages all energy trading operations, the P2P smart contract is used to control the relevant information between producers and consumers. The prosumer-to-grid (P2G) smart contract is used to manage the data related to the main power grid. AlSkaif et al. [18] constructed a full P2P energy trading model based on blockchain for residential energy systems with different distributed energy types. In this scheme, two new trading strategies are proposed, one is the matching strategy of supply and demand, and the other is the matching strategy based on distance. In the supply and demand matching strategy, the transaction coefficients are calculated according to the power demand and excess power. In the distance-based matching strategy, P2P trading between nearby participants is encouraged to reduce the loss of long-distance transmission, thus improving energy efficiency.

3 Blockchain-Based Smart Medical

In current city life, medical data is basically stored and managed by each hospital separately. However, the centralized storage of medical data will cause certain security problems. As a distributed ledger, blockchain provides distributed and secure medical data storage.

Blockchain provides an efficient and low-latency method for data storage. For example, Gaetani et al. [19] use blockchain to ensure data integrity. They design a database based on blockchain to solve the low throughput and high latency in the original cloud computing environment. However, the medical data of patients are stored separately in each hospital, resulting in the fragmentation of the medical data storage of patients, which brings inconvenience for patients to seek medical treatment across hospitals.

To build a full life cycle of medical data storage management for patients, Yue et al. [20] proposed a Healthcare Data Gateway based on blockchain, which guarantees patients to control their private information. The complete data storage framework was built through the blockchain, which built a lifelong traceable and tamper-proof data record for patients. Since all data in a single blockchain system has been stored. However, the data storage capacity of the current blockchain is limited. To improve the scalability of blockchain, Sun et al. [7] proposed a hybrid storage mode based on blockchain pluses IPFS. Therefore, in the face of massive medical data, the hybrid storage model can safely and irreversibly store medical data in blockchain and IPFS without violating the concept of blockchain distributed and trusted storage, which satisfies big data storage requirements to a certain extent.

In addition, there is a problem with privacy disclosure in the process of sharing medical data. To realize the safe and reliable sharing of medical data under the premise of ensuring user privacy, Kumar et al. [21] proposed a patient-centered framework and a data access control mechanism for PHRs stored in semi-trusted servers. This scheme uses ABE technology to encrypt the PHR files of each patient. However, the ABE system needs additional computing overhead to perform attribute revocation and encrypt the data again. To reduce costs, Gu et al. [22] proposed a more effective ABS scheme using monotone predicates, but it cannot solve the problems caused by access control modifications. In this regard, Guo et al. [23] introduced a signature scheme with multiple permissions to ensure the effectiveness of medical data in the blockchain.

In the smart medical system, blockchain can effectively solve the problems of data isolation. In addition, it can tamper with the current medical data storage, and relieve the pressure on data storage caused by massive medical data.

4 Blockchain-Based Smart Transportation

In this section, several of the newly developed blockchain applications that relate to the transportation industry will be thoroughly discussed. Transportation serves as both the urban development's infrastructure and the heartbeat of the city. Promoting the establishment of smart cities requires the development of smart transportation. Smart transportation scenario refers to the industry developed by using cutting-edge technologies (e.g., blockchain etc.) combined with the traditional transportation industry.

Traditional transportation system has two main problems. First, the level of intelligence and digitalization of traditional transportation infrastructure is low, and the capability of controlling information is weak. Second, some data-centric transportation applications, such as vehicular ad-hoc networks (VANETs), suffer from the lack of a fixed infrastructure and dynamic information communication. Blockchain can assist in building a distributed, secure and trustworthy smart transportation system. Applying blockchain technology to the transportation industry can achieve borderless connectivity and flat sharing of transportation system data.

Currently, many related research [8,24,25] are applying blockchain technology to various aspects of transportation. In the aspect of big data management. To solve the serious road safety problem caused by the easy modification of car odometer data, Preikschat et al. [24] proposes a blockchain-based distributed database. This distributed database enables data sharing of odometers and ensures that data is protected from tampering. This scheme enhances the public transparency of odometer data to a certain extent.

In the aspect of transport detection management, Yang et al. [8] give an implementation of an IoV trust management system using blockchain. The scheme focuses on verifying the received information by the Bayesian model and calculating the trust offset value of the received information. It realizes the combination of data and blockchain. Additionally, Liu et al. [25] assert that blockchain technology might be used to plan the charging and discharging of electric vehicles. In this scheme, electric vehicles firstly place charging and discharging orders to the public blockchain trading platform of the smart grid. Then, matching orders are processed and authenticated by peer nodes within the blockchain network. Finally, both organizations save the confirmed requests in a distributed manner. All the above researches prove that blockchain provide some technical support to ITS, thus improving the distributed capability as well as the security of the system.

Some related research has shown that blockchain can help build vehicular VANETs and manage vehicle communications. Conventional VANETs suffer from the lack of fixed infrastructure and dynamic information communication. It has been demonstrated that dynamic VANETs can be safely managed using software-defined VANETs [26]. Following that, it is incorporated with blockchain technology to offer distributed control over the entire network. Moreover, To improve the interactive security of electric vehicles in VANETs, data coins and energy coins stimulated by blockchain can be defined as a new type of cryptocurrency and can be applied to the vehicles [27]. This literature focuses on workload proofs using data contribution frequencies and energy contribution amounts to achieve distributed consensus among vehicles. In addition, a significant problem that needs to be resolved is how to store and transmit data securely in VANETs environment. It has been demonstrated [28] that the issue of safe data storage on VANETs can be resolved by implementing a distributed transaction storage method based on a blockchain.

Further, to address the consistency and non-tamperability of transmitted data, Zhang [29] proposed a data sharing and storage system based on blockchain. The system allows having digital signatures in a self-organizing on-board network. Roadside units (RSU) can also execute smart contracts to define the parameters of data exchange and store replica sensor data in a distributed method.

All the above applications prove that blockchain technology can bring certain technical support to the transportation field, thus making transportation under smart cities more efficient and safer.

5 Blockchain-Based Smart Manufacturing

Smart manufacturing achieves production optimization and improves production efficiency by utilizing new-generation information technologies, including but not limited to the Industrial Internet of Things (IIoT) and automation [30,31]. However, smart manufacturing is currently facing challenges such as security, interoperability, privacy protection, and traceability [32,33]. As a decentralized and trustless distributed data ledger, blockchain provides new ideas for solving the problems existing in smart manufacturing. Next, the applications of blockchain in IIoT and supply chain management are discussed below.

IIoT is an application of the IoT in industry and one of the core technologies of smart manufacturing. However, the traditional IIoT has problems of high cost, high communication overhead, low security, and high latency. The distributed storage of blockchain can effectively solve the problems of a single point of failure in centralized IIoT and improve its robustness of it [34].

A variety of solutions have been proposed by researchers in response to the problems existing in IIoT. Huo et al. [35] proposed a trusted identifier co-governance architecture with the fusion of blockchain and Handle technology. The architecture is divided into three levels. In the second level, blockchain is introduced to provide efficient and stable identity resolution services. The proposed solution not only further improves the performance of identity resolution services, but also ensures data security and reliability. Although blockchain can make IIoT securer, there are still shortcomings in communication efficiency. Wang et al. [36] attempted to replace the Merkle tree in the blockchain with incremental aggregator subvector commitment, which realizes the aggregation proof of multiple data blocks. It effectively solves the problem of low communication efficiency caused by verifying data. In addition, the combination of blockchain and other technologies (e.g., cryptography, machine learning, artificial intelligence) can better ensure the security of IIoT. Tan et al. [37] proposed an IIoT key data protection scheme based on blockchain, where the private key is split and encrypted through the Shamir secret sharing algorithm. They also publish it on the blockchain. Besides, Mansour [38] proposed an Intrusion

Detection System for IIoT, which not only improves the accuracy of intrusion detection, but also uses blockchain to achieve secure data transmission. The above applications show that the deep integration of blockchain and IIoT can solve the problems of insecure data transmission, low communication efficiency, which improves the security and stability of the IIoT equipment.

Supply chain is an important part of smart manufacturing and is moving towards digitization. However, it faces challenges and problems in information sharing, data security, traceability and enterprise interconnection [9,39]. Blockchain provides new ideas for optimizing the management of smart manufacturing supply chain.

Many studies focus on blockchain-based solutions to the supply chain. For example, Wu and Zhang [40] investigated the problem of supply chain trust management. They proposed a framework for trust management utilizing blockchain to make the interactions among supply chain entities more reliable. The improved EigenTrust algorithm used in it optimizes the trust management model. Specifically, the malicious nodes are identified by calculating the trust value of the node. Thus, the influence of malicious nodes, who may give a negative evaluation, can be reduced.

To achieve supply chain transparency and traceability, Xu et al. [9] put forward a supply chain management system based on Ethereum. The combination of traditional database technology and blockchain enables producers, managers, and customers to obtain the required information according to their needs, and ensures the correctness and credibility of the information. In addition, Weaterkamp et al. [41] designed a supply chain system with traceability utilizing smart contracts. In contract, the ingredients of products are defined as recipes. Every ingredient is a token, which represents a batch of real goods. The token is unique and unforgeable. When products are manufactured, the input tokens are consumed and new tokens are created. It is more convenient to trace the conversion process of products.

The above applications show that blockchain effectively solves problems in the supply chain. It makes the smart manufacturing supply chain more secure, reliable and transparent and improves its efficiency.

6 Discussions

While blockchain brings convenience to smart cities, we also hope to protect citizens' data security and privacy. However, there are some privacy leakage issues with the blockchain. For example, users participating in the construction of the blockchain cannot guarantee complete anonymity, resulting in the disclosure of user identity data. This is because the blockchain is open and transparent, and all network participants can see all the information stored on the blockchain. Although the blockchain maps network participants to pseudonymous addresses in order to ensure anonymity. However, attackers can use the combination of public information stored in the blockchain and external information to track the actions of users, so as to obtain the real identity of the user, resulting in the disclosure of user privacy. Therefore, it is crucial to study the anonymity of the blockchain to ensure the privacy of users.

In addition, the scalable storage of smart blockchains is a prerequisite for future urban applications. In smart cities, various data collection devices will generate a large amount of data, which requires blockchain technology to process. However, each node in the traditional blockchain contains complete transaction information, which cannot store all transaction information due to the limited storage space of the blockchain. Therefore, it is impossible to directly apply the blockchain to smart cities. We need to study the scalability of the blockchain to meet the application needs of massive data in smart cities. In addition, we can use decentralized storage methods in the blockchain system to increase its storage, thereby providing the possibility for a wide range of applications in smart cities.

7 Conclusion

This survey explored the application of blockchain technology in smart cities. First, we discussed in detail the relevant applications of blockchain technology in smart cities from the perspectives of medical, transportation, energy and manufacturing. Further, we also discussed the privacy and scalability of blockchain-based applications in smart cities. To sum up, the application of blockchain to smart cities improve people's quality of life. Moreover, we also believe that this survey provides new ideas for the development of smart cities.

Acknowledgements. This work is partially supported by the National Key Research and Development Program of China (Grant No. 2021YFB2701300), Natural Science Foundation of Shandong Province (Grant No. ZR2020ZD01).

References

1. Gai, K., Wu, Y., Zhu, L., Choo, K.R., Xiao, B.: Blockchain-enabled trustworthy group communications in UAV networks. IEEE Trans. Intell. Transp. Syst. **22**(7), 4118–4130 (2021)
2. Gai, K., Qiu, M.: Optimal resource allocation using reinforcement learning for IoT content-centric services. Appl. Soft Comput. **70**(1), 12–21 (2018)
3. Zhang, Y., Gai, K., Xiao, J.: Blockchain-empowered efficient data sharing in internet of things settings. J-SAC **40**(12), 3422–3436 (2022)
4. Gai, K., Tang, H., Li, G.: Blockchain-based privacy-preserving positioning data sharing for IoT-enabled maritime transportation systems. IEEE Trans. Intell. Transp. Syst. (9), 1–15 (2022)
5. Gai, K., Guo, J., Zhu, L., Yu, S.: Blockchain meets cloud computing: a survey. IEEE Commun. Surv. Tutor. **22**(3), 2009–2030 (2020)
6. Liang, G., Weller, S., Luo, F.: Distributed blockchain-based data protection framework for modern power systems against cyber attacks. IEEE Trans. Smart Grid **10**(3), 3162–3173 (2018)
7. Sun, J., Yao, X., Wang, S., Wu, Y.: Blockchain-based secure storage and access scheme for electronic medical records in IPFS. IEEE Access **8**, 59389–59401 (2020)

8. Yang, Z., Yang, K., Lei, L., et al.: Blockchain-based decentralized trust management in vehicular networks. IEEE Internet Things J. **6**(2), 1495–1505 (2018)
9. Xu, Z., Zhang, J., Song, Z., et al.: A scheme for intelligent blockchain-based manufacturing industry supply chain management. Computing **103**(8), 1771–1790 (2021)
10. Gai, K., Qiu, M., Zhao, H., Tao, L., Zong, Z.: Dynamic energy-aware cloudlet-based mobile cloud computing model for green computing. J. Netw. Comput. Appl. **59**, 46–54 (2016)
11. Bao, J., He, D., Luo, M., Choo, K.K.R.: A survey of blockchain applications in the energy sector. IEEE Syst. J. **15**(3), 3370–3381 (2020)
12. Dileep, G.: A survey on smart grid technologies and applications. Renewable Energy **146**, 2589–2625 (2020)
13. Wang, Y., Su, Z., Zhang, N.: SPDS: a secure and auditable private data sharing scheme for smart grid based on blockchain. IEEE Trans. Industr. Inf. **17**(11), 7688–7699 (2020)
14. Bera, B., Saha, S., Das, A.K.: Designing blockchain-based access control protocol in IoT-enabled smart-grid system. IEEE Internet Things J. **8**(7), 5744–5761 (2020)
15. Gai, K., Wu, Y., Zhu, L., Qiu, M., Shen, M.: Privacy-preserving energy trading using consortium blockchain in smart grid. IEEE Trans. Industr. Inf. **15**(6), 3548–3558 (2019)
16. Ali, F.S., Bouachir, O., Özkasap, Ö.: Synergychain: blockchain-assisted adaptive cyber-physical P2P energy trading. IEEE Trans. Industr. Inf. **17**(8), 5769–5778 (2020)
17. Khalid, R., Javaid, N., Javaid, S., Imran, M., Naseer, N.: A blockchain-based decentralized energy management in a P2P trading system. In: IEEE International Conference on Communications, pp. 1–6. IEEE (2020)
18. AlSkaif, T., Crespo-Vazquez, J.L., Sekuloski, M., van Leeuwen, G., Catalão, J.P.S.: Blockchain-based fully peer-to-peer energy trading strategies for residential energy systems. IEEE Trans. Industr. Inf. **18**(1), 231–241 (2021)
19. Gaetani, E., Aniello, L., Baldoni, R.: Blockchain-based database to ensure data integrity in cloud computing environments (2017)
20. Yue, X., Wang, H., Jin, D., et al.: Healthcare data gateways: found healthcare intelligence on blockchain with novel privacy risk control. J. Med. Syst. **40**(10), 1–8 (2016)
21. Kuma, M.R., Fathima, M.D., Mahendran, M.: Personal health data storage protection on cloud using MA-ABE. Int. J. Comput. Appl. **75**(8), 11–16 (2013)
22. Gu, K., Jia, W., Wang, G., Wen, S.: Efficient and secure attribute-based signature for monotone predicates. Acta Informatica **54**(5), 521–541 (2017)
23. Guo, R., Shi, H., Zhao, Q., Zheng, D.: Secure attribute-based signature scheme with multiple authorities for blockchain in electronic health records systems. IEEE Access **6**, 11676–11686 (2018)
24. Preikschat, K., Böhmecke-Schwafert, M., Buchwald, J., Stickel, C.: Trusted systems of records based on blockchain technology-a prototype for mileage storing in the automotive industry. Concurr. Comput. **33**(1), e5630 (2021)
25. Liu, C., Chai, K., Zhang, X., et al.: Adaptive blockchain-based electric vehicle participation scheme in smart grid platform. IEEE Access **6**, 25657–25665 (2018)
26. Zhang, D., Yu, F., Yang, R.: Blockchain-based distributed software-defined vehicular networks: a dueling deep Q-learning approach. IEEE Trans. Cogn. Commun. Netw. **5**(4), 1086–1100 (2019)
27. Liu, H., Zhang, Y., Yang, T.: Blockchain-enabled security in electric vehicles cloud and edge computing. IEEE Network **32**(3), 78–83 (2018)

28. Zheng, D., Jing, C., Guo, R., Gao, S., Wang, L.: A traceable blockchain-based access authentication system with privacy preservation in vanets. IEEE Access **7**, 117716–117726 (2019)
29. Zhang, X., Chen, X.: Data security sharing and storage based on a consortium blockchain in a vehicular ad-hoc network. IEEE Access **7**, 58241–58254 (2019)
30. Leng, J., Ye, S., Zhou, M.: Blockchain-secured smart manufacturing in industry 4.0: a survey. T-SMC **51**(1), 237–252 (2021)
31. Thoben, K.D., Wiesner, S., Wuest, T.: "Industrie 4.0" and smart manufacturing - a review of research issues and application examples. Int. J. Autom. Technol. **11**(1), 4–16 (2017)
32. Phuyal, S., Bista, D., Bista, R.: Challenges, opportunities and future directions of smart manufacturing: a state of art review. Sustain. Futures **2**, 100023 (2020)
33. Zhou, K., Liu, T., Zhou, L.: Industry 4.0: towards future industrial opportunities and challenges. In: 2015 12th International Conference on Fuzzy Systems and Knowledge Discovery, pp. 2147–2152 (2015)
34. Hassan, M.U., Rehmani, M.H., Chen, J.: Privacy preservation in blockchain based IoT systems: integration issues, prospects, challenges, and future research directions. Futur. Gener. Comput. Syst. **97**, 512–529 (2019)
35. Huo, R., Zeng, S., Di, Y., Cheng, X., et al.: A blockchain-enabled trusted identifier co-governance architecture for the industrial internet of things. IEEE Commun. Mag. **60**(6), 66–72 (2022)
36. Wang, J., Chen, J., Ren, Y., Sharma, P.K., Alfarraj, O., Tolba, A.: Data security storage mechanism based on blockchain industrial internet of things. Comput. Ind. Eng. **164**, 107903 (2022)
37. Yu, K., Tan, L., Yang, C., Choo, K.R., et al.: A blockchain-based shamir's threshold cryptography scheme for data protection in industrial internet of things settings. IEEE Internet Things J. **9**(11), 8154–8167 (2022)
38. Mansour, R.F.: Blockchain assisted clustering with intrusion detection system for industrial internet of things environment. Expert Syst. Appl. **207**, 117995 (2022)
39. Zuo, Y.: Making smart manufacturing smarter - a survey on blockchain technology in industry 4.0. Enterprise Inf. Syst. **15**, 1323–1353 (2021)
40. Wu, Y., Zhang, Y.: An integrated framework for blockchain-enabled supply chain trust management towards smart manufacturing. Adv. Eng. Inform. **51**, 101522 (2022)
41. Westerkamp, M., Victor, F., Küpper, A.: Blockchain-based supply chain traceability: token recipes model manufacturing processes. In: 2018 IEEE International Conference on Internet of Things, Halifax, NS, Canada, pp. 1595–1602 (2018)

Topic-Aware Model for Early Cascade Population Prediction

Chunyan Tong[1], Zhanwei Xuan[1], Song Yang[1], Zheng Zhang[1], Hongfeng Zhang[2], Hao Wang[2], Xinzhuo Shuang[3], and Hao Sun[4(✉)]

[1] State Key Laboratory of Communication Content Cognition, People's Daily Online, Beijing 100733, China
{tongchunyan,xuanzhanwei,yangsong,zhangzheng}@people.cn
[2] Wuhan Second Ship Design and Research Institute, Wuhan 430064, China
[3] Fuxin Higher Vocational College, Liaoning 123000, China
[4] Electronic Information School, Wuhan University, Wuhan 430072, China
2021202120053@whu.edu.cn

Abstract. This paper introduces an early content propagation popularity prediction model based on graph neural network and variational inference topic dependent dynamic variational autoencoder model (CD-VAE). CD-VAE captures the dynamics in the content propagation process, aggregates the topological information in the information diffusion process using GraphSAGE, approaches the uncertainty in terms of time and node from the perspective of probability by introducing two variational autoencoders, considers the changes in semantic characteristics in the process by integrating natural language processing methods into the model, and therefore significantly improves its prediction performance.

Keywords: Graph Attention Network · Attention Mechanism · Topic-dependent · Variational Autoencoder · Artificial Neural Network

1 Introduction

With the growing prevalence of mobile devices, online social networks have become one of the most important service of the Internet [1]. A large number of famous social APP, such as Twitter, Facebook and Youtube, have appeared all over the world. This study is conducted on the Weibo, a famous microblogging network in China.

In Weibo network, users are connected by directed following relationships and the information can diffuse through users' reposting behavior from their followees. When users retweet or post information, the platform pushes the information to his/her fans, which will accelerate its diffusion.

Models that predict the popularity of content can be roughly divided into three categories. Over the past few years, feature-based methods are used to

M. Qiu et al. (Eds.): SmartCom 2022, LNCS 13828, pp. 495–504, 2023.
https://doi.org/10.1007/978-3-031-28124-2_47

predict popularity, which emphasize devising effective festures and adopt classical machine-learning models [2–5], such as Naive Bayes, Support Vector Machine (SVM) and Multi-Layer Perceptron (MLP), for prediction. However, those methods rely heavily on the devised features. Then, generative models based on Hawkes process [6,7] or Possion process [8,9] are used in predictive model Generative models are used to model the intensity of the information diffusion. However, those models assume that later retweeting is only affected by the previous publishing or retweeting, which is not satisfied in real diffusion process. Recently, deep learning models are developed to predic diffusion process [10–12]. In a deep learning model, a diffusion process is embedd into a vector, which can be used to predict user behavior or diffusion. However, the existing deep learning models only rely on time-series information, while other features such as user preference, text content and user relationship are not extracted.

In summary, the difficulties of information diffusion models include: (1) Insufficient use of structure, time, and text features, ignoring joint modeling at the text-structure-temporal level; (2) There is little information, and the uncertainty of users and time levels is large, making early prediction difficult; (3) The prediction model has poor migration and generalization ability.

Our work can be concluded as follows: a deep generative model-Topic-dependent Dynamic Variation-al Autoencoder (CD-VAE) is proposed to achieve early content propagation popularity prediction. Specifically, the model learns structural features using an inductive graph neural network GraphSAGE, uses Transformer network that is based on attention mechanism to learn time series information. Finally, CD-VAE uses two variational autoencoders to model the node-level and temporal-level uncertainty to achieve propagation prediction.

2 Preliminaries

2.1 Problem Definition

The popularity prediction discussed in this paper is defined in Definition.

Definition. Information Popularity Prediction. For information C_i, the observation time window is $[t_0, T]$, and the prediction time window is $(T, T+\tau)$. The diffusion process can be expressed as $G_i = G(E, V)$, in which $e_{ij} \in E$ means user u_j retweet the information from u_i. Based on the diffusion process G_i, text content of the tweet and the time information in the observation window, the target is to predict the popularity increase size ΔP_i in the prediction time window.

2.2 Topic-Wise Propagation Scale

Consider the conditions of time delay, time, user historical activity, edge historical contact, published text, and forwarded generated text, the content cascade is divided into different scale intervals to obtain the distribution Y. The user's historical activity is expressed by the number of contents published and forwarded by users in history, while the historical contact degree is measured by the number of contents forwarded between two users in history, that is, the frequency of side occurrence. They are divided into different intervals according to the frequency to get their distribution. In this section, published texts and forwarded generated texts are divided into 10 categories, namely sports, finance and economics, real estate, home furnishing, education, science and technology, fashion, politics, games and entertainment, and their distribution is obtained. According to different X, the obtained joint distribution is as shown in Fig. 1:

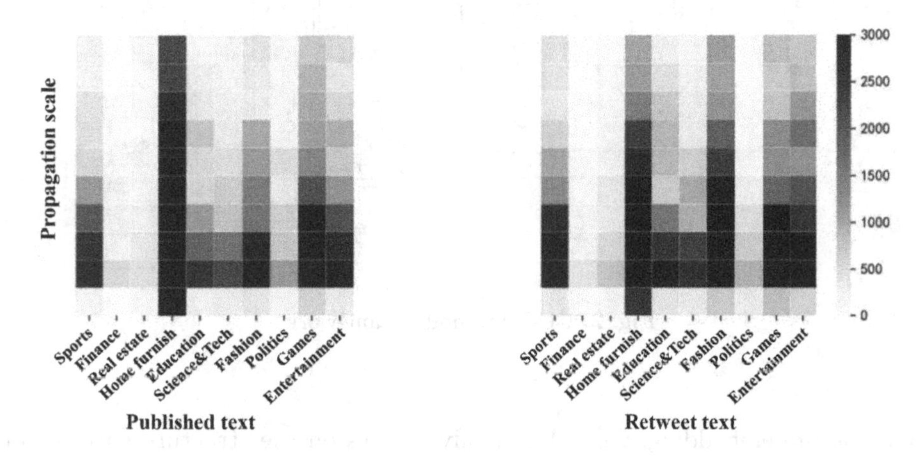

Fig. 1. Joint distribution of propagation scale of different topics

3 Model

3.1 Popularity Prediction Model

Framework. CD-VAE tackles the challenges of existing prediction models by considering the following characterstics of information diffusion on social networks.

- Topic dependence of user behavior. User social relationship and User Generated Content (UGC) are factors that influencing information diffusion and are topic dependent.
- Structure dependence. When the retweeting structure changes, some users' retweeting behavior of the information will be changed.
- Realtime dependence: The retweeting sequence is composed of a time sequence. In addition, information retweeting will decay over time. Information has a greater popularity scale during the day than at night, which can cause the time series to be unstable.

These three characteristics make our problem more complicated, and our model is to deal with the difficulties above.

We propose an end-to-end neural network framework to deal with these three characteristics. The popu-larity prediction model is mainly composed of 5 parts as is shown in Fig. 2:

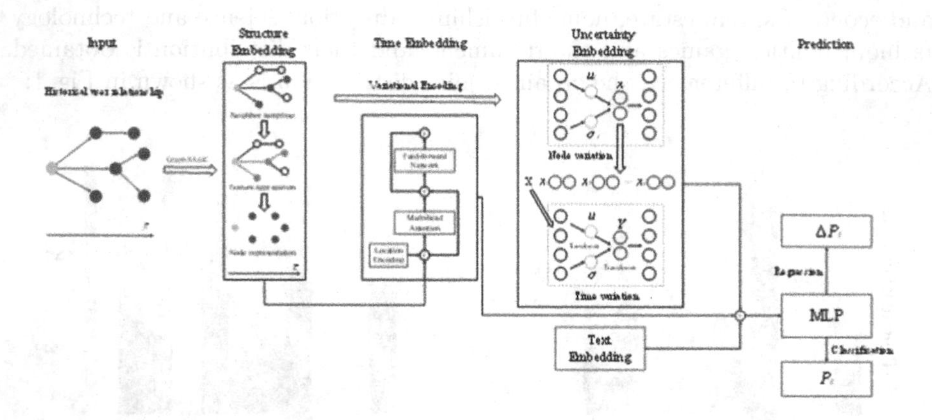

Fig. 2. CD-VAE model framework

- Structure embedding: CD-VAE mainly focuses on the structure patterns in the content propagation process and the potential relationships in the network. Specifically, the model uses an inductive graph neural network method GraphSAGE to embed nodes to learn the network structure.
- Time embedding: CD-VAE adopts Transformer network based on attention mechanism to embed time series information in the propagation process.
- Uncertainty embedding: CD-VAE uses the Variable Autoencoder (VAE) to model the node level and time level uncertainties in the content propagation process.
- Text embedding: CD-VAE uses natural language processing methods to represent non node attributes in text, and uses GraphSAGE to embed an undirected acyclic graph with attributes to obtain text representation.
- Prediction: CD-VAE combines Transformer, variational inference, and text embedding to get the final representation, and inputs it into the MLP for classification and regression prediction tasks.

Diffusion Dynamic Graph. For a piece of information C_i, the first step is to jointly model the user, content, time, and structure characteristics in the observation time window into the diffusion dynamic graph G_i. The N retweeting sequences in the observation time window are represented as N subgraphs $G_i = \{g_i(t_0), g_i(t_1), \cdots, g_i(t_{N-1})\}$. The sequence set from t_0 to t_{N-1}describes the time characteristics. $g_i(t_j) = (V_i^{t_j}, E_i^{t_j}, D_i^{t_j}, F_i^{t_j})$ contains the set of nodes and edges, that is, describes the structural characteristics. User characteristics are described by attribute $D_i^{t_j}$, and content characteristics are described by attribute $F_i^{t_j}$. The user's behavioral characteristics and social characteristics are a process that depends on long-term formation, so we assume that the user's characteristics remain unchanged during the observation period. Social characteristics are presented in the form of edge attributes, so there is $D_i^{t_j} = S_i = \{s_i(0), s_i(1), \cdots, s_i(K_e - 1)\}$. User behavior characteristics and content characteristics belong to the node level, next we combine content characteristics and user behavior characteristics to node attributes.

Compared with the traditional method, we take retweeting generated text by each user as its text feature to form the user text content set $Text_i = \{text_i(0), text_i(1), \cdots, text_i(K_v - 1)\}$, where $text_i(k)$ represents a word sequence. Then we use the text embedding method to embed the user text content. We use the self-attention content embedding model (SACE) shown in Fig. 3 to learn text representation.

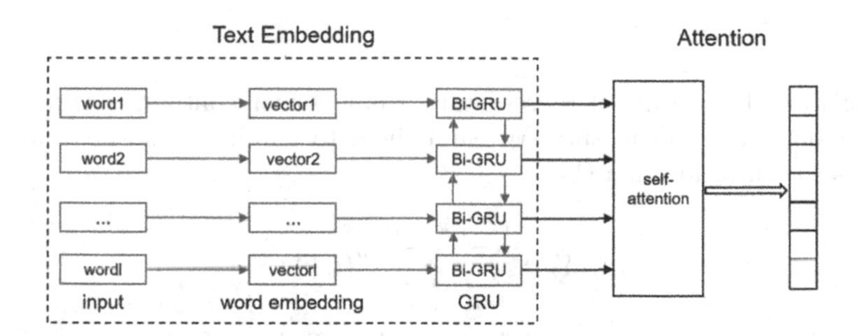

Fig. 3. SACE text embedding model

Spatio-Tempooral Dependence. We model and represent information diffusion dynamics in a structure-time mode. In order to obtain structural information, we need to use graph embedding technology to learn the representation of diffusion dynamic graph G_i. User B retweeting information of user A appears to be a one-way process, and direction needs to be considered. Therefore, g is a directed acyclic graph with node attributes and edge attributes, and it is difficult to embed it effectively. Therefore, g is a directed acyclic graph with node attributes and edge attributes, and it is difficult to embed it effectively. We use graph attention network to achieve this, as shown in Fig. 4. The GAT action process is divided into 3 steps: 1) Segmentation: divide the graph into 4 subgraphs according to direction, node and edge; 2) Aggregation: Use the attention mechanism to learn features for the graphs where the nodes and edges are located; 3) Combination: aggregate the sub-graphs representations.

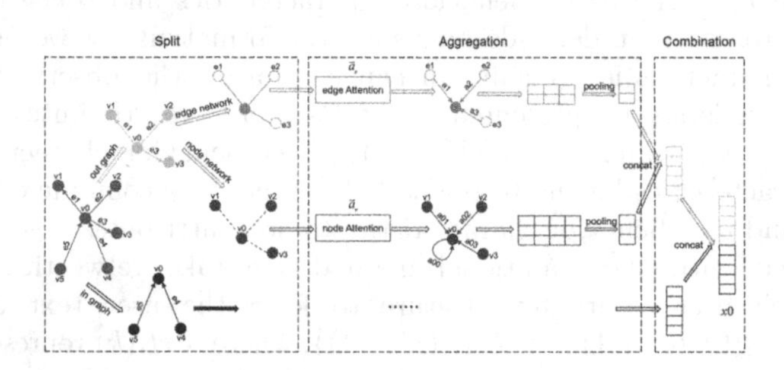

Fig. 4. Improved graph attention network: uses three layers of segmentation, aggregation, and combination to learn directed acyclic graphs with node attributes and edge at-tributes.

Prediction. The result of representation is a tensor containing three dimensions: node, time, and feature. We sum them to eliminate the node and time scale, as shown in formula (1).

$$Y_i = \sum_{t=t_0}^{t_{N-1}} \sum_{k=0}^{K_v-1} x''(t,k) \tag{1}$$

Y_i is the final representation of information C_i through the proposed model, then we connect it to a multilayer perceptron (MLP) for prediction tasks.

$$\Delta P_i = f(C_i(User, Content, Structure, Time)) = MLP(Y_i) \tag{2}$$

Our goal is to obtain the popularity increment ΔP_i within the predicted time τ according to the user, content, structure, and time characteristics of the information C_i within the observation time T. The loss function is defined as formula (3):

$$loss = \frac{1}{c}\sum_{i}^{j}(\log \Delta P_i - \log \overline{\Delta P_i})^2 + \mu L_{reg} \tag{3}$$

Among them, c is the quantity of information, (ΔP_i) represents the predicted increment of information C_i, and $(\overline{\Delta P_i})$ represents the true increment. L_{reg}

represents the L2 regularization term, which is to prevent overfitting during training, and μ is a hyperparameter.

4 Experiments

4.1 Dataset

We used Weibo, China's largest Weibo platform to obtain real-world data. Specifically, we tested CD-VAE on the dataset in the 2021 Artificial Intelligence Challenge contest hosted by People's Daily Online. These Weibo data are sampled. The posts with less than 10 retweets are filtered and we get a total of 18000 posts finally. For contrast, we also tested our model on the dataset that DeepHawkes used to compare the performance of CD-VAE and DeepHawkes. Figure 5 shows the relationship between the number of posts and the popularity of posts.

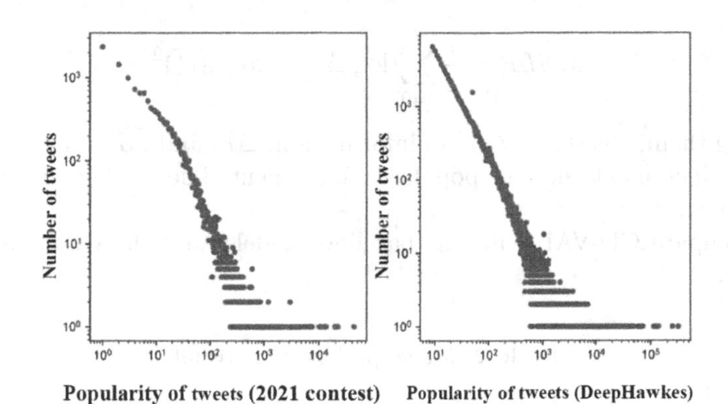

Fig. 5. Blog post popularity distribution relationship

User behavior characteristics and social characteristics will change over a relatively long period. Therefore, we consider user characteristics to remain unchanged in the short-term, we use the data to extract user-related priors characteristics.

Node and edge information of the dataset are listed in Table 1.

Table 1. Dataset statistics

Dataset	2021 Contest	DeepHawkes
Content count	18000	119313
Avg. nodes	63.51	142.24
Avg. edges	63.67	143.90

4.2 Baselines

In this part, we mainly refer to some of relevant models in the 3 types of information popularity prediction mentioned in the Introduction as the baseline model:
Feature-deep: A method based on feature design.
TopoLSTM:A method based on diffusion model [13].
DeepHawkes: A method based on generative models [14].
CasCN: A model based on deep learning [15].
NPP: A model based on deep learning [16].
CasFlow: A model based on deep learning and generative process [17].

4.3 Experimental Result

For popularity prediction, we define the information popularity prediction problem as a regression problem, and use MSLE for evaluation.

$$MSLE = \frac{1}{c}\sum_{i=0}^{c}(\log \Delta P_i - \log \overline{\Delta P_i})^2 \tag{4}$$

Among them, c is the amount of information, ΔP_i and $\overline{\Delta P_i}$ are the predicted popularity increment and true popularity increment. The smaller the MSLE, the better the effect.

We compare CD-VAE with the baseline models, and the results are shown in Table 2.

Table 2. Model performance result

	2021 Contest Data		DeepHawkes Data	
Model	Acc	MSLE	Acc	MSLE
Feature-deep	52.03%	2.36	57.58%	3.58
DeepHawkes	64.84%	1.74	65.67%	2.67
CasCN	65.57%	1.72	66.27%	2.58
NPP	67.19%	1.62	66.52%	2.55
CasFlow	65.98%	1.64	65.12%	2.39
CD-VAE	74.02%	1.39	69.42%	2.23

From Table 2, we can see that Feature-deep model requires manual design of structure, time, content, and user characteristics, which is inefficient, and its performance is not as good as other methods. As described in Sect. 2, the method based on feature design relies heavily on feature selection and design. CD-VAE also uses structure, content, time, and user characteristics to achieve better results.

Although DeepHawkes method introduces a deep learning method on the basis of Hawkes, the focus is still on the learning of temporal characteristics. CasCN fully obtains temporal and structural information, but it does not consider the characteristics of users and content. In addition, CasCN learns network characteristics through the GCN layer. In contrast, our CD-VAE performs local graph learning during structural dependence modeling, which has better results. It can also handle directed acyclic graphs with attributes (nodes, edges). NPP fully mixes user, content, time characteristics, and establishes the relationship between content and user characteristics, showing better results. But in the process of information diffusion, the network topology changes with time, and NPP does not consider the structure-dependent information. From the above experimental results, we conclude that CD-VAE comprehensively considers user, content, structure, and time information, models user topic dependence and spatiotemporal dependence, and shows excellent performance on the Weibo datasets.

5 Conclusion and Future Work

We propose a topic-dependent dynamic variational autoencoder model (CD-VAE) to predict information popularity. CD-VAE is a model based on deep learning and Bayesian learning. It considers the characteristics of early propagation of Weibo content, and learns structural information via GraphSAGE, learns time information using Transformer, and deploys two variation autoencoders to learn the uncertainty of users and time in the content propagation process. The text information of users is embedded as node attributes using NLP methods, and text features are learned in a graph-dependent way. We use Weibo data to conduct experiments to predict the information popularity. The results show that CD-VAE has better results, and existing deep learning methods on different datasets.

Acknowledgement. This work was supported by the Open Funding Projects of the State Key Laboratory of Communication Content Cognition (No. 20K05 and No. A02107).

References

1. Wang, J., Qiu, M., Guo, B.: Enabling real-time information service on telehealth system over cloud-based big data platform. J. Syst. Architect. **72**, 69–79 (2017)
2. Szabo, G., Huberman, B.A.: Predicting the popularity of online content. Commun. ACM **53**(8), 80–88 (2010)
3. Ma, Z., Sun, A., Cong, G.: On predicting the popularity of newly emerging hashtags in twitter. J. Am. Soc. Inform. Sci. Technol. **64**(7), 1399–1410 (2013)
4. Bakshy, E., Hofman, J.M., Mason, W.A., Watts, D.J.: Everyone's an influencer: quantifying influence on twitter. In: Proceedings of the Fourth ACM International Conference on Web Search and Data Mining, pp. 65–74 (2011)

5. Cheng, J., Adamic, L., Dow, P.A., Kleinberg, J.M., Leskovec, J.: Can cascades be predicted? In: Proceedings of the 23rd International Conference on World Wide Web, pp. 925–936 (2014)
6. Matsubara, Y., Sakurai, Y., Prakash, B.A., Li, L., Faloutsos, C.: Rise and fall patterns of information diffusion: model and implications. In: Proceedings of the 18th ACM SIGKDD International Conference on Knowledge Discovery and Data Mining, pp. 6–14 (2012)
7. Mishra, S., Rizoiu, M.A., Xie, L.: Feature driven and point process approaches for popularity prediction. In: Proceedings of the 25th ACM International on Conference on Information and Knowledge Management, pp. 1069–1078 (2016)
8. Shen, H., Wang, D., Song, C., Barabási, A.-L.: Modeling and predicting popularity dynamics via reinforced poisson processes. In: Proceedings of the AAAI Conference on Artificial Intelligence, vol. 28 (2014)
9. Gao, S., Ma, J., Chen, Z.: Modeling and predicting retweeting dynamics on microblogging platforms. In: Proceedings of the Eighth ACM International Conference on Web Search and Data Mining, pp. 107–116 (2015)
10. Defferrard, M., Bresson, X., Vandergheynst, P.: Convolutional neural networks on graphs with fast localized spectral filtering. In: Advances in Neural Information Processing Systems, vol. 29 (2016)
11. He, K., Zhang, X., Ren, S., Sun, J.: Deep residual learning for image recognition. In: Proceedings of the IEEE Conference on Computer Vision and Pattern Recognition, pp. 770–778 (2016)
12. Vaswani, A., et al.: Attention is all you need. In: Advances in Neural Information Processing Systems, vol. 30 (2017)
13. Wang, J., Zheng, V.W., Liu, Z., Chang, K.C.C.: Topological recurrent neural network for diffusion prediction. In: 2017 IEEE International Conference on Data Mining (ICDM), pp. 475–484. IEEE (2017)
14. Cao, Q., Shen, H., Cen, K., Ouyang, W., Cheng, X.: Deephawkes: bridging the gap between prediction and understanding of information cascades. In: Proceedings of the 2017 ACM on Conference on Information and Knowledge Management, pp. 1149–1158 (2017)
15. Chen, X., Zhou, F., Zhang, K., Trajcevski, G., Zhong, T., Zhang, F.: Information diffusion prediction via recurrent cascades convolution. In: 2019 IEEE 35th International Conference on Data Engineering (ICDE), pp. 770–781. IEEE (2019)
16. Chen, G., Kong, Q., Nan, X., Mao, W.: NPP: a neural popularity prediction model for social media content. Neurocomputing **333**, 221–230 (2019)
17. Xu, X., Zhou, F., Zhang, K., Liu, S., Trajcevski, G.: Casflow: exploring hierarchical structures and propagation uncertainty for cascade prediction. IEEE Trans. Knowl. Data Eng. (2021)

GeoNet: Artificial Neural Network Based on Geometric Network

Xiangyang Cui[1], Zhou Yan[1], Song Yang[1], Zheng Zhang[1], Hongfeng Zhang[2], Hao Wang[2], Xinzhuo Shuang[3], and Qi Nie[4(✉)]

[1] State Key Laboratory of Communication Content Cognition, People's Daily Online, Beijing 100733, China
{cuixiangyang,yanzhou,yangsong,zhangzheng}@people.cn
[2] Wuhan Second Ship Design and Research Institute, Wuhan 430064, China
[3] Fuxin Higher Vocational College, Liaoning 123000, China
[4] Electronic Information School, Wuhan University, Wuhan 430072, China
nieqi@whu.edu.cn

Abstract. Artificial neural network has achieved great success in many fields. Considering the unique advantages of naturally generated networks, we combine the geometric complex network model with the existing neural network model to build a neural network with geometric space structure characteristics. We proposes a GeoNet neural network model based on a random geometric network structure and finds that the neural network with a natural structure has good classification performance, and the classification accuracy is higher than the widely used neural network structure.

Keywords: Complex Network · Geometric Network Model · Network Generation · Residual Structure · Artificial Neural Network

1 Introduction

Artificial neural network (ANN) has achieved great success in machine learning such as image recognition, object detection, computer vision, natural language processing, and so on, which drives deep learning to become a very popular research topic. As an important hyperparameter, the structure of a network is a significantly vital factor promoting the development of the neural network. Currently, ANNs have evolved from simple chain-like models to structures with multiple wiring paths such as ResNets [1], DenseNets [2], etc. In deep learning, how computational networks are wired is crucial for building intelligent machines.

A large number of studies show that the naturally formed network structure has many advantages such as high information transmission efficiency [3,4], high economy [3,5], and good robustness [6]. Therefore, we have reason to doubt the superiority of the traditional structure: whether the naturally formed network structure has more advantages than the current artificially designed structure.

M. Qiu et al. (Eds.): SmartCom 2022, LNCS 13828, pp. 505–514, 2023.
https://doi.org/10.1007/978-3-031-28124-2_48

In this paper, the network structure with natural characteristics is combined with the neural network, and the naturally formed complex network is applied to the structural design of the neural network. By comparing the artificial residual model with the real network structure, this paper believes that the structural features of efficient training naturally exist in the natural network. In this paper, stochastic geometric networks are used to describe the topological connection of artificial neural networks, so that the neural networks have the key characteristics of natural networks, such as small-world effect, powerlaw degree distribution, and high aggregation. According to the existing neural network technology, this paper proposes a GeoNet neural network model based on a random geometric network structure and finds that the neural network with a natural structure has good classification performance, and the classification accuracy is higher than the widely used neural network structure.

2 Related Work

In 2006, the concept of deep learning accelerated the research of neural networks, in which convolutional neural networks are widely used. In 2012, AlexNet [7] convolutional neural network won the championship in the ImageNet image recognition competition. The network uses a Rectified Linear Unit (ReLU) instead of the traditional sigmoid activation, which alleviates the problem of gradient disappearance in the backpropagation of the neural network, and uses dropout technology to prevent the neural network from overfitting. Subsequently, the vision group of Oxford University proposed the VGG network [8] in 2014, which uses 3×3 The small-size convolution kernel replaces the large-size convolution kernel and improves the network representation ability by cascading. The research also explores the relationship between the depth of the convolution neural network and its performance, showing that the depth of the neural network is a key factor in the network performance. Because of VGG's excellent feature extraction ability, the VGG network has been widely used in image processing tasks such as style migration. In the same year, Google proposed GoogLeNet [9–11], which uses the multibranch Inception structure to increase the adaptability of the network to multi-scale features. In 2015, He Kaiming and others proposed ResNet [?], aiming to alleviate the problem of gradient disappearance of network backpropagation training, so that networks can be stacked to a deeper depth. Although the residual structure of ResNet reduces the training difficulty of the network, there is evidence that the design will result in only a small number of residual blocks learning useful information, and this problem is described as dividing feature reuse [12]. To alleviate this problem, WideResNet [13] studied the impact of increasing the residual block width on the performance of the residual network. This study shows that increasing the number of convolution cores is beneficial to the performance of the residual neural network. Pyramid-Net [14] studied the scale change of network width and depth on this basis and achieved a good balance between them. DenseNet model [2] uses a dense connection structure to enable all layers in the network to accept the features of all previous convolution layers. This structure can effectively alleviate the problem of gradient disappearance and promote feature reuse.

The performance of the convolutional neural network is improved with the growth of model size. Generally, better network performance can be obtained by increasing the depth and expansion width. However, the high computational complexity, memory consumption, and delay limit the practical application of deep networks. MobileNet [15] uses depth separable convolution to reduce the redundant parameters of traditional 3D convolution and the computation required for convolution. At the same time, the network uses the RELU6 activation function to maintain the robustness of the network under low floating point numerical accuracy. MobileNet's low computational complexity and low latency make it suitable for mobile device deployment scenarios. In 2018, Google proposed MobileNet V2 [16], which uses expansion convolution compression reverse residuals and Linear Bottlenecks to retain feature information as much as possible. Recently, Google proposed MobileNet V3 [17]. By introducing lightweight attention [18] and the h-swish activation function, the network has significantly improved its performance in classification, semantic segmentation, and target detection tasks. In addition to manually designing the neural network structure, the neural network architecture search [19–23] (Neural Architecture Search, NAS) uses deep learning to automatically screen the neural network structure, which is also widely used in the current neural network design.

3 Geometric Neural Network Model

Real networks are usually characterized by high concentration and small-worldness. Among them, the high aggregation is reflected in the triangular connection relationship in the network, which forms the same connection form as the shortcut connection of the residual network. At the macro scale, the small-world feature of the network makes the path length between network nodes shorter, which is conducive to slowing down the gradient attenuation phenomenon of neural network training. Figure 1 shows the structural similarity between the residual network and the real network. The real network naturally has structural characteristics that are easy to train.

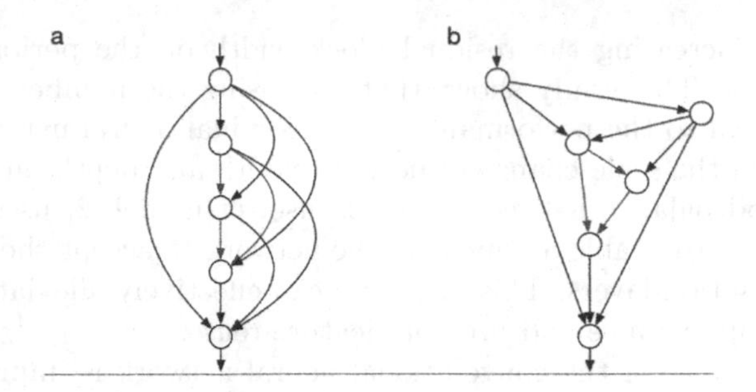

Fig. 1. Schematic diagram of residual network and real network. (a) Resnet network structure; (b) Real network structure. In the figure, red lines are used to identify a group of triangular connections in the network. The triangle connection in the real network has the same connection form as the shortcut structure in the residual network. (Color figure online)

3.1 Network Connection Structure and Function

Connection Structure of the Network. The generation network of the S^1 model has the characteristics of scale-freeness, high aggregation and small-worldness, and the network nodes have the characteristics of spatial distribution. Its model features are highly consistent with the structural features of the brain neural network [24,25], so this paper uses S^1 model to simulate the real network structure, and the connection relationship of the neural network is generated according to the following model. The n nodes of the network are uniformly distributed in the ring with a radius of $R = \frac{n}{2\pi}$. Each node in the network has a parameter κ which follows that $\kappa \sim c\kappa^{(-\gamma)}$. The connection probability between two nodes follows that

$$p(x) = p(\kappa, \kappa', d) = \frac{1}{1 + \frac{d}{\mu\kappa\kappa'}^{\beta}} \quad (1)$$

$$x = \frac{d}{\mu\kappa\kappa'} \quad (2)$$

where $\mu = \frac{1}{2I\langle\kappa\rangle}$; $I = \int p(x)dx$; $\langle\kappa\rangle$ represents node variable κ Expectation of distribution. The expectations for generating node degrees in the network are $\bar{k}(\kappa) = \kappa$. $\beta(>1)$ is the parameter that controls the network aggregation coefficient.

Function of Nodes and Edges. In this paper, network edge connection describes the data flow in the network, and the edge connection transfers the output characteristics of the node to its neighbors. The edge connection direction reflects the direction of data transmission; The network node realizes the feature processing function of data, specifically including three stages of feature aggregation, feature transformation and feature distribution [26].

3.2 Network Nodes Model

Based on the neural network technologies such as batch normalization [27], deeply separable convolution [28], inverse residual structure [16] and channel attention mechanism [18], this paper considers the following node feature transformation model, as depicted in Fig. 2.

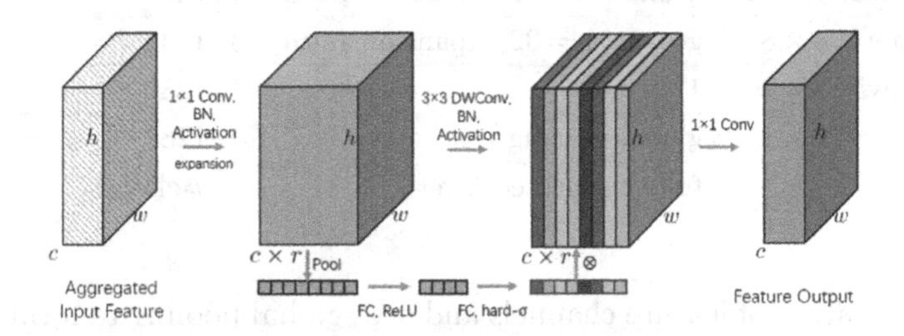

Fig. 2. Characteristic transformation of nodes.

The feature transformation of nodes includes three convolution layers. Aggregation characteristics of nodes are used in the first convolution layer. 1×1 convolution is used for improving the number of channels of the feature. The second convolution layer is 3×3 depth convolution, which is used to extract the spatial features of the image and reduce the number of parameters of the model. The third layer is a 1×1 convolution for restoring the channel dimension of the original feature. After the first convolution layer, this paper uses a lightweight channel attention mechanism to weigh the importance of different channels to improve the network performance. The first 1×1 convolution layer and depth convolution adopt batch normalization and nonlinear activation to improve the training speed of the network and the nonlinearity of feature transformation; In the node model, the last 1×1 convolution layer is used to realize linear weighting between characteristic channels [29].

3.3 Geometric Neural Network Architecture

According to the connection structure and node model of the neural network, this paper constructs the image classification convolution neural network GeoNet model with the network structure shown in Table 1.

GeoNet model is mainly divided into three parts. The first part consists of two 3×3 Convolution layer conv1 and conv2, which is used to extract low-level features of images. The second part is composed of random geometric networks conv3 and conv4. In order to realize feature reuse in neural networks, this paper uses two 32-node random geometric networks stacked to replace a single 64-node random geometric network. For the input nodes of conv3 and conv4, the stripe of the depth convolution is set to 2 to reduce the size of the feature. Finally, the GeoNet uses 1280 dimensional 1×1 convolution layer conv5 to realize the

Table 1. GeoNet network architecture.

Stage	Output	Layer	Output Channel
input	32×32	-	3
conv1	32×32	3×3 conv	36
conv2	32×32	3×3 conv	36
conv3	16×16	geonet, n = 32 expansion ratio = 4	72
conv4	8×8	geonet, n = 32 expansion ratio = 3	144
conv5	8×8	1×1 conv	1280
gp	1×1	global pooling	1280
fc		fully connected layer	#classes

linear combination of feature channels and uses global pooling to form the global features of the image. The logit vector of global features that distinguishes the category is output through the full connection layer. In order to reduce the risk of overfitting, the dropout layer [30] with a probability of 0.5 is used to regularize the network after the full connection layer.

4 Performance Evaluation

4.1 Experimental Data Set

The experiments in this paper use the public CIFAR-10 and CIFAR-100 datasets. Because of their rich categories and appropriate image size, CIFAR-10 and CIFAR-100 datasets are suitable for model verification under complex connection structures.

4.2 Model Training

For image classification, this paper uses the cross-entropy loss function and the Minibatch Gradient Descent method for training. The initial learning rate is set to $\eta = \frac{B}{256}\eta_{base}$, where B is batch size set to 64; basic learning rate is set as $\eta_{base} = 0.1$. The momentum parameter is set to 0.9.

This paper adopts the half cycle cosine attenuation adjustment strategy [31] and the linear warm-up strategy [32] to gradually adjust the learning rate in the training process. For the linear warm-up strategy, the learning rate is linearly adjusted from 0 to the initial learning rate in the first five epochs.

In this paper, label smoothing [10,33], regularization and data enhancement techniques are used to avoid the overfitting problem of the network. In which, the label smoothing parameter is set to 0.1. The regularization method of weight attenuation is used for the training of weight parameters in the network, and the weight attenuation parameter is set to 5e–5. In this paper, data enhancement techniques such as whitening, random translation (the translation distance is less than or equal to 4 pixels), and horizontal random flipping are used to process the training image and serve as the input of the neural network.

4.3 Image Classification Performance

In this paper, the convolution step of the benchmark model is adjusted to adapt to the image input size of CIFAR-10 and CIFAR-100 datasets, while the network connection mode remains unchanged. In order to achieve the comparison of network structures, this paper constructs the Resnetlike-GeoNet network, which uses the node model proposed in this paper, but uses the connection mode of the residual network to compare the performance of natural network structure and residual network structure. Through experiments, Table 2 shows the classification performance of this model and the benchmark model.

Table 2. Image classification performance

Model	Number of parameters	accuracy(%)	
		CIFAR-10	CIFAR-100
Resnet-50 [1]	23.74M	93.36	75.74
Resnet-50 v2 [13]	23.72M	94.31	77.21
Densenet-121 [2]	7.14M	94.33	75.85
Mobilenet [15]	3.33M	92.43	72
Mobilenet v2 [16]	2.38M	93.19	73.22
GeoNet	9.39M	95.59	79.73
GeoNet-Reslike	9.39M	95.20	78.39

The parameters of GeoNet: $\gamma = 9$. $\beta = 5.5$, $\langle \kappa \rangle = 3$.

The experimental results show that the GeoNet model in this paper achieves good classification accuracy in CIFAR-10 and CIFAR-100 datasets, with accuracy rates of 95.59% and 79.73%. Compared with other models, this paper achieves the best classification performance; Compared with the results of other models, the accuracy of the GeoNet model in this paper has been greatly improved. The results show that the network structure based on the geometric network model in this paper is superior to the classical model with residual structure in classification performance. To verify the advantages of network structure, the GeoNet model in this paper is superior to the Resnetlike-GeoNet model in classification performance. The network structure with natural characteristics is conducive to the improvement of neural network performance. Moreover, we draw the accuracy change curves of different models in the training process, as shown in Fig. 3. Compared with other models, GeoNet network has a faster convergence speed.

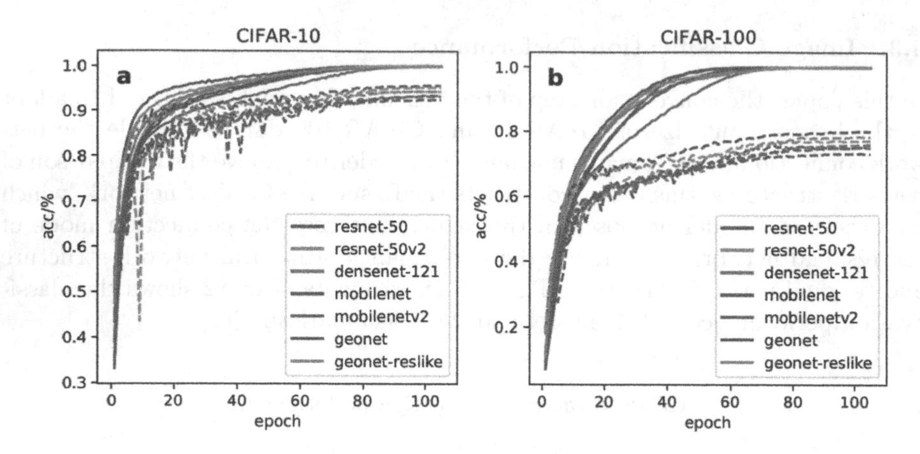

Fig. 3. Accuracy curve. The solid line in the figure represents the accuracy of the training set, and the dotted line represents the accuracy of the test set. (a) CIFAR-10 data set training curve; (b) CIFAR-100 dataset training curve.

5 Conclusion

The realistic biological neural network is distributed in a certain geometric space. This paper combines the geometric complex network model with the existing neural network model to build a neural network with geometric space structure characteristics. This chapter clarifies that due to the high aggregation characteristics and small world effect of the real network, the real network structure has some similarities with the current artificially designed residual network. The shortcut connection naturally exists in the real network structure, meeting the characteristics of efficient training of the backpropagation algorithm. With the help of current neural network components, this paper designs GeoNet neural network and tests the classification performance of the neural network with geometric space structure. Experiments show that GeoNet network has good classification performance, and its performance exceeds the current commonly used network structure.

Acknowledgement. This work was supported by the Open Funding Projects of the State Key Laboratory of Communication Content Cognition (No. 20K05 and No. A02107).

References

1. He, K., Zhang, X., Ren, S., Sun, J.: Deep residual learning for image recognition. In: Proceedings of the IEEE Conference on Computer Vision and Pattern Recognition, pp. 770–778 (2016)
2. Huang, G., Liu, Z., Van Der Maaten, L., Weinberger, K.Q.: Densely connected convolutional networks. In: Proceedings of the IEEE Conference on Computer Vision and Pattern Recognition, pp. 4700–4708 (2017)

3. Gulyás, A., Bíró, J.J., Kőrösi, A., Rétvári, G., Krioukov, D.: Navigable networks as nash equilibria of navigation games. Nat. Commun. **6**(1), 1–10 (2015)
4. Latora, V., Marchiori, M.: Efficient behavior of small-world networks. Phys. Rev. Lett. **87**(19), 198701 (2001)
5. Latora, V., Marchiori, M.: Economic small-world behavior in weighted networks. Eur. Phys. J. B-Condensed Matter Complex Syst. **32**(2), 249–263 (2003)
6. Albert, R., Jeong, H., Barabási, A.-L.: Error and attack tolerance of complex networks. Nature **406**(6794), 378–382 (2000)
7. Krizhevsky, A., Sutskever, I., Hinton, G.E.: Imagenet classification with deep convolutional neural networks. Commun. ACM **60**(6), 84–90 (2017)
8. Simonyan, K., Zisserman, A.: Very deep convolutional networks for large-scale image recognition. arXiv preprint arXiv:1409.1556 (2014)
9. Szegedy, C., et al.: Going deeper with convolutions. In: Proceedings of the IEEE Conference on Computer Vision and Pattern Recognition, pp. 1–9 (2015)
10. Szegedy, C., Vanhoucke, V., Ioffe, S., Shlens, J., Wojna, Z.: Rethinking the inception architecture for computer vision. In: Proceedings of the IEEE Conference on Computer Vision and Pattern Recognition, pp. 2818–2826 (2016)
11. Szegedy, C., Ioffe, S., Vanhoucke, V., Alemi, A.A.: Inception-v4, inception-resnet and the impact of residual connections on learning. In: Thirty-First AAAI Conference on Artificial Intelligence (2017)
12. Srivastava, R.K., Greff, K., Schmidhuber, J.: Highway networks. arXiv preprint arXiv:1505.00387 (2015)
13. Zagoruyko, S., Komodakis, N.: Wide residual networks. arXiv preprint arXiv:1605.07146 (2016)
14. Han, D., Kim, J., Kim, J.: Deep pyramidal residual networks. In: Proceedings of the IEEE Conference on Computer Vision and Pattern Recognition, pp. 5927–5935 (2017)
15. Howard, A.G., et al.: Mobilenets: efficient convolutional neural networks for mobile vision applications. arXiv preprint arXiv:1704.04861 (2017)
16. Sandler, M., Howard, A., Zhu, M., Zhmoginov, A., Chen, L.-C.: Mobilenetv 2: inverted residuals and linear bottlenecks. In: Proceedings of the IEEE Conference on Computer Vision and Pattern Recognition, pp. 4510–4520 (2018)
17. Howard, A., et al.: Searching for mobilenetv3. In: Proceedings of the IEEE/CVF International Conference on Computer Vision, pp. 1314–1324 (2019)
18. Hu, J., Shen, L., Sun, G.: Squeeze-and-excitation networks. In: Proceedings of the IEEE Conference on Computer Vision and Pattern Recognition, pp. 7132–7141 (2018)
19. Zoph, B., Vasudevan, V., Shlens, J., Le, Q.V.: Learning transferable architectures for scalable image recognition. In: Proceedings of the IEEE Conference on Computer Vision and Pattern Recognition, pp. 8697–8710 (2018)
20. Zoph, B., Le, Q.V.: Neural architecture search with reinforcement learning. arXiv preprint arXiv:1611.01578 (2016)
21. Pham, H., Guan, M., Zoph, B., Le, Q., Dean, J.: Efficient neural architecture search via parameters sharing. In: International Conference on Machine Learning, pp. 4095–4104. PMLR (2018)
22. Liu, C., et al.: Progressive neural architecture search. In: Proceedings of the European Conference on Computer Vision (ECCV), pp. 19–34 (2018)
23. Liu, H., Simonyan, K., Yang, Y.: Darts: differentiable architecture search. arXiv preprint arXiv:1806.09055 (2018)
24. Bassett, D.S., Bullmore, E.T.: Small-world brain networks revisited. The Neuroscientist **23**(5), 499–516 (2017)

25. Chialvo, D.R.: Critical brain networks. Physica A Stat. Mech. Appl. **340**(4), 756–765 (2004)
26. Xie, S., Kirillov, A., Girshick, R., He, K.: Exploring randomly wired neural networks for image recognition. In: Proceedings of the IEEE/CVF International Conference on Computer Vision, pp. 1284–1293 (2019)
27. Ioffe, S., Szegedy, C.: Batch normalization: accelerating deep network training by reducing internal covariate shift. In: International Conference on Machine Learning, pp. 448–456. PMLR (2015)
28. Sifre, L., Mallat, S.: Rigid-motion scattering for texture classification. arXiv preprint arXiv:1403.1687 (2014)
29. He, K., Zhang, X., Ren, S., Sun, J.: Identity mappings in deep residual networks. In: Leibe, B., Matas, J., Sebe, N., Welling, M. (eds.) ECCV 2016. LNCS, vol. 9908, pp. 630–645. Springer, Cham (2016). https://doi.org/10.1007/978-3-319-46493-0_38
30. Srivastava, N., Hinton, G., Krizhevsky, A., Sutskever, I., Salakhutdinov, R.: Dropout: a simple way to prevent neural networks from overfitting. J. Mach. Learn. Res. **15**(1), 1929–1958 (2014)
31. He, T., Zhang, Z., Zhang, H., Zhang, Z., Xie, J., Li, M.: Bag of tricks for image classification with convolutional neural networks. In: Proceedings of the IEEE/CVF Conference on Computer Vision and Pattern Recognition, pp. 558–567 (2019)
32. Goyal, P., et al.: Accurate, large minibatch SGD: training imagenet in 1 hour. arXiv preprint arXiv:1706.02677 (2017)
33. Müller, R., Kornblith, S., Hinton, G.E.: When does label smoothing help? In: Advances in Neural Information Processing Systems, vol. 32 (2019)

Research on Blockchain-Based Smart Contract Technology

Hongze Wang(✉) and Qinying Zhang

Department of Information Engineering, Wuhan Institute of City, Wuhan, China
1540509443@qq.com

Abstract. With the continuous development of blockchain technology, smart contract has become an important research object among the achievable technologies on blockchain technology. Based on the characteristics of decentralization, tamper-proof and transparency of blockchain, it provides a reliable technical support for the implementation of smart contract. Based on blockchain smart contract technology, this paper aims to design a smart contract management engine with higher versatility, security, and feasibility to develop a smart contract from the joint participation of multiple users. The smart contract is proliferated through the P2P network and deposited into the blockchain. And the blockchain is designed to automatically execute the smart contract, providing a new solution to the problems of opacity, easy tampering, and low efficiency of the traditional contract. It is safer and more reliable. By specifying the treaty and trigger conditions through the program code, once the conditions are met, the contract will be automatically executed, which greatly reduces the time and space costs. Using Ether and smart contracts to develop distributed applications to realize this technology, the feasibility of applying blockchain technology in this field is explored, and a new technical implementation is provided for traditional contract signing.

Keywords: Blockchain · Smart contracts · Decentralization · Byzantine fault-tolerant algorithm

1 Introduction

As digital assets have become more and more important in our lives, the protection of digital assets, transactions, etc. have also become more and more important in our lives. At present, our control of digital assets is not secure enough, our operation of digital assets are dependent on the third party, our digital assets are stored in the database of the third party, the security and privacy of digital assets are realized through the third party, that is to say, the preservation of our digital assets need to rely on the trustworthiness of the third party.

Since 2016, smart contract technology, represented by Ethernet, has suddenly become a hot topic in all walks of life. Smart contracts are essentially programs that digitize contracts and then run them on a computer. The concept was

M. Qiu et al. (Eds.): SmartCom 2022, LNCS 13828, pp. 515–524, 2023.
https://doi.org/10.1007/978-3-031-28124-2_49

first proposed by Nick Szabo back in 1994, but because there was no suitable platform for the development of smart contracts, they remained at a conceptual stage until the rise of blockchain technology, which provided a suitable platform for smart contracts.

Based on further optimization of the Byzantine fault-tolerant algorithm, this paper analyzes and designs an improved S-PBFT based on this algorithm to achieve a more stable smart contract system by optimizing the algorithmic mechanism in it. The experimental data in this paper illustrates that the performance metrics of the smart contract system in terms of latency and resource utilization are significantly improved after the optimization of S-PBFT.

The main contents of this paper are as follows:

- Introduce smart contract technology from the traditional contracting system for smart deployment and execution of contracts.
- Further modularize and analyze the body of smart contracts, loading methods, consensus algorithms used, and deployment environment.
- Combine the optimized Byzantine fault-tolerant algorithm S-PBFT with the practical application of the system, and verify the optimized algorithm in the application. The results show that the method of this paper largely improves the efficiency of system processing in practical applications.

2 Background

Blockchain technology is a new application model of computer technology such as distributed architecture data storage, P2P real-time transmission, and secrecy algorithm, which solves the core problem of Decentralization through distributed architecture database, digital encryption technology and unique consensus algorithm without the premise of third-party credit institutions, and realizes a decentralized and credible system without third parties. The theoretical basis of the consensus algorithm is Byzantine fault tolerance algorithm, common consensus algorithms such as proof of stake (PoS) [16].

And smart contract is based on the digital asset control program in blockchain 2.0. In layman's terms, smart contract is to code the relevant business logic and algorithm involved, and program the complex relationship between the legal agreement, business agreement and network in reality, which contains four core parts: one is digital asset and smart property; two is digital identity, building digital identity authentication service; three is digital asset custody, relying on blockchain technology to keep all kinds of property; fourth is contract arbitration, the arbitration platform to execute delivery contracts [3].

In a broad sense this technology is not only applicable to business, but also has great application prospects in distributed computing, Internet of Things, artificial intelligence and other fields.

3 Related Work of the Smart Contract Operation Mechanism

3.1 Contract Subject

The operation mechanism of a smart contract is shown in Fig. 1. The smart contract, which has both value and state attributes, is pre-programmed with What-If and If-Then statements in the code to provide trigger scenarios and response rules for the contract terms. The user will be provided with the returned contract address and contract interface information and can invoke the contract by initiating a transaction [18]. When miners receive a contract to create or invoke a transaction, they create or execute the contract code in a sandbox execution environment. The contract code automatically determines whether the current scenario meets the contract trigger conditions based on trusted external data sources and world state checks to strictly enforce the response rules and update the world state [11]. After the transaction is validated and packaged into a new data block, the new block will be certified by consensus algorithm and linked to the main chain of the blockchain, and all updates will take effect [21].

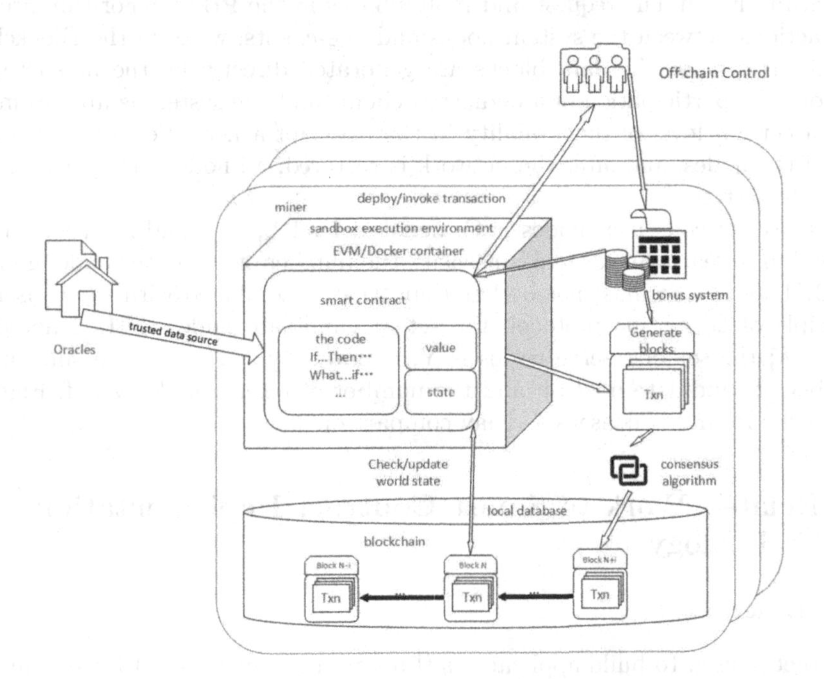

Fig. 1. The operation mechanism of smart contracts

3.2 Data Loading Method

The data layer includes state data, transaction data, application data, contract code, etc. State data and transaction data are generally stored on-chain for verifiable and observable purposes [25]. The majority of blockchains adopt the on-chain method to publish the application data and code to the chain, and then load the data and code from the chain and execute them, but there are disadvantages. The storage burden will make the efficiency low [7].

3.3 Consensus Mechanism

The consensus mechanisms currently used are Proof of Stake (PoS), Proof of WorkProof of workload mechanism (PoW), (DPoS), and Proxy proof-of-stake mechanism (DPoS), [24] and the state machine Byzantine system requires all nodes to perform a uniform operation to maintain the consistency of the system state [15]. Most of the existing state machine Byzantine systems are based on the PBFT algorithm, whose mechanism increases exponentially with the number of system nodes, thus lacking practical application value, which requires optimization of the Byzantine protocol in conjunction with practical application scenarios. An improved Byzantine algorithm, S-PBFT, is designed for the possible communication instability in distributed networks [12]. With the help of its algorithm, it can The request and reply phases in the PBFT algorithm are the interactions between the system nodes and the clients, while in the Blockchain consensus process the data blocks are generated directly by the master node without the participation of a dedicated client, and the system is able to maintain a certain level of functionality in the event of a network outage in up to 1/3 of the nodes, and after the network is restored, all nodes can quickly reach Consistent state [5].

The set of consensus nodes is G, denoted by $1, 2, ..., g$, and in order for the algorithm to reach an effective consensus, the number of consensus nodes g in the S-PBFT algorithm must not be less than $2f + 1$, combined with the consensus principle of Byzantine protocol; the set of candidate nodes is H, denoted by $1, 2, ..., h$; the set of reserve nodes is Y, denoted by $1, 2, ..., y$. The sum of the number of candidate nodes h and the number of reserve nodes y is f. Figure 2 shows the results of consensus delay comparison [1].

4 Related Work to Smart Contract Implementation Technology

4.1 Ether

Ethernet is used to build applications through a complete set of Ethereum Virtual Machinecode scripting language, which is similar to assembly language [2]. We know that programming directly in assembly language is very painful, but programming in Etherium does not require the direct use of the EVM language, but rather something likeC language It is similar to a high-level language like Python, Python, etc. Lisp and other high-level languages, and then converted to EVM language by compiler [20].

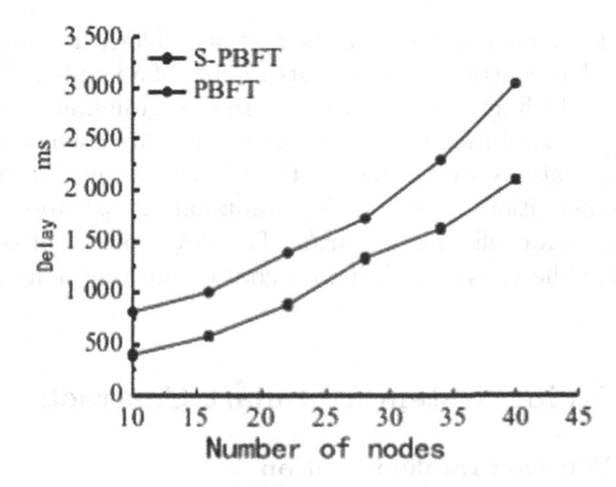

Fig. 2. Consensus Latency Comparison Results

4.2 Core Technologies

The general model of blockchain is shown in Fig. 3. It usually has six layers, including the application layer, contract layer, incentive layer, consensus layer, network layer, and data layer. Among them, consensus layer and incentive layer are usually bound together, the reason is that in a large P2P network, a certain incentive is needed to reach most consensus [8], so the design of consensus algorithm also needs to take the rules of incentive distribution into account. The data layer is the most critical part of the blockchain, it is the root of the blockchain and will determine the organization based on the connectivity of individual blocks [19]. It can be considered as a composite data structure. The network layer is mainly responsible for the validation and distribution of data throughout the dis-

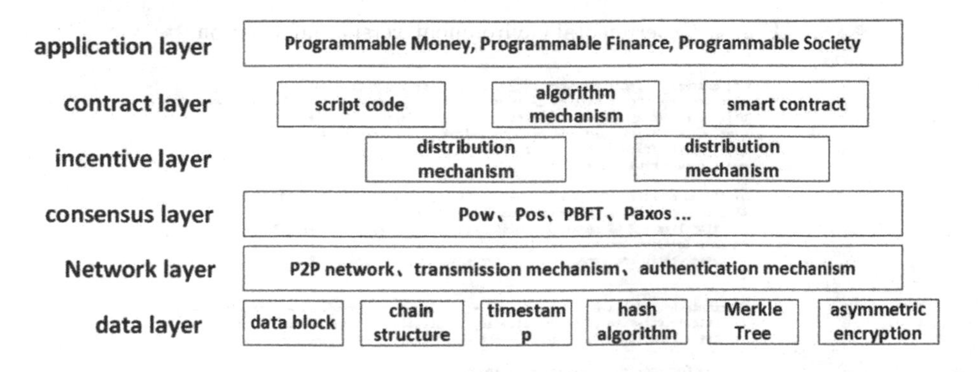

Fig. 3. Blockchain general model

tributed network. Smart contracts can be seen as a blockchain-based specificity application [14]. The contract layer consists mainly of a number of scripts, called smart contracts, which are prerequisites for the programmable functionality of the blockchain [23]. And the application layer is the layer where the blockchain development applications are located. In the blockchain model, each layer has a variety of implementations, and complex combinations produce a wide variety of blockchain types for different scenarios. The development of smart contracts has also liberated the use of blockchain scenarios, making it more flexible and diverse [9,22].

5 Smart Signing System for Smart Contracts

5.1 System Function Implementation

The IDE used is STS (Spring Tool Suite). Part of the experimental environment is shown in Fig. 4.

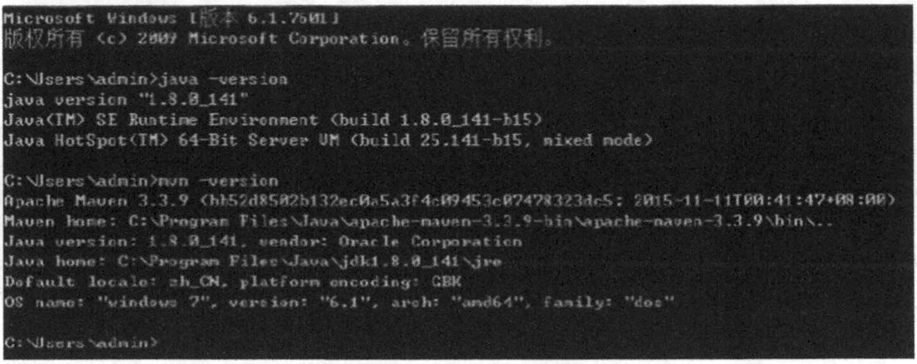

```
Microsoft Windows [版本 6.1.7601]
版权所有 (c) 2009 Microsoft Corporation。保留所有权利。

C:\Users\admin>java -version
java version "1.8.0_141"
Java(TM) SE Runtime Environment (build 1.8.0_141-b15)
Java HotSpot(TM) 64-Bit Server VM (build 25.141-b15, mixed mode)

C:\Users\admin>mvn -version
Apache Maven 3.3.9 (bb52d8502b132ec0a5a3f4c09453c07478323dc5; 2015-11-11T00:41:47+08:00)
Maven home: C:\Program Files\Java\apache-maven-3.3.9-bin\apache-maven-3.3.9\bin\..
Java version: 1.8.0_141, vendor: Oracle Corporation
Java home: C:\Program Files\Java\jdk1.8.0_141\jre
Default locale: zh_CN, platform encoding: GBK
OS name: "windows 7", version: "6.1", arch: "amd64", family: "dos"

C:\Users\admin>
```

Fig. 4. Experimental environment version information

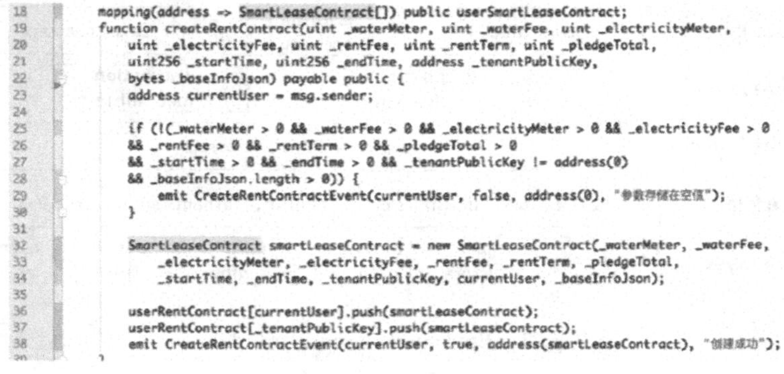

```
18  mapping(address => SmartLeaseContract[]) public userSmartLeaseContract;
19  function createRentContract(uint _waterMeter, uint _waterFee, uint _electricityMeter,
20      uint _electricityFee, uint _rentFee, uint _rentTerm, uint _pledgeTotal,
21      uint256 _startTime, uint256 _endTime, address _tenantPublicKey,
22      bytes _baseInfoJson) payable public {
23      address currentUser = msg.sender;
24
25      if (!(_waterMeter > 0 && _waterFee > 0 && _electricityMeter > 0 && _electricityFee > 0
26      && _rentFee > 0 && _rentTerm > 0 && _pledgeTotal > 0
27      && _startTime > 0 && _endTime > 0 && _tenantPublicKey != address(0)
28      && _baseInfoJson.length > 0)) {
29          emit CreateRentContractEvent(currentUser, false, address(0), "参数存储在空值");
30      }
31
32      SmartLeaseContract smartLeaseContract = new SmartLeaseContract(_waterMeter, _waterFee,
33          _electricityMeter, _electricityFee, _rentFee, _rentTerm, _pledgeTotal,
34          _startTime, _endTime, _tenantPublicKey, currentUser, _baseInfoJson);
35
36      userRentContract[currentUser].push(smartLeaseContract);
37      userRentContract[_tenantPublicKey].push(smartLeaseContract);
38      emit CreateRentContractEvent(currentUser, true, address(smartLeaseContract), "创建成功");
```

Fig. 5. Create contract source code

A contract system based on smart contracts and blockchain is proposed for software implementation. 1) This system digitizes traditional contracts and then deploys them to the blockchain environment, and utilizes the enforceability of programmed contracts and the decentralized features of blockchain to achieve a reliable smart contract behavior that is de-trusted between users [4]; 2) The Truffle framework, Testrpc test environment, and MetaMask wallet are organically combined together to form a complete system development and testing environment [17]. After rigorous testing, the intelligent distributed application of the contract system designed in this paper has the characteristics of robustness, simplicity of operation and practicality, and the system part of the implementation code is shown in Fig. 5.

5.2 System Functional Framework Diagram

Unlike the traditional application architecture, this system adopts the Ethernet decentralized application architecture. Smart contracts are executed on the Ethernet virtual machine EVM, and the client only needs to make RPC calls through the web browser to interact with the instance, and the client does not need to request access to the centralized server [6].

The framework of each function of the system is shown in Fig. 6.

Various equipment requirements
Various equipment systems
iOS client
android client
windows client
mac client
Follow-up technical support
Contract technical support
forensics
Technical Reports
apply for arbitration
....
Legal Information Supplement
Contract Type Analysis
Real-time legal and regulation enquiry
busines s
travel
Work
...
natural facts
Behavio r
Intelligent data analysis
key information extraction
Secondary data comparison
the amount
time
ratio
...
User Information Module
encrypted storage
personal identity verification
Account public key pair/private key pair

Fig. 6. System function diagram

By simulating and maintaining the operation of the whole Blockchain network through two computers together, each computer acts as a node, and these participating nodes prove the information and reach consensus through the improved S-PBFT mechanism based on the one proposed in Sect. 3.3. After consensus is reached, the management node broadcasts and stores the transaction information in the supply chain, and the other consensus nodes make relevant queries and traceability of the transaction information. Through a specific case implementation, the technical feasibility and the actual demand in the society, once again proves that the potential of smart contract technology in the future is not limited to this, and its technical maturity will be slowly shown in the future [10,13] (Fig. 7).

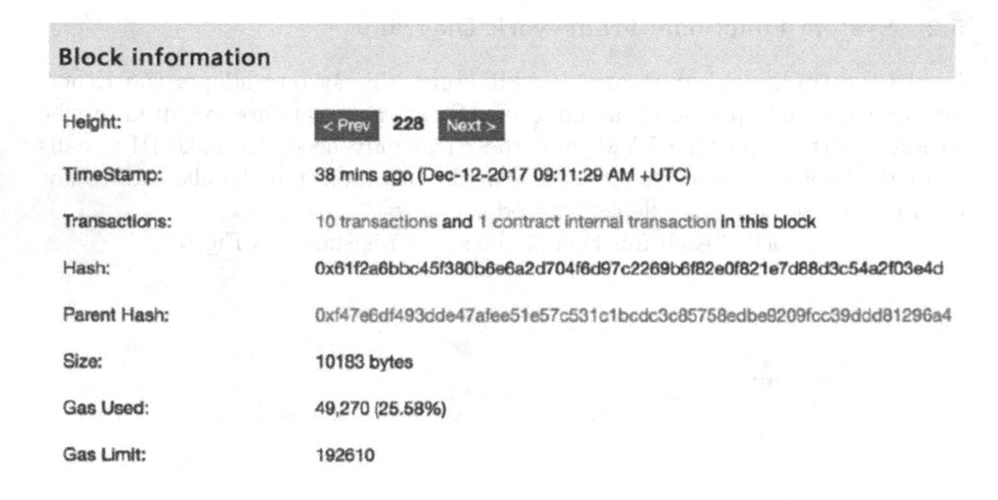

Fig. 7. Block information in the application backend

6 Conclusion

With the development of the information age, intelligent and digital related technologies are realized, and Smart contracts based on Blockchain are one of the most promising technologies among the currently achievable technologies with the help of this development. Smart contracts can not only be applied to more complex application scenarios and more advanced functional requirements, but also Smart contracts can provide an effective way for disciplines such as big data.

Smart contracts are still in a preliminary stage, and there is a lack of further exploration of some key technologies and theoretical knowledge, as well as some social and legal challenges brought about by them in reality are not taken into account, this paper only introduces and summarizes the smart contract technology itself, and verifies the actual effect of smart contract technology in application through a specific case, aiming to provide some reference for scholars in related fields. This paper only introduces and summarizes the smart contract technology itself, and verifies the practical effect of smart contract technology in application through a specific case, aiming to provide some reference and reference for scholars in related fields.

References

1. Zhang, Y., Guo, K.W., Hu, K.: Research on data sharing incentive mechanism based on smart contract. Computer Engineering (048-008) (2022)
2. Das, D., Banerjee, S., Chatterjee, P., Biswas, M., Biswas, U., Alnumay, W.: Design and development of an intelligent transportation management system using blockchain and smart contracts. Cluster Comput. 1–15 (2022). https://doi.org/10.1007/s10586-022-03536-z
3. Fang, L.: A new interpretation of smart contracts for contracts under blockchain technology. J. Chongqing Univ. (Soc. Sci. Ed.) **27**(5), 14 (2021)
4. Gai, K., Wu, Y., Zhu, L., Zhang, Z., Qiu, M.: Differential privacy-based blockchain for industrial internet-of-things. IEEE Trans. Ind. Inform. **16**(6), 4156–4165 (2019)
5. Gurgun, A.P., Koc, K.: Administrative risks challenging the adoption of smart contracts in construction projects. Eng. Constr. Archit. Manag. **29**(2), 989–1015 (2021)
6. Hauck, R.: Blockchain, smart contracts and intellectual property. Using distributed ledger technology to protect, license and enforce intellectual property rights. Legal Issues Digital Age **1**(1), 17–41 (2021)
7. He, H.W., Yan, A., Chen, H.Z.: A review of blockchain-based smart contract technologies and applications. Comput. Res. Dev. **55**(11), 15 (2018)
8. Hewa, T.M., Hu, Y., Liyanage, M., Kanhare, S.S., Ylianttila, M.: Survey on blockchain-based smart contracts: technical aspects and future research. IEEE Access **9**, 87643–87662 (2021)
9. Huanhuan, T.: Research on smart contract technology and application based on blockchain. Computer Programming Skills and Maintenance (003) (2022)
10. Hunhevicz, J.J., Motie, M., Hall, D.M.: Digital building twins and blockchain for performance-based (smart) contracts. Autom. Constr. **133**, 103981 (2022)
11. A formal verification method for smart contracts. Inf. Secur. Res. **2**(12), 10 (2016)
12. Kirli, D., et al.: Smart contracts in energy systems: a systematic review of fundamental approaches and implementations. Renew. Sustain. Energy Rev. **158**, 112013 (2022)
13. Kõlvart, M., Poola, M., Rull, A.: Smart contracts. In: Kerikmäe, T., Rull, A. (eds.) The Future of Law and eTechnologies, pp. 133–147. Springer, Cham (2016). https://doi.org/10.1007/978-3-319-26896-5_7
14. Liu, Y., Zhou, Z., Yang, Y., Ma, Y.: Verifying the smart contracts of the port supply chain system based on probabilistic model checking. Systems **10**(1), 19 (2022)
15. Qiu, H., Qiu, M., Memmi, G., Ming, Z., Liu, M.: A dynamic scalable blockchain based communication architecture for IoT. In: Qiu, M. (ed.) SmartBlock 2018. LNCS, vol. 11373, pp. 159–166. Springer, Cham (2018). https://doi.org/10.1007/978-3-030-05764-0_17

16. Shiyi, L., Lei, Z., Desheng, L.: A review of blockchain-based smart contract application research. Comput. Appl. Res. **38**(9), 12 (2021)
17. Sigalov, K., et al.: Automated payment and contract management in the construction industry by integrating building information modeling and blockchain-based smart contracts. Appl. Sci. **11**(16), 7653 (2021)
18. Tian, Z., Li, M., Qiu, M., Sun, Y., Su, S.: Block-def: a secure digital evidence framework using blockchain. Inf. Sci. **491**, 151–165 (2019)
19. Wu, Y., Li, J., Zhou, J., Luo, S., Song, L.: Evolution process and supply chain adaptation of smart contracts in blockchain. J. Math. **2022** (2022)
20. Research and implementation of blockchain-based smart contracts. Ph.D. thesis, Southwest University of Science and Technology
21. Xin, Y., Ran, C.: A review of blockchain smart contract technology applications. Modern Information Technology
22. Xiong, X.: Research on blockchain smart contract technology. Ph.D. thesis, University of Electronic Science and Technology
23. Feiyue, W.: Status and outlook of blockchain technology development. J. Autom. **42**(4), 14 (2016)
24. Yuan, X.: Blockchain smart contract technology application. China Finance (6), 2 (2018)
25. Zhao, F., Tan, J.C.: Smart contract security issues and research status. Inf. Technol. Netw. Secur. **40**(5), 7 (2021)

Cross-Chain-Based Distributed Digital Identity: A Survey

Tianxiu Xie[1,3], Hong Zhang[2,3], Yiwei Feng[2,3], Jing Qi[1], Chennan Guo[4], Gangqiang Yang[5], and Keke Gai[1](✉)

[1] School of Cyberspace Science and Technology, Beijing Institute of Technology, Beijing 100081, China
{3120215672,3120221282,gaikeke}@bit.edu.cn

[2] School of Computer Science and Technology, Beijing Institute of Technology, Beijing 100081, China
{3220211142,3220211128}@bit.edu.cn

[3] Yangtze Delta Region Academy, Beijing Institute of Technology, Jiaxing 314000, China

[4] School of Software, Dalian University of Technology, Dalian 116081, China
guochennan@dlut.edu.cn

[5] School of Information Science and Engineering, Shandong University, Jinan 266237, Shandong, China
g37yang@sdu.edu.cn

Abstract. Due to the high degree of privacy and sensitivity, it is difficult to share distributed digital identity with multiple parties. Blockchain-based distributed digital identities could address the issues of data sharing. However, the blockchain systems contain a wide variety of heterogeneous autonomous systems, and each blockchain system performs independent identity authentication. To be specific, it is difficult to realize mutual recognition and trust in digital identity. In this survey, we introduce the technical architecture of distributed digital identity and research the technical principle and development application of cross-chain technology. In addition, we expound the research status of cross-chain technology and the cross-chain-based distributed digital identity mechanism to realize unified and trusted cross-chain identity mutual recognition.

Keywords: Distributed digital identity · Cross-chain technology · Blockchain · Relay chain

1 Introduction

With the continuous enrichment and complexity of the application scenarios of the Internet, activities in the physical world are increasingly dependent on digital identities. Digital identity can support users to obtain different network services. However, in the traditional digital identity situation, a user has different digital identity accounts in multiple application systems, which will lead to decentralized information management and complicated identity authentication. In order to

M. Qiu et al. (Eds.): SmartCom 2022, LNCS 13828, pp. 525–534, 2023.
https://doi.org/10.1007/978-3-031-28124-2_50

make digital identity more secure and complete, the *World Wide Web* (W3C) proposes *Distributed Identities* (DID). Due to the characteristics of security, credibility and decentralization [1,2], the blockchain-based DID can adapt to the new Internet system Web3.0 [3]. In addition, DID fundamentally addresses the issues of large-scale and messy user identities and limited data sharing in the Internet. As the key to trusted interaction, the blockchain-based DID is of great significance in interconnection and privacy identity management.

Distributed digital identity not only facilitates digital services, but also makes the demand for data sharing and application collaboration among blockchains increasingly apparent. The interoperability and scalability have always limited the large-scale application of blockchain [4]. Specifically, cross-chain technology facilitates the wider application of DID. At present, due to the independence of blockchains in different application fields [5,6], distributed identity information is in an isolated state, which violates the original intention of distribution. Therefore, cross-chain technology is introduced into distributed identity. Cross-chain-based DID can realize mutual identity recognition, identity transfer and multi-party authentication, thereby promoting mutual trust and efficient collaboration of different blockchain systems.

Currently, the cross-chain-based DID faces an important challenge of ensuring the authenticity, validity, and legitimacy of the information source. Cross-chain identity authentication requires secure access to identity data and user permissions, making the process highly secure and reliable. Specifically, distributed digital authentication requires identity authentication and two-way verification. Therefore, the cross-chain-based DID technology needs to provide authentication services for different certificate domains, so as to realize the authentication of various digital certificates. The application of cross-chain technology in DID will bring about the security and credibility of identity data cross-chain.

In order to deal with these challenges, we researched a relay chain-based distributed digital identity. We proposed a *Cross-chain Channel-based DID* (C3-DID) model that enabled distributed storage of DIDs and was resistant to multiple threats [7]. In addition, we proposed a relay chain-based cross-chain system that achieved asynchronous consensus and improved the defensiveness and scalability of the cross-chain system [8].

The rest of the paper consists of the following sections. Section 2 gives the architecture and functionality of distributed digital identities and shows related research and landing projects. Section 3 introduces cross-chain related technologies. In addition, in Sect. 4, we present the existing research and project of the cross-chain-based distributed digital identity. Finally, Sect. 5 gives a summary and outlook.

2 Distributed Digital Identity

Distributed Identity (DID), as defined by the W3C distributed digital identity specification, is a new globally unique identifier. This identifier can be used not only for people, but also for everything, including a car, an animal or even a machine [9]. The core components of DID technology consist of three parts: the DID, the DID Document, the Verifiable Credential and the Verifiable Expression. Figure 1 shows the architecture of DID. A DID identification is a string of a specific format used to represent the numerical identity of an entity, which in this case can be a person, machine or object [10,11].

The *DID Document* (DID Doc) contains all the information related to the DID identity and is linked to the DID by a generic data structure *Uniform Resource Locator* (URL). Moreover, the DID controller is used to write or change data in the DID Doc. The DID document itself cannot represent the real identities of the users, so we provide the *Verifiable Credentials* (VCs) to support identity authentication. The binding of a user-centric identity to other identifiers issued by the *Certificate Authority* (CA) is known as the VC. It also provides a system similar to a PKI [12,13].

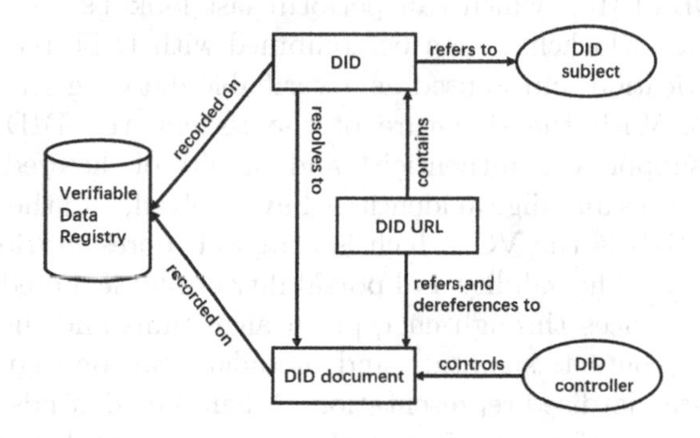

Fig. 1. Distributed Digital Identity Architecture

The distributed identity system has attracted extensive research from scholars. Zhou et al. [14] propose the EverSSDI framework, a distributed identity management system that relies on smart contracts, blockchain and the *Interplanetary File System* (IFS). In this framework, users first encrypt personal information and store them in the IPFS system, then use smart contracts for data verification. Another blockchain-based distributed identity management scheme, DNS-Idm, is proposed in [15]. In DNS-Idm, users and service providers can authenticate and declare identity attributes using real identities, while smart contracts protect the relevant operations among the authenticators and the users in this scheme.

Efat et al. [16] have proposed DT-SSIM, a distributed and trusted framework for self-sovereign identity management. DT-SSIM combines blockchain-based smart contract technology with a secret sharing scheme that can provide a reliable self-sovereign identity-based service for the IoT and can prevent tampering of IoT identity credentials. DT-SSIM uses the secret sharing method and smart contract technology to manage the generated identity shares and verify the user's identity declaration. In addition, DT-SSIM has proposed a series of novel algorithms to ensure the information security of externally stored identity credentials. It has realized identity verification without disclosing the original identity data. For this framework, future improvements could be directed towards replacing dynamic active secret sharing with a moving target defence mechanism, in which the external sharing of identities could be updated periodically and the storage location dynamically refreshed, a modification and integration that could significantly improve the stability of the framework and increase the cost of attack for attackers.

Deepak et al. [17] have proposed CanDID, a user-friendly and usable distributed identity implementation platform in which users can manage their own certificates. CanDID uses a decentralized committee of nodes to provide a robust solution for users' real identities, while preventing users from generating multiple identities and allowing for the identification of sanctioned users.

One effective technique to address distributed authentication is *Distributed Hash Tables* (DHT) [18], which can perform fast lookups and load identifiers on key-value pairs. Blockchain can be combined with DHT technology for distributed authentication and is used as a verifiable data registry that can store public key DID. With the signature of the issuers, the DID stored on the blockchain can support the authenticity and validity of the credentials.

Digital credentials and digital identities have evolved, and the latest evolution of digital credentials is the VC, which is a digital representation of real-world physical credentials. The validity and portability of physical credentials is transferred to digital devices through encryption algorithms and digital signatures, and the declared content, signature and metadata can be digitally verified in seconds. It is a standardised representation of digital credentials and is designed to bring the benefits of physical credentials to the digital world by referring to real-world physical credentials in terms of usage scenarios and core model design. Typical features of VCs are cryptographic security, privacy protection and machine readability.

Figure 2 shows the VC model proposed by the W3C. There are three main user roles in this model. Issuer is the certificate issuer who can be an individual or an organization. The certificate issuer actually issues a certificate to the public key of the person who will hold the certificate and digitally signs it with his or her private key. In this way, when the certificate holder presents the certificate, it also needs to use a digital signature method to prove that it is indeed the owner of the public key of the person to whom the certificate is issued (a very common method of digital signature verification). The certificate is digitally signed by the certificate issuer, so using the digitally signed verification method, anyone can instantly verify that the certificate is valid, and any tampering with the certificate will result in the invalidation of the digital signature. As the certificate issuer needs to know the holder's public key, it is generally the case that the certificate holder first applies to the issuer, providing his or her public key and proof of digital signature with this public key, as well as other evidence and data required by the issuer when applying. The issuer will verify these data and after confirming that they are correct, the certificate can be generated and sent to the certificate holder.

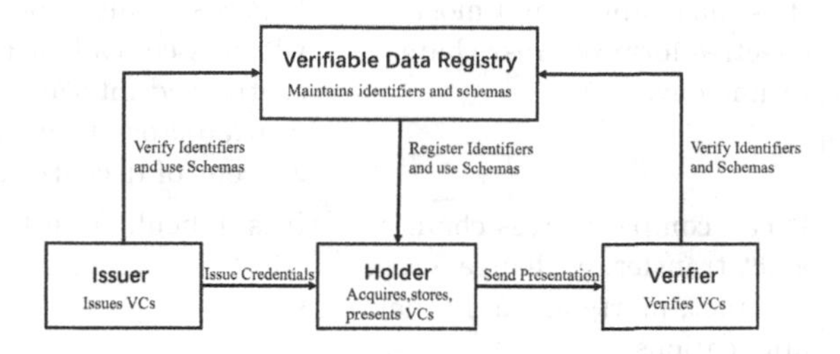

Fig. 2. Validatable Credential Data Model

The certificate holder is responsible for keeping his or her own certificate after the above process has been applied for. Usually these certificates are in digital form and can be downloaded and saved, and some advanced digital wallet systems have full digital certificate storage and management capabilities. In most cases, the responsibility for keeping the certificate rests with the certificate holder and the certificate issuer can decide for itself whether or not to keep a copy of the issued certificate. If the certificate holder loses the certificate, or the private key used to prove their identity, they will have to reapply for a new certificate. When a certificate needs to be verified, the certificate holder is responsible for providing a digital certificate and providing a digital signature to prove that it has the public key of the person to whom the certificate was issued.

The certificate verifier only needs to use the digital signature algorithm to verify the identity of the certificate holder and verify the correctness of the digital certificate to basically determine the authenticity of the certificate and the legitimacy of the certificate holder. Usually the certificate also contains some meaningful information inside, such as the expiry date, which is also used as a basis for determining the authenticity of the certificate.

3 Cross-Chain Technology

The interoperability of blockchains limits the large-scale application of blockchains. The cross chain technology is a key technology to achieve the Internet of Value and a bridge for blockchain expansion. Three cross chain technologies were mentioned, notary mechanism, side chain and relay chain and hash locking. In this section, We briefly introduce and compare these cross chain technologies, as shown in Table 1.

Notary mechanism is essentially a way of mediation by introducing a third-party intermediary to verify and transmit cross chain messages between two chains that cannot directly interoperate. By introducing one or more trusted third parties as credit endorsements, notary mechanism continuously monitors events on the chain, and is responsible for verifying and forwarding cross chain

Table 1. Comparison of Different Cross-chain Technologies

Cross Chain	Advantage	Disadvantage
Notary mechanism	It is the simplest and most effective form of cross chain technology.	It needs to introduce notaries, which is equivalent to adding a centralized intermediary, and is contradictory to its own concept of decentralization.
Side chain and relay chain	It can complete cross chain asset transfer, exchange, cross contract, mortgage and other applications.	It is difficult from technology.
Hash locking	It can also be used to exchange cross chain assets without mutual trust.	It does not realize cross chain transfer of assets, nor so-called cross chain contracts, but only cross chain exchange.
Distributed private key technology	Users always have control over their assets.	The contract is incomplete and needs further optimization.

messages on other chains according to the obtained event information. It is a simple solution to achieve blockchain interoperability. It does not require complex proof of workload or proof of equity, and is easy to interface with existing heterogeneous chains. However, its disadvantage is that it can cause asset exchange and has the problem of over concentration.

As for side chain and relay chain technology, the side chain is referred to the blockchain which is parallel to the main blockchain. The side chain implementation is to lock the temporary digital currency on the main chain and release the equivalent digital assets on the side chain through the two-way pegging technology. Relay chain is the role of "intermediary", which is the integration and expansion of notary mechanism and side chain mechanism [19]. From the perspective of form, relay chain is a way, and side chain is a result. Side chain expresses the relationship between two chains, not a cross chain technology or scheme.

Due to the unidirectional and low collision of hash function, Hash locking is a mechanism that takes advantage of the delayed execution of transactions in the blockchain. In order to realize the asset security on each blockchain, for the asset transaction process, hash locking needs to ensure the transaction atomicity, that is, asset transactions are either completed or not completed [20]. A contract with a hash locking mechanism is used to lock assets to achieve pledge effect, providing a trust basis for transactions between different assets. The advantage of hash lock is that if the transaction fails due to various reasons, the time lock can enable all parties involved in the transaction to recover their own funds to avoid losses caused by fraud or transaction failure.

Distributed private key technology [21] is a technology based on multi-party computation and threshold keys in cryptography. It separates the use right and ownership of digital assets, so as to realize the decentralization of assets control rights. The assets on the original blockchain are safely mapped to each blockchain system, thus realizing the cross chain circulation and value transfer of assets among multiple blockchain systems.

Early cross chain technologies such as BTCRelay [22], they focus more on asset transfer. Interleader protocol first proposed by Ripple Lab [23] allows two different accounting systems to freely transmit currency to each other through a third-party "connector" or "verifier". It is applicable to the accounting systems of various blockchains and can accommodate their differences, so it can be used as a unified payment standard. BTC Relay connects Ethereum network and Bitcoin network through Ethereum smart contract, enabling users to verify Bitcoin transactions on Ethereum [24]. However, the existing cross chain technologies, represented by Polkadot and Cosmos, pay more attention to cross chain infrastructure. Polkadot integrates a variety of blockchain technologies of different public blockchains, and various public blockchains can communicate with each other by any message and exchange tokens. Cosmos provides convenience for application developers to use their own public blockchains by providing modular blockchains. This mode is very suitable for public blockchains focusing on vertical fields. The emerging FUSION realizes multi currency smart contracts, which is a public blockchain with great application value and can generate rich cross chain financial applications.

A large number of blockchain systems with different characteristics have formed a large number of value islands, and direct value circulation cannot be carried out among blockchains. It limits the functional expansion and development space of the blockchain. The security concerns observed on centralized cryptocurrency exchanges motivated the design of atomic swaps, which apply to the exchange of money between any two users. However, there is currently no atomic swap protocol that is both universal and multi-asset, that is, compatible with all cryptocurrencies and supports the exchange of multiple coins in a single atomic swap. [25] introduced a general currency exchange protocol for securely exchanging multiple coins from any currency to multiple currencies of any other currency, for any kind of currency, in addition to the ability to verify signatures, the protocol no need to rely on any special scripts in the blockchain.

4 Relay Chain-Based Identity Cross-Chain Technology

The relay chain is to establish an additional chain between the source chain and the target chain to complete the verification and execution of cross-chain transactions. The nodes in the relay chain are deployed in various blockchain networks, and the information of initiating cross-chain transactions is synchronized at all times. After the user initiates a cross-chain transaction in the source chain, the relay node will forward the information to the relay chain and verify the transaction data. After the verification is completed, the corresponding transaction will be constructed through the consensus node in the relay chain, and complete the signature, and finally transport it to the target chain through the relay node.

BitXHub adopts the relay chain to provide safe and efficient cross-chain services, and solves the core problems of capturing, transmitting and verifying cross-chain transactions. It focuses on the interoperability of ledgers between heterogeneous and homogeneous consortium chains, and supports asset exchange, data interoperability and service complementarity. XuperChain has implemented a set of trusted digital identity solutions, which aims to connect multiple industries and promote cross-institution and multi-scenario identity authentication and data cooperation. *TencentCloud Decentralized Identity* (TDID) provides infrastructure for trusted digital identity and data exchange services across systems and institutions. Based on the blockchain, TDID provides a mechanism for distributed generation, holding and verification of identity identifiers and VC that carry identity data to encrypt security, protect data privacy. It can reliably express various types of identities and credentials in the real world on the Internet in a way that can be machine-verified by a third party.

The use process of digital identity in BitXHub in practical application scenarios is as follows.

1. Apply for a DID: Initiate a request to the management relay chain and send the DID name;
2. Approval the DID: The administrator approves the user's application request, and the result is "approved" or "rejected";
3. Register the DID: After approval, the user can register the DID. First, organize the relevant information of the chain identity into DID Doc for storage to obtain the storage address docAddr and content hash docHash. Then initiate a registration request to the highest relay chain with docAddr and internal docHash as parameters;
4. Parsing the DID: The information of the DID is synchronized on all relay chains, so DID parsing can be initiated to any relay chain. First, the client initiates a resolution request to any relay chain, and obtains the relevant information of the identity. Then obtain the actual DID Doc through the storage address, perform hash verification, and if it passes, it proves that the DID Doc is credible.

5 Conclusion

The scalability of blockchain has always been a research focus. The early blockchain technology was developed with an independent single chain, but due to the limitations of the efficiency and performance of the single chain, it could not support the application requirements of Web3.0. Therefore, the single blockchain technology was gradually extended to multi-chain collaborative development. Many cross-chain approaches have emerged. As a solution to maximize the scalability and interoperability of blockchains, cross-chain technology has received widespread attention since the birth of Bitcoin. In this survey, we researched the recent project on DID and cross-chain technology and analyzed cross-chain-based DID methods and business processes. In the future, we will conduct more in-depth research on the interoperability and robustness of cross-chain technology, the secure transmission of cross-chain data, and cross-chain efficiency.

Acknowledgements. This work is partially supported by the National Key Research and Development Program of China (Grant No. 2021YFB2701300), Natural Science Foundation of Shandong Province (Grant No. ZR2020ZD01), and Graduate Interdisciplinary Innovation Project of Yangtze Delta Region Academy of Beijing Institute of Technology (Jiaxing) (No. GIIP2022-019).

References

1. Gai, K., Wu, Y., Zhu, L., Zhang, Z., Qiu, M.: Differential privacy-based blockchain for industrial internet-of-things. IEEE Trans. Industr. Inform. **16**(6), 4156–4165 (2019)
2. Gai, K., Wu, Y., Zhu, L., Xu, L., Zhang, Y.: Permissioned blockchain and edge computing empowered privacy-preserving smart grid networks. IEEE Internet Things J. **6**(5), 7992–8004 (2019)
3. Gai, K., Zhang, Y., Qiu, M., Thuraisingham, B.: Blockchain-enabled service optimizations in supply chain digital twin. IEEE Trans. Serv. Comput. (2022)
4. Gai, K., Wu, Y., Zhu, L., Choo, K., Xiao, B.: Blockchain-enabled trustworthy group communications in UAV networks. IEEE Trans. Intell. Transp. Syst. **22**(7), 4118–4130 (2020)
5. Gai, K., Wu, Y., Zhu, L., Qiu, M., Shen, M.: Privacy-preserving energy trading using consortium blockchain in smart grid. IEEE Trans. Industr. Inform. **15**(6), 3548–3558 (2019)
6. Gai, K., Tang, H., Li, G., Xie, T., Wang, S., et al.: Blockchain-based privacy-preserving positioning data sharing for IoT-enabled maritime transportation systems. IEEE Trans. Intell. Transp. Syst. (2022)
7. Xie, T., Zhang, Y., Gai, K., Xu, L.: Cross-chain-based decentralized identity for mortgage loans. In: Qiu, H., Zhang, C., Fei, Z., Qiu, M., Kung, S.-Y. (eds.) KSEM 2021. LNCS (LNAI), vol. 12817, pp. 619–633. Springer, Cham (2021). https://doi.org/10.1007/978-3-030-82153-1_51
8. Zhang, S., Xie, T., Gai, K., Xu, L.: ARC: an asynchronous consensus and relay chain-based cross-chain solution to consortium blockchain. In: 2022 IEEE 9th International Conference on CSCloud, pp. 86–92. IEEE (2022)
9. Naik, N., Jenkins, P.: Governing principles of self-sovereign identity applied to blockchain enabled privacy preserving identity management systems. In: 2020 IEEE ISSE, pp. 1–6. IEEE (2020)
10. Li, K., Ren, A., Ding, Y., Shi, Y., Wang, X.: Research on decentralized identity and access management model based on the oidc protocol, pp. 252–255 (2020)
11. Rupa, C., Patan, R., Al-Turjman, F., Mostarda, L.: Enhancing the access privacy of IDaaS system using SAML protocol in fog computing. IEEE Access **8**, 168793–168801 (2020)
12. Zhou, H., Zhu, L.: Research and design of CAS protocol identity authentication. In: 2020 International Conference on CVIDL, pp. 384–387. IEEE (2020)

13. Hu, X., Tan, W., Ma, C.: Comment and improvement on two aggregate signature schemes for smart grid and VANET in the learning of network security. In: 2020 ICISE-IE, pp. 338–341. IEEE (2020)
14. Zhou, T., Li, X., Zhao, H.: EverSSDI: blockchain-based framework for verification, authorisation and recovery of self-sovereign identity using smart contracts. Int. J. Comput. Appl. Technol. **60**(3), 281–295 (2019)
15. Jamila, A., Sarwar, S., Hector, M., Zeeshan, P., Keshav, D.: DNS-idM: a blockchain identity management system to secure personal data sharing in a network. Appl. Sci. **9**(15), 2953 (2019)
16. Samir, E., Wu, H., Azab, M., Xin, C., Zhang, Q.: DT-SSIM: a decentralized trustworthy self-sovereign identity management framework. IEEE Internet Things J. **9**(11), 7972–7988 (2022)
17. Deepak, M., Harjasleen, M., Zhang, F., Jean-Louis, N., Alexander, F., et al.: CanDID: can-do decentralized identity with legacy compatibility, sybil-resistance, and accountability. In: 2021 IEEE SP, pp. 1348–1366. IEEE (2021)
18. Alizadeh, M., Andersson, K., Schelén, O.: Comparative analysis of decentralized identity approaches. IEEE Access **10**, 92273–92283 (2022)
19. Kannengießer, N., Pfister, M., Greulich, M., Lins, S., Sunyaev, A.: Bridges between islands: cross-chain technology for distributed ledger technology (2020)
20. Deng, L., Chen, H., Zeng, J., Zhang, L.-J.: Research on cross-chain technology based on sidechain and hash-locking. In: Liu, S., Tekinerdogan, B., Aoyama, M., Zhang, L.-J. (eds.) EDGE 2018. LNCS, vol. 10973, pp. 144–151. Springer, Cham (2018). https://doi.org/10.1007/978-3-319-94340-4_12
21. Li, D.W., Yu, J., Gao, X., Al-Nabhan, N.: Research on multidomain authentication of IoT based on cross-chain technology. Secur. Commun. Netw. (2020)
22. Frauenthaler, P., Sigwart, M., Spanring, C., Sober, M., Schulte, S.: ETH relay: a cost-efficient relay for ethereum-based blockchains. In: 2020 IEEE International Conference on Blockchain, Rhodes, Greece, pp. 204–213. IEEE (2020)
23. Armknecht, F., Karame, G., Mandal, A., Youssef, F., Zenner, E.: Ripple: overview and outlook. In: International Conference on Trust and Trustworthy Computing, Crete, Greece, pp. 163–180, 2015. Springer
24. Zhang, J., Liu, Y., Zhang, Z.: Research on cross-chain technology architecture system based on blockchain. In: Liang, Q., Wang, W., Liu, X., Na, Z., Jia, M., Zhang, B. (eds.) CSPS 2019. LNEE, vol. 571, pp. 2609–2617. Springer, Singapore (2020). https://doi.org/10.1007/978-981-13-9409-6_318
25. Thyagarajan, S.A., Malavolta, G., Moreno-Sanchez, P.: Universal atomic swaps: secure exchange of coins across all blockchains. In: 2022 IEEE Symposium on SP, pp. 1299–1316. IEEE (2022)

Research on Power Border Firewall Policy Import and Optimization Tool

Chen Zhang[1(✉)], Dong Mao[1], Lin Cui[2], Jiasai Sun[1], Fan Yang[1], and Cong Cao[3(✉)]

[1] State Grid Zhejiang Electric Power Corporation Information and Telecommunication Branch, Hangzhou, China
chzhangxd@163.com

[2] Beijing Guowang Xintong Accenture Information Technology Company, Beijing, China

[3] Hangzhou Innovative Institute, Beihang University, Hangzhou, China
caocong0419@163.com

Abstract. At present, the security strategy of the lower boundary protection equipment of the power system can no longer meet the needs of the current business growth. A large number of redundant strategies cause the protection performance of the boundary firewall to decline. At the same time, the large number of business growth causes the network boundary order of the power grid system to be blurred. In order to prevent the paralysis and partial collapse of the network and ensure the reliability and integrity of the power business data and enterprise information, this paper develops a smart border firewall optimization tool. This tool can not only integrate the security device policies of different manufacturers through *Simple Policy Specification Description Language* (SPSDL), but also prioritize security rules according to the frequency of use through keyword filtering algorithms and rule optimization decision trees, then realize the classification, streamlining, optimization and upgrading of firewall security rules. The research results show that the power system firewall can achieve an accuracy rate of more than 90% when the strategy is imported. The rule optimization part can reduce the unique correlation addition index of this paper to about 0.2, which solves the problem of firewall security strategy import language diversification. It further eases the pressure of firewall policy redundancy under the power system.

Keywords: Blockchain · Food Safety · Traceability technology · Information dentification · Consensus mechanism

1 Introduction

With the upgrading and the improvement of the software [1, 2], hardware [3, 4], and network [5–7], the security demand for power grid business is diversified and explosive, and the relevant security hardware equipment, software systems, and alarm emergency protection measures are constantly upgraded and improved. The traditional network

This work was supported by the State Grid Zhejiang Electric Power Co., Ltd. Technology Project (No. 5211XT22000D).

M. Qiu et al. (Eds.): SmartCom 2022, LNCS 13828, pp. 535–545, 2023.
https://doi.org/10.1007/978-3-031-28124-2_51

architecture mainly considers the communication requirements of business applications [8], simply implements static security policies [9] in the aggregation layer and access layer and does not consider the security requirements of terminal devices for multi-source requests in different operating environments. To solve this problem, the security protection of firewall boundary devices uses an ECA (*Enforcement Object/Controlled Object/Action*) based policy description language to uniformly normalize the policies of various types of firewall network security devices into a standard format, and realizes real-time control of terminal device traffic through policy account building, grouping classification, log analysis, etc. It has become an effective security policy management mode [10].

At present, the research on firewall boundary devices mainly includes filtering optimization of security policies, quantification of firewall performance indicators and application of distributed firewalls. As early as 1997, Lupu and Sloman published a conflict optimization analysis on firewall security policies at the IFIP/IEEE International Symposium. Based on this research [11], Bartal et al. developed the Firmato Firewall Management Toolkit [12] in 1999, realized the ternary separation management of security policies, network topology, and firewall hardware devices, and automatically generated firewall configuration files for multiple gateways and implemented advanced debugging. However, on the rule set with more than 20 host groups, firewall security policy rules are as dense as "spaghetti"; Moreover, Firmato does not allow users to control the order of rules, which makes it impossible for users to achieve finer control. To enable users to sort out firewall security rules, Hu et al. proposed a firewall framework based on rule segmentation, which effectively improved nearly 70% of rule conflicts through the network packet space segmentation and redundancy elimination mechanism defined by the firewall [13].

In addition, Han Guolong, Wang Wei and Sheng Honglei took the firewall of Tianjin Electric Power Company as an example in 2018 to propose an optimization scheme for expired rules, duplication and unreasonable configuration, and introduced practical application scenarios to expand the way of strategy optimization [14]. In 2021, Michigan State University proposed a cooperative firewall security policy framework VGuard based on *Virtual Private Networks* (VPNs), which processes data packets through Xhash (XOR and Secure Hash functions are parallel) and decision graph, and the processing efficiency is much higher than the linear search mode. The privacy and security of data packets are also guaranteed through VGuard. However, VGuard is a product based on VPC and lacks certain adaptability [15]. With the development of artificial intelligence and deep learning, in 2022, Journal of Shaanxi University of Science and Technology proposed the application of a security policy tool based on *Improved Mayfly Algorithm* (IMA) in the firewall [16] and optimized the parameters in the *Support Vector Machine* (SVM) firewall configuration model by training and testing the firewall dataset.

2 Power System Boundary Firewall Security Architecture

In order to realize the multidimensional analysis of the security strategy of the multi-node boundary firewall under the distributed power network information system and to clearly and accurately grasp the real-time state of the boundary equipment, this paper constructs the security boundary firewall model according to Fig. 1.

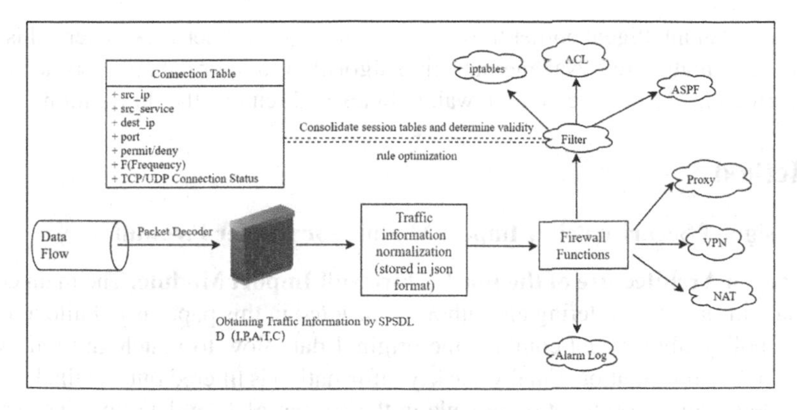

Fig. 1. Border Security Firewall Architecture

Considering the intelligent filtering and standardized parsing of external network traffic information and the merging and de-duplication of border device security policies [17], this paper uses Simple Policy Specification Description Language (Simple Policy Specification Description Language, SPSDL and DFA algorithms filter key information, and combine with decision tree algorithm to merge, de-duplicate and classify firewall security policies. Finally, key information in the security policy (SRC_IP, DEST_IP, PORT, and ACTION) is selected to determinc the functional dimensions of the firewall security policy architecture, including the import accuracy, correlation addition coefficient, rule scanning time, and coverage rate of key information.

Before the policy management of the firewall at the security boundary of the power network information system, it is necessary to interpret the extensibility and formalize the traffic data of the A-node and use the multivariate group theory to reify the traffic data as the set D(I, P, A, T, C), where *I* represents the source IP address set of the traffic data, *P* indicates the source port set of the traffic data, *A* indicates the protocol (TCP/UDP) carried by the traffic data, *T* indicates the sending time of the traffic data, and *C* indicates the specific content of the traffic data. Based on this, the multiple sets of traffic data are transferred to the firewall of each security boundary node under the power network system. The boundary firewall combines the existing policy library to compile the traffic set into an extensible language for formal processing. Different from the traditional firewall protection system, the intelligent boundary firewall under the power network system can monitor the network traffic in real time through the log storage detection system (Splunk, ELK, etc.) during the use of the application platform [18]. Combined with the intelligent security policy differentiation mode under the boundary firewall, the data packet can be filtered more efficiently. The boundary firewall architecture adopted in this paper will regularly apply:

$$BOOL(E, T, L) \tag{1}$$

Verify the effectiveness of the firewall policy. Where, E indicates whether the policy belongs to the policy library, T indicates the policy life cycle, and L indicates whether the policy log exists. Any Boolean value in the three rules is false, and the overall Boolean value is false. The firewall policy will no longer be effective. Aiming at the classification of intelligent border firewall strategies in power network system, this paper uses key information retrieval and matching algorithm to filter data packets and extract key information. Then match the firewall policies and remove the duplication.

3 Method

3.1 Design of Security Policy Import Module for Border Firewall

The Overall Architecture of the Border Firewall Import Module. The main content of the key information filtering algorithm constructed in this paper is to build a firewall security policy library, and combine the original data flow to match and retrieve the security policy information. Finally, the key information is filtered out and finally output to the rule optimization function module in the format of Json data flow. The specific construction mode is shown in Fig. 2:

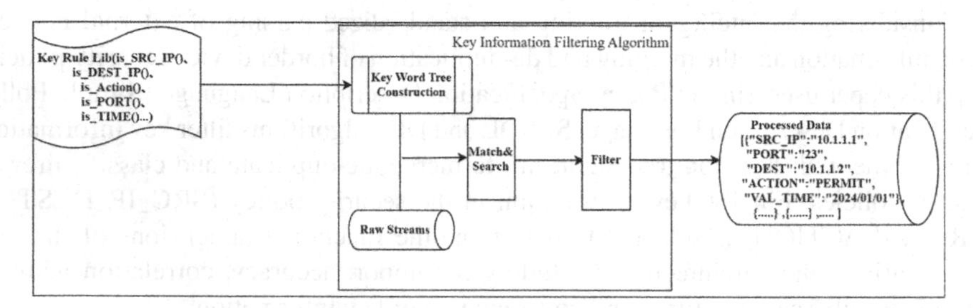

Fig. 2. Security Boundary Firewall Import Module

As there are many types of firewall network security devices, the current policy description language cannot meet the description of the policy. If a new firewall device type is added, its policy cannot be quickly imported and analyzed, resulting in poor flexibility and scalability [19–21]. Therefore, through the simple policy description language SPSDL and the key information matching algorithm, a Json format based formalization mechanism is proposed, which can realize the unified formalization of various types of firewall network security device policies into the standard format under the condition of the complex and changeable firewall network devices, so as to be imported into the system, laying a foundation for policy analysis.

Policy Standardization of Border Firewall. In order to solve the problem of policy description, this paper designs a Simple Policy Specification Description Language (SPSDL) based on ECA (*Enforcement Object/Controlled Object/Action*).

SPSDL Morphology. Morphology is the basic grammatical unit of language, which has definite meaning and plays various roles in policy compilation. How the words of a language are classified, divided into several categories, depends mainly on the convenience of processing. In SPSDL language will be divided into four categories:

1) key words: also called the reserved word. These words have a fixed meaning in the SPSDL.

2) operator: including logical operators, assignment operator, etc.
3) constants: such as digital constant, Boolean constants, character constants, etc.
4) delimiter: such as ";", "{";"}", etc.

Application of Key Information Matching Algorithm in Security Policy Import Module of Border Firewall. Many accurate key information retrieval and matching algorithms are based on deterministic finite automation (DFA). However, the key information of the security policy is changeable, and there are many key information dimensions to be retrieved, and the algorithm and time complexity are high, which cannot be completed through the retrieval sensitive word matching algorithm. This article is based on the following strategies:

$$S : \{\{S1 : r1, r2.., r_i\}, \{S2 : r_{i+1}, r_{i+2}.., r_{i+j}\}...\} \tag{2}$$

Key information such as IP address, port, action and time is found by combining lexical analysis, syntax analysis and function judgment such as IP and PORT, and finally exported to the policy optimization module in JSON format. The algorithm simplifies the complexity of the model and improves the running speed (Fig. 3).

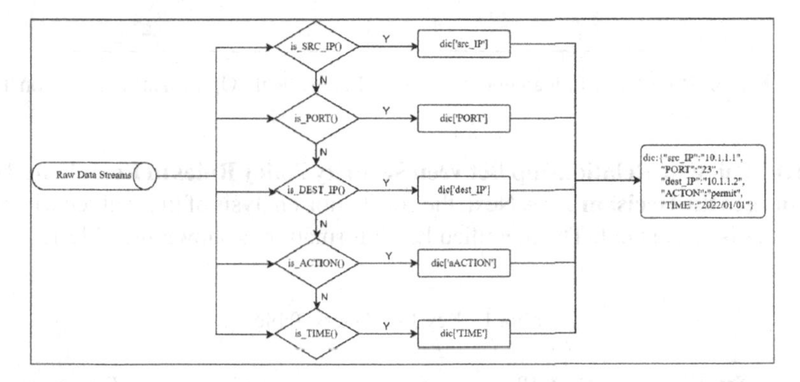

Fig. 3. Key information matching algorithm implementation model

3.2 Rule Optimization of Firewall Strategy in Power System

The security policy of the boundary firewall under the power system integrates a series of rules and information, which define the operations to be performed on the specified data packets and are finally presented in the form of < Condition", "Action">. The conditions in the rule consist of a set of fields used to identify the matching of specific types of packets (such as src_ip, dest_ip, src_port, dest_port, protocol, TCP/UDP, Expiration, Explanation, Log, etc.).

Judgment on the Validity of Rules for Flow Control. In the process of matching and optimizing traffic rules of firewall security policy, the unilateral relationship of rules is analyzed first. You only need to scan the rule column once, and then scan the rule's time parameter, comment parameter, generation log flag, and effective flag in turn. For example, check whether the comment is empty, and whether the calculation time parameter is expired. For each rule, the corresponding analysis results can be obtained without relying on the information of other rules. Therefore, context related parameters such as priority need not be considered. The specific process is shown in Fig. 4.

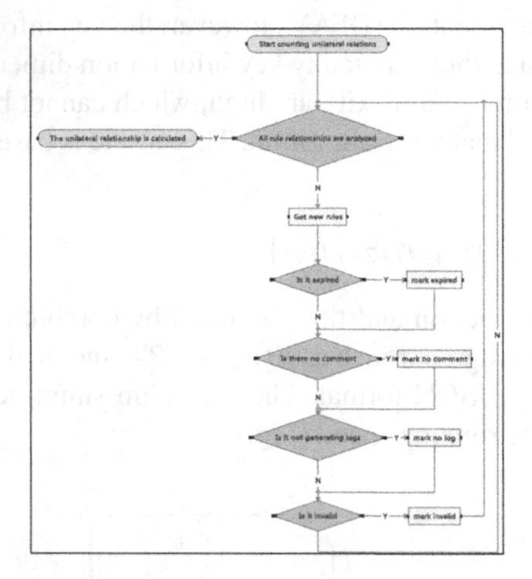

Fig. 4. Rule Valid Judgement

Fig. 5. Rule Optimization Decision Tree

Judge the Pairwise Relationship Between Security Policy Rules in Real-Time Traffic According to the Decision Tree. Next, the correlation analysis of firewall security policy traffic rules is carried out. The identified key information is shown in Table 1.

Table 1. Key Info Query Table

Id	*Src_IP*	*Dest_IP*	*Service*	*Action*	*Frequency*
1	*A*	*B*	*C*	permit	***f1***
2	any	any	any	permit	f_2

The relationship analysis of rules is generally divided into irrelevant, intersecting, subset or exclusive relationships. If there is no association between the two rules, the next step will not be processed; If the two rules are intersected or included, the union is used; If the relationship between the two rules is mutually exclusive, select the one with higher priority for the next action after judging the priority. Due to the existence of priority, if the actual matching traffic of a high priority rule is small, it will result in a large number of matching attempts. Therefore, moving the rules with high matching frequency to the high priority direction as far as possible can reduce the cumulative matching frequency. Obviously, this also requires the matching information of the traffic and rules previously analyzed. Due to the existence of the rule security zone, in order to ensure that moving the rule location will not cause changes in the actions of the security device when filtering packets, the movable location of the rule will be limited within its security zone. The rule optimization decision tree is shown in Fig. 5.

Regularly Identify and Update Flow Control Rules. For the formed policy rule base, it is necessary to regularly check the relationship and effectiveness of rules. At first, judge the validity of all rules in the firewall policy rule library, delete invalid rules, and repeat the previous operations for valid policy rules, as shown in the following figure (Fig. 6):

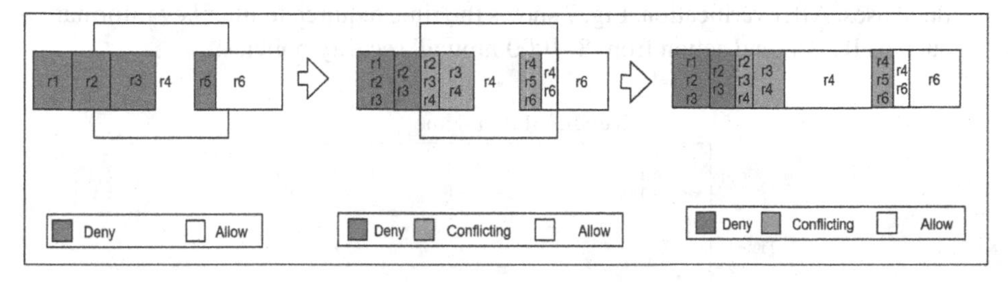

Fig. 6. Rule Optimization Flow Chart

Finally, the correlation coefficient *Cor ()* between two pairs of rules is used to determine whether all rules have been processed. If all correlation coefficients are zero, the processing is completed; otherwise, continue to process the rules with non-zero correlation coefficients until the correlation coefficients cannot be reduced. The processing table is shown below. If the following conditions are achieved, the scanning and updating will be completed (Table 2).

Table 2. Rule Validation Judgement Table

Regular_update_time_T	*Rule_Validation_BOOL (E, T, L)*	*Is_Cor (r1 & r2, r1 & r3...ri&rj)*
Date	True	False

4 Calculation

Throughput, number of concurrent connections, new connection speed, delay, packet loss rate, data backup speed, system recovery time, etc. can all be used as the main performance test indicators of the security boundary firewall. This paper focuses on the normalization import of firewall security boundary policy information and the optimization of security policy rules. Therefore, the accuracy rate of key information imported by security boundary firewall policy, the running time of policy import, the coverage rate of policy key information, the rule scanning time, the rule priority and correlation addition indicators are selected as the experimental evaluation results.

Border Firewall Security Policy Import Module. The import effect of key information is affected in many ways. This paper mainly evaluates the import effect of the security policy import module through the import time of key information, the coverage of key information in all policies, and the accuracy of relevant information when importing into the rule base.

1) Key Information Import Time
   The import duration of key information is mainly aimed at the time when the firewall boundary filters out IP, port number, action and other related information in all policy databases. After verification, Fig. 7 shows the time required to filter key information such as IP, port and action from 8–1000 firewall security policies.

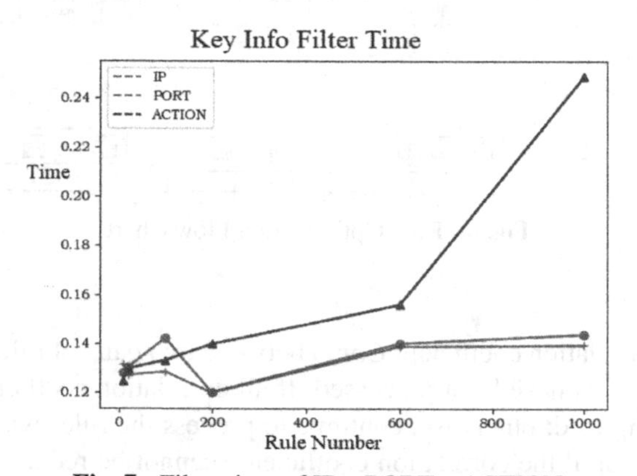

Fig. 7. Filter time of IP、PORT、ACTION

2) Key Information Coverage
   To ensure the efficiency of the border firewall security policy import module, this paper also counts the frequency of key information, that is, key information coverage. The calculation formula and process are as follows (Fig. 8)

$$F = \frac{1}{N}\sum_{i=1}^{N}\frac{k}{n_i} \tag{3}$$

   Through the calculation of the above process, the frequency of key information such as IP, port and action of the security boundary firewall is calculated as shown in Fig. 9. It can be seen that the frequency of the ACTION field is relatively high, and the error of the ACTION field in the statistical process is minimal.
3) Key Information Import Accuracy
   In the border firewall based security policy import module, the key information matching algorithm also counts the matching accuracy of key information. The experimental results show that more than 90% of the key information can be matched successfully, and the identification error is mainly due to the confusion of subnet mask and IP address. In addition, according to the statistical results, the source IP address and actions appear frequently in the security policy, so they have high priority. The specific results are shown in Table 3:

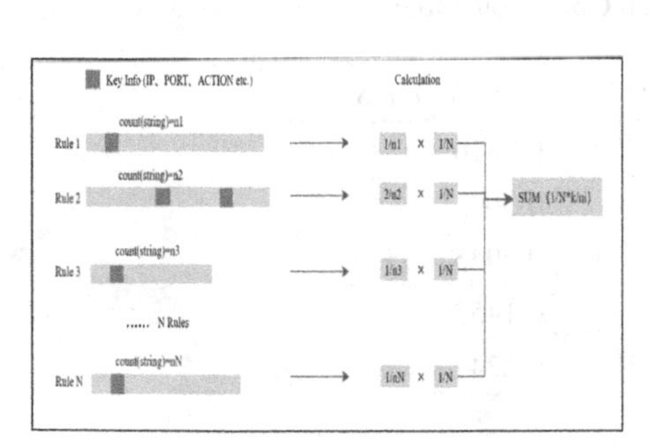

Fig. 8. Key info. Coverage Rate

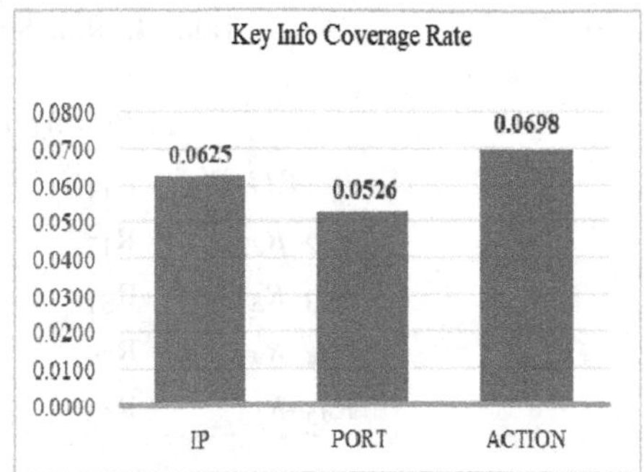

Fig. 9. Key Info. Coverage Rate Result

Table 3. Key Info. Match Accuracy

	SRC_IP	*DEST_IP*	*PORT*	*ACTION*	*TOTAL*
Match_Accuracy	0.9091	0.9091	0.9231	1.00	0.9353
Error_Rate	0.0909	0.0909	0.0769	0.00	0.0647
Frequency_f	0.7273	0.2727	0.5455	0.8182	

Evaluation Index of Policy Rule Optimization Module of Boundary Firewall.

1) Correlation Bonus Indicator.
   To judge the performance of firewall policy rules based on traffic control, it is necessary to consider whether the correlation indicators in the firewall policy rule library meet the standards. Therefore, the correlation plus indicator S is introduced:

$$S = Cor1(r_1, r_2) + Cor2(r_1, r_3) + \ldots + Cor_{\frac{n\times(n-1)}{2}}(r_n, r_{n-1}) \tag{4}$$

   When S is close to 0, it indicates that the rules in the firewall policy rule base are uncorrelated. At this time, the simplification of the firewall policy rule base reaches the optimal state.
2) Rule Priority Indicator.
   The priority of rules in the rejected domain is often higher than that of rules in the accepted domain. Based on this, the rule priority needs to be determined according to frequency f and action A. The specific calculation is as follows:

$$P = k_1 \times f + k_2 \times A + \varepsilon \tag{5}$$

3) Performance Analysis of Rule Optimization.
   In this paper, 55 rules in the security policy library are randomly selected for performance optimization analysis and the final results are shown in Table 4:

Table 4. Rule Scan Condition Table

	Priority_P	*Correlation_S*
R1–R11	R_{11}	0.2310
R_{12}–R_{22}	R_{17}	0.3001
R_{23}–R_{33}	R_{31}	0.2945
R_{34}–R_{44}	R_{39}	0.1453
R_{45}–R_{55}	R_{53}	0.2247

The results show although the rules are deduplicated and optimized, some rules cannot be completely irrelevant, so the optimization algorithm needs to be further improved.

5 Conclusion

In order to solve the problem of identification and standardization of diversified security policy description languages, this paper adds a security policy description language based on policy knowledge base and completes the policy normalization import. In addition, aiming at the redundancy, invalidity and all pass problems of firewall security policies, a group of firewall policy analysis and optimization algorithms are designed to help administrators automatically analyze redundant, conflicting, overlapping, loose and other policy problems, solve the boundary leak caused by firewall configuration problems, reduce the security operation and maintenance costs, and improve the border protection capability.

References

1. Tao, L., Golikov, S., et al.: A reusable software component for integrated syntax and semantic validation for services computing. In: IEEE Symposium on SOSE, pp. 127–132 (2015)
2. Li, J., Ming, Z., et al.: Resource allocation robustness in multi-core embedded systems with inaccurate information. JSA **57**(9), 840–849 (2011)
3. Qiu, M., Ming, Z., et al.: Three-phase time-aware energy minimization with DVFS and unrolling for chip multiprocessors. J. Syst. Archit. **58**(10), 439–445 (2012)
4. Qiu, M., Li, H., Sha, E.: Heterogeneous real-time embedded software optimization considering hardware platform. In: ACM Symposium on Applied Computing, pp. 1637–1641 (2009)
5. Qiu, M., et al.: Efficent algorithm of energy minimization for heterogeneous wireless sensor network. In: Sha, E., et al. (eds.) EUC 2006. LNCS, vol. 4096, pp. 25–34. Springer, Heidelberg (2006). https://doi.org/10.1007/11802167_5
6. Niu, J., Gao, Y., Qiu, M., Ming, Z.: Selecting proper wireless network interfaces for user experience enhancement with guaranteed probability. JPDC **72**(12), 1565–1575 (2012)
7. Hu, F., Lakdawala, S., et al.: Low-power, intelligent sensor hardware interface for medical data preprocessing. IEEE Trans. Inf. Technol. Biomed. **13**(4), 656–663 (2009)

8. Jian-bing, L.I.U., Xu-yan, M.A., Xiao-hong, W.A.N.G., Zhen-xin, W.A.N.G.: Security policy of active security network architecture. Inf. Secur. Res. **7**(11), 998–1006 (2021)
9. Ming, H.A.N.: Internal network security policy. Inf. Comput. **04**, 157–158 (2018)
10. Yan, Z.: Optimization of communication network firewall strategy. Comput. Knowl. Technol. **17**(07), 46–47+53 (2021). https://doi.org/10.14004/j.cnki.ckt. 2021.0724
11. Lupu, E., Sloman, M.: Conflict analysis for management policies. In: Lazar, A.A., Saracco, R., Stadler, R. (eds.) Integrated Network Management V. IM 1997. IFIP — The International Federation for Information Processing, pp. 430–443. Springer, Boston (1997). https://doi.org/10.1007/978-0-387-35180-3_32
12. Bartal, Y., Mayer, A., Nissim, K., et al.: Firmato: a novel firewall management toolkit. ACM Trans. Comput. Syst. (TOCS) **22**(4), 381–420 (2004)
13. Hu, H., Ahn, G.J., Kulkarni, K.: Detecting and resolving firewall policy anomalies. IEEE Trans. Dependable Secure Comput. **9**(3), 318–331 (2012)
14. Han, G., Wang, W., Sheng, H.: Research on firewall strategy sorting and optimization method. Electr. Power Inf. Commun. Technol. **16**(06), 31–35 (2018)
15. Liu, A.X., Li, R.: Collaborative enforcement of firewall policies in virtual private networks. In: Liu, A.X., Li, R. (eds.) Algorithms for Data and Computation Privacy, pp. 139–170. Springer, Cham (2021). https://doi.org/10.1007/978-3-030-58896-0_6
16. Gao, Z., Zhang, Y., et al.: Improved mayfly algorithm and its application in firewall policy configuration. J. Shaanxi Univ. Sci. Technol. (Nat. Sci. Ed.) **38**(02), 41–48 (2022)
17. Liu, K.: Research and Implementation of Firewall Deep Packet Detection Technology. Beijing University of Posts and Telecommunications (2013)
18. Ren, Z.: Research on Key Technologies of Firewall Security Policy Configuration. National University of Defense Science and Technology (2011)
19. Chen, X.: Analysis of security policy conflicts in multi device firewalls. Comput. CD Softw. Appl. (02), 104+102 (2012)
20. Deng, W., Liang, Y.: Semantic analysis method of firewall security policy. Comput. Eng. Appl. (26), 135–137 (2007)
21. Wang, B.: Active Security Policy Firewall Based on Honeynet. Beijing University of Posts and Telecommunications (2010)

Block-gram: Mining Knowledgeable Features for Smart Contract Vulnerability Detection

Tao Li[1,2,3,4], Haolong Wang[2], Yaozheng Fang[2], Zhaolong Jian[2], Zichun Wang[2], and Xueshuo Xie[2,3,4](✉)

[1] Tianjin Key Laboratory of Network and Data Security Technology, Tianjin, China
[2] College of Computer Science, Nankai University, Tianjin, China
{litao,wzc,xueshuoxie}@nankai.edu.cn,
{whlong,fyz,jianzhaolong}@mail.nankai.edu.cn
[3] Key Laboratory of Blockchain and Cyberspace Governance of Zhejiang Province, Hangzhou, China
[4] State Key Laboratory of Computer Architecture, Institute of Computing Technology, Chinese Academy of Sciences, Beijing, China

Abstract. Effective vulnerability detection of large-scale smart contracts is critical because smart contract attacks frequently bring about tremendous economic loss. However, code analysis requiring traversal paths and learning methods requiring many features training is too time-consuming to detect large-scale on-chain contracts. This paper focuses on improving detection efficiency by reducing the dimension of the features, combined with expert knowledge. We propose a feature extraction method *Block-gram* to form low-dimensional knowledgeable features from the bytecode. We first separate the metadata and convert the runtime code to opcode sequence, dividing the opcode sequence into segments according to some instructions (*jump*, etc.). Then, we mine extensible *Block-gram* features for learning-based model training, consisting of 4-dimensional block features and 8-dimensional attribute features. We evaluate these knowledge-based features using seven state-of-the-art learning algorithms to show that the average detection latency speeds up 25 to 650 times, compared with the features extracted by *N-gram*.

Keywords: Smart Contract · Bytecode · Opcode · Knowledgeable Features · Vulnerability Detection

1 Introduction

Smart contracts are widely deployed on blockchain to implement complex transactions, such as decentralized applications on Ethereum [1]. As of August 9, 2022, the number of smart contracts exceeded 51.1 million on Ethereum[1]. These smart contracts are written in a domain specific language (e.g., *Solidity*), compiled into bytecodes, executed as opcodes in EVM after being deployed on-chain by the

[1] https://explore.duneanalytics.com/.

M. Qiu et al. (Eds.): SmartCom 2022, LNCS 13828, pp. 546–557, 2023.
https://doi.org/10.1007/978-3-031-28124-2_52

consensus mechanism [2]. Due to running on distributed nodes of blockchain, once vulnerabilities are found, they are difficult to upgrade and repair [3,4]. For example, hackers exploited a re-entrancy vulnerability in the DAO contract to steal 3.6 million ETH[2]. According to SlowMist Hacked, a huge economic loss of more than $10 billion has been caused due to the security issues of smart contracts[3]. With the rapid increase in smart contracts, an efficient smart contract vulnerability detection method is particularly important for the blockchain [5].

Nowadays, the mainstream smart contract vulnerability detection methods include code analysis and machine learning. For code analysis, we can analyze the types or causes of vulnerabilities with expert knowledge through formal verification, symbolic execution, fuzz testing, etc. [6–10]. But these methods need to traverse more paths of code or complexity mathematical proofs, the detection is time-consuming and labor-intensive. For machine learning, they primarily capture code features by training machine learning models to infer whether it is vulnerable. Some smart contract vulnerability detection algorithms are based on the combination of text information and neural network [5,11,12] or based on the combination of smart contract graph information and neural network [13,14]. But all of them need a large feature space for training, and the dimensional of features will influence the model performance and detection latency.

In this paper, we focus on detection efficiency of large-scale smart contracts and face the following challenges: (1) how to improve detection performance combined with vulnerability features through expert knowledge; (2) how to reduce the dimension of feature space for optimizing the detection latency without influencing the model performance. To tackle the first challenge, we preprocess the bytecode to opcode sequences according to the disassembling rules of Ethereum and divide the opcode sequences into flow graphs through some instructions (*jump*, etc.) for extracting 4-dimensional block features. For the second challenge, we mine other 8-dimensional attribute features of vulnerabilities through expert knowledge, to construct extensible 12-dimensional *Block-gram* features. The *Block-gram* features have lower dimensional than thousand dimensions features by *N-gram*, and will significantly reduce the detection latency without influencing the model performance. In summary, the contributions:

- We mine the extensible low-dimensional *Block-gram* features from bytecode, including 4-dimensional block features, and 8-dimensional attribute features by smart contract vulnerabilities analysis.
- We introduce expert knowledge when constructing opcode sequence flow graphs and extracting attribute features, and combine vulnerability analysis to improve the performance of the above low-dimensional features.
- We validate the efficiency of *Block-gram* features on seven state-of-the-art machine learning algorithms. The evaluation shows that the above low-dimensional features can flexibly support multiple detection algorithms and significantly reduce detection latency.

---

[2] http://www.coindesk.com/daoattacked-code-issue-leads-60-million-ether-theft, 2016.

[3] https://hacked.slowmist.io/en/.

2 Preliminaries

2.1 Ethereum Smart Contract

Ethereum Virtual Machine. EVM provides 142 opcodes or bytecodes with 10 functions, such as *STOP* and *PUSH* operations, etc. There are only 256 opcodes at maximum, but some instructions are not defined now, only for future expansion. EVM has a simple stack structure with a maximum stack size of 1024. Each opcode is allocated one byte (for example, *STOP* is *0x00*), pushes or pops a certain number of elements from the stack, and can obtain information about the execution environment, or interact with other blockchains smart contracts [15]. During execution, the bytecode is split into bytes (1 byte equals 2 hex characters). Bytes in region *0x60–0x7f* (*PUSH1–PUSH32*) are treated differently because they contain data that needs to be pushed into the stack. If the call count is over 1024, the call-stack attack may take place. EVM explains how to change the system state given a series of the above instructions and a small part of environmental data.

Smart Contract Compilation. As shown in Fig. 1, the developers write the source code in a high-level language (e.g., *Solidity*). The source codes are compiled into byte arrays encoded by hexadecimal digits with a compiler as bytecodes [16]. Then, the bytecodes are uploaded to EVM with an Ethereum client and can be translated into EVM instructions or opcodes [17]. After the source code of the contract is compiled into EVM bytecode and ABI, it can be deployed using the *Web3.js* interface. For contract deployment, it is essential to execute a transaction, which has no destination address but the data field is EVM bytecode [18]. When processing this transaction, the EVM executes the input data as code. The bytecode is divided into deployment code, runtime code, and metadata. After the contract deployment, EVM will store the runtime code and metadata on the blockchain, and then match their storage addresses to the contract account to complete the deployment of the contract. It would be easier to analyze smart contracts with bytecodes or opcodes, because: (1) bytecodes or opcodes are not had man-made variables that are defined in source codes; (2) bytecodes or opcodes are easy to collect from the public blockchain.

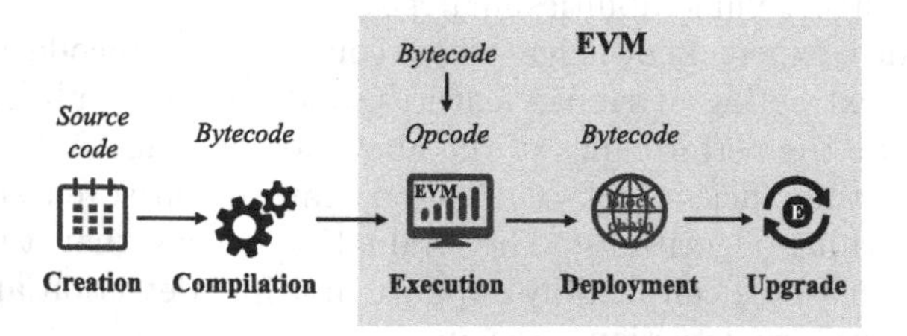

Fig. 1. Smart Contract Compilation Process.

Table 1. Comparisons among smart contract vulnerability detection methods.

Type	Name	Base Model	Detection Source	Platform
Code analysis	KEVM	Formal Verification	Bytecode	EVM
	Oyente	Symbolic Execution	Bytecode & ETH Condition	EVM
	Contractfuzzer	Fuzz Testing	EVM ABI & EVM Log	EVM
	Ethir	Intermediate Representation	Bytecode	EVM
Learning methods	DR-GCN	GCN	Source code	ETH & VNT
	TMP	TGNN	Source code	ETH & VNT
	Deescvhunter	DNN	Source code	ETH
	Eth2vec	DNN	Bytecode	ETH
	Escort	DNN	Bytecode	ETH
	Rechecker	BLSTM	Sourcecode	ETH
	ContractWard	XGBoost	Bytecode	ETH
	Vscl	DNN	Bytecode	ETH
	EtherGIS	GNN	Bytecode	ETH
	SafeSC	LSTM	Opcode	ETH

2.2 Smart Contract Security Analysis

Code Analysis. As shown in Table 1, most of the traditional smart contract vulnerability detection methods are based on program code analysis and program path analysis. Hildenbrandt et al. [6] present KEVM based on formal verification and provide an executable formal specification for EVM's program language using the K framework. Luu et al. [8] use symbolic execution to implement Oyente which traverses smart contract execution paths based on control flow graphs to detect vulnerability. Jiang et al. [9] propose Contractfuzzer that sets up test cases and analyzes smart contract behavior logs to detect vulnerabilities based on fuzz testing. Albert et al. [10] implement Ethir based on intermediate representation and analyze the security properties of bytecode by converting Oyente's control flow graph into a rule-based representation. Due to the need to traverse most paths of the code, the detection is time-consuming and labor-intensive, and difficult to use for large-scale contract detection.

Learning Method. As shown in Table 1, Wang et al. [5] propose ContractWard that can extract bigram features from smart contract opcodes and use multiple machine learning algorithms for vulnerability detection. Mi et al. [12] apply novel feature vector generation techniques from bytecode and metric learning-based deep neural network to detect vulnerability. Zhuang et al. [13] use DR-GCN to convert source code into contract graph and use graph convolutional neural network to build a vulnerability detection model. Based on DR-GCN, TMP considered the time sequence information in the contract graph and used the time sequence graph neural network to build a vulnerability detection model. Zeng et al. [14] use graph neural network and expert knowledge to build the control flow graph with attribute and input graph attribute features into graph neural networks to detect vulnerabilities. However, due to the lack of expert knowledge, most learning methods use high-dimensional feature spaces for training, such as *N-gram* extracting thousands of dimensional features. The detection is still time-consuming and may not be suitable for batch vulnerability detection. Therefore, combining the expert knowledge used in code analysis with the learning methods is precisely the problem that our work mainly solves.

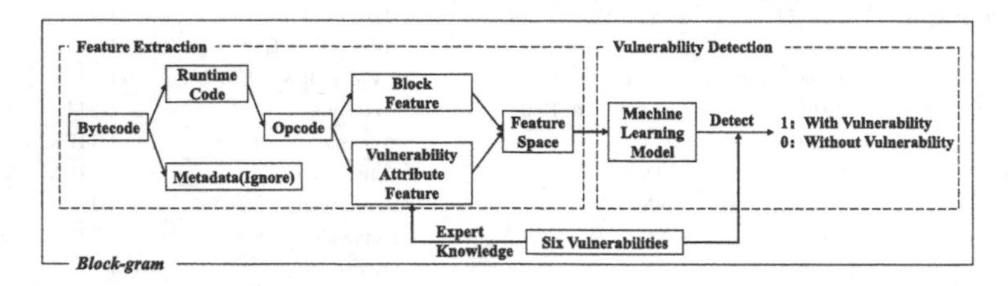

Fig. 2. The *Block-gram* feature extraction method overview.

3 Detailed Design

3.1 Overview

As shown in Fig. 2, we present a detailed description of the features extraction method *Block-gram* using smart contract bytecode. We first extract the 4-dimensional bytecode block features from the opcode sequence flow graph according to EVM disassembling rules. Then, we divide the opcodes into eight categories and count their ratios as 8-dimensional attribute features through six vulnerabilities analysis by expert knowledge. Finally, we use these 12-dimensional *Block-gram* features by seven state-of-the-art machine learning algorithms to detect six vulnerabilities. The *Block-gram* features include in:

- **Rule-based bytecode block features.** We first collect the bytecode of the metadata header of various solidity versions. The metadata in the bytecode is separated from the runtime code by string matching, and convert the runtime code into opcode according to Ethereum disassembling rules. Then we divide the opcode sequence into blocks through `JUMP` opcode, and generate the opcode sequence flow graph, denoted by the adjacency matrix. We extract 4-dimensional bytecode block features from the adjacency matrix. These features can preserve the relationship between different bytecode blocks.
- **Attribute Features with Expert Knowledge.** We first divide the opcodes of Ethereum into eight categories, such as block opcodes, system opcodes, stack opcodes, etc. Each type of opcode corresponds to several smart contract vulnerabilities according to the vulnerability analysis with expert knowledge. We count their ratios as 8-dimensional extensional attribute features. These features can preserve the expert knowledge on vulnerability, and extend with the new vulnerabilities.

3.2 Rule-Based Bytecode Block Features

The key to bytecode preprocessing are metadata separation and bytecode conversion. In the metadata separation module, we separate metadata and runtime-code according to metadata header(*e.g., 0x65 'bzzr0' 0x58 0x20 <32 bytes swarm* *hash> 0x00 0x29*). Then we convert the runtimecode to the opcode according to Ethereum conversion rules. The bytecode of each non-PUSH class opcodes is converted into the corresponding opcode(for example, *STOP* is *0x00*, *ADD* is *0x01*). But for the PUSH class opcodes, we get the length of their parameters according to the type of the PUSH opcode, so as to convert the bytecode.

The existing processing methods of opcode sequence are divided into block-sequence method and natural language processing method. Block-sequence method divides the opcode sequence into blocks and edges. Natural language processing method uses *N-gram* method to process the opcode sequence. We choose block-sequence method to process opcode sequence and build adjacency matrix. First, we divide the opcode into blocks according to the jump class instruction, and then determine the edges between blocks according to the jump type at the end of each block. Each blocks has the following boundary.

- **Starting point:** `JUMPDEST...`
- **End point:** `JUMP, JUMP I, STOP, REVERT, RETURN, SELFDESTRUCT, INVALID`

In order to construct the edges between blocks, we divide edges into three categories according to the end point of blocks:

- `JUMP`: If there is `PUSHn` before `JUMP`, the parameter of `PUSHn` is the destination address of `JUMP`. If there is not `PUSHn` before `JUMP`, we using stack execution algorithm provided by EtherSlove [19] to calculate the destination address.
- `JUMP I`: `JUMP I` is a conditional jump. True edge's target is the parameter of the PUSH opcode; false edge's target is the offset of the following block. This means that if a basic block ends with JUMP I, then there will be two edges starting from this block.
- `REVERT, SELFDESTRUCT, RETURN, INVALID, STOP`: These opcodes mean the interruption of control flow, so they have no subsequent basic blocks.

When we get the blocks and edges, we can build an opcode sequence flow graph. Because the opcode sequence flow graph is a directed graph, we use the adjacency matrix to represent the opcode sequence flow graph and extract sequence features from the adjacency matrix. Then, we extract four-dimensional smart contract block features according to the adjacency matrix. These features represent the sequence attribute of the opcode.

- **Number of nodes.** The number of rows or columns of the adjacency matrix is the number of nodes. The number of nodes represents how many basic blocks are in the opcode sequence flow graph.
- **Number of edges.** The sum of the elements in the adjacency matrix is the number of edges. The number of edges represent how many basic edges are in the opcode sequence flow graph.
- **Maximum out-degree.** The maximum value of the sum of the elements in row i is the maximum out-degree. Out-degree is the sum of the times when a block of the opcode sequence flow graph is used as the starting point.

- **Maximum in-degree.** The maximum value of the sum of the elements in row j is the maximum in-degree. In-degree is the sum of the times when a block of the opcode sequence flow graph is used as the end point.

Table 2. *Block-gram* Knowledgeable Features.

Feature		Opcode	Value	Knowledge
Block Feature	Node	None	>0	Feature of block
	Edge		>0	
	Maxout		>0	
	Maxin		>0	
Attribute Feature	Unary-arithmetic ratio	`ISZERO, NOT...`	(0, 1)	Feature of integer overflow
	Binary-arithmetic ratio	`ADD, AND, SHA3...`	(0, 1)	
	Block ratio	`NUMBER, BLOCKHASH, COINBASE...`	(0, 1)	Feature of timestamp dependency
	Control-flow ratio	`JUMP, JUMP I, JUMPDEST...`	(0, 1)	Feature of re-entrancy
	Environment ratio	`CALLER, CALLDATASIZE...`	(0, 1)	Feature of TOD
	System ratio	`CALL, RETUREN, REVERT...`	(0, 1)	Feature of callstack depth attack
	Stack ratio	`POP, PUSH, SWAP...`	(0, 1)	
	Invalid ratio	Others	(0, 1)	Feature of invalid opcodes

3.3 Attribute Features with Expert Knowledge

There are some common vulnerabilities in Ethereum, such as integer overflow vulnerabilities, integer underflow vulnerabilities, callstack depth attack vulnerability, transaction-ordering dependence vulnerability, timestamp dependency vulnerability, and re-entrancy vulnerability. We analyze above vulnerabilities and opcodes related to them and extract attribute features. We divide the opcodes into eight categories, such as unary arithmetic opcodes, binary arithmetic opcodes, block opcodes, control-flow opcodes, environmental opcodes, system opcodes, stack opcodes, and invalid opcodes. As shown in Table 2, each type of opcode corresponds to several smart contract vulnerabilities according to the vulnerability analysis with expert knowledge. These features can preserve the expert knowledge on vulnerability, and extend with the new vulnerabilities. We define *count (opcode, i)* as the number of all opcodes of the smart contract i and *count (j, i)* as the number of opcodes corresponding to feature j in the smart contract i. For example, *count (1, 1)* is the number of unary arithmetic opcodes in smart contract 1. $count(1, i)/count(opcode, i)$ is ratio of feature 1.

- **Unary and Binary arithmetic opcodes ratio:** In Ethereum, some opcodes are responsible for arithmetic operations, including unary, binary, and ternary arithmetic opcodes. Only these arithmetic opcodes cause integer overflow. In our research on smart contract, we find that almost no contract contain ternary arithmetic codes, so we only consider unary(*e.g.,* `ISZERO`, `NOT`) and binary(*e.g.,* `ADD`, `AND`, `SHA3`) arithmetic opcodes.
- **Block opcodes ratio:** In Ethereum, some opcodes are related to the block information. `BLOCKHASH` shows the hash value of the block, `COINBASE` is the address of the miner. At Ethereum system layer, block information is often used as a seed for generating random numbers. However, the block timestamp, number, and other information are often used by attackers. Block information opcodes are related to timestamp dependency vulnerabilities.

- **Control-flow opcodes ratio:** Some opcodes are used to change the control flow. `JUMP` is an unconditional jump that takes the top element of the stack as the destination address. `JUMP I` is a conditional jump that has the same destination address with `JUMP` if the top element of the stack is not zero; otherwise, EVM will execute the opcodes following `JUMP I`. The re-entrancy vulnerability stems from the attacker's cyclic call changing the control flow. Control flow opcodes are related to this vulnerability.
- **Environmental opcodes ratio:** Some opcodes are responsible for interacting with contracts and message calls and transactions. `ADDRESS` is the address of the currently executed contract. `ORIGIN` is the address of the initiator of the transaction. `CALLER` is the address of the caller of the message. Since the environmental opcodes can obtain transaction information, they are related to Transaction-Ordering Dependence vulnerabilities that also rely on transaction information.
- **System opcodes ratio:** The system opcodes are responsible for calling between smart contracts. `CALL` calls the function in other contracts. `RETURN` returns from the contract calls. `REVERT` reverts transaction and return data.
- **Stack opcodes ratio:** EVM is a stack machine, and all calculations are performed on a data area called the stack. Some opcodes deal with the elements of the stack. `POP` pops the top element of the stack and discards it. `SWAP1` exchanges the top two members of the stack. They are stack opcodes. System opcodes and stack opcodes act on calls and stack operations, and these opcodes may cause callstack depth attack vulnerabilities.
- **Invalid opcodes ratio.** Invalid opcodes refer to the opcodes irrelevant to the six vulnerabilities detected.

3.4 Low-Dimensional Knowledgeable Features

During feature extraction, to make the trained model suitable for all smart contracts, we first select 4-dimensional block features to highlight the relationship between different opcode blocks. The 4-dimensional block features are the number of nodes, the number of edges, the maximum out-degree, and the maximum in-degree of the control flow graph. These 4-dimensional features represent the complexity of the smart contract, emphasizing the role of the features of the smart contract itself in vulnerability detection. Then, we investigate the causes of vulnerabilities in smart contracts and mine 8-dimensional attribute features for opcodes associated with them. For example, unary-arithmetic and binary-arithmetic opcodes modify integers in Ethereum, and improper arithmetic operations can lead to integer overflow vulnerabilities; block information opcodes are closely related to timestamp vulnerabilities and block parameter dependency vulnerabilities; control flow opcodes are related to the re-entrancy vulnerability stems from the attacker's cyclic call changing the control flow. As shown in Table 2, we combined the 4-dimensional block features and 8-dimensional attribute features together for efficient vulnerability detection. When constructing the opcode sequence flow graph, we used depth-first-search algorithm, and the time complexity is $O(N)$. When constructing the adjacency matrix, the time complexity is $O(N^2)$. Therefore, the time complexity of feature extraction is $O(N + N^2)$, where N represents the nodes amount of the above graph.

Features Normalization. Due to the difference in feature extraction, the first four-dimensional features are large integers, and the last eight-dimensional features are decimals between 0 and 1. Therefore, if these 12-dimensional features are directly used as the input of machine learning models, some machine learning models (such as K-Nearest Neighbors) will only focus on the first four-dimensional features while ignoring the last eight-dimensional features during training. In addition, some machine learning models also have requirements for the format of input data. To make *Block-gram* features suitable for most mainstream machine learning models, we use linear normalization to process these features. The normalization method is defined in Eq. (1).

$$x^*_{(n,f)} = \frac{x_{(n,f)} - \min\limits_{0<i<r} x_{(i,f)}}{\max\limits_{0<i<r} x_{(i,f)} - \min\limits_{0<i<r} x_{(i,f)}} \tag{1}$$

where $x_{(n,f)}$ represents the value of the feature f in the n row.

4 Evaluation

4.1 Experimental Setup

Configuration. We perform experiments on a Windows 10 machine with 12th Gen Intel Core 2.10 GHz CPUs and 32 GB RAM and use the GPU of a 1060ti graphics card to train the model and predict the results. To verify the efficiency and the validity of the above 12-dimensional feature, we choose seven state-of-the-art machine learning algorithms, such as eXtreme Gradient Boosting (XGBoost), Random Forest (RF), K-Nearest Neighbors (KNN), Logistic Regression (LR), Decision Tree (DT), Naive Bayes, and Long short-term memory (LSTM), as the training and detection model. We use the sklearn library in python3.6.8 to build the machine learning algorithms. We also select accuracy, recall, F1-score, and latency as the measured metrics of the model.

Datasets. We select 3000 smart contracts as the dataset for performance analysis from Contractward [5], 70% for training, and 30% for testing. The size of the dataset is 62.7 MB. There are 871 contracts with vulnerabilities and 2179 contracts without vulnerabilities. The vulnerabilities of the dataset include integer overflow and integer underflow vulnerabilities, callstack depth attack vulnerability, transaction-order dependence vulnerability, timestamp dependency vulnerability, and re-entrancy vulnerability.

4.2 Performance Analysis

Detection Performance. As shown in Table 3, in terms of accuracy, the maximum value of the seven models is 82.22% and the minimum value of the seven

Table 3. Model Performance by *Block-gram* Features.

Model	Feature	Accuracy (%)	Recall (%)	F1-score (%)	Latency (ms)
XGBoost	*Block-gram*	**82.22**	93.27	88.16	**0.2**
	N-gram	80.4	95.3	87.33	15
Random Forest	*Block-gram*	**81.77**	95.00	88.10	**0.3**
	N-gram	70.55	98.27	82.58	12
K-Nearest Neighbors	*Block-gram*	**75.44**	**84.51**	**83.01**	**0.004**
	N-gram	51.55	36.62	51.77	0.1
Logistic Regression	*Block-gram*	75.44	98.28	85.04	**0.01**
	N-gram	75.88	87.17	83.70	0.4
Decision Tree	*Block-gram*	77.44	84.66	84.20	**0.004**
	N-gram	76.11	83.72	83.27	2.6
Naive Bayes	*Block-gram*	**76.66**	**87.17**	**84.13**	**0.0003**
	N-gram	60.33	58.37	67.63	0.09
LSTM	*Block-gram*	**73.88**	66.28	59.55	1
	N-gram	69	75.10	58.42	90

models is 73.88% when they use *Block-gram* features. When they use features extracted from opcodes by *N-gram*, almost all models' accuracy drops. *Block-gram* features perform better than features extracted by *N-gram* in terms of accuracy. In terms of recall and F1-score, the performance of K-Nearest Neighbors and Naive Bayes drops significantly when they use features extracted by *N-gram*. And other measured metrics of K-Nearest Neighbors and Naive Bayes also dropped significantly. The two models poor perform when dealing with *N-gram* features. The reason is that we have considered the real jump relationship when the smart contract runs and the expert knowledge of six vulnerabilities but the *N-gram* method only extracts the combination of the opcode sequence.

Detection Latency. As shown in Fig. 3, the detection latency of all models is greatly improved when they use *Block-gram* features. Among the seven models, the detection latency of the decision tree model has reduced 650 times when using *Block-gram* features compared with using *N-gram*. In addition, compared with the traditional method Oyente and other machine learning methods (for example, VSCL and EtherGIS), the latency of using *Block-gram* features is also

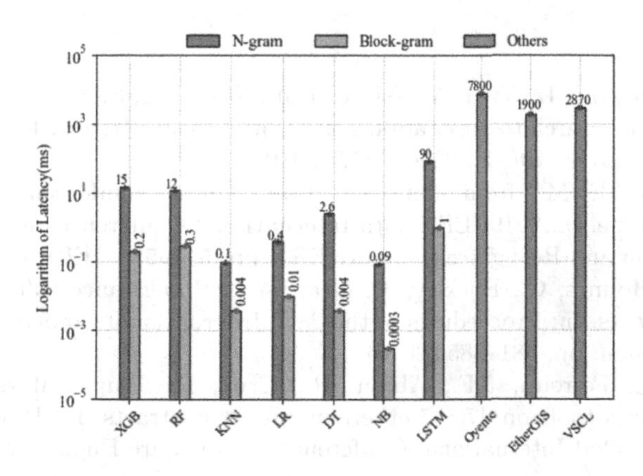

Fig. 3. Comparisons among smart contract detection latency.

significantly reduced. Since VSCL and EtherGIS did not publish open-source code, we directly cited the detection delay published in the paper. Their experimental environment is far better than our work. The reason is that the dimension of the feature space extracted by *N-gram* is too high, as high as tens of hundreds of dimensions during the initial extraction. In training, the features will be expanded up to tens of thousands of dimensions. When using *Block-gram* features, the initially extracted feature space is only 12-dimensional. Even if it needs to expand during the training process, it is only 15-dimensional at most. Significant differences in feature space dimensions lead to differences in latency. The experimental results demonstrate that *Block-gram* features are efficient and can be used by mainstream machine learning models.

5 Conclusion

This paper addressed improving the detection efficiency for the quick detection requirements of large-scale smart contracts. We only used extensible 12-dimensional features mining from bytecode and opcode. The low-dimensional features will speed up the detection time 25 to 650 times without influencing the model performance (*accuracy* etc.). We can also extend these features to support more vulnerability detection and security analysis in the future. Compared with the existing thousand-dimensional feature space, the features improve the detection efficiency and extend the detection range. The evaluation based on seven state-of-the-art learning-based methods has shown the effectiveness of *Block-gram* features and can significantly improve detection efficiency.

Acknowledgment. This work is partially supported by the National Natural Science Foundation (62272248), the Open Project Fund of State Key Laboratory of Computer Architecture, Institute of Computing Technology, Chinese Academy of Sciences (CARCH201905, CARCHA202108), the Natural Science Foundation of Tianjin (20JCZDJC00610) and Sponsored by Zhejiang Lab (2021KF0AB04).

References

1. Wang, S., Ouyang, L., Yuan, Y., Ni, X., Han, X., Wang, F.-Y.: Blockchain-enabled smart contracts: architecture, applications, and future trends. IEEE Trans. Syst. Man Cybern. Syst. **49**(11), 2266–2277 (2019)
2. Ma, F., et al.: EVM*: from offline detection to online reinforcement for ethereum virtual machine. In: 2019 IEEE 26th International Conference on Software Analysis, Evolution and Reengineering (SANER), pp. 554–558. IEEE (2019)
3. Sai, A.R., Holmes, C., Buckley, J., Gear, A.L.: Inheritance software metrics on smart contracts. In: Proceedings of the 28th International Conference on Program Comprehension, pp. 381–385 (2020)
4. Durieux, T., Ferreira, J.F., Abreu, R., Cruz, P.: Empirical review of automated analysis tools on 47,587 ethereum smart contracts. In: Proceedings of the ACM/IEEE 42nd International Conference on Software Engineering, pp. 530–541 (2020)

5. Wang, W., Song, J., Xu, G., Li, Y., Wang, H., Su, C.: ContractWard: automated vulnerability detection models for ethereum smart contracts. IEEE Trans. Netw. Sci. Eng. **8**(2), 1133–1144 (2020)
6. Hildenbrandt, E., et al.: KEVM: a complete formal semantics of the ethereum virtual machine. In: 2018 IEEE 31st Computer Security Foundations Symposium (CSF), pp. 204–217. IEEE (2018)
7. Krupp, J., Rossow, C.: teEther: gnawing at ethereum to automatically exploit smart contracts. In: 27th USENIX Security Symposium (USENIX Security 2018), pp. 1317–1333 (2018)
8. Luu, L., Chu, D.-H., Olickel, H., Saxena, P., Hobor, A.: Making smart contracts smarter. In: Proceedings of the 2016 ACM SIGSAC Conference on Computer and Communications Security, pp. 254–269 (2016)
9. Jiang, B., Liu, Y., Chan, W.K.: ContractFuzzer: fuzzing smart contracts for vulnerability detection. In: 2018 33rd IEEE/ACM International Conference on Automated Software Engineering (ASE), pp. 259–269. IEEE (2018)
10. Albert, E., Gordillo, P., Livshits, B., Rubio, A., Sergey, I.: EthIR: a framework for high-level analysis of ethereum bytecode. In: Lahiri, S.K., Wang, C. (eds.) ATVA 2018. LNCS, vol. 11138, pp. 513–520. Springer, Cham (2018). https://doi.org/10.1007/978-3-030-01090-4_30
11. Gai, K., Qiu, M.: Reinforcement learning-based content-centric services in mobile sensing. IEEE Netw. **32**(4), 34–39 (2018)
12. Mi, F., Wang, Z., Zhao, C., Guo, J., Ahmed, F., Khan, L.: VSCL: automating vulnerability detection in smart contracts with deep learning. In: 2021 IEEE International Conference on Blockchain and Cryptocurrency (ICBC), pp. 1–9. IEEE (2021)
13. Zhuang, Y., Liu, Z., Qian, P., Liu, Q., Wang, X., He, Q.: Smart contract vulnerability detection using graph neural network. In: IJCAI, pp. 3283–3290 (2020)
14. Zeng, Q., et al.: EtherGIS: a vulnerability detection framework for ethereum smart contracts based on graph learning features. In: IEEE 46th Annual Computers, Software, and Applications Conference (COMPSAC), pp. 1742–1749. IEEE (2022)
15. Li, T., et al.: SmartVM: a smart contract virtual machine for fast on-chain DNN computations. IEEE Trans. Parallel Distrib. Syst. **33**(12), 4100–4116 (2022)
16. Qiu, H., Qiu, M., Memmi, G., Ming, Z., Liu, M.: A dynamic scalable blockchain based communication architecture for IoT. In: Qiu, M. (ed.) SmartBlock 2018. LNCS, vol. 11373, pp. 159–166. Springer, Cham (2018). https://doi.org/10.1007/978-3-030-05764-0_17
17. Gai, K., Wu, Y., Zhu, L., Zhang, Z., Qiu, M.: Differential privacy-based blockchain for industrial Internet-of-Things. IEEE Trans. Ind. Inf. **16**(6), 4156–4165 (2019)
18. Tian, Z., Li, M., Qiu, M., Sun, Y., Su, S.: Block-DEF: a secure digital evidence framework using blockchain. Inf. Sci. **491**, 151–165 (2019)
19. Contro, F., Crosara, M., Ceccato, M., Dalla Preda, M.: EtherSolve: computing an accurate control-flow graph from ethereum bytecode. In: 2021 IEEE/ACM 29th International Conference on Program Comprehension (ICPC), pp. 127–137. IEEE (2021)

Enhanced 4A Identity Authentication Center Based on Super SIM Technology

Renjie Niu[1], Zixiao Jia[2], Yizhong Liu[2](✉), Jianhong Lin[1], Xiaoli Huang[1], and Min Sun[1]

[1] China Mobile Information Technology Co., Ltd., Shenzhen, China
{niurenjie,linjianhong,huangxiaoliit,sunmin}@chinamobile.com
[2] School of Cyber Science and Technology, Beihang University, Beijing, China
{jiazixiao,liuyizhong}@buaa.edu.cn

Abstract. This paper explains how large-scale IT systems balance security and enabling under the pressure of supporting complex business models, constantly introducing new technologies, maintaining unified identity management, and meeting the business diversification of digital transformation. The paper introduces the necessity and basic idea of using China Mobile Super SIM technology to realize the enhanced Authentication, Authorization, Accounting, and Audit (4A) identity authentication center. We then describe the system's design principles, references, and architecture and detail several key processes and capabilities. Finally, the implementation and evolution of the enhanced identity authentication system are described.

Keywords: Super SIM card · Identity authentication · 4A · Treasury mode · Operation review

1 Introduction

With the continuous deepening of the development of the digital economy around the world, thanks to the rapid progress in cloud system [1,2] and computer capability [3,4], various companies have accelerated the pace of digital transformation [5,6]. In this context, it is a considerable challenge for all communication industry businesses and security personnel to manage the identity, authentication, authorization, and access control of various operation and maintenance activities so as to ensure the security of crucial systems and sensitive information [7]. On November 19, 2020, China Mobile released the Super SIM card at the Global Partner Conference. It is an integrated security product based on the SIM card security chip and uses different functions to achieve different levels of authentication [8,9]. It contains an encryption chip and NFC function, which can be used as a meal card, access control card, transportation card, and key to a car offline, or can perform financial security authentication, 5G electronic signature [10], and large-value transfers online.

M. Qiu et al. (Eds.): SmartCom 2022, LNCS 13828, pp. 558–569, 2023.
https://doi.org/10.1007/978-3-031-28124-2_53

This paper combines the super SIM authentication technology with the 4A system, adopts the concept of zero trust and multi-factor authentication, takes the 4A system as the core, and uses the super SIM to achieve the security capability of EAL5+, so as to build a convenient authentication process that supports hundreds of millions of times a day. If an identity authentication center is to be implemented based on super SIM technology, the following basic requirements must be met:

Independent Computing Requirements. Use the independent and removable key computing entity to achieve chip-level security assurance and achieve the self-security capability of EAL5+; Implement centralized remote management of the chip to avoid the risk of man-in-the-middle attacks and data replay attacks, and increase the complexity and time of cracking.

No Perception/Less Perception Requirement. Perform multiple automatic authentications in scenarios such as user login, resource login, multi-factor authentication, secondary authentication, and vault mode, and combine zero-trust technology to achieve a non-perceptual, non-disturbing automated authentication process [11], which improves security and reduces users' number of passive authentications;

Centralized Management Requirements. Realize the management of subject, object, authorization relationship, trust degree, and operation review; Allow different business operation and maintenance modes to adopt different authentication strengths and authorization relationships and simultaneously perform remote manual intervention control on abnormal subjects, access relationships, devices, and SIM cards.

In summary, the contributions of this paper include the following points:

- We put forward some design principles based on the super SIM technology, thus completing the design of the enhanced 4A identity authentication center so that it can well meet the three basic requirements mentioned above.
- Based on the design architecture of the enhanced 4A identity authentication center, we have realized the four capabilities of the identity authentication center, namely, no perception/low perception authentication capability, multi-factor authentication capability, secondary authentication capability, and operation review (vault mode) capability.
- Considered and designed the security of the enhanced 4A certification center, improved its security performance in the process of authentication and transmission, and enhanced the robustness of the certification center so that it can automatically detect and process abnormal behaviors in operation.

The remainder of the work is organized as follows. Section 2 introduces the design principles of the enhanced 4A identity authentication center and its overall process. Section 3 analyzes the unique capabilities of the enhanced 4A identity authentication center. The security of the system and its ability to handle abnormal behavior are discussed in Sect. 4. Finally, concluded remarks of the conducted work are discussed in Sect. 5.

2 Design of Enhanced 4A Identity Authentication Center

By designing an identity authentication center that combines super SIM and 4A, system security can be further improved, user perception can be optimized, information systems and physical systems can be connected, and refined management can be achieved. The core design principles are as follows:

The principle of dual-core certification. Two cores are formed by super SIM and 4A, combined through a secure and open channel. The core of the super SIM is the authentication capability of the EAL5+ level to realize the security root on the user side. The mobile phone and the super SIM integrate biometrics, geographical location, behavioral characteristics, and zero-trust capabilities of the mobile terminal to achieve financial-level authentication. It has become a universal ID card and pass card for users in the physical and information worlds. As the core capability of identity management and control of various business systems, the 4A system has gradually evolved into an identity middle platform to integrate current new, and future advanced identity-related capabilities.

User-perceived optimization principle. Achieving "safety" and "convenience" has always been challenging. It is necessary to use intelligent analysis, environmental perception, and chip-level security technologies to judge the security certification status through technical means. Only when certification is required are users required to authenticate, so that to improve security capabilities. At the same time, it optimizes user perception and improves refined identity management and control, reducing the "repetitive" and "tedious" experience of users in the authentication process.

The principle of advanced security management and control. In the face of critical systems and sensitive data operations, security controls must be improved to achieve real-time authentication and real-time audit of people, data, and behavior.

The principle of minimum authorization and convenient permission application. Due to the complex business scale of operators and the complex roles and permissions of operation and maintenance and operation personnel, in order to ensure access to the gateway system and sensitive data, it is necessary to adopt an on-demand, and per-event authorization mode, which is intuitively reflected in that users can see the function entry. However, it is necessary to apply for permission before use. The approval action is automatically performed by the work order or authorized by the superior. The user obtains the temporary permission and withdraws it immediately after the work is completed or after the timeout. In order to simplify the authorization work of the superior supervisor, it is necessary to combine the super SIM computing power of the supervisor to complete the authorization through a simple click operation on the mobile phone.

2.1 Identity Authentication Logic Combining Super SIM and 4A

In order to realize the integration of super SIM and 4A, it is necessary to incorporate the "5G super SIM card authentication open platform" into the identity security infrastructure connected with 4A to implement the overall integrated identity authentication logic. The enhanced 4A identity authentication center logic based on super SIM technology is shown in Fig. 1.

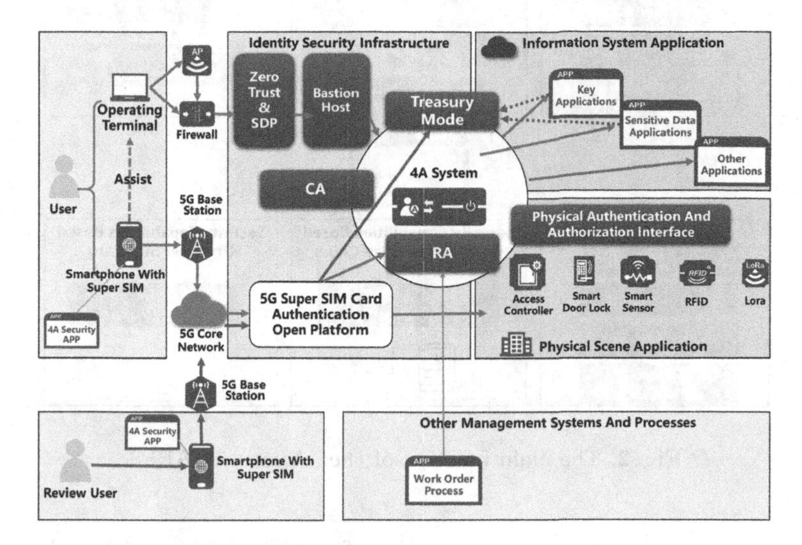

Fig. 1. Logic diagram of enhanced 4A identity authentication center based on super SIM technology

2.2 The Main Functions of 4A Security APP

Due to super SIM technology's emergence, smartphones' security has been dramatically improved, and mobile phone authentication, operation, and approval have been greatly improved. It is necessary to redesign a 4A security APP based on super SIM cards and smartphones. Its main functional structure is shown in Fig. 2.

4A security APP is to transform the 4A capabilities from the traditional PC field to the mobile Internet field. It utilizes the security features of current smartphones and super SIMs. Based on the characteristics of mobile phones being carried around and closely related to natural persons, the APP program is used to convert the identity of natural persons into the information world. Its main features are divided into two fields, one is the functional field, including security authentication function, operation review function, and zero trust function; the other is its own security field, including software-based security capabilities, security capabilities based on mobile phone chips, and security capabilities based on super SIM card.

2.3 Certification Process of Enhanced 4A Authentication Center

Authentication Process of Enterprise-Level Application System. The main principle of enterprise-level application system authentication is based on the 4A system to achieve "one point authentication, everywhere authentication;

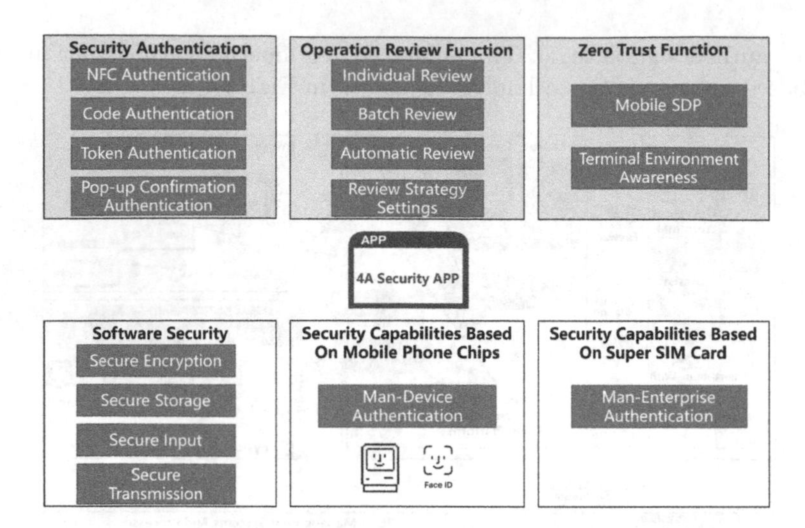

Fig. 2. The main function of the 4A security APP

one point exit, everywhere exit", that is, after the user completes the "man-device" verification and "man-identity authentication" in combination with the Super SIM and the 4A system, the authentication of the system, operation and data within its authority is also completed at the same time. It is only necessary to realize the simplified single sign-on process from the 4A system to other application systems. Once a login terminal is withdrawn from the 4A system, other login rights on the login terminal will also be withdrawn.

As shown in Fig. 3, the user authentication process in the information system is as follows:

1. The user first logs in to the 4A portal and sends an authentication request to the 4A system;
2. 4A system sends an authentication challenge to the login portal, such as a username and password or a QR code that needs to be scanned with a mobile phone;
3. In the process of combining the super SIM, the user needs to use the 4A login APP on the mobile phone to scan the code;
4. While using the mobile phone, user must use fingerprint, face, encryption, and other technologies to pass the "man-device" authentication. Most of these authentication processes are completed by mobile phone software or exclusive security chips (such as Apple's T2 chip);
5. After the "man-device" authentication is completed, the "man-enterprise identity" authentication needs to be completed again. At this time, it needs to be completed by the authentication computing power of the super SIM;
6. The 5G-based super SIM security application will be connected to 4A through the base station, core network, and super SIM card authentication open platform to complete the authentication of "man-enterprise identity";

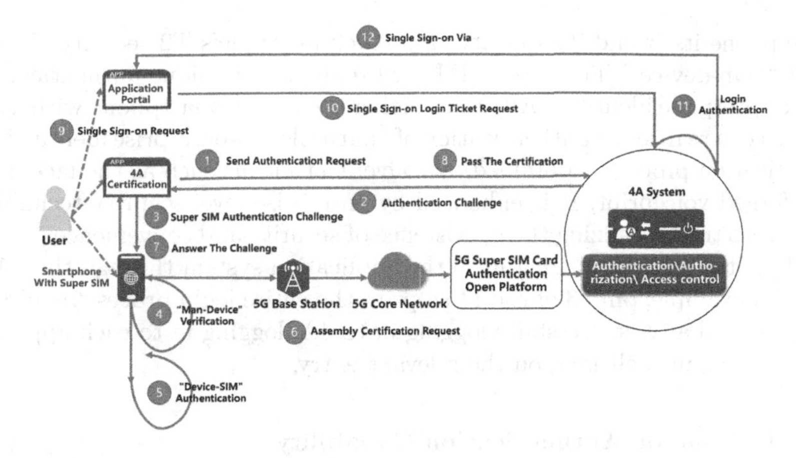

Fig. 3. Authentication application of super SIM and 4A technology in information system

7. After the certification is completed, the mobile phone super SIM application will complete the challenge response required for 4A certification, manifesting as passing the code on the 4A certification portal.
8. The 4A system judges the authentication process to be passed, and a complete set of enterprise identity authentication from natural persons, mobile devices, and PC terminals has been completed at this time;
9. In the second stage, the user requests to log in to a specific application at this time, which can be jumped through 4A or directly logged in to the application;
10. The modified applications will use the 4A single sign-on function to initiate a ticket login verification request to 4A;
11. 4 System A will re-verify factors such as people, PC terminals, mobile devices, enterprise applications, and related policies to determine whether the single sign-on process is released;
12. After the application login determination is completed, 4A will notify the application to release the user, thus completing the authentication login of the user to the application system.

In the above process, the user's recognition perception is the process of logging in to the 4A portal. In the subsequent use process, unless the security policy, timeout, user offline, secondary authentication, vault mode, and other processes are triggered, the user can use it within the adequate time. Log in to use authorized applications.

3 Enhanced Capability of 4A Certification Center

3.1 No Perception/Low Perception Authentication Capability

In the identity authentication logic, a smartphone with a super SIM card acts as a convenient and credible man-machine trust consistency system, in which the mobile phone itself and its security chip (such as Apple's T2 security chip) can realize "man-device." The super SIM card realizes the unique authentication of "device-enterprise identity." After combining the two, a smartphone with a super SIM card can realize the authentication of "natural man-enterprise identity." This authentication process is obtained by convenient means such as the face, fingerprint, forged voiceprint. It is enhanced by enterprise-level security technologies such as zero trust, realizing the coexistence of security and convenience.

Taking the steps of logging in to the application system through the 4A system as an example, only 3 of the 11 steps in the entire login process require user responses, and after successfully logging in to 4A, logging in to each application again only requires clicking on the relevant entry.

3.2 Multi-factor Authentication Capability

Multi-factor authentication is a method of computer access control. Users must pass two or more authentication mechanisms before being authorized to use computer resources. For example, the user wants to enter a PIN, insert a bank card, and finally, through fingerprint comparison, these three authentication methods can obtain authorization. This authentication method can improve security.

Smartphones with a Super SIM card are equipped with a security token function, which can form two-factor authentication with a username and password in certain situations. This specific situation is based on the evaluation of the zero-trust mechanism. When the security score of the user's device, network, behavior, and other factors is too low, the two-factor authentication will be turned on. The security score will be restored and returned under single-factor authentication if the two-factor authentication is passed.

3.3 Secondary Authentication Capability

The secondary authentication capability will be activated when the user's operation behavior triggers the policy. These behaviors include dangerous operations, access to sensitive data, behavior deviations from the baseline, and changes in the security status of the registrant's terminal. The action of the secondary authentication is to change the implicit authentication judgment to the explicit one and require the user to re-enter the user name, password, or super SIM card token code.

After the secondary authentication is passed, there will be no secondary authentication until the following secondary authentication action is triggered.

3.4 Operation Review (Vault Mode) Capability

Vault mode will be activated when user actions trigger policies. These actions include pre-determined actions involving dangerous operations and accessing sensitive data. The action of operation review is to suspend the operation process first and introduce a reviewer into the review process. The reviewer is usually the superior leader of the user or a post with management responsibility in the relevant operation field.

The operation reviewer will use a smartphone with a super SIM card to complete most of the operation review actions. The reviewer first needs to log in to the 4A security APP on the mobile phone and complete the "man-device" authentication and "man-enterprise identity." Due to the high frequency of participation in operation review, the 4A security APP will support interface click and batch confirmation methods, allowing reviewers to submit review results to ensure safety quickly.

4 The Security of the Enhanced 4A Certification Center

A smartphone with a super SIM card is a security-increasing computer with independent computing and storage capabilities. Combined with some smartphones' existing independent security chips, the architecture of dual independent security computers is formed. The independent secure computer architecture formed by the super SIM card is shown in Fig. 4.

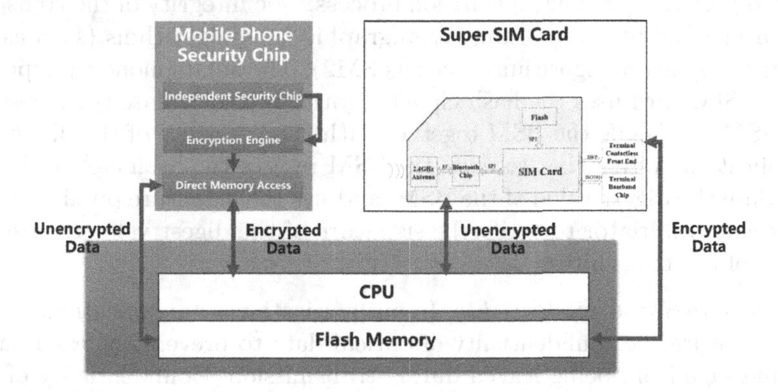

Fig. 4. Independent secure computer architecture formed by super SIM cards

4.1 Security Enhancement During Authentication and Transmission

Because the super SIM card has storage and computing capabilities, it supports domestic SM2, SM3, SM4, and SM9 security algorithms and supports RSA 2048, SHA 256, TDES, AES, ECC (Secp256r1), X 25519, ED 25519 algorithms. It can realize the encryption ability of the root key without the SIM so that the mobile phone carrying the SIM card, related applications, and network links cannot unilaterally steal or tamper with the transmission and storage containers. The transmitted data can be decrypted only through the internal program of the super SIM card and the super SIM authentication center. Code protection is enabled on the super SIM card, the long-term key K stored in the SIM card can only be accessed by the SIM card itself and the card-issuing operator, while most smartphones are used in network access and data transmission. The keys are all derived based on the key K, which shows that the communication process of the smartphone has exceptionally high security.

Authenticity of Data Transmission: In the communication of the authentication request, it is indispensable to ensure the authenticity of the identity of the person transmitting the data to prevent the user authentication data from being counterfeited during the transmission process. A smartphone with a Super SIM card first digitally signs the basic security messages (BSM) it advertises using the private key corresponding to the pseudonymous certificate. Then the smartphone sends that signed message along with the pseudonymous certificate or the digest value of the pseudonymous certificate. The application system that receives the message first uses the certificate of the certificate authority (CA) that issued the pseudonym certificate to verify whether the signed certificate in the message is valid and then uses the public key in the pseudonym certificate to verify whether the signature in the signed message is correct. Finally, the receiving initiator uses the verified BSM content to determine the sent data content.

Data Transmission Integrity: In authentication communication, it is necessary to ensure the integrity of the transmitted data and prevent the message from being tampered with during the transmission process. The integrity of the transmitted data can be guaranteed by using cryptographic hash algorithms (such as SM3) and digital signature algorithms (such as SM2). The smartphone equipped with the super SIM card uses the hash cipher algorithm to calculate the digest value of the BSM and sends the BSM together with the signature of the digest value; the application system that receives the BSM re-uses the hash cipher algorithm to calculate the digest value of the BSM, and use the signature public key of the authentication initiator to verify the signature of the digest value to ensure the integrity of the transmitted data.

Data Transmission Confidentiality: In authentication communication, it is necessary to ensure the confidentiality of critical data to prevent the essential data in the message from being leaked during transmission. Confidentiality of transmitted data can be guaranteed by using a symmetric cryptographic algorithm such as SM4. The authentication initiator, 4A system, and application system can obtain the symmetric encryption key through the key distribution platform or preset methods, use the symmetric key to encrypt the important information in the BSM, and then the main certificate initiator sends the BSM out. The application or 4A system that receives the message uses the symmetric key to decrypt the important information encrypted in the BSM.

Anti-replay: Multiple parties in the authentication process can use timestamps or cache queues to prevent anti-replay attacks. The authentication initiator, application, and 4A should all judge whether the timestamp in the received message has expired. Both communication parties need accurate time synchronization and guarantee the time source from a legitimate authority, such as GNSS (Global Navigation Satellite System) time.

Behavioral Non-repudiation: The authentication initiator uses the private key to sign the BSM digitally, and the applications and 4A systems that receive the data use the sender's public key to verify the signature data, preventing the sender from denying its behavior of sending the BSM.

Discovery and Disposal of Abnormal Behavior: The enhanced 4A identity authentication capability based on super SIM technology also has security risks, mainly from the following directions: 1) On the mobile terminal side, the user's mobile phone and the super SIM card may be lost, fraudulently used, or replaced with the super SIM card. 2) On the communication network side, there may be a network configuration error, which may cause it to fail to connect with the right SIM card authentication center or fail to connect with the 4A system. 3) On the application side, there may be security issues in the 4A system itself, leading to risks such as leakage, tampering, and unavailability of user identity data.

Therefore, the project must have various abnormal behavior detection mechanisms and handling plans during the operation process. For security issues on the terminal side, it is necessary to conduct data analysis on the behavior of terminals, SIM cards, and users and analyze possible physical intrusion, network hijacking, and user impersonation. If the secondary authentication fails, the device and SIM card will be temporarily locked and confirmed by the security operator through dialogue with the direct user. When addressing network-side security issues, the 5G network evaluation process should be combined to minimize network infrastructure's hidden dangers in advance. When addressing the security risks on the application side, it is indispensable to improve its security and simultaneously design a bypass plan for the system's unavailable risks.

The super SIM could be used in 5G, which is a research hotspot. 5G could be combined with the emerging blockchain technology [12] to realize better performance [13], scalability [14], and higher lever of security [15,16]. Blockchain consensus [17,18] is the most important fundamental, which decides the the system performance and security [19]. Sharding blockchain [20,21] can be used in 5G authentication protocol, combined with the super SIM as a trusted hardware, to process large-scale user requests. Moreover, the 5G and blockchain technology can be used to process the cloud storage, access control problems, etc. [22–24].

5 Conclusion

The enhanced 4A identity authentication center based on Super SIM technology was an important innovation that combined the 4A system covering the entire network with Super SIM technology. The system provided a new method of identity control for the problems of expanding business scale and complex team composition faced by enterprises in digital transformation. In the future, when identity control needs are more diversified, the identity control system must achieve higher security, smarter perception, and finer management and control. At the same time, the system will also need to balance the contradiction between security and convenience, and unify the contradiction between identity and business complexity. The main evolution direction of this project is to create a safe, convenient, highly compatible identity authentication system. This system will use the capabilities of Super SIM fully and current smartphones, making mobile phones a powerful identification tool for natural persons, thereby creating a modern identity control solution.

Acknowledgment. This paper is supported by the National Natural Science Foundation of China (U21B2021, 62202027, 61972018, 61932014), Yunnan Key Laboratory of Blockchain Application Technology (202105AG070005-YNB202206).

References

1. Qiu, M., Chen, Z., et al.: Energy-aware data allocation with hybrid memory for mobile cloud systems. IEEE Syst. J. **11**(2), 813–822 (2014)
2. Gai, K., Qiu, M., Elnagdy, S.: A novel secure big data cyber incident analytics framework for cloud-based cybersecurity insurance. In: IEEE BigDataSecurity, pp. 171–176 (2016)
3. Qiu, M., Xue, C., Shao, Z., Zhuge, Q., Liu, M., Sha, E.H.-M.: Efficent algorithm of energy minimization for heterogeneous wireless sensor network. In: Sha, E., Han, S.-K., Xu, C.-Z., Kim, M.-H., Yang, L.T., Xiao, B. (eds.) EUC 2006. LNCS, vol. 4096, pp. 25–34. Springer, Heidelberg (2006). https://doi.org/10.1007/11802167_5
4. Niu, J., Gao, Y., et al.: Selecting proper wireless network interfaces for user experience enhancement with guaranteed probability. JPDC **72**(12), 1565–1575 (2012)
5. Gai, K., Zhang, Y., et al.: Blockchain-enabled service optimizations in supply chain digital twin. IEEE TSC, pp. 1–12 (2022)
6. Qiu, M., Qiu, H., et al.: Secure data sharing through untrusted clouds with blockchain-enabled key management. In: 3rd SmartBlock Conference, pp. 11–16 (2020)
7. Huang, S., Guo, B., Liu, Y.: 5G-oriented optical underlay network slicing technology and challenges. IEEE Commun. Mag. **58**(2), 13–19 (2020)
8. Li, Y., Gai, K., et al.: Intercrossed access controls for secure financial services on multimedia big data in cloud systems. ACM TMCCA **12**(4), 1–18 (2016)
9. Gao, X., Qiu, M.: Energy-based learning for preventing backdoor attack. In: Memmi, G., Yang, B., Kong, L., Zhang, T., Qiu, M. (eds.) KSEM 2022. LNAI, vol. 13370, pp. 706–721. Springer, Cham (2022). https://doi.org/10.1007/978-3-031-10989-8_56
10. Vinel, A.V., Breu, J., Luan, T.H., Hu, H.: Emerging technology for 5G-enabled vehicular networks. IEEE Wirel. Commun. **24**(6), 12 (2017)
11. Wang, T., Zhang, Y., et al.: An effective edge-intelligent service placement technology for 5G-and-beyond industrial IoT. IEEE TII **18**(6), 4148–4157 (2022)
12. Liu, Y., Liu, J., Zhang, Z., Yu, H.: A fair selection protocol for committee-based permissionless blockchains. Comput. Secur. **91**, 101718 (2020)
13. Liu, Y., Liu, J., Yin, J., Li, G., Yu, H., Wu, Q.: Cross-shard transaction processing in sharding blockchains. In: Qiu, M. (ed.) ICA3PP 2020. LNCS, vol. 12454, pp. 324–339. Springer, Cham (2020). https://doi.org/10.1007/978-3-030-60248-2_22
14. Liu, Y., Liu, J., et al.: SSHC: a secure and scalable hybrid consensus protocol for sharding blockchains with a formal security framework. IEEE TDSC **19**(3), 2070–2088 (2022)
15. Liu, Y., Liu, J., et al.: A secure shard reconfiguration protocol for sharding blockchains without a randomness. In: IEEE TrustCom, pp. 1012–1019 (2020)
16. Liu, Y., Liu, J., Hei, Y., Xia, Yu., Wu, Q.: A secure cross-shard view-change protocol for sharding blockchains. In: Baek, J., Ruj, S. (eds.) ACISP 2021. LNCS, vol. 13083, pp. 372–390. Springer, Cham (2021). https://doi.org/10.1007/978-3-030-90567-5_19
17. Liu, Y., Liu, J., Zhang, Z., Xu, T., Yu, H.: Overview on consensus mechanism of blockchain technology. J. Cryptol. Res. **6**(4), 395–432 (2019)

18. Hu, B., Zhang, Z., et al.: A comprehensive survey on smart contract construction and execution: paradigms, tools, and systems. Patterns **2**(2), 100179 (2021)
19. Liu, Y., Xia, Y., Liu, J., Hei, Y.: A secure and decentralized reconfiguration protocol for sharding blockchains. In: IEEE HPSC 2021, pp. 111–116. IEEE (2021)
20. Liu, Y., Liu, J., et al.: Building blocks of sharding blockchain systems: concepts, approaches, and open problems. Comput. Sci. Rev. **46**, 100513 (2022)
21. Kokoris-Kogias, E., Jovanovic, P., et al.: OmniLedger: a secure, scale-out, decentralized ledger via sharding. In: IEEE SP 2018, pp. 583–598 (2018)
22. Hei, Y., Liu, J., et al.: Making MA-ABE fully accountable: a blockchain-based approach for secure digital right management. Comput. Netw. **191**, 108029 (2021)
23. Hei, Y., Li, D., Zhang, C., Liu, J., Liu, Y., Wu, Q.: Practical AgentChain: a compatible cross-chain exchange system. Future Gener. Comput. Syst. **130**, 207–218 (2022)
24. Hei, Y., Liu, Y., Li, D., Liu, J., Wu, Q.: Themis: an accountable blockchain-based P2P cloud storage scheme. Peer-to-Peer Netw. Appl. **14**(1), 225–239 (2021)

Verifiable, Fair and Privacy-Preserving Outsourced Computation Based on Blockchain and PUF

Jiayi Li[1], Xinsheng Lei[2](✉), Jieyu Su[3], Hui Zhao[4], Zhenyu Guan[3], and Dawei Li[3]

[1] Datang Microelectronics Technology Co., LTD., Beijing, China
[2] China Information and Communication Technology Group Co., LTD., Beijing, China
leixsh@cict.com
[3] School of Cyber Science and Technology, Beihang University, Beijing 100191, China
{sujieyu,guanzhenyu,lidawei}@buaa.edu.cn
[4] Henan University, Zhengzhou, Henan, China
zhh@henu.edu.cn

Abstract. With the increasing maturity of *Industrial Internet of Things* (IIoT) technology, resource-constrained devices are widely applied to segments of the factory, which puts significant pressure on their computational capacity. In order to address this issue securely, we propose a verifiable and privacy-preserving outsourced computation system that employs SRAM PUF to safeguard the hardware security of devices and blockchain to achieve public verifiability and data privacy, thereby greatly guaranteeing the security of outsourced computation in the IIoT environment. Additionally, we protect the rights of calculators using a mechanism that identifies malicious calculators. Finally, compared with other existing schemes, the experimental results demonstrate that our scheme provides more efficient and secure outsourced computation services for IIoT devices.

Keywords: IIoT · Blockchain · Outsourced computation · PUF · Verifiability · Privacy

1 Introduction

The *Industrial Internet of Things* (IIoT) is the extension of the *Internet of Things* (IoT) in industry [15,17,19]. At the advent of the fourth industrial revolution, applying IIoT technology will become comprehensive, eventually realizing the transformation of manufacturing processes. In the IIoT environment, massive data [3,8,13] makes it impracticable for resource-constrained devices [20–22] to process it locally. This practical necessity is effectively addressed by the emergence of outsourced computation.

As shown in Fig. 1, Edge outsourced computation is a new computing paradigm [32] of providing real-time computational services, which originates

M. Qiu et al. (Eds.): SmartCom 2022, LNCS 13828, pp. 570–580, 2023.
https://doi.org/10.1007/978-3-031-28124-2_54

from outsourcing customers' computing overhead to the cloud [7]. The advancement of edge outsourced computation benefits several applications in the IIoT environment, including data processing in supply chain optimization, smart grids and intelligent industrial parks. However, due to the distributed network environment, security and accuracy of outsourced computation cannot be protected [6]. Further, hardware security of IIoT devices is not fully guaranteed, which gives more weight to device authentication. PUF is a novel hardware primitive utilized in authentication, whose unclonability and low-power dissipation meet the needs of authentication in the IIoT system.

To address the aforementioned problems, we present a blockchain-based outsourced computation system with PUF and implement outsourced linear regression computation. *Linear regression* (LR) is a simple but efficient machine learning algorithm utilized in the IIoT environment [2,23]. A standard model of linear regression is: $y = X\beta$, where y and β are $m \times 1$ and $n \times 1$ vectors in $\mathbb{R}^m$ respectively and X is an $m \times n$ matrix in $\mathbb{R}^{m\times n}$, for $m > n$ [1]. We apply the least squared error method to find an approximation of β using the formula $\beta = (X^TX)^{-1}X^Ty$ [1]. Then, in order to verify the results, we introduce blockchain [4,5] into our system, which is a decentralized ledger using cryptography to host applications, store data, and share information securely [18,25].

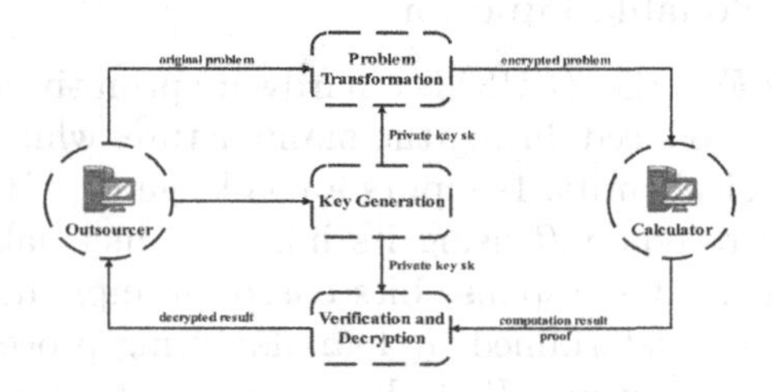

Fig. 1. The model of edge outsourced computation

The main contributions of our paper can be concluded as follows: (1) We propose a verifiable and privacy-preserving outsourced computation system using PUF and blockchain, which achieves data privacy and public verifiability and allows IIoT devices to access computing resources securely. (2) Our proposed scheme guarantees the hardware security of IIoT devices with low security levels utilizing SRAM-PUF-based authentication scheme, which only allows PUF-authenticated outsourcers to initiate outsourced tasks. (3) We introduce a mechanism to identify malicious calculators, allowing the selection of calculators on the blockchain. The experimental results indicate that our scheme not only is in milliseconds level with high verification success rate, which is more efficient than other related schemes, but also achieves privacy, verifiability, and fairness.

In the rest of this paper, Sect. 2 reviews the related work. Fundamental knowledge is interpreted in Sect. 3. Section 4 presents the system model and the threat model. We detail our scheme in Sect. 5. We implement our proposal and analyze the experimental results in Sect. 6. Section 7 concludes our work.

2 Related Work

Researches in outsourced computation cover diverse directions, the majority of which focus on data privacy and verifiability. In the study of data privacy, Liu et al. [16] presented a privacy-preserving outsourced computing scheme that employs Intel SGX as a trusted execution environment to guarantee data privacy. Compared with them, Zhang et al. [30] avoided introducing a trusted third party and proposed an IoT RSA outsourced computation scheme based on the Chinese residual theorem that simultaneously assures the privacy and security of the scheme. In the study of verifiability, Li et al. [12] devised a secure distributed outsourced computation system for modulo power operations, which combines logical partitioning and segmentation methods to verify the results. However, the verification can only be performed by users. Further, Li et al. [14] and Zhang et al. [29] achieved public verifiability. Li et al. [14] validated the results based on the Elgamal algorithm and protected data privacy. Zhang et al. [29] created a safe, equitable, and verifiable method to outsource linear regression computation to unreliable cloud, which enables public verifiability.

3 Preliminaries

3.1 Physical Unclonable Function

Physical Unclonable Function (PUF) is a hardware primitive employing intrinsic random deviations introduced during the manufacture, which provides a unique fingerprint for a physical entity. It inputs a challenge C and outputs an unpredictable and unique response R using its inherent unclonability [11]. Particularly, SRAM PUF generates response bits based on unpredictable initial values of SRAM cells that are determined by manufacturing process, which makes it convenient to be integrated into IIoT devices without additional changes [10]. Therefore, we employ SRAM PUF in the authentication phase of our system.

3.2 Sparse Matrix Masking

The core concept of *Sparse matrix masking* (SMM) technique is to mask sensitive data using a randomly selected sparse matrix, which can be utilized in matrix-related calculations to avoid expensive encryption operations and secure data privacy effectively [31]. In order to reduce computational consumption while protecting information related to zero elements, we introduce the definitions of χ–sparse matrix and adaptive adjustment function in [31], which adjusts placements of non-zero elements according to non-zero elements of the input matrix, thereby concealing zero-element-related information of the original matrix.

4 Problem Formulation

4.1 System Model

Five entities compose our system as illustrated in Fig. 2: task outsourcers, task calculators, certificate authority(CA), authentication committee, and blockchain.

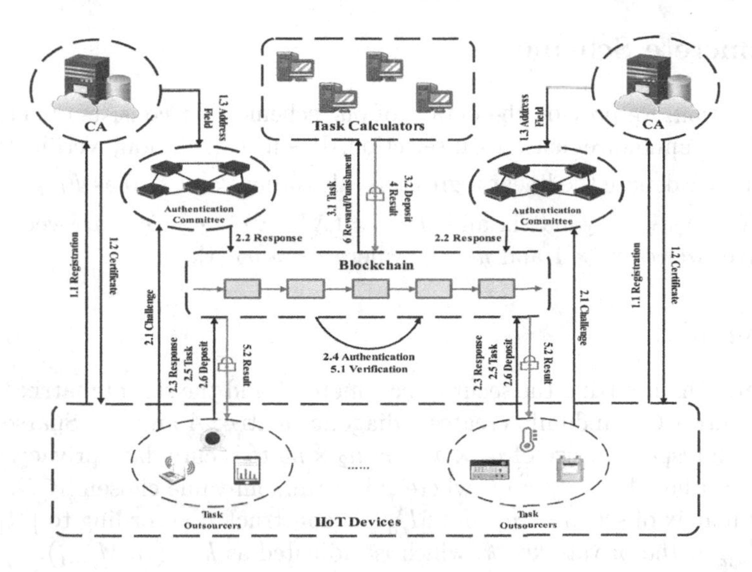

Fig. 2. System model

- Task Outsourcers and Task Calculators. The task outsourcers are resource-constrained devices initiating the outsourced tasks and the task calculators are resource-rich devices accomplishing the outsourced tasks.
- CA and Authentication Committee. CA is responsible for issuing certificates for task outsourcers and distributing the information of CRPs to nodes in the P2P network, which comprise the authentication committee. In our system, each participant in authentication committee holds a part of CRPs.
- Blockchain. Blockchain is a recorder of all transactions in the system that shows the outcomes of authentication and outsourced tasks.

4.2 Threat Model

We present our threat model in this part. The primary security threats originate from calculators and adversaries in the P2P network, therefore, we define identity unforgeability, privacy-preserving, and verifiable.

- Identity unforgeability. The adversary cannot forge the identity of low-security-level devices from the physical layer to steal their assets, and the forged authentication information of the adversary cannot be authenticated.
- Privacy-preserving. The outsourced computation should protect the privacy of the outsourced data that belongs to outsourcers, even if the openness and transparency of the blockchain-based system introduce the issue of privacy disclosure, which makes private data susceptible to eavesdropping.
- Verifiable. Under the premise of guaranteeing data privacy, the verification algorithm does not validate a calculator if the calculator submits inaccurate results in an attempt to obtain rewards.

5 Concrete Scheme

In this section, we provide the details of our scheme that consists of six phases: Setup, Task announcement, Task selection, Result calculation, Verification and Payment. We denote the linear regression calculation $\overrightarrow{y} = X \cdot \beta$ as $\Phi_{LR} = (X, \overrightarrow{y})$, where X is a $n_1 \times n_2$ matrix and $\overrightarrow{y}$, $\beta = (X^t \cdot X)^{-1} \cdot X^t \cdot \overrightarrow{y}$ are vectors with respective sizes of $n_1 \times 1$ and $n_2 \times 1$, where $n_1 > n_2$ [1].

5.1 Setup

KeyGen. On inputting the security parameter λ and the size of matrix (n_1, n_2), the outsourcer O_i randomly creates a diagonal matrix A and a χ-Sparse matrix M_{post}^{χ} with respective size of $n_1 \times n_1$ and $n_2 \times n_2$ to secure data privacy. Specifically, A satisfies $A^t \cdot A = \varphi^2 \cdot I$, where φ is a random value chosen by O_i, I is an identity matrix of size $n_1 \times n_1$ and M_{post}^{χ} is constructed according to [31]. Set A and M_{post}^{χ} as the private key k, which is indicated as $k = (A, M_{post}^{\chi})$.

ProbGen. On inputting the private key k, O_i produces an encrypted LR problem $\Phi'_{LR} = (X', \overrightarrow{y}')$. Firstly, O_i executes the function $M_{adj-post}^{\chi} \leftarrow adj(X, M_{post}^{\chi})$ in [31] to obtain matrix $M_{adj-post}^{\chi}$. Then, O_i calculates $X' = A \cdot X \cdot M_{adj-post}^{\chi}$ and $\overrightarrow{y}' = A \cdot \overrightarrow{y}$, which are outputted as encrypted inputs of the LR problem.

Registration. This phase is performed jointly by CA, authentication committee and outsourcer which is embedded in a SRAM PUF. Firstly, O_i sends its identity to CA, who issues a certificate to O_i in response. Then CA randomly assigns the memory address fields of SRAM PUF to nodes in authentication committee [10].

5.2 Task Announcement

Authentication. According to [10], first, O_i randomly chooses CRPs on the basis of the distribution of memory address fields and submits the challenges to authentication committee. Given the challenges, each node delivers the hash value of the response $H(R_i)$ to the smart contract based on its possession of the partial CRPs. After powering on the SRAM PUF, O_i concatenates the responses $R_1', R_2', \ldots, R_n'$ into a string and signs it as $\sigma = Sig_{sk}(H(R_1' \ldots R_n', r))$, where r is randomly chosen. Subsequently, O_i calculates the value of $R_{e_1}, R_{e_2}, \ldots, R_{e_n}$, where $R_{e_i} = g^{H(R_i')}(1 \leq i \leq n)$. Then, O_i sends σ and R_{e_i} to the smart contract, which checks the validity of the signature and the equation $g^{H(R_i)} \stackrel{?}{=} R_{e_i}(1 \leq i \leq n)$. The authentication of O_i will be successful if the equation and the signature are validated, otherwise, it fails.

Initialization. On the premise that authentication is successful, O_i initiates the outsourced task $T_i = (ID, input(X^{'}, \overrightarrow{y}^{'}), \Phi^{'}_{LR} = (X^{'}, \overrightarrow{y}^{'}), reward, deposit, Time)$. Six parameters characterize T_i: ID denotes the task identifier; $input(X^{'}, \overrightarrow{y}^{'})$ is the input data; $\Phi^{'}{}_{LR} = (X^{'}, \overrightarrow{y}^{'})$ represents the description of the problem; *reward* indicates the task reward; *deposit* denotes the deposit required to be submitted; $Time$ is the deadline for the task. After initialization, O_i submits T_i to the smart contract.

5.3 Task Selection

In this phase, calculators select public tasks on the blockchain according to their computational resources by signing tasks and submitting the deposit. The selections will be confirmed if calculators pass the related validation.

5.4 Result Calculation

First, C_j calculates of $\beta^{'} = (X^{'t} \cdot X^{'})^{-1} \cdot X^{'t} \cdot \overrightarrow{y}^{'}$ locally, which guarantees data privacy. After that, C_j calculates the hash value $H = H(\beta^{'}||address)$ and submits it to the smart contract. Eventually, C_j submits $\beta^{'}$ and its public key address after the confirmation of H.

5.5 Verification

First, the smart contract validates the identity of C_j by comparing the hash value $H^{'} = H(\beta^{'}||address)$ with H. Then, it checks whether the equation $\Delta\overrightarrow{y}^{'} = \overrightarrow{y}^{'} - X^{'} \cdot \beta^{'}$ holds. If $\forall i \in \{1, \ldots, n\}$, $\Delta\overrightarrow{y}^{'}(i, 1)$ is within the standard error interval, the result will be accepted, otherwise, rejected. After validation, the result $\beta^{'}$ is recorded on the blockchain, which can be decrypted using the private key k of O_i according to the formula $\beta = M^{X}_{adj-post} \cdot \beta^{'}$.

5.6 Payment

In this phase, all honest calculators that submit correct results on time are rewarded, while the ones that submit inaccurate results or fail to complete the task before deadline are deemed malicious. Calculators marked as malicious are forced to submit more deposits than honest ones and their deposits are fully deducted for the purpose of protecting the rights of all calculators. As all nodes are considered rational, we believe that this strategy can effectively decrease malicious calculators in the system.

6 Performance Evaluation

In this section, We first compare our scheme with other existing works, and then we evaluate the time cost and the predictive performance of our scheme. The scheme is performed using the Combined Cycle Power Plant Data Set from the UCI Machine Learning Repository [9,26]. Finally, we prove that our scheme achieves the security requirements.

6.1 Performance Comparison

We compare the performance of our scheme with some existing works, including verifiability, fairness, privacy, authentication, and security. Based on the comparison results displayed in Table 1 using "Yes" and "No", we can observe that our scheme is more advanced than other related works.

Table 1. Performance comparison with existing schemes.

Ref.	Verifiable	Fairness	Privacy	Authentication	Security
[27]	No	Yes	Yes	No	Yes
[24]	No	Yes	Yes	No	Yes
[28]	Yes	Yes	Yes	No	Yes
[14]	Yes	Yes	Yes	No	Yes
Our scheme	Yes	Yes	Yes	Yes	Yes

6.2 Prototype Evaluation

Our scheme is implemented on a consortium blockchain using Hyperledger Fabric v2.3. We deploy Go-language-based smart contracts on a desktop with a quad-core processor, 4 GB RAM and 20 GB memory, which is running Ubuntu 22.04.1. The implementation can be separated into three parts. In the first part, we evaluate the time cost of each stage multiple times and average the results. From Fig. 3, we can clearly see that the overhead of all stages is milliseconds, which is efficient and applicable in the IIoT environment. Compared with the scheme mentioned above, our scheme is more advanced in that it achieves identity unforgeability, data privacy and public verifiability with acceptable time consumption. Then, we analyze the time complexities of our algorithms as shown in Table 2.

Table 2. Time complexities of our proposed algorithms.

Setup	Task announcement	Task selection	Result calculation	Verification	Payment
$O(n_1^2)$	$O(n)$	$O(1)$	$O(n_2^3)$	$O(n_2^2)$	$O(1)$

In the second part, we evaluate the ratio of verification that successfully identifies malicious calculators and show the resistance of our scheme to malicious calculators in Fig. 4. With the percentage of malicious calculators varying from 10% to 50%, a graceful increase in the ratio of verification that identifies incorrect results can be observed, which proves the effectiveness of the verification.

In the third part, we evaluate the predictive performance of the linear regression model. The comparison between measured and predicted data is presented

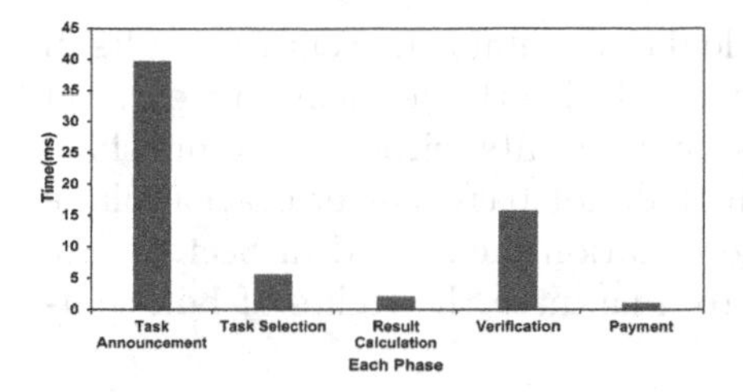

Fig. 3. The time consumption of each stage

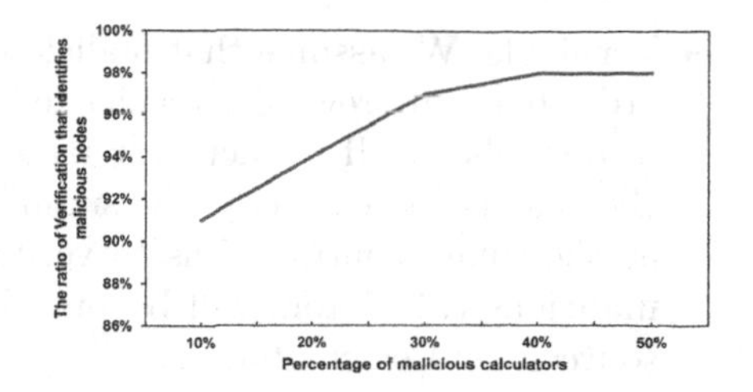

Fig. 4. The success ratio of verification that identifies malicious calculators

in Fig. 5. As we can see, the majority of the predicted results fall in the standard error interval indicating that the predictive performance satisfies the requirements of the IIoT environment, which demonstrates the practical application value of our proposal.

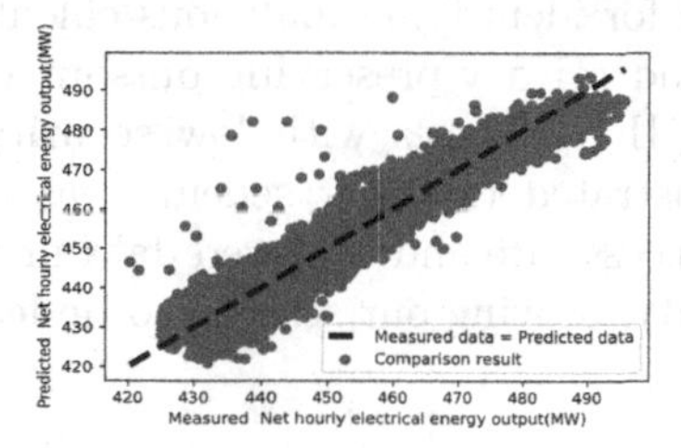

Fig. 5. Comparison between measured and predicted data

6.3 Security Analysis

We analyze the security requirements in this section, including identity unforgeability, privacy-preserving, and verifiable.

- Identity unforgeability. We assume that adversaries attack the hardware of devices in the system from the physical layer to forge their identity, which can be detected through authentication. Due to the unclonability of PUF, this attack changes the CRPs of PUF and consequently leads to the failure of authentication. Devices that fail to pass the authentication can not publish outsourced tasks in our system, as a result of which the assets of devices are protected.
- Privacy-preserving. In our system, a malicious entity can eavesdrop private and sensitive data contained in the outsourced tasks from the transmission channel and blockchain. Therefore, we use the SMM technique to encrypt the original input data so that malicious eavesdroppers can not decrypt private information from public task-related data in the P2P network and the privacy of data is protected.

- Verifiable. We assume that malicious calculators submit inaccurate results in order to obtain rewards, which can be detected with the verification algorithm in our scheme. If a calculator tries to submit results without accomplishing the calculation correctly, the submission shall not pass the verification phase as the smart contract fails to verify the equation mentioned in Sect. 5. The malicious calculators will be punished to guarantee the rights of both outsourcers and calculators.

7 Conclusion

In this paper, we proposed a new edge outsourced computation scheme for the IIoT environment. Our scheme adopted SRAM PUF to accomplish device authentication and guarantee their hardware security. To the best of our knowledge, our scheme was the first outsourced computation scheme that achieved hardware security of IIoT devices. In addition, to verify the accuracy of results and protect sensitive data, the scheme utilized blockchain and SMM technologies to achieve public verifiability and data privacy. Furthermore, the scheme introduced a mechanism for identifying malicious calculators. Our scheme provided a verifiable, fair and privacy-preserving outsourced computation method for resource-constrained IIoT devices with low security level. Eventually, the experimental results illustrated that our scheme was at the millisecond level with high verification success rate and achieved data privacy, public verifiability and identity unforgeability, proving our scheme to be feasible and fair.

References

1. Chen, F., Xiang, T., Lei, X., Chen, J.: Highly efficient linear regression outsourcing to a cloud. IEEE Trans. Cloud Comput. **2**(4), 499–508 (2014)
2. Ciulla, G., D'Amico, A.: Building energy performance forecasting: a multiple linear regression approach. Appl. Energy **253**, 113500 (2019)
3. Gai, K., et al.: A novel secure big data cyber incident analytics framework for cloud-based cybersecurity insurance. In: IEEE BigDataSecurity (2016)
4. Gai, K., Fang, Z., et al.: Edge computing and lightning network empowered secure food supply management. IEEE IoT J. **9**(16), 14247–14259 (2022)
5. Gai, K., Hu, Z., Zhu, L., Wang, R., Zhang, Z.: Blockchain meets DAG: a BlockDAG consensus mechanism. In: Qiu, M. (ed.) ICA3PP 2020. LNCS, vol. 12454, pp. 110–125. Springer, Cham (2020). https://doi.org/10.1007/978-3-030-60248-2_8
6. Gai, K., Wu, Y., et al.: Permissioned blockchain and edge computing empowered privacy-preserving smart grid networks. IEEE IoT J. **6**(5), 7992–8004 (2019)
7. Gai, K., Wu, Y., et al.: Differential privacy-based blockchain for industrial Internet-of-Things. IEEE TII **16**(6), 4156–4165 (2020)
8. Hu, F., Lakdawala, S., et al.: Low-power, intelligent sensor hardware interface for medical data preprocessing. IEEE TITB **13**(4), 656–663 (2009)
9. Kaya, H., Tüfekci, P., Gürgen, F.S.: Local and global learning methods for predicting power of a combined gas & steam turbine. In: Conference on Emerging Trends in Computer and Electronics Engineering (ICETCEE), pp. 13–18 (2012)

10. Li, D., Chen, R., et al.: Blockchain-based authentication for IIoT devices with PUF. J. Syst. Archit. **130**, 102638 (2022)
11. Li, D., et al.: PUF-based intellectual property protection for CNN model. In: Memmi, G., Yang, B., Kong, L., Zhang, T., Qiu, M. (eds.) KSEM 2022. LNAI, vol. 13370, pp. 722–733. Springer, Cham (2022). https://doi.org/10.1007/978-3-031-10989-8_57
12. Li, H., Yu, J., Zhang, H., Yang, M., Wang, H.: Privacy-preserving and distributed algorithms for modular exponentiation in IoT with edge computing assistance. IEEE Internet Things J. **7**(9), 8769–8779 (2020)
13. Li, J., Ming, Z., et al.: Resource allocation robustness in multi-core embedded systems with inaccurate information. J. Syst. Archit. **57**(9), 840–849 (2011)
14. Li, T., Tian, Y., Xiong, J., Bhuiyan, M.Z.A.: FVP-EOC: fair, verifiable, and privacy-preserving edge outsourcing computing in 5G-enabled IIoT. IEEE Trans. Ind. Inf. **19**(1), 940–950 (2023)
15. Li, Y., Gai, K., et al.: Intercrossed access controls for secure financial services on multimedia big data in cloud systems. ACM TMCCA **12**(4s), 1–18 (2016)
16. Liu, Z., et al.: A privacy-preserving outsourcing computing scheme based on secure trusted environment. IEEE Trans. Cloud Comput. 1–12 (2022)
17. Qiu, H., Dong, T., et al.: Adversarial attacks against network intrusion detection in IoT systems. IEEE Internet Things J. **8**(13), 10327–10335 (2020)
18. Qiu, H., Qiu, M., Memmi, G., Ming, Z., Liu, M.: A dynamic scalable blockchain based communication architecture for IoT. In: Qiu, M. (ed.) SmartBlock 2018. LNCS, vol. 11373, pp. 159–166. Springer, Cham (2018). https://doi.org/10.1007/978-3-030-05764-0_17
19. Qiu, M., Chen, Z., et al.: Energy-aware data allocation with hybrid memory for mobile cloud systems. IEEE Syst. J. **11**(2), 813–822 (2014)
20. Qiu, M., Jia, Z., et al.: Voltage assignment with guaranteed probability satisfying timing constraint for real-time multiproceesor DSP. JSPS **46**, 55–73 (2007). https://doi.org/10.1007/s11265-006-0002-0
21. Qiu, M., et al.: Energy minimization with soft real-time and DVS for uniprocessor and multiprocessor embedded systems. In: IEEE DATE, pp. 1–6 (2007)
22. Qiu, M., et al.: Dynamic and leakage energy minimization with soft real-time loop scheduling and voltage assignment. IEEE TVLSI **18**(3), 501–504 (2009)
23. Qiu, M., Thuraisingham, B., et al.: Special issue on robustness and efficiency in the convergence of artificial intelligence and IoT. IEEE IoT J. **8**(12), 9460–9462 (2021)
24. Shao, J., Wei, G.: Secure outsourced computation in connected vehicular cloud computing. IEEE Netw. **32**(3), 36–41 (2018)
25. Tian, Z., Li, M., Qiu, M., Sun, Y., Su, S.: Block-DEF: a secure digital evidence framework using blockchain. Inf. Sci. **491**, 151–165 (2019)
26. Tüfekci, P.: Prediction of full load electrical power output of a base load operated combined cycle power plant using machine learning methods. Int. J. Electr. Power Energy Syst. **60**, 126–140 (2014)
27. Yang, H., Su, Y., et al.: Privacy-preserving outsourced inner product computation on encrypted database. IEEE TDSC **19**(2), 1320–1337 (2022)
28. Yu, X., Yan, Z., Zhang, R.: Verifiable outsourced computation over encrypted data. Inf. Sci. **479**, 372–385 (2019)
29. Zhang, H., Gao, P., et al.: Machine learning on cloud with blockchain: a secure, verifiable and fair approach to outsource the linear regression. arXiv preprint arXiv:2101.02334 (2021)

30. Zhang, H., Yu, J., et al.: Efficient and secure outsourcing scheme for RSA decryption in Internet of Things. IEEE IoT J. **7**(8), 6868–6881 (2020)
31. Zhao, L., Chen, L.: Sparse matrix masking-based non-interactive verifiable (outsourced) computation, revisited. IEEE TDSC **17**(6), 1188–1206 (2018)
32. Zhao, M., Hu, C., et al.: Towards dependable and trustworthy outsourced computing: a comprehensive survey and tutorial. JNCA **131**, 55–65 (2019)

Architecture Search for Deep Neural Network

Xiangyu Gao[1], Meikang Qiu[2(✉)], and Hui Zhao[3]

[1] New York University, New York, NY, USA
xg673@nyu.edu
[2] Dakota State University, Madison, SD 57042, USA
Meikang.Qiu@dsu.edu
[3] Henan University, Kaifeng, China
zhh@henu.edu.cn

Abstract. Deep learning has become a popularly used tool in large amount of applications. Given its ability to explore the input and output relationship, deep learning can perform well in terms of prediction. However, one important drawback in this framework is that the model cannot be easily trained due to huge search space and large number of parameters. In response to such problem, this paper proposes a novel way to build a deep neural network. Specifically, it tries to consider an unbalanced structure of deep neural network by expanding the number of nodes in the beginning to extract as much information as possible and then shrinking quickly to converge to the final result. The experiment results show that our proposed structure can output equally good output with much faster time compared with traditional methods.

Keywords: Deep Learning · Machine Learning · Neural Architecture Search · Deep Neural Network · Information Set · Back Propagation

1 Introduction

Deep learning has become an essential tool across different fields including cyber security [31,32,37], stock trading [49] and Go playing [46]. Specifically, as for complex problems, deep learning tries to build a layer-by-layer structure so as to precisely capture the relationship between the input features and output results. Most of these relationships are nonlinear, so cutting it into different layers and learning the relationship through deep neural network model is a good option. For example, Gao and Qiu [10] consider the energy-based learning as the tool to prevent backdoor in various situations.

Due to the good performance of deep learning techniques, people are always trying to explain the reason why deep learning can do better than traditional machine learning methods. Artzai et al. [28] attribute this to the fact that the multi-level structure can provide the system a way to learn complex functions to map input to the output. This also reflects the motivation of the invention of deep learning, which tries to simulate the human brain structure and behave like the cell-to-cell connection. Therefore, with the development of computation

M. Qiu et al. (Eds.): SmartCom 2022, LNCS 13828, pp. 581–596, 2023.
https://doi.org/10.1007/978-3-031-28124-2_55

technology, one famous deep learning machine AlphaGo [46] can leverage deep neural network models to easily beat top-level human being Go players.

However, there are still several problems existing in the deep learning framework [23]. First of all, it is usually hard to explain the result of deep learning because the complex structure, including many layers and a lot of nodes, makes itself impossible to clearly show the correlation between the input and the output. In addition, even with Hecht-Nielsen's back propagation [14] method, it still takes too long to train the deep learning model.

Recently, Kung [20] proposed an effective neural architecture search algorithm to find useful DNN structures for training. Specifically, it comes up with an *X-learning Neural Architecture Search* (XNAS) to automatically train the network's structure and parameters, which seems to be more competitive than state-of-the-art approaches. However, different from the previous approaches, we think if it is possible for the user decouple structure exploration and model training, this can save a lot of time to do deep learning model training because we only need to focus on one concrete model to train the deep neural network.

In order to speed up the training process of deep learning model, we claim that setting up the concrete and effective structure of *Deep neural network* (DNN) beforehand should be helpful for the future model training process. Accordingly, in this paper, we come up with a novel technique to build the structure of DNN in order to speed up the training speed. To be specific, we consider the DNN as a structure to extract useful information from the input features so as to quickly build the connection between input and output. Therefore, we want to include as much information as possible in the initial several layers.

However, too many nodes within the DNN will increase the training difficulties a lot. Our algorithm aims at reducing the number of nodes dramatically layer by layer until the final one. To be specific, we will set the number of nodes in the first layer to be equal to the number of input features. Then the number of nodes in the next layer would be half of the number of nodes in the previous layer. We will also guarantee that each layer (except the output layer) should have more than one node.

There are a lot of benefits of such DNN architecture design. In the beginning, this design can help the structure avoid losing any important information. As the data goes through layers, more important information has been extracted, which means there is no need to remain the number of nodes the same as layers before. In order to reduce the training complexity, it is reasonable to reduce the number of nodes in the following layers. We believe this should be an effective way to find a trade-off between fitting accuracy and training speed in DNN. There are three main contributions of this paper:

- We give a detailed analysis of deep learning to find some possible reason of its low training speed.
- We propose an deep neural network structure that can quickly capture the input and output relationship.
- We compare our proposed structure with several existing structures to illustrate the benefits of our proposal.

The remainder of this paper has the following structure. Section 2 summarizes the background of deep neural network and overviews several up-to-date structure learning techniques in the domain of DNN. Then, Sect. 3 starts from a simple example and formally presents the algorithm design of our proposed architecture search algorithm for DNN. Furthermore, Sect. 4 gives the implementation details together with several experiment results. Finally, conclusions are shown in Sect. 5.

2 Background

In this section, we will provide a brief description of the deep neural network and explain the pros and cons of DNN. Then, several related structure learning techniques will also be mentioned to show some of the current trials to find the close-to-optimal structure of the DNN.

2.1 Deep Neural Network

With the rapid development of computer capability [39,43,44], new algorithm design [26,41,42], and cloud computing [22,30,36], large amounts of data [9,16,21] can be generated in a very short time period. Hence, this demands fast processing capability of big data. Machine learning [33–35] and artificial intelligence are the results of this big demand. There are many different learning models. DNN [4] is one important part of a broader family of machine learning methods based on artificial intelligence. It has become a widely used technique across several fields. The general structure of DNN is shown in the Fig. 1.

At a really high level, one DNN framework consists of two main components: layer and node. The first layer is called the input layer while the last layer is called the output layer. All layers between input and output layer are all regarded as hidden layers. The data would come from the input layer and then pass through all layers before arriving at the last layer, which stores the output information. Each node will represent a value and be used as the inputs to all nodes in the next layer. Usually, the nodes' values in the next layer will be a linear combination of all nodes in the layer before and their output will be determined by the activation functions chosen by the users. Popularly used activation functions include linear combination, sigmoid, tanh and relu. Selecting the best activation function [13] in different scenarios remains an interesting research topic in DNN fields.

In fact, DNN achieve good performance in many areas, such as various software [45,50,52], complicated systems [29,40], and parallel structures [17,38,53]. Even though people do not have official proof of why DNN can work well under a lot of scenarios, there are several intuitive explanation of such phenomenon.

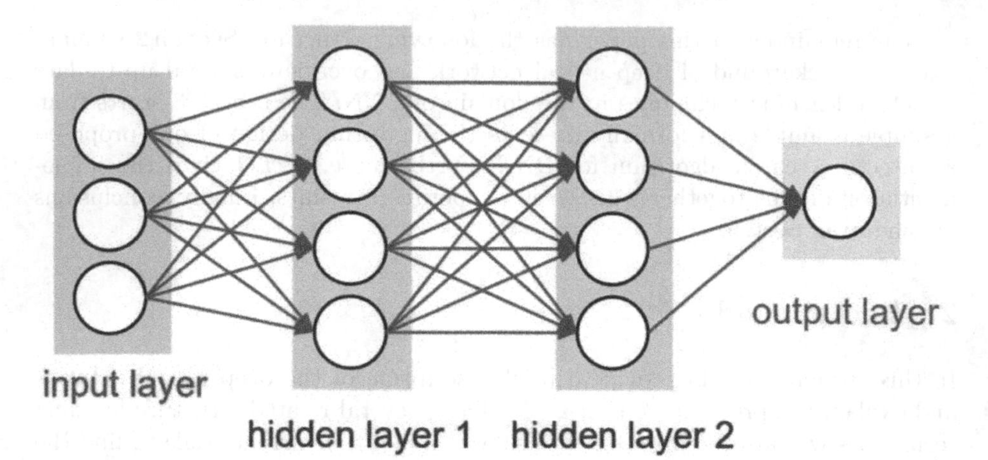

Fig. 1. The visualization of deep neural network.

After all, the motivation to develop the DNN structure originates from people's idea to simulate the function in human's brain. They want to find a way to let the machine understand new knowledge similar to how people learn new stuff.

In reality, most of the input/output relationship cannot be easily captured by a simple function. Therefore, the DNN is able to divide such complex relationship into multiple stages, each of which can be determined by one simple linear relationship. Then, the combination of multiple layers' simple function can precisely understand the complex input/output relationship if sufficient data is provided to train the model.

Therefore, as is shown in Fig. 2, when the amount of training data is small, deep learning and old machine learning techniques will generate quite similar outcomes. However, as the amount of data increases, the deep learning will significantly out-perform the old machine learning techniques. Hence, We can draw the conclusion that in the current era of big data, deep learning will be a better choice.

However, DNN has several drawbacks [1]. Here, We will list two most important ones. First of all, the training time of DNN structure would be quite long due to a large number of nodes and complex layer-to-layer connection relationship. Some techniques [19], including back propagation [14], have been proposed to speed up the process, but they are still not sufficient. In addition, DNN, similar to many other machine learning techniques, is hard to explain. This brings problems that people sometimes cannot differentiate between scenarios where

DNN is a suitable model or scenarios that overfitting [6] happens. Therefore, even if the fitting accuracy over the training set is good for DNN, people are not willing to totally rely on DNN for new datasets, which strictly restrict the number of application fields of DNN.

2.2 Structure Searching Techniques

An important improvement in DNN is related to the structure learning techniques development. To be specific, when selecting DNN as our target model structure, people usually want to do the model selection together with model training. In other words, they can only provide a general model, but more details (e.g., number of layers, number of nodes per layer) will be finalized during the training process. This increase the difficulty-level of the training process.

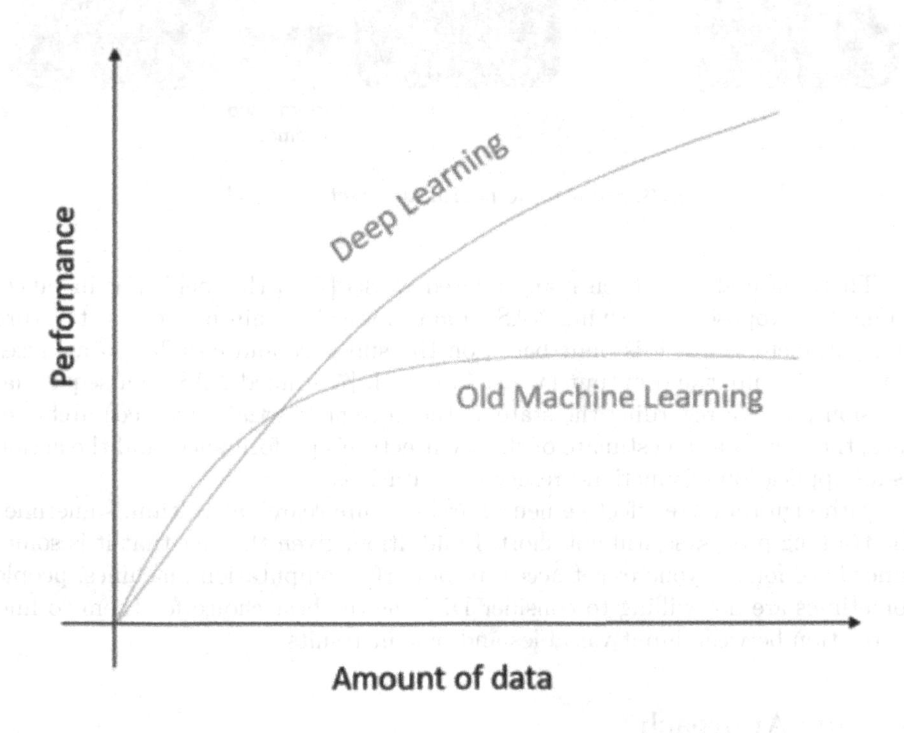

Fig. 2. deep learning vs traditional machine learning.

Just as shown in Fig. 3, *Neural Architecture Search* (NAS) is a technique for automating the design of *Artificial Neural Networks* (ANN), a widely used model in the field of machine learning. It has three main components: the search space defines the type(s) of neural networks that can be designed and optimized; the search strategy defines the approach used to explore the search space; the performance estimation strategy evaluates the performance of a possible neural network from its design. By now, NAS methods have outperformed several manually designed architectures on some tasks including image classification [54] and semantic segmentation [5].

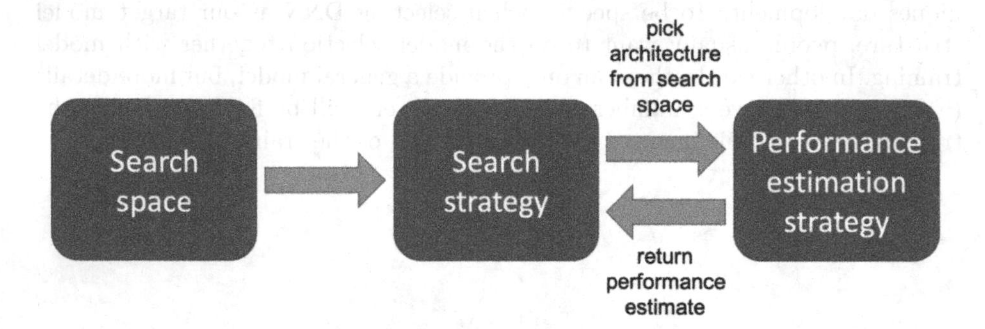

Fig. 3. Steps to do neural architecture search.

There have already been many related works [7] in this field. For instance, Kung [20] proposed X-learning NAS to automatically train network's structure and parameters, which is built based on the subspace and correlation analyses between the input and output layers. Cai et al. [7] framed NAS as a sequential decision process: regarding the state as the current (partially trained) architecture, the reward as an estimate of the architecture's performance, and the action as an application of function-preserving mutations.

Although they are effective neural architecture search algorithm, sometimes the training process is still not short. In addition, given the fact that it is sometimes hard for everyone to get access to powerful computation machines, people sometimes are not willing to consider DNN as the best choice for them to find correlation between input variables and output results.

3 Our Approach

In response to the problems shown in the previous section, we will present our approach in this section. To be specific, in order to deal with the problem that it takes too long to train a DNN model, we start from one motivation example which offers a modified DNN model with smaller search space. Then, a formalized algorithm will be provided in the following subsection together with our analysis of such algorithm.

3.1 A Small Example

General DNN structure is illustrated in the LHS of Fig. 4. Assume we have four layers in total, the number of nodes in each layer is the same across layers. In addition, all nodes of two adjacent layers are fully connected by linear functions. The assumption behind is that it will collect all information from the input set and then find the correlation between input and output nodes.

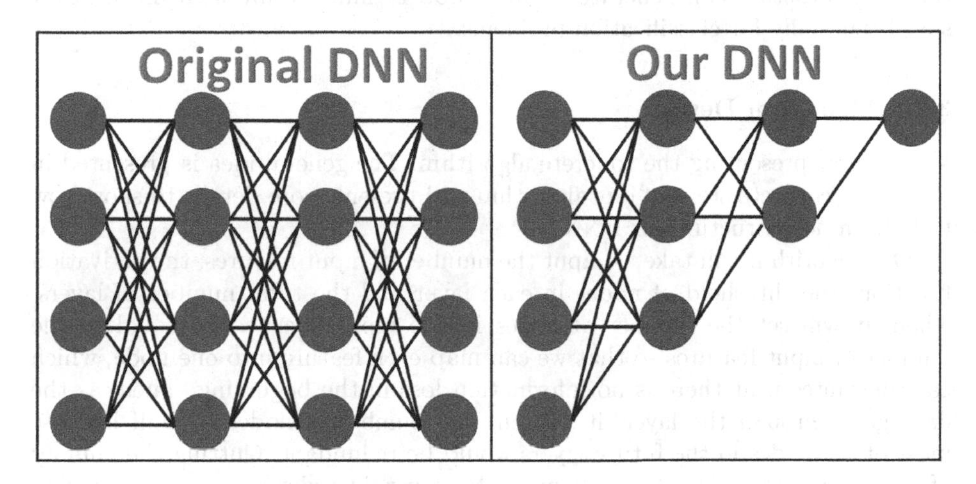

Fig. 4. Comparison of our approach and original DNN (suppose the total number of layers is four).

However, this might easily cause a problem: when we need to train the DNN, it might require too many computation resources and long training time. Specifically, in some scenarios where multiple layers are needed, the total number of nodes will increase linearly, causing the training complexity [25] to increase exponentially.

We believe it is easy to reduce the training complexity by considering another better DNN structure. If we retrospect the DNN structure shown on the LHS of Fig. 4, it is usually unnecessary for us to put so many nodes in the last layer. Usually, if people use DNN to do classification problem, the output is just one node with binary values. Giving too many nodes to the last layer is redundant. In addition, it is highly possible that most of the useful information can be stored with only a few number of nodes, making it unnecessary to keep the number of nodes the same across layers.

Therefore, we give a new DNN structure shown on the RHS of Fig. 4. In other words, given the situation that there are four layers in total, it keeps the number of nodes equal to the number of input features in the first layer. Afterwards, we reduce the number of nodes in each layer and leave with only one node in the last layer. Given this is just a motivation example, there are some slight difference between this example and the proposed algorithm. But the main idea behind is that we want to remove the redundant nodes in DNN structure so as to speed up the training process. Since we also want to keep the fitting accuracy of our framework, we believe the structure with decreasing number of nodes as the layer increase will be the good solution.

Compared with original DNN, our proposed structure reduces the number of nodes by almost a half (16 vs 10) and simplify the training process. We claim that this structure can generate equally good training result with much faster speed especially for classification problems.

3.2 Algorithm Design

We are now presenting the concrete algorithm. The general idea is presented in Fig. 4 but concrete steps of our algorithm will present more details to show how to build a new structure for DNN.

Our algorithm will take as input the number of input features, the activation function, the threshold of nodes in each layer and the total number of layers. Then, it will set the number of nodes in the input layer to be equal to the number of input features so that we can map each feature into one node, which can guarantee that there is no information loss in the beginning. Then, as the data goes through the layer, it will cut the number of nodes by half because most of the nodes in the future layers would be redundant. Cutting the number of nodes can save a lot of search space. However, in order to prevent removing too many nodes, this algorithm will force the number of nodes in each layer to be at least as large as the threshold given by the users. This can guarantee that we do not over-reducing necessary information.

3.3 Algorithm Analysis

The intuition behind our proposed algorithm can be explained as follows. In the beginning, it wants to collect as much information as possible, which forces its number of nodes in the input layer to be equal to the total number of input features. However, it would be a waste if we keep the number of nodes the same across layers because the goal of DNN is to collect useful information but drop useless one. Therefore, in order to strike a balance between fitting effect and the complexity of DNN model, we reduce the number of nodes by half layer by layer so as to reduce the number of necessary nodes exponentially. If we compare our proposed DNN structure against the original DNN structure, our search space is much smaller, leading to the fact that our training time will be much shorter than original DNN model.

We believe this algorithm can have good performance in scenarios where the final output is only a binary variable (e.g., whether the price is over or below the average). As for the classification problem, we only need one nodes in the output layer to determine the final output, reducing the nodes in each layer will not have too many negative effects on the fitting results. Extending such structure design

---

Algorithm 1. ***DNN structure generation*** Algorithm

Require: M input features, activation function of DNN F, total number of layers of DNN N, accuracy level L, and threshold of nodes K per layer.
Ensure: The threshold K must be smaller than the dimension M of input features.
Ensure: The threshold K must be greater than 2.

1: Set up a Deep Neural Network structure with only N layers (one input layer, one output layer and N - 2 hidden layer).
2: Set the number of node in the input layer equal to M.
3: Set the number of node in the output layer equal to 1.
4: Set the number of nodes in each hidden layer
for i = 1; i < M - 1; i++:
Set the number of node in the layer i to be max(number of nodes in previous layer/2, K).
5: Train the DNN on the input data set for several epochs until the accuracy is greater than or equal to accuracy level L.
6: Output the DNN training result together with the training time (number of epochs) required to generate such result.
Output results: The final results include all parameters of the trained DNN framework together with the fitting score.

---

from classification problem to non-classification problem will be an interesting direction for future work.

4 Implementation and Experiments

This section is mainly about the implementation details and the experiment results of our proposed algorithm through showing its comparison against other model benchmarks. In the traditional DNN training process, people would directly feed the data into the neural network pipeline and train such model using feedforward [47] and backpropagation methodology. We want to see how much improvement our algorithm can provide. We are concerned of three main metrics: accuracy [18], loss [48] and training time [3].

Accuracy is an important determinant since it directly decides how good the correlation between input variables and output variables could be learnt by DNN. Besides, we also want to get the reasonably good result as soon as possible; hence, the training time is another dimension people pay attention to. In our experiment setup, we regard the number of epochs required to reach some specific level as the metric to measure the length of training time. Since usually the fitting result of DNN will keep improving as the number of epochs increase, we want to see whether it is possible for our proposed structure to quickly learn the correlation between input features and output results.

4.1 Data Description

In our experiment, we leverage our model to fit the open source data of house price [2]. This is because the data set is popularly used by researchers to test

DNN model. Usually the price of a house depends on many features, so within the data set, the output Y variable is binary, where 1 means the price is above the median and 0 means the price is below the median. Ten features (e.g., overall quality, overall condition, distance to surrounding utilities) are considered as important determinants to the final price. There are 1,460 data points in total within the data set.

We utilize related packages in python (e.g., sklearn [27], keras [12]) to implement our algorithm. Concretely speaking, house price data is stored in the csv format. We read the data from csv file and store it into the pandas data frame. Then, we implement several preprocessing techniques into the data with inserted packages such as standardization. The standardized data is split into training set and testing set. In our setting, we put the proportion of training data to be 70% of the whole data set while the remaining 30% are used as the testing set [11]. The DNN model from keras package will be trained in the training data and we test its performance over testing data set.

4.2 Benchmark Selection

As for benchmarks, we mainly consider the original DNN with the number of nodes the same across layers as the candidates. Since our approach is an improvement based on the original DNN model, this should be the best candidate we want to compare against. To be specific, we let the total number of layers of DNN to be four. As for the number of nodes, we set it to be ten because of ten features in total for the input data. The number of nodes is the same across all layers for the original DNN model while as for our proposed DNN model, it has five, two and one node for the second, third and last layer.

There are several dimensions we can do to increase the number of benchmarks. For instance, we can set the number of layers to be a variable and generate a bunch of different DNN models. In addition, we can also randomize the number of nodes per stage. In this paper, we mainly focus on the comparison of our deterministic algorithm against the original DNN structure.

In the future, we also want to extend our approach to other deep learning models [8,24] to show the benefits of such proposal.

4.3 Results

When it comes to the results comparison, we compare our method against original DNN model because we want to see how much improvement such proposed algorithm can bring. As for the metrics, we take into consideration accuracy rate [18], loss value [48] together with required time spent training the model.

The comparison results are illustrated in Fig. 5, Fig. 6, Fig. 7, and Fig. 8. In general, given the high accuracy and low loss value, we can see that DNN should be a suitable candidate model for house price prediction. It means that DNN is a suitable model for people to use to predict the house price over this given data set. In reality, given the fact that whether the house price is above or below the

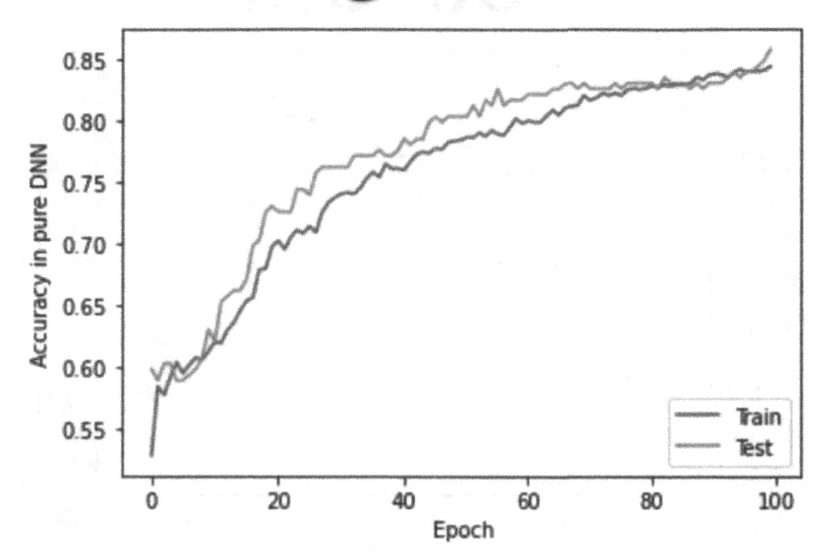

Fig. 5. Accuracy value of original DNN.

median should be determined by a series of factors, DNN should be better than other simple models.

We can also find other interesting phenomena from the figures. First of all, it is easy to witness that the big trend is that the fitting results get better and better as the number of epochs increases, which conforms to the common sense that longer training time will lead to better fitting result [15]. In addition, we can find that the improvement speed of the accuracy decreases as the training time increases. It means that in the beginning, a fixed number of training epochs can improve the accuracy a lot but further improvement will take much longer training time.

Therefore, when using DNN as a candidate structure, people do not want to reach 100% accuracy but will stop whenever the accuracy reaches a threshold. Otherwise, they have to wait too long for the training process to stop. Spending too much time increasing the accuracy might have some side effects such as overfitting [51].

Compared with the original DNN, our DNN model achieve the similar (even better) fitting effect from Fig. 5 and Fig. 6. Specifically, the accuracy of the testing set is always higher than the training set and after 100 epochs, the accuracy increases to a level greater than 85%. In comparison, the accuracy of traditional accuracy is always lower than 85%. When it comes to the loss value in Fig. 7 and Fig. 8, our approach is slightly lower than original DNN structure but they are very close to each other (45% vs 50%). How to reduce the loss rate for our

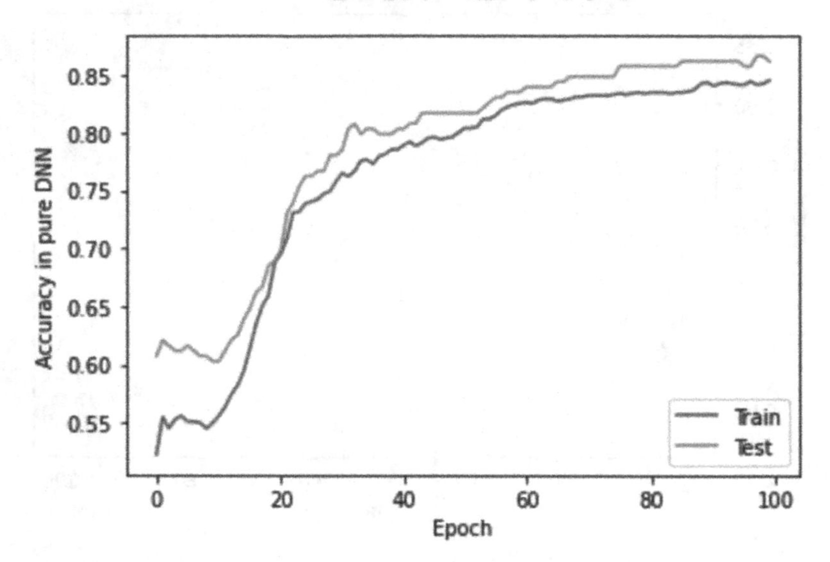

Fig. 6. Accuracy value of our DNN structure.

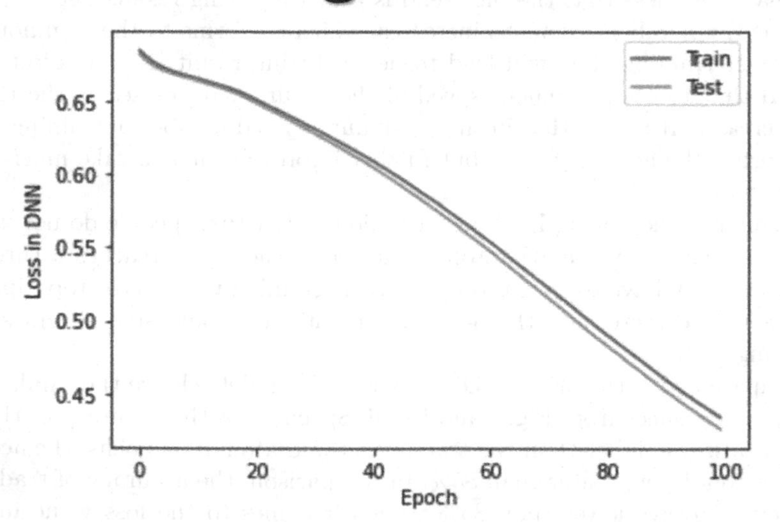

Fig. 7. Loss value of original DNN.

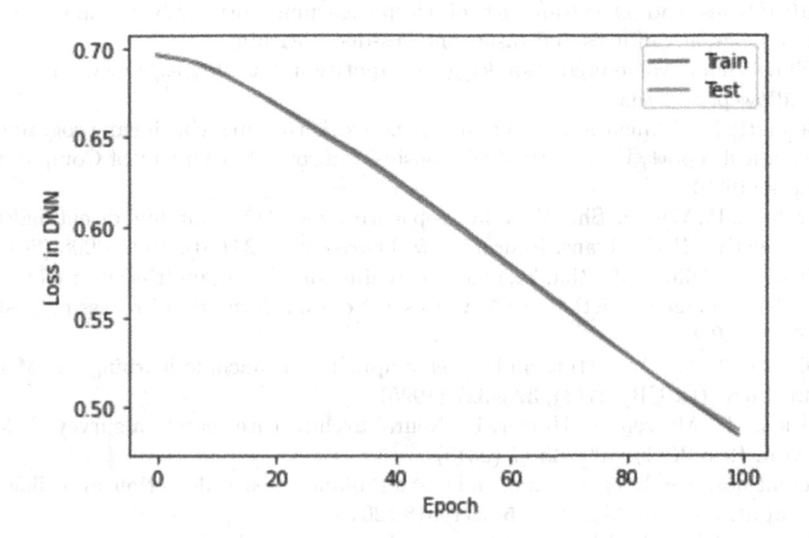

Fig. 8. Loss value of our DNN structure.

proposed DNN to a level that is equivalent to original DNN should be a next step worthwhile to explore.

In addition, we also measure the training time of both original DNN and our DNN, which is the main part to show the benefits of our proposal. It is easy to see that our approach converges much faster than original DNN when setting the accuracy threshold to be 80%. For instance, it only takes around 40 epochs for our DNN model to reach accuracy level of over 80% while original DNN requires more than 60 epochs. We can see that our proposed DNN can be trained more quickly to reach good accuracy result. This means that our model is simpler to train and faster to reach a level with relatively high accuracy.

5 Conclusions

In this paper, we presented a new structure for deep neural network, which decreases the number of nodes in each layer exponentially. Our approach is an extension of existing DNN structure but aims at speeding up the training time of the model. The experiment results shown that in terms of house price prediction, our model's fitting result is similar to the currently popularly used one but with faster training time. We hope this proposal can motivate more deep learning users to implement in more application fields.

References

1. Advantages and Disadvantages of Deep Learning. https://www.analyticssteps.com/blogs/advantages-and-disadvantages-deep-learning
2. Zillow's home value prediction kaggle competition data. https://www.kaggle.com/c/zillow-prize-1/data
3. Bennett, K., Mangasarian, O.: Neural network training via linear programming. Technical report, University of Wisconsin-Madison, Department of Computer Sciences (1990)
4. Brahma, P., Wu, D., She, Y.: Why deep learning works: a manifold disentanglement perspective. IEEE Trans. Neural Netw. Learn. Syst. **27**(10), 1997–2008 (2015)
5. Chen, L., Collins, M., Zhu, Y., et al.: Searching for efficient multi-scale architectures for dense image prediction. In: Advances in Neural Information Processing Systems, vol. 31 (2018)
6. Dietterich, T.: Overfitting and undercomputing in machine learning. ACM Comput. Surv. (CSUR) **27**(3), 326–327 (1995)
7. Elsken, T., Metzen, J., Hutter, F.: Neural architecture search: a survey. J. Mach. Learn. Res. **20**(1), 1997–2017 (2019)
8. Ferentinos, K.: Deep learning models for plant disease detection and diagnosis. Comput. Electron. Agric. **145**, 311–318 (2018)
9. Gai, K., Qiu, M., Elnagdy, S.: A novel secure big data cyber incident analytics framework for cloud-based cybersecurity insurance. In: IEEE BigDataSecurity Conference (2016)
10. Gao, X., Qiu, M.: Energy-based learning for preventing backdoor attack. In: Memmi, G., Yang, B., Kong, L., Zhang, T., Qiu, M. (eds.) KSEM 2022. LNAI, vol. 13370, pp. 706–721. Springer, Cham (2022). https://doi.org/10.1007/978-3-031-10989-8_56
11. Gholamy, A., Kreinovich, V., Kosheleva, O.: Why 70/30 or 80/20 relation between training and testing sets: a pedagogical explanation. Technical report: UTEP-CS-18-09 (2018)
12. Gulli, A., Pal, S.: Deep Learning with Keras. Packt Publishing Ltd. (2017)
13. Hayou, S., Doucet, A., Rousseau, J.: On the selection of initialization and activation function for deep neural networks. arXiv preprint arXiv:1805.08266 (2018)
14. Hecht-Nielsen, R.: Theory of the backpropagation neural network. In: Neural Networks for Perception, pp. 65–93. Elsevier (1992)
15. Hoffer, E., Hubara, I., Soudry, D.: Train longer, generalize better: closing the generalization gap in large batch training of neural networks. In: Advances in Neural Information Processing Systems, vol. 30 (2017)
16. Hu, F., Lakdawala, S., et al.: Low-power, intelligent sensor hardware interface for medical data preprocessing. IEEE TITB **13**(4), 656–663 (2009)
17. Huang, H., Chaturvedi, V., et al.: Throughput maximization for periodic real-time systems under the maximal temperature constraint. ACM TECS **13**(2s), 1–22 (2014)
18. Johnson, V., Rogers, L.: Accuracy of neural network approximators in simulation-optimization. J. Water Resour. Plan. Manag. **126**(2), 48–56 (2000)
19. Kamarthi, S., Pittner, S.: Accelerating neural network training using weight extrapolations. Neural Netw. **12**(9), 1285–1299 (1999)
20. Kung, S.: XNAS: a regressive/progressive NAS for deep learning. ACM Trans. Sens. Netw. (TOSN) **18**(4), 1–32 (2022)

21. Li, J., Ming, Z., et al.: Resource allocation robustness in multi-core embedded systems with inaccurate information. J. Syst. Archit. **57**(9), 840–849 (2011)
22. Li, Y., Gai, K., et al.: Intercrossed access controls for secure financial services on multimedia big data in cloud systems. ACM TMCCA **12**(4s), 1–18 (2016)
23. Marcus, G.: Deep learning: a critical appraisal. arXiv preprint arXiv:1801.00631 (2018)
24. Miglani, A., Kumar, N.: Deep learning models for traffic flow prediction in autonomous vehicles: a review, solutions, and challenges. Veh. Commun. **20**, 100184 (2019)
25. Mikolov, T., Deoras, A., Povey, D., Burget, L., Černocký, J.: Strategies for training large scale neural network language models. In: 2011 IEEE Workshop on Automatic Speech Recognition and Understanding, pp. 196–201. IEEE (2011)
26. Niu, J., Gao, Y., et al.: Selecting proper wireless network interfaces for user experience enhancement with guaranteed probability. JPDC **72**(12), 1565–1575 (2012)
27. Pedregosa, F., Varoquaux, G., Gramfort, A., Michel, V., Thirion, B., Grisel, O., et al.: Scikit-learn: machine learning in Python. J. Mach. Learn. Res. **12**, 2825–2830 (2011)
28. Picon Ruiz, A., Alvarez Gila, A., Irusta, U., Echazarra Huguet, J., et al.: Why deep learning performs better than classical machine learning? Dyna Ingenieria E Industria (2020)
29. Qi, D., Liu, M., et al.: Exponential synchronization of general discrete-time chaotic neural networks with or without time delays. IEEE Trans. Neural Netw. **21**(8), 1358–1365 (2010)
30. Qiu, H., Dong, T., et al.: Adversarial attacks against network intrusion detection in IoT systems. IEEE Internet Things J. **8**(13), 10327–10335 (2020)
31. Qiu, H., Qiu, M., Liu, M., Memmi, G.: Secure health data sharing for medical cyber-physical systems for the healthcare 4.0. IEEE J. Biomed. Health Inform. **24**(9), 2499–2505 (2020)
32. Qiu, H., Qiu, M., Lu, R.: Secure V2X communication network based on intelligent PKI and edge computing. IEEE Netw. **34**(2), 172–178 (2019)
33. Qiu, H., Zeng, Y., et al.: DeepSweep: an evaluation framework for mitigating DNN backdoor attacks using data augmentation. In: ACM Asia CCS (2021)
34. Qiu, H., Zheng, Q., et al.: Deep residual learning-based enhanced JPEG compression in the Internet of Things. IEEE Trans. Ind. Inform. **17**(3), 2124–2133 (2020)
35. Qiu, H., Zheng, Q., et al.: Topological graph convolutional network-based urban traffic flow and density prediction. IEEE Trans. ITS **22**(7), 4560–4569 (2020)
36. Qiu, M., Chen, Z., et al.: Energy-aware data allocation with hybrid memory for mobile cloud systems. IEEE Syst. J. **11**(2), 813–822 (2014)
37. Qiu, M., Gai, K., Xiong, Z.: Privacy-preserving wireless communications using bipartite matching in social big data. FGCS **87**, 772–781 (2018)
38. Qiu, M., Guo, M., et al.: Loop scheduling and bank type assignment for heterogeneous multi-bank memory. JPDC **69**(6), 546–558 (2009)
39. Qiu, M., Jia, Z., et al.: Voltage assignment with guaranteed probability satisfying timing constraint for real-time multiproceesor DSP. JSPS **46**, 55–73 (2007). https://doi.org/10.1007/s11265-006-0002-0
40. Qiu, M., Khisamutdinov, E., et al.: RNA nanotechnology for computer design and *in vivo* computation. Philos. Trans. Roy. Soc. A (2013)
41. Qiu, M., Li, H., Sha, E.: Heterogeneous real-time embedded software optimization considering hardware platform. In: ACM SAC, pp. 1637–1641 (2009)

42. Qiu, M., Xue, C., Shao, Z., Zhuge, Q., Liu, M., Sha, E.H.-M.: Efficent algorithm of energy minimization for heterogeneous wireless sensor network. In: Sha, E., Han, S.-K., Xu, C.-Z., Kim, M.-H., Yang, L.T., Xiao, B. (eds.) EUC 2006. LNCS, vol. 4096, pp. 25–34. Springer, Heidelberg (2006). https://doi.org/10.1007/11802167_5
43. Qiu, M., Xue, C., Shao, Z., Sha, E.: Energy minimization with soft real-time and DVS for uniprocessor and multiprocessor embedded systems. In: IEEE DATE Conference, pp. 1–6 (2007)
44. Qiu, M., Yang, L., et al.: Dynamic and leakage energy minimization with soft real-time loop scheduling and voltage assignment. IEEE TVLSI **18**(3), 501–504 (2009)
45. Qiu, M., Zhang, K., Huang, M.: Usability in mobile interface browsing. Web Intell. Agent Syst. **4**(1), 43–59 (2006)
46. Silver, D., et al.: Mastering the game of go without human knowledge. Nature **550**(7676), 354–359 (2017)
47. Svozil, D., Kvasnicka, V., Pospichal, J.: Introduction to multi-layer feed-forward neural networks. Chemom. Intell. Lab. Syst. **39**(1), 43–62 (1997)
48. Takase, T., Oyama, S., Kurihara, M.: Effective neural network training with adaptive learning rate based on training loss. Neural Netw. **101**, 68–78 (2018)
49. Takeuchi, L., Lee, Y.: Applying deep learning to enhance momentum trading strategies in stocks. Technical report, Stanford University, Stanford, CA, USA (2013)
50. Tao, L., Golikov, S., Gai, K., Qiu, M.: A reusable software component for integrated syntax and semantic validation for services computing. In: IEEE Symposium on Service-Oriented System Engineering, pp. 127–132 (2015)
51. Ying, X.: An overview of overfitting and its solutions. In: Journal of Physics: Conference Series, vol. 1168, no. 2 (2019)
52. Zhang, K., Kong, J., et al.: Multimedia layout adaptation through grammatical specifications. Multimedia Syst. **10**(3), 245–260 (2005)
53. Zhang, L., Qiu, M., et al.: Variable partitioning and scheduling for MPSoC with virtually shared scratch pad memory. J. Sig. Process. Sys. **58**(2), 247–265 (2018). https://doi.org/10.1007/s11265-009-0362-3
54. Zoph, B., Vasudevan, V., Shlens, J., Le, Q.: Learning transferable architectures for scalable image recognition. In: Proceedings of the IEEE Conference on Computer Vision and Pattern Recognition, pp. 8697–8710 (2018)

Blockchain Development

Siqi Xie[1,2], Jiahong Cai[1,2](✉), Hangyu Zhu[1,2,3], Ce Yang[1,2], Lin Chen[1,2], and Weidong Xiao[4]

[1] School of Computer Science and Engineering, Hunan University of Science and Technology, Xiangtan 411201, China
{jiahongcai,yangce}@hnust.edu.cn
[2] Hunan Key Laboratory for Service Computing and Novel Software Technology, Xiangtan 411201, China
[3] Guangdong Financial High-Tech Zone "Blockchain +" Fintech Research Institute, Foshan 528253, China
[4] School of Software Engineering, Xiamen University of Technology, Xiamen 361024, China
xiaoweidong@xmut.edu.cn

Abstract. In recent years, blockchain research has set off an upsurge in academia, and it is called the next generation of value Internet. Because of its decentralization, anonymity, security, immutability, traceability and other characteristics, blockchain is gradually accepted and developed by people. With the deepening of research and the integration of technologies such as deep learning, blockchain has gradually been applied to various fields such as credit reporting, government, medical care, and industrial Internet of Things, not just the initial virtual currency field. This article mainly discusses the three important stages of blockchain public chain development, namely Bitcoin, Ethereum, and meta-verse, and introduces some basic supporting technologies of blockchain, as well as the research status and future trends of blockchain. Simple Analysis. By vertically introducing the development history of the blockchain, researchers can have a more concrete understanding of the status quo of the blockchain, and provide ideas for blockchain-related research.

Keywords: Blockchain · Bitcoin · Ethereum · Metaverse · Smart contracts

1 Introduction

Since the birth of blockchain, most countries had a positive attitude towards its research and industry development, with the United States legislating support, Canada having a positive view, and the Asia-Pacific region watching to explore the development of blockchain technology. In fact, blockchain is a database placed in a non-secure environment, which uses cryptography [1] to ensure that existing data cannot be changed, and consensus algorithms [2] to reach a consensus on new data. Only after Satoshi Nakamoto published "Bitcoin: A Peer-to-Peer Electronic Cash System" [3] blockchain received widespread attention because the core idea of implementing Bitcoin is the blockchain, so Bitcoin is also called the blockchain 1.0 era by scholars. With the continuous research and development of applications, some problems gradually appeared in

M. Qiu et al. (Eds.): SmartCom 2022, LNCS 13828, pp. 597–606, 2023.
https://doi.org/10.1007/978-3-031-28124-2_56

the Bitcoin system, and to solve these problems, Ethereum was created, and this means that the blockchain officially entered the 2.0 era. The metaverse as the latest concept [4] has caused another lively discussion in the academic world [5, 6], and blockchain technology is one of the basic technologies to realize the meta-verse, and people's vision of blockchain 3.0 is that it can realize assets on the chain and construct various applications based on blockchain as the underlying framework to promote the large-scale creation in the fields of science, health [7], education, and images [8], so the metaverse whether the development of blockchain can bring it into the 3.0 era, everything is possible.

The remainder of this paper is organized as follows. Section 2 focuses on the birth of the blockchain and the basic components of the Bitcoin blockchain. Section 3 focuses on the features of the Ethereum blockchain and its improvement on the Bitcoin blockchain. Section 4 introduces the metaverse concept and the application of blockchain in the metaverse. Finally, Sect. 5 concludes the paper.

2 Blockchain 1.0 - Bitcoin

Scholars in different fields give different definitions of blockchain, which according to the literature [9] is a decentralized technology for transactions and data management. For users of the technology, blockchain represents a great improvement in the field of information collection, distribution, and governance [10]. The literature [11] compares blockchain to a tamper-proof digital ledger, implemented in a distributed manner. The literature [12] considers blockchain as a technology that makes the concept of a shared registry in distributed systems a reality in many application domains.

The concept of blockchain can be traced back to "How to Time-Stamp a Digital Document" written by Stuart Haber and W. Scott Stornetta in 1991 [13]. However, it was not until 2008 that the concept of blockchain appeared in the form of "block" and "chain" in Satoshi Nakamoto's article, attracting the attention of scholars from a new perspective of the application. This article focuses on Bitcoin, a virtual currency that he proposed. Bitcoin is not the earliest attempt at currency in the digital world. Virtual currencies such as Goofycoin and Zainucoin emerged before this, but they all ultimately failed. Until the emergence of Bitcoin, the problem of value and reliability was balanced with simple logic. A synthesis of the failures of previous virtual currencies sums up the following ideas.

1. Original currency transactions are simple and unlimited because they are made directly person-to-person, without the need for a bank-like third party. In fact, virtual currencies can also take advantage of this feature to simplify transactions, so the idea of decentralization emerged, directly turning the currency payment process into a "transaction" to a "transaction" form.
2. Digital products are replicable in nature and if they appear in monetary transactions, they are prone to double-spending problems, so to prevent "double-spending", it is better for everyone to witness it.
3. Since there is no third-party intervention and no concept of a balance wallet, a distributed consensus system is necessary to reach a consensus on the transaction witnessed by all. In order to prevent someone from being evil in the consensus

process, a penalty and reward mechanism was introduced to implement a consensus mechanism for many people—Proof of Work.

It was these seemingly simple, yet logical structures that led others to agree that the mechanism by which this currency operated was valid and reliable. Since the basis for Bitcoin's implementation of distributed consensus logic is the blockchain, there are great similarities between the two structures. Zhang et al. [14] summarize the structure of Bitcoin, dividing it into a data layer, a network layer, a consensus layer, a contract layer, and an application layer.

2.1 Data Layer

Block. A blockchain can be viewed as a distributed database that consists of individual blocks linked to each other. There are two main parts of the block, which are the block header and the block body. The block header stores the hash of the previous block, the hash of the content of the block body of the current block (the root of the Merkle tree), and the padding data Nonce. The block body contains the branch nodes of the Merkle tree, i.e., the specific transactions encapsulated in the block. As miners continue to experiment and over time, a continuously growing chain of blocks is created, and the constant updating of the chain represents the updating of the state of the Bitcoin ledger.

Timestamp. Blockchain is a chain structure. The process of chain formation is related to time stamps. The timing of each transaction is in order to prevent some illegal transactions. The timestamp identifies the time of each transaction and forms a chain relationship, its structure in Bitcoin is shown in Fig. 1.

Fig. 1. Blockchain timestamp structure.

Hash Algorithm. The hashing algorithm used in the blockchain is SHA256. There is a probability of hash collision in the hashing algorithm, i.e., it is possible that there are two different numbers that have the same value obtained after calculation by the hash function [15]. However, since the output space of SHA256 is 2^{256}, with the existing computer computing power to forge a number and artificially create a hash collision unlikely, SHA256 can be used to verify whether a block in the blockchain has been tampered with. Hashing the transaction information in the block can form a summary of the block information, and when you need to verify the authenticity of the data, you can hash the transaction information again to determine whether there is a problem with the data.

2.2 Consensus Layer

Proof-of-Work. Bitcoin uses a Proof-of-Work (PoW) mechanism. Proof-of-Work is the mining process, which is simply the search for a number (Nonce) that will satisfy the target value by increasing the number of random bits in the block so that the zero bits when SHA256 is performed. The amount of work required is exponentially related to the number of required zero bits. The process of searching for Nonce is a continuous trial-and-error process, which requires CPU and power consumption. Finding a Nonce that meets the requirements represents the creation of a valid block. The target value is typically reset every 2016 blocks to ensure a frequency of one block generated every ten minutes.

Consensus Mechanism. In the traditional BFT consensus algorithm, the identity of each node must be known, but the nature of blockchain is fully public, everyone can participate in the system and can't reveal their identity, so BFT is no longer applicable. Blockchain is a timestamped server on a peer-to-peer basis, and proof-of-work solves the problem of identifying representatives in consensus decisions, in addition to preventing witch attacks and malicious chain generation. Since proof of work is essentially CPU arithmetic, a CPU represents one vote in a consensus decision. Most decisions are represented by the longest chain since a longer chain represents more workload proofs invested. As long as most of the CPU arithmetic is controlled by honest nodes, then honest chains will grow faster and outpace other competing chains.

2.3 Network Layer

Broadcast Mechanism. To give a macro summary of the Bitcoin transaction process. When a transaction is initiated, the initiator broadcasts it to all nodes, which receive the transaction in a block and perform a proof of work, and do not stop until one node finds the proof and broadcasts the block it is into all nodes. When the transactions in the block are verified by the nodes as all valid and unspent, the node accepts the block, and creates the next chain of acceptances to the block and uses the hash of the accepted block as the previous hash stored in the block header.

Validation Mechanism. In Bitcoin, since there is no concept of a wallet, this type of cryptocurrency ledger can also be seen as a state transition system, and the "state" refers to the current unused currency (UTXO), each UTXO contains a denomination and an owner, and the UTXO is updated after each transaction. UTXO is also one of the steps to verify the validity of the transaction. If two nodes include a block at the same time in this process, the remaining nodes must make a choice and continue mining. If the chain where the block is located is no longer the longest, the mined block will be invalidated.

2.4 Contract Layer

Incentives. All nodes are profit-driven, and the honesty of most nodes is critical to blockchain growth, so there must be incentives to ensure that nodes remain honest and gain more than they would from a malicious attack. Node mining consumes CPU time and power, while the revenue is mainly through transaction fees. If the input value of a transaction is less than the output value, the difference is the transaction fee, which is included in the block containing the transaction. Also, every miner who successfully generates a block has the ability to include in the block a fee issued to itself, worth 12.5 BTC. if a malicious node, wants to get a fee by forging a transaction, then it will re-mine more blocks to make it more than the longest chain, thus making the forged transaction legitimate, which will undoubtedly consume more resources. Therefore, most nodes will still choose to be honest in order to gain revenue.

The final summary of the bitcoin transactions and the blockchain generation process is shown in Fig. 2.

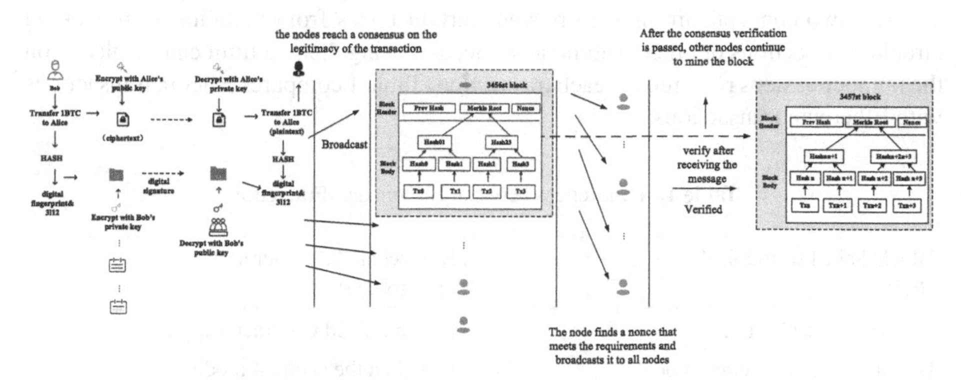

Fig. 2. Blockchain generation process.

3 Blockchain 2.0 - Ethereum

3.1 EThereum's Improvements to BItcoin's Logic

Bitcoin, as a virtual currency, is based on the core idea of a numerical exchange system. And while decentralization, consensus, hashing, and proof-of-work are all joint technologies adopted to allow this currency to be recognized and circulated. After the emergence of Bitcoin's decentralized consensus system, many currencies developed using this model have emerged, such as domain coins, colored coins, and meta-coins. However, Butlerin, the founder of Ethereum, argues that there are some problems with the scripts that implement this process of UTXO for verifying transactions in Bitcoin [16]. For example, lack of Turing completeness, lack of state, etc., so he proposed the Ethereum platform to address these problems and make some improvements.

Account. The "state" in Bitcoin consists of UTXO, while the "state" in Ethereum consists of accounts. It is divided into a contract account and an external account. Both of them contain transaction counters, Ethereum balances, storage for the account, and contract accounts additionally contain a contract code for the account. The external account has no code, the account holder can send messages to the contract account by creating and signing transactions, and when the contract account receives the message, its code will be activated to perform the corresponding operation.

Trading. Since Ethereum is based on an account model, if a cryptocurrency transaction is initiated, then only the sufficiency of the balance on the Ethereum blockchain account needs to be verified, and the source of the Ethereum on the Ethereum blockchain does not need to be stated. This is also an improvement to the transaction-based model in Bitcoin. It is suitable for cases where transactions are large and complex. A transaction in Ethereum needs to contain the message recipient, a signature identifying the sender, the amount of Ethereum transferred, an optional data field, the maximum number of steps allowed for the transaction to execute, and a fee paid by the sender for each computational step. All but the last two concepts are fields that must be included in a cryptocurrency. The last two concepts are used to prevent certain nodes from maliciously provoking circular transactions or other arithmetic waste, and using them, a limit can be placed on the number of steps required for each transaction. Table 1 compares Bitcoin transactions with Ethereum transactions.

Table 1. Differences between Bitcoin and Ethereum.

Blockchain 1.0 Bitcoin (P2P)	Blockchain 2.0 Ethereum (end-to-end)
Decentralized Currency	Decentralized Contract Support
Based on the transaction model No concept of balance	Based on the account model Concept of balance
Go and check if the currently used bitcoin has already been used	Natural protection against a double-spend attack

3.2 Other Improvements in Ethereum

Ethereum not only improves on the Bitcoin blockchain but also establishes a protocol that can create decentralized applications that can be effectively used for small and micro applications and improve the interaction between programs. The proposal of smart contracts has brought the application of blockchain technology to a new level. Based on it, Ethereum is more like a development platform where users can write different applications through different smart contracts and allows everyone to write smart contracts through built-in Turing completeness, so Ethereum is a black box close to a Turing machine.

Smart Contracts. Smart contracts appeared before the birth of Bitcoin and almost simultaneously with the birth of the Internet. It was conceptualized by SZABO in 1995 [9], published on the website of the Extropy Institute. But the Internet environment at the time was not suitable for the development of smart contracts, they were never well used. Until the successful emergence of Bitcoin and the continuous development of blockchain technology provided good underlying support for the development of smart contracts.

Smart contracts applied on the blockchain are defined: smart contracts are event-driven, stateful programs that run on a replicable and shared ledger and are capable of holding the capital on the ledger [17]. In blockchain systems, smart contracts can run in three environments, namely embedded, virtual machine-based, and container-based [18].

Ethereum smart contracts are executed inside the Ethereum Virtual Machine EVM. The creation and use of a contract can be seen as a transaction process. Among them, creating a contract can be seen as a special kind of transaction process, where the creation function implements the creation of a new contract using a set of fixed parameters that produce a new set of states. The process is as follows.

$$(\sigma', g', A) \equiv \Lambda(\sigma, s, o, g, p, v, i, e) \tag{1}$$

Where σ' is the latest status, g' is the available gas value, A is the sub-state, σ is the system status, s is the transaction sender, o is the source account body, g is the available gas value, p is the gas price, v is the account balance, i is the initialization EVM code, e is the depth of the created contract stack.

Message. Another special feature of Ethereum is messages, which are executed in a similar but different way to "transactions". A transaction is initiated by a message sent from an external account and is real-name. A message is initiated by a contract to other contracts and is anonymous. It contains the sender of the message, the receiver of the message, the amount of Ethereum to be transferred with the message, an optional data field, and the maximum number of computation steps allowed for the message. The emergence of blockchain 2.0 - Ethereum has expanded the application of blockchain technology from cryptocurrency transactions to savings wallets [19], crop insurance [20], intelligent multi-signature escrow [21], cloud computing [22], and many more areas [23, 24].

4 Blockchain 3.0 - Metaverse

4.1 Introduction to the Concept of Metaverse

The metaverse as the latest buzzword has attracted a wide range of attention from industry and academia. The metaverse seamlessly merges the real and virtual worlds while allowing computers to perform many complex activities, including creation, presentation, entertainment, social networking, and trade, thus promising an exciting digital world. In exploring the metaverse, a better real world can also be transformed [25]. The concept of metaverse was first mentioned in a science fiction novel called Snow Crash [26], and the development of blockchain has made it possible to realize this world that exists only in science fiction. Some technology giants, such as Facebook [27], Epic [28], and Jingteng Tech, are working on the integration of the metaverse into their lives.

The image of users in the metaverse is a projection of humans in the real world, and as the metaverse develops, the image, creation, and consumption of humans in it will refine and influence the real world. In an idealized metaverse, transactions between virtual goods, such as clothes, cars, and real estate, can be realized, as well as the exchange of virtual goods for real substances, in either form, it will affect the regular economy in the real world. The economic system in the metaverse can be divided into four parts: digital creation, digital assets, digital market, and digital currency. The article [25] summarizes them and compares their differences with the conventional economic system, as shown in Fig. 3.

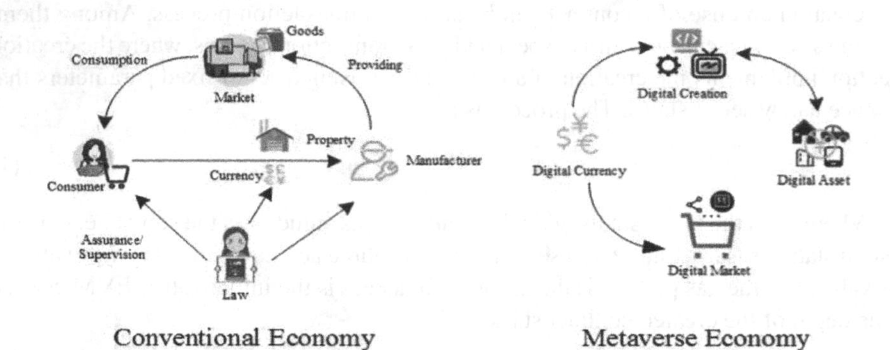

Fig. 3. Comparison of Traditional Economy and Metaverse Economy.

4.2 The Role of Blockchain in the Metaverse

If Ethereum is a black box close to a Turing machine, then by the 3.0 era blockchain will evolve into a Turing machine. Blockchain does three main things in this Turing machine. One is to ensure that most nodes are good [29]. Two is to maintain data security from being tampered with [30–32]. The third is to ensure that the transactions work correctly. The metaverse is a virtual world with an operation mechanism similar to the real world [33], a complete and self-consistent economic system, a complete industrial chain for producing and consuming digital products, and various transactions linking this virtual world together. The metaverse is positioned to be fair, notarized, and self-organizing, then the centralized economic system in the real world cannot function well in the metaverse due to the high transaction volume involved and the complexity of transactions. Thanks to the aid of big data techniques [34–36], the emergence of blockchain has broken the transaction barriers between regions in the circulation of money and opened the barriers in production, life, learning, and work so that the metaverse economic system is decentralized and the transactions using virtual images and virtual assets in the metaverse are legal and effective.

5 Conclusion

The development of blockchain technology from 1.0 to 3.0 is both the continuous improvement of technology and the continuous expansion of application scenarios, but there are also some problems during the development process, such as how to effectively communicate between different chains, the emergence of mining machines leading to the concentration of arithmetic power, how to establish a perfect cross-chain protocol when one chain represents one currency, and how to use blockchain to handle massive transactions. For any technology, it will go through the process of gradually increasing the heat, reaching the peak, and then gradually decreasing and finally leveling off. At present, the research on selfish mining, fragmentation technology, and micropayment channel in blockchain had gradually matured, while the research on blockchain economy, decentralized finance, the integration of blockchain and 5G/6G, blockchain and edge computing were gradually rising in popularity. This also shows that blockchain technology is and will be changing various fields such as finance and communication. In addition to research on blockchain application areas, many scholars are also exploring the improvement aspects of blockchain based technologies, such as side-chain, lightning network, cross-chain, etc. are still in the popular stage of research.

References

1. Huang, Y., Liang, W., Long, J., et al.: A novel identity authentication for FPGA based IP designs. In: 17th IEEE TrustCom/BigDataSE, pp. 1531–1536 (2018)
2. Bach, L.M., Mihaljevic, B., Zagar, M.: Comparative analysis of blockchain consensus algorithms. In: 41st IEEE MIPRO, pp. 1545–1550 (201)
3. Nakamoto, S.: Bitcoin: a peer-to-peer electronic cash system. Decent. Bus. Rev., 21260 (2008)
4. Wang, Y., Su, Z., Zhang, N., et al.: A survey on metaverse: fundamentals, security, and privacy. IEEE Commun. Surv. Tutor. (2022)
5. Liang, W., Tang, M., et al.: SIRSE: a secure identity recognition scheme based on electroencephalogram data with multi-factor feature. Comput. Electr. Eng. **65**, 310–321 (2018)
6. Xu, Z., Liang, W., Li, K.C., et al.: A time-sensitive token-based anonymous authentication and dynamic group key agreement scheme for industry 5.0. IEEE TII (2021)
7. Li, Y., Liang, W., Peng, L., et al.: Predicting drug-target interactions via dual-stream graph neural network. IEEE/ACM Trans. Comput. Biol. Bioinform. (2022)
8. Liang, W., Li, Y., Xie, K., et al.: Spatial-temporal aware inductive graph neural network for C-ITS data recovery. IEEE Trans. Intell. Transp. Syst. (2022)
9. Szabo, N.: Smart contracts: building blocks for digital markets. EXTROPY J. Transhumanist Thought (16) **18**(2), 28 (1996)
10. Bambara, J.J., Allen, P.R.: Blockchain. A practical Guide to Developing Business, Law and Technology Solutions. McGraw-Hill Professional, New York (2018)
11. Yaga, D., Mell, P., Roby, N., et al.: Blockchain technology overview. arXiv preprint arXiv: 1906.11078 (2019)
12. Belotti, M., Božić, N., Pujolle, G., et al.: A vademecum on blockchain technologies: when, which, and how. IEEE Commun. Surv. Tutor. **21**(4), 3796–3838 (2019)
13. Haber, S., Stornetta, W.S.: How to time-stamp a digital document. In: Menezes, A.J., Vanstone, S.A. (eds.) CRYPTO 1990. NCD, vol. 537. Springer, , Heidelberg (1990). https://doi.org/10.1007/3-540-38424-3_32
14. Zhang, F., Shi, B., Jiang, W.: A review of key technologies and applications of blockchain. J. Netw. Inf. Secur. **4**(4), 22–29 (2018)
15. Rachmawati, D., Tarigan, J.T., Ginting, A.B.C.: A comparative study of message digest 5 (MD5) and SHA256 algorithm. J. Phys. Conf. Ser. **978**(1), 012116 (2018)
16. Buterin, V.: A next-generation smart contract and decentralized application platform. White Pap. **3**(37), 2–1 (2014)
17. Osterland, T., Rose, T.: Model checking smart contracts for ethereum. Pervasive Mob. Comput. **63**, 101129 (2020)

18. Jili, F., Xiaohua, L., Tiezheng, N., et al.: Overview of smart contract technology in blockchain system. Comput. Sci. **46**(11), 1–10 (2019)
19. Praitheeshan, P., Pan, L., Doss, R.: Security evaluation of smart contract-based on-chain ethereum wallets. In: Kutyłowski, M., Zhang, J., Chen, C. (eds.) NSS 2020. LNCS, vol. 12570, pp. 22–41. Springer, Cham (2020). https://doi.org/10.1007/978-3-030-65745-1_2
20. Jha, N., Prashar, D., Khalaf, O.I., et al.: Blockchain based crop insurance: a decentralized insurance system for modernization of Indian farmers. Sustainability **13**(16), 8921 (2021)
21. Yang, X., Liu, M., Au, M.H., et al.: Efficient verifiably encrypted ECDSA-like signatures and their applications. IEEE TDSC **17**, 1573–1582 (2022)
22. Gai, K., Guo, J., Zhu, L., et al.: Blockchain meets cloud computing: a survey. IEEE Commun. Surv. Tutor. **22**(3), 2009–2030 (2020)
23. Liang, W., Yang, Y., Yang, C., et al.: PDPChain: a consortium blockchain-based privacy protection scheme for personal data. IEEE Trans. Reliab. (2022)
24. Liang, W., Xiao, L., Zhang, K., et al.: Data fusion approach for collaborative anomaly intrusion detection in blockchain-based systems. IEEE Internet Things J. (2021)
25. Yang, Q., Zhao, Y., Huang, H., et al.: Fusing blockchain and AI with metaverse: a survey. IEEE Open J. Comput. Soc. **3**, 122–136 (2022)
26. Joshua, J.: information bodies: computational anxiety in Neal Stephenson's snow crash. Interdiscip. Lit. Stud. **19**(1), 17–47 (2017)
27. Meta, I.: A social technology company. Meta **12**(11), 2021 (2021)
28. Games, E.: Fortnite. Epic Games (2017)
29. Liang, W., Tang, M., Long, J., et al.: A secure fabric blockchain-based data transmission technique for industrial Internet-of-Things. IEEE TII **15**(6), 3582–3592 (2019)
30. Li, Y., Gai, K., et al.: Intercrossed access controls for secure financial services on multimedia big data in cloud systems. ACM Trans. Multimedia Comput. Commun. Appl. (2016)
31. Gai, K., Qiu, M., Elnagdy, S.: A novel secure big data cyber incident analytics framework for cloud-based cybersecurity insurance. In: IEEE BigDataSecurity (2016)
32. Qiu, H., Dong, T., Zhang, T., Lu, J., Memmi, G., Qiu, M.: Adversarial attacks against network intrusion detection in IoT systems. IEEE IoT J. **8**(13), 10327–10335 (2020)
33. Kumar, P., Kumar, R., et al.: PPSF: a privacy-preserving and secure framework using blockchain-based machine-learning for IoT-driven smart cities. IEEE Trans. Netw. Sci. Eng. **8**(3), 2326–2341 (2021)
34. Hu, F., Lakdawala, S., et al.: Low-power, intelligent sensor hardware interface for medical data preprocessing. IEEE Trans. Inf. Technol. Biomed. **13**(4), 656–663 (2009)
35. Niu, J., Gao, Y., et al.: Selecting proper wireless network interfaces for user experience enhancement with guaranteed probability. JPDC **72**(12), 1565–1575 (2012)
36. Qiu, M., Xue, C., Shao, Z., Zhuge, Q., Liu, M., Sha, E.H.M.: Efficent algorithm of energy minimization for heterogeneous wireless sensor network. In: Sha, E., et al. (eds.) EUC 2006. LNCS, vol. 4096, pp. 25–34. Springer, Heidelberg (2006). https://doi.org/10.1007/11802167_5

Research on Blockchain-Based Food Safety Traceability Technology

Qinying Zhang(✉) and Hongze Wang

Department of Information Engineering, Wuhan Institute of City, Wuhan, China
446618463@qq.com

Abstract. Food traceability can be used to quickly pinpoint problematic links and minimise the risks associated with food safety incidents by viewing the trajectory of food circulation after a food safety incident has occurred. Blockchain, as an emerging technology with characteristics such as decentralisation, asymmetric encryption, tamper-proof and traceability, can be used to generate link traceability technology, which provides great convenience for us to effectively regulate food safety. In response to the need for multiple parties to share information in the traditional food safety traceability system there are many problems such as non-uniformity of the traceability chain, user cultivation and high traceability costs. This paper introduces Blockchain technology to improve the food safety traceability system, using the decentralised and fully distributed DNS service provided by Blockchain to achieve domain name query and resolution through peer-to-peer data transfer services between various nodes in the network, which can be used to ensure that the operating system and firmware of the infrastructure of the food production process have not been tampered with, introducing QR code technology, RFID, ZigBee.Web Web server and other key IoT technologies to achieve information identification and coding for identifying production objects, and identify the bottlenecks that limit the performance of the system by analysing and studying the consensus mechanism of the super ledger Fabric, and use smart contracts and consensus mechanisms as support to build food safety data assurance and food supervision methods to improve overall traceability efficiency.

Keywords: Blockchain · Food safety · Traceability technology · Information dentification · Consensus mechanism

1 Introduction

As a country with a large population, Food Safety is a matter of great importance to people's livelihood. The recurring Food Safety problems in recent years warrant our consideration, and how to make consumers eat with greater peace of mind is an issue that we urgently need to address. In fact, any issue concerning Food Safety can be solved by discussing Traceability systems. With globalisation, the food supply chain is becoming more and more complex. Food Safety

M. Qiu et al. (Eds.): SmartCom 2022, LNCS 13828, pp. 607–616, 2023.
https://doi.org/10.1007/978-3-031-28124-2_57

Traceability systems mainly address safety issues in the information field, reducing the risk of hazards occurring and giving consumers a real and tangible sense of security. Blockchain technology is a new technology in the Internet era, which has features such as decentralisation and tamper-proof, and the whole process can achieve anonymous machine autonomy and procedural auditing, which can be applied to Food Safety Traceability systems to address the trust crisis in a targeted manner and alleviate resource wastage.

Based on the idea of consensus mechanism optimization, this paper analyzes and studies the existing consensus framework and transaction consensus process of Blockchain Fabric, and makes Fabric support Byzantine fault tolerance by adopting the parallel plus cache scheme strategy for sorting services, so as to realize the optimization of Blockchain Fabric consensus mechanism. The experimental data in this paper shows that the optimized Fabric system makes the system Byzantine fault-tolerant while the system processing efficiency is also significantly improved.

The main points of this paper are as follows.

- Starting from the traditional traceability system itself, blockchain technology is introduced for the traceability of the information base and the decentralisation of the information processing.
- Use RFID, QR code, Zigbee and Web server technologies in different fields to realize the information identification function of food products.
- Combining the idea of consensus algorithm optimization with the practical application of the system, the method of this paper is tested and analyzed in the Hyperledger Caliper program, and the experimental results show that the method of this paper makes the Fabric system fault-tolerant and greatly improves the system processing efficiency and increases the system throughput to a certain extent.

2 Background

The continuous high rate of economic and social development in China is accompanied by a continuous improvement in people's living standards. The constant Food Safety accidents and other incidents in the market have caused a great impact on government departments, the stability of enterprises and the safety of people's lives. By describing the current situation in the field of Food Safety and analysing the shortcomings of existing Food Safety Traceability systems, this paper attempts to use Blockchain technology as a core technology and carry out the design of Food Safety Traceability methods by utilising the decentralised advantages of this technology.

Blockchain technology is essentially a decentralised distributed ledger concave, which uses cryptography to make the data on the ledger tamper-proof and uses Consensus mechanisms to ensure data consistency between distributed nodes. The decentralised storage mechanism, data immutability and data Traceability are well suited to the food Traceability application scenario.

3 Related Work on Food Safety Analysis

In practice, the current food traceability system adopts a segmented supervision system, and the traceability system is obviously "fragmented", with a wide range of participation and huge coverage [11]. The existing food quality traceability system itself has the following main problems. Poor data security, the central database can be tampered with, it is difficult to identify the responsible parties for food quality problems [16], the cost of traceability is high, consumer participation is low, the information acquisition and sharing capability of the traceability system is insufficient, and the management of the quality and safety traceability system is difficult and inefficient [6].

4 The Framework of Food Safety Traceability System

Blockchain technology has been introduced to ensure the traceability of food safety production information by decentralising the information through the traceability of the food information database, thus ensuring the traceability of all aspects of production [22]. In this process, food safety production information is entered into a database that integrates blockchain technology, which actively scans and decomposes the data after entry, dividing a complete base data into data consisting of "data headers + data blocks". This process is illustrated in Fig. 1.

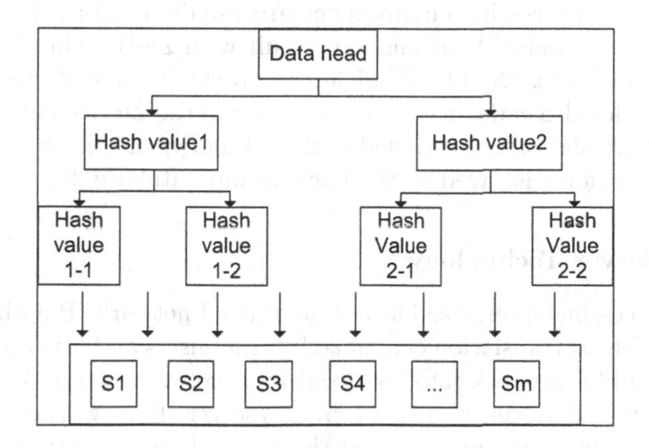

Fig. 1. Blockchain technology-based decomposition of entry data processing diagram

5 Related Work of Food Safety Traceability System

5.1 RFID Technology

The use of RFID radio frequency identification technology can be equipped with electronic tags from the raw material production, processing and purchase of food to the product sales stage, while using the global unified article coding technology to determine a unique ID number for each food product, and then using radio frequency identification technology to read or write the food process information, while using wireless transmission technology to release all the information in the food chain to the network management platform in a timely manner [19].

5.2 Two-Dimensional Code Technology

The introduction of two-dimensional code technology allows for the marking of food safety production information [21].

PDF417 Barcode. This code has a high error correction capability, with each line of the code involving both basic information about the line and also recording some information reflecting the characters used for error correction at the location [24].

Maxi Code. The main purpose is to enable the tracking and searching of packages [17].

5.3 ZigBee Technology

The intelligent sensors such as temperature, humidity, acidity and alkalinity distributed in the processing environment are usually composed of three parts: acquisition circuit, control transmission circuit with ZigBee chip CC2530 as the core, and antenna, etc. [2]. The CC2530 can transmit the collected data to the aggregation node, also called the gateway, through the ZigBee network, which is the management platform mentioned earlier. Finally, the client program of the management platform is saved to the background database [9].

5.4 Web Server Technology

Each food processing enterprise has a unique fixed network IP address, through the network address translation can make Internet users and other ordinary users through the public network into the local area network of the food processing enterprise [3]. The data centre uses a browser/server (B/S) working mode, which allows users to view data in real time through a browser without the need to install any client software [18].

6 Related Work to Research on Consensus Algorithms for Hyperledger

6.1 Overview of the Consensus Algorithm

Consensus ensures that the nodes in a distributed system can reach a final agreement even if a part of the nodes fails arbitrarily. In order to address the problem of non-determinism in blockchain transactions where there is a certain probability that the consensus outcome will be different, many blockchain platforms address the problem of non-determinism in blockchain transactions by writing smart contracts in specific languages [1].

6.2 Blockchain Fabric Consensus Mechanism

In order to improve the operational efficiency of the system for large-scale application scenarios, Fabric adopts the transaction model of executing first and then sequencing and verifying last (as shown in Fig. 2 [20]). Compared with sequential execution, executing transactions first avoids the problems of low performance, poor scalability and flexibility. The transaction steps: initially verifying the transaction and endorsing it [14], the consensus sorting service sorts the endorsed transactions and generates blocks, verifies the endorsement strategy of the transaction and submits it to the ledger [13].

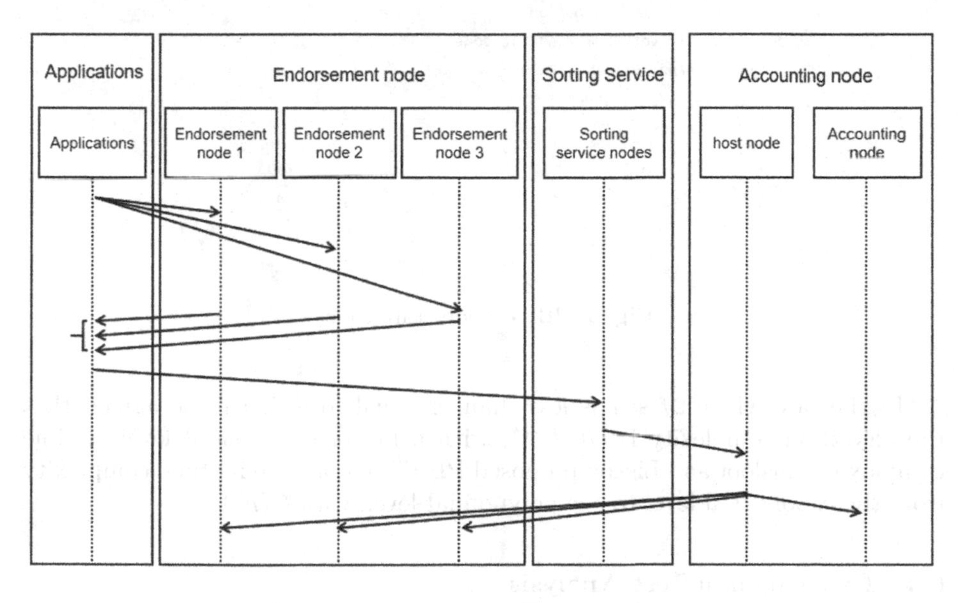

Fig. 2. Hyperledger transaction flow

6.3 Sorting Service

The following improvements are made to the sorting service.

Divide the sorting nodes into groups, taking into account the complexity of PBFT consensus and the fact that PBFT consensus efficiency decreases rapidly as the number of consensus nodes increases [7]. Continue to combine the ideas of POS and VRF. The nodes involved in the sorting service are divided into multiple parallel groups or organizations by means of a random lottery, but in order to increase the execution efficiency and throughput of the system [10]. A Java-based implementation of PBFT, such as Fig. 3, shows the PBFT node state code and tests its performance under different node states [14].

```
1  public enum VoteEnum {
2    PREPREPARE("节点生成区块", 10),
3    PREPARE("节点收到区块，进入准备状态，并对外广播", 20),
4    COMMIT("节点收到超过2f+1个不同节点的commit消息后，" +
5          "进入committed状态，并将其持久化到区块链数据库中", 40);
6
7    // 投票情况描述
8    private String msg;
9    // 投票情况状态码
10    private int code;
11
12    // 根据状态码返回对应的Enum
13    public static VoteEnum find(int code) {
14      for (VoteEnum ve : VoteEnum.values()) {
15        if (ve.code == code) {
16          return ve;
17        }
18      }
19      return null;
20    }
```

Fig. 3. PBFT node state code

In the case where $2f + 1$ is less than or equal to n, it can be agreed that the algorithm complexity is $O(n^{f+1})$, with a higher exponential level of time complexity. Kasloot and Liskov proposed PBFT to optimize its time complexity from the exponential level to the polynomial level, with $O(n^2)$.

6.4 Experimental Test Analysis

The Hyperledger Caliper program was chosen as the performance benchmarking framework [12]. Hyperledger Caliper allows users to measure the performance of the blockchain platform through a customised test environment and corresponding parameters for the test tool that needs to measure the performance of the blockchain platform [8]. Caliper's simplified architecture is shown in Fig. 4.

6.5 Analysis of Experimental Results

Optimized throughput of the Fabric system can reach 6000 tps with a single block transaction count of 200 at a node thread count of 30, with latency below 500 ms.

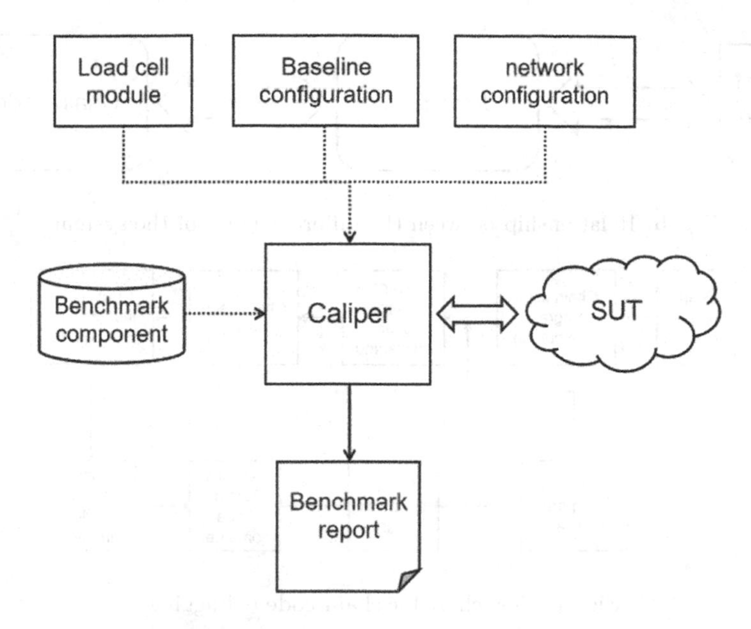

Fig. 4. Sketch of Caliper test architecture

This is a significant improvement in throughput over the Fabric system and reduces transaction latency at high performance [23]. This performance meets the performance requirements of food traceability system scenarios (Table 1).

Table 1. Comparison of Fabric performance before and after optimization

Arithmetic	Handing capacity/tps	Delay/ms	CPU (avg)/%
Fabric v1.4	2876	680	50.43
Optimized Fabric	5962	496	78.19

7 Blockchain-Based Food Traceability System Implementation

7.1 Functional Implementation of the System

The implementation of the food traceability system is divided into two parts, namely the web application and the functional implementation of the chain code [17]. The web application is further divided into the back-end business based on the java SDK of the Super Ledger Fabric and the front-end page based on the front-end architecture [4]. The front-end web page application interacts with the back-end business through a RESTful interface, and the back-end business interacts with the blockchain platform through a gRPC interface, as shown in Fig. 5.

Fig. 5. Relationship between the different parts of the system

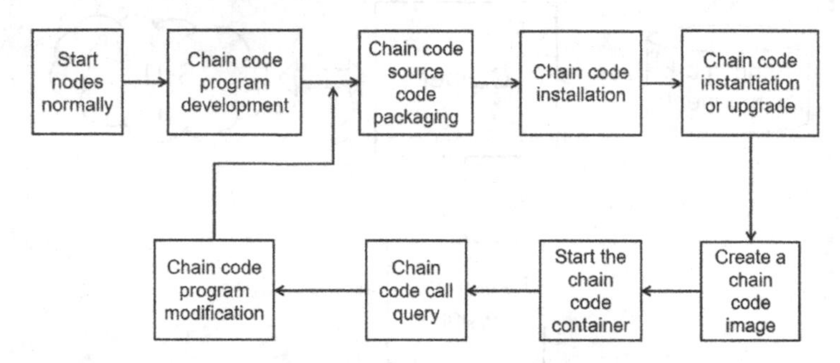

Fig. 6. Flow chart for chain code debugging

7.2 Chain Code Development

Smart contracts, the core of the blockchain, define the executable logic that can be generated to add to the distributed ledger data [15]. Smart contracts act as a set or series of common contracts defined by the system business prior to the transaction, including common terms, data, rules, concept definitions and processes. The Superledger Fabric currently supports chain code development in Go, Java and Node.js development languages [5]. The specific debugging process for chain code is shown in Fig. 6.

Fig. 7. Instantiation results diagram

A script to quickly start the blockchain network and initialize it is available from the Fabric's project on GitHub. First start the blockchain network and front-end services with the script, then create the channel and install the instantiated chaincode. The instantiation is completed as shown in Fig. 7.

Once the chain code has been deployed, the system can be accessed via a browser to verify the relevant functionality.

8 Conclusion

The food safety traceability system has been deeply integrated with blockchain technology. Blockchain ensures the data and transparency of food-related information with its decentralized service, and provides a reliable traceability system for consumers. At the same time, the framework of food traceability system based on blockchain technology as the core and Fabric's optimized consensus algorithm made relative improvements to initially realize a demonstration system of blockchain food traceability.

References

1. Bao, S.: Research on quality and safety assurance system of food supply chain. Ph.D. thesis, Chang'an University (2015)
2. Fei, C., Chunming, Y., Tao, C.: Blockchain-based food traceability system design. Comput. Eng. Appl. **57**(2), 10 (2021)
3. Furong, L., Sha, C.: Research on building China's agricultural products quality and safety traceability system based on blockchain technology. Rural Finance Res. (12), 5 (2016)
4. Gai, K., Wu, Y., Zhu, L., Zhang, Z., Qiu, M.: Differential privacy-based blockchain for industrial Internet-of-Things. IEEE Trans. Ind. Inf. **16**(6), 4156–4165 (2019)
5. Gajo, M.: Valereum blockchain übernimmt mehrheit an börse von gibraltar. Die Aktiengesellschaft **67**(1–2), r8 (2022)
6. He, H.W., Yan, A.: A review of blockchain-based smart contract technologies and applications. Comput. Res. Dev. **55**(11), 15 (2018)
7. Leibfried, P., Petry, H.: Blockchain in der Finanzberichterstattung. Controlling Manag. Rev. **66**(1), 54–59 (2022). https://doi.org/10.1007/s12176-021-0440-3
8. Liu, Q., X.C.: Construction of agricultural products quality traceability system based on blockchain technology. High Technol. Commun. **29**(3), 9 (2019)
9. M., L.T., X., Y., N., Z.Z., Y., T., X., W., Chaoqun., L.: Research on the application of blockchain+Internet of Things in agricultural products traceability. Comput. Eng. Appl. **57**(23), 11 (2021)
10. Müller, M.: Unternehmen halten blockchain für wichtig. Die Aktiengesellschaft **66**(17), r253–r254 (2021)
11. Shao, Q., Jin, C., Zhang, Z.: Blockchain technology: architecture and progress. J. Comput. Sci. **41**(5), 20 (2018)
12. Qiu, H., Qiu, M., Memmi, G., Ming, Z., Liu, M.: A dynamic scalable blockchain based communication architecture for IoT. In: Qiu, M. (ed.) SmartBlock 2018. LNCS, vol. 11373, pp. 159–166. Springer, Cham (2018). https://doi.org/10.1007/978-3-030-05764-0_17
13. Shengyan, L.: Research on blockchain-based food safety traceability technology. Modern Food **28**(17), 3 (2022)
14. Sun, Z., Xu, Q., Shi, B.: Price and product quality decisions for a two-echelon supply chain in the blockchain era. Asia-Pac. J. Oper. Res. **39**(01), 2140016 (2022)
15. Tian, Z., Li, M., Qiu, M., Sun, Y., Su, S.: Block-DEF: a secure digital evidence framework using blockchain. Inf. Sci. **491**, 151–165 (2019)
16. Wang Hongmei, Y.Y.: Research on blockchain-based food safety traceability technology. Electron. Des. Eng. **27**(13), 6 (2019)

17. Xiong, L.B., Na, Z.X., Xin, X.: Supply chain management in the era of e-commerce. China Manag. Sci. (03), 1–7 (2000)
18. Yang, D.R., Wenhui, Z., Snap, W., Long, W.: Design and implementation of hyperledger fabric-based food traceability system. Electron. Technol. Appl. **47**(3), 6 (2021)
19. YG, Y., SHuxin, Z.: A review of blockchain consensus mechanisms. Inf. Secur. Res. **4**(4), 11 (2018)
20. Ying, L.: Research on the mechanism of realizing the key links of sustainable green food supply chain. Ph.D. thesis, Dalian University of Technology (2014)
21. Yong, Y., Xiao-Chun, N.I., Zeng, S., Fei-Yue, W.: The development status and outlook of blockchain consensus algorithm. Zidonghua Xuebao/Acta Automatica Sin. **44**(11), 2011–2022 (2018)
22. Youlin, Z.: Exploration of food safety in China. North China Trade Econ. (10), 4 (2009)
23. Zhao, Z., Ma, J.: Application of blockchain in trusted digital vaccination certificates. China CDC Weekly **4**(6), 106 (2022)
24. Zhihong, D., Jinwang, L.: Exploring the privacy protection mechanism of super ledger. Sci. Technol. Innov. (17), 2 (2021)

Few-Shot Learning for Medical Numerical Understanding Based on Machine Reading Comprehension

Xiaodong Zeng, Wenhui Hu(✉), Xueyang Liu, Yuhang Chen, Wenyu Shao, and Lizhuang Sun

Peking University, No. 5 YiHeYuan Road, Haidian District, Beijing, China
{zengxiaodong,swy1798,sunlizhuang}@stu.pku.edu.cn, {huwenhui, liuxueyang,2101210553}@pku.edu.cn

Abstract. Numerical understanding relies on some content understanding techniques, which can be based on rules, entity extraction, and machine reading comprehension. Traditional methods often require a large number of regular expressions or a large number of data annotations, and often do not have a deep understanding of numerical values, lacking the ability to distinguish similar numerical values. In this paper, we propose a few-shot learning framework for numerical understanding tasks in Chinese medical texts, and through dynamic negative sampling of the training data, the model's ability to discriminate similar numerical values is enhanced. We use patient text data provided by 13 hospitals in Beijing to conduct experiments. The results show that our newly proposed method is superior to training the baseline pretrained language model directly, the EM increases by 38% and the F1 increases by 27.59%.

Keywords: Numerical Understanding · Few-shot Learning · Negative Sampling

1 Introduction

With the development of computer [1–3] and network [4–6], hospitals have increasingly collected and stored text data [7–9] related to the patient's condition through the electronic medical record technology [10] to better help analyze and predict the patient's condition and better serve the patient health management work in recent years. Among these text data, there are relatively regular structured texts, such as physical examination reports, inspection reports, etc. However, there are also a large number of unstructured data produced by doctors and nurses through hand-writing and keyboard typing, such as ward round records, patient complaints, admission records, discharge records, etc. These unstructured data often contain very important numerical information and disease entity information. Numerical information is helpful for diagnosis. Considering the diagnosis of diarrhea, 3 times a day and 10 times a day are different.

M. Qiu et al. (Eds.): SmartCom 2022, LNCS 13828, pp. 617–628, 2023.
https://doi.org/10.1007/978-3-031-28124-2_58

1.1 Numerical Understanding

The numerical understanding task was first systematically proposed by Corey A Harper et al. [11, 12], which mainly proposed five dimensions for numerical understanding, including numerical unit, modifier, measured entity, measured property, and qualifier. The English training data provided by it includes fields of agriculture, biology, chemistry, computer science, earth science, engineering, materials science, mathematics, and medicine, with a total of 448 documents, and the average number of numerical values in per document is about 5. Early numerical understanding systems were often constructed by rule-based methods. By defining a series of complex grammar rules, the numerical value and its corresponding five understanding dimensions were extracted according to lexical analysis tree, syntax analysis tree, and regular expressions [13]. Later, some scholars adopted NER (*Named Entity Recognition*) methods to define numerical entities and related entities through BIO or BIOES tags, which used LSTM, Bi-LSTM, CRF, ELMo, BERT, GPT-3 to train an entity extraction model, then extracted the entities and matched pairs based on a set of predefined distance rules [14, 15]. Some scholars used the methods based on machine reading comprehension for numerical understanding [16, 17]. The above-mentioned NER method first extracts numerical entities, and then performs machine reading comprehension on the extracted numerical entities to find their corresponding references. The question-answering method has gradually become a mainstream numerical understanding technology. However, these methods either rely on complex rules design or require a relatively large training data set to achieve better results.

1.2 Formula Definition

We analyze Chinese medical texts, especially patient discharge summaries and ward round records. A large number of meaningful values consist of a number and a unit. We focus on finding the reference, which is the combination of measured entity and measured property, as shown in the following Table 1:

Table 1. Examples of Chinese Medical Numerical Understanding.

Quantity	Digit	Unit	Reference
36.5 °C	36.5	°C	T
83次/分 (83 times/min)	83	次/分 (times/min)	P
20次/分 (20 times/min)	20	次/分 (times/min)	R
11月 (11 months)	11	月 (months)	反复便血 (Recurrent blood in the stool)
3月 (3 months)	3	月 (months)	乏力 (Weak)
4天 (4 days)	4	天 (days)	腹胀 (Abdominal distension)

The numerical extraction part can complete high-accuracy extraction through a simple rule-based method, such as regular expressions. The output is one of the inputs of our entire system.

In the numerical understanding part, the mainstream and effective method is to ask questions about the extracted numerical values, fill in the numerical values into a slot to get the question sentence: "{quantity}prompt?", and then ask the model with the context and questions to get the answers as the numerical references.

Specifically, we can define the problem as the following mathematical formula: Suppose the question is Q:

$$Q = \{q_1, q_2, q_3, q_4 \ldots q_m\}$$

Suppose the text is C:

$$C = \{c_1, c_2, c_3, c_4 \ldots c_n\}$$

Then the input is I:

$$I = \{[cls], q_1, q_2, \ldots q_m, [sep], c_1, c_2, \ldots c_n, [sep]\}$$

After going through the deep transformer encoder network, we get the vector of the last hidden layer as H:

$$H = \{h_0, h_1, h_2, \ldots h_m, h_{m+1}, h_{m+2}, h_{m+3}, \ldots h_{m+n+1}, h_{m+n+2}\}$$

The output of the model is:

$$P_{start(i)} = softmax_i(W_{start}|H)$$

$$P_{end(i)} = softmax_i(W_{end}|H)$$

Finally, the loss of the model is:

$$loss = -log\, p_{start}(s^*) - log\, p_{end}(e^*)$$

1.3 Overview of Our Work

We propose a few-shot learning framework and a dynamic negative sampling method for numerical understanding tasks to address the above two problems. We assume that there is a special coreference relationship between numerical value and reference, and use the data of antecedent and anaphoric in the ontoNotes_Release_5.0 [18] coreference resolution task to pretrain the current numerical understanding as a large-scale middle task. We suppose the language model to be knowledgeable [19–21], let the language model do the fill-mask task to generate the most suitable question for the current scene, and select the most effective question from a large number of generated questions. We use negative sampling to generate (context, quantity', no answer) as one unanswerable triple for each positive (context, quantity, reference) triple. We use a training method similar to SQuAD2.0 [22]. After comparing and assuming the experimental results, we design a dynamic negative sampling method that increases the difficulty of negative samples as the number of the training epoch increases.

In summary, the main contributions of this paper are as follows:

1. A few-shot learning framework for Chinese medical numerical understanding is proposed, and the evaluation indicators are greatly improved.
2. A dynamic negative sampling method is designed to generate unanswerable training samples, thereby enhancing the model's ability to discriminate similar values.

The results show that our newly proposed method is superior to training the baseline pretrained language model directly, the EM increases by 38% and the F1 increases by 27.59%.

2 Related Work

2.1 Regular Expressions

We compare our method with "Microsoft Recognizers Text". This is a very simple method, through the enumeration of the rules of expression for templated extraction, this method has advantages in time efficiency, but cannot meet the richness of natural language expression in the extraction effect.

2.2 BERT + CRF

Some scholars adopted NER (Named Entity Recognition) methods to define numerical entities and related entities through BIO or BIOES tags, which used LSTM, Bi-LSTM, CRF, ELMo, BERT, GPT-3 to train an entity extraction model, then extracted the entities and matched pairs based on a set of predefined distance rules [14, 15].

2.3 QuAnt

Some scholars used the methods based on machine reading comprehension for numerical understanding [16, 17]. The above-mentioned NER method first extracts numerical entities, and then performs machine reading comprehension on the extracted numerical entities to find their corresponding references. The question-answering method has gradually become a mainstream numerical understanding technology. However, these methods either rely on complex rules design or require a relatively large training data set to achieve better results.

Due to the lack of numerical understanding datasets in the Chinese medical field, we found that training with only a small amount of annotated data could not achieve the desired accuracy, however, a large number of annotations brought a large cost. On the other hand, when we used mainstream numerical understanding techniques for numerical understanding in the Chinese medical field, the model is not able to discriminate similar numerical values. Consider the quantity "38 °C", which is similar to the quantity "37 °C" extracted from "the temperature of patient is 37 °C", when we ask questions about 38 °C, the model still gives "temperature".

3 Our Approach

We present a few-shot learning framework for numerical understanding, and a dynamic negative sampling method for unanswerable question for numerical understanding (FSLUA). We will describe this algorithm in detail in this section.

Our FSLUA designed an intermediate task for numerical understanding to improve with only a small number of downstream samples. We use the knowledge characteristics of the language model to generate questions. We use the idea of negative sampling to generate unanswerable samples, improving the model's ability to understand similar values (Fig. 1).

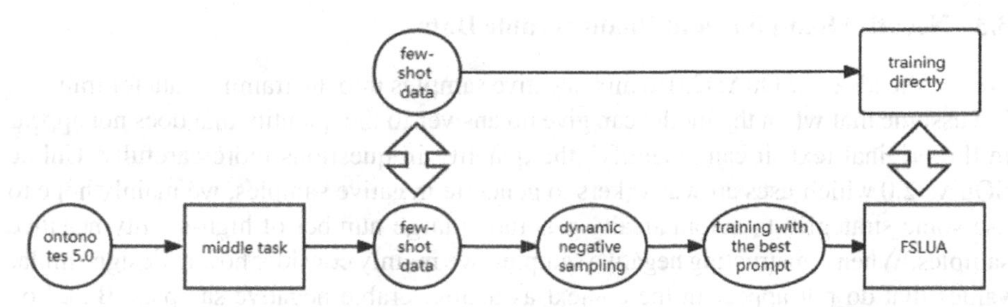

Fig. 1. The framework of our FSLUA.

3.1 Design of Middle Task

Since both a value and its reference describe the same thing, we hypothesize that a value and a reference can be viewed as a special coreference relationship. Specifically, it can correspond to two types of coreference. The QR value is in front of the reference, which can be regarded as the antecedent. The RQ value is after the reference, which can be regarded as the anaphoric (Table 2).

Table 2. Basic Types Divided by Relative Position.

Type	Context	Quantity	Reference
QR	8%的病人不容乐观 (8% of patients are not optimistic)	8%	病人(patients)
RQ	病人的体温为37度 (The patient's temperature is 37°)	37度 (37°)	体温 (temperature)

We use the coreference resolution dataset of ontonotes 5.0 [18] and train it under our existing machine reading comprehension framework. This process can be regarded as a middle task for numerical understanding tasks or large-scale pre-training to improve the model in the current framework.

3.2 Searching the Best Question Template

We use the same BERT [23–25] model as the subsequent machine reading comprehension for the fill mask task. Based on some inspirations for using BERT to do question and answer tasks [19–21], the language model is knowledgeable after pre-training. Since the language model can give answers based on its own knowledge, we assume that the language model can also use its own knowledge to give a question that is most suitable for the current scene. The sentence we design for the input model is "original text + numerical value + [MASK]*n + reference to numerical value", and the [MASK] part is filled in by the BERT model.

3.3 Negative Sampling and Unanswerable Data

We use the idea of SQuAD2.0 to mix negative samples into the training data for training. We assume that when the model can give no answer to the quantity that does not appear in the original text, it can "identify" the quantity in questions more carefully. Unlike SQuAD2.0 which uses crowdworkers to generate negative samples, we mainly hope to use some strategies to automatically generate a large number of high-quality negative samples. When constructing negative samples, we mainly consider how to design similar values that do not appear in the context as unanswerable negative samples. Based on numerical analysis, we consider the modification strategy in terms of two independent parts, including digit and unit.

For digit type, we replace the original n bits with the n bits random digits in replace (Table 3).

Table 3. Numeric Types and Modification Strategies.

Example	Quantity type	Modify	Replace
37	Int	n bits	n bits random digit
15.3	Float	n bits, except "."	n bits random digit
8–10, 159/87	With special token	n bits	n bits random digit

For the unit type, we consider two negative sampling methods. The first one is random sampling. For a certain unit, one negative unit is randomly sampled from all the unit library, except for the original unit. The other is sampling based on edit distance. We calculate the edit distance between a certain unit and all units in the unit library, and sort according to the edit distance, then randomly select a unit in the top k as a negative sample each time (Table 4).

Table 4. Unit Negative Sampling Strategies and Examples.

Unit negative sampling method	Original	Candidate	Negative
Random sampling	mg	元, L, mol, ... (Yuan-RMB, L, mol)	元 (Yuan-RMB)
Top k sampling based on edit distance	mg	kg, g, m, ml......	kg

4 Experiment

In this section, we evaluate our FSLUA on one dataset collected from 13 hospitals in Beijing and compare it with some previous baselines [13, 14, 16, 17].

4.1 Dataset

We collected data on 200,000 patients from 13 hospitals in Beijing, with a total of 19,730,000 medical records. We randomly selected 100 discharge summary texts for annotation, and selected 145 < context, quantity, reference> as the training set, the other 559 <context, quantity, reference> as the validation set.

4.2 Evaluation Metric

Following previous work, we mainly evaluate the numerical understanding performance on our dataset with EM (exact match) and F1 score. For Chinese, the F1 score is calculated based on the character level split.

4.3 Implementation

Pre-trained language models based on continuous word masking such as SpanBERT, WWM (whole word masking) [24, 25] have been shown to have good results on the machine reading comprehension task and other natural language understanding tasks. According to the characteristics of Chinese text, we choose chinese-roberta-wwm-ext-large as our baseline model. For a fair comparison with our model, the baseline model is also used in the NER tasks in the experiments.

We train all the models on 8 NVIDIA SXM4 A100 GPUs with 40 GB memory. The max sequence length is 512, and we will do padding to the max length. We set the batch size to 64. We use AdamW optimizer with the learning rate of 3e−5, linear decay with 5% warm up steps.

4.4 Results and Analysis

Table 5 illustrates the evaluation result of our model and the previous baselines.

Table 5. Evaluation Result.

Method	EM (%)	F1 (%)
Regular Expressions	14.67	21.88
SciBERT + CRF	36.12	44.26
QuAnt	48.20	61.63
FSLUA (Ours)	86.20	89.22

It can be seen that the accuracy of the rule-based method is not good, largely because we did not manually design all the relevant rules. What's more, complex rules are also difficult to deal with casually styled natural language. SciBERT + CRF reduces the enumeration of a large number of rules, but is not completely rule-free, so the accuracy is also not ideal. The mainstream QA method is better than the ruled-related methods. However, it does not meet business needs due to lacking enough annotation. The EM of our FSLUA increases by 38% and the F1 of our FSLUA increases by 27.59%, compared with QuAnt.

4.5 Ablation Study

In this section, we evaluate the EM and F1 score of our FSLUA, along with analysis on how its output differs from training directly, the effects of different training schemes (Table 6).

Table 6. Experiment of Pretraining with Different Amount of Middle Task.

The amount of data of coreference elimination pre-training	EM (%)	F1 (%)
0	48.20	61.63
1000	52.34	69.34
2000	66.73	76.23
3000	72.41	84.70
4000	72.62	83.22
5000	72.51	83.17
6000	70.12	81.23

We first use different amounts of coreference resolution data for pre-training, then use the training set for finetune, and perform evaluation on the validation set, finally get the following results. After experimental verification, we find that with the increase of coreference resolution data, the final accuracy of the model gradually increases. When the coreference resolution data reaches 3000, the model accuracy basically reaches its peak value. Adding the data continuously will not improve the accuracy of the model, however, there will be a certain decline, as shown in the table above. In this process, the model learns to find the related pairs according to coreference resolution, which helps the model to find numerical values and their corresponding references to a certain extent. However, the two tasks are not exactly the same after all. Excessive data can easily cause the model to over-fit the coreference resolution task and fail to migrate to the numerical understanding task after training with small samples. Therefore, adding an appropriate amount of coreference resolution data as an intermediate task before the training of the numerical understanding task is helpful with only a small number of samples. Compared with training directly, the EM increases by 24.42% and the F1 increases by 23.07%.

We search the question and get 191 templates in total. We have selected some templates with better accuracy and put them in the paper for comparison (Table 7).

Table 7. Experiment of Searching Templates.

Templates: {quantity}prompt	EM (%)	F1 (%)
是什么? (shi shen me)	78.11	84.63
是甚么? (shi shen me)	85.01	87.84
是什事? (shi shen shi)	85.01	88.02
都是? (dou shi)	81.56	87.02
就是? (jiu shi)	78.11	80.18
是什点? (shi shen dian)	85.01	85.01
份是? (fen shi)	85.01	85.91
钟是? (zhong shi)	81.56	84.72

Note that the meanings of the templates in the table are all "what is it" with different special characters, so we just indicate the Chinese pinyin in the table.

Overall, language expressions that are more human-like will have higher EM and F1 scores, for example, "是什么? (shi shen me)", "是甚么? (shi shen me)". However, we also found that the best template "是什事? (shi shen shi)" is not absolutely coherent in terms of language, which is also contrary to our initial assumption. But we believe that these templates are completed by the model itself, so these prompts are the most suitable ways for the model to do this process, which will be slightly different from the paradigm learned by human. The best EM value is up to 85.01%, and the F1 value is up to 88.02%.

We compared the effects of different negative sampling methods on the experimental results. Training with 1 or 2 or 3 or 4 directly, the negative samples are very similar to the positive quantities from start to end, which will have an undesirable impact on the positive samples. The final result is worse than not using the strategy, because the model will also give no answer to some positive data. However, training with 5 or 6 or 7 or 8 directly, the negative samples are relatively simple for the model, thus the training for the model is insufficient (Table 8).

Table 8. Experiment of Unanswerable Negative Sampling.

id	Strategy	EM (%)	F1 (%)
1	Only quantity, n = 2	81.51	83.26
2	Only quantity, n = 1	79.23	82.13
3	Only unit, top k = 20	84.54	87.31
4	Only unit, top k = 10	82.31	85,12
5	Both quantity and unit, n = 2 top k = 20	85.05	88.06
6	Both quantity and unit, n = 1 top k = 20	85,82	88.93
7	Both quantity and unit, n = 2 top k = 10	85.45	88.57
8	Both quantity and unit, n = 1 top k = 10	85.23	88.38
9	Dynamic Negative Sampling	86.2	89.22

After some experiments, we try a training order "5-6-7-8-3-4-1-2", whose degree of confusing the model increases as the epoch increases. And we call this method dynamic negative sampling. After experimental comparison, it is found that the accuracy is better than directly using 1, 2, 3, 4, 5, 6, 7, and 8. By increasing the noise gradually, the model can learn from simple negative samples to difficult negative samples, which results in a large boosting.

5 Conclusion

We present a few-shot learning framework for numerical understanding, and a negative sampling method for unanswerable question for numerical understanding. The middle task part of few-shot learning framework boosts the EM from 48.20% to 72.62%, F1 from 61.63% to 84.70%, compared with training directly. Then the question searching part of few-shot learning framework boosts the EM to 85.10%, F1 to 88.02%. Although there is not much confusing data for similarity numerical understanding in the validation dataset, the confusing questions get the right answer after our dynamic negative sampling. In order to illustrate our conclusion, we do some test case, and model can distinguish the difference between the similar quantities. Finally, the negative sampling boosts the EM to 86.2%, F1 to 89.22%.

References

1. Qiu, M., Li, H., Sha, E.: Heterogeneous real-time embedded software optimization considering hardware platform. In: ACM symposium on Applied Computing, pp. 1637–1641 (2009)
2. Li, J., Ming, Z., et al.: Resource allocation robustness in multi-core embedded systems with inaccurate information. J. Syst. Archit. **57**(9), 840–849 (2011)
3. Qiu, M., Chen, Z., Ming, Z., Qin, X., Niu, J.: Energy-aware data allocation with hybrid memory for mobile cloud systems. IEEE Syst. J. **11**(2), 813–822 (2014)

4. Niu, J., Gao, Y., et al.: Selecting proper wireless network interfaces for user experience enhancement with guaranteed probability. JPDC **72**(12), 1565–1575 (2012)
5. Qiu, M., Xue, C., Shao, Z., Zhuge, Q., Liu, M., Sha, E.-M.: Efficent algorithm of energy minimization for heterogeneous wireless sensor network. In: Sha, E., Han, S.-K., Xu, C.-Z., Kim, M.-H., Yang, L.T., Xiao, B. (eds.) EUC 2006. LNCS, vol. 4096, pp. 25–34. Springer, Heidelberg (2006). https://doi.org/10.1007/11802167_5
6. Qiu, H., Dong, T., et al.: Adversarial attacks against network intrusion detection in IoT systems. IEEE Internet Things J. **8**(13), 10327–10335 (2020)
7. Qiu, H., Zheng, Q., et al.: Topological graph convolutional network-based urban traffic flow and density prediction. IEEE Trans. ITS **22**(7), 4560–4569 (2020)
8. Li, Y., Gai, K., et al.: Intercrossed access controls for secure financial services on multimedia big data in cloud systems. ACM Trans. Multimedia Comput. Commun. Appl. **12**(4s), 1–18 (2016)
9. Gai, K., Qiu, M., Elnagdy, S.: A novel secure big data cyber incident analytics framework for cloud-based cybersecurity insurance. In: IEEE BigDataSecurity (2016)
10. Hu, F., Lakdawala, S., et al.: Low-power, intelligent sensor hardware interface for medical data preprocessing. IEEE Trans. Inf. Technol. Biomed. **13**(4), 656–663 (2009)
11. Harper, C., Cox, J., Kohler, C., et al.: SemEval-2021 task 8: MeasEval–extracting counts and measurements and their related contexts. In: 15th International Workshop on Semantic Evaluation (SemEval-2021), pp. 306–316, August 2021
12. Therien, B., Bagherzadeh, P., Bergler, S.: CLaC-BP at SemEval-2021 task 8: SciBERT plus rules for MeasEval. In: SemEval-2021, pp. 410–415, August 2021
13. Foppiano, L., Romary, L., et al.: Automatic identification and normalization of physical measurements in scientific literature. In: ACM Symposium on Document Engineering, pp. 1–4 (2019)
14. Gangwar, A., Jain, S., Sourav, S., Modi, A.: Counts@IITK at SemEval-2021 task 8: SciBERT based entity and semantic relation extraction for scientific data. arXiv preprint arXiv:2104.01364 (2021)
15. Kohler, C., Daniel Jr, R.: What's in a measurement? Using GPT-3 on SemEval 2021 task 8–MeasEval. arXiv preprint arXiv:2106.14720 (2021)
16. Davletov, A., Gordeev, D., Arefyev, N., Davletov, E.: LIORI at SemEval-2021 task 8: ask transformer for measurements. In: 15th SemEval-2021, pp. 1249–1254, August 2021
17. Avram, A.M., Zaharia, G.E., Cercel, D.C., Dascalu, M.: UPB at SemEval-2021 task 8: extracting semantic information on measurements as multi-turn question answering. arXiv preprint arXiv:2104.04549 (2021)
18. Hovy, E., Marcus, M., Palmer, M., et al.: OntoNotes: the 90% solution. In: Human Language Technology Conference of the NAACL, Companion Volume: Short Papers, pp. 57–60, June 2006
19. Jiang, Z., Xu, F.F., Araki, J., Neubig, G.: How can we know what language models know? Trans. Assoc. Comput. Linguist. **8**, 423–438 (2020)
20. Roberts, A., Raffel, C., Shazeer, N.: How much knowledge can you pack into the parameters of a language model? arXiv preprint arXiv:2002.08910 (2020)
21. Zhang, Z., Han, X., Liu, Z., Jiang, X., Sun, M., Liu, Q.: ERNIE: enhanced language representation with informative entities. arXiv preprint arXiv:1905.07129 (2019)
22. Rajpurkar, P., Jia, R., Liang, P.: Know what you don't know: unanswerable questions for SQuAD. arXiv preprint arXiv:1806.03822 (2018)
23. Devlin, J., Chang, M.W., Lee, K., Toutanova, K.: BERT: pre-training of deep bidirectional transformers for language understanding. arXiv preprint arXiv:1810.04805 (2018)

24. Joshi, M., Chen, D., Liu, Y., et al.: SpanBERT: improving pre-training by representing and predicting spans. Trans. Assoc. Comput. Linguist. **8**, 64–77 (2020)
25. Cui, Y., Che, W., Liu, T., Qin, B., Yang, Z.: Pre-training with whole word masking for Chinese BERT. IEEE/ACM Trans. Audio Speech Lang. Process. **29**, 3504–3514 (2021)

A Feature Extraction Algorithm Based on Blockchain Storage that Combines ORB Feature Points and Quadtree Division

Yawei Li[1], Yanli Liu[1](✉), Heng Zhang[1], and Neal Xiong[2]

[1] Shanghai Dianji University, P.O. Box 201306, Shanghai, China
216003010223@st.sdju.edu.cn, {liuyl,zhangheng}@sdju.edu.cn
[2] Department of Mathematics and Computer Science, Sul Ross State University, Alpine, TX 79830, USA

Abstract. In the process of 5G power grid inspection robot moving for a long time, the sensor constantly collects the feature information of the substation. Due to the limited memory capacity, this feature information must be stored on the network, which requires the storage network must have strong security. The storage network based on blockchain can better solve the problem of feature data encryption, and a good feature extraction method can also relieve the pressure of network storage. As the input information of the whole SLAM, feature points play a crucial role in the detection performance and accuracy of the whole SLAM. When the extracted feature points are few or evenly distributed, they cannot express the information of the whole environment, which will make the mapping and localization error of SLAM system larger, and seriously lead to the loss of tracking. In this paper, we first analyze the standard ORB algorithm and the Qtree_ORB algorithm. Aiming at the problems existing in the two algorithms, an improved ORB feature extraction algorithm is developed. For the problem that the feature points extracted by the Qtree_ORB algorithm are too uniform, the maximum division depth of the quadtree is limited according to the number of feature points required for each layer of the image pyramid, which improves the problem that the feature points are too uniform. Finally, we evaluate the performance of the improved algorithm, and analyze the uniformity of feature points to verify the performance and robustness of the improved algorithm.

Keywords: Visual SLAM · Blockchain · ORB Feature Points · Quadtree · Feature extraction

1 Introduction

Feature point extraction and matching is the most basic front-end step of visual SLAM (*Simultaneous Localization and Mapping*), which directly determines the effect of SLAM localization and mapping [1–3]. Feature points can express the essence of the image and analyze the information in the environment. They play

M. Qiu et al. (Eds.): SmartCom 2022, LNCS 13828, pp. 629–638, 2023.
https://doi.org/10.1007/978-3-031-28124-2_59

an important role in the algorithm based on feature matching [4,5]. High-quality feature points will have a better contribution to the system performance and robustness of SLAM, and it is not easy to lose environmental information and tracking loss. In addition, image feature points are also an important representation of scene information. If the extracted image feature points are not uniform, they cannot fully reflect the information in the environment and also affect the system accuracy of SLAM. For mobile robots, robotic SLAM is generally required to have low cost [6], low power consumption [7,8], and high accuracy [9,10]. Connecting the robot to the web and cloud would expose the robot SLAM process to possible hacking, which could lead to loss of map data and or tampering with the SLAM process [11–13]. Blockchain is an alternative to build a transparent and secure environment for data storage [14–16], which can be well protected against data tampering in transmission [17–19]. The storage network based on blockchain can better solve the problem of feature data encryption [20–22], and a good feature extraction method can also relieve the pressure of network storage [23].

There are many algorithms for the extraction of environmental feature points, each with its own characteristics. The standard ORB(Oriented FAST and Rotated BRIEF) [24,25] algorithm is widely used because of its high real-time performance [26]. However, the standard ORB feature point extraction algorithm [27] also has some unavoidable problems. For example, most of the feature points will be concentrated in the regions with strong texture information, while the number of feature points extracted in some regions with weak texture information is relatively small [28]. Yang et al. [29] proposed a 3D reconstruction method for weakly textured scenes. They use calibrated cameras to take photos from multiple views, and then use these images to complete high-precision reconstruction through SFM (*Structure from Motion*) and MVS (*Multiple View Stereovision*) algorithms. Zhou et al. [30] add line features on the basis of ORB-SLAM, and construct an improved point-line integrated monocular VSLAM(Visual SLAM) algorithm to solve the problem of missing feature points in regions with weak textures. An algorithm for extracting ORB feature points based on quadtree structure (Qtree_ORB) is proposed in ORB-SLAM2 [31]. The Qtree_ORB algorithm divides the quadtree structure for each layer of the image pyramid, and retains the point with the largest Harris response value in each node. Although the algorithm can evenly extract the feature points in the environment, the feature points are too uniform in this algorithm, and the operation efficiency is slow due to too many quadtree nodes, and the algorithm retains some low-quality features point. Therefore, it is necessary to study a feature point extraction algorithm with feature point homogenization, better real-time performance and higher efficiency [32].

This paper improves the feature point extraction algorithm in ORB-SLAM based on the above problem, reduces the impact on real-time performance due to too deep division of quadtrees [33], and limits the division depth of quadtrees to reduce the impact of division depth on feature point extraction when processing feature point homogenization, which effectively solves the above problem.

The main contributions of this paper are as follows: (1) This paper improves the feature point extraction algorithm based on quadtree in ORB-SLAM2 algorithm, limits the division depth of quadtree based on the number of feature points required by each layer of the image pyramid, reduces the number of iterations of quadtree partition and the extraction of low-quality feature points. (2) In this paper, the meshing of each image layer is performed adaptively according to the number of feature points required. The number of divisions is then reduced by limiting the quadtree segmentation depth. The problem of uniform extraction of feature points is improved, and the feature matching time is reduced.

The rest of this paper is organized as follows. The standard ORB algorithm and the Qtree_ORB algorithm are described in Sect. 2. The improved Qtree_ORB feature point algorithm is described in Sect. 3. Some experiment results are shown in Sect. 4, and Sect. 5 sets out the conclusions.

2 Standard ORB and Qtree_ORB Feature Point Algorithm

2.1 Introduction to Standard ORB Algorithms

In the research process, it has been found that the corner and edge points in the image are more representative of the information in the scene than the grayscale values of the image pixels directly [34]. Researchers define these points as key points, which describe information such as the position, direction and scale of feature points in the scene. But how to describe a feature point is also a problem that needs to be solved, so the concept of the descriptor is proposed to describe that feature point. ORB algorithm combines the two concepts and proposes an ORB feature point extraction algorithm [35].

2.2 Standard Qtree_ORB Feature Point Algorithm

In order to extract feature points in the ORB-SLAM2 algorithm, the ORB algorithm is used for extraction, and the quadtree structure is used to scree the extracted feature points [36]. It mainly includes the following processes: (1) The process of feature point extraction at different scales is solved by constructing image pyramid. (2) Mesh the images constructed in the first step at different scales. (3) Limit the extraction of feature points, set a minimum threshold, extract feature points from the grid set in the second step. If feature points are not enough to extract, lower the threshold and continue to extract feature points until enough feature points are extracted. (4) Based on the quadtree, the obtained FAST corner points are selected uniformly.

3 Improved Qtree_ORB Feature Point Algorithm

3.1 Construction of Image Pyramids

By constructing an n-layer image pyramid to increase the scale information of the image, the problem that ORB feature points do not have scale invariance is solved. Marking the initial image as I_0, the latter $I_1\tilde{}I_{n-1}$ layers can be obtained by down sampling the image from the previous layer. The scaling scale of each layer is:

$$S_i = ScaleFactor^i (i = 0, 1, ..., n - 1) \tag{1}$$

Where n is the number of layers of the pyramid (n = 8), and $ScaleFactor$ is the scale reversal of each layer of the image pyramid. S_i is the scaling scale of the image pyramid for each layer set. Define $Invs = 1/ScaleFactor$, the width and height of the original image are W and H, and the corresponding relationship between the width and height of images in each layer of the image pyramid is:

$$w_i = w \times InvS^i \tag{2}$$

$$h_i = h \times InvS^i \tag{3}$$

Where W_i is the width of the image at layer i, h_i is the height of the image at layer i, $i = 0, 1, \ldots, 7$.

Through the construction of image pyramids, different scale information is added to the image to satisfy the problem of scaling in the image scene, which solves the problem that ORB feature extraction algorithm does not have scale invariance.

3.2 Division of the Grid

After obtaining the image pyramid of each layer, we divide the image pyramid of each layer into grids. The original ORB-SLAM2 algorithm uses a fixed aspect ratio to mesh the image, but this does not adapt well to the environment in the scene. In this paper, we improve it by using an adaptive form of meshing, which adaptively meshes each layer of the image according to the number of feature points needed, and determines the meshing according to the area.

$$\begin{cases} width = maxBordeX - minBordeX \\ height = maxBordeY - minBordeY \\ W = \sqrt{width \times height \times \epsilon / N} \end{cases} \tag{4}$$

Where $width$ is the width of each pyramid image, $height$ is the height of each pyramid image, ϵ is the scale factor, $maxBordeX$, $minBordeX$, $maxBordeY$, $minBordeY$ is the boundary coordinates of each image layer, and the experimentally verified ϵ value is 1.8. N is the required number of feature points. W is the width and height of the divided grid, which is rounded if it is not an integer.

The actual width and height of the grid are: where $wCell$ is the actual width of the divided grid, $hCell$ is the actual height of the divided grid, and $round$ is the rounding function.

$$\begin{cases} wCell = round(width/W) \\ hCell = round(height/W) \end{cases} \tag{5}$$

The actual divided rows and columns are calculated by the width and height of the image pyramid of each layer. In the case that they are not integers, they are rounded upward, as shown in the following formula:

$$\begin{cases} nCols = width/W \\ nRows = height/W \end{cases} \tag{6}$$

After meshing the images of each layer, feature points are extracted from the images in the mesh, and the feature points of each layer are calculated according to the following criteria:

$$N_i = N \times \frac{1 - InvS^i}{1 - InvS} \tag{7}$$

Define N_i as the number of feature points per layer, and $InvS$ is set as each layer's reciprocal of the image pyramid scale factor.

After the grid division in the feature point extraction process, a minimum threshold is set and the feature points are extracted in the grid. After feature extraction, if enough feature points cannot be extracted from the grid, the threshold is lowered until enough feature points are finally extracted. Finally the non-maximal value suppression algorithm is used to eliminate the overlapping feature points.

3.3 Division of Quadtrees

After the information extraction of the scene by the ORB algorithm extraction process, there are many redundant and duplicate feature points, which are eliminated and selected by the quadtree in this paper. Extracted feature points are divided into four nodes using a quadtree structure, and feature points in the nodes are filtered according to the Harris response values.

The standard quadtree filters feature points by splitting nodes at each level. However, in some scenarios, the division depth of the quadtree is too deep, which has an impact on the efficiency of the algorithm and thus on the real-time performance of SLAM, and extracted feature points are too uniform when the depth of the quadtree is too deep. To address these problems, this paper limits the division depth of quadtrees. Firstly, the maximum partition depth D_{max} of the pyramid is set according to the number of feature points required by each layer of image pyramid. The relationship between the division depths in a quadtree can be expressed as follows:

$$d \leqslant D_{max} \tag{8}$$

Where d is the current division depth of the quadtree, D_{max} is the maximum division depth. The expression of the relationship between the number of nodes per layer and the depth of division is:

$$Nodes_i = 4^d \leqslant 4^{D_{max}} \tag{9}$$

Where $Nodes_i$ is the number of nodes at the current division depth.

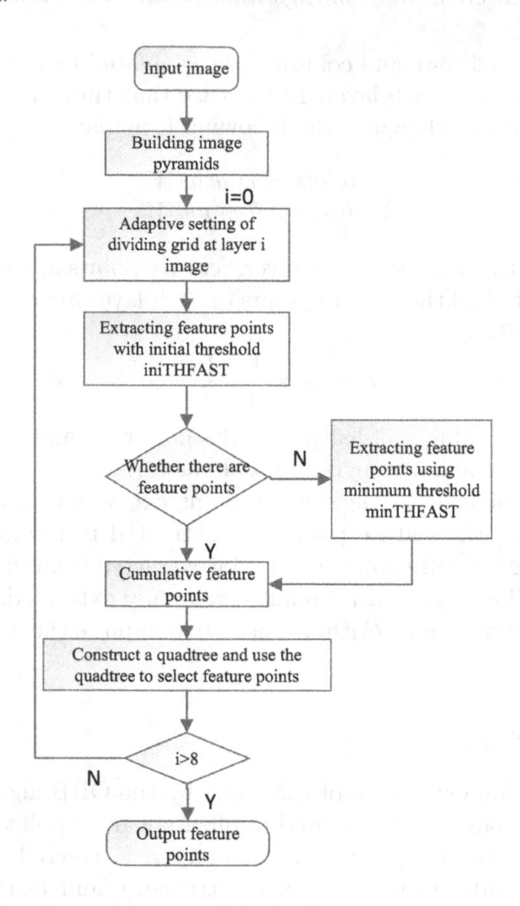

Fig. 1. Flowchart of improved Qtree_ORB feature point algorithm

Figure 1 shows the overall flow chart of the improved algorithm in this section. Firstly, feature points are extracted through adaptive thresholds in each layer of image pyramid, then feature points are screened according to the improved quadtree structure, and finally feature points are retained according to Harris response values [37].

4 Experimental Verification

4.1 Experimental Evaluation Index

The improved feature point extraction algorithm in this paper uses the uniformity of distribution of image feature points for evaluation [38].

$$\sigma = \sqrt{\frac{(m_1 - \overline{m})^2 + (m_2 - \overline{m})^2 + ... + (m_{10} - \overline{m})^2}{10}} \tag{10}$$

Where σ denotes the uniformity of the feature point distribution, $m_1, m_2, .., m_{10}$ indicates the ten divided regions, $\overline{m}$ is the average of ten areas. The time taken to extract feature points in the image is measured between the improved algorithm and the original algorithm, and the cost time T is compared to reflect the operation efficiency of the algorithm.

4.2 Comparison of Feature Point Uniformity

In this experiment, the images in the Mikolajczyk [39] dataset are used to compare ORB-SLAM2 with the improved feature point extraction algorithm in this paper. Each group of images in the Mikolajczyk dataset contains six images, which are divided into one standard image and five images in different changing environments. In this experiment, eight datasets are selected for testing, where boat is the rotation change dataset, bark is the scale change dataset, leuven is the illumination change dataset, bikes and trees are the blur change dataset, wall and graf are the viewpoint change dataset, and ubc is the image compression change dataset. In this paper, the first two images in the dataset are selected as samples for the experiments of feature point extraction by different algorithms in the above six variations, respectively. The hardware configuration of all experiments in this paper is as follows: GPU NVIDIA 2080Ti, CPU: Inter I5-10300H and Ubuntu18.04 operating system.

All the data in the experiments in this paper have different errors in different hardware devices, so the data measured in the experiments are taken 10 times to avoid random mistakes. After experimentally determining the depth of the quadtree division in the image pyramid the maximum depth of the first to the fourth level is taken as 3, the maximum depth of the fifth to the sixth is 2, and the maximum depth of the seventh to the eighth level is taken as 1. In order to balance the algorithm between the calculation speed and the uniformity of feature points, the minimum threshold of the Harris response value is 15. In this paper, the first benchmark image and the second matching image in 8 sets of datasets are selected. According to the feature point evaluation criteria set in this paper, the uniformity of image feature points is tested in different scenes in the dataset. In order to quantify the distribution of feature points extracted from images by different algorithms, the evaluation criteria in this paper are selected to calculate the uniformity of feature points in images, as shown in Table 1.

Table 1. Comparison of the uniformity of image feature point distribution.

Algorithm	boat	bark	leuven	tress	wall	ubc	graf	bikes	Uniformity
Improved_Qtree_ORB	85.27	17.85	35.62	18.39	52.68	37.04	25.36	38.26	38.81
Qtree_ORB	90.49	54.72	43.65	88.74	97.85	78.91	48.64	25.75	66.09
ORB	179.72	267.74	174.58	179.72	134.72	183.55	186.12	176.23	185.30

In Table 1, the improved Qtree_ORB algorithm extracts feature points with less uniformity compared to the standard ORB and Qtree_ORB. Since the standard deviation of the number of feature points is used as the uniformity standard in the standard in this section, it can be seen that the feature points extracted by the improved Qtree_ORB algorithm in this paper are more uniform and have a great improvement compared to the other two. This is due to the restriction on the division depth of the quadtree in the Improved_Qtree_ORB algorithm so that it extracts essentially the same number of feature points in each quadtree node. The uniformity of Improved_Qtree_ORB is 38.81, which is 79.1% higher than ORB algorithm and 41.2% higher than Qtree_ORB's 66.09.

Table 2 shows the time spent by the three algorithms in extracting feature points as a reflection of the operational efficiency of the three algorithms. In this table, ORB algorithm consumes the shortest time and runs with the highest efficiency. The least efficient operation is the Qtree_ORB algorithm, mainly because its algorithm needs to add a quadtree to manage the feature points when extracting them. Then an image pyramidal grid division is used for each layer to extract feature points, which is not limited by the depth of its division in the quadtree division process, resulting in a longer time to extract feature points. Compared to Qtree_ORB, the average running time of our method tested with multiple datasets is improved by 5.27%. The Improved_Qtree_ORB algorithm running efficiency is 8.93% higher than the original Qtree_ORB algorithm on the bot dataset. Our method saves running time while ensuring good uniformity of feature points. In this paper, the efficiency of the algorithm is improved by limiting the depth of the quadtree division.

Table 2. Feature point extraction time for Mikolajczyk dataset.

Algorithm	bark	wall	graf	bikes	leuven	boat	ubc	trees	Average
ORB	130.27	195.85	152.62	190.39	160.68	164.04	188.52	215.82	174.77
Improved_Qtree_ORB	168.49	278.72	196.65	225.74	214.85	240.91	235.73	316.62	234.71
Qtree_ORB	178.72	284.74	206.58	231.72	236.72	264.55	242.36	336.74	247.77

5 Conclusion

The standard visual SLAM algorithm uses a quadtree form for feature point extraction in the scene. However, when the depth of the quadtree is too deep, the problem of extracting feature points too uniformly is observed when extracting feature points. In this paper, the algorithm for extracting feature points was improved, and the division principle of each layer of the quadtree was connected with the image pyramid, which solves the problem of low quality and too uniform feature points. It can better extract environmental information, reduce the redundancy of extracting environmental information, and relieve the storage pressure on the blockchain network. Finally, the uniformity of the extracted feature points was experimentally compared to analyze the advantages and disadvantages of the improved algorithm. The experimental results showed that our method can meet the real-time requirements of the system.

Acknowledgements. This work was supported in part by the National Natural Science Foundation of China under Grant 61963017; in part by National Natural Science Foundation of China, under Grant (No. 62177019, F0701); in part by Shanghai Educational Science Research Project, China, under Grant C2022056; in part by the Outstanding Youth Planning Project of Jiangxi Province, China, under Grant 20192BCBL23004.

References

1. Jia, Y., et al.: A survey of simultaneous localization and mapping for robot. In: IEEE 4th Advanced Information Technology, Electronic and Automation Control Conference (IAEAC), vol. 1, pp. 857–861 (2019)
2. Yao, Y., et al.: Privacy-preserving max/min query in two-tiered wireless sensor networks. Comput. Math. Appl. **65**(9), 1318–1325 (2013)
3. Guo, J., Ni, R., Zhao, Y.: DeblurSLAM: a novel visual slam system robust in blurring scene. In: IEEE 7th ICVR, pp. 62–68 (2021)
4. Li, Y., et al.: Research and improvement of feature detection algorithm based on FAST. Rendiconti Lincei. Scienze Fisiche e Naturali **32**(4), 775–789 (2021). https://doi.org/10.1007/s12210-021-01020-1
5. Szeliski, R.: Feature detection and matching. In: Szeliski, R. (ed.) Computer Vision. TCS, pp. 333–399. Springer, Cham (2022). https://doi.org/10.1007/978-3-030-34372-9_7
6. Qiu, M., et al.: Energy minimization with soft real-time and DVS for uniprocessor and multiprocessor embedded systems. In: IEEE DATE, pp. 1–6 (2007)
7. Qiu, M., et al.: Dynamic and leakage energy minimization with soft real-time loop scheduling and voltage assignment. IEEE TVLSI **18**(3), 501–504 (2009)
8. Wu, C., Luo, C., Xiong, N., Zhang, W., et al.: A greedy deep learning method for medical disease analysis. IEEE Access **6**, 20021–20030 (2018)
9. Li, J., Ming, Z., et al.: Resource allocation robustness in multi-core embedded systems with inaccurate information. J. Syst. Archit. **57**(9), 840–849 (2011)
10. Zhou, Y., Zhang, Y., et al.: A bare-metal and asymmetric partitioning approach to client virtualization. IEEE Trans. Serv. Comput. **7**(1), 40–53 (2012)
11. Qiu, H., Qiu, M., Memmi, G., Ming, Z., Liu, M.: A dynamic scalable blockchain based communication architecture for IoT. In: Qiu, M. (ed.) SmartBlock 2018. LNCS, vol. 11373, pp. 159–166. Springer, Cham (2018). https://doi.org/10.1007/978-3-030-05764-0_17
12. Zhihui, L., Gai, K., et al.: Machine learning empowered content delivery: status, challenges, and opportunities. IEEE Netw. **34**(6), 228–234 (2020)
13. Zhang, Y., Gai, K., et al.: Blockchain-empowered efficient data sharing in Internet of Things settings. IEEE J. Sel. Areas Commun. **40**(12), 3422–3436 (2022)
14. Qiu, H., Qiu, M., Lu, R.: Secure V2X communication network based on intelligent PKI and edge computing. IEEE Netw. **34**(2), 172–178 (2019)
15. Qiu, H., Zeng, Y., et al.: DeepSweep: an evaluation framework for mitigating DNN backdoor attacks using data augmentation. In: ACM Asia CCS (2021)
16. Gai, K., et al.: A novel secure big data cyber incident analytics framework for cloud-based cybersecurity insurance. In: IEEE BigDataSecurity (2016)
17. Gai, K., Yulu, W., et al.: Differential privacy-based blockchain for industrial Internet-of-Things. IEEE Trans. Ind. Inform. **16**(6), 4156–4165 (2019)
18. Tian, Z., Li, M., et al.: Block-DEF: a secure digital evidence framework using blockchain. Inf. Sci. **491**, 151–165 (2019)

19. Zhang, Y., Gai, K., Wei, Y., Zhu, L.: BS-KGS: blockchain sharding empowered knowledge graph storage. In: Qiu, H., Zhang, C., Fei, Z., Qiu, M., Kung, S.-Y. (eds.) KSEM 2021. LNCS (LNAI), vol. 12817, pp. 451–462. Springer, Cham (2021). https://doi.org/10.1007/978-3-030-82153-1_37
20. Kumar, P., Kumar, R., et al.: PPSF: a privacy-preserving and secure framework using blockchain-based machine-learning for IoT-driven smart cities. IEEE Trans. Netw. Sci. Eng. **8**(3), 2326–2341 (2021)
21. Cheng, H., et al.: Multi-step data prediction in wireless sensor networks based on one-dimensional CNN and bidirectional LSTM. IEEE Access **7**, 117883–117896 (2019)
22. Fu, A., Zhang, X., et al.: VFL: a verifiable federated learning with privacy-preserving for big data in industrial IoT. IEEE TII **18**(5), 3316–3326 (2020)
23. Huang, S., Zeng, Z., et al.: An intelligent collaboration trust interconnections system for mobile information control in ubiquitous 5G networks. IEEE TNSE **8**(1), 347–365 (2020)
24. Bin, Z., Xiaohu, Z., Zhishuai, Y.: Image feature matching algorithm based on improved ORB. Laser Optoelectron. Prog. **58**(2), 0210006 (2021)
25. Gao, Y., et al.: Human action monitoring for healthcare based on deep learning. IEEE Access **6**, 52277–52285 (2018)
26. Xia, F., et al.: Adaptive GTS allocation in IEEE 802.15.4 for real-time wireless sensor networks. J. Syst. Archit. **59**(10), 1231–1242 (2013)
27. Dong, H., Song, W., et al.: Autonomous recognition technology of carrier robot on various terrain environment. Inst. Mech. Eng. Part D J. Autom. Eng. **235**(9), 2568–2584 (2021)
28. Yang, S., et al.: MGC-VSLAM: a meshing-based and geometric constraint VSLAM for dynamic indoor environments. IEEE Access **8**, 81007–81021 (2020)
29. Yang, X., Jiang, G.: A practical 3D reconstruction method for weak texture scenes. Remote Sens. **13**(16), 3103 (2021)
30. Zhou, F., Zhang, L., et al.: Improved point-line feature based visual SLAM method for complex environments. Sensors **21**(13), 4604 (2021)
31. Mur-Artal, R., Tardós, J.: ORB-SLAM2: an open-source SLAM system for monocular, stereo, and RGB-D cameras. IEEE Trans. Robot. **33**(5), 1255–1262 (2017)
32. Lin, C., He, Y.-X., Xiong, N.: An energy-efficient dynamic power management in wireless sensor networks. In: 2006 Fifth International Symposium on Parallel and Distributed Computing, pp. 148–154. IEEE (2006)
33. Chen, Y., et al.: KNN-BLOCK DBSCAN: fast clustering for large-scale data. IEEE Trans. Syst. Man Cybern. Syst. **51**(6), 3939–3953 (2019)
34. Wu, C.: UAV autonomous target search based on deep reinforcement learning in complex disaster scene. IEEE Access **7**, 117227–117245 (2019)
35. Liu, X., Yuan, X., Kang, H., Kang, H., Yu, C.: An image registration method based on improved TLD and improved ORB for mobile augmented reality. In: 2021 IEEE International Conference on Systems, Man, and Cybernetics (SMC), pp. 2127–2132. IEEE (2021)
36. Zhang, D., Yang, T.: Visual object tracking algorithm based on biological visual information features and few-shot learning. Comput. Intell. Neurosci. **2022** (2022)
37. Zhao, J., et al.: An effective exponential-based trust and reputation evaluation system in wireless sensor networks. IEEE Access **7**, 33859–33869 (2019)
38. Yao, J., Zhang, P., et al.: An adaptive uniform distribution ORB based on improved quadtree. IEEE Access **7**, 143471–143478 (2019)
39. Mikolajczyk, K., Tuytelaars, T., et al.: A comparison of affine region detectors. Intl. J. Comput. Vis. **65**(1), 43–72 (2005). https://doi.org/10.1007/s11263-005-3848-x

Smart Contract Vulnerability Detection Model Based on Siamese Network

Weijie Chen[1], Ran Guo[2(✉)], Guopeng Wang[3(✉)], Lejun Zhang[1,2(✉)], Jing Qiu[2], Shen Su[2], Yuan Liu[2], Guangxia Xu[2], and Huiling Chen[4]

[1] College of Information Engineering, Yangzhou University, Yangzhou 225127, China
{MZ120200850,zhanglejun}@yzu.edu.cn

[2] Cyberspace Institute Advanced Technology, Guangzhou University, Guangzhou 510006, China
{guoran,qiujing,yuanliu,xugx}@gzhu.edu.cn

[3] Engineering Research Center of Integration and Application of Digital Learning Technology, Ministry of Education, Beijing 100039, China
wangguopeng@sohu.com

[4] Department of Computer Science and Artificial Intelligence, Wenzhou University, Wenzhou 325035, China

Abstract. Blockchain is experiencing the transition from the first generation to the second generation, and smart contract is the symbol of the second generation blockchain. Under the background of the explosive growth of the second-generation blockchain platform and applications represented by smart contracts, frequent smart contract vulnerability events seriously threaten the ecological security of the blockchain, reflecting the importance and urgency of smart contract vulnerability detection. In this paper, we proposed a smart contract vulnerability detection method based on a Siamese network. We combined the Siamese network with Long Short-Term Memory (LSTM) Network neural network to complete the task of smart contract vulnerability detection. The Siamese network used in this paper consists of two subnetworks that share the same parameters onto a low dimension and easily separable feature space. Siamese network is now widely used in the field of image similarity and target tracking. In this paper, we improve the Siamese network so that it can be used for smart contract vulnerability detection. By comparing with previous research results, the model has better vulnerability detection performance and a lower false-positive rate.

Keywords: Smart Contract · Deep learning · Siamese Network

1 Introduction

The idea of smart contract was proposed by Nick Szabo in the 1990s [1], but due to the lack of a trusted execution environment at the time, smart contracts were not used and developed. Blockchain has the characteristics of openness, transparency, decentralization and invariance. Due to its characteristics, blockchain is now used in the Internet of Things [2], finance and other fields. So blockchain can be used as a vehicle for smart contracts

M. Qiu et al. (Eds.): SmartCom 2022, LNCS 13828, pp. 639–648, 2023.
https://doi.org/10.1007/978-3-031-28124-2_60

[3]. Compared to traditional contracts, smart contracts are themselves participants and executors of the contract, so the process of contract execution does not require third-party involvement.

Due to the convenience, smart contracts are beginning to be widely used in finance, energy, smart cities and other fields. Ethereum [4] is one of the most popular blockchain platforms and has deployed tens of thousands of smart contracts controlling billions of dollars' worth Ether (Cryptocurrency of Ethereum). In 2016, hackers exploited the reentrancy vulnerability to attack The Dao, resulting in the loss of over 60 million USD [5]. Security incidents like this can create a serious trust crisis in the blockchain, so we need to build an efficient and smart contract vulnerability detection tool. In this paper, we propose a smart contract vulnerability detection model based on Siamese networks named SCVSN model, which performs vulnerability detection by calculating similarity and has the advantages of simple structure, high correct rate and low false positive rate. Through experiments, we demonstrate that our proposed SCVSN model has better performance in vulnerability detection compared to previous smart contract vulnerability detection tools. The main contributions of this paper are described as follows:

1) Applying the natural language processing side to smart contract processing, and proposing a smart contract embedding method that reduces the impact of irrelevant content on smart contract vulnerability detection.
2) This is the first model that applies Siamese network to smart contract vulnerability detection at the source code level. SCVSN improves the accuracy of smart contract vulnerability detection and reduces the false positive rate of smart contract vulnerability detection by calculating the similarity. And to ensure that the number of positive and negative sample pairs is balanced during training, we also propose a classification algorithm for positive and negative sample pairs.
3) Many experiments are performed to confirm the performance of SCVSN model, and this is compared with the recently proposed deep learning based smart contract vulnerability detection method.

The paper is organized as follows: We discuss our methodology for our study in Section 2. We present the experimental procedure which includes dataset processing and performs performance comparisons in Section 3. Finally, we conclude the whole paper and future work in Section 4.

2 Smart Contract Vulnerability Detection Model Based on Siamese Network

From the perspective of model structure, the SCVSN model has the following main steps: (1) The first step is to input the sol files. (2) The second step is to process the sol files and use word2vec for word embedding. (3) The third step SCVSN model performs feature extraction on the word vectors, calculates the Euclidean distance, and obtains the final result by comparing it with the boundary value(m). The architecture of the SCVSN model is shown in Fig. 2.

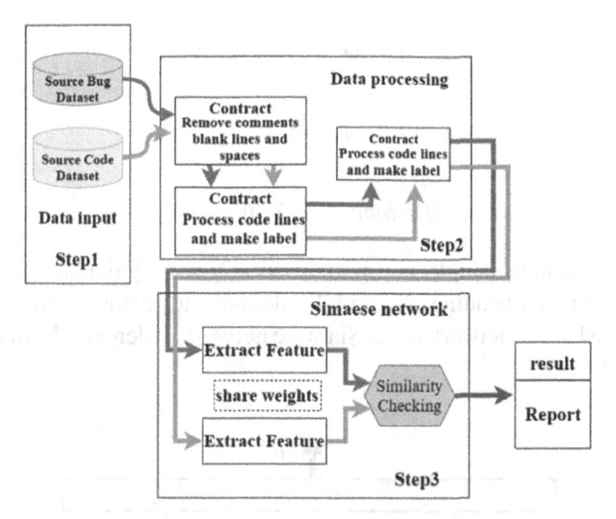

Fig. 2. SCVSN model structure

2.1 Siamese Network Structure in SCVSN Model

In the third step of our proposed model, feature extraction using the SCVSN model is required to calculate the Euclidean distance d. Then the Euclidean distance d is compared with the edge value margin to finally derive the result. The Siamese network structure used in our study is the combination of the Siamese network and LSTM neural network and the structure of the Siamese network used in this paper is shown in Fig. 3. It can be seen that two subnetworks have the same structure and the weights are shared between the two subnetworks. The subnetworks in the Siamese network used in SCVSN model consisted of only the input layer, LSTM neural network layer, Dropout layer, ReLU layer and fully connected layer, the structure of the subnetwork is simple. The input data goes through the LSTM neural network layer to extract the features and then passes through the Dense layer, which nonlinearized the previously extracted features and extracts the association between them, and then passes through the Dropout layer to avoid overfitting, the data are then processed using the nonlinear ReLU activation function. For large-scale data, ReLU has a better fitting ability compared to tanh activation function and sigmoid activation function, and also better enhances the nonlinearity of the model, making the neural network more discriminative. The data goes through the Dropout layer and ReLU layer once more to further increase the stability of the model, followed by the vector transformation through the Dense layer, and finally the feature extraction results are output. After the two subnetwork models extract the feature values, the Siamese network determines whether the two smart contracts are similar by calculating the Euclidean distance between the output values of the two subnetworks. The processing of the Siamese network structure used in this paper can be represented by the following equation.

$$word2vec(C_1) \Rightarrow C_1 \quad (1)$$

$$word2vec(C_2) \Rightarrow C_2 \quad (2)$$

$$LSTM\,(C_1) \Rightarrow X_1 \tag{3}$$

$$LSTM\,(C_2) \Rightarrow X_2 \tag{4}$$

$$|Distance(X_1, X_2)| \Rightarrow d \tag{5}$$

where C_1 and C_2 denote sample smart contracts, C_1 and C_2 denote the input matrices obtained after word embedding, X_1 and X_2 denote the feature values obtained after processing by the LSTM network in the Siamese network, d denotes the distance between two smart contracts.

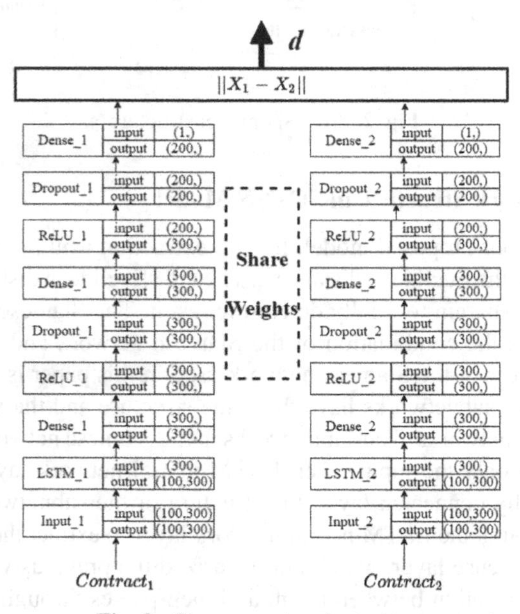

Fig. 3. Siamese network structure

LSTM can make full use of temporal relationship and semantic information among data. The benefits of using LSTM for our subnetwork are that LSTM is able to establish connections between units in a loop and remember previous inputs through their internal states, LSTM neural networks have specialized gate control structures capable of controlling the memory and forgetting of data [6], and LSTM neural networks do not suffer from gradient disappearance and gradient explosion. So, the LSTM neural network is used as the subnetwork of the Siamese network structure in the research.

2.2 The Core Idea of SCVSN Model

Our proposed approach uses Siamese network model to compute the similarity of two smart contracts, and based on comparing the similarity, thus performing vulnerability detection of smart contracts. The core idea of our proposed model can be briefly described as two subnetworks of SCVSN model, one of which extracts the feature of a smart contract *A* and other extracts the feature of a smart contract *B*. Smart contract *A* contains the vulnerability and is used as a reference sample, smart contract *B* may or may not contain vulnerability. First of all, the features extracted from the two subnetworks are subjected to Euclidean distance calculation, and then the Euclidean distance is compared with the threshold value. If the value of Euclidean distance is lower than the threshold, it means that the two smart contracts are similar and smart contract *B* contains vulnerability, if the Euclidean distance is higher than the threshold, it means that the two contracts are not similar and thus it is judged that smart contract *B* does not contain the vulnerability.

In order to implement the idea, we proposed, the following issues need to be addressed: 1). How to implement smart contract vulnerability detection via SCVSN model. 2). How to train SCVSN model to improve their feature extraction capabilities.3). How to reduce rate of false positives.

First Question. Our proposed SCVSN model performs features extraction through subnetworks, After the training of the SCVSN model is completed (the training process described in detail in second question), the subnetwork that specifically extracts features of the smart contract containing the vulnerability saves the trained parameters. In testing or practical use, the sample to be tested is passed through another subnetwork of SCVSN model to extract features, then the Euclidean distance between the two-feature value is calculated, followed by a comparison of the Euclidean distance with a threshold value to determine whether the sample to be tested contains vulnerabilities. The structure of our proposed SCVSN model is simple, using only 2 LSTM neural network layers and 4 ReLU layers, but achieves good detection results and small computational effort. After the network is trained, the features of the reference samples (smart contracts containing vulnerabilities) can be extracted in advance, and the samples to be tested only need to extract their features and then calculate the Euclidean distance and compare it with the threshold value. The vulnerability detection process is shown in Fig. 4.

Second Question. Our approach is to first use the 50 smart contract samples that have been processed in the Source Bug Dataset as the reference samples, since these 50 smart contract samples are certain to contain vulnerabilities, then 50 smart contract reference samples containing vulnerabilities are combined with 2918 smart contract samples already processed in Source Code Dataset to form positive and negative sample pairs respectively. The positive sample is composed of a sample of smart contracts containing vulnerabilities in the Source Bug Dataset and a sample of smart contracts containing vulnerabilities in the Source Code Dataset, and this sample pair corresponds to a label of 1, negative sample is a sample pair consisting of a sample of smart contracts with vulnerabilities in the Source Bug Dataset and a sample of smart contracts without vulnerabilities in the Source Code Dataset, and this sample pair corresponds to a label of 0. After constructing a large number of positive and negative samples, the samples are input into the network model, which is used to train the network model and improve the ability of the SCVSN model to extract features.

Third Question. Since smart contract vulnerability can appear anywhere in a smart contract, the fact that two smart contracts are similar does not necessarily mean that the tested smart contract contains the smart contract vulnerabilities. As shown in Fig. 4,

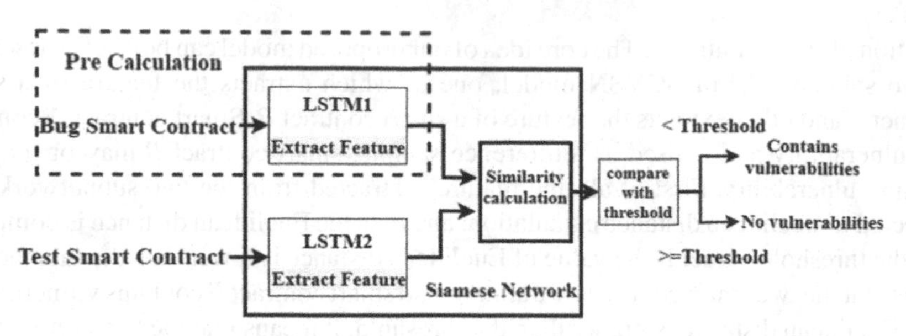

Fig. 4. Vulnerability detection process

where Contract1 is the one containing reentrancy vulnerability When the attacker uses reentrancy vulnerability to attack, Contract1 will not execute according to the logic intended by the contract creator, where the vulnerability appears in line 12. Contract2 does not contain reentrancy vulnerability. These two smart contracts are similar, but one contains reentrancy vulnerability and the other does not. If Contract1 is used as the reference sample, the result must be that the two smart contracts are similar by the usual method of calculating similarity, thus proving that Contract2 contains reentrancy vulnerability, but this is not the actual case. By training the SCVSN model, the false positive rate of smart contract vulnerabilities in this case can be reduced. The previous subsection illustrates that SCVSN model uses two subnetworks together to extract features, and the parameters are shared between the subnetworks, providing a stronger feature extraction capability than a single network, so the loss function used in SCVSN model needs to be able to handle the relationship among sample pair labels, Euclidean distance and threshold. Therefore, the loss function used in our proposed SCVSN model is Contrastive Loss which shown in Eq. (6).

$$loss = \frac{1}{2N}\sum\nolimits_{n=1}^{N} yd^2 + (1-y)\max(m-d,0)^2 \quad (6)$$

In the formula, d denotes the Euclidean distance between the features of two smart contract samples, which can also indicate the degree of similarity between two smart contract samples, with the expression $d = ||a_n - b_n||^2$, where a and b denote two different smart contracts. y indicates the label of weather two smart contract samples match, y = 1 means they are similar, y = 0 means they are not similar, and this labelling rule is also used when construct the labels of positive and negative sample pairs, m indicates threshold. The advantage of our choice to use Contrastive Loss as the loss function is that the overall loss function ensures that sample pairs that are already similar remain similar when SCVSN model performs feature extraction, while sample pairs that are not similar remain dissimilar after feature extraction. When the sample pair label y = 1, the loss function is shown in Eq. (7).

$$loss = \frac{1}{2N}\sum_{n=1}^{N} yd^2 \quad (7)$$

The sample pair label y = 1 indicates that both smart contracts composing the sample pair contain vulnerabilities, and if the Euclidean distance between the two smart sample features is too large, it indicates that SCVSN model is not good at extracting critical information and needs to increase the loss. When the sample pair label y = 0, the loss function is shown in Eq. (8).

$$loss = \sum (1 - y) \max(margin - d, 0)^2 \tag{8}$$

The sample pair label y = 0 indicates that one of the two smart contract samples is not containing the smart contract vulnerability. If the Euclidean distance between the two smart contract sample pairs is too small, the same indicates that SCVSN model is not good at extracting critical information and needs to increase the loss. If the Euclidean distance between sample pairs is too large over the threshold margin, it is a normal sample pair, but it does not help the model's ability to extract features, so we choose to ignore these sample pairs, which can greatly reduce the amount of computation and increase the training speed.

3 Experiment

3.1 Experimental Procedure

During the experiments, the evaluation criteria for model performance is the accuracy of the model on the test set. We recorded the training data for each epoch and calculated the average of the accuracy and loss values based on the recorded data, as shown in Fig. 5.

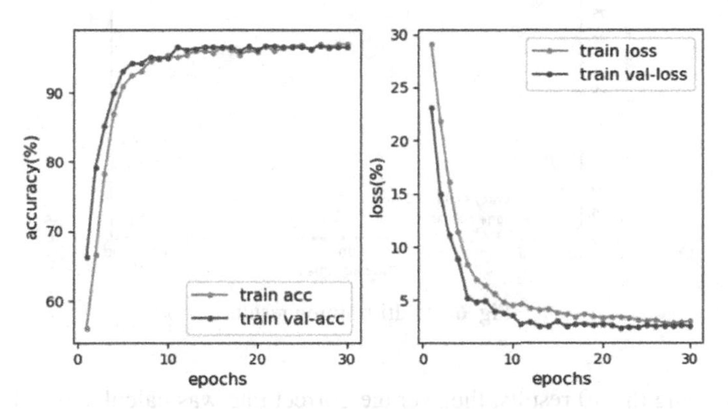

Fig. 5. Training results

In the research, SCVSN model uses two datasets, the first one is a recently released large-scale dataset named Smart Bugs Dataset-wild [7]. It contains 47398 real and unique sol files, which together have about 203,716 smart contracts in total, this is because a sol file may contain more than one smart contract. The Source Code Dataset used for training is extracted from this dataset. In addition, the SCVSN model used another small dataset SolidiFI Benchmark [8], which contains 7 different vulnerabilities, each vulnerability type contains 50-sol files, and this dataset serves as Source Bug Dataset.

The plot on the right shows the relationship between epoch and loss function, and the plot on the left shows the relationship between epoch and accuracy. In this paper, the data of each epoch is recorded during the training to facilitate the observation and optimization of the model. From the change of the curve, it can be seen that the loss value and the correct rate of the model on both the training and test sets are changing rapidly in the first 10 epochs. After 10 epochs, the loss values and accuracy rates of the model on the training and test sets fluctuate up and down regularly, and the overall trend tends to level off, indicating that the model starts to converge gradually at this point. In the right plot, it can be seen that the curve of loss declines in the test set and training set is gradually easing, and there is no violent jitter, indicating that the value of batch size in the experiment is correct. The loss curves of the training and test sets crossed during the training process, and the distance between the two curves was moderate, without overlap or excessive distance, indicating that there was no overfitting or underfitting in the training process.

In order to correctly represent the data of the model, we generate a graph of the data from multiple training of the model, and the specific results are shown in Fig. 6. The straight line in the figure indicates the average accuracy calculated from the test results after several training times. The results of 40 times of the model on the test set are recorded in Fig. 6, with the worst accuracy of 94.35% and the best accuracy of 97.14%.

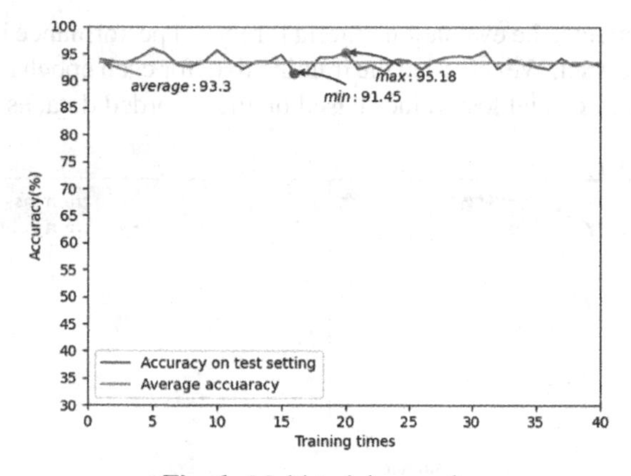

Fig. 6. Multi-training results

By averaging the 40 results, the average correct rate was calculated to be 96.01%, and it can also be seen that the results of all 40 tests fluctuate around 96.01%. So, we conclude that our proposed SCVSN model has a correct detection rate of 96.01% for smart contract reentrancy vulnerability.

3.2 Experimental Performance Comparison

The performance metrics are those introduced above, including ACC, REC, FPR and F1. In the same experimental setting, the DeeSCV Hunter, the Peculiar, the BLSTM-ATT, and the TMP were selected for comparison. SCVSN data is the average of multiple test results. The results of the comparison are shown in the Table2. The measurements we chose were accuracy (ACC), precision (PRE), F1-score (F1), and recall rate (REC). They can be calculated by thc following formulas:

$$\mathrm{ACC} = \frac{TP + TN}{TP + TN + FP + FN} \tag{9}$$

$$\mathrm{TPR} = \frac{TP}{TP + FP} \tag{10}$$

$$\mathrm{Recall} = \frac{TP}{TP + FN} \tag{11}$$

$$\mathrm{FPR} = \frac{FP}{FP + TN} \tag{12}$$

$$\mathrm{F1} = \frac{2*PRE*TPR}{PRE + TPR} \tag{13}$$

TP denotes positive samples predicted by the model as positive class, TN denotes negative samples predicted by the model as negative class, FP denotes negative samples predicted by the model as positive class, and FN denotes positive samples predicted by the model as negative class.

It can be seen from Table 1 that our proposed SCVSN model has an excellent performance in smart contract vulnerability detection. The SCVSN model has relatively stable values in all four-performance metrics, where the SCVSN model has a correct rate of 96.01%, which is better than the models involved in the comparison. The Recall of the SCVSN model proposed in this paper is 96.64%, which exceeds the rest of the models, this shows that the SCVSN approach to detecting smart contract vulnerabilities can reduce the false positive rate in vulnerability detection.

Table1. Comparison of experimental results.

Model	ACC (%)	REC (%)	PRE (%)	F1(%)
SCVSN	96.01	96.04	94.25	94.78
DeeSCVHunter [9]	93.02	83.46	90.70	86.87
Peculiar [7]	92.37	92.4	91.80	92.10
BLSTM-ATT [10]	88.47	88.48	88.5	88.26
TMP [11]	84.48	82.63	74.06	83.82
DR-GCN [11]	81.47	80.89	72.36	76.39
LSTM	81.91	91.43	76.80	83.48

4 Conclusions

We apply Siamese networks to smart contract vulnerability detection for the first time, and we perform smart contract vulnerability detection by comparing similarity. Our proposed SCVSN model has an excellent performance in smart contract vulnerability detection due to the simple structure of Siamese networks, powerful similarity computation capability and feature extraction ability. In the work of this paper, we apply a natural language processing approach to the processing of smart contracts for two realistic datasets and apply the processed dataset to the training and testing of the model. We demonstrate through extensive experiments that our proposed SCVSN model can achieve good results in smart contract reentrancy vulnerability detection tasks. Our research also provides a new way of thinking about vulnerability detection, specifically by calculating the similarity to detect code vulnerabilities, and our code will be open-sourced afterwards.

SCVSN can detect whether the smart contract contains vulnerabilities, but cannot detect what kinds of vulnerabilities, so our future work is to study how to make the SCVSN model detect the types of vulnerabilities contained in the smart contract.

References

1. Giancaspro, M.: Is a 'smart contract' really a smart idea? Insights from a legal perspective. Comput. Law Secur. Rev. **33**(6). 825–835(2017)
2. Qiu, H., Qiu, M., Memmi, G.: A dynamic scalable blockchain based communication architecture for IoT. In: International Conference on Smart Blockchain, pp. 159–166 (2018)
3. Khan, S.N., Loukil, F., Ghedira-Guegan, C., Benkhelifa, E., Bani-Hani, A.: Blockchain smart contracts: applications, challenges, and future trends. Peer-to-Peer Networking Appl. **14**(5), 2901–2925 (2021). https://doi.org/10.1007/s12083-021-01127-0
4. Oliva, G.A., Hassan, A.E., Jiang, Z.M.: An exploratory study of smart contracts in the Ethereum blockchain platform. Empir. Softw. Eng. **25**(3), 1864–1904 (2020). https://doi.org/10.1007/s10664-019-09796-5
5. Samreen, N.F., Alalfi, M.H.: Reentrancy vulnerability identification in ethereum smart contracts. In: 2020 IEEE International Workshop on Blockchain Oriented Software Engineering (IWBOSE), pp. 22–29 (2020)
6. Karevan, Z., Suykens,A. Transductive lstm for time-series prediction: An application to weather forecasting. Neural Networks **125**, 1–9 (2020)
7. Wu, H., et al.: Peculiar: Smart contract vulnerability detection based on crucial data flow graph and pre-training techniques. In: 2021 IEEE 32nd International Symposium on Software Reliability Engineering (ISSRE), pp. 378–389 (2021)
8. Ghaleb, A., Pattabiraman, K.: How effective are smart contract analysis tools? evaluating smart contract static analysis tools using bug injection. In: Proceedings of the 29th ACM SIGSOFT International Symposium on Software Testing and Analysis, pp. 415–427 (2020)
9. Yu, X., Zhao, H. Hou, B., Ying, Z., Wu, B. Deescvhunter: a deep learning-based framework for smart contract vulnerability detection. In: 2021 International Joint Conference on Neural Networks (IJCNN), pp. 1–8 (2021)
10. Qian, P., Liu, Z., He, Q., Zimmermann, R., Wang X. Towards automated reentrancy detection for smart contracts based on sequential models. IEEE Access **8**, 119685–19695(2020)
11. Zhuang, Y., Liu, Z., Qian, P., Liu, Q., Wang, X., He, Q.: Smart contract vulnerability detection using graph neural network. In: IJCAI 2020, pp. 3283–3290 (2020)

Context-User Dependent Model for Cascade Retweeter Prediction

Tong Chunyan[1], Zhang Kai[1], Yang Song[1], Zhang Zheng[1], Zhang Hongfeng[2], Wang Hao[2], Shuang Xinzhuo[3], and Liu Yerui[4](✉)

[1] State Key Laboratory of Communication Content Cognition, People's Daily Online, Beijing 100733, China
{tongchunyan,zhangkai,yangsong,zhangzheng}@people.cn
[2] Wuhan Second Ship Design and Research Institute, Wuhan 430064, China
[3] Fuxin Higher Vocational College, Liaoning 123000, China
[4] Electronic Information School, Wuhan University, Wuhan 430072, China
1093252514@qq.com

Abstract. This paper proposes a retweeter predction model based on attention model and Tranformer encoding–Forwader Prediction Model based on User Preference (FPM-UP). Considering the impact of release time, content, and current external context information in the forwarding process, FPM-UP integrates user attribute embedding and context-user dependency into a temporal and text attention model for the prediction of the next forwarding user. Compared with the existing methods, FPM-UP significantly improves the prediction accuracy.

Keywords: Social Network · Forwarding Prediction · Attention Mechanism · Graph Neural Network · Artificial Neural Network

1 Introduction

In major online social networking platforms such as Twitter or Weibo, users are connected by their following status and the information can diffuse via users' reposting behavior from their followees. Such a retweeting behavior pushs the original information to more users thus accelerating the diffusion of the original information. Such diffusion paths are also refered to as cascade structures, which has been widely studied [1–6].

Generally, retweeter prediction models can be divided into two categories: social network prediction meth-ods and diffusion path prediction methods.

The social network prediction method describes users and their relationships as a complex network, in which retweeting behavior is carried on the edges. This method originally originated from two theories: Independent Cascade (IC) and Linear Threshold (LT) [7]. Recently, many papers have revised the theory [8,9]. The basic problem of this method is that the assumptions are usually difficult to meet in realworld data.

M. Qiu et al. (Eds.): SmartCom 2022, LNCS 13828, pp. 649–658, 2023.
https://doi.org/10.1007/978-3-031-28124-2_61

The diffusion path prediction method expresses the retweeting process as a user sequence. The goal of the model is to predict the next or all future users participating in the retweeting of the information given the histor-ical diffusion path of a piece of information.

RNN model is used in the SNIDSA model proposed by Wang [10] to extract the currently observed retweeting information. Attention mechanism is used to learn the relationship between users before RNN. CYAN-RNN model uses a two-layer RNN structure to learn global information and local information separately. Also, attention mechanism [11] is used to correct the impact of current global information on local information. In the DCE model pro-posed by Zhao [12], the joint embedding method is used to learn the influence of the diffusion sequence to learn the user embedding vector. These models have had some success. However, due to the inherent defect that RNN cannot learn long-term feature when extracting sequence features, none of these models can achieve better results.

Many scholars have relied on the attention mechanism in diffusion sequence prediction to achieve good results. In the HiDAN model proposed by Wang [13], the traditional RNN model architecture was abandoned, and a redesigned non-sequential model based on the attention mechanism was used to predict the diffusion sequence. Ma [14] discussed the influence of followers' hot topic discussion on users with self-attention model.

In addition, some scholars try to describe the attractiveness of other information to users. In the HMCF model proposed by Yang [15], text and visual information are used to describe the attractiveness of graphic information in different places of brochures to users. Wu [16] used user learning model and object learning mod-el in its model to learn user preference characteristics and information content characteristics. Wang [17] pro-posed to design corresponding description vectors for different roles of users, discussing their different charac-teristics as publishers and receivers.

In summary, current works ignore the underlying users graph and text information in the information. However, it is possible to extract enough features to pre-dict both popularity and retweeters.

In this paper, a model is proposed for different features based on the results of the analysis of forwarding behavior. For user relationship network, this paper embeds user attributes through Graph Attention Network. For the user's forwarding process, this paper designs an attention model to extract the dependency information between context and users. Subsequently, the propaga-tion process relies on the Transformer encoder to further extract user relationship information. Finally, this paper designs a temporal and text attention model to model the influence of release time, the content, and the current external context information on the forwarding process.

The rest of this paper is organized as follows. Data analysis is illustrated in Sect. 2. Models are defined in Sect. 3. In Sect. 4, we analyze the experimental results. Finally, we briefly review the work in Sect. 5.

2 Data Analysis and Discoveries

2.1 Data Analysis and Discoveries

Many studies have shown that, time, user, structure, and content characteristics are effective for information popularity prediction. We analyze the effectiveness of different features in predicting information popularity and retweeters from the perspective of information entropy.

2.2 Text Content and Publish Time and Delay Time

Text content is the main information described by microblog, which is the content that users read before retweeting.

For each user, the retweeting frequency of different content is analyzed. The counters are normalized for eliminate the unbalance of categories' counter and user's retweeting times. In order to investigate whether users' preferences are stable, we calculated the information entropy of all users' retweeting probablities, which is shown in Eq. (1).

$$E_u = -\sum_{c \in C} P_u(c) \log P_u(c) \tag{1}$$

where C means all possible text content categories, and $P_u(c)$ means the probability that user u retweets a microblog of category c.

Time mainly affects the activeness of users, which indirectly affects the probability of retweeting. Time features can be divided into two categories: release time features and delay time features.

In order to quantitatively describe the impact of the release time feature on the user's willingness to retweet, we use Shannon entropy of information published by the user when the release time of the information is not given.

$$H_{pt}^{sh} = -\sum_{u \in G, p \in P} P(u,p) \log P(u,p) \tag{2}$$

where P represents the set of all possible publish times. H_{pt}^{sh} means the information entropy of users' retweeting. P(u,p) is the retweeting probability of user u at release time p.

To distinguish the influence of publish time, the Condition entropy of user retweeting with the same publish time was calculated as in Eq. (3).

$$H_{pt}^{con} = -\sum_{p \in P} P(p) \sum_{u \in G} P(u,p) \log P(u,p) \tag{3}$$

The second type of time feature is the interval between the retweeting behavior and the release time of the content which mainliy affects the popularity of the content. With the extension of time, the popularity decay is approximately exponential.

3 Model

3.1 Framework

To solve the above problems, we propose a Forwader Prediction Model based on User Preference (FPM-UP). In FPM-UP, in order to deal with the observed retweeting information, we use Transformer encoders as sequence feature extractors. Transformer encoders are suitable for dealing with such problems where the relationship between nodes appears unclear. At the same time, there is a delay in each step of the retweeting process. This pa-per uses the attention mechanism to consider the influence of the delay time on each step of the retweeting process. For the release time and text content of the mi-croblog, we embedded them into vectors respectively, and again used the attention mechanism to consider their influence on the retweeting process, so as to get the final retweeting cascade embedded vector. Finally, we consider the similarity between the obtained sequence expression vector and the user expression vector, and select the user with the highest similarity as the prediction result. The overall frame of the model is shown in Fig. 1.

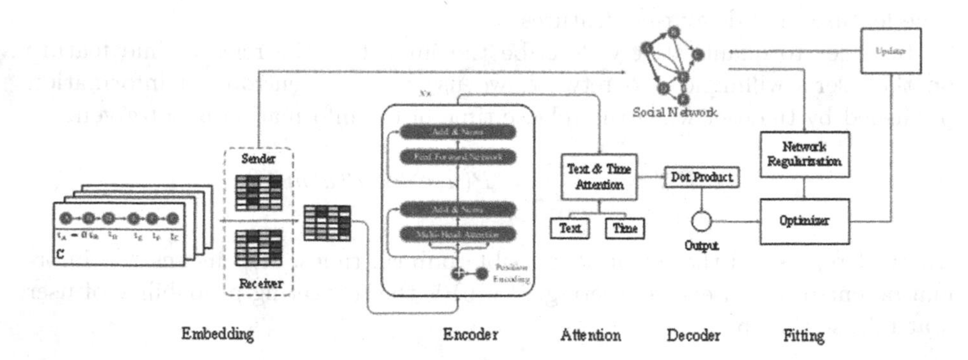

Fig. 1. FPM-UP model framework

3.2 Tweet Embedding

Content and release time belong to the information of microblogs themselves, which mainly affect users' pref-erence for different types of microblogs.

For text information, FPM-UP uses the standard Bert model to embed text content as vectors. In order to include the influence of external information on content preference, we extracted the content of "Hot Search List" from Weibo, converted it into vector expression, and calculated its similarity with the content currently pro-cessed.

$$\beta_i = \frac{exp(f_t(W_t x_t + b_t) \cdot f_h(W_h x_h + b_h))}{\sum_{i=0}^{N_h} exp(f_t(W_t x_t + b_t) \cdot f_h(W_h x_h + b_h))} \tag{4}$$

The external influence vector finally obtained as Eq. (5):

$$e_h = \sum_{i=0}^{N_h} \beta_i x_h \tag{5}$$

where, x_t represents the embedded information of the text content, and x_h represents the embedded information of the top search list content.

In addition, the release time of a microblog will affect the user's activity, thus affecting the user's retweeting intention. Cycle information is used to describe the publishing time characteristics of Weibo. The cycle information represents the time period of the day when the release time is located. One-hot coding is converted to vector representation using the full connection layer.

$$e_p^t = f_d(W_{td} x_w + b_w) \tag{6}$$

The $W_{td} \in R^{d_t \in N_d}$, $b_w \in R^{d_t}$ is the model parameters, N_d is the number of segments of one day, whose time length is $24/N_d$ hours. The final embedded vector is

$$e^w = f_x(W_w[x_t, e_h, e_p^t] + b_w) \tag{7}$$

in which $[\cdot]$ is concatenation operation.

3.3 Retweeting Sequence Embedding

To embed the retweeting sequence accurately, each user is transformed into a user expression vector using two full connection layers, one vector representing the extent to which the user influences other users as a pub-lisher and the other vector representing the extent to which the user prefers different content as a receiver.

$$e_i^{up} = f_x(W_u p x_i + b_u p) \tag{8}$$

$$e_i^{ur} = f_x(W_u r x_i + b_u r) \tag{9}$$

where $W_{up} \in R^{d_{up} \times N}$,$W_{ur} \in R^{d_{ur} \times N}$,$b_p \in R^{d_{up}}$,$b_r \in R^{d_{ur}}$is model parameters, d_{up} and d_{ur} represent publisher embedded dimension and receiver embedded dimension respectively, and f_x represents nonlinear activation function.

The embedding of retweeting sequence mainly uses the encoder module in Transformer module. In each layer of Transformer, the input data is an expression of the user's influence, arranged in the order in which it is retweeted, denoted as $C_o \in R^{d_{up} \times l}$, where l represents the sequence length. The embedding results are mainly achieved through the self-attention mechanism.

$$SA(C_o) = f_x\left(\frac{(C_o \cdot W^Q) \cdot (C_o \cdot W^K)^T}{\sqrt{d_{up}}} \cdot M\right) \cdot (C_o \cdot W^V) \tag{10}$$

In order to analyze the influence of delay information on retweeting intention, the corresponding vector representation $\lambda_i \in R^{d_d}$ is determined according to different delay times. Then, the vector representation is spliced with the vector representation of microblog content and publishing time to obtain $e^f = [\lambda_i, e^w]$. The coefficient of attention mechanism was calculated as Eq. (11):

$$\alpha_i = \frac{\exp\left(w^f \cdot \left(e^f \odot e_i^c\right)\right)}{\exp\left(w^f \cdot \left(e^f \odot e_i^c\right)\right)} \tag{11}$$

where $\omega^f \in R^d$ is model parameter. Finally, the overall embedding result is shown in Eq. (12):

$$e^c = \sum_{i=1}^{n} \alpha_i \cdot e_i^c \tag{12}$$

3.4 Prediction

Given the cascade embedding result e^c at the time of t_i, the estimated probability of user u_i retweeting the Weibo is shown in Eq. (13):

$$\hat{p}(u_{i+1}|c_i) = softmax(e_c \cdot e_{ur} \tag{13}$$

The objective of model optimization is to maximize the retweeting probability of real retweeters, and the objective function is shown in Eq. (14):

$$\mathcal{L} = -\sum_{m=1}^{M} \sum_{i=1}^{n_m - 1} \log \hat{p}\left(u_{i+1} \mid c_i\right) \tag{14}$$

4 Experiments

4.1 Dataset

We used Douban, China's largest online platform for book and movie sharing, where users can share their current reading status.

In this dataset, each book is regarded as a piece of content. This dataset contains the encrypted user ID and reading time, as well as the relationship network between users.

Finally, we extract dataset for retweeter predition through sampling. The statistical data of the dataset we use in this paper is shown in Table 1.

Table 1. Douban dataset for retweeter prediction

Item	Information
Number of Users	3796
Number of Tweets	3349
Number of Edges	79356
Average Retweeting Times	21.66
Average Occurrences of Users	19.99

4.2 Experimental Result

For retweeter prediction, the performance of the model refers to the success rate of predicting the next retweeting user. Two methods are usually used to quantify the model performance: MRR index (Mean Reciprocal Rank) and A@k accuracy.

MRR uses probability ranking to measure model performance, and the specific calculation method is as Eq. (15):

$$MRR = \frac{1}{N}\sum_{i=1}^{N}\frac{1}{rank_i} \tag{15}$$

where $rank_i$ represents the probability ranking of the correct result in the ith prediction result. The larger the MRR value, the better the model performance.

A@k represents the probability of real retweeting users ranking top k in probability.

$$A_k = \frac{\sum_{i\in U}|\{u_i \mid u_i \in U_{ik}\}|}{|U|} \tag{16}$$

The higher A@k is, the better the model performance is, and the smaller k is, the more important the negative index is.

Here we use map@k, which comprehensively considers the effect of A@k and MRR. It considers the probability ranking of the model and the ranking of the model itself by applying an average operation on the ranking result.

$$map_k = \frac{1}{N}\sum_{i=1}^{N} s\left(r_i, k\right) \tag{17}$$

$$s\left(r_i, k\right)=\begin{cases}\frac{1}{N}, r_i \leq k \\ 0, r_i > k\end{cases} \quad (18)$$

The experimental results are shown in Table 2 and Table 4.

Table 2. Experimental result

Model	MRR	map@1	map@10	map@20	map@50	map@100
FOREST	0.03400	1.74	3.01	3.15	3.26	3.30
HiDAN	0.04783	1.87	3.97	4.27	4.51	4.62
DeepDiffuse	0.04283	1.90	3.66	3.87	4.04	4.13
TopoLSTM	0.03751	1.34	3.03	3.29	3.49	3.59
SNIDSA	0.04621	1.93	4.03	4.23	4.41	4.49
FPM-UP	0.05210	2.05	4.22	4.60	4.90	5.04

By observing the data in the experimental results, the following rules can be found.

- The FPM-UP model significantly improves the pre-diction accuracy on the same dataset. For instance, The FPM-UP model's accuracy rate is about 9.1% higher in comparison with HiDAN in terms of map@100, proving that using FPM-UP on cascade modeling can improve the prediction accuracy.
- The nonsequential models are more suitable for the modeling of forwarding cascades. In the com-parison methods, HiDAN, SNIDSA and the FPM-UP method proposed in this paper are all non-sequential models.

5 Conclusion

This article summarizes the current processing se-quence prediction problem of the main difficulties and verifies the impact of a variety of characteristics for re-tweeting process. A diffusion sequence prediction model based on Transformer encoder is proposed. This model uses relevant methods of representation learning to learn the influence of upstream users on subsequent retweeting. In the end, this paper verifies the performance of the model on the real data set. The results show that the accuracy of the proposed model has about 10% higher accuracy than that of the current models.

Acknowledgement. This work was supported by the Open Funding Projects of the State Key Laboratory of Communication Content Cognition (No. 20K05 and No. A02107).

References

1. Gong, H., Wang, P., Ni, C., Cheng, N.: Popularity prediction in microblogging network: a case study on sina weibo. In: Proceedings of the 22nd International Conference on World Wide Web, pp. 177–178 (2013)
2. Cheng, J., Adamic, L., Dow, P.A., Kleinberg, J.M., Leskovec, J.: Can cascades be predicted? In: Proceedings of the 23rd International Conference on World Wide Web, pp. 925–936 (2014)
3. Gao, S., Ma, J., Chen, Z.: Effective and effortless features for popularity prediction in microblogging network. In: Proceedings of the 23rd International Conference on World Wide Web, pp. 269–270 (2014)
4. Zhang, B., Qian, Z., Lu, S.: Structure pattern analysis and cascade prediction in social networks. In: Frasconi, P., Landwehr, N., Manco, G., Vreeken, J. (eds.) ECML PKDD 2016. LNCS (LNAI), vol. 9851, pp. 524–539. Springer, Cham (2016). https://doi.org/10.1007/978-3-319-46128-1_33
5. Gai, K., Qiu, M., Ming, Z., Zhao, H., Qiu, L.: Spoofing-jamming attack strategy using optimal power distributions in wireless smart grid networks. IEEE Trans. Smart Grid **8**(5), 2431–2439 (2017)
6. Gai, K., Yulu, W., Zhu, L., Lei, X., Zhang, Y.: Permissioned blockchain and edge computing empowered privacy-preserving smart grid networks. IEEE Internet Things J. **6**(5), 7992–8004 (2019)
7. Kempe, D., Kleinberg, J., Tardos, É.: Maximizing the spread of influence through a social network. In: Proceedings of the ninth ACM SIGKDD International Conference on Knowledge Discovery and Data Mining, pp. 137–146 (2003)
8. Gao, S., Pang, H., Gallinari, P., Guo, J., Kato, N.: A novel embedding method for information diffusion prediction in social network big data. IEEE Trans. Ind. Inf. **13**(4), 2097–2105 (2017)
9. Qiu, J., Tang, J., Ma, H., Dong, Y., Wang, K., Tang, J.: Deepinf: social influence prediction with deep learning. In: Proceedings of the 24th ACM SIGKDD International Conference on Knowledge Discovery & Data Mining, pp. 2110–2119 (2018)
10. Wang, Z., Chen, C., Li, W.: A sequential neural information diffusion model with structure attention. In: Proceedings of the 27th ACM International Conference on Information and Knowledge Management, pp. 1795–1798 (2018)
11. Vaswani, A., et al.: Attention is all you need. In: Advances in Neural Information Processing Systems, vol. 30 (2017)
12. Zhao, Y., Yang, N., Lin, T., Philip, S.Y.: Deep collaborative embedding for information cascade prediction. Knowl.-Based Syst. **193**, 105502 (2020)
13. Wang, Z., Li, W.: Hierarchical diffusion attention network. In: IJCAI, pp. 3828–3834 (2019)
14. Ma, R., Hu, X., Zhang, Q., Huang, X., Jiang, Y-G.: Hot topic-aware retweet prediction with masked self-attentive model. In: Proceedings of the 42nd International ACM SIGIR Conference on Research and Development in Information Retrieval, pp. 525–534 (2019)
15. Yang, Y., Duan, Y., Wang, X., Huang, Z., Xie, N., Shen, H.T.: Hierarchical multi-clue modelling for poi popularity prediction with heterogeneous tourist information. IEEE Trans. Knowl. Data Eng. **31**(4), 757–768 (2018)

16. Wu, Q., Gao, Y., Gao, X., Weng, P., Chen, G.: Dual sequential prediction models linking sequential recommendation and information dissemination. In: Proceedings of the 25th ACM SIGKDD International Conference on Knowledge Discovery & Data Mining, pp. 447–457 (2019)
17. Wang, Z., Chen, C., Li, W.: Information diffusion prediction with network regularized role-based user representation learning. ACM Trans. Knowl. Disc. Data (TKDD) **13**(3), 1–23 (2019)

Heterogeneous System Data Storage and Retrieval Scheme Based on Blockchain

Ni Zhang[1], BaoQuan Ma[1](✉), Peng Wang[1], XuHua Lei[1,2], YeJian Cheng[1,2], JiaXin Li[1], XiaoYong Huai[1], ZhiWei Shen[1], NingNing Song[1], and Long Wang[1]

[1] National Computer System Engineering Research Institute of China, Beijing 102200, China
15501210877@163.com

[2] School of Computer Science and Technology, Xidian University, Xi'an 710071, China

Abstract. In the field of information interaction, when a project involves a large amount of heterogeneous information, it is difficult to transmit and update the required information timely, accurately, reliably and securely in such a complex environment to maintain synchronization. At present, blockchain itself has problems such as high storage pressure of nodes, low access efficiency, and simple query. Therefore, this paper takes this as a starting point and proposes a data mapping method of physical resources based on node attributes and heterogeneous nodes, which provides a general method for the data mapping from actual physical resources to information domain. This method can define the attributes of heterogeneous physical information nodes and support unified expression of various physical information resources on the same platform in the real physical world, then connect heterogeneous physical information nodes and improve the sharing ability of data resources between nodes. When accessing data, the corresponding content can be found in the off-chain database by obtaining the index information of the off-chain location stored on the chain. This method takes advantage of the large space and high access efficiency of the off-chain storage system to share the pressure of on-chain data storage. This paper expounds the design idea of the system, introduces the design objective and method in detail, gives the flow chart of the system operation, and carries out a simple software test and verification. Finally, this paper summarizes the work and prospects the development direction of future work, hoping to provide inspiration for solving such problems.

Keywords: Blockchain · Data Query · Heterogeneous Information · Data Storage · Retrieval Scheme

1 Introduction

Cyber-Physical Systems (CPS) is a complex system with new capabilities, where computing [1–3], physical elements [4–6] and network environment [7–9] are tightly coupled. With the integration of CPS and technologies such as 5G and blockchain, the security problems [10–12] of CPS have been exposed. In 2017, Ouaddahet proposed a perceptive security framework for CPS, which laid the foundation for the security mechanism of CPS [10–12], but the different security requirements of each layer brought complexity to the solution. Meanwhile, the growing scale of CPS also brought hidden dangers to the confidentiality and integrity [12]. System framework as Fig. 1 shown.

M. Qiu et al. (Eds.): SmartCom 2022, LNCS 13828, pp. 659–668, 2023.
https://doi.org/10.1007/978-3-031-28124-2_63

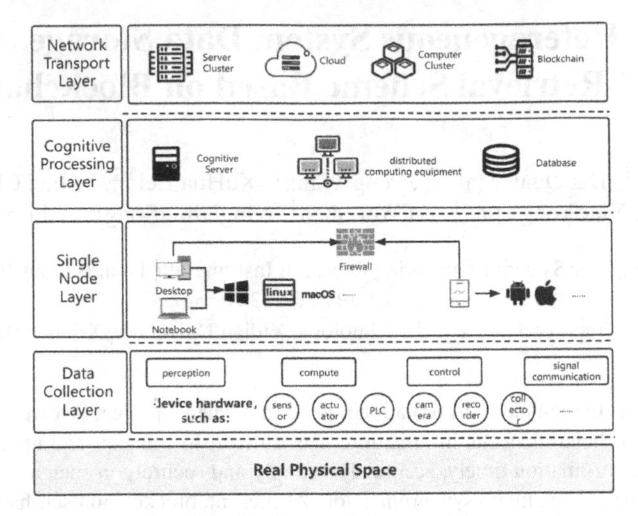

Fig. 1. System Framework

Blockchain-related technologies have brought new solutions to this problem. Blockchain has been widely applied in many fields due to its characteristics of decentralization, immutable, whole-process retention, traceable, collective maintenance, openness and transparency [13], from the initial Bitcoin project to the present application in how to provide abundant query functions to massive applications. At present, many scholars have proved through research and testing that blockchain technologies can make the system have a higher level of security and better privacy protection ability [14].

Due to the strong real-time, dynamic and massive nature of these resources [15, 16], it brings a huge challenge to the query [17]. How to effectively integrate heterogeneous platforms and heterogeneous data, complete the association between human society and the physical resource information world, build a unified information system, manage data, build a storage system, and set up an efficient query mechanism [18] are important issues that need to be solved at present. Therefore, this paper proposes a method of resource index construction and query in CPS to solve the query problem in resource management under CPS environment [19].

The structure of the paper is as follows: Sect. 2 the Blockchain Architecture Design for Task Management Business; Sect. 3 the Software Design; Sect. 4 Work Summarization.

2 Related Work

Jinshan Shi et al. [20] proposed an IoT access control framework based on blockchain. Siyuan Wang et al. [21] proposed the power token ring network based on blockchain to solve the unauthorized access of the Internet of Things [22, 23]. Guanjie Cheng et al. [24] proposed a data management architecture of IoT based on blockchain and edge computing to achieve data security management. However, they assume that the edge nodes are safe and reliable, ignore that the edge nodes should be confirmed by the trusted mechanism when accessing the IoT under the expansion of the scale of the IoT, and also ignore the computing resources brought by thc edge nodes. That is, the role of the edge nodes in the process of access control and storage query is not considered under the current new paradigm of the Internet of Things system integrating edge computing.

3 Preliminaries

3.1 Design Goals

1. Aiming at the problem of heterogeneous physical resource mapping, this paper proposes a mapping method based on node attributes and node data, which provides a general method for actual physical resources to be mapped into the information domain. This method can achieve the definition of attributes and characteristics of heterogeneous physical information nodes and nodes access, support the unified representation of various physical information resources in the real physical world on the same platform, and further improve the ability of data resource sharing among nodes.
2. This paper proposes a storage scheme that combines the information of the on-chain index table with the off-chain database. The blockchain data storage problem is solved by the sharing mode combining the on-chain and off-chain storage of data blockchain, where the original data is stored off-chain and the data description and data sharing log are stored on-chain. The data attributes are graded according to different sensitivities to meet the requirement of flexible data sharing.
3. Aiming at the complex data information generated by various heterogeneous resources, this paper proposes an index construction method based on abstract. This method supports fast query based on attributes through index query processing method of classification. It uses the on-chain and off-chain collaborative way to share data to release a large amount of space on the blockchain, ensure the efficient transmission of data on the chain, and improve the efficiency of information query and sharing.

3.2 Blockchain Technology

Blockchain provides a decentralized distributed data system [25]. The trusted or semi-trusted nodes of the participating system jointly maintain a growing chain through a consensus mechanism, which eliminates the need for centralized control and uses cryptography mechanisms such as Hash, digital certificates, and signatures to ensure that records cannot be forged or destroyed. In a distributed storage management system based on blockchain, data requesters must obtain data access permission from the blockchain before accessing specific data. The system uses the aggregation of data access rights as an incentive mechanism in the blockchain to encourage institutions to participate in building the blockchain.

Blockchain technology can improve the trust between the sharing parties, which has gradually become a consensus in the current data sharing mode. Leveraging blockchain's de-neutralization and immutable nature, data providers can use it as a tool for logging data usage. At present, most applications based on blockchain use the characteristics of it to achieve functions such as trusted depository and traceability query. In the above applications, blockchain system can be regarded as a new type of secure and trusted distributed database system. However, blockchain itself has shortcomings in storage and query, such as low storage efficiency, huge storage cost, slow query speed and single query function on the chain, etc. These problems have been restricting the development of blockchain, and become the bottleneck of blockchain application landing. Although scholars at home and abroad have done a lot of research to solve above questions, most of them only analyze the similarities and differences of data management technology between traditional database and blockchain database [26].

4 Heterogeneous Data Storage and Retrieval Method Based on Blockchain

4.1 Procedure of Operation

1) Data generation: a large amount of data will be generated in production activities. Upon the consent of the organization, the institutions will collect, clean and process these data.
2) Data storage: The institutions grades the data according to data grading criteria, encrypts and stores the data using the algorithm proposed in this paper [27].
3) Data registration: Institutions will upload data information (including data description, encryption key, institution information, relevant individuals, access control policies and data addresses, etc.) to the data sharing platform and store the data information on the blockchain [28].
4) Key application: Users can apply for keys to the data sharing platform according to their own attributes. After authenticating their identity, attribute authorization agencies can generate corresponding keys using the key generation method and send them to users.
5) Data request: the users retrieve data on the shared platform and send a data request to the platform. After receiving the request, the platform returns data information to users.
6) Query: Participants can query data usage by sending requests to the platform, as Fig. 2 shown.

4.2 Storage Methods

This scheme combines the on-chain index table information with the off-chain database for storage. On the one hand, it can release a large amount of space on the blockchain. On the other hand, it can improve the efficiency of information sharing. The index table on the chain stores the index information (index category and the address of the encrypted file) and forms the index block and stores it on the blockchain. The index table corresponds the information category queried by the queriers to the storage address value and occupies a small portion of memory on the chain. The off-chain database stores the encrypted data files uploaded by the data owner to ensure the security of the data. The data storage process as Fig. 3 shown.

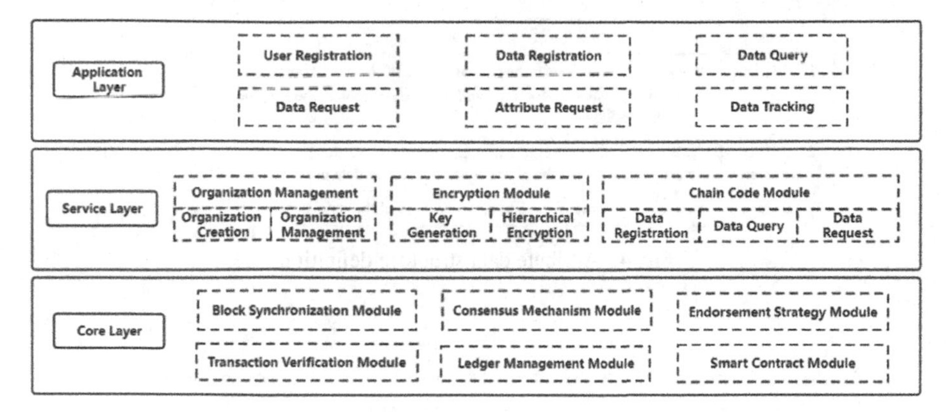

Fig. 2. Hierarchy diagram

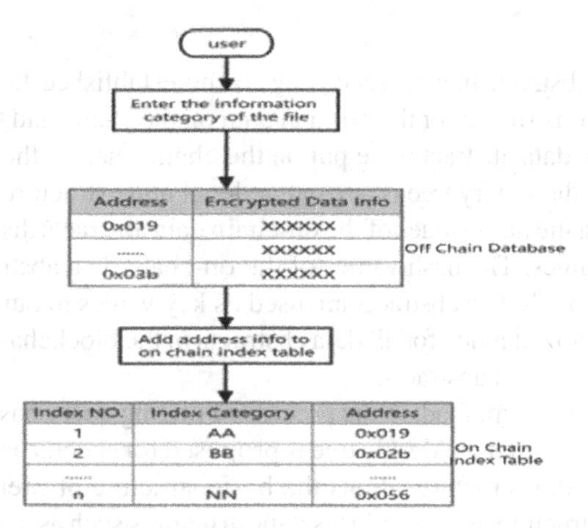

Fig. 3. Storage process

4.3 Index Build

The composition structure of data attributes is complex and diverse, and the access control policies should be different for data from different sources and with different levels of sensitivity. Aiming at the high flexibility privacy protection of data, the data is divided into the following three levels: high sensitivity, medium sensitivity and low sensitivity. Different access control policies can be implemented for different levels of data. The specific data grading criteria are as follows: (1) Highly sensitive data: information that can be identified or that will cause significant impact if exposed. (2) Medium sensitive data: the information that cannot identify personal identity and still has important significance after being blurred can retain the fuzzy result. (3) Low sensitive data: other non-important information. At the same time, in order to realize data access control, data users need to be divided according to their identity information, including user type and user permission level. The attribute and relational data structures are shown in Fig. 4 and Fig. 5.

Filed	Description
key	Index Key, whose value is the full hash value of the data digest.
preVersion	Pointer Field, whose value is the address of data node corresponding to the previous version of this data digest.
UserType	Value, whose value is the data user type.
UserLevel	Value, whose value is the professional level of the data user.
AbAddr	Value, whose value is the data digest address of the data digest hash.

Fig. 4. Attribute data structure definition

Field	Description
key	Index Key, whose value is the public prefix of the data digest hash.
next	Pointer Array

Fig. 5. Relational data structure definition

Construct data abstract, index. According to the established field extraction rules, the data abstract is constructed for the original data on the chain, and the abstract index is constructed for each data abstract to be put on the chain, such as the abstract dictionary tree index. Abstract dictionary tree is a resource level node structure definition which is constructed by using the hash values of the on-chain data abstracts distributed in different blocks as the key values. The hash values of the on-chain data abstracts are calculated, and the hash values of all data abstracts are used as key values to build a dictionary tree, so as to build a centralized index for all data abstracts in the blockchain, thus accelerating the query efficiency of data abstracts.

Hybrid index building method. In the process of tracing query based on blockchain, a hybrid index structure based on Merkle tree is proposed for specific multiple information query. The hybrid index structure adopts the basic structure of Merkle tree, combined with Merkle index structure, and introduces data structures such as hash table into Merkle tree to improve the efficiency of transaction query. It supports for member existence queries, that is, whether the collection contains the element. During the construction of Merkle tree nodes, the corresponding transaction keyword information of the node is stored in each node. When querying the transaction information corresponding to the keyword key, this paper will first calculate from the root node of the Merkle tree. If the key exists in Bloom Filter (BF) node, it will successively judge whether the queried keyword exists in BF of the subtrees around the node. If so, the query will continue; otherwise, *none* will be returned. Multiple pruning operations can be performed to improve the search efficiency during searching and traversing the Merkle tree. For the data that does not exist in the blockchain, the previous block is directly searched because there is no index stored on the key in the root node Bloom filter. In the block header, this paper introduces a Hash Table. The hybrid index structure records the transaction information corresponding to the ID in the Hash Table and stores it in the leaf node of the Merkle tree, so that the leaf node index where the transaction information is located can be quickly located according to the ID in the tracing process.

Based on the hybrid index structure of Merkle tree, the algorithm is constructed as follows: ① Input transaction set is sorted according to the numerical attributes; ② After the maximum and minimum attributes are obtained, they are placed in the block header; Traverse the transactions, Hash the transaction information and place it on the leaf node of the Merkle tree. Take out the ID of the transaction information and the corresponding location index stored in the leaf node to construct the Hash Table and store it in the block header. ③ Hash the Hash value of leaf nodes in pairs, and calculate the keywords in the transaction information of corresponding nodes in the calculation process. The Bloom filter (BF) is stored in the node, and then the Merkle tree is constructed. The Merkle root Hash is placed in the block header, and the hybrid index structure based on the Merkle tree is constructed. The output hMerkleTree is the blockchain structure, and the hybrid index structure is shown in Fig. 6.

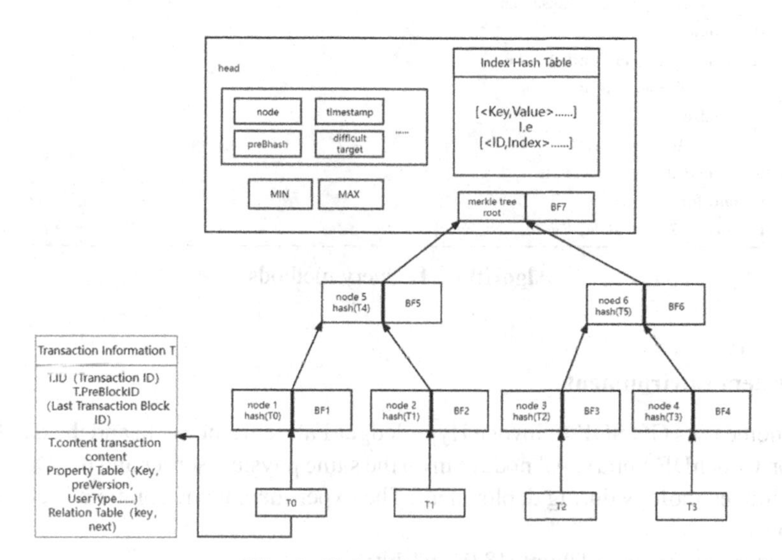

Fig. 6. Hybrid index structure

4.4 Query Methods

In the algorithm, as long as the input tracing source code can be in the alliance chain to take out the corresponding transaction address, and then take out the specific value from the database according to the address. In line 1, from the newly generated block to the Genesis block, the information of the products stored in the block from production to sales can be obtained, so there may be multiple transactions in one tracing source. In line 2, judge whether proCode is in the hash value of Merkle root. If not, directly search the next block; if so, enter the Merkle tree for search. In lines 2 to 12, the hash values of the left and right child nodes are used to determine which subtree the transaction exists in, and then several pruning operations are carried out until back to the source code to find the corresponding designated trading under the chain in the address database, and then root address queries specific transaction from the database under the chain, and adds the transaction to the list. After traversing over all blocks, return the transaction list.

Algorithm : On-chain and Off-chain Collaborative Query Algorithm

Input: Data index identification : proCode

Output: A collection of transactions that meet the requirements:$T^p=\{t_1^p.data, t_2^p.data, ..., t_m^p.data\}$

1: **for** block in blockchains **do**
2:     **if** proCode in block.Merkle.root.bloom **then**
3:         **while** root.leftchild root.rightchild **do**
4:             **if** proCode in root.leftchild.bloom **then**
5:                 root = root.leftchild;
6:             **elseif** proCode in root.rightchild.bloom
7:                 root = root.rightchild;
8:             **end if**
9:         **end while**
10:         Address=root.transaction;
11:     **end if**
12:     **if** Address in CouchDB **then**
13:         T^p.add.(data);
14:     **else**
15:         Address is err!
16:     **end if**
17: **end for**
18: **return** $T^p=\{t_1^p.data, t_2^p.data, ..., t_m^p.data\}$

Algorithm. 1. Query methods

4.5 System Environment

This scheme uses GO SDK to invoke Hyperledger Fabric framework, and the database is based on CouchDB library. All nodes run on the same physical server and use Docker virtualization technology during deployment. The experimental environment is configured as follows:

1) Operating system: Ubuntu18.04, 64-bit;
2) CPU and memory: 4 cores (vCPU) 16 GiB.

In this paper, Hyperledger Fabric V2.3 licensed blockchain is used to implement data storage and retrieval testing. Network nodes include each node of blockchain and each node of attribute authority. Among them, the blockchain node, as the underlying core of the system, provides support for storage and retrieval testing; Attribute authority node provides encryption and decryption support for access control algorithm. There are two organizations in the blockchain network, namely, Institution Org and Requester Org, where the institution organization corresponds to the data collector in the data sharing, and the requester organization corresponds to the individual and the consumer in the data sharing.

5 Conclusion

In this paper, we propose a mapping method based on node attributes and node data to solve the heterogeneous physical resource mapping problem, which provides a general method for mapping actual physical resources to the information domain. On this basis, a blockchain-oriented storage, index construction and efficient query method is proposed.

This method can realize the definition of attributes and characteristics of heterogeneous physical information nodes and node access, and support the unified expression of various physical information resources in the real physical world on the same platform. A storage scheme combining the on-chain index table information with the off-chain database is adopted, in which the original data is stored under the chain, and the data description and data sharing log are stored on the chain. The attributes of data are graded according to different sensitivities to meet the flexibility of data sharing. Aiming at the complex data information generated by various heterogeneous resources, an index query processing method based on attribute classification is proposed, which supports the fast query based on attribute. This scheme adopts the on-chain and off-chain collaborative method to share data, which can release a large amount of space on the blockchain and ensure the efficient transmission of data on the chain. While improving data security and removing information barriers, it also greatly improves the query efficiency. However, the method proposed in this paper has slightly increased spatial complexity, and the current society has increasingly higher requirements for data privacy sharing. Therefore, in the future, on the basis of ensuring storage and query efficiency, the storage burden of the system will be reduced as much as possible, and further in-depth research will be made on how to share traceable data securely on the blockchain.

References

1. Qiu, M., Jia, Z., et al.: Voltage assignment with guaranteed probability satisfying timing constraint for real-time multiproceesor DSP. J. Signal Proc. Sys. **46**, 55–73 (2007). https://doi.org/10.1007/s11265-006-0002-0
2. Qiu, M., Yang, L., et al.: Dynamic and leakage energy minimization with soft real-time loop scheduling and voltage assignment. IEEE TVLSI **18**(3), 501–504 (2009)
3. Qiu, M., Xue, C., Shao, Z., Sha, E.H.M.: Energy minimization with soft real-time and DVS for uniprocessor and multiprocessor embedded systems. In: IEEE DATE Conference, pp. 1–6 (2007)
4. Hu, F., Lakdawala, S., et al.: Low-power, intelligent sensor hardware interface for medical data preprocessing. IEEE Trans. Inf. Technol. Biomed. **13**(4), 656–663 (2009)
5. Qiu, M., Li, H., Sha, E.H.M. : Heterogeneous real-time embedded software optimization considering hardware platform. In: Proceedings of the ACM Symposium on Applied Computing, pp. 1637–1641 (2009)
6. Qiu, H., Zheng, Q., et al.: Topological graph convolutional network-based urban traffic flow and density prediction. IEEE Trans. ITS **22**(7), 4560–4569 (2020)
7. Qiu, M., Chen, Z., et al.: Energy-aware data allocation with hybrid memory for mobile cloud systems. IEEE Systems J. **11**(2), 813–822 (2014)
8. Niu, J., Gao, Y., Qiu, M., Ming, Z.: Selecting proper wireless network interfaces for user experience enhancement with guaranteed probability. JPDC **72**(12), 1565–1575 (2012)
9. Qiu, M., Xue, C., Shao, Z., et al.: Efficient algorithm of energy minimization for heterogeneous wireless sensor network. In: Sha, E., Han, S.K., Xu, C.Z., Kim, M.H., Yang, L.T., Xiao, B. (eds.) Embedded and Ubiquitous Computing. EUC 2006. Lecture Notes in Computer Science, vol. 4096, pp. 25–34. Springer, Berlin (2006). https://doi.org/10.1007/11802167_5
10. Li, Y., Gai, K., et al.: Intercrossed access controls for secure financial services on multimedia big data in cloud systems. ACM TMCCA **12**(4s), 1–18 (2016)
11. Qiu, H., Dong, T., et al.: Adversarial attacks against network intrusion detection in IoT systems. IEEE Internet Things J. **8**(13), 10327–10335 (2020)

12. Gai, K., Qiu, M., Elnagdy, S.: A novel secure big data cyber incident analytics framework for cloud-based cybersecurity insurance. In: IEEE BigDataSecurity (2016)
13. Qiu, H., Qiu, M., Memmi, G., Ming, Z., Liu, M.: A Dynamic scalable blockchain based communication architecture for IoT. In: Qiu, M. (eds.) Smart Blockchain. SmartBlock 2018. Lecture Notes in Computer Science(), vol. 11373, pp. 159–166. Springer, Cham (2018). https://doi.org/10.1007/978-3-030-05764-0_17
14. Gai, K., Wu, Y., Zhu, L., et al.: Differential privacy-based blockchain for industrial internet-of-things. IEEE Trans. Industr. Inf. **16**(6), 4156–4165 (2020)
15. Li, Y., Gai, K., Qiu, L., et al.: Intelligent cryptography approach for secure distributed big data storage in cloud computing. Inf. Sci. **387**, 103–115 (2017)
16. Li, J., Ming, Z., et al.: Resource allocation robustness in multi-core embedded systems with inaccurate information. J. Syst. Arch. **57**(9), 840–849 (2011)
17. Qiu, M., Qiu, H.: Review on image processing based adversarial example defenses in computer vision. In: IEEE 6th BigDataSecurity, pp. 94–99, Baltimore, MD, USA (2020)
18. Tian, Z., Li, M., Qiu, M., et al.: Block-DEF: a secure digital evidence framework using blockchain. Inf. Sci. **491**, 151–165 (2019)
19. Tang, S., Du, X., Lu, Z., et al.: Coordinate-based efficient indexing mechanism for intelligent IoT systems in heterogeneous edge computing. JPDC **166**, 45–56 (2022)
20. Du, X., Tang, S., et al.: A novel data placement strategy for data-sharing scientific workflows in heterogeneous edge-cloud computing environments. In: IEEE ICWS, pp. 498–507 (2020)
21. Jinshan, S., Li, R., Tingting, S.: Internet of Things access control framework based on blockchain. J. Comput. Appl. **40**(04), 931–941 (2020)
22. Siyuan, W., Shihong, Z.: Access control mechanism based on block chain and power in Multi-domain Internet of Things. Chin. J. Appl. Sci. **39**(01), 55–69 (2021)
23. Qiu, M., Qiu, H., et al.: Secure data sharing through untrusted clouds with blockchain-enabled key management. In: 3rd International Conference on SmartBlock, Zhengzhou, China, pp. 11–16 (2020)
24. Gai, K., Zhang, Y., et al.: Blockchain-enabled service optimizations in supply chain digital twin. IEEE Trans. Serv. Comput. (2022)
25. Guanjie, C., Zhengjie, H., Shuiguang, D.: Internet of things data management based on blockchain and edge computing. J. Internet Things **4**(02), 1–9 (2020)
26. XinG, X., ChEn, Y., Li, T., et al.: A blockchain index structure based on subchain query. J. Cloud Comput. **10**(1), 1–11 (2021)
27. Hegde, P., Streit, R., Georghiades, Y., Ganesh, C., Vishwanath, S.: Achieving almost all blockchain functionalities with polylogarithmic storage. In: Eyal, I., Garay, J. (eds.) Financial Cryptography and Data Security. FC 2022. Lecture Notes in Computer Science, vol. 13411, pp. 642–660. Springer, Cham (2022). https://doi.org/10.1007/978-3-031-18283-9_32
28. Validi, A., Kashansky, V., Khiari, J., et al.: Hybrid on/off blockchain approach for vehicle data management, processing and visualization exemplified by the adapt platform. arXiv: 2208.06432 (2022)
29. Sagirlar, G., Sheehan, J.D., Ragnoli, E.: On the design of co-operating blockchains for IoT. In: 3rd IEEE International Conference on Information and Computer Technologies (ICICT), pp. 548–552 (2020)

Fbereum: A Novel Distributed Ledger Technology System

Dylan Yu[1], Ethan Yang[2], Alissa Shen[3], Dan Tamir[4](✉), and Naphtali Rishe[5]

[1] Dulles High School, Sugar Land, Texas, USA
[2] Westwood High School, Austin, Texas, USA
[3] St. Stephen's Episcopal School, Austin, Texas, USA
[4] Texas State University, San Marcos, Texas, USA
dt19@txstate.edu
[5] Florida International University, Miami, FL, USA
ndr@acm.org

Abstract. Over the past several years, due to the progression toward data-driven scientific disciplines, the field of Big Data has gained significant importance. These developments pose certain challenges in the area of efficient, effective, and secure management and transmission of digital information. This paper presents and evaluates a novel Distributed Ledger Technology (DLT) system, Fibereum, in a variety of use-cases, including a DLT-based system for Big Data exchange, as well as the fungible and non-fungible exchange of artwork, goods, commodities, and digital currency. Fibereum's innovations include the application of non-linear data structures and a new concept of Lazy Verification. We demonstrate the benefits of these novel features for DLT system applications' cost performance and their added resilience towards cyber-attacks via the consideration of several use cases.

Keywords: Distributed Ledger Technology · Blockchain · Bitcoin · Ethereum · Hyper Ledger · Consensus Verification · Cybersecurity · Proof of Work · Proof of Elapsed Time

1 Introduction

This paper presents Fibereum, a Distributed Ledger Technology (DLT) system that applies novel methods to address several issues with current Blockchain [1–5] data structures and storage mechanisms, including deficiencies with respect to defending against blockchain attacks. A typical implementation of conventional DLT systems is via Blockchain. That is, conventional DLT systems, such as the permissionless Bitcoin [3] and Ethereum [4], as well as the permission-based Hyper Ledger Technology [5], use a linear data structure to manage blocks of data. Furthermore, conventional DLT systems often require a complex time- and energy-consuming process for the verification of the DLT system integrity. The DLT system presented here, Fibereum, utilizes novel methods for Big Data exchange, as well as for the fungible and non-fungible exchange of artwork, goods, commodities, and digital currency.

M. Qiu et al. (Eds.): SmartCom 2022, LNCS 13828, pp. 669–684, 2023.
https://doi.org/10.1007/978-3-031-28124-2_64

The novel features of the proposed DLT system include: **(i)** the enablement of the use of non-linear data structures-based systems; **(ii)** employing a procedure of lazy verification, where the verification of the DLT system integrity is delayed indefinitely and applied only on a need-to-do basis; and **(iii)** the enablement of permissionless as well as permission-based implementations. The use of non-linear data structures for the storage of transactions in blocks and storage of blocks within the DLT system enhances the efficiency of the overall system. For example, in one implementation, the proposed DLT system can use cryptographic trees for intra-block and inter-block management in conjunction with lazy verification. This approach significantly improves the management and security of Big Data and other types of digital data-driven systems. Furthermore, the use of lazy verification along with non-linear data structures, as well as the utilization of a time/energy consumption-efficient consensus mechanism, can provide a significant saving in energy consumption.

Fibereum introduces several innovative modifications to Blockchain technology, providing a more general framework for DLT systems. The modifications extend the utility of DLT systems to several new applications, including business-to-business data governance and data exchange. The Fibereum DLT system has various applications in the fields of Big Data, including Healthcare, Transportation, Smart Cities, the Internet of Things, and process control. The Fibereum DLT system is also suitable for applications such as digital currency, smart contracts, licensing [6], inventory management [7], supply chain management [8], counterfeit detection [9], and the exchange of copyrighted material, e.g., Non-Fungible Tokens [10].

We have developed and implemented an event-based simulation for Fibereum use cases. The simulation and theoretical analysis show that, due to the option to use non-liner data structures and the concept of lazy verification, in most cases Fibereum computational complexity is lower than other DLT systems such as the Bitcoin Blockchain.

The rest of this paper is organized as follows. Section 2 provides background information and definitions. Section 3 includes a literature review. Section 4 presents the main features of the Fibereum DLT system. Section 5 presents several Fibereum use-cases, and Sect. 6 includes a conclusion and directions for further research.

2 Background and Definitions

2.1 Definitions

A blockchain is a peer-to-peer network that stores transactions between multiple parties, organized as a cryptographic linked list – i.e., a chain of nodes. Blockchains attempt to guarantee decentralization, transparency, and immutability [1, 2].

A **Merkle tree** is a tree in which all the leaves contain the cryptographic hash of a block of data, potentially along with the data, and every non-leaf node contains the hash of its child nodes' data [11]. The notion of the basic Merkle tree can be extended to Merkle Heaps (Min and Max heaps) [4], Merkle binary search trees [4], Merkle Hash tables [4], and Merkle cyclic and acyclic graphs [4].

A **permissionless** DLT system is open to the public. Any user can create or access data or smart contracts in the DLT system, and all the transactions made on the DLT system are displayed to all the users, making the permissionless DLT system completely transparent [1–3]. In general, blocks are mined by users, referred to as miners, onto the ledger, in exchange for incentives for the miners [1–3]. Furthermore, users may be engaged in establishing the DLT system's integrity (potentially with verification incentives) [1–3]. For specific use-cases, Fibereum offers a permissionless version of a DLT system, where the system construction is extremely simple and does not require significant incentives for miners. Additionally, Fibereum offers an alternative approach for verifying the integrity and incentivizing the verification process. This approach is referred to as 'Lazy Verification.' In this case, initially, the DLT system is in a "verifiable" state. The verification and its incentives are enacted only on a "need to do" basis.

Lazy verification is a form of consensus term-setting, first introduced in the context of Fibereum, where the consensus verification process (as well as verification incentives) is/are delayed as much as possible and only performed when an immediate urgent need, e.g., taking care of an exception, arises. While continuous and prompt verification, which may require ample incentives, computational resources, and a high amount of energy consumption, is mandatory in certain use-cases and applications (e.g., digital currency), lazy verification allows for an efficient method of verifying transactions in a DLT system as it removes the unnecessary steps of verifying every block before an exception has occurred. The verification algorithm may apply the same concept as the standard blockchains' verification procedures, such as the Proof-of-Work (PoW)-based Byzantine consensus [3, 12], but with more scalability for Big Data when there are large amounts of data entering the DLT system at a high rate.

A **permission-based** DLT system is a private network where only certain users are authorized to access the DLT system. The network users are identifiable and complete anonymity is not possible [13, 14]. Hence, access control and encryption may be implemented as a part of the permission mechanism [13]. **Hyperledger Fabric Technology** (HFT) is the most commonly used framework for permission-based blockchains [5]. Fibereum offers a permission-based version of a DLT system.

A **Consensus Algorithm** replaces a centralized authority to preserve the security and fault tolerance of a DLT system. Two consensus algorithms, Proof-of-Work (PoW) [3, 13] and Proof-of-Elapsed-Time (PoET) [14] are most relevant for the Fibereum use-cases. Other commonly used consensus algorithms include Proof-of-Inclusion [15] and Proof-of-Stake [16]. The Practical Byzantine Fault Tolerance is often used as a part of consensus algorithms [12]. We elaborate on the PoET mechanism, which is less known to many DLT system practitioners and is advantageous in terms of cost performance over other consensus mechanisms, especially with respect to some Fibereum use-cases.

PoET is a consensus algorithm in which all nodes "sleep" for an arbitrary amount of time, with the first node to wake up receiving authorization/rewards for verification and mining. PoET is more energy efficient and less resource costly than PoW. Nevertheless, this algorithm must resolve "collisions" in a way that may be similar to the collision detection, avoidance, and resolution of the Carrier Sense, Multiple Access, with Collision Detection (CSMACD) procedures that govern many of the commonly used communication protocols [17, 18].

Notably, Fibereum's use of lazy verification, along with providing the option to use non-linear data structures as well as time/energy consumption efficient DLT system construction and consensus verification mechanisms, can provide improved resilience against attacks, as well as a significant saving in operational costs.

2.2 Resilience and Security

In this sub-section, we list several of the common attacks applied to existing blockchain DLT systems and related security concerns. In Sect. 4, we will refer to these items in the context of Fibereum.

Some of the common attacks on the Bitcoin Blockchain are Eclipse/Sybil attacks [20] and double-spending attacks [19]. Among other attack types are the Vector76 attack [20], the Blockchain reorganization attack [3, 21], and Denial of Service (DoS) attacks [22]. Additionally, careless management of passwords and security measures might jeopardize the anonymity of the DLT system users. The **Majority Attack / 51% attack,** is one of the most commonly discussed attacks in the context of digital coins [3, 21]. In this type of attack, the attacker controls more than 50% of the network's computation power and thus is able to successfully perform bogus blockchain modifications and reorganizations and obtain [temporary] consensus for the bogus blocks [3, 21].

Many of the attacks on the Bitcoin blockchain listed above are applicable to the Ethereum blockchain. An additional set of attacks on Ethereum exploits its smart contract functionality, specifically the computational complexity of the embedded "Turing Complete" [23] functions, which is quantified in terms of "gas," reflecting the cost of computation [4]. The major Ethereum attacks include Reentrancy [24], Front running [25], Integer Overflow and Underflow attacks [26], Unexpected Revert attacks [26], Gas Limit attacks [27], Block Stuffing attacks [27], and Multi-Signature attacks [28].

Common Attacks on HFT Blockchain deal with the centralized and permission-based aspects of the HFT DLT system, particularly attacking the membership service provider that authorizes and provides permissions for entrance into the blockchain and blockchain transactions [29]. The major HFT attacks are the Insider Threat attacks [30] and the Certificate of Authority attack [31].

3 Literature Review

DLT systems provide a decentralized platform. Hence, its potential usage in the field of big data exchange may have significant benefits. Nevertheless, to the best of our knowledge, Fibereum is the first DLT system that provides an optimal solution for that purpose while offering significant benefits in other use-cases. Since Fibereum is a unique DLT system using lazy verification and non-linear data structures, its composition and computation processes are especially efficient. An extensive literature review performed resulted in very few publications that specifically address the issues that Fibereum addresses. Three papers are listed below.

Cäsar et al. have developed a DLT system named Cerberus that focuses on the particular ordering of State Machine Replication (SMR) across a network of unreliable machines [32]. This DLT's consensus mechanism is based on a leader-based Byzantine fault-tolerant consensus approach [12]. In contrast, we propose consensus mechanisms such as the lazy verification mechanism, which minimizes the need for prompt and incentivized consensus, and enhance fault tolerance at lower computational resources.

Snow has developed Factom, a general-purpose data layer that creates a consensus system to ensure that entries are quickly recorded [33]. Comparatively, Fibereum offers the lazy verification approach, which is more suitable for numerous use-cases (see Sub-Sect. 4.2 and Sect. 5). Additionally, Fibereum offers the use of other non-linear data structures, such as Merkle-heaps, for the Inter-block DLT system's construction and maintenance. Thus, Fibereum enables a more efficient method of verification and DLT operation, especially for Big Data exchange.

Parachain [34] uses chains that are processed in parallel, thereby has the potential to improve throughput. Parachain DLT has not addressed certain issues related to overseeing blockchain creation. Fibereum provides better support of "safe" parallelism via the mechanism of using Merkle Heaps for inter-block construction; at the same time, the mechanism is highly efficient and provides $O(log(n))$ [4] complexity for DLT consensus and construction.

Other relevant papers that are not completely overlapping Fibereum concepts and targeted use cases. Examples include work Gay et al. [35], and by Zhu et al. [36].

4 The Fibereum DLT

The Fibereum DLT introduces the following novel features: (i) Options for storing information/transactions blocks in data structures, including Merkle trees, Merkle heaps, or Merkle hash tables, rather than as a linear list in the form of a blockchain; (ii) Enablement of low complexity algorithms and parallel processing; (iii) Lazy Verification – minimizing the need for incentivized DLT systems' construction and consensus verification; (iv) Enablement of permission-based and permissionless modes of access and operations; (v) Enablement of encryption and compression of the data; (vi) Enabling improved cyber security, protection against attacks, and fault tolerance; (vii) Providing additional layers of encryption and digital signatures (in addition to cryptographic hash functions referred to as the digests [1, 2]); (viii) Enablement of Embedded Turing Complete [23], static and/or dynamic, code, which provides efficient management of smart contracts [4] and End User License Agreements (EULA) [35, 36]; (ix) Enablement of efficient management of static, dynamic, and ad-hoc federated data, including terms and policy management for monetization (please see the section on use-cases); (x) Enablement of systems for data governance and currency exchange; (xi) Providing an option for using more than one DLT system in tandem. The latter is referred to as multi-plan implementations. For example, in the exception maintenance use-cases (4.2 and 5.1), we introduce three DLT system plans: one for data transactions governance, one for data exchange, and one for smart contracts – defining data ownership, usage policies, rights, management, and governance.

4.1 A Merkle-Based Verification System

Implementations of Merkle trees-based non-linear DLT systems offer several key advantages over linear blockchains. These advantages include: (i) Merkle trees-based implementations maintain the integrity by cascading any change to the cryptographic hash.

Pointing from the previous node in the tree back to the Merkle root would invalidate the changed block. (ii) Merkle trees-based implementations are typically efficient for the construction and verification of DLT systems, offering the complexity of $O(log(n))$ (or, in some cases, $O(n \times log(n))$) rather than $O(n)$ (or, in some cases, $O(n^2)$). Hence, Merkle tree-based implementations can reduce the temporal and spatial computational complexity. Moreover, without a Merkle tree, the data would need to be sent across the network for verification. Hence, Merkle trees reduce the data transfer delay. (iii) The Merkle tree structure of the local blockchain can use Proof of Inclusion [15], a method of verifying the validity of data without needing to move data across all parts of the network. This algorithm can work in conjunction with consensus mechanisms, such as PoW and PoET. The joint Merkle heap, proposed in some of the Fibereum use-cases, is beneficial for efficient traversal without revealing all portions of the structure. To further demonstrate the principles of Fibereum's operation and its novelty, we present an important use-case here, and the rest of the use-cases are presented in Sect. 5.

4.2 Exception Maintenance 1

This use-case considers the situation that an airplane manufacturing company X buys an engine from an engine manufacturing company Z, and the engine is installed on an airplane of an airline company Y. In other words, this is a business-to-business-to-business (B2B2B) scenario. To simplify the example, we assume that the engine is in X's possession, and the use-case is a typical B2B use-case between X and Y. Generally, Y owns the data. However, in some scenarios, the ownership of the data might be shared between X and Y. It is assumed that according to a licensing agreement between X and Y, Y collects and owns the engine's sensor data. The proposed Fibereum DLT system is designed to be used for managing data usage, ownership, and storage used by the companies and their affiliates in a way that enables dealing with exceptions in the regular operation of the airplane.

DLT System Operation Procedures

(i) Verifiable sensor data is collected in a joint heap, accessible by both X and Y. The data is "verifiable" in the sense that it includes means for verifying its authenticity, e.g., cryptographic hashes of time stamps and sensor IDs. (ii) Each heap node stores specific components of the sensor data (e.g., temperature, pressure, and position) in the storage area. In some implementations, the heap storage is based on a string pool [38]. (iii)Similarly to blockchains, such as the Bitcoin blockchain, the data is stored in blocks and organized in blocks via an internal Merkle tree (the Intra-Merkle Tree). These blocks are maintained by an external Merkle heap (the Inter-Merkle heap). (iv) Inter-block and Intra-block storage via Merkle heaps or trees simply imply that both parties can traverse each piece of sensor data as well as the entire collection of sensor data efficiently and "quickly." (v) If legitimately requested, timely verification is used to ensure data integrity. The fact that the verification is done only on a need-to-do basis and at the time of the need-to-do verification is the origin of the name "lazy verification." The lazy verification process can reduce the cost of operations.

Lazy Verification

The lazy verification process is activated when the need arises (e.g., dealing with an exception in the engine's operation). At this time, a new DLT system based on a data structure such as a sorted Merkle (tree in this example) or a Blockchain (in use-case 5.1) may be constructed. If there is no malicious activity, valid blocks are fetched from the joint heap of the first DLT, and each valid block is appended to the second DLT system in the order of the recency of activities.

Our current implementation of the system has the following components: permissionless or permission-based application of the Fibereum DLT system used for validated exchanges, min-heap DLT system for exchanges that still have to be validated, and license agreement programmed into smart contracts.

To elaborate: data continuously flows from Company X's engine to the joint Merkle heap-based DLT system (DLT System-1) shared by both X and Y. On exception (e.g., overheating, mis-assembly, or exhaustion), X or Y can choose to request lazy verification regarding the exception. In this case, X and Y can nominate one or more proxies and assign the task of verification to the proxies. At this point, the process might resemble verification on blockchains, such as the Bitcoin blockchain. The proxies act like miners and assemble a second DLT system (DLT System-2), which is a Merkle Tree-based DLT or a blockchain-based DLT. DLT System-2 contains only blocks that have been verified by the proxies. Following the request, Companies X and Y (or their proxies) check the sensor data stored in DLT System-2 to determine whether there is a legitimate error, such as engine exhaustion. If there is a nonfunctional component, the lazy verification tags the exception as valid, and Company Y can take steps to fix the issue with the engine. Otherwise, the lazy verification deems the exception invalid, implying that there has been a human error in maintaining the engine and that the engine is functioning properly.

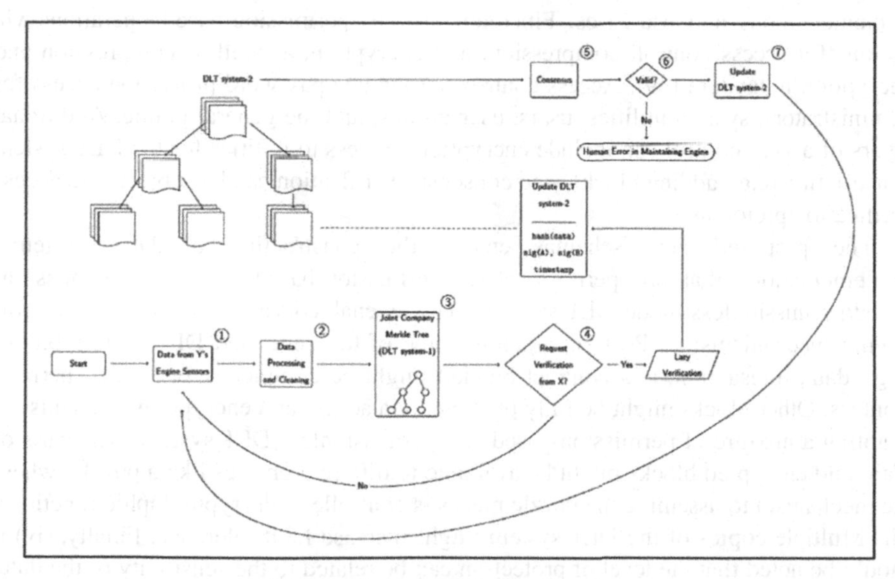

Fig. 1. Big Data Exchange Use-Case

Figure 1 depicts the process and the related scenario of lazy verification. As shown in the figure, sensor data from Company Y's engine is transported from X's plane to the shared heap, demonstrating Fibereum's ability to function in B2B relationships.

As Fig. 1 shows, in a regular mode of operation: (i) Company Y collects data from the engine. (ii) Y might perform data pre-processing and conditioning. (iii) Y places the data in a verifiable form into DLT System-1. This process (i, ii, and iii) continues as long as there is no valid request for verification, e.g., Company X wishing to terminate the contract with Company Y and replace the engine with a new engine from Company W.) (iv) In case that X, Y, or an authorized third party initiates a legitimate request for verification, X and Y nominate miners (e.g., proxies). (v) The proxies generate local copies of DLT System-1. (vi) A consensus algorithm, such as PoET, is applied and used to construct and/or extend a second DLT system (DLT System-2). For clarity, DLT system-2 is referred to as the global DLT system – that is, a DLT system that has been verified or extended according to a consensus algorithm. If the data is valid, DLT System-2 is updated. Otherwise, an exception is raised. (vii) DLT System-2, the global DLT system, is used to address the exception.

If some blocks do not pass consensus verification due to a malicious act or negligence by one of the parties breaching the agreement, then the matter may be further pursued, e.g., brought to courts or arbitration. Note that in order to save storage space, DLT System-2 might only contain cryptographic pointers to nodes of DLT System-1, thereby serving as a transaction management DLT system.

4.3 Fibereum Permissionless and Permission-Based DLT System Applications

The general mode of operation of Fibereum is permissionless. Nevertheless, in some implementations and use-cases, Fibereum enables permission-based operations via options for access control, compression, and encryption, as well as compression and encryption in tandem [39]. Access control can include password protection access for administrators, system utilities, users, user groups, and the general public. Additional layers of access control can include encryption. Access to utilities for the DLT system construction (e.g., adding blocks) and consensus verification can be subjected to access verification protocols.

The proposed new schemas enable the construction of DLT systems' implementations that are permissionless, permission-based or mixed permission-based/permissionless mode DLT systems. This is enabled via one or more of the following mechanisms: (i) Post verification, some of the generated DLT system blocks (e.g., data, transaction, and contract blocks) might be completely open, i.e., permissionless. Other blocks might be fully protected via access and encryption mechanisms, enabling a mixture of permission-based and permissionless DLT systems. (ii) Parts of plain and encrypted blocks might be available to different entities like a puzzle, where the mechanism to assemble the puzzle pieces is controlled via cryptographic functions. (iii) Multiple copies of the DLT system might increase fault tolerance. Finally, (iv) it should be noted that the level of protection can be related to the sensitivity of the data, where sensitive data might be encrypted and subject to access control.

4.4 Cyber Security and Fault Tolerance Enablement

The proposed Fibereum DLT system schema provides additional security layers for permission-based and permissionless DLT systems. The concept of lazy verification enables storing the data in local storage, where access can be controlled and protected in several ways. The data can be digitally signed, compressed, and encrypted – potentially using methods for tandem compression and encryption. In some implementations, the digests, digital signatures, as well as public and private keys used for encryption can be associated with the physical devices used to generate, transmit, or process the data and with the time that the data was generated. Furthermore, the verification stage can be limited to "trusted parties" and proxies that have protected and potentially permission-based access to the data. Consequently, Fibereum provides better protection against commonly used Blockchain attacks listed in Sect. 3. The following subsection provides further details concerning potential attacks on Fibereum and the resilience of Fibereum implementations to such attacks. This resilience is referred to as "counterattacks."

4.5 Counterattacks by Fibereum

The main threats to Fibereum (and many other DLT systems) are Sybil [20], Majority (51%) attack [3, 21], Denial of Service (DoS) [22, 28], and insider attacks [30]. Several components of the Fibereum DLT system can reduce or completely eliminate the risk emanating from the above and other DLT system attacks. First, when applicable, the option for lazy verification provides ample time to detect and prevent those attacks. Second, some implementations may use the PoET verification protocol for lazy or prompt verification. This increases the resilience of Fibereum to DLT system attacks. Finally, the utilization of permission-based or mixed permission-based and permissionless systems' components can be a paramount counterattack method.

It has been established that without a centralized authority, a system might be susceptible to Sybil attacks [29]. Consensus algorithms (e.g., PoW and PoET) mitigate the effects of Sybil attacks, and permission-based implementations of Fibereum can completely prevent such attacks. Similarly, a Majority Attack would fail with permission-based Fibereum utilizing PoET. In particular, data passing the lazy verification process would enter the second Merkle-based DLT system (System-2), where it would then need to pass through PoET in order for the first and/or the second DLT system to be updated.

Many forms of DoS counterattack methods in general networks exist; some of these methods are applicable to Blockchain DoS [22, 28]. Fibereum can be more resilient to DoS since it is possible that for long periods of time, the only DLT system activity is updating DLT System-1 with new verifiable data, which provides ample time to detect and mitigate a DoS attack.

Insider attacks are the most difficult to prevent, but they often only affect permission-based DLT systems. Measures to reduce the threat of insider attacks have been proposed. Some of these measures are common to many other permission-based systems. Other measures use a blockchain traceability system with a differential traceability algorithm, both of which can be implemented into Fibereum [30].

4.6 Fibereum Smart Contracts

Fibereum Smart contracts are irreversible contracts enforced by the program code embedded in the Fibereum blocks, which are not controlled by users. These contracts are limited by their inability to send HTTP requests and access off-chain data directly. Ethereum can circumvent this limitation via oracles, but this leaves transactions susceptible to attacks that manipulate data and price values [26]. In some implementations, in order to resolve this issue, Fibereum maps the smart contract onto a specific version of an End User License Agreement (EULA) [37]. Hence, the EULA might guide the lazy verification procedure.

5 Additional Fibereum Use-Cases

In this section, several Fibereum use-case examples are presented, concentrating on B2B Applications of the Fibereum DLT system for [Big] Data Exchange use-cases. The concerns related to the data exchange use-cases are data ownership, data rights, data use agreements, managing survivability and termination clauses for contracts, data integrity, liability, monetary value, and responsibility for disclosing the data and its use to third parties, governments, and governing authorities. In these types of use-cases, our objectives include creating a framework for policies governing data exchange in a B2B environment, where data exchange transactions are bounded by a legal contract, potentially in the form of license agreements or subscriptions (signed or click-through), specifying certain terms, such as ownership, usage policies, rights, management, and governance. Often, these agreements take the form of End User License Agreements, Developer License Agreements, and Data License Agreements. Given a predetermined legal contract, Fibereum aims to minimize the computational burden of consensus verification. It should be noted that Fibereum also provides efficient mechanisms for support of other use-cases, including digital currency exchange, B2C data exchange, as well as services related to data exchange. This section includes examples of these use-cases as well.

Some of the use-cases might include additional Fibereum-based DLT systems, e.g., a DLT system for transaction management that records the process of data exchange and a smart contract DLT system that is used to dictate data usage, rights, and termination. The use-case examples, however, do not elaborate on the internals of the smart contract DLT systems and their operation. Finally, all the use cases may deploy a permissionless version of Fibereum, a permission-based version, or a combination.

5.1 Exception Maintenance 2: Data Networks

In some Fibereum implementations, DLT System-2 is a Blockchain DLT. As an example of such an implementation, one can consider a use-case where company U is a process control firm that has sensors installed in an oil refinery that belongs to company V. In this implementation of the Fibereum DLT system, the data is stored locally by the data stakeholders (e.g., by companies U, V, and their affiliates/proxies) in Merkle heaps. Additionally, a Fibereum DLT system for transaction management may record the process of data exchange, and a Fibereum smart contract DLT system may be employed to dictate data usage, rights, and termination. When new sets of sensor data are available, they are aggregated into blocks, and the blocks are inserted into the Merkle heaps. At the same time, the transaction management DLT system is updated. At the time that consensus verification is mandated (e.g., a dispute between Company X and Company Z or a discovery subpoena by local authorities due to an accident), the integrity of the data stored in the heaps and proxies is assessed.

The following is a flowchart of the of DLT system-2 construction and lazy verification procedure applied in the Exception Handling use cases:

The Lazy Verification Procedure

Require: LazyVerification(*DLT system-1*, *DLT system-2*)
1: **while** *DLT system-1* is non-empty **do**
2: *d* ← POP(*DLT system-1*)
3: **if** *d* is valid **then**
4: APPEND(*DLT system-2*, *d*)
5: **else if** *d* is invalid **then**
6: raise legal issue
7: **end if**
8: **end while**

5.2 Digital Currency

This use-case considers digital currency applications that are similar to Bitcoin and Ethereum digital coins exchange. In contrast to most other digital coin DLT systems, both the intra-block and the inter-block may be managed via Merkle trees. Due to the nature of the application and potential attacks, the verification may be prompt and incentivized using fees or digital coin mining rewards. A PoET consensus mechanism may be employed to reduce operational complexity and energy consumption and improve counterattack capabilities.

5.3 Targeted Advertising

This use-case may be a B2B or a B2C use-case. For example, assume that Company P manufactures autonomous vehicles and Company Q, or a consumer R, uses these vehicles, which collect federated data along with sensor data. Specifically, suppose that a consumer X buys a car manufactured by Company Y, and Company Z wishes to access parts of the sensor data from the car. The process is similar to the process described in the above exception maintenance use-cases. However, it might utilize two Merkle heaps or one heap with access control to heap elements. Both classified and unclassified information is accessible to X and Y via one heap, but, for the protection of X's privacy, the second heap contains only unclassified information for Z.

5.4 Patient Medical History

This use-case considers situations where patient X wishes to switch from care provider Y to care provider Z and then transfer their medical history to Z. Due to stringent confidentiality requirements, it is most likely that the DLT system implementation would be permission-based, preventing unauthorized access to medical records. Any new medical data that goes into the records by the care provider must be verifiable and potentially include encryption, digests, and the digital signature of the patient before it is added as a block to the DLT. Provider Y uses its own encryption, digests, and digital signature to securely access medical records. If the patient wishes to share their information with other care providers, e.g., Z, then Provider Y might require a digital signature of Patient X and Provider Z for consent to release information. The DLT system includes a network of care providers, as medical records may need to be transferred from one care provider to another care provider. In this case, the medical records and other information are encrypted in a Merkle heap (DLT system-1) and, following verification, sent to other care providers through the Merkle tree of DLT System-2. Lazy verification, initiated on a "need to do" basis, e.g., switching a care provider, is used to check permissions and verify information correctness.

5.5 Digital Cartography

This use-case considers a situation where the system includes satellites, e.g., X1, X2, X3, and X4, a ground station Y, a user Z, and an object of interest W. The information generated by the remote sensing satellites and gathered by the ground station (e.g., GPS locations of Object W) is stored in the Merkle heap-based DLT System-1 and is accessible to User Z. The system allows User Z to request a legitimate verification of certain parts of the information. This triggers lazy verification and the creation of DLT System-2. The verification may include proxies. Since storage is placed within a heap, the location may be constantly updated, and User Z can constantly update the positions of Object W by requesting lazy verification. If an exception occurs, the data stored in the heap can be used to track previous locations with precise timestamps from the satellites' atomic clock in order to help figure out what may have happened.

5.6 Non-fungible Tokens

This use-case considers situations where a DLT system is used for the exchange and management of Non-Fungible Tokens (NFT) [10]. In contrast to most other NFT DLT systems, both the intra-block and the inter-block may be managed via Merkle trees. Due to the nature of this application and potential attacks, the verification may be prompt and incentivized using fees[1].

5.7 Smart Contracts and Licensing Agreements

This pertains to cases where a DLT system is used for smart contract management. In contrast to most other smart contract DLT systems (e.g., Ethereum-based smart contracts), both the intra-block and the inter-block may be managed via Merkle trees. Due to the nature of the application and potential attacks, the verification may be prompt and incentivized using fees[1].

USCG is interested in exploring, along with our team, the utility of DLT systems for these use-cases and may utilize the DLT systems for licensing, e.g., licensing of fishing companies and vessels. Given that the DLT users are not necessarily USCG staff members, the system may have to tighten security measures with respect to access to the DLT and the construction of DLT system blocks.

5.8 Copyrighted Material

This use-case considers situations where Consumer X wishes to access copyrighted material produced by Company Y. In a possible DLT system implementation, the operation procedures are similar to the operating procedures of the DLT system described in Use-case 4.1. However, the amount of data stored in DLT System-1 is not as big as the amount of data expected in Use-case 4.1. Furthermore, the DLT system may be permission-based so that only Consumer X and Company Y can access the material. Release of the material requires the consent of both parties, i.e., Consumer X and Company Y must both sign in order to sell content to a third party. Note that this use case has some overlap with NFT and can be used as a DLT system for NFT.

5.9 Commerce, Supply Chain Management, and Inventory Management Systems

This use-case considers situations where Company X wishes to transport goods to a warehouse owned by Company Y. In this case, a comprehensive database accessible by both X and Y can be used. This DLT system is likely to be permission-based so that only the two parties and their affiliates have access to it. The original owner of the DLT system, Company X, shares parts of the database of verifiable transactions with Company Y in a Merkle heap-based DLT System-1. Lazy verification may be used to generate DLT System-2 in order to mine data, verify transactions, manage inventory, and validate the integrity of the data and the underlining supply chain. Inventory management can be implemented in a similar way.

The US Coast Guard (USCG) is interested in exploring, along with our team, the utility of DLT systems for these use-cases and may utilize DLT systems under the assumption that the users are internal to the organization. Hence, they may have "some" level of trust by the system (e.g., after supplying credentials that associate them with the USCG).

5.10 Weather Broadcasting

This use-case considers a situation where two or more weather stations (e.g., X and Y) wish to share data regarding weather conditions in a certain region. The DLT system may be permission-based so that only Stations X and Y can access the data. Sensor data created by X and Y is verifiable and stored in DLT System-1. Lazy verification and the

[1] A PoET consensus mechanism may be employed to reduce operation complexity and energy consumption and improve counterattack capabilities.

creation/update of DLT System-2 may take place when there are discrepancies between X and Y as to the data relied upon or in their weather prediction.

6 Conclusion and Future Research

The DLT systems and methods discussed above include novel modifications to blockchain technology, providing a superior framework for DLT systems. Those modifications can improve the cost/performance of DLT systems in current applications and extend the utility of current DLT systems to several new applications and use-cases. The Fibereum DLT system has various applications in the fields of Big Data, including Transportation, Smart Cities, Healthcare, Process Control, and Internet of Things. Fibereum DLT systems are also suitable for other applications, such as Digital Currency, Smart Contracts and licensing, and Supply Chain Management.

Future work can include: (i) Implementations for other use-cases and data exchange applications; (ii) Monetization and control of federated data; (iii) Enhancements to B2C applications where concerns may include tight privacy constraints, as well as consumer rights protection; (iv) Further exploration of the utility of additional cryptographic data structures and other non-linear data structures, e.g., directed acyclic graphs for transactions and/or data storage; (v) Appending new data transaction information to a concurrent transaction data structure rather than directly appending it to the DLT; (vi) Appending new sensor data to a concurrent sensor data structure rather than directly appending it to the DLT; (vii) Further exploring the management of federated data where different parts of copies of the data reside in the DLT systems of individual parties; (viii) Exploring implementations where the data is compressed and encrypted, potentially for enabling permission-based access.

Acknowledgment. This material is based in part upon work supported by the Department of Homeland Security grants TXST83938 and E2055778 and by the National Science Foundation under Grant CNS-2018611 and CNS-1920182.

References

1. Narayanan, A.: Bitcoin and Cryptocurrency Technologies: A Comprehensive Introduction. Princeton University Press, Princeton (2016)
2. Lipton, A., Treccani, A.: Blockchain and Distributed Ledgers: Mathematics, Technology, and Economics. World Scientific, Singapore (2022)
3. Nakamoto, S.: Bitcoin: A Peer-to-Peer Electronic Cash System. http://www.bitcoin.org/bitcoin.pdf
4. Wood, G.: Ethereum: a secure decentralised generalised transaction ledger. Ethereum Proj. Yellow Pap. **151**, 1–32 (2014)
5. Horowitz, E., Sahni, S.: Fundamentals of Data Structures. Sung Kung Computer Book Co. (1987)
6. Androulaki, E., et al.: Hyperledger fabric. In: Proceedings of the Thirteenth EuroSys Conference. ACM (2018)
7. Introduction to smart contracts. https://ethereum.org/en/developers/docs/

8. Muller, M.: Essentials of Inventory Management. HarperCollins Leadership, Nashville (2019)
9. Chopra, S., Meindl, P.: Supply chain management. Strategy, planning & operation. In: What's New in Operations Management. Pearson (2018)
10. Fraga-Lamas, P., Fernandez-Carames, T.M. Leveraging distributed ledger technologies and blockchain to combat fake news (2019). CoRRabs/1904.05386. arXiv:1904.05386. http://arxiv.org/abs/1904.05386
11. Wang, Q., et al.: Non-Fungible Token (NFT): Overview, Evaluation, Opportunities and Challenges (2021). arXiv:2105.07447
12. Merkle, R.: A certified digital signature. In: Crypto, vol. 89, pp. 218–238 (1989)
13. Castro, M., Liskov, B.: Practical byzantine fault tolerance and proactive recovery. ACM Trans. Comput. Syst. **20**(4), 398–461 (2002)
14. Dabbagh, M., et al.: A survey of empirical performance evaluation of per-missioned blockchain platforms. Comput. Secur. **100**, 102078 (2021)
15. Pal, A., Kant, K.: DC-PoET: proof-of-elapsed-time consensus with distributed coordination for blockchain networks. In: 2021 IFIP Networking Conference (IFIP Networking), pp. 1–9 (2021)
16. Peng, K.: A general, flexible, and efficient proof of inclusion and exclusion (1970)
17. Proof-of-stake (POS). https://ethereum.org/en/developers/docs
18. Liu, Y.C.: Performance of a CSMA/CD protocol for Local Area Networks. https://ieeexplore.ieee.org/document/1146621
19. Tenenbaum, A.S.: Computer Networks. Prentice Hall, Indore (2009)
20. Heilman, E., Kendler, A., Zohar, A., Goldberg, S.: Eclipse attacks on bitcoin's peer-to-peer network. In: usenixsecurity15 (2015)
21. Iqbal, M., Matulevi˘cius, R.: Exploring sybil and double-spending risks in blockchain systems. IEEE Access **9**, 76153–76177 (2021)
22. Joshi, J., Mathew, R.: A survey on attacks of bitcoin. In: Proceeding of the International Conference on Computer Networks, Big Data and IoT (ICCBI - 2018), pp. 953–959 (2020)
23. Frankenfield, J.: 51% attack: definition, who is at risk, example, and cost (2022). https://www.investopedia.com/
24. Raikwar, M., Gligoroski, D.: DoS Attacks on Blockchain Ecosystem (2022)
25. Jansen, M.: Do smart contract languages need to be turing complete?" In: Blockchain and Applications, pp. 19–26 (2020)
26. Ethereum smart contracts. In: 2020 IEEE International Workshop on Blockchain Oriented Software Engineering (IWBOSE). IEEE (2020)
27. Torres, C.F., Camino, R., State, R.: An empirical study of frontrunning on the ethereum blockchain. In: Usenixsecurity21 (2021)
28. Ali Khan, Z., Siami Namin, A.: A Survey on Vulnerabilities of Ethereum Smart Contracts (2020). ArXiv 2012.1448
29. Chen, H., et al.: A Survey on Ethereum Systems Security: Vulnerabilities, Attacks and Defenses (2019). ArXiv.1908.04507
30. Samreen, N.F., Alalfi, M.H.: SmartScan: an approach to detecting denial of service vulnerability in ethereum smart contracts. CoRR abs/2105.02852 (2021). arXiv: 2105–02852. https://arxiv.org/abs/2105.02852
31. Gojka, E.E., et al.: Security in distributed ledger technology: an analysis of vulnerabilities and attack vectors. In: Intelligent Computing, pp. 722–742 (2021)
32. Putz, B., Pernul, G.: Trust factors and insider threats in permissioned distributed ledgers - an analytical study and evaluation of popular DLT Frameworks. https://core.ac.uk/outputs/232204350
33. Davenport, A., Shetty, S., Liang, X.: Attack surface analysis of permissioned blockchain platforms for smart cities. In: 2018 IEEE International Smart Cities Conference (ISC2), pp. 1–6 (2018)

34. Cäsar, F.: Cerberus: A parallelized BFT consensus protocol for radix (1970). https://api.semanticscholar.org/CorpusID:221297416
35. Snow, P., et al.: Factom Ledger by Consensus (2015). https://cryptochainuni.com/wp-content/uploads/Factom-Ledger-by-Consensus
36. Kaur, G., Gandhi, C.: Chapter 15 - scalability in blockchain: challenges and solutions. In: Handbook of Research on Blockchain Technology, pp. 373–406 (2020)
37. Gai, K., Wu, Y., Zhu, L.Q., Shen, M.: Privacy-preserving energy trading using consortium blockchain in smart grid. IEEE Trans. Industr. Inf. **15**(6), 3548–3558 (2019)
38. Zhu, L. Wu, Y. Gai, K., Kwang, K., Choo. R.: Controllable and trustworthy blockchain-based cloud data management. Future Gener. Comput. Syst. **91**, 527–535 (2019)
39. Cotton, H., Bolan, C.: User perceptions of end user license agreements in the smartphone environment. In: Australian Information Security Management Conference, pp 235–244 (2018)
40. Bubel, R., Hähnle, R., Geilmann, U.: A formalisation of java strings for program specification and verification. In: Software Engineering and Formal Methods
41. Tamir, D., Bruck, D.: Compression and Decompression Engines and Compressed Domain Processors, US Patent 10404277 (2019)

Indistinguishable Obfuscated Encryption and Decryption Based on Transformer Model

Pengyong Ding[1], Zian Jin[2], Yizhong Liu[2(✉)], Min Sun[1], Hong Liu[1], Li Li[1], and Xin Zhang[1]

[1] China Mobile Information Technology Company Limited, Shenzhen, China
{dingpengyong,sunmin,liuhong,liliit,zhangxinit}@chinamobile.com
[2] Beihang University, Beijing, China
{JinZiAn,liuyizhong}@buaa.edu.cn

Abstract. To solve the problem in secure encryption in cryptography, *indistinguishability Obfuscation* (iO) was born. It is a crypto-complete idea, based on which we can build many cryptographic construction. The implementation of it can hide both the dataset and the program itself. In this paper, we use the idea of translation in the (*Natural Language Processing*) NLP-like language model to realize the conversion between plaintexts and ciphertexts with the help of hints. We trained a self-attention transformer model, successfully hiding the dataset as well as the encryption and decryption programs. The input of the encryption model is a plaintext prefixed with a hint and the output is the result of encryption using one of the specified algorithms. The input and output of the decryption model are the opposite of the encryption one.

Keywords: Information Security · Indistinguishable Confusion · Transformer · Blockchain

1 Introduction

With the informationnization of the whole society, new information technologies such as cloud computing [1–3], new computer hardware [4–6], the Internet of Things [7–9], blockchain [10–12], etc., have been gradually applied to various industries in society, leading to the exponential growth of all kinds of data [13–15]. As the core asset of information systems, data are of great value, so data security [16] is becoming more and more important. In the information era, data assets have become one of the most important assets of each enterprise, thus data security [17–19] has become a key concern for enterprises. However, transmitting data in public is inevitable, so the security of data encryption and decryption is the top priority.

Secure encryption has a crypto-complete construction in cryptography, also known as indistinguishable Obfuscation(iO) [20], which could hide both the

M. Qiu et al. (Eds.): SmartCom 2022, LNCS 13828, pp. 685–694, 2023.
https://doi.org/10.1007/978-3-031-28124-2_65

dataset and the program itself. Based on iO, we could build an encryption algorithm that can implement almost all other encryption protocols, including but not limited to public key encryption, function encryption, digital watermarking system, etc. Computer scientists proved that indistinguishable obfuscation can be the basis for almost all cryptographic protocols (except black-box obfuscation), including classical cryptographic tasks (e.g., public-key encryption) and emerging tasks (e.g., fully homomorphic encryption). It also covers cryptographic protocols that no one knows how to construct, such as deniable encryption and function encryption. But until now there is no practical iO implementation. This paper provides a viable solution for implementing it.

Blockchain [21] is a distributed system with multiple participants, secret communication [22], transaction processing [23], and committee reconfiguration [24,25] among the participants often requires public-key cryptographic systems. Therefore, iO-based public-key cryptographic protocols can significantly improve the security of communication between the participants of a blockchain system [26]. The iO-based public-key cryptographic protocols require shorter keys to achieve the same security strength, which can effectively reduce the communication complexity of blockchain consensus protocols [27] and accelerate the agreement of the consensus [28]. It can be seen that iO could be well coupled with blockchain and has a wide application prospect.

This paper analyzed and discussed the combination of iO and blockchain [29,30]. Applying iO to blockchain can improve the security of the system [31,32] or reduce the time needed for consensus, which has a wide application prospect in various field [33–36]. The combination of blockchain and iO cryptography need to be studied further [37,38].

The rest of the paper is organized as follows. In Sect. 2, we introduce the basic ideas of the method, followed by the description of the algorithm. Then, we present the experiment and give the evaluation of the results in Sect. 3. Finally, we conclude the paper in Sect. 4.

2 Constructing Indistinguishable Obfuscated Encryption and Decryption Based on Transformer Model

2.1 Basic Ideas

In this paper, we implement the conversion of plaintext and ciphertext by combining hints through a translation behavior similar to that in the language model [39]. The ciphertext output from the model plus some noise (encrypted data from public transmissions, etc.) is then converted to plaintext as input to the model.

For the linguistic autoregressive model, we build the following conditional language model, which predicts a word y from the first n known words, or phrases.Known $X(x_1, x_2, ..., x_n)$, according to the conditional language model,

$$p(y_t \mid y_1, y_2, ..., y_{t-1}, X) \tag{1}$$

to output the corresponding $y = (y_1, y_2, ..., y_T)$.

The Transformer [40] self-attentive architecture, on the other hand, has a great improvement in the expressiveness of the model for computing these conditional probabilities. The above unconditional language model could be transformed into a conditional language model when the specific encryption algorithm is known. The model in this paper is controlled based on the following 2 methods:

1) When encrypting the plaintext or decrypting the ciphertext, we can add hints (numbers or other representations of encryption algorithm categories) in front of the input.
2) In the normalization of the Transformer model's translation behavior, the main method layer is normalized by the following formula:

$$\mu^l = \frac{l}{H}\sum_{i=1}^{H} a_i^l \quad \sigma^l = \sqrt{\frac{l}{H}\sum_{i=1}^{H}\left(a_i^l - \mu^l\right)^2} \tag{2}$$

l denotes the Lth hidden layer, H denotes the number of nodes in the layer, and a denotes the value of a node before activation. Suppose the current input is x^t and the previous hidden state is h^{t-1} , then the weighted input vector (input of the nonlinear cell) is

$$a^t = W_{hh}h^{t-1} + W_{\text{xh}}x^t \tag{3}$$

Layer normalization for weighted input vectors, followed by scaling and translation (for recovering nonlinearity)

$$y = g \cdot \hat{a}^t + b \quad \hat{a}^t = \frac{a^t - u^t}{\sigma^t} \tag{4}$$

where g is denoted as the gain and b denotes the bias parameter. The above is the basic normalization, while the condition c (the representation of cryptographic algorithm classes) is turned into the same dimensions as g and b in equation (4) by different matrix transformations (i.e., linear mapping), respectively, and then the transformations are mapped to g and b.

$$g' = w_g * c + g \quad b' = w_b^* c + b \tag{5}$$

The input is processed by the transformations mentioned above, and the translation behavior of the Transformer model is controlled by the method in 2), so that the conversion of plaintext and ciphertext are realized. At the same time, the ciphertext transmitted to the public is obfuscated and only the corresponding decoding module can convert it to plaintext, so that there is no fear of interception. The model can also generate an obfuscated ciphertext with a specified encryption algorithm on demand, which can only be decrypted into plaintext by the trained decryption model.

To be able to train more layers (e.g., 100, 1000 or even 10,000 layers), the scaling multiplier associated with the number of layers needs to be adjusted during layer normalization to prevent gradient explosion. The scaling multiplier is given by the following formula, where N is the number of layers:

$$\alpha = (2N)^{1/4} \quad \beta = (8N)^{-1/4} \tag{6}$$

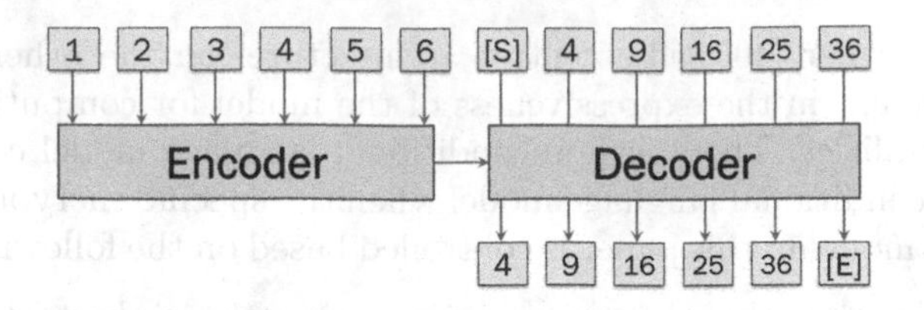

Fig. 1. Simple architecture for encryption and decryption model

2.2 Algorithm Description

With the rapid development of multicore [41–43] and large memory [44–46], large amounts of data and big software [47–49] can be implemented quickly. Hence, privacy and security becomes a big concern in data storage, processing, and transmission [50–52]. The whole system architecture is divided into an encryption module and a decryption module, which form the transformer model [53–55]. Figure 1 is a schematic diagram of the encryption model and the decryption training model.

Algorithm1 is the description of our specific algorithm (encryption and decryption processes are almost the same, only the source and the target are reversed).

Algorithm 1.

Input: source data, *src_seq*; target with label, *target_seq*; algorithm category, C; number of self-attentive computation block layers, *num_layers*; batch, *epoch*

1: $num\_layers = 200, \alpha = 1599.75,\ \beta = 4.472,\ initialweights(including\ w_g, w_b$, etc.)
2: **for** each $i \in [0, epoch]$ **do**
3:     **for** each $j \in [0, num\_layers]$ **do**
4:         $new\_src\_seq \leftarrow LN(src\_seq_{i,j} + C_{i,j})$
5:         $out\_projection \leftarrow Self\_Attention(new\_src\_seq)$
6:         $loss \leftarrow loss + (out\_projection, target\_seq_{i,j})$
7:     **end for**
8: **end for**
9: MIN($loss$)

As can be seen from Algorithm1, we treat plaintext and ciphertext as sequences that can be translated into each other, add hints and obfuscation before the input data, and make the algorithm or obfuscation a condition to be added during the layer normalization of the deep self-attentive computation. We control the translation behavior of the transformer by the method described above.

2.3 Algorithm Implementation

The implementation steps of the indistinguishable obfuscation and decryption algorithm based on the transformer model are as follows.

1) Determine the number of model layers, calculate the parameters that can prevent gradient explosion for each layer normalized by the number of layers, i.e., equation (6), and initialize the individual weights.
2) Input source sample data and target data, the input data can be prefixed to indicate which encryption algorithm is used, and it can also be used as obfuscated data. The model is trained to predict the ciphertext after encryption.
3) At each layer of layer normalization, i.e., equation (4), algorithms or obfuscated data are added to the normalization as conditions in order to control the translation behavior, i.e., equation (5).
4) input the source sequence into the model, and then perform the self-attentive calculation after the above process, and the model outputs the predicted target sequence, which is subjected to cross-entropy loss calculation with the real target sequence, and the gradient is calculated according to the loss value to update each parameter.
5) Repeat steps 2 to 4 until the loss converges to 0, and the training of the encryption model is completed.
6) Train the corresponding decryption model and repeat the above steps 1 to 5, the difference is that the input of step 2 is the ciphertext predicted by the encryption model, i.e., the source sequence and the target sequence of the above encryption model are switched.

The entire data flow for model training can be briefly summarized as follows.

(i) Encryption model training process: input plaintext or plaintext with prefix added in front, label is the encrypted ciphertext, and the model training is mainly to predict the encrypted ciphertext.
(ii) Decryption model training process: the input is the ciphertext predicted by the above encryption model, the label is the decrypted plaintext, and the model training is mainly to predict the decrypted plaintext.

Figure 2 describes the algorithm flow, in which we can find that there are hints (indicating the type of cryptographic algorithm) at the data input and at the normalization of each layer, similar to the "noise perturbation" added to the neural network, thus allowing the model to learn better.

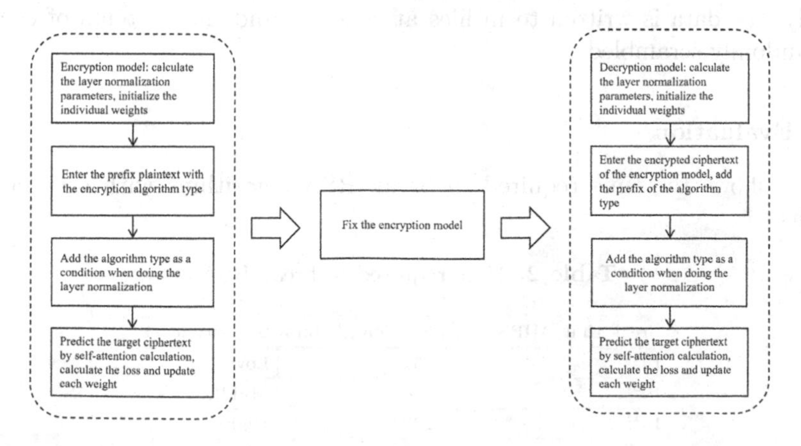

Fig. 2. Algorithm flow

3 Experiment and Analysis

3.1 Experimental Environment

In order to verify and analyze the algorithm proposed in this paper, we conducted simulation experiments using the following hardware (arithmetic) and data.

Table 1. Hardware configuration

Name	Quantity	Configuration
Calculation Cards	1	A100 80G
CPU	2	64 cores 128 threads intel Xeon
Memory	3	32*3 total 96G
Hard Disk	1	2T

3.2 Experiment

Randomly generate 600,000 pieces of plaintext, each of which is a number of length 11. The 600,000 plaintext are equally divided into 6 data sets, and the data in different data sets are encrypted differently. Then the transformed data sets are combined and a global random disruption is performed. The file is stored by rows, and each row represents a sample, what we want to do is to randomly disrupt the file by rows:

(a) Assuming that the file has a total of $m * n$ lines, the original file is divided equally into m files of n lines each.
(b) Randomly breaking up the file by line for each n lines, since n is arbitrarily specified, so this step can be done in memory.
(c) Read the first line of each file (to get m lines of data) and write these m lines of data randomly to an output file.
(d) Read the 2nd,..., n^{th} line of each file in turn, and repeat the operation in step 3 to generate m output files.

Simply, the data is written to m files at random, and the contents of each file are randomly scrambled.

3.3 Evaluation

Table 2 shows the time required to break RSA algorithms with different key lengths.

Table 2. Time required to break RSA

Break time/MIPS year	Key length/bits	Security level
10^4	512	Low
10^8	768	Middle
10^{11}	1024	High
10^{20}	2048	High

Table 3 shows the time required to break the iO algorithm proposed in this paper. Observing the data, we can see that the security level of iO algorithm is high when the encryption algorithm type is greater than 6 and the key length is greater than or equal to 512 bits.

Table 3. Security analysis of iO

Num of encrytion algorithms	n=6	n=7
Break time 512(Security Level)	10*7! = 10 (Middle)	10*8! = 10 (High)
Break time 768(Security Level)	10*7! = 10 (High)	10*8! = 10 (High)
Break time 1024(Security Level)	10*7! = 10 (High)	10*8! = 10 (High)

4 Conclusion

This paper introduced a feasible algorithm implementation for iO with information security protection as the core. By performing security encryption and decryption in a way similar to language translation, it provided a new scheme for data security transmission, and provides a research idea for the feasibility of iO implementation, which is of great significance to improve data security transmission and information protection level. This paper also analyzed and discussed the combination of iO and blockchain. Our study showed that applying iO to blockchain can improve the security of the system or reduce the time needed for consensus, which has a wide application prospect in various field. In future, we will continue to study the combination of blockchain and iO cryptography.

Acknowledgement. This paper is supported by the National Natural Science Foundation of China (U21B2021, 62202027, 61972018, 61932014).

References

1. Kaufman, L.M.: Data security in the world of cloud computing. IEEE Secur. Priv. **7**(4), 61–64 (2009)
2. Li, Y., Gai, K., et al.: Intercrossed access controls for secure financial services on multimedia big data in cloud systems. In: ACM TMCCA (2016)
3. Qiu, M., Chen, Z., et al.: Energy-aware data allocation with hybrid memory for mobile cloud systems. IEEE Syst. J. **11**(2), 813–822 (2014)
4. Qiu, M., et al.: Energy minimization with soft real-time and DVS for uniprocessor and multiprocessor embedded systems. In: IEEE DATE, pp. 1–6 (2007)
5. Qiu, M., Jia, Z., et al.: Voltage assignment with guaranteed probability satisfying timing constraint for real-time multiproceesor DSP. In: JSPS (2007)
6. Qiu, M., et al.: Dynamic and leakage energy minimization with soft real-time loop scheduling and voltage assignment. IEEE TVLSI **18**(3), 501–504 (2009)

7. Qiu, M., Xue, C, et al.: Efficient algorithm of energy minimization for heterogeneous wireless sensor network. In: IEEE EUC, pp. 25–34 (2006)
8. Niu, J., Gao, Y., et al.: Selecting proper wireless network interfaces for user experience enhancement with guaranteed probability. JPDC **72**(12), 1565–1575 (2012)
9. Qiu, H., Zheng, Q., et al.: Deep residual learning-based enhanced JPEG compression in the internet of things. IEEE Trans. Ind. Inf. **17**(3), 2124–2133 (2020)
10. Qiu, M., Qiu, H., et al.: Secure data sharing through untrusted clouds with blockchain-enabled key management. In: 3rd SmartBlock Conference, pp. 11–16 (2020)
11. Gai, K., Zhang, Y., et al.: Blockchain-enabled service optimizations in supply chain digital twin. In: IEEE TSC (2022)
12. Qiu, M., Qiu, H.: Review on image processing based adversarial example defenses in computer vision. In: IEEE 6th International Conference BigDataSecurity, pp. 94–99 (2020)
13. Hu, F., Lakdawala, S., et al.: Low-power, intelligent sensor hardware interface for medical data preprocessing. IEEE TITB **13**(4), 656–663 (2009)
14. Li, J., Ming, Z., et al.: Resource allocation robustness in multi-core embedded systems with inaccurate information. J. Syst. Arch. **57**(9), 840–849 (2011)
15. Qiu, H., Zheng, Q., et al.: Topological graph convolutional network-based urban traffic flow and density prediction. IEEE Trans. ITS **22**(7), 4560–4569 (2020)
16. Denning, D.E.R.: Cryptography and Data Security, vol. 112. Addison-Wesley Reading, Boston (1982)
17. Qiu, H., Qiu, M., Lu, R.: Secure V2X communication network based on intelligent PKI and edge computing. IEEE Network **34**(2), 172–178 (2019)
18. Qiu, H., Zeng, Y., et al.: Deepsweep: an evaluation framework for mitigating DNN backdoor attacks using data augmentation. In: ACM AsiaCCS (2021)
19. K. Gai et al. A novel secure big data cyber incident analytics framework for cloud-based cybersecurity insurance. In IEEE BigDataSecurity (2016)
20. Jain, A., Lin, H., Sahai, A.: Indistinguishability obfuscation from well-founded assumptions. In: Proceedings of the 53rd Annual ACM SIGACT Symposium on Theory of Computing, pp. 60–73 (2021)
21. Liu, Y., Liu, J., Zhang, Z., Yu, H.: A fair selection protocol for committee-based permissionless blockchains. Comput. Secur. **91**, 101718 (2020)
22. Liu, Y., Liu, J., Qianhong, W., Hui, Yu., Hei, Y., Zhou, Z.: SSHC: a secure and scalable hybrid consensus protocol for sharding blockchains with a formal security framework. IEEE Trans. Dependable Secur. Comput. **19**(3), 2070–2088 (2022)
23. Liu, Y., Liu, J., Yin, J., Li, G., Yu, H., Wu, Q.: Cross-shard transaction processing in sharding blockchains. In: Qiu, M. (ed.) ICA3PP 2020. LNCS, vol. 12454, pp. 324–339. Springer, Cham (2020). https://doi.org/10.1007/978-3-030-60248-2_22
24. Liu, Y., Liu, J., Hei, Y., Tan, W., Wu, Q.: A secure shard reconfiguration protocol for sharding blockchains without a randomness. In: TrustCom 2020, pp. 1012–1019. IEEE (2020)
25. Liu, Y., Xia, Y., Liu, J., Hei, Y.: A secure and decentralized reconfiguration protocol for sharding blockchains. In: IEEE HPSC 2021, pp. 111–116. IEEE (2021)
26. Pass, R., Seeman, L., Shelat, A.: Analysis of the blockchain protocol in asynchronous networks. In: Coron, J.-S., Nielsen, J.B. (eds.) EUROCRYPT 2017. LNCS, vol. 10211, pp. 643–673. Springer, Cham (2017). https://doi.org/10.1007/978-3-319-56614-6_22
27. Liu, Y., Liu, J., Zhang, Z., Tongge, X., Hui, Yu.: Overview on consensus mechanism of blockchain technology. J. Cryptologic Res. **6**(4), 395–432 (2019)

28. Kokoris-Kogias, E., Jovanovic, P., Gasser, L., Gailly, N., Syta, E., Ford, B.: Omniledger: a secure, scale-out, decentralized ledger via sharding. IEEE SP **2018**, 583–598 (2018)
29. Bin, H., Zhang, Z., Liu, J., Liu, Y., Yin, J., Rongxing, L., Lin, X.: A comprehensive survey on smart contract construction and execution: paradigms, tools, and systems. Patterns **2**(2), 100179 (2021)
30. Liu, Y., Qiu, M., Liu, J., Liu, M.: Blockchain-based access control approaches. In: IEEE CSCloud 2021, pp. 127–132. IEEE (2021)
31. Liu, Y., Liu, J., Salles, M.A.V., Zhang, Z., Li, T., Bin, H., Henglein, F., Rongxing, L.: Building blocks of sharding blockchain systems: Concepts, approaches, and open problems. Comput. Sci. Rev. **46**, 100513 (2022)
32. Liu, Y., Liu, J., Hei, Y., Xia, Yu., Wu, Q.: A secure cross-shard view-change protocol for sharding blockchains. In: Baek, J., Ruj, S. (eds.) ACISP 2021. LNCS, vol. 13083, pp. 372–390. Springer, Cham (2021). https://doi.org/10.1007/978-3-030-90567-5_19
33. Hei, Y., Liu, J., Feng, H., Li, D., Liu, Y., Qianhong, W.: Making MA-ABE fully accountable: a blockchain-based approach for secure digital right management. Comput. Networks **191**, 108029 (2021)
34. Hei, Y., Li, D., Zhang, C., Liu, J., Liu, Y., Qianhong, W.: Practical agentchain: a compatible cross-chain exchange system. Future Gener. Comput. Syst. **130**, 207–218 (2022)
35. Hei, Y., Liu, Y., Li, D., Liu, J., Qianhong, W.: Themis: an accountable blockchain-based P2P cloud storage scheme. Peer-to-Peer Netw. Appl. **14**(1), 225–239 (2021)
36. Xie, J., et al.: A survey of blockchain technology applied to smart cities: research issues and challenges. IEEE Commun. Surv. Tutorials **21**(3), 2794–2830 (2019)
37. Gai, K., Guo, J., Zhu, L., Shui, Yu.: Blockchain meets cloud computing: a survey. IEEE Commun. Surveys Tutorials **22**(3), 2009–2030 (2020)
38. Beck, R., Avital, M., Rossi, M., Thatcher, J.B.: Blockchain technology in business and information systems research. Bus. Inf. Syst. Eng. **59**(6), 381–384 (2017). https://doi.org/10.1007/s12599-017-0505-1
39. Vaswani, A., et al.: Attention is all you need. In: Advances in Neural Information Processing Systems, vol. 30 (2017)
40. Kitaev, N., Kaiser, Ł., Levskaya, A.: Reformer: the efficient transformer. arXiv preprint arXiv:2001.04451 (2020)
41. Qiu, M., Li, H., Sha, E.: Heterogeneous real-time embedded software optimization considering hardware platform. In: ACM SAC, pp. 1637–1641 (2009)
42. Qiu, M., Sha, E., Liu, M., Lin, M., Hua, S., Yang, L.: Energy minimization with loop fusion and multi-functional-unit scheduling for multidimensional DSP. JPDC **68**(4), 443–455 (2008)
43. Qi, D., Liu, M., Qiu, M., Zhang, S.: Exponential synchronization of general discrete-time chaotic neural networks with or without time delays. IEEE Trans. Neural Networks **21**(8), 1358–1365 (2010)
44. Li, J., Qiu, M., et al.: Thermal-aware task scheduling in 3D chip multiprocessor with real-time constrained workloads. ACM Trans. Embedded Comput. Sys. (TECS) **12**(2), 1–22 (2013)
45. Shao, Z., Wang, M., et al.: Real-time dynamic voltage loop scheduling for multi-core embedded systems. IEEE Trans. Circuits Syst. II **54**(5), 445–449 (2007)
46. Qiu, M., Ming, Z., et al.: Three-phase time-aware energy minimization with DVFS and unrolling for chip multiprocessors. J. Syst. Archit. **58**(10), 439–445 (2012)

47. Tao, L., Golikov, S., Gai, K., Qiu, M.: A reusable software component for integrated syntax and semantic validation for services computing. In: IEEE Symposium on Service-Oriented System Engineering, pp. 127–132 (2015)
48. Zhang, K., Kong, J., Qiu, M., Song, G.: Multimedia layout adaptation through grammatical specifications. Multimedia Syst. **10**(3), 245–260 (2005)
49. Qiu, M., Zhang, K., Huang, M.: Usability in mobile interface browsing. Web Intell. Agent Syst. **4**(1), 43–59 (2006)
50. Zhang, L., Qiu, M., Tseng, W., Sha, E.: Variable partitioning and scheduling for mpsoc with virtually shared scratch pad memory. J. Signal Proc. Syst. **58**(2), 247–265 (2018). https://doi.org/10.1007/s11265-009-0362-3
51. Qiu, M., et al.: RNA nanotechnology for computer design and in vivo computation. Philos. Trans. Roy. Soc. A **371**(2000), 20120310 (2013)
52. Qiu, M., Gai, K., Xiong, Z.: Privacy-preserving wireless communications using bipartite matching in social big data. FGCS **87**, 772–781 (2018)
53. Qiu, M., Su, H., Chen, M., Ming, Z., Yang, L.: Balance of security strength and energy for a PMU monitoring system in smart grid. IEEE Commun. Mag. **58**(5), 142–149 (2012)
54. Qiu, H., Dong, T., et al.: Adversarial attacks against network intrusion detection in IoT systems. IEEE Internet Things J. **8**(13), 10327–10335 (2020)
55. Gao, X., Qiu, M.: Energy-based learning for preventing backdoor attack. KSEM **3**, 706–721 (2022)

An Investigation of Blockchain-Based Sharding

Jiahong Xiao[1,2(✉)], Wei Liang[1,2], Jiahong Cai[1,2], Hangyu Zhu[1,2,3], Xiong Li[4], and Songyou Xie[5]

[1] School of Computer Science and Engineering, Hunan University of Science and Technology, Xiangtan 411201, China
xiaoiaong@163.com, wliang@hnust.edu.cn
[2] Hunan Key Laboratory for Service Computing and Novel Software Technology, Xiangtan 411201, China
[3] Guangdong Financial High-Tech Zone "Blockchain +" Fintech Research Institute, Foshan 528253, China
[4] School of Computer Science and Engineering, University of Electronic Science and Technology of China, Chengdu 611731, China
[5] College of Computer Science and Electronic Engineering, Hunan University, Changsha 410082, China

Abstract. Nowadays, blockchain distributed ledger technology is becoming more and more prominent, and its decentralization, anonymization, and tampering obvious features have been widely recognized. These excellent technical features of blockchain have also made it a hot issue for global research. With the wide application of blockchain technology in various industries, some defects are gradually exposed, and more prominently, the blockchain system is unable to meet the current demand of explosive growth of data volume and frequent data interaction. As one of the key technologies to solve this problem, sharding technology is gaining attention. This article introduces common blockchain scaling schemes and focuses on an overview of blockchain sharding. Sharding technology is introduced from two perspectives of intra-slice consensus and inter-slice consensus. The current mainstream slicing technology is summarized according to three different slicing methods: network sharding, transaction sharding, and state sharding. Finally, the challenges faced by current blockchain sharding technology are analyzed and the full text is summarized.

Keywords: Blockchain · Scaling · Sharding · Consensus Algorithm · Capacity Extension

1 Introduction

The digital economy has become a major component of today's world, and blockchain as new technology is gaining increasingly widespread attention worldwide and has penetrated into various aspects of people's production life, such as the Internet of Things [1, 2], finance [3–5], energy [6–8], health care [9, 10], and many other fields [11]. Blockchain [12] is a new application technology with technical features such as distributed data storage [13–15], decentralization, peer-to-peer transactions, and dynamic

M. Qiu et al. (Eds.): SmartCom 2022, LNCS 13828, pp. 695–704, 2023.
https://doi.org/10.1007/978-3-031-28124-2_66

encryption algorithms [16, 17]. However, the performance [18–20] of current decentralized systems based on blockchain technology is low and cannot afford the data processing under big data. Blockchain sharding technology is proposed as an important solution for blockchain sharding.

In 2016, Luu et al. [21] proposed an Elastico algorithm to combine the sharding method of expansion in the database with blockchain technology to improve the efficiency of transaction processing and reduce the cost of transactions. Since then, researchers from all walks of life have conducted a lot of research on sharding technology in terms of consensus mechanisms, distributed ledgers, and sharding methods. What we call sharding is not an innovative concept, but the essence of the idea is to divide the data in the database into small data shards [22], and then store these shards in different locations. This will avoid generating a large number of data access requests in a short period of time, thus overprocessing the transactions of one server. In traditional blockchain networks, transactions must be confirmed by each node in the network so as to ensure the security of transactions, but this is one of the difficulties in improving the speed of transactions. In contrast, the application of sharding technology in blockchain systems is to divide the blockchain network into several sub-networks, each of which will contain some nodes, and new transactions will be randomly assigned to individual shards for processing. Based on the sharding problem in the blockchain sharding scheme, this article introduces the research progress of blockchain sharding technology and gives a comprehensive introduction to sharding-related technologies.

The rest of this paper is as follows: Sect. 2 introduces blockchain technology and common blockchain scaling schemes; Sect. 3 categorizes blockchain slicing schemes based on consensus algorithms; Sect. 4 introduces three current mainstream slicing approaches and provides theoretical references for further research and development of blockchain slicing technology; Sect. 5 outlines the current challenges faced by blockchain slicing technology; Sect. 6 summarizes the paper.

2 Related Technology Introduction

Blockchain technology is increasingly developing into a mature digital economy infrastructure [23], which is essentially a bookkeeping method, usually using a general ledger with peer-to-peer storage [24], and has a wide range of applications in many fields [25, 26]. In this article, we present blockchain in a six-layer model, which are the application layer, contract layer, incentive layer, consensus layer, network layer, and data layer.

According to the six-layer architecture model of blockchain, its sharding schemes are divided into two main categories: on-chain scaling and off-chain scaling [27, 28]. On-chain expansion is the optimization and improvement of the basic structure, model, and algorithm of the blockchain from the data layer, network layer, consensus layer, and incentive layer of the blockchain, including changing the size of the block, sharding, changing the consensus mechanism, isolating the witness, increasing the capacity of the blockchain, and other aspects. The off-chain expansion is to build a transaction network on another layer other than the main chain, which is an adjustment of the contract and application layers of the blockchain. Placing some complex work under the chain and returning the result to the chain reduces the workload on the chain and improves the performance of the blockchain, including state channels [29], side chaining [30], cross-chain, and off-chain computing. Blockchain scaling techniques are shown in Table 1.

3 Consensus Algorithm-Based Sharding Technology

The essence of a blockchain is a distributed application, and the first problem in solving a distributed system is to reach a consensus among multiple independent nodes, i.e., all of them have to be consistent. The most common consensus algorithms are the proof-of-work (PoW) mechanism [31], proof-of-stake (PoS) mechanism [32], and delegated proof-of-stake (DPoS) mechanism [33]. However, high-performance consensus protocols used in distributed databases cannot be applied to blockchains, and secondly, blockchains rely on BFFs, which have been shown to lead to low scalability performance [34].

Table 1. Blockchain System Architecture.

Heading level	Example	Font size and style
On-chain expansion	Data Layer	Change block size
		Isolation testimonial
	Network layer	Sharding
	Consensual level	Pos, PBFT
Off-chain expansion	Tier 2 improvements	Status channel
		Side chain
		Cross-chain
		Below the chain calculation

3.1 Intra-slice Consensus Protocol

The consensus protocol is divided into intra-slice consensus and cross-slice consensus. Intra-slice consensus requires each node in the same slice to agree and broadcast according to the slice's protocol and finally results in a consensus result for the whole slice. In 1999, The PBFT proposed by Castro et al. [35] is the earliest BFT algorithm, in which the whole system works if more than 2/3 of the nodes are normal. In 2016, Miller et al. [36] proposed the HoneyBadgerBFT algorithm, which is the first asynchronous BFT algorithm that can guarantee its effectiveness without time constraints.

Figure 1 presents two different on-slice consensus flows. The consensus protocol used for blockchain sharding is based on the Practical Byzantine Consensus Protocol, and Fig. 1(a) depicts the flow of the intra-slice consensus mechanism based on PBFT, describing how a request is then passed between nodes in a distributed system running the PBFT protocol. PBFT is widely used in many blockchain systems, such as Zilliqa, and

Hyperledger, where each consensus of consensus nodes for a block requires receiving and forwarding multiple blocks. Figure 1(b) gives the intra-slice consensus phase under the two-layer slice consensus protocol under the federated chain, in which the task is to gather the intra-slice and cross-slice transactions forwarded to the slice into a single block and submit it to the blockchain after completing the consensus.

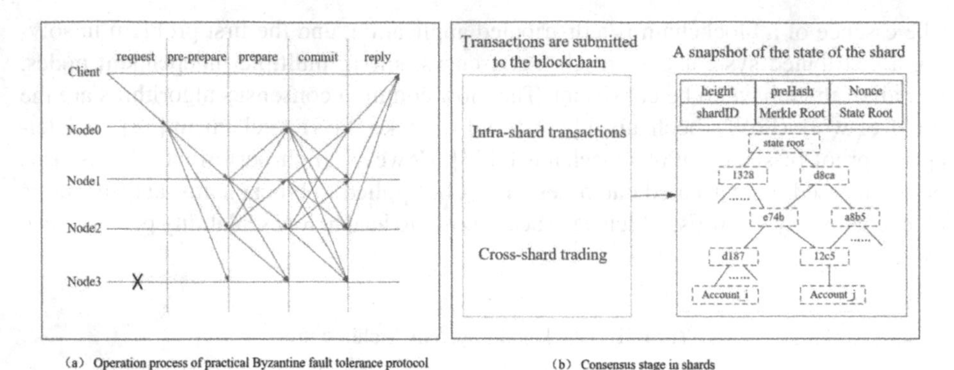

Fig. 1. Intra-slice consensus.

3.2 Cross-Slice Consensus Protocol

In traditional blockchain systems, communication between all nodes is required to maintain identical copies of the blockchain, which leads to relatively low performance of the blockchain system. Now all the slices are facing the challenge of cross-slice transaction processing and need to handle data from multiple slices. To prevent the problem of double spending [37], all the slices that involve cross-slice transactions need to execute a multi-stage protocol to confirm the transactions.

There are three main approaches to cross-slice consensus, which are transaction atomization, transaction centralization, and the use of class routing protocols. Figure 2 describes three different cross-slice consensus processes. Figure 2(a) depicts the RSTP protocol, which efficiently handles intra-slice and cross-slice transactions by invoking FBFT [38]. In RSTP, the operations of input and output slices are separated. Figure 2(b) depicts the two-layer slice consensus protocol mechanism [39], which is based on the principle of committing transactions to the blockchain before updating them using state snapshots, which in turn enables the final atomicity of cross-slice transactions. In the intra-slice public phase, all transactions are submitted to the blockchain, and the intra-slice submitted blocks generate a new snapshot of the partition state. In the cross-slice consensus phase, the snapshots of the slice state generated by all the slices are synchronized to the DAG global state snapshot, and then the global state snapshot is updated for the cross-slice transactions. The consensus flow of the cross-slice consensus phase is shown in Fig. 2(c).

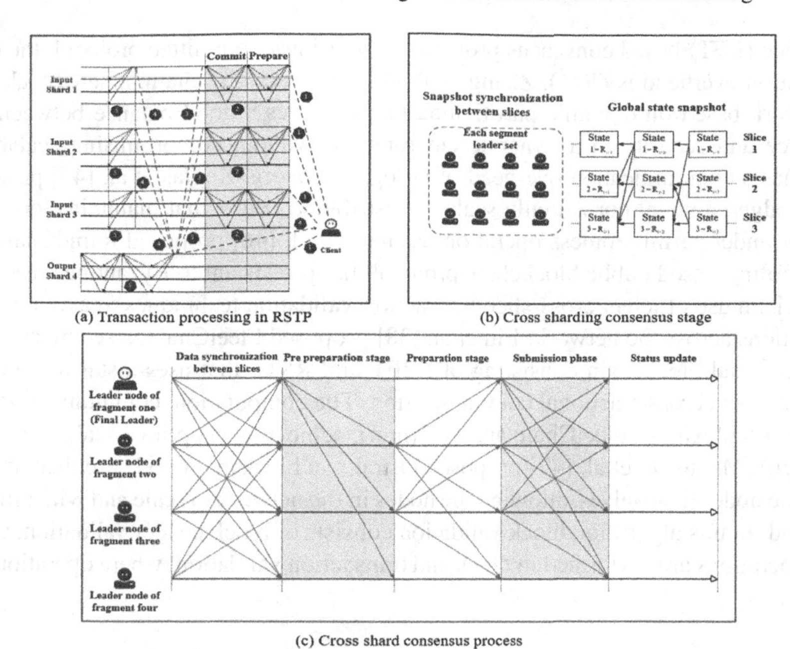

(a) Transaction processing in RSTP

(b) Cross sharding consensus stage

(c) Cross shard consensus process

Fig. 2. Cross-slice consensus.

4 Mainstream Sharding Technology

4.1 Network Sharding

The sharding technique was first proposed by Wang et al. [40]. The major challenge of the blockchain sharding protocol is that the vast majority of transactions in the blockchain system are cross-slicing, and these cross-slicing transactions not only reduce the throughput of the system but also increase the cost of transaction processing. Network sharding is the most basic sharding method, which divides the nodes in a blockchain network into individual shards, and each network independently performs consensus and processes transactions, thus improving the throughput of the blockchain. Network sharding involves the problem of how to ensure the security of on-chain transactions after slicing [41], and network slicing usually takes into account factors such as slicing size, slicing security, and slicing method, and the process of slicing is usually implemented using random functions. Sharding divides the entire blockchain network into several small subnetworks, and each slice can independently and simultaneously build consensus and process different transactions in the network in parallel to increase the throughput and reduce the latency of the blockchain.

Zhou [42] proposed an NRSS sharding scheme to reduce the difference in performance among slices through a node scoring strategy; Cai et al. [43] proposed a multi-objective optimization algorithm (MaOEA-DRP) based on a dynamic reward and penalty mechanism to optimize the slice validation effectiveness model and then obtain an optimal blockchain slice scheme; Since most shard systems use a Byzantine Fault

Tolerance (BFT) based consensus protocol as their intra-committee protocol, the communication overhead is $O(n^2)$. Zhang et al. [44] proposed Skychain, a new blockchain framework based on dynamic partitioning that achieves a good balance between performance and security in a dynamic environment without compromising scalability, and it has a communication overhead of $O(\log n)$; Kokoris-Kogias et al. [45] proposed OmniLedger, a novel horizontally scalable distributed ledger that maintains long-term security under permissionless operation; Zamani et al. [46] proposed RapidChain, the first sharding-based public blockchain protocol that is resistant to Byzantine errors, and RapidChain use efficient cross-slice transaction validation techniques to avoid passing transactions across the network; Liu et al. [38] proposed FleetChain, a secure and scalable fractional blockchain consisting of FBFT and RSTP that uses a star network for both intra- and cross-fractional communication. The computation, communication, and storage complexity of FleetChain are all $O(n/m)$, achieving an optimal scaling factor of $O(n/\log n)$; Manuskin et al. [47] proposed Ostraka, a blockchain node architecture that slices the nodes themselves, allowing the nodes in the network to scale and with minimal overhead. In this algorithm, block validation consists of block header validation, where most operations are performed in $O(1)$, and transaction validation, where operations are in $O(n)$.

4.2 Transaction Sharding

Transaction sharding is based on network sharding, where different transactions are divided into different slices based on certain rules to reach an agreement. All the sliced networks are able to perform consensus and process transactions at the same time and assign some highly related transactions to the same slice, thus reducing the transaction processing overhead. The idea is to allocate transactions in the slices according to specific rules, both to achieve parallel processing and to avoid double-spending attacks. Transaction sharding focuses on which transactions are to be assigned to which slices. Transaction sharding allows individual network slices to have greater processing power over transactions. Although transaction sharding can improve the operational efficiency of the network to a certain extent, it cannot fundamentally solve the resource deficiencies in the network.

Nguyen et al. [48] proposed a new sharding paradigm-Optchain based on OmniLedger, which utilizes a lightweight dynamic transaction placement method that assigns related transactions to the same slice, minimizing transactions across slices, significantly reducing the confirmation time required for transactions, and increasing throughput. Its average computation time takes only $O(n)$. Castro et al. [35] proposed Zilliqa, which runs consensus mechanisms at a higher frequency within each slice to significantly increase throughput by processing transactions in parallel, but the scheme does not slice the blockchain data and is vulnerable to attacks; Liu et al. [49] proposed a secure and scalable hybrid consensus (SSHC) and fair slice selection (FSS) with the strict proof scheme that designs a transaction batching mechanism for the corresponding slice to handle transactions across the slice, thus reducing the number of calls to the BFT algorithm.

4.3 State Sharding

In blockchain systems, state sharding is a common solution, also known as data sharding. It classifies the data of the whole network according to the state, and each slice keeps only part of the data, not the whole blockchain. State sharding is the most ideal sharding method, and only by implementing state sharding can we essentially solve the problem of public chain scalability. At the same time, state sharding is the most challenging sharding scheme so far, which stores completely different ledgers in each shard, and the whole sharding network forms a complete ledger.

Chen et al. [50] proposed SSChain, a new non-reconfiguration structure that supports transaction slicing and state slicing; Wang et al. [51] proposed Monoxide, which introduces a human-specific asynchronous consensus that enables full slicing; Huang et al. [52] proposed BrokerChain, a state slicing design based on the account \ balance of cross-slicing blockchain protocol. In the success case, the computing complexity of cross-shard verification is O(1); Zilliqa can realize transaction slicing and state slicing, based on Zilliqa, Harmony [53] can not only split transaction validation like Zilliqa but also split data state, and its slicing process ensures high security and makes up for some of the shortcomings of Zilliqa.

5 The Challenges of Sharding Technology

The current development of blockchain has entered a bottleneck period, unable to meet the needs of large-scale application scenarios in the era of big data, and a large part of the bottleneck faced by blockchain technology today is due to the need to go to distributed consensus, while resource scarcity also makes blockchain systems have significant performance defects. The introduction of sharding can fundamentally reduce the resources required by nodes and is the most effective on-chain scaling solution. However, although the sharding technology has solved the problem of blockchain scalability [54] and improved the performance of transaction processing, there are still many areas for improvement, and still face many challenges.

The main challenges within the sharding are as follows: the security threat is by far the most important issue facing the blockchain. When sharding divides the tasks of the whole network into n different shards, the arithmetic power is also allocated to the corresponding shards, so for a single shard, only 1/n of the original arithmetic power is available, and at this time the difficulty of launching a 51% attack on a single shard is reduced to 1/n of the original one. This leads to a significant decrease in the security of the system. Under the PoW consensus, the blockchain mainly faces the 51% attack problem. That is, the blockchain can be tampered with and forged as long as it has more than 51% of the computing power on the network.

PBFT allows less than (n-1)/3 number of failed nodes and does not require multiple blocks to determine, so there are multiple sharding projects that have chosen the PBFT consensus mechanism within sharding. However, because PBFT consensus has the fault tolerance of (n-1)/3 nodes, 100 nodes cannot fully secure the slice. A witch attack is when a node disguises itself as n nodes and falsely claims that it stores n copies of data, thus weakening the redundant backup role of the data. The PBFT itself cannot prevent the witch attack, so it needs to be prevented with other means.

The challenges between slices are as follows: Because digital information is easily replicated, double-splash attacks must be taken into account regardless of decentralization. Double-splash attacks within slices can be prevented using traditional methods, but in the UTXO model, an attacker can create multiple transactions with the same input but different outputs to perform a double-splash attack. In order to avoid "double spend attacks", the system must communicate a lot across slices, which increases the complexity of the system and also degrades its performance of the system.

6 Conclusion

Blockchain scalability is a hot topic right now, and sharding technology is one of the effective ways to solve this problem. Without reducing decentralization, sharding is the most likely solution to achieve high-performance on-chain scalability. This article firstly gives an overview of blockchain slicing-related technologies based on blockchain sharding consensus protocol, and classifies blockchain consensus algorithms from two aspects: intra-chip consensus and cross-chip consensus; secondly, it summarizes and outlines the existing blockchain sharding technologies from three aspects: network sharding, transaction sharding, and state sharding, and to help readers quickly understand blockchain sharding. Finally, with the development of blockchain sharding technology, more efficient blockchain slicing schemes will emerge in the future, and these results will further improve the blockchain technology ecology based on the sharding mechanism.

References

1. Xu, Z., et al.: A time-sensitive token-based anonymous authentication and dynamic group key agreement scheme for industry 5.0. IEEE TII **18**(10), 7118–7127 (2021)
2. Xie, Y., Liang, W., Li, R.F., et al.: An in-vehicle CAN signal packing algorithm for connected vehicle environment. J. Softw. **27**(09), 2365–2376 (2016)
3. Chen, Y., Bellavitis, C.: Blockchain disruption and decentralized finance: The rise of decentralized business models. J. Bus. Vent. Insights **13**, e00151 (2020)
4. Li, Y., Gai, K., et al.: Intercrossed access controls for secure financial services on multimedia big data in cloud systems. ACM Trans. Multi. Comput. Commun. Appl. **12**(4s), 1–18 (2016)
5. Gai, K., Qiu, M., Elnagdy, S.: A novel secure big data cyber incident analytics framework for cloud-based cybersecurity insurance. In: IEEE BigDataSecurity (2016)
6. Zuo, C., et al.: Trust-aware and low energy consumption security topology protocol of wireless sensor network. J. Sens. (2015)
7. Qiu, M., Xue, C., Shao, Z., Sha, E.H.M.: Energy minimization with soft real-time and DVS for uniprocessor and multiprocessor embedded systems. In: IEEE DATE Conference, pp. 1–6 (2007)
8. Qiu, M., Xue, C., Shao, Z., Zhuge, Q., Liu, M., Sha, E.H.M.: Efficent algorithm of energy minimization for heterogeneous wireless sensor network. In: Sha, E., Han, S.K., Xu, C.Z., Kim, M.H., Yang, L.T., Xiao, B. (eds.) Embedded and Ubiquitous Computing. EUC 2006. Lecture Notes in Computer Science, vol. 4096, pp. 25–34. Springer, Berlin, Heidelberg (2006).https://doi.org/10.1007/11802167_5
9. Peng, L., et al.: Improved low-rank matrix recovery method for predicting miRNA-disease association. Sci. Rep. **7**(1), 1–10 (2017)

10. Hu, F., Lakdawala, S., et al.: Low-power, intelligent sensor hardware interface for medical data preprocessing. IEEE TITB **13**(4), 656–663 (2009)
11. Qiu, H., Zheng, Q., et al.: Topological graph convolutional network-based urban traffic flow and density prediction. IEEE Trans. ITS **22**(7), 4560–4569 (2020)
12. Gai, K., Zhang, Y., Qiu, M., Thuraisingham, B.: Blockchain-enabled service optimizations in supply chain digital twin. IEEE Trans. Serv. Comput. (2022)
13. Jing, W., Chong, Z., Wei, L., Xiangyang, L.: Localized restoration coding based on Pyramid codes in distributed storage systems. J. Electr. Meas. Instrum. **31**(09), 1481–1487 (2017). https://doi.org/10.13382/j.jemi.2017.09.020
14. Qiu, M., Chen, Z., Ming, Z., Qin, X., Niu, J.: Energy-aware data allocation with hybrid memory for mobile cloud systems. IEEE Syst. J. **11**(2), 813–822 (2014)
15. Li, J., Ming, Z., et al.: Resource allocation robustness in multi-core embedded systems with inaccurate information. J. Syst. Arch. **57**(9), 840–849 (2011)
16. Conoscenti, M., Antonio, V., De Martin, J.C.: Blockchain for the internet of things: a systematic literature review. In: IEEE/ACS AICCSA (2016)
17. Qiu, H., Dong, T., et al.: Adversarial attacks against network intrusion detection in IoT systems. IEEE Internet Things J. **8**(13), 10327–10335 (2020)
18. Niu, J., Gao, Y., Qiu, M., Ming, Z.: Selecting proper wireless network interfaces for user experience enhancement with guaranteed probability. JPDC **72**(12), 1565–1575 (2012)
19. Qiu, M., Li, H., Sha, E.: Heterogeneous real-time embedded software optimization considering hardware platform. In: Proceedings of the ACM Symposium on Applied Computing, pp. 1637–1641 (2009)
20. Qiu, M., Jia, Z., et al.: Voltage assignment with guaranteed probability satisfying timing constraint for real-time multiproceesor DSP. J. Signal Proc. Syst. **46**, 55–73 (2007)
21. Luu, L., et al.: A secure sharding protocol for open blockchains. In: Proceedings of the 2016 ACM SIGSAC Conference on Computer and Communications Security (2016)
22. Liu, Y., Wang, Y., Jin, Y.: Research on the improvement of MongoDB auto-sharding in cloud environment. In: 7th IEEE ICCSE (2012)
23. Liang, W., et al.: Secure data storage and recovery in industrial blockchain network environments. IEEE Trans. Ind. Inf. **16**(10), 6543–6552 (2020)
24. Wang, S., et al.: Blockchain-enabled smart contracts: architecture, applications, and future trends. IEEE Trans. SMC: Syst. **49**(11), 2266–2277 (2019)
25. Gai, K., et al.: Permissioned blockchain and edge computing empowered privacy-preserving smart grid networks. IEEE Internet Things J. **6**(5), 7992–8004 (2019)
26. Kumar, P., Kumar, R., et al.: PPSF: a privacy-preserving and secure framework using blockchain-based machine-learning for IoT-driven smart cities. IEEE Trans. Netw. Sci. Eng. **8**(3), 2326–2341 (2021)
27. Yu, H., Zhang, Z., Liu, J.: Research on bitcoin blockchain scaling technology. Comput. Res. Develop. **54**(10), 2390–2403 (2017)
28. Zeng, S., et al.: Blockchain scaling for bitcoin: key technologies, constraints and derived problems.J. Auto. **45**(06) 1015–1030 (2019). https://doi.org/10.16383/j.aas.c180100
29. Zhang, F., et al.:Federated learning meets blockchain: state channel based distributed data sharing trust supervision mechanism. IEEE IoT J. (2021)
30. Singh, A., et al.: Sidechain technologies in blockchain networks: An examination and state-of-the-art review. J. Netw. Comput. Appl. **149**, 102471 (2020)
31. Vukolić, M.: The quest for scalable blockchain fabric: proof-of-work vs. BFT replication. In: Camenisch, J., Kesdoğan, D. (eds.) Open Problems in Network Security. iNetSec 2015. Lecture Notes in Computer Science, vol. 9591, pp. 112–125 . Springer, Cham (2016). https://doi.org/10.1007/978-3-319-39028-4_9
32. Kang, J., et al.: Incentivizing consensus propagation in proof-of-stake based consortium blockchain networks. IEEE Wirel. Commun. Lett. **8**(1), 157–160 (2018)

33. Luo, Y., et al.: A new election algorithm for DPos consensus mechanism in blockchain. In: 2018 7th International Conference on Digital Home (ICDH). IEEE (2018)
34. Dang, H., et al.:Towards scaling blockchain systems via sharding. In: Proceedings of the 2019 International Conference on Management of Data (2019)
35. Castro, M., Liskov, B.: Practical byzantine fault tolerance. OsDI. **99**, 1999 (1999)
36. Miller, A., et al.: The honey badger of BFT protocols. In: Proceedings of the 2016 ACM SIGSAC Conference on Computer and Communications Security (2016)
37. Karame, G.O., et al.: Misbehavior in bitcoin: A study of double-spending and accountability. ACM Trans. Inf. Syst. Secur. (TISSEC) **18**(1), 1–32 (2015)
38. Liu, Y., Liu, J., Li, D., Yu, H., Wu, Q.: FleetChain: a secure scalable and responsive blockchain achieving optimal sharding. In: Qiu, M. (eds) Algorithms and Architectures for Parallel Processing. ICA3PP 2020. Lecture Notes in Computer Science, vol. 12454, pp. 409–425. Springer, Cham (2020). https://doi.org/10.1007/978-3-030-60248-2_28
39. Nankun, L.: Research and system implementation of blockchain sharding technology in federated chain scenario. Univ. Electr. Sci. Technol. (2021). https://doi.org/10.27005/d.cnki.gdzku.2021.005445
40. Wang, G., et al.: Sok: sharding on blockchain. In:Proceedings of the 1st ACM Conference on Advances in Financial Technologies (2019)
41. Zhang, S., et al.: A novel blockchain-based privacy-preserving framework for online social networks. Connect. Sci. **33**(3), 555–575 (2021)
42. Yang, Y.Z.: Research on the improvement of blockchain slicing strategy. Tianjin Univ. (2019). https://doi.org/10.27356/d.cnki.gtjdu.2019.002400
43. Cai, X., et al.: A sharding scheme-based many-objective optimization algorithm for enhancing security in blockchain-enabled industrial internet of things. IEEE Trans. Ind. Inf. **17**(11), 7650–7658 (2021)
44. Zhang, J., et al.: Skychain: a deep reinforcement learning-empowered dynamic blockchain sharding system. In: 49th IEEE ICPP (2020)
45. Kokoris-Kogias, E., et al.: Omniledger: a secure, scale-out, decentralized ledger via sharding. In: 2018 IEEE Symposium on Security and Privacy (SP) (2018)
46. Zamani, M., Movahedi, M., Raykova, M.: RapidChain: a fast blockchain protocol via full sharding. IACR Cryptol. ePrint Arch. **2018**, 460 (2018)
47. Manuskin, A., Mirkin, M., Eyal, I.: Ostraka: secure blockchain scaling by node sharding. In 2020 IEEE European Symposium on Security and Privacy (EuroS&PW) (2020)
48. Nguyen, L.N., et al.: Optchain: optimal transactions placement for scalable blockchain sharding. In: IEEE 39th International Conference on Distributed Computing Systems (ICDCS) (2019)
49. Liu, Y., et al.: SSHC: A secure and scalable hybrid consensus protocol for sharding blockchains with a formal security framework. IEEE Trans. Depend. Secure Comput. **19**(3), 2070–2088 (2020)
50. Chen, H., Wang, Y.: Sschain: A full sharding protocol for public blockchain without data migration overhead. Pervasive Mob. Comput. **59**, 101055 (2019)
51. Wang, J., Hao, W.: Monoxide: scale out blockchains with asynchronous consensus zones. In: 16th USENIX NSDI (2019)
52. Huang, H, et al.: BrokerChain: a cross-shard blockchain protocol for account/balance-based state sharding. In: IEEE INFOCOM (2022)
53. Harmony Team: Technical Whitepaper. Harmony. https://harmony.one/whitepaper.pdf. Accessed11 Feb 2020
54. Croman, K.: On scaling decentralized blockchains. In: Clark, J., Meiklejohn, S., Ryan, P.Y.A., Wallach, D., Brenner, M., Rohloff, K. (eds.) FC 2016. LNCS, vol. 9604, pp. 106–125. Springer, Heidelberg (2016). https://doi.org/10.1007/978-3-662-53357-4_8

Author Index

M. Qiu et al. (Eds.): SmartCom 2022, LNCS 13828, pp. 705–708, 2023.
https://doi.org/10.1007/978-3-031-28124-2

Printed in the United States
by Baker & Taylor Publisher Services

Printed in the United States
by Baker & Taylor Publisher Services